The physics of vibration

The physics of vibration

A. B. PIPPARD, FRS
Emeritus Professor of Physics, University of Cambridge

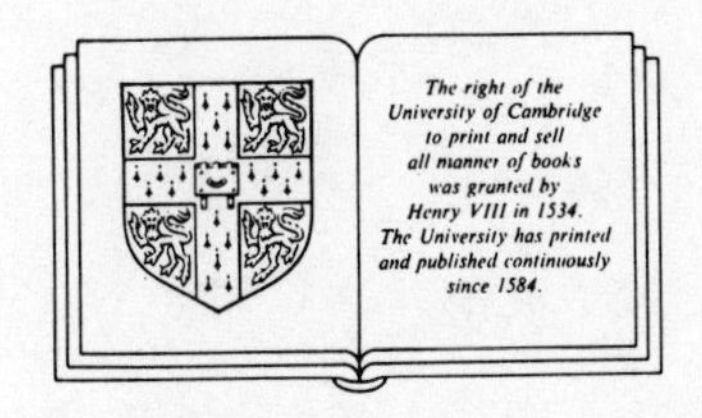

CAMBRIDGE UNIVERSITY PRESS
Cambridge
New York Port Chester
Melbourne Sydney

Published by the Press Syndicate of the University of Cambridge
The Pitt Building, Trumpington Street, Cambridge CB2 1RP
40 West 20th Street, New York, NY 10011, USA
10 Stamford Road, Oakleigh, Melbourne 3166, Australia

Part 1 first published 1979
Part 2 first published 1983
This volume (Parts 1 and 2 combined) first published 1989

Printed in Great Britain at Bath Press, Bath

British Library cataloguing in publication data

Pippard, A. B. (Alfred Brian), 1920–
The physics of vibration. – Combined ed.
1. Vibration
I. Title
531′.32

Library of Congress cataloguing in publication data

Pippard, A. B.
The physics of vibration / A. B. Pippard.
p. cm.
Includes index.
ISBN 0 521 37200 3
1. Vibration. I. Title
QC136.P56 1989
531′.32–dc20 89-7369 CIP

ISBN 0 521 37200 3

JWA

Contents

9 The driven anharmonic vibrator; subharmonics; stability

10 Parametric excitation

11 Maintained oscillators

12 Coupled vibrators

Note: Equations, diagrams and references are numbered serially in each chapter, and the chapter number is omitted if it is something in the same chapter that is referred to. Thus (4.23) in chapter 7 means equation 23 of chapter 4; (23) in chapter 7 means equation 23 in that chapter.

Cross-references are given as page-numbers in the margin.

Preface to Part 1

The plan and purpose of this book are outlined in chapter 1, and what is said there need not be repeated here. A preface does, however, offer the opportunity of acknowledging the help I have received during its preparation, and indeed over the years before it was even conceived. I cannot begin to guess how far my understanding and opinions have benefited from the good fortune that has allowed me to spend so much of my working life in the Cavendish Laboratory, surrounded by physics and physicists of the highest quality, and by students who at their best were at least the equal of their teachers. For the resulting gifts of learning, casually offered and accepted for the most part, and now untraceable to their source, I extend belated thanks. More specifically, in the preparation of the material I have relied heavily on Douglas Stewart for the construction of apparatus, Christopher Nex for programming the computer to draw many of the diagrams (and incidentally for two notes on his bassoon in fig. 5.8), Gilbert Yates for help in the theory and practice of electronics, and Shirley Fieldhouse for typing and retyping of the manuscript. My thanks to each of them for much willing and expert help, and to the following who by discussion and criticism, by assistance in carrying out experiments, or by providing material for diagrams have given support and comfort on the way: Dr J. E. Baldwin, Mr W. E. Bircham, Prof. J. Clarke, Dr M. H. Gilbert, Prof. O. S. Heavens, Dr R. E. Hills, Dr I. P. Jones, Mr P. Jones, Mr P. Martel, Prof. D. H. Martin, Mr P. D. Maskell, Dr P. J. Mole, Mr E. Puplett, Dr G. P. Wallis and Mr D. K. Waterson.

I am not the first scientific author to have struggled with the limitations of alphabets and the restrictions imposed by convention in the choice of symbols, nor am I the first to have to confess in the end that perfect consistency has eluded me. The failure would have been even more obvious without the intervention of Cambridge University Press whose attention to detail, scrupulous yet kindly, I acknowledge with gratitude.

A. B. PIPPARD

Cambridge 1978

Preface to Part 2

Ten years ago, when I started writing on the physics of vibration, I had in mind a single volume. Four years ago I was reconciled to the need for two, and now must confess that the complexity of the subject has made it advisable to get the second of the projected three parts into print without waiting for the third to accompany it between the same covers. It is still my hope to do justice to the vibrations of extended systems, but the difficulties are considerable and not made easier by the vigour with which some of the central topics are being pursued at present.

Of all the encouragement I have enjoyed I particularly wish to record with the warmest thanks the help of Dr Edmund Crouch and Dr John Hannay who long ago, as research students, derived for me some solutions of Schrödinger's equation which provided a stout anchor for my thoughts: Dr Bob Butcher who has firmly guided me in my brief excursions into structural chemistry and molecular spectra: and Dr Andrew Phillips who devoted more time than he could have been expected to spare to a critical reading of much of the typescript. If the state of that typescript as delivered to the printers did not achieve even a modest standard of tidiness, the fault is entirely the consequence of copious afterthoughts on my part, and in no way to be blamed on Mrs Janet Thulborn whose patient and faithful typing deserved better respect, and has certainly earned my gratitude. I am happy to acknowledge the generous provision by Dr Brian Petley of the original of fig. 19.5. The typography of Cambridge University Press continues to give me great satisfaction, but an oversight of mine in volume 1 has been criticized – without chapter numbers at the head of each page it is unnecessarily difficult to locate an equation. This fault has been avoided in the present volume.

A. B. PIPPARD

Cambridge 1981

Preface to the combined volume

In combining Parts 1 and 2 into a single volume only minor changes have been made, apart from the correction of errors. I have not attempted to bring the treatment up to date even for such rapidly expanding fields as the study of chaos in non-linear systems, but have been content to add a small number of references. Where an argument could be corrected, clarified or extended in the space of a few sentences, these are signposted in the margin and placed at the end of the chapter. The system of marginal cross-references has met with critical approval and I have taken the opportunity of adding to their number.

My hopes of ever completing a further volume on the vibrations of extended systems have by now grown faint. There are too many exciting new ideas that are not yet ready for the type of exposition that suited the present work. Fortunately the development, by both classical and quantal methods, of the physics of simple vibrators produces a reasonably self-contained argument, and I am grateful, as always, to Cambridge University Press for making possible this synthesis of concepts which in modern physics should be regarded always as complementary, never as antagonistic.

A. B. PIPPARD

Cambridge 1988

Preface to the combined volume

In combining Parts 1 and 2 into a single volume only minor changes have been made, apart from the correction of errors. [illegible]

1 General introduction – author to reader

The writing of this book has occupied several years, and now that it is finished it is time to ask what sort of a book it is. For all its length, it turns out to be only the first volume out of two, any hope of covering the topic in a single volume having vanished as the project developed. It must be obvious, therefore, that it is not a textbook in the sense of an adjunct to a conventional course of lectures (there are not even any questions at the ends of the chapters). On the other hand, it is certainly not an advanced treatise, for many of the more difficult topics are treated at a much lower level than is to be found in the specialist works devoted to them. Moreover, I can claim no professional skill in most of them, and this is both a confession and an advertisement. For by writing about them in a way that illuminates for me the essential physical thought underlying what is often a very complicated calculation, I hope to have provided a treatment that will enlighten others in the same unlearned state. The field is very wide, ranging from applied mathematics (non-linear vibration and stability theory) to electrical engineering (oscillators), and taking in masers, nuclear magnetic resonance, neutron scattering and many other matters on the way. Nobody could hope to make himself a master of all these, and few advanced treatises dealing with one topic think fit to mention the analogies with others. Yet it is both useful and enjoyable to recognize that unities lie behind so many diverse phenomena, and I have written for the reader who shares my pleasure in these things. At the same time, I hope no one will be so high-minded as to take offence if occasionally a page or two are devoted to some rather trivial example; there is room in physics for magpies as well as for more systematic collectors, and I am not ashamed to take delight in things for their own sake as well as for their relationship to other things. Besides, you can always skip the boring bits. The cross-references in the margin are there, among other reasons, to make skipping easy – you will be directed back to the passages you have missed if it turns out later that they are relevant to another discussion.

The purpose of the book, then, is to bridge the gap between the normal undergraduate curriculum and detailed treatments of special topics at the research level, and I have aimed at a level of treatment that will be accessible to a student who has mastered the elementary course text and is hungry for more. The emphasis throughout is on the physical processes, and very little is expected in the way of mathematical skill. Complex variable theory at the level of contour integration is only used on those rare occasions when it significantly shortens an argument. For the most part, elementary algebra and calculus provide all that is needed, but in saying this it is only fair to draw attention to the need to think graphically. A differential equation is a specification, such as even a computer can be made to understand, of how to draw a curve, and I take for granted the reader's willingness to think in these terms. Formal solutions are fine if they exist, but exact mathematics, except (possibly even) in the hands of an expert, is a very limited tool. The scientist who scorns the qualitative condemns himself to the quantitative, and his imagination will remain earthbound. The manipulation of ideas with the help of graphs is part of the technique of imaginative physical thinking, without which exact mathematical analysis can easily degenerate into sterile manipulation.† It is with this in mind that I offer the book to the applied mathematician as well as to the physicist and the engineer – in so far as we are all concerned with the real world we need to make a connection in our minds between it and the abstract processes of analysis.

Consistent with this approach is an emphasis on experiment, and not only by reference to accounts in learned journals; too often these resemble the offerings of a supermarket, the final result hygienically wrapped to discourage uncomfortable thoughts of the original raw material. In this volume, however, we shall meet many problems that can be investigated with simple equipment, and I have described a number of these and presented the untreated results of my own measurements with the aim of encouraging you to carry them out for yourself. There is hardly a single one from which I did not learn something that finds an echo in my account of the phenomenon it illustrates; sometimes, indeed, my first thoughts were proved so mistaken by the experiment that the first draft had to be scrapped. This is not to say that a better physicist or a better mathematician would not have got it right first time, without the aid of experiment; but most of us find it easier to get the answer if we already know what answer we want, and if a quick experiment starts us off on the right lines the time spent on it is time well spent.

It may be helpful to say something about the background knowledge of physics that I have assumed. It is not constant throughout the book, but rather I have tried to keep in mind, as I wrote each chapter, the sort of reader who might wish to understand what I wanted to say. Thus chapter 2 is a revision course for someone who is reasonably familiar with simple harmonic motion, with the aim of deepening understanding and at the same

† Graphical methods must not be confused with the formal processes of Euclidean or cartesian geometry. In a few places I choose to use geometry to demonstrate a particular exact result; others might prefer to do it by algebra, but the choice is merely a matter of taste, not of principle. I find it helpful to represent the manipulations of complex algebra on an Argand diagram, but have no fault to find with those who get on perfectly well without this aid. But anyone who tackles differential equations without sketching solutions handicaps himself severely.

time indicating some of the problems that are not described by linear differential equations; but it assumes no broad knowledge of other branches of physics. By the time chapter 6 is reached I take for granted a grounding in wave-motion and the general character of such special forms as acoustic and electromagnetic waves. And chapter 8 is written in the expectation that only someone with a background of elementary solid-state and quantum physics will wish to come to grips with the rather difficult physical concepts involved in the calculation of the relaxation time in proton resonance. The justification for this approach is that, although I have written in the style of a continuous narrative, the argument is not a closed integrated logical progress. There is indeed an underlying unity, illustrated by many varied examples, but the sense of unity need not be destroyed by selecting a few and neglecting the rest. My earlier advice to skip what seems trivial or dull applies equally to the too-difficult.

As for the subject matter itself, 'vibration' is taken to encompass almost anything that fluctuates in time more or less periodically. I have not included wave-propagation as an independent branch of the subject, except in so far as the vibrations of strings and solid bodies, and the oscillations of electric fields in cavities, can be considered as standing waves; and a few examples bring in the scattering of waves by resonant obstacles. No attempt, however, is made to treat such matters as wave-groups or diffraction except as incidentals to the main study. The only defence I can offer is that one must draw the line somewhere, and to explore wave-propagation in the same spirit which I have brought to vibration would require another volume at least. The size of the present venture is already regrettable, if only because it has led to a separation of classical and quantal treatments, something that happens all too often in the teaching of physics. Almost inevitably it fosters the feeling that quantum mechanics is intrinsically better than classical mechanics, and that classical mechanics is something real physicists ought to grow out of. This is a disputable proposition. Of course, if they give different answers when applied to a given problem one must expect the quantal treatment to be more nearly correct. There are, however, many problems in vibration where they give the same answer, or where the quantal treatment shows the behaviour to be the same as that of an equivalent classical system. It is then usually easier to think about the classical equivalent, especially if it is complicated. For classical physics is the mode of description appropriate to direct observation, and the imaginative faculties develop through the contemplation of the world around. It is easier to guess the answer to a difficult classical problem than to a difficult quantal problem – indeed, the more sophisticated the intellectual framework the harder it is to extend one's understanding beyond those problems that can be solved exactly. When it is a matter of formulating fundamental laws, quantum mechanics must be accepted as having superseded classical mechanics; but, when specific problems are to be solved, quantum mechanics can prove unconscionably intractable. The physicist or engineer concerned with specific problems must therefore develop his classical skills, analytical and imaginative, to the highest degree, but he must also develop a reliable appreciation of how far he dare thrust his classical argument into quantal territory.

Part 1 of this work, which occupies the rest of this volume, is essentially classical, with only an occasional sortie into problems that are normally regarded as quantal but which yield satisfactorily to classical analysis. In the second volume I propose to take up the quantal theme as Part 2, with examples of strictly quantal phenomena and a careful exposition of the safe limits of the classical approach to the same problems. This will leave the field clear for a treatment, in Part 3, of the vibrations of complex systems, including normal mode analysis, by whatever techniques seem appropriate. Finally, I intend to present accounts of a varied collection of maintained oscillations, especially those with a complicated spatial structure such as aerodynamic vibrations and Gunn oscillators.

But this is in the future, and already I have shown an unbecoming personal touch in revealing my aims and aspirations. It is time to disappear from the scene. But though, following custom, *I* adopt the cloak of invisibility and simultaneously cease to acknowledge the existence of *You*, my reader, there will still be found, as *We*, the assumption of collaboration between writer and reader without which a book might as well remain unwritten.

Part 1

The simple classical vibrator

2 The free vibrator

Introduction: the special role of harmonic vibrators

This chapter is mainly concerned with simple vibrators executing periodic, or nearly periodic, motion undisturbed by external influences. In this context the term *simple* refers to systems which can be characterized by a single variable – the angular displacement of a pendulum, the height of a bouncing ball, the charge on a capacitor connected in series with an inductor. The behaviour of such systems may be displayed as a graph of the value of this variable (which we shall call the *displacement*) against time, as in fig. 1 which shows the behaviour of the three examples cited. The first

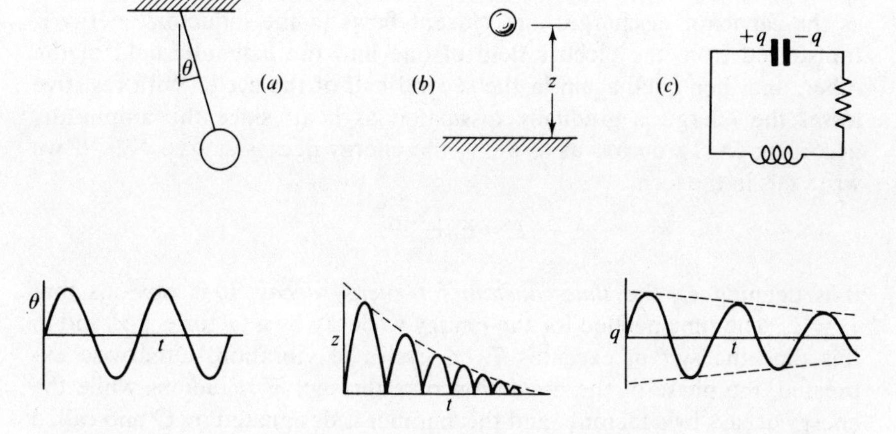

Fig. 1. Three vibrators, (*a*) simple pendulum, (*b*) bouncing ball, (*c*) series resonant circuit, with graphs of the variation of displacement with time.

and third are very similar, the oscillations being simple harmonic, i.e. sinusoidal in form, decaying slowly for a massive pendulum swinging in air, but rather more rapidly for a typical LC circuit at room temperature on account of the resistance of the inductor. The decay of the oscillations in the circuit follows an exponential law, so that the charge on one capacitor plate may be expressed as

$$q = q_0 \cos(\omega' t + \phi)\, e^{-t/\tau_a}, \qquad (1)$$

q_0, ω', ϕ and τ_a being constants. If the *time-constant for amplitude decay*, τ_a, is so large that the decay may be neglected, e^{-t/τ_a} being put equal to unity, the oscillation of q is strictly periodic with period $T_0 = 2\pi/\omega'$, since adding T_0 to t in (1) adds 2π to the argument of the cosine and leaves q unchanged; ω' is the *angular frequency* and is the reciprocal of the time required to change the argument of the cosine by unity. The argument, $(\omega' t + \phi)$, is called the *phase* of the oscillation. The moment from which time is measured may usually be chosen at will, and ϕ, the *phase-constant*, is introduced to indicate what point in the cycle of oscillation is reached at the instant $t = 0$. The *amplitude* of the oscillation is denoted by q_0; during a complete cycle the displacement ranges between $\pm q_0$. If the damping is not negligible, but is exponential as in (1) (this is not a universal form of damping, as later examples will show), the oscillations are not strictly periodic, for it is no longer true that $q(t+T_0) = q(t)$. They now obey the modified periodicity relation, $q(t+T_0) = \alpha q(t)$, where α is a constant equal to e^{-T_0/τ_a} and represents the factor by which the amplitude is changed in one period. Between one maximum deflection and the next, on the opposite side of the mean position, a time $\frac{1}{2}T_0$ elapses, and the ratio of the maxima is $\alpha^{-\frac{1}{2}}$, or e^{Δ} where $\Delta = \frac{1}{2}T_0/\tau_a$; Δ is called the *logarithmic decrement.*

34

At the moment of maximum displacement, when $q = \pm q_0$, no current flows and the energy wholly resides in the capacitor, $E = q_0^2/2C$. If resistive losses are negligible during one cycle the total energy does not change, but, as the capacitor discharges and current flows in the inductor, energy is transferred from the electric field of one into the magnetic field of the other, and then back again in the second half of the cycle. With resistive losses the energy is gradually dissipated as heat; since the amplitude, according to (1), decays as $q_0 e^{-t/\tau_a}$, the energy decays as $q_0^2 e^{-2t/\tau_a}$. If we write this in the form

$$E = E_0 e^{-t/\tau_e},$$

thus defining τ_e, the *time-constant for energy decay*, it is obvious that $\tau_e = \frac{1}{2}\tau_a$. The time needed for the energy to decay by a factor e is τ_e and in this time the system executes T_0/τ_e cycles of vibration. Otherwise expressed, the phase of the oscillation runs through $\omega'\tau_e$ radians while the energy decays by a factor e, and this number is designated by Q and called the *quality factor* or *selectivity*. Systems of high Q vibrate for many cycles before coming to rest.

One more symbol may be introduced at this point, though its utility will not be obvious yet: we define ω'' as $1/\tau_a$. Then, since $\tau_e = 1/2\omega''$,

53

$$Q = \tfrac{1}{2}\omega'/\omega''. \tag{2}$$

The LCR circuit is the archetypal vibrator, serving as a model for the enormous variety of systems whose behaviour may be more or less adequately represented by a single variable oscillating according to (1). Exponentially damped simple harmonic oscillations play so fundamental a role in so many branches of physics that there is a danger of overlooking, or treating as relatively unimportant, systems which exhibit different forms of oscillation, for example a bouncing ball. To be sure, one could hardly contend that the bouncing ball is important in itself, but one does not have to look far to find examples of non-sinusoidal vibrations that deserve to be treated seriously – the vibrations of the reed of a woodwind instrument and the beating of the heart are examples. The reason for the tendency to neglect such systems of which, as we proceed, we shall meet many other instances, is that, unlike harmonic vibrators, they cannot be accommodated in a single formal treatment; almost every anharmonic system demands its own analysis. One therefore gets a much less tangible return for a given amount of mathematics. Moreover the analysis of anharmonic vibrations is almost always more troublesome.

118
253
280
341
351
353

To appreciate the reason for the peculiar position occupied by harmonic vibrators, let us compare the equation governing the LCR circuit with that for the bouncing ball, as elementary a vibratory system as one may hope to find in the whole menagerie of phenomena that refuse to conform to a common calculus. The ball offers a rare case of easy analysis. So long as it is in flight, its height z obeys the differential equation

$$\ddot{z} = -g. \tag{3}$$

Each flight, between one bounce and the next, is described by the general solution of (3):

$$z = v_0(t-t_0) - \tfrac{1}{2}g(t-t_0)^2, \tag{4}$$

in which t_0 is the instant at which it leaves the ground ($z = 0$) with initial velocity v_0. Each loop of the curve in fig. 1(b) is a parabola with a maximum height, z_m, of $\frac{1}{2}v_0^2/g$, and the time of flight between bounces is $2v_0/g$. If Newton's law of impact is obeyed, each bounce reduces v_0 by a factor ε; thus successive values of z_m form a geometrical progression $1:\varepsilon^2:\varepsilon^4$ etc., and the successive 'periods' are in the ratios $1:\varepsilon:\varepsilon^2$ etc. It is easily seen that with a very bouncy ball having ε nearly unity, the envelope of the loops is parabolic. The successive values of z_m would lie on an exponential decay curve if the period remained constant, but the shortening of the period with diminishing amplitude leads to a well-defined terminal point at the bottom of the parabola, a form of behaviour markedly different from exponential decay which in principle goes on for ever.

The apparently elementary formulation of the problem in (3) and the statement of its solution (4) is deceptive, since a wealth of complicated physics is hidden in the rule connecting the value of v_0 in each flight to that which follows, as summed up in the law of impact. One can appreciate this if one attempts to replace (3) by a differential equation which shall govern the whole history as a continuous narrative, and not as a succession of separate episodes. The reader is earnestly advised not to waste any time on so fruitless an exercise; it is enough to recognize that very strange mathematical functions would be required, while by contrast the LCR circuit is governed by a straightforward differential equation possessing, unlike the bouncing ball, the property of *linearity*. This point will be explained presently, but first we must derive the equation for the circuit, starting with the specially easy case where the resistance, R, is small enough to be ignored.

If the capacitor plates hold charges $\pm q$ the potential difference across them is q/C, and this is only supportable by the inductor if the current, i, through it is changing according to the law of induction that $L\,\mathrm{d}i/\mathrm{d}t$ is the magnitude of the induced e.m.f. Since $i=\dot{q}$ the equation for q follows immediately:

$$L\ddot{q}+q/C=0. \qquad (5)$$

Substituting $q=q_0\cos(\omega' t+\phi)$ as a trial solution (this is (1) without the decay), one sees that (5) is satisfied provided that

$$\omega'=(LC)^{-\frac{1}{2}}. \qquad (6)$$

The value of q_0 is irrelevant, and this is where the property of linearity reveals itself with its important consequences. In a linear equation, of which (5) is an example, each term is of such a form that if it is evaluated for trial functions $q(t)$ and $aq(t)$, a being a constant, the results differ only in being scaled by the same multiplier a. Consequently if the equation is satisfied by $q(t)$ it is also satisfied by $aq(t)$. The solutions shown in fig. 1 reflect the difference between a linear and a non-linear system. In the decay of the sinusoidal oscillation of the LCR circuit (which remains a linear system when R is introduced) the magnitude of the oscillations diminishes but their frequency is unchanged – the solution at a later stage is the same as earlier except for a scaling down of the amplitude. On the other hand, the frequency of the bouncing ball increases as its amplitude drops – the time variation is affected along with the magnitude. This is enough to show that if we were to derive a complete equation describing its motion, it would be non-linear. It is worth noting that the converse cannot be assumed without further thought; we shall meet cases later in the chapter, and elsewhere, of systems whose period is independent of amplitude but which are nevertheless non-linear. **15**

Equation (5) is a particular case of a linear equation, containing two different operators: the operation of multiplication by a constant, such as converts q to q/C, and that of double differentiation, such as converts q to $\ddot{q}$. There are many other linear operators, of which examples will be given later. Those in (5) belong to a special class of particular concern to the present argument, with two cardinal properties – *superposition* and *translational invariance*. Equations of motion possessing these properties generate harmonically oscillatory solutions, so that many different physical systems, governed by a wide variety of equations, have certain common features in their behaviour. It is this that gives such general application to the theory of harmonic vibrations, in contrast to systems governed by non-linear equations, where the solutions bear little family resemblance to each other. **44**

Superposition. If $\mathscr{L}(q)$ is a linear operator and $q_1(t)$ and $q_2(t)$ are sufficiently regular functions that $\mathscr{L}(q_1)$ and $\mathscr{L}(q_2)$ are meaningful,

$$\mathscr{L}(aq_1+bq_2)=a\mathscr{L}(q_1)+b\mathscr{L}(q_2), \tag{7}$$

a and b being constants. For example the operator $(L\mathrm{d}^2/\mathrm{d}t^2+1/C)$ in (5) possesses the property of superposition expressed by (7). If $\mathscr{L}$ is used to designate this operator, (5) may be written in the form $\mathscr{L}(q)=0$. It is obvious that if $q_1(t)$ and $q_2(t)$ are solutions, so also is $aq_1(t)+bq_2(t)$. Because of this all possible solutions can be written in terms of a limited number of special solutions, two in the case of second-order equations like (5). All that is needed is that q_1 and q_2 shall not be multiples of the same function, and then any solution may be generated by giving appropriate values to a and b. Nothing like this holds for a non-linear equation. A second order non-linear equation has as its general solution a functional form containing two arbitrary constants, but different choices of these constants normally generate a family of curves of greater variety than anything that can be resolved into the superposition of two basic forms.

Translational invariance. Any solution $q(t)$ can be shifted bodily along the time axis, becoming $q(t+t_0)$, and still remain a solution whatever the value of t_0. This property is shared by the bouncing ball and the LCR circuit, and by any differential equation in which the coefficients do not involve the independent variable (t in this case). It is an obvious physical necessity in a circuit whose components do not change with time, or in a mechanical system where the forces between constituent parts depend on where they are and what they are doing, but not on the exact instant in time at which the configuration occurs. On the other hand, a circuit whose parameters are being altered (e.g. parametric amplifiers), or a simple pendulum whose string is steadily shortened, may be a linear system yet not possessed of translational invariance in time. The solutions are additive but their character changes with time; the period of the simple pendulum, for example, does not stay constant. Each such system demands special treatment, but there is such a wealth of interest in invariant systems that we must defer this problem till chapter 10. **285**

Granted the invariance, it is still necessary to indicate why linear systems are to be singled out, and the reason is not far to seek. In equilibrium the potential energy, V, of a body is at a minimum with respect to small displacements, and the usual form of V will be a quadratic function of displacement. This is not to say that other variations are excluded – a ball resting on a horizontal table has V independent of position, the bouncing ball has V a linear function of height, z, so long as z is positive, but something much stronger for negative z. The norm for V, however, is a smooth function that can be expanded as a power series about its minimum; the linear terms must vanish but only rare circumstances cause the quadratic terms to be absent. So long as the motion to be studied is confined to the range of displacement within which the quadratic expansion is adequate, the total energy, E, of the body may be expressed as a quadratic function of position and velocity. Thus, for motion in one dimension

$$E = T + V, \tag{8}$$

where $T = \frac{1}{2}m\dot{x}^2$, the kinetic energy, and $V = \frac{1}{2}\mu x^2$, the potential energy. For an isolated body E is invariant, so that on differentiating (8) with respect to x we find, since $\mathrm{d}\dot{x}/\mathrm{d}x = \ddot{x}/\dot{x}$,

$$m\ddot{x} + \mu x = 0, \tag{9}$$

a linear equation of the same form as (5). All that this argument says is that if V is a quadratic function of displacement, the force $-\partial V/\partial x$ is a linear function. It is convenient, however, to express this fact in terms such as (8), the quadratic dependence of both kinetic and potential energies on velocity (more generally, momentum) and position coordinates, since a very general development of the theory of linear vibrators, involving many coordinates, may start from here. One can see, for example, how electromagnetic oscillations fit into the common framework; in the LC circuit the electric field $\mathscr{E}$ is proportional to q, the magnetic field $\mathscr{B}$ to $\dot{q}$, and the total energy which has contributions from $\mathscr{E}^2$ and from $\mathscr{B}^2$ takes the form of (8), with electrical energy playing the part of V, magnetic energy the part of T. And this argument does not depend in any marked degree on the possibility of assigning different energy terms to specific circuit components. Any field configuration in free space has energy quadratic in $\mathscr{E}$ and $\mathscr{B}$, and it is therefore readily understandable that electromagnetic oscillations should commonly conform to the harmonic pattern. So long as the fields are not so powerful as to distort the material structures that determine their configuration, or to cause dielectric materials etc. to respond non-linearly, the quadratic dependence is very accurately followed; indeed, among the most perfect examples of harmonic motion electromagnetic oscillations occupy a high place. So important is electromagnetism in the structure of the physical world that this alone would justify paying the greatest attention to harmonic oscillations, even if they were not equally fundamental to small-amplitude mechanical and acoustical vibrations, as well as providing an essential basis for describing transitions of quantized systems.

372

510

Anharmonic simple vibrators

There will be time enough to develop a uniform treatment of linear systems, starting with the next chapter, and we may briefly luxuriate in the richer world of non-linear vibrators. Consider a particle moving along a straight line, without dissipation, under the influence of a position-dependent force, so that its equation of motion takes the form:

$$m\ddot{x} = F(x). \tag{10}$$

Obviously (5) is a special case of this equation. There is no essential difficulty about step-by-step integration of (10). If the time scale is divided into equal short steps of length θ, and the value of x at the nth stage denoted by x_n, then $(x_{n+1}-2x_n+x_{n-1})/\theta^2$ is the first approximation to the value of $\ddot{x}$ at the nth stage. We therefore write as an approximation

$$m(x_{n+1}-2x_n+x_{n-1})/\theta^2=F(x_n),$$

i.e.

$$x_{n+1}=2x_n-x_{n-1}+\theta^2F(x_n)/m.$$

This equation allows x_{n+1} to be computed from the two preceding values, x_n and x_{n-1}; once the process has been started by assuming two neighbouring values, x_0 and x_1, say, it can be continued indefinitely. By choosing a small enough interval, θ, any desired accuracy may be achieved at the cost of computing time.

If the form of $F(x)$ is such as to constrain the particle within certain limits, it is likely to return ultimately to x_0, and the cycle will then repeat itself. In making this assertion we assume that it returns to x_0 with the same velocity, $\dot{x}$, as it had originally; this, of course, is necessary to conserve energy. Since V is determined by x alone, so too must T and $|\dot{x}|$ be unique functions of x once the total energy has been fixed. The cycle, therefore, repeats itself exactly. In addition each half-cycle, from one limiting point to the other, is a mirror image of the preceding half-cycle, with the velocity at each point reversed in sign. This is an elementary consequence of the time-reversal symmetry of (10) – if t is replaced by $-t$ the equation is unaltered, for only t^2 appears in it. A record of the solution, $x(t)$, can be read equally well forwards or backwards, and is indeed, because of the symmetry of the half-cycles, the same curve either way. Fig. 2(a) is a possible oscillation obeying an equation of the form (10) – fig. 2(b) is not, and when we find such behaviour in a periodic system we must look for breakdown of time-reversal symmetry in the system, usually due to instabilities or other irreversible phenomena (but see also chapter 8).

41
209

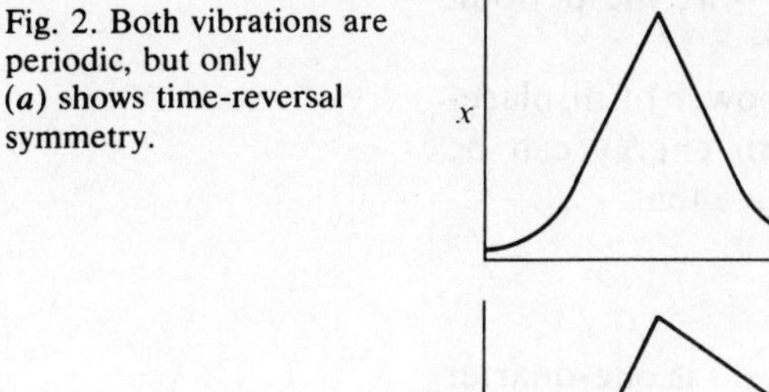

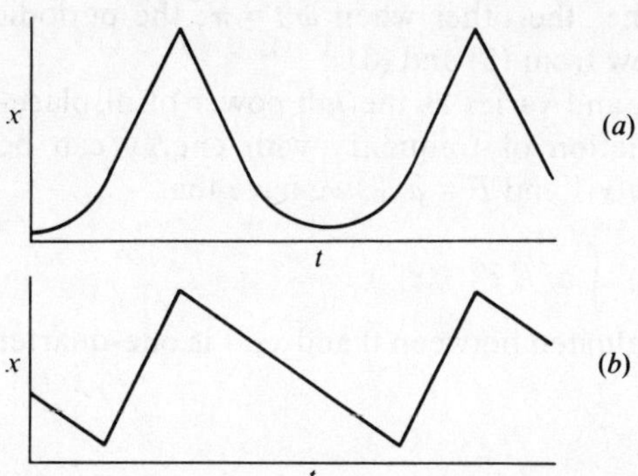

Fig. 2. Both vibrations are periodic, but only (a) shows time-reversal symmetry.

The energy equation (8) often provides a better starting point for integrating the equation of motion, being in fact the first integral of (10). Instead of step-by-step integration the formal solution may be written as an indefinite integral which is then evaluated, by numerical quadrature if necessary. For, since $T=\frac{1}{2}m\dot{x}^2$, (8) may be written in the form

$$\dot{x}=[2(E-V)/m]^{\frac{1}{2}},$$

from which it follows that

$$t=\int\frac{\mathrm{d}x}{[2(E-V)/m]^{\frac{1}{2}}}. \tag{11}$$

Two features of the solution are obvious from (11) – once the total energy E is fixed the form of the solution is also fixed, and because an arbitrary integration constant may be added the solution is invariant with respect to shifts in time.

A particle vibrating in a potential well is confined to a region in which $E-V$ is everywhere positive or zero, the limits of its motion being the points where $E=V$. If the integration is started from one of these limits, $t=0$ at the outset and the value of the integral when the other limit is reached is half the period. Because the integrand is infinite at both limits, a little care is needed, but, this aside, (11) provides a clear procedure for determining the period, as may be illustrated by performing the integral for one of the few cases that is analytically straightforward, the linear vibrator described by (9). Here the force $F(x)$ is $-\mu x$, the minus sign indicating an attractive force towards a point chosen as the origin of x. The potential energy is $\frac{1}{2}\mu x^2$, and

$$t=\left(\frac{m}{\mu}\right)^{\frac{1}{2}}\int\frac{\mathrm{d}x}{(x_0^2-x^2)^{\frac{1}{2}}}=\left(\frac{m}{\mu}\right)^{\frac{1}{2}}\cos^{-1}(x/x_0)+t_0,$$

where x_0 is $2E/\mu$ and t_0 is constant. With ω' written for $(\mu/m)^{\frac{1}{2}}$ the general solution takes the form,

$$x=x_0\cos[\omega'(t-t_0)].$$

If t_0 is chosen to be either 0 or π/ω', so as to give x one of its limiting values, $\pm x_0$, when $t=0$, it reaches the other when $\omega' t=\pi$; the periodic time is $2\pi/\omega'$, as we already know from (5) and (6).

If the potential is symmetrical and varies as the nth power of displacement, so that $V\propto|x|^n$, the variation of frequency with energy can be derived from (11). Writing $V=a|x|^n$ and $E=ax_0^n$, we have that

$$t=(\tfrac{1}{2}m/a)^{\frac{1}{2}}\int\mathrm{d}x/(x_0^n-|x|^n)^{\frac{1}{2}}.$$

In particular, if the integral is evaluated between 0 and x_0, t is one-quarter of the period, so that

$$T_0=(8m/a)^{\frac{1}{2}}\int_0^{x_0}\mathrm{d}x/(x_0^n-x^n)^{\frac{1}{2}}$$

$$=(8m/a)^{\frac{1}{2}}x_0^{1-\frac{1}{2}n}\int_0^1\mathrm{d}\xi/(1-\xi^n)^{\frac{1}{2}},\quad \text{with } \xi\equiv x/x_0.$$

The integral is a function of n only, not x_0, so that for any given value of n we have that

$$T_0 \propto x_0^{1-\frac{1}{2}n} \propto E^{1/n-\frac{1}{2}}.$$

The only power law that makes T_0 independent of E is the parabolic potential, $n=2$. The bouncing ball has $n=1$ and $T_0 \propto E^{\frac{1}{2}}$. At the other extreme, as $n \to \infty$, T_0 tends to become proportional to $E^{-\frac{1}{2}}$; this is the case of a ball moving on a flat surface and bouncing back and forth between vertical walls – the frequency is proportional to the speed.

The determination of the period may be represented graphically, as in fig. 3. If a horizontal line at the energy E is drawn across the graph of $V(x)$, cutting it at the limiting points of the vibration, the vertical distance between the lines is $E-V$, and $(E-V)^{-\frac{1}{2}}$ is readily constructed from this.

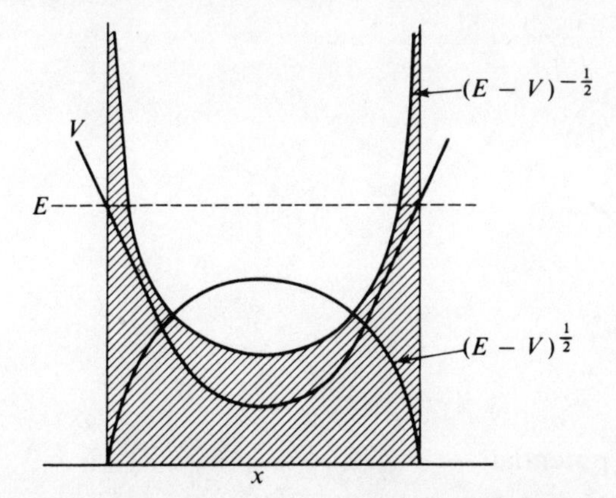

Fig. 3. Graphical construction for the period.

The area under the curve, shown shaded, then represents the period (apart from constants). With a parabolic potential an increase of E increases the width of the shaded area, but decreases its height so as to keep the area constant; the vibrator is *isochronous*, with frequency independent of energy. There is a continuous range of forms for $V(x)$ which lead to isochronous vibrations, for any potential curve $V(z)$ may be 'sheared' without altering the frequency.[7] In fig. 4, the curve B is a sheared version

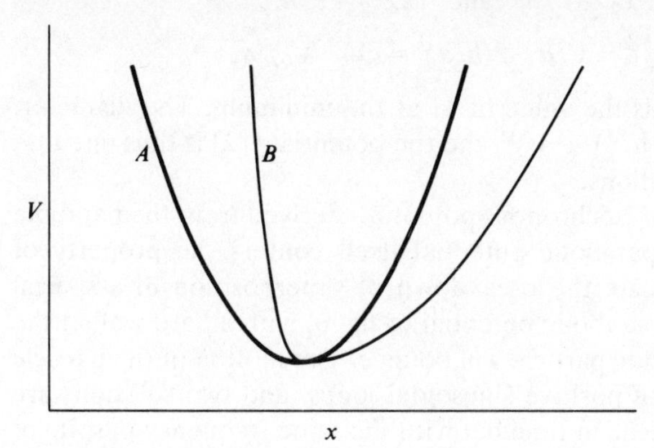

Fig. 4. B is a sheared version of A; both potentials yield the same variation of frequency with energy.

of A, both having the same horizontal diameters at the same value of V. The horizontal shift is different for different V, and may be chosen to vary arbitrarily provided the resulting $V(x)$ is single-valued. Both these potential curves yield the same variation of period T_0 with E, for on carrying through the graphical construction of fig. 3 the curve for $(E-V)^{-\frac{1}{2}}$ is itself only sheared, leaving the area under it unchanged. It is clear from this that a specification of how the frequency varies with energy is not enough to fix the form of the potential. In particular, if ω' is independent of E the potential need not be parabolic; any curve derived from a parabola by shearing is a potential in which vibrations are isochronous. The essential property, apart from single-valuedness, is that the diameter must vary as $V^{\frac{1}{2}}$ when V is measured from the lowest point.

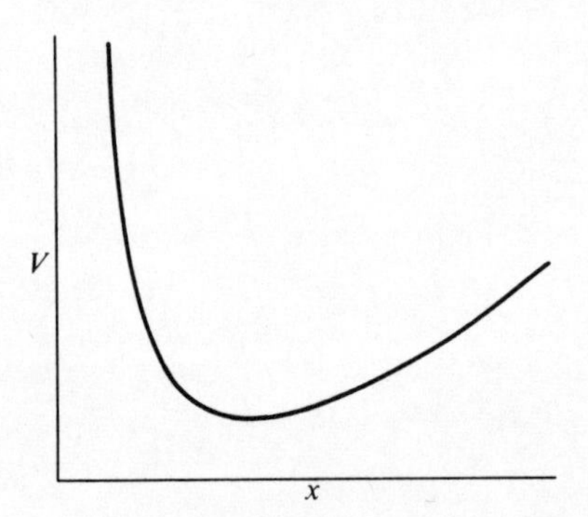

Fig. 5. The isochronous potential ax^2+b/x^2.

As an example consider the potential,

$$V=ax^2+b/x^2, \tag{12}$$

29

shown in fig. 5. To find the separation of the two positive values of x at which V takes a given value, rearrange (12) as a quadratic equation in x^2:

$$ax^4-Vx^2+b=0,$$

with solutions $\pm x_1$ and $\pm x_2$. The equation shows directly that

$$x_1^2+x_2^2=V/a \quad \text{and} \quad x_1^2x_2^2=b/a.$$

Hence $$(x_1-x_2)^2=V/a-2(b/a)^{\frac{1}{2}}=(V-V_0)/a,$$

where $V_0=2(ba)^{\frac{1}{2}}$ and is the value of V at the minimum. The diameter, $|x_1-x_2|$, is proportional to $(V-V_0)^{\frac{1}{2}}$ and the potential (12) is thus one that yields isochronous vibrations.

Of the whole family of isochronous potentials derived from the parabola by shearing, only the parabolic potential itself confers the property of superposition. To illustrate the breakdown of superposition in a special case, consider the half-parabolic potential of fig. 6, with a hard wall at the left, off which the vibrating particle will bounce. The motion of the particle consists of a sequence of positive sinusoidal loops, and two solutions are shown in fig. 7(a) displaced in time but with the same frequency in spite of

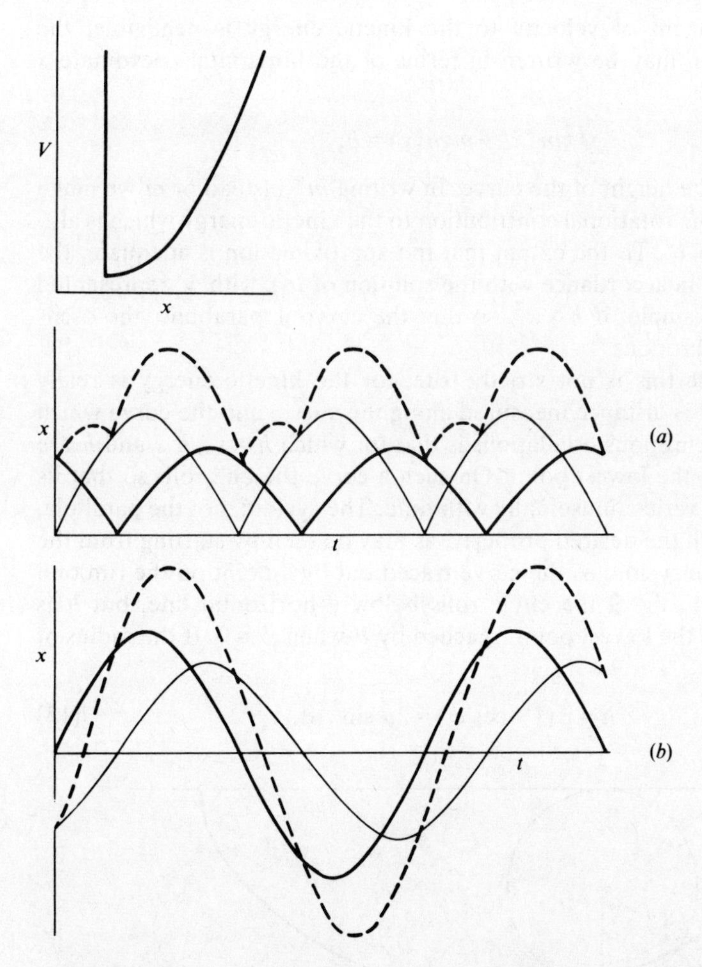

Fig. 6. An isochronous potential generating vibrations that are not superposible.

Fig. 7. (*a*) Failure of superposition with the potential of fig. 6; (*b*) success of superposition with a parabolic potential. In both diagrams the broken curve is the sum of the two continuous curves.

having different amplitudes. The sum of the two solutions, with its abrupt changes of slope, is obviously no solution in its own right. By contrast, the fully parabolic potential yields solutions which are strictly sinusoidal as well as isochronous, as in fig. 7(*b*). Since any two sinusoids of the same period add to give another sinusoid of the same period, the sum is also a true solution of the equation of motion, and superposition applies.

A convenient way to demonstrate vibrations in a variety of potentials is to roll a steel ball on a curve of the desired form, as in fig. 8; by the use of a groove or of rails the ball may be kept on the curve without excessive friction. So long as the curve is very shallow, and the contribution of the

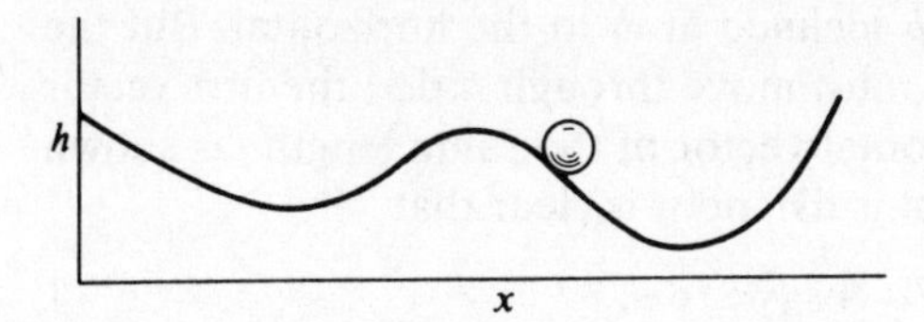

Fig. 8. Ball rolling on a curve.

vertical component of velocity to the kinetic energy is negligible, the energy equation may be written in terms of the horizontal coordinate x alone:

$$\tfrac{1}{2}m^*\dot{x}^2 + mgh(x) = E,$$

where $h(x)$ is the height of the curve. In writing m^* instead of m we make allowance for the rotational contribution to the kinetic energy which is also proportional to $\dot{x}^2$. To the extent that the approximation is adequate, the model behaves in accordance with the solution of (8), with V represented by mgh; for example, if $h \propto x^2$, so that the curve is parabolic, the oscillations are isochronous.

But of course this is not strictly true, for the kinetic energy is really $\tfrac{1}{2}m^*\dot{s}^2$, where s is distance measured along the curve, and the curve which yields truly isochronous oscillations is that for which $h \propto s^2$, if s and h are measured from the lowest point. On such a curve the ball rolls so that its displacement s varies sinusoidally with time. The cycloid, not the parabola, is the curve with the desired property, as may be seen by starting from the definition of the cycloid as the curve traced out by a point on the rim of a rolling wheel. In fig. 9 the circle rolls below a horizontal line, but h is measured from the lowest point, reached by P when $\phi = 0$. If the radius of the circle is a,

$$h = a(1 - \cos\phi) = 2a\sin^2\tfrac{1}{2}\phi. \tag{13}$$

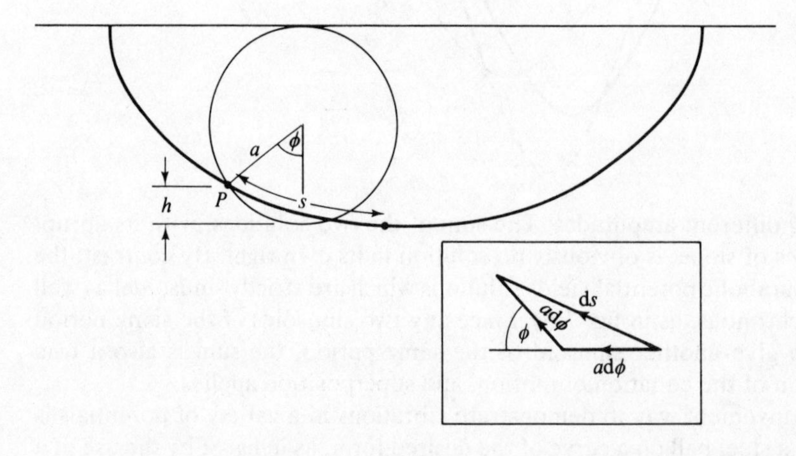

Fig. 9. Geometry of a cycloid.

To find how s depends on ϕ, consider what happens to the point P when the circle rolls through $d\phi$. If its centre remained fixed, the movement of P would be a vector of length $a\,d\phi$ inclined at ϕ to the horizontal. But the circle does not slip and its centre must move through $a\,d\phi$; the first vector therefore has added to it a horizontal vector of the same length, as shown in the inset to fig. 9. The resultant is ds and it is clear that

$$ds = 2a\cos(\tfrac{1}{2}\phi)\,d\phi,$$

so that, on integrating,

$$s = 4a \sin(\tfrac{1}{2}\phi) = (8ah)^{\frac{1}{2}} \quad \text{from (13).}$$

The relation between s and h is what is required for isochronous oscillation. This property of the cycloid was discovered by Huygens[1] who also showed how a simple pendulum could be constrained by cycloidal cheeks so that its bob described a similar cycloid (fig. 10). The result is amusing and elegant, rather than useful, for an unconstrained pendulum with a small amplitude of oscillation is very nearly isochronous, and it is easier to keep the amplitude constant than to devise practicable tricks to render variations of amplitude harmless.

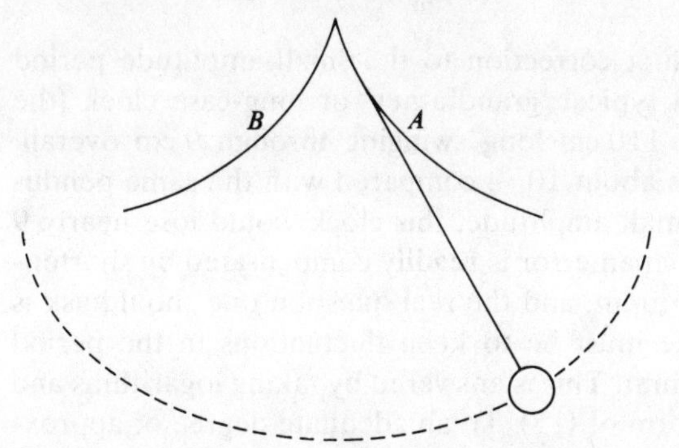

Fig. 10. Huygens' cycloidal pendulum. With the string constrained by the cheeks, *A* and *B*, the bob executes the cyloidal path shown as a broken line.

To expand on this last point let us calculate how frequency depends on amplitude for a simple pendulum whose bob moves in a circle. The displacement is described by the angle θ between the string and the vertical, and the energy equation takes the form

$$\tfrac{1}{2}ml^2\dot{\theta}^2 + mgl(1-\cos\theta) = E,$$

which may be rearranged in the form

$$\dot{y}^2 = \omega_0^2(\sin^2 y_0 - \sin^2 y),$$

in which $y = \frac{1}{2}\theta$, $\omega_0^2 = g/l$ and E is written as $2(g/l)\sin^2 y_0$; y_0 measures the amplitude of swing, the pendulum being at rest, with $\dot{y} = 0$, when $y = \pm y_0$. From this the period follows, by the same argument as yielded (11), and with the symmetry of V used to divide the cycle into four equal parts:

$$T_0 = \frac{4}{\omega_0}\int_0^{y_0} \frac{\mathrm{d}y}{(\sin^2 y_0 - \sin^2 y)^{\frac{1}{2}}}. \tag{14}$$

The integrand is infinite at the upper limit, but this awkwardness can be removed by a change of variables. Put $\sin\phi = \sin y/\sin y_0$, so that

$$\sin y_0 \cos\phi = (\sin^2 y_0 - \sin^2 y)^{\frac{1}{2}};$$

also $$\sin y_0 \cos\phi \, \mathrm{d}\phi = \cos y \, \mathrm{d}y,$$

so that $\mathrm{d}y/(\sin^2 y_0 - \sin^2 y)^{\frac{1}{2}} = \mathrm{d}\phi/\cos y$. As y ranges from 0 to y_0, ϕ ranges from 0 to $\frac{1}{2}\pi$, and (14) takes the form

$$T_0 = \frac{4}{\omega_0}\int_0^{\frac{1}{2}\pi} \frac{\mathrm{d}\phi}{(1-\sin^2 y_0 \sin^2\phi)^{\frac{1}{2}}},$$

with an integrand that is finite and expandable in series:

$$T_0 = \frac{4}{\omega_0}\int_0^{\frac{1}{2}\pi} (1+\tfrac{1}{2}\sin^2 y_0 \sin^2\phi + \cdots)\,\mathrm{d}\phi$$

$$= \frac{2\pi}{\omega_0}(1+\tfrac{1}{4}\sin^2 y_0 + \cdots) = \frac{2\pi}{\omega_0}(1+\tfrac{1}{4}y_0^2+\cdots). \qquad (15)$$

The second term is the first correction to the small-amplitude period $2\pi/\omega_0$, i.e. $2\pi(l/g)^{\frac{1}{2}}$. For a typical 'grandfather' or long-case clock (the author's) with a pendulum 110 cm long swinging through 9 cm overall, $y_0 = \frac{1}{4}\times\frac{9}{110} \doteqdot 0.02$, and $\frac{1}{4}y_0^2$ is about 10^{-4}; compared with the same pendulum swinging with a very small amplitude, this clock would lose nearly 9 seconds a day. Of course such an error is readily compensated by shortening the pendulum by about $\frac{1}{4}$ mm, and the real question one should ask is how constant the amplitude must be to keep fluctuations in the period below a certain prescribed limit. This is answered by taking logarithms and differentiating the second form of (15). To an adequate degree of approximation,

$$\delta T_0/T_0 = \tfrac{1}{2}y_0\delta y_0 = \tfrac{1}{2}y_0^2(\delta y_0/y_0).$$

For the particular clock just mentioned, $\delta T_0/T_0 = 2\times 10^{-4}\,(\delta y_0/y_0)$; time-keeping to better than 1 second a day means that the amplitude must not vary by more than 8% overall. This is not too stringent a requirement and if most domestic pendulum clocks are not nearly as regular as this it is probably for quite other reasons. Astronomical pendulum clocks have been made to keep time to something better than 1 second a year.[2] Apart from anything else in the design (and these instruments are marvels of craftsmanship) the amplitude of the swing must not only be small but very well stabilized for such precision to be attained.

To go from this nearly harmonic system to the opposite extreme, consider the potential curve with two minima in fig. 11. So long as $E < E_1$, the vibrations are confined to one or other of the minima, being nearly harmonic when the amplitude is very small but being increasingly affected by the maximum at E_1 as E approaches this value. At higher energies the oscillations become symmetrical, with fast sections of the cycle as the body passes over the minima and slow sections both at the ends and in the middle. Just at the critical point, defined by $E = E_1$, the frequency drops to zero, since it takes an infinite time for the body to cross the peak when its velocity momentarily vanishes. To see how the frequency varies in this energy range consider E to be marginally greater than E_1, so that the

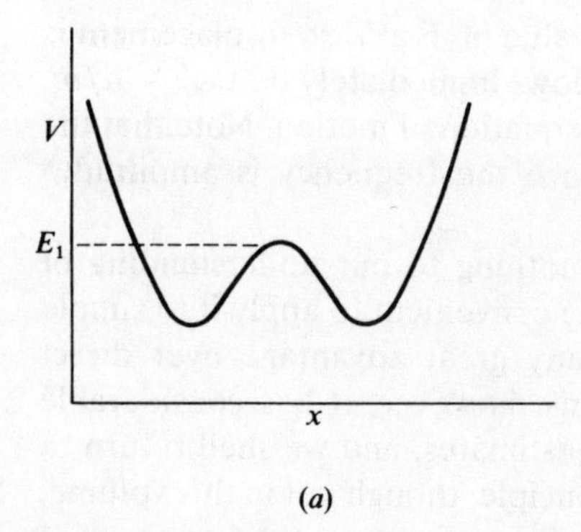

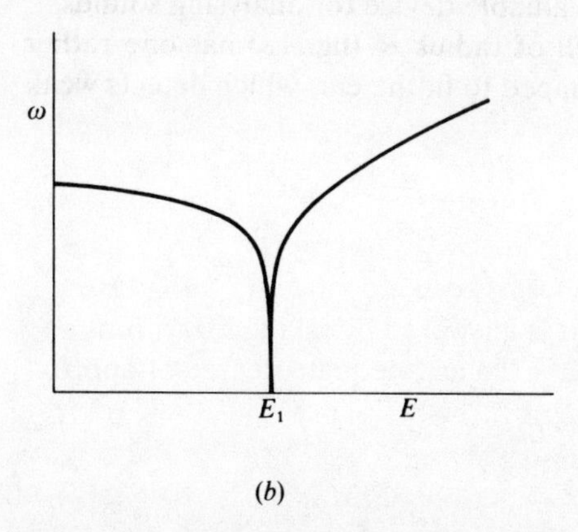

Fig. 11. (*a*) Potential curve with two minima; (*b*) schematic variation of frequency with energy in this potential.

integral (11) takes the form

$$t \sim \int_0^x \frac{dx}{(x^2+\varepsilon)^{\frac{1}{2}}},$$

with x^2 representing the form of the peak when x is small, and ε the excess energy $E-E_1$. The time taken to travel from the centre to a certain small displacement x_1 is therefore proportional to $\sinh^{-1}(x_1/\varepsilon^{\frac{1}{2}})$, which tends to $\frac{1}{2}\ln(4x_1^2/\varepsilon)$ as ε tends to zero. The time spent near the peak thus ultimately controls the period, and the frequency tends to zero as $(-\ln\varepsilon)^{-1}$. This is an extremely precipitate fall; the same behaviour is found as E tends to E_1 from below.

Using energy conservation to calculate vibration frequency

We return now to harmonic vibrators which are at rest, with their energy wholly potential, when their displacement is greatest, and are moving at their greatest speed, with their energy wholly kinetic, when their displacement is zero. If T and V are given by (8), and the displacement is represented by $a\cos\omega' t$, the velocity is $-a\omega'\sin\omega' t$. With a maximum displacement of a and a maximum velocity of $a\omega'$, the value of V at maximum displacement is $\frac{1}{2}\mu a^2$, and the value of T at zero displacement is $\frac{1}{2}m(a\omega')^2$. Since these must be equal it follows immediately that $\omega'^2 = \mu/m$, as already found by direct solution of the equation of motion. Note that the amplitude, a, drops out, as expected since the frequency is amplitude-independent.

This method of determining ω' adds nothing to our understanding of what is happening, and although it may be convenient to apply it to simple vibrators it cannot be said to possess any great advantage over direct methods. With more complicated systems, however, it has considerable merit, especially in making approximate estimates, and we shall return to its application in discussing Rayleigh's principle, though not in this volume. Meanwhile let us illustrate its use by calculating the resonant frequency of a Helmholtz resonator, formerly a valuable device for analysing sounds.[3] A more-or-less spherical metal shell of radius R (fig. 12) has one rather large hole A and a small outlet B shaped to fit the ear, which detects weak

See note on p. 43.

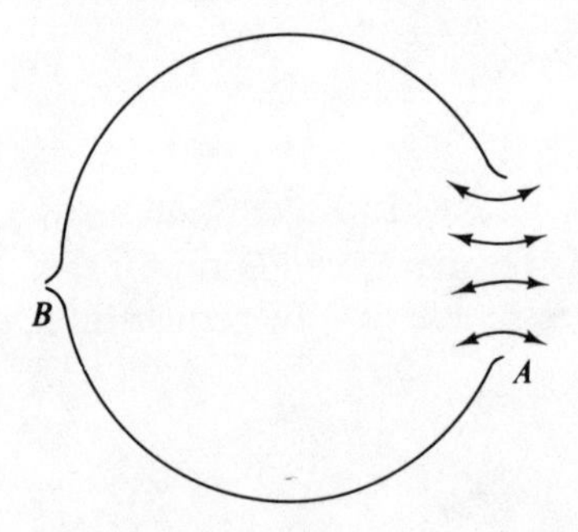

Fig. 12. Section of Helmholtz resonator.

vibrations of pressure within the shell. When exposed to a sound of the right frequency the air within oscillates, flowing to and fro through A as indicated by the arrows. The resonant frequency is essentially determined by the volume of the shell and the area of the hole.† At the maximum inward displacement of the air everything within the shell is uniformly compressed, and the energy is wholly potential; similarly at the maximum outward displacement, when the air within is expanded below atmospheric pressure, which also increases its potential energy. Midway between these two phases the air is at atmospheric pressure, but that part lying close to A is flowing at its greatest speed and carries the energy of vibration in kinetic form. Just how much the kinetic energy is when the mean speed across the aperture is $\dot{z}$ demands some knowledge of the flow pattern, a matter we return to in a moment. For the present, however, it is enough to recognize that the flow speed drops off rapidly away from the aperture, so that the kinetic energy is the same as would be possessed by the mass of gas contained in a fairly short cylinder whose radius r is defined by the hole A,

† That humble musical instrument, the ocarina, is essentially a Helmholtz resonator with a variety of holes to allow different notes to be sounded, and a mouthpiece like that of the recorder to excite it into vibration. The following theory shows why the note rises in pitch as more holes are opened.

and whose length l is as yet undetermined. If this cylinder be imagined vibrating harmonically with small amplitude, a ($\ll l$), its maximum speed is $a\omega'$ and its kinetic energy takes the form:

$$T = \tfrac{1}{2}\pi r^2 l\rho a^2 \omega'^2, \tag{16}$$

where ρ is the density of the gas.

To calculate the potential energy at the moment of maximum inward displacement, we note that the volume of the gas in the shell, originally $\frac{4}{3}\pi R^3$, has been reduced by $\pi r^2 a$ and has therefore suffered a fractional change $\frac{3}{4}r^2 a/R^3$. If the adiabatic bulk modulus of the gas is K (its actual value γP need not be used), the excess pressure within the shell at this instant is $\frac{3}{4}Kr^2 a/R^3$, and the work that would be needed to create this strain by slowly forcing the gas in is one-half the product of the pressure and the volume of gas moved. Hence, when the energy is wholly potential,

$$V = \tfrac{1}{2} \times (\tfrac{3}{4}Kr^2 a/R^3) \times \pi r^2 a. \tag{17}$$

Equating (16) and (17) yields the frequency:

$$\omega'^2 = \tfrac{3}{4}Kr^2/\rho l R^3 = \tfrac{3}{4}c_s^2 r^2/lR^3,$$

where c_s is the velocity of sound, $(K/\rho)^{\frac{1}{2}}$. Now the wavelength, λ, in open air of a sound of frequency ω' is $2\pi c_s/\omega'$. The resonance condition for the Helmholtz resonator can therefore be expressed in purely geometrical terms by the formula

$$\lambda = \frac{4\pi(lR^3)^{\frac{1}{2}}}{\sqrt{3r}}. \tag{18}$$

One of a 100-year-old set of resonators in the Cavendish Laboratory museum was confirmed with a signal generator and loudspeaker as resonating at its stated pitch (g below middle C, with a wavelength of 1.77 m); it had $R = 9$ cm and $r = 1.9$ cm, and (18) is satisfied if $l = 3$ cm, a very reasonable-looking value for a hole 3.8 cm in diameter. A good theoretical estimate is easy if R is large enough for the flow through the hole to follow the pattern of flow through a hole in a plane sheet, for this is the same as the electric field radiating from a charged metal disc of radius r, a standard problem in electrostatics. By following through the analogy between field lines and flow lines one can see that $\frac{1}{2}l$ is the spacing of plates in an idealized capacitor (see fig. 13) that would have the same capacitance, $8\varepsilon_0 r$, as the isolated disc. In this way l is found to be $\frac{1}{2}\pi r$, which is very close to 3 cm in the present example, so that theory and experiment correspond satisfactorily.

It may be noted that determination of vibration frequency by the energy argument is not in principle confined to harmonic vibrations, since in any force field the vibrating object carries its energy alternately in potential and kinetic form. But in practice the relationship between maximum deflection, maximum speed and frequency ($v_{max} = \omega' a$ for harmonic vibrations) is not known without a full solution of the equation of motion, and the technique

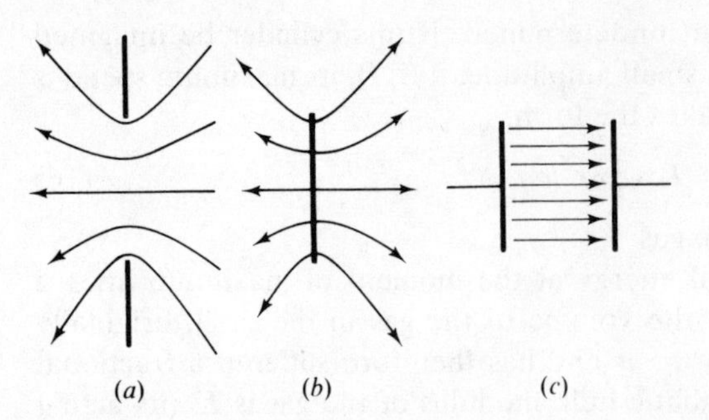

Fig. 13. (*a*) Streamline flow of air through circular hole in a flate plate; (*b*) electric field lines around a charged plate the same size as the hole in (*a*); (*c*) an idealized parallel plate capacitor (no fringing field) having the same capacitance as the capacitance to earth of (*b*).

is therefore virtually useless except in this one case. To illustrate the danger of applying the method naively, consider a simple pendulum swinging up to a horizontal position, so that at the moment of rest its potential energy is mgl. If the bob were to move strictly harmonically along its arc, one would write the amplitude as $\frac{1}{2}\pi l$ and the kinetic energy as it passes its lowest point as $\frac{1}{8}m\omega'^2\pi^2l^2$. Equating these two energies, and remembering that the small-amplitude frequency, ω_0 is $(g/l)^{\frac{1}{2}}$, one estimates the period to be increased by a factor $\frac{1}{2}\pi/\sqrt{2}$, i.e. by 11%, though the true answer is 18%.

Systems with more than one mode

The systems considered so far have been simple in the special sense of having only one significant variable, and simple in the colloquial sense also. There are many others which at first sight are of a complexity that would make one expect their description also to be complex, yet which on investigation turn out to be reducible to simple systems. Take, for example, the case of two identical pendulums hung from a slack line, as in fig. 14.

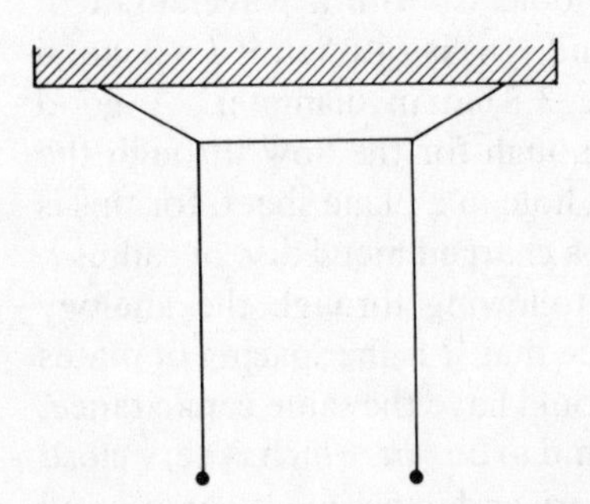

Fig. 14. Two pendulums coupled by hanging from a slack string.

When one is displaced in a direction normal to the paper and set swinging, after a few swings the other joins in, being excited by the movement of the line supporting it, and the first pendulum begins to lose amplitude; there comes a time when the first is at rest and only the second swinging, and then the process repeats in reverse until all the energy is back where it started – and so on until the vibrations are damped out. Clearly this is not described by a single variable executing sinusoidal vibrations of constant amplitude. But now displace both pendulums by the same amount in the same direction, and they will be found to vibrate together without any exchange of energy; and the same thing happens when they are displaced equally in opposite directions, though the frequency is slightly higher now that they are in antiphase than when they were in phase. The system is capable of executing simple harmonic vibrations in two different modes, and the peculiar exchange of energy first described is a consequence of exciting both simultaneously.

365 This is a phenomenon to which we shall devote much attention later, and it is enough to note, to clarify the issue in a preliminary way, that if x_1 and x_2 are the displacements of the two pendulums, (x_1+x_2) and (x_1-x_2) both behave like the coordinate of a simple vibrator and can execute only harmonic oscillations:

$$(x_1+x_2)=A_1\cos(\omega_1 t+\phi_1), \tag{19}$$

$$(x_1-x_2)=A_2\cos(\omega_2 t+\phi_2). \tag{20}$$

These can coexist independently, and any motion of the system can be represented by a superposition of the two modes with arbitrary choice of the constants $A_{1,2}$ and $\phi_{1,2}$ (always subject to the amplitude being small enough for the linear approximation to hold). Thus the two *normal modes* which undergo harmonic motion are the symmetric mode, in which $x_2=x_1$ and A_2 but not A_1 vanishes, and the antisymmetric mode, in which $x_2=-x_1$ and A_1 but not A_2 vanishes. If both modes are excited equally, as when $A_1=A_2=1$, and if $\phi_1=\phi_2=0$, adding and subtracting (19) and (20) show that

$$x_1=\tfrac{1}{2}(\cos\omega_1 t+\cos\omega_2 t)=\cos\bar{\omega}t\cos\delta t$$

and

$$x_2=\tfrac{1}{2}(\cos\omega_1 t-\cos\omega_2 t)=\sin\bar{\omega}t\sin\delta t,$$

where

$$\bar{\omega}=\tfrac{1}{2}(\omega_1+\omega_2)\quad\text{and}\quad\delta=\tfrac{1}{2}(\omega_2-\omega_1).$$

If the two modes have nearly the same frequency, $\delta\ll\bar{\omega}$, so that x_1 and x_2 both oscillate with mean frequency $\bar{\omega}$ but with amplitudes that vary slowly, that of x_1 as $\cos\delta t$, that of x_2 as $\sin\delta t$, the peaks of amplitude alternating between one pendulum and the other.

The two pendulums represent a physical system whose configuration is determined by two coordinates, x_1 and x_2, and which has two normal modes of vibration. A stretched string needs a continuous function, effectively an infinite number of discrete variables, to describe its shape at any instant, and is capable of executing harmonic vibrations at a series of different frequencies. In this particular case the normal mode frequencies are all multiples of the lowest, *fundamental*, frequency as can be readily demonstrated on a piano. Depress the key of C, two octaves below c′ (middle C), gently so as to raise the damper without sounding the note. Then strike the octave above, c, and release it – the low string will be heard sounding an octave above its own natural pitch. Similarly with the g next above, when the low C-string will be excited to vibrate in its third harmonic mode and sound the note g; and so on through c′ and the e′, g′ and c″ above it. Intermediate notes excite no response from the low string. In its fundamental mode the string vibrates as a simple loop with a single antinode at its centre (fig. 15), and the other modes excited in the way described have 2, 3, 4, 5, 6 and 8 loops (the seventh harmonic sounds about halfway between a′ and b′ and is not represented in the diatonic scale; it

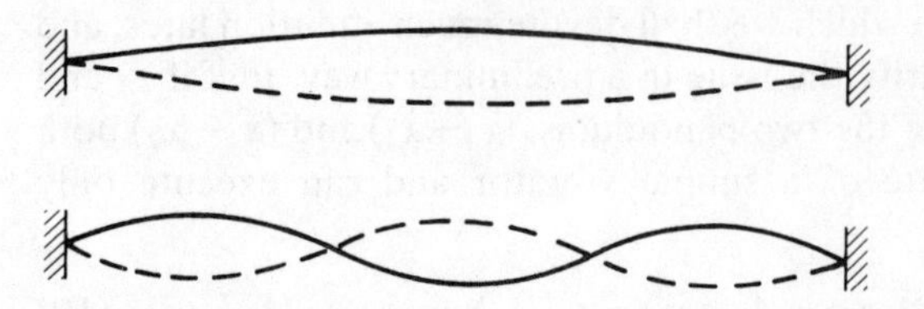

Fig. 15. Fundamental and third harmonic vibrations of a stretched string. The full and broken lines show the extremes of displacement half a cycle apart.

cannot be excited by any string on the piano, but if you have a steady tenor or higher voice you may be able to excite it by singing loudly at the C-string). The string has a virtually unlimited series of modes, each of which by itself confers strictly harmonic vibrations on each element of the string. When several are simultaneously excited, as by the piano hammer, the superposition of all the modes with their different frequencies and standing-wave patterns causes each element of the string to execute a different non-sinusoidal motion. Even more complicated are the vibrations of drums and bells, but the same resolution into normal modes is possible, and each normal mode has the character of a simple vibration and can be represented as the harmonic motion of a single variable. All these systems obey linear equations of motion, on which the resolution into normal modes depends. Non-linear systems with many variables are quite different and do not conform to any simple generalized behaviour, but each has to be treated separately. We shall discuss a few examples of this extremely difficult field at appropriate places later.

225, 385

A system with two independent normal modes that deserves some consideration is the simple pendulum swinging not in a plane but in any direction and arranged, for example, as in fig. 16, so that the frequency depends on the direction of its swing. The normal modes can be seen by inspection – if it swings in and out of the paper it is effectively of length l_1, if

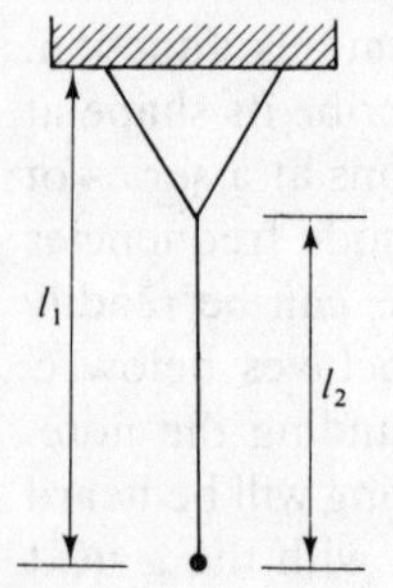

Fig. 16. A simple pendulum supported on a loop so as to vibrate with different frequency in two normal planes.

in the plane of the paper its effective length is l_2. With these directions as x- and y-axes, motion in each of the two normal modes follows one of the equations:

$$\left.\begin{aligned} x &= A_1 \cos(\omega_1 t + \phi_1), \\ y &= A_2 \cos(\omega_2 t + \phi_2), \end{aligned}\right\} \tag{21}$$

where $\omega_1^2 = g/l_1$ and $\omega_2^2 = g/l_2$. The combination of these two motions gives rise to the well-known Lissajous figures, of which examples are shown in fig. 17. If the ratio ω_1/ω_2 is a rational fraction of the form N/M (N and M having no common factor) the figure is closed after N cycles in the x-direction and M in the y-direction; the whole pattern is contained within a rectangle of sides $2A_1$ and $2A_2$. If ω_1/ω_2 is irrational the figure never closes and ultimately every point within the rectangle is traversed. These figures are most easily demonstrated with a cathode-ray tube, by applying separate alternating voltages of variable frequency to the X and Y plates.

The same behaviour is shown by a body confined to motion in two dimensions, and acted upon by springs so that its potential energy is a positive definite quadratic function of its displacement from equilibrium; any arrangement of springs leads to this, for small displacements at least. If the axes are chosen arbitrarily the potential energy takes the form $ax^2 + by^2 + cxy$, but by choosing the orientation of the axes appropriately the cross-term, cxy, can be eliminated to leave V in the form

373

$$V = a'x^2 + b'y^2.$$

The force acting on the displaced body is a two-dimensional vector $F(x, y)$ having components

$$F = (-\partial V/\partial x, -\partial V/\partial y) = (-2a'x, -2b'y).$$

With this coordinate system the variables are separated in the equations of motion:

$$m\ddot{x} = F_x = -2a'x$$

and

$$m\ddot{y} = F_y = -2b'y,$$

so that the normal mode frequencies are given by $\omega_1^2 = 2a'/m$, $\omega_2^2 = 2b'/m$. The Lissajous figures result from the simultaneous excitation of both.

In the special case of an isotropic two-dimensional vibrator (e.g. the conical pendulum), $a = b$ and $\omega_1 = \omega_2$. Then F is in the direction of the displacement and the system can oscillate in a straight line in any direction:

$$\mathbf{r} = \mathbf{A} \cos(\omega t + \phi),$$

where $\mathbf{A}$ is any constant vector in the plane. But any combination of x and y oscillations with arbitrary amplitude and phase difference is equally possible, giving rise to motion in an elliptical path. Let the phase difference

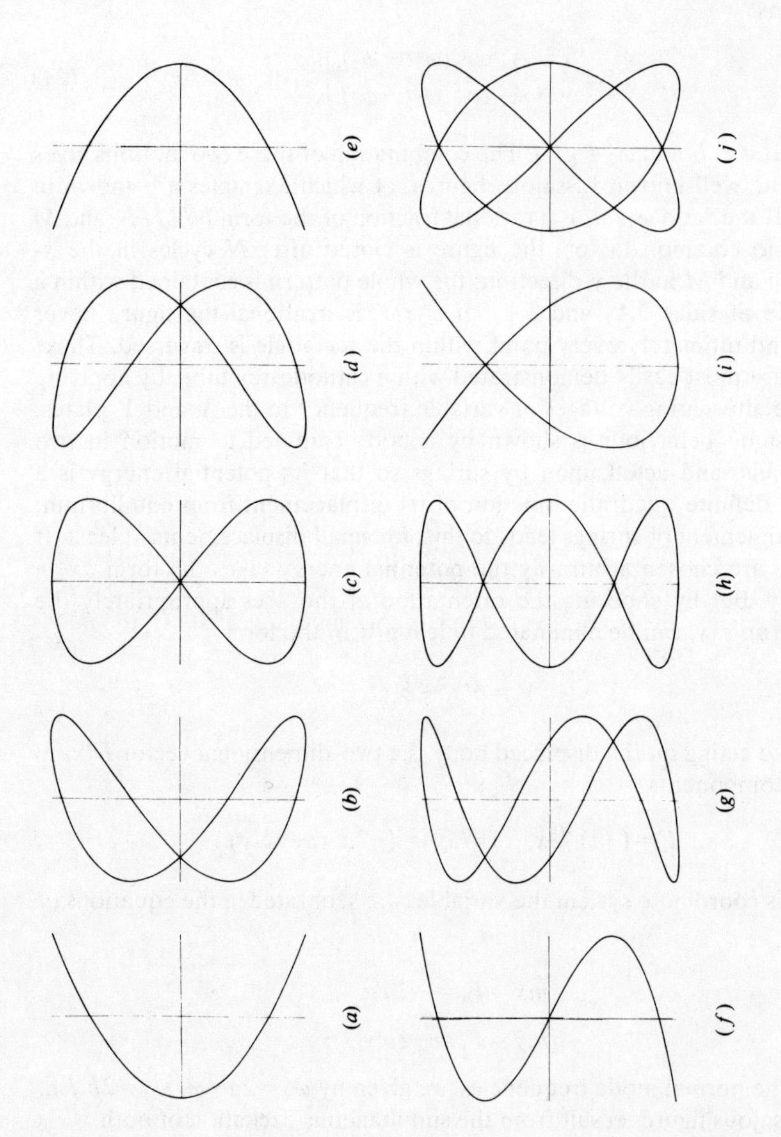

Fig. 17. Lissajous figures. Top row (*a*)–(*e*), $\omega_1 = 2\omega_2$ and $\phi = 0, \frac{1}{4}\pi, \frac{1}{2}\pi, \frac{3}{4}\pi, \pi$. Bottom row (*f*)–(*h*), $\omega_1 = 3\omega_2$ and $\phi = 0, \frac{1}{4}\pi, \frac{1}{2}\pi$; (*i*) and (*j*) $\omega_1 = 1.5\omega_2$ and $\phi = 0, \frac{1}{4}\pi$.

be 2ϕ and choose the origin of time so that the motion can be written in the form

$$x = A_1 \cos(\omega t - \phi) = A_1(\cos \omega t \cos \phi + \sin \omega t \sin \phi),$$

$$y = A_2 \cos(\omega t + \phi) = A_2(\cos \omega t \cos \phi - \sin \omega t \sin \phi).$$

Then $x/A_1 + y/A_2 = 2 \cos \omega t \cos \phi$ and $x/A_1 - y/A_2 = 2 \sin \omega t \sin \phi$;

hence $(x/A_1 + y/A_2)^2 \sin^2 \phi + (x/A_1 - y/A_2)^2 \cos^2 \phi = \sin^2 2\phi.$

The time-variable has been eliminated to yield the equation of the trajectory, which is quadratic and readily shown to be an ellipse. The reader may work out for himself the relationship between A_1, A_2 and ϕ on the one hand, and the shape and orientation of the ellipse on the other. The formulae are rather complicated and not particular useful.†

The conical pendulum has the property that for small displacements all trajectories are closed and have the same period. The oscillatory radial motion is thus isochronous, though it is not harmonic. Let us separate this radial motion from the rotation, examining the bob from a frame of reference that rotates in step with it. The angular momentum $L = mr^2\Omega$ when the radial displacement is r and the instantaneous angular velocity Ω, and since L is a constant of the motion, $\Omega = L/mr^2 \propto 1/r^2$. To the observer in the frame which also rotates at the variable angular velocity Ω, the bob is subject to a centrifugal force $mr\Omega^2$ as well as the restoring force $-\mu r$ due to the string, and therefore appears to be moving in a straight line under the influence of the force field

$$F(r) = mr\Omega^2 - \mu r = L^2/mr^3 - \mu r,$$

which is derivable from a potential

$$V(r) = \tfrac{1}{2}\mu r^2 + L^2/2mr^2.$$

This has the same form as (12) which was earlier cited as an example of an isochronous potential. We are now in a position to write down the form of the vibration in this particular potential, for it is just the radial motion of a conical pendulum moving in an ellipse. By choosing the x- and y-axes to coincide with the axes of the ellipse we arrange that the two oscillations in

† The analysis of circular or elliptical motion into two harmonic oscillations along orthogonal axes is only useful in a linear system. A simple example shows how the argument may fail for a non-linear vibrator; a conical pendulum swinging in a circle with the string making an angle θ to the vertical has a frequency $(l\sec\theta/g)^{\frac{1}{2}}$, becoming greater as θ increases, while oscillations in a plane have lower frequencies at larger amplitudes. The former obviously is not a simple composition of two of the latter oscillations. In the same way if we were to set a body moving on a smooth bowl formed as a figure of revolution from a cycloid, we might naively expect the isochronous property to give rise to simple closed orbits whatever the shape or amplitude; but we should be wrong.

(21) are in phase-quadrature – the x-displacement is greatest as the bob crosses the x-axis ($y=0$), and similarly for the y-displacement:

$$x = C_1 \cos \omega t, \qquad y = C_2 \sin \omega t.$$

Hence
$$r^2 = C_1^2 \cos^2 \omega t + C_2^2 \sin^2 \omega t,$$

which can be written in the form

$$r = C_1(1 + B \sin^2 \omega t)^{\frac{1}{2}},$$

where C_1 can take any positive value and $B(=C_2^2/C_1^2 - 1)$ any value between -1 and $+\infty$. The same result may of course be reached directly by integrating (11).

Dissipation

The simple vibrators discussed so far have been idealized to the extent of neglecting the inevitable losses that in practice cause the amplitude of a passive vibrating system to decay. The use of the term 'passive' distinguishes one class of vibrating systems from another, 'active' vibrators which are supplied by an energy source. In a passive system the energy of vibration may be converted into heat but not vice versa, and therefore decay is the only mode of change;† by contrast the active system may build up in amplitude until the dissipation matches the power input or the character of the system is changed by some catastrophe. The dependence of dissipation rate on amplitude of motion in a passive vibrator may take different forms, of which the simplest from a mathematical point of view is that exemplified by an ohmic resistor in an oscillating circuit or viscous damping of a mechanical vibrator. In both cases the force responsible for damping is proportional to the first derivative of the displacement. For the circuit of fig. 1(c) the displacement is measured by q, $\dot{q}$ is the current through R and $R\dot{q}$ is the potential difference across R; if q and $\dot{q}$ are positive, the electric field between the capacitor plates and the field in the resistor have the same sign, so that (5) must be supplemented with another positive term:

$$L\ddot{q} + R\dot{q} + q/C = 0, \tag{22}$$

which may be written in what will henceforth be taken as the standard form:

$$\ddot{\xi} + k\dot{\xi} + \omega_0^2 \xi = 0. \tag{23}$$

In this particular case the generalized displacement, ξ, stands for q, $k = R/L$ and $\omega_0^2 = (LC)^{-1}$. Similarly the pendulum bob in fig. 1(a), when

† For the present we ignore the interchange between random thermal motions and vibration of the system, which is of great importance in small systems and in delicate instruments, but rarely noticeable otherwise. This question will be discussed in chapter 4 and subsequently.

89, 156, 361

displaced through a positive angle θ and moving outwards with a positive velocity $l\dot{\theta}$, experiences a restoring force proportional to θ and a force from any viscous fluid surrounding it proportional to $\dot{\theta}$ and in the same sense as the restoring force. Its equation of motion therefore has the form (23) also.

The popularity of (23) as an expression of damped vibratory motion arises partly from the great importance of electrical circuits for which it really is a very good representation, but partly also from the linearity of the equation which makes it mathematically tractable. There is something of a tendency among physicists to try to reduce everything to linearity, and this may be defended if the linear approximation is not too fanciful and if the alternative (of tackling the real problem) involves analysis of a different order of difficulty. Since we shall be discussing linear or linearized systems so much in this book let us recognize at the outset, by means of a few examples, that the reality may not always conform to what we might wish, rather more so with damping forces than with the restoring force in small-amplitude vibrations. Thus a mechanical vibrator immersed in a viscous fluid, and moving so slowly that the flow round it is laminar, will indeed suffer a retarding force proportional to its velocity. But if it is moving fast enough to create a turbulent wake the force will be more nearly proportional to v^2; while if it is rubbing against some fixed surface the frictional force may be independent of the speed and merely directed oppositely to the motion. These are elementary examples of non-linear damping effects, and more complicated cases are by no means unusual; for example the frictional force may be substantially higher when there is no slipping and fall as soon as slipping begins.

41, 336

Each case demands its own special treatment if a complete analytical solution is to be found, but a very adequate approximation may be derived from energy considerations if the rate of damping is not too great. Before giving this we shall find complete solutions for two models which will serve for comparison with the approximate result. The first, which is important in its own right, and therefore deserves rather full discussion, is the linear equation (23) whose solution is the exponentially damped harmonic oscillation expressed by (1). Substituting $\xi = \xi_0 \cos(\omega' t + \phi)\,\mathrm{e}^{-t/\tau_a}$ in (23), one finds that the terms in $\cos(\omega' t + \phi)$ and $\sin(\omega' t + \phi)$ both vanish and the equation is satisfied when

$$\text{either} \quad \left.\begin{cases} \omega'^2 = \omega_0^2 - \tfrac{1}{4}k^2, \\ \tau_a = 2/k, \end{cases}\right\} \text{ if } k < 2\omega_0, \tag{24}$$

$$\text{or} \quad \left.\begin{cases} \omega' = 0, \\ \tau_a = 2/[k \pm (k^2 - 4\omega_0^2)^{\frac{1}{2}}], \end{cases}\right\} \text{ if } k > 2\omega_0. \tag{25}$$

The non-oscillatory solution (25) arises when the damping is too high (e.g. $R^2 > 4L/C$) to permit vibrations, and the system is said to be overdamped. The complete solution with two adjustable constants then takes the form

$$\xi = \xi_1 \,\mathrm{e}^{-t/\tau_1} + \xi_2 \,\mathrm{e}^{-t/\tau_2}, \tag{26}$$

where τ_1 and τ_2 are the two values appearing in (25). A typical curve of the time-variation of ξ in an overdamped system is shown in fig. 18, where $\xi(0)$ has been made equal to zero by choosing $\xi_2=-\xi_1$.

In the limit of very heavy overdamping, τ_1 approaches $1/k$ and τ_2 approaches k/ω_0^2. For the LCR circuit $\tau_1=L/R$, the time-constant for L and R in series, while $\tau_2=RC$, the much longer time-constant for C and R in series. When an impulse is applied to the system at rest, $\xi_2=-\xi_1$ in (26); the amplitudes of the two terms match initially, and the initial rise is therefore dominated by ξ_1, with the much larger slope. After one or two times τ_1, however, ξ_2 takes over and the fall back to zero is very slow, being governed by τ_2. It is easy to see the physical processes at work here. The initial impulsive e.m.f. which establishes a current has finished before any significant charge has appeared on the capacitor. Immediately afterwards, therefore, the capacitor plays no part and the decay of the current is controlled by L and R, with time-constant L/R. A little later, when the current is small, the capacitor, which by now has a charge, will begin to take over, and by the time the current has fallen to zero it will be dominant. For at this moment the charge on the capacitor reaches its peak value, and thereafter discharge occurs; and when R is large the discharge rate, governed by the time-constant RC, is slow so that inductive effects are negligible.

The underdamped system, obeying (24), is oscillatory with a Q-value that is determined by use of (2). Since $\omega''=1/\tau_a=\frac{1}{2}k$,

$$Q=\omega'/k=(\omega_0^2/k^2-\tfrac{1}{4})^{\frac{1}{2}}. \tag{27}$$

Thus the LCR circuit, for which $\omega_0^2=(LC)^{-1}$ and $k=R/L$, has a Q-value given by $(L/R^2C-\frac{1}{4})^{\frac{1}{2}}$ or approximately (when $Q\gg 1$) $\omega_0 L/R$ or $(\omega_0 RC)^{-1}$.

A particular case arises at the point of critical damping, $k=2\omega_0$, at which $Q=0$ and (24) gives place to (25). If an underdamped system is initially at rest and is struck sharply at $t=0$ so as to initiate vibration, the subsequent motion is described by the equation

$$\xi=\xi_0 \sin\omega' t\cdot \mathrm{e}^{-t/\tau_a}, \tag{28}$$

since this choice of phase ($\phi=-\pi/2$) ensures that ξ vanishes at the beginning. Now as k is increased towards $2\omega_0$, τ_a decreases to $1/\omega_0$ and ω' to zero, so that an ever smaller fraction of a cycle of $\sin\omega' t$ is accomplished before the exponential term kills the vibration. In the limit, as the critical condition is reached, it is permissible to replace $\sin\omega' t$ by $\omega' t$, so that

$$\xi=\dot{\xi}_0 t\,\mathrm{e}^{-\omega_0 t},$$

the constant $\dot{\xi}_0$ being the initial value of $\dot{\xi}$ due to the impulse. The form of this function is very similar to fig. 18.

With an instrument like a moving coil galvanometer, whose damping is largely provided by induced currents flowing in an external resistor, and is therefore adjustable, it is customary to arrange that it is close to critical damping, for it then responds most rapidly. If the damping is less it

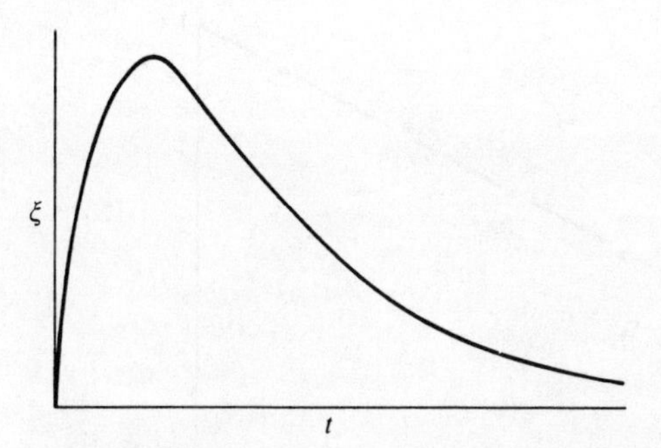

Fig. 18. Displacement of an overdamped vibrator struck while at rest, at $t=0$; the two time-constants here differ by a factor of two.

oscillates several times before reaching a steady reading; if it is more the longer of the two time-constants in (25) causes it to be sluggish. The galvanometer incidentally is a very useful device for submitting the theory of linear vibrators to experimental check, for the decrement is easily measured, and the damping is adjustable in a controlled way by connecting different resistors across the coil. If the coil is deflected through an angle θ, the e.m.f. induced by its motion is proportional to $\dot{\theta}$ (the constant of proportionality, depending on the strength of the magnetic field and the dimensions of the coil, does not concern us for this purpose), and the current in the circuit of total resistance R is therefore proportional to $\dot{\theta}/R$, as is the consequent torque on the coil. The equation of angular motion therefore takes the form

$$I\ddot{\theta}+(\lambda_0+A/R)\dot{\theta}+\mu\theta=0, \tag{29}$$

where I is the moment of inertia of the coil and μ the torsion constant of the suspension; λ_0 takes care of the air-damping present at all times, which leads to dissipation even on open circuit ($R=\infty$). This equation has the same form as (23) and we make the following identifications:

$$k=(\lambda_0+A/R)/I,$$

$$\omega_0^2=\mu/I.$$

8 The logarithmic decrement $\Delta=\pi/\omega'\tau_a=\frac{1}{2}\pi k/(\omega_0^2-\frac{1}{4}k^2)^{\frac{1}{2}}$ from (24);

hence $$(\lambda_0+A/R)/(\mu I)^{\frac{1}{2}}=2\Delta/(\Delta^2+\pi^2)^{\frac{1}{2}}.$$

This is a convenient formula to compare with experiment, since a graph of $\Delta/(\Delta^2+\pi^2)^{\frac{1}{2}}$ against $1/R$ should be linear over the whole range of damping up to the critical condition. The result of a test is shown in fig. 19 which substantiates the theory very satisfactorily. It is worth noting that the measurements have been taken to very high damping conditions, with the ratio of successive swings on opposite sides as high as 600. But although this ratio cannot be measured very accurately, errors become much less significant when the logarithm of a large number is taken to form Δ, and still less significant when $\Delta/(\Delta^2+\pi^2)^{\frac{1}{2}}$ is formed, since this quantity tends to a constant, unity, as the decrement rises without limit while k approaches its critical value $2\omega_0$.

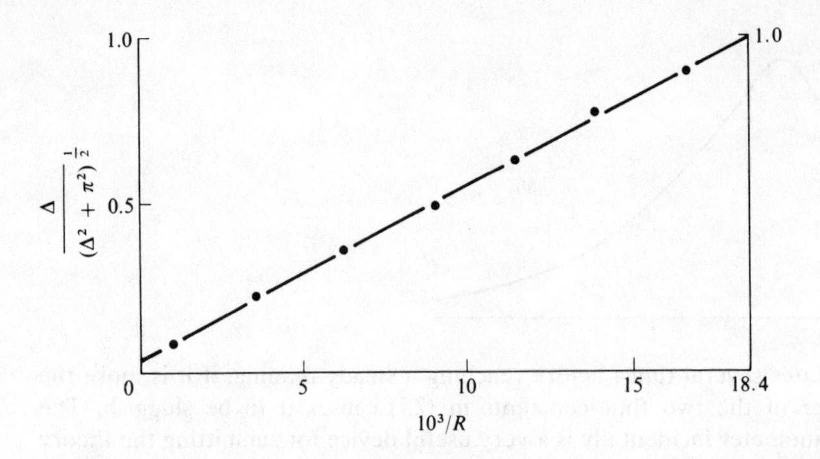

Fig. 19. Experimental verification of variation of damping with circuit resistance in a moving-coil galvanometer. The instrument has a resistance of 10 Ω, and the graph shows that critical damping ($\Delta = \infty$) occurs when the total circuit resistance is 54 Ω (i.e. 44 Ω connected across the terminals). To obtain these points it was necessary to keep the deflection of the light spot (at 1 m) less than 10 cm; with larger deflections the air damping increased very rapidly.

The second dissipative model, this time non-linear, can be dealt with quickly. If the frictional force is independent of speed, the vibrating body is subject to a constant retarding force so long as it continues moving in the same direction; as soon as it reverses the force also reverses. With constant frictional force F, and with restoring force $-\mu x$, proportional to displacement, as the only other force acting, the equation of motion for a body of mass m takes the form

$$m\ddot{x} + \mu x \pm F = 0, \tag{30}$$

the sign in front of F being the same as the sign of the velocity $\dot{x}$. By moving the origin of x to $\mp F/\mu$ (i.e. writing $x' = x \pm F/\mu$), (30) is cast into the standard form for undamped sinusoidal oscillations, analogous to (5):

$$m\ddot{x}' + \mu x' = 0.$$

The complete solution now follows immediately – each half-cycle between positive and negative extremes is of sinusoidal form and of the same frequency, but alternate half-cycles are centred on different values of x, as shown in fig. 20; the vibration stops abruptly at a turning point where the

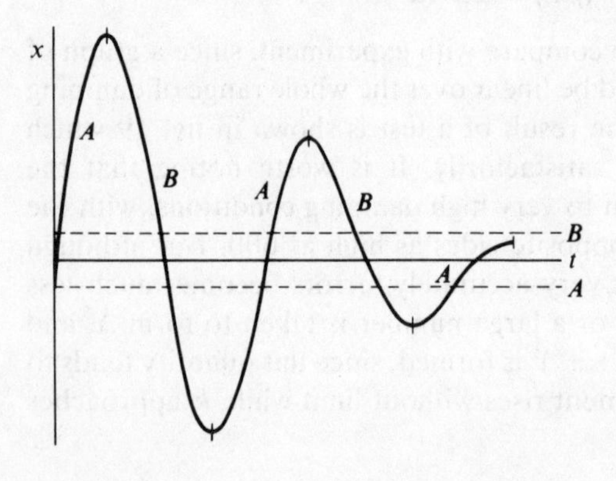

Fig. 20. Frictional damping. The segments labelled A are sinusoids vibrating about the horizontal line A, and similarly for B.

restoring force is no longer as large as F. Until that moment is reached, the turning points are successively $2F/\mu$ nearer the origin, and in one complete cycle the amplitude decreases by $4F/\mu$. In contrast to the first model (damping force $\propto$ velocity) which gives exponential decay, the amplitude now decays in a linear fashion.

The difference between the two models is readily demonstrated with a rigid pendulum, about 1 m long. If the damping is provided by fixing a stiff wire onto the pendulum, and letting it dip into a can of treacle, the exponential decay obtained (fig. 21) shows the force to be proportional to velocity. But if, instead, a long wire be stretched horizontally so that the

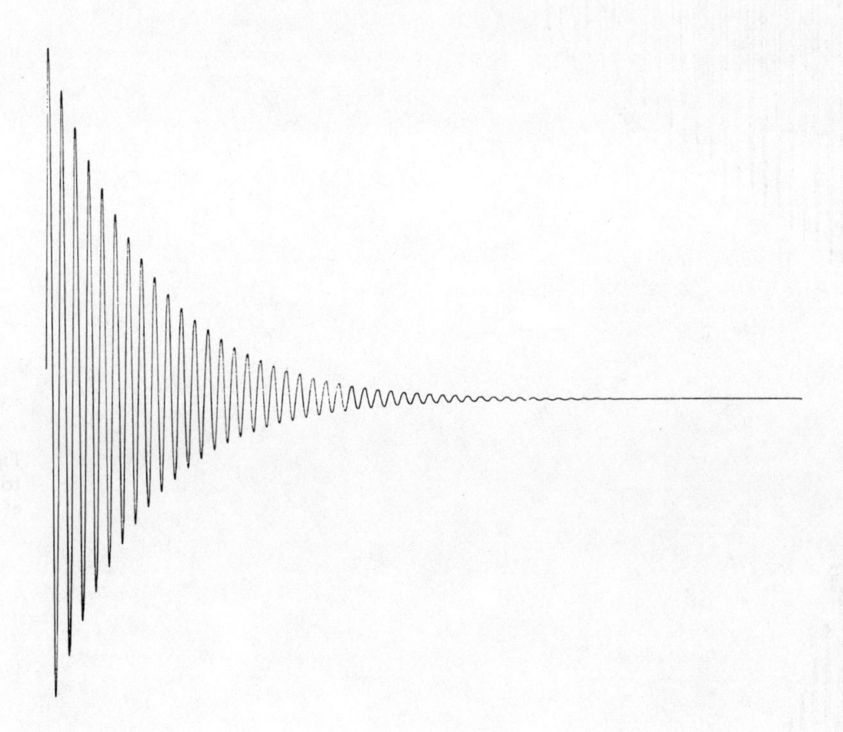

Fig. 21. Viscous damping of a rigid pendulum.

pendulum rod rubs lightly against it, the resulting frictional damping gives the almost linear decay of fig. 22. The behaviour recorded in fig. 23 was obtained by letting a rather stiff wire attached to the pendulum rub against the fixed horizontal wire. When the amplitude is large slipping occurs, with rapid damping due to friction, but at a smaller amplitude the two wires stick together, and there is enough flexibility in the wire attached to the pendulum to permit vibration to continue with little damping, though at a frequency that is 8% higher because of the extra restoring force of the wire. Exactly the same effect has been observed with superconducting discs

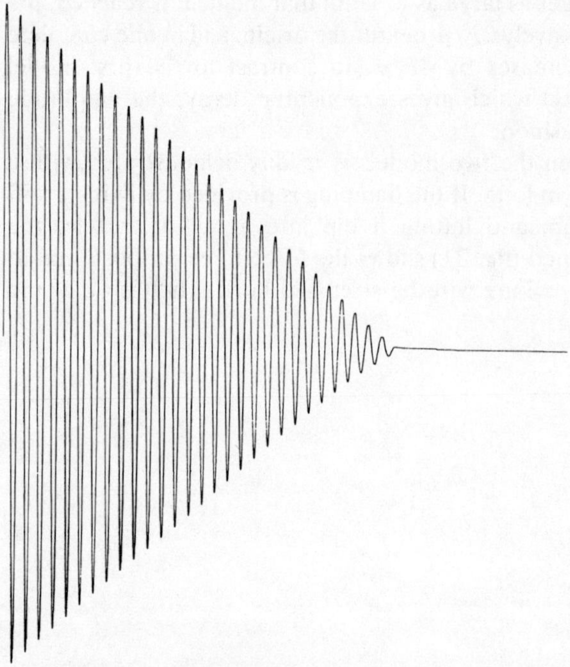

Fig. 22. Frictional damping of a rigid pendulum.

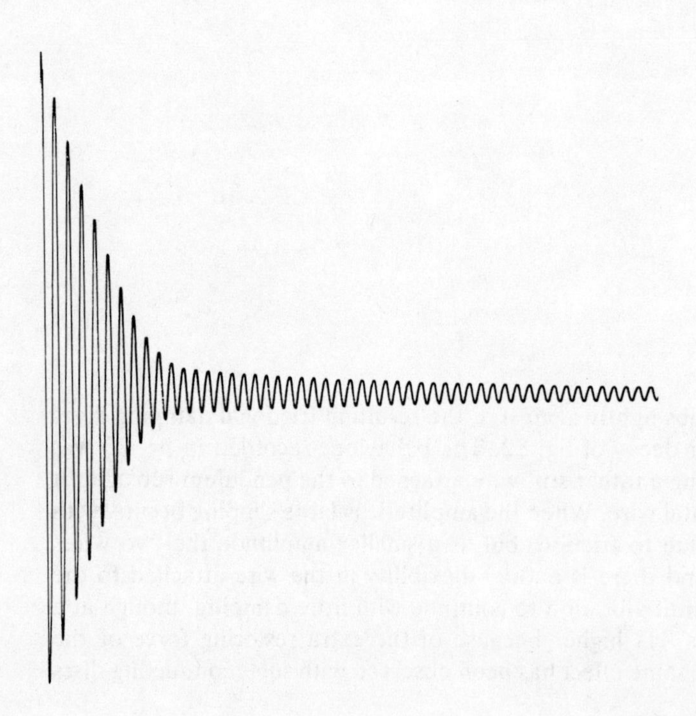

Fig. 23. Pendulum with frictional damping only at large amplitude.

oscillating in a magnetic field; so long as the magnetic flux can move relative to the metal a frictional force is observed, but when the disc oscillates with only small amplitude the flux is pinned and provides an elastic restoring force. Damping curves like fig. 24 are obtained and may be analysed to give a measure of the pinning and frictional forces involved when magnetic fields penetrate and move within a superconductor.

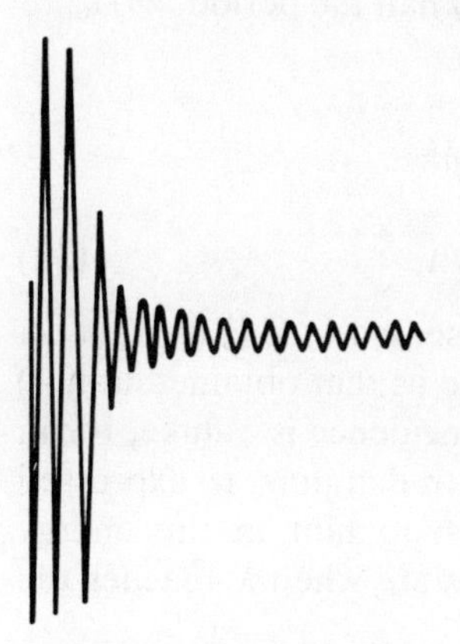

Fig. 24. Oscillation of a superconducting disc in a non-uniform field (P. C. Wraight[4]) exhibiting the same behaviour as fig. 23.

The behaviour represented by figs. 21 and 22 may now be compared with the predictions of an approximate analysis based on the rate of dissipation of energy. If the damping force is weak (though varying with velocity) it will not greatly perturb the sinusoidal form of the vibrations, and we may take $x = x_0 \sin \omega' t$ still to represent them, while bearing in mind that x_0 now decreases slowly. If x_0 is taken as constant over one cycle the velocity, and hence the damping force, $F(x)$, are determined everywhere. The work done in a cycle is proportional to $\oint F(x)\,\mathrm{d}x$, where x is taken through a complete cycle between $\pm x_0$; alternatively we may calculate how F varies with time and write the work done as $\oint F(t)\dot{x}(t)\,\mathrm{d}t$. It is this that determines the decrease in the energy stored in the vibrator, and hence the change in x_0 during the cycle. Clearly if x_0 changes substantially in a cycle the argument can be only approximate.

Let us apply the argument to the two models already treated. First, when damping is viscous the mechanical version of (23) takes the form

$$m\ddot{x} + \lambda\dot{x} + \mu x = 0. \tag{31}$$

If the oscillation be assumed to have the form $x_0 \cos \omega_0 t$, where $\omega_0^2 = \mu/m$, the velocity $\dot{x}$ is $-\omega_0 x_0 \sin \omega_0 t$, and the damping force F is $\lambda\omega_0 x_0 \sin \omega_0 t$. The rate of dissipation of energy is $-F\dot{x}$, i.e. $\lambda\omega_0^2 x_0^2 \sin^2 \omega_0 t$, and the energy dissipated in one cycle is therefore given by

$$\Delta E = -\int_0^{2\pi/\omega_0} \lambda\omega_0^2 x_0^2 \sin^2 \omega_0 t\,\mathrm{d}t = -\pi\lambda\omega_0 x_0^2.$$

This may be compared with the stored energy, $E=\frac{1}{2}\mu x_0^2$, to give the decay of energy per cycle in the form

$$\Delta E/E=-2\pi\lambda\omega_0/\mu.$$

Since $E\propto x_0^2$, we may take the logarithmic decrement (per half-cycle), $-\frac{1}{2}\Delta x_0/x_0$, to be $-\frac{1}{4}\Delta E/E$, i.e. $\frac{1}{2}\pi\lambda\omega_0/\mu$. Alternatively, the rate of decay of amplitude with time is obtained by dividing this by half the period, π/ω_0, to yield

$$\frac{1}{x_0}\frac{\mathrm{d}x_0}{\mathrm{d}t}=-\tfrac{1}{2}\lambda\omega_0^2/\mu=-\tfrac{1}{2}\lambda/m.$$

Hence $$x_0\propto \mathrm{e}^{-t/\tau_\mathrm{a}} \quad \text{where } \tau_\mathrm{a}=2m/\lambda. \tag{32}$$

Now the parameter k introduced in (23) is represented in (31) by λ/m, so that the value of τ_a given by (32) is $2/k$, the same as that obtained in (24) by direct solution of the equation. The exact coincidence is a fluke, for in the energy calculation variations of frequency with damping, as expressed in (24), were neglected; and certainly there is no hint in the energy calculation of the transition to the overdamped state when λ reaches the critical value $2(m\mu)^{\frac{1}{2}}$.

In the second model, with frictional damping, the energy loss is most readily computed by integration over x rather than t, for during one cycle the particle moves through $4x_0$ against a constant retarding force F, and therefore loses energy $\Delta E=-4Fx_0$. But $E=\frac{1}{2}\mu x_0^2$, so that $\Delta E=\mu x_0\Delta x_0$. Hence $\Delta x_0=-4F/\mu$, once again in coincidence with the exact treatment. These two examples give us confidence in using the energy argument even when the decrement is not particularly slow, and indeed though not exact it is a most useful first attack on what may otherwise be a tedious analytical problem.[5] It may be especially valuable for hysteretic systems, in which the force is not determined at each instant by the position and velocity, but depends on the whole previous history. Whatever the underlying physical process, the dissipation per cycle is given by the area enclosed by a graph of force versus displacement as the latter runs through a complete cycle.

It may be observed that when F is independent of v, the energy loss per cycle $\Delta E\propto x_0$, while when $F\propto v$, $\Delta E\propto x_0^2$. It is easily seen by writing down the integral for the energy loss that if $F\propto v^n$, n being positive, $\Delta E\propto x_0^{n+1}$. The form of decay when $n>1$ follows immediately, for we write

$$\Delta x_0/x_0\propto\Delta E/E\propto x_0^{n-1};$$

hence $$\mathrm{d}x_0/\mathrm{d}t\doteqdot\Delta x_0/T_0\propto x_0^n,$$

and $$x_0\propto(t-t_0)^{1/(1-n)}, \quad \text{where } t_0 \text{ is a constant.} \tag{33}$$

For example, a vane moving in a light mobile fluid experiences a force proportional to v^2 when moving fast enough to set up a turbulent wake, and a vibrator so damped will decay as $1/(t-t_0)$; if the system is started when $t=0$ and in the first period, of duration T_0, drops to an amplitude α

times the initial amplitude, t_0 must be given the value $-\alpha T_0/(1-\alpha)$. One would not expect (33) to hold as the amplitude gets small, since turbulent flow will give place to laminar, with a force proportional to v; the subsequent decay will then be exponential.

Maintained vibrations

We have already drawn a distinction between passive and active vibrators, the latter being supplied with energy from an external source. This may be insufficient to overcome completely the internal dissipation, in which case the system need not differ in any marked degree from a passive vibrator, though its decay time may be longer than could easily be achieved without external aid. If the provision of energy is steadily increased, τ_a will also increase until, at a critical rate of supply that exactly balances the internal dissipation, the vibrations will maintain themselves at constant amplitude. Beyond this critical condition the amplitude grows, with τ_a in (1) negative so that in a linear system the increase is exponential. This growth cannot, of course, continue indefinitely; something must break or change, or the energy source prove insufficient. Non-linearity is therefore an ultimately inevitable feature of a spontaneously oscillating system. Yet until non-linearity prevails there need be no mathematical discontinuity to distinguish active from passive systems. That is to say, the standard form (23) has the same type of solution whether k is positive or negative – changing the signs of k and t together leaves the equation unaltered, and mathematically time-reversal is no great matter. In real life it is of the greatest consequence, however, and we are bound to regard as altogether different in kind systems whose equations of motion can nevertheless all be treated in a common framework. A great variety of oscillators will be discussed in more or less detail in later chapters and in the next volume. Let us note here some of the types with which we shall be concerned, cataloguing them under different headings in order to put some rough order into what could become a bewildering list. The first two headings comprise systems which can be brought into formal correspondence with simple passive vibrators.

(*a*) *Negative resistance devices.* If the current through a circuit element, such as a semiconducting tunnel diode, does not increase steadily with the potential difference across it, but has a region of negative slope as in fig. 25, it is possible to bias the device at N and incorporate it in a resonant circuit where it will counteract the lossy effects of the circuit resistance. Related to this type of device are those in which other mechanisms of loss are reversed in sign; in masers and lasers, for instance, artificial dielectrics are created by rearranging the populations of different energy levels of molecules, so as to turn the normal dielectric loss into dielectric gain.

330

345
603

314

(*b*) *Feedback oscillators.* A signal derived from an oscillating system is amplified and fed back into the system with such a phase as to encourage the oscillation to grow (*positive feedback*). Fig. 26 illustrates such an arrangement. Ignoring all but the essentials, we may think of a current i in

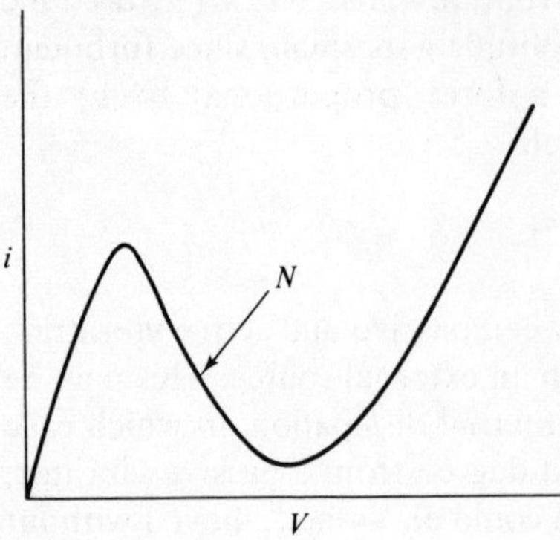

Fig. 25. Current–voltage characteristic of a tunnel diode, showing negative slope region, N.

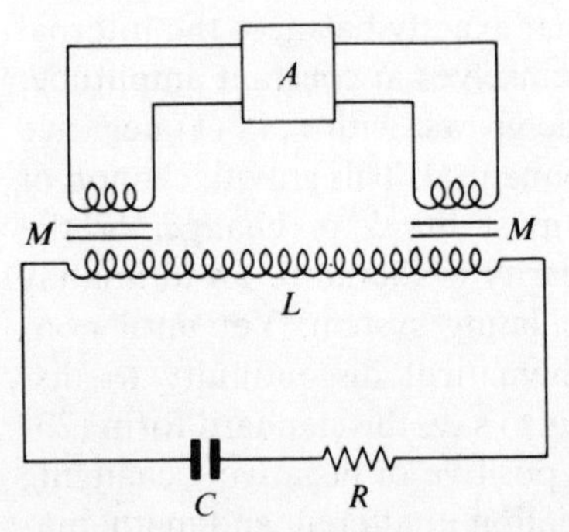

Fig. 26. Use of amplifier (A) and feedback to maintain oscillation in a resonant circuit.

the resonant circuit inducing an e.m.f. $M\,\mathrm{d}i/\mathrm{d}t$ in the input circuit of the amplifier. If this generates an output current $\mu M\,\mathrm{d}i/\mathrm{d}t$, the e.m.f. induced back into the resonant circuit is $\mu M^2\,\mathrm{d}^2i/\mathrm{d}t^2$, which must be added to the e.m.f. $L\,\mathrm{d}i/\mathrm{d}t$ due to the inductance itself. Corresponding to (22) we now have

$$\mu M^2\dddot{q}+L\ddot{q}+R\dot{q}+q/C=0. \qquad (34)$$

Without going into the solution of a cubic equation we may see how this can lead to spontaneous oscillation very much in the same way as introducing a negative resistance. For if q be supposed to oscillate sinusoidally as $q_0 \sin\omega' t$,

$$R\dot{q}=R\omega' q_0\cos\omega' t$$

and $$\mu M\dddot{q}=-\mu M^2\omega'^3 q_0\cos\omega' t.$$

Both terms have the same phase (i.e. cosine rather than sine), and if the connections to the amplifier are made so that μ is positive the feedback term counteracts the resistive term.

A related system, though one that is essentially non-linear, is the escapement mechanism of pendulum clocks and balance-wheel watches. The drive here is automatically synchronized with the oscillatory system, as in the feedback oscillator, but the driving force is not proportional to the oscillatory amplitude; instead, a certain minimum amplitude is needed to

323

actuate the escapement, and above this amplitude the energy supplied is roughly independent of amplitude.

It is at about this point in the catalogue that we discover classes of maintained vibrators which cannot usefully be thought of as resembling passive vibrators running backwards in time.

359 (*c*) *Relaxation oscillators.* This is a rather ill-defined term which is used to cover apparently very different systems. Perhaps the nearest we can get to characterizing the type is by observing that there are intrinsically unstable systems which are prevented from reaching a stable configuration by a sudden change in the behaviour of one element of the system, having the effect of returning them to their original unstable state; each subsequent attempt to reach equilibrium suffers the same fate and the system therefore oscillates. It might be contended that a bouncing ball is a **9** relaxation oscillator in this sense, but a more convincing example is the 'linear time-base' circuit of fig. 27. The capacitor charges through the resistor until the breakdown voltage of the gas tube is reached, whereupon the capacitor discharges through the ionized gas; once the discharge is over, the ionized molecules recombine and the charging starts again. The only stable state for the circuit containing the capacitor and the cell is that in which the capacitor is fully charged to the cell voltage, V_0, and this state is barred by the breakdown of the tube.

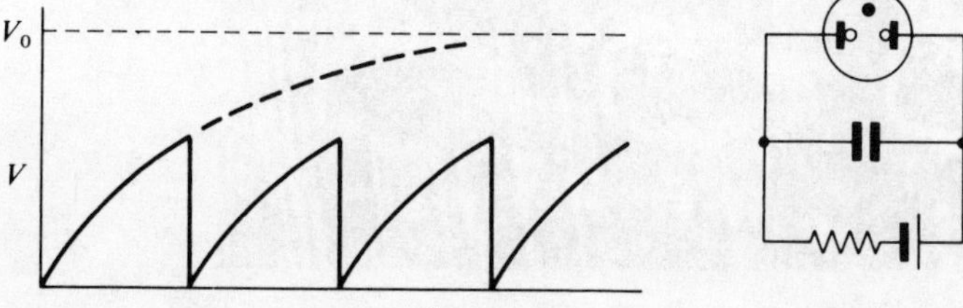

Fig. 27. The 'linear time-base' circuit using breakdown in a gas-filled tube.

The on–off switch of a thermostat, as in a domestic electric oven or a tank for tropical fish, can be considered as constituting a relaxation oscillator. There are two different equilibrium conditions, too cold (when the switch is off) and too hot (when it is on), and both are precluded through the action of the switch. These are examples of what we shall call *astable* systems, in contrast to the *unstable* systems of, say, class (*b*) above.† A feedback oscillator circuit can in principle be set in a condition of unstable equilibrium (e.g. no current in the resonant circuit of fig. 26 or in the input circuit of the amplifier, and such steady current in the output circuit of the amplifier as results from zero input current); but any chance fluctuation sets it oscillating with ever-increasing amplitude. With the astable system, on the other hand, no such condition of even transient rest can be found.

The stick–slip phenomenon may also be put in the class of relaxation **336** oscillators, though it may not represent a strictly astable situation. This is a

† This is not quite the normal use of the word 'astable', but vibrators that have been called astable have enough in common with these systems for the application not to be misleading.

subtlety that need not be pursued here – it is the discontinuous character of the phenomenon itself that matters. This arises when the frictional force on a sliding body, though more-or-less independent of speed once sliding has started, is less than the force required to get it moving from rest. If such a body is dragged along a surface by a spring it tends to stick until the spring has stretched enough to dislodge it, when it jumps forward, the tension of the spring being now greater than the sliding friction; but as the spring tension is reduced by this accelerated movement the body slows down and eventually sticks again. The process is then repeated. It is this effect that causes a door-hinge and a blackboard chalk to squeak, a stainless steel spoon to judder along the bottom of an aluminium saucepan, and, more beneficially, a violin bow to set the strings in vibration. The photograph of fig. 28 shows a magnified image (in a scanning electron microscope) of the pits left by a steel scriber drawn swiftly and lightly across an aluminium sheet. Here the sticking is associated with plastic deformation of the aluminium creating a ridge which holds the scriber until the force on it is enough to make it jump and begin to gouge out the next pit.

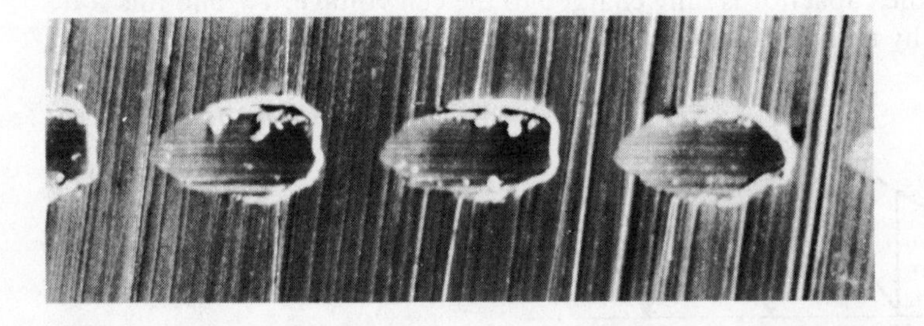

Fig. 28. Scanning electron micrograph of the track left by a steel scriber moving from left to right on an aluminium plate. The pits are $\frac{1}{4}$ mm apart, and the bright edges show piled-up aluminium scraped from the troughs.

(*d*) *Aerodynamic oscillations.* We use this term to cover more than aerodynamic effects, but these are best known. Wind blowing past a tall chimney sheds a stream of vortices alternately from each side, as in fig. 29 which shows the phenomenon on the laboratory scale with oil rather than air. So long as the chimney is immovable the frequency of vortex formation

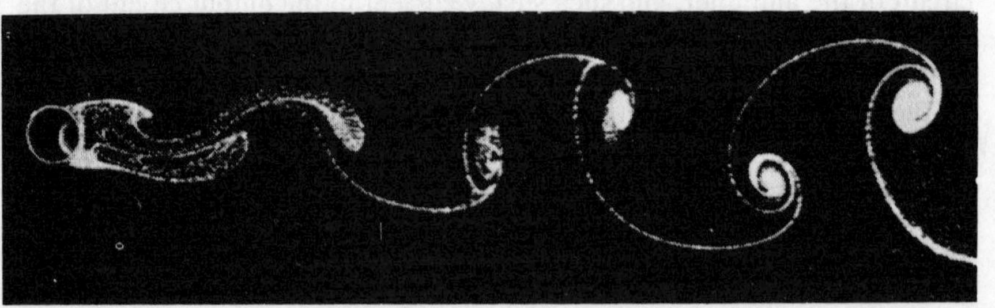

Fig. 29. Vortex pattern behind a cylindrical obstacle (F. Homann[6]).

is not very well defined, but the vortices exert a lateral force and if the chimney can move they may set it swaying at its own natural frequency. In its turn the movement of the chimney tends to control the formation of vortices which, on becoming synchronized with the sway, are able to make it grow still further, possibly in the end causing collapse. A similar effect set the Tacoma bridge into such oscillations as to break it, and on a much gentler scale the singing of telegraph wires and the production of sound in a flute, recorder and flue pipe of an organ are all of the same basic nature.

Additional note (see p. 22)

Rayleigh's principle, developed in detail in his classic *Theory of Sound*,[8] allows the frequency of the lowest mode of a complex linear system to be estimated rather well, even if the pattern of vibration cannot be found with any precision. It rests on two propositions: (i) if there are no velocity-dependent forces, a linear system possesses normal modes in which every part executes harmonic vibrations at the same frequency and in phase. There are instants when every part is at rest, at maximum displacement, and the energy is wholly potential; at instants, one-quarter of a cycle later, when every part is at its equilibrium position, the energy is wholly kinetic. If the system consists of mass elements m_i whose displacement in the first case is a_i, the kinetic energy at the latter instant is $\Sigma\,\frac{1}{2}m_i\omega^2 a_i^2$. The potential energy at the former instant is calculated from the a_i and the spring constants. Equating the two determines ω. (ii) If some arbitrary assumption is made concerning the relative magnitudes of the a_i, the value of ω so determined is not less than the frequency of the lowest mode. If, then, one imagines ω to be represented on a multi-dimensional plot, with relative values of all the a_i as coordinates, the point representing the true form of the lowest mode is at an absolute minimum of ω. It follows that ω is rather constant in the neighbourhood of this point, so that poor guesses of the a_i may still yield good estimates of ω.

209

3 Applications of complex variables to linear systems

Complex exponential solutions to equations of motion

Vibrators having the properties of linearity and translational invariance may obey a considerable variety of equations. We have already met in chapter 2 a few examples of linear differential equations with constant coefficients; (2.22) and (2.34) are of second and third order respectively, and there is no difficulty in devising circuits with equations of higher order than this. In other systems there may be effects not so easily described by differential coefficients; by the use of delay lines it is possible to introduce forces determined by what the system was doing at some earlier time, yet still linear in these quantities. What we expect of a linear system whose parameters do not change with time is that its equation of motion will be expressible in the form

30, 40

$$\int_{-\infty}^{\infty} \xi(t')K(t'-t)\,\mathrm{d}t' = 0. \qquad (1)$$

which includes all the cases so far cited, if we allow the function K to take appropriate forms, including the Dirac δ-function and its derivatives. The δ-function itself, $\delta(t'-t)$, with its sharp peak of unit area at the moment when $t'-t$ vanishes, generates the value $\xi(t)$ from the integral. Putting $K(t'-t)=\dot{\delta}(t'-t)$, the first derivative of the δ-function, generates $\dot{\xi}(t)$; and so on for higher derivatives. Similarly if K is given the form $\delta(t'-t+t_0)$, the integral generates the value of ξ at $t-t_0$, and in this way a delay can be introduced. In general $K(t'-t)$ can take a suitable form to construct from $\xi(t)$ a summation (or, if K is a smooth function, an integral) over values of ξ and its derivatives both at the instant t and at any other instants. The linear property is clearly exhibited and so is the time-independence of the physical parameters, for nothing in (1) is related to any absolute time but only to what is occurring at the moment t and at other moments specified in relation to t.

The form of (1) is such that exponential solutions are to be looked for; on substituting e^{pt} for $\xi(t)$, we have immediately

$$\mathrm{e}^{pt}\int_{-\infty}^{\infty} \mathrm{e}^{px}K(x)\,\mathrm{d}x = 0, \quad \text{where } x = t'-t, \qquad (2)$$

and any value of p that causes the integral to vanish allows (1) to be satisfied at all times. The peculiar property that singles out the exponential function is this – the exponential curve is similar to itself everywhere, in the sense that shifting it bodily through t_0 has the same effect as multiplying it at each point by a constant, e^{pt_0}. As a special case of this property of similarity, the nth derivative bears a constant relationship at every point to the value of the function itself: $d^n\xi/dt^n = p^n\xi$. If the derivatives, delayed values etc. combine at any instant, t, to satisfy (1), they are bound to do so at all times since every term is scaled by the same amount in moving to another value of t.

It is one thing to recognize that self-similar exponential solutions to (1) are likely, but one must also consider other possibilities. It may be that no value of p exists which causes the integral in (2) to vanish; or that there are other solutions which are not self-similar. It may be shown, however, that all solutions of (1) may be cast in exponential form, provided p is permitted complex values.[1] For example, the sinusoidal solutions we have met in simple cases, of the form $\xi = \cos(\omega t + \phi)$, may be constructed by adding exponential solutions in which $p = \pm i\omega$:

$$\cos(\omega t+\phi) = A\,e^{i\omega t} + B\,e^{-i\omega t},$$

if

$$A = \tfrac{1}{2}e^{i\phi} \quad \text{and} \quad B = \tfrac{1}{2}e^{-i\phi}. \tag{3}$$

Since the displacement, ξ, is a real quantity any expression of it in terms of complex values of p must also include the complex conjugate; for example, in (3) $B\,e^{-i\omega t}$ is the complex conjugate of $A\,e^{i\omega t}$. The values of p, if complex, always appear in pairs, p and p^* together. For if e^{pt} is a solution of (1), e^{p^*t} is a solution of the complex conjugate equation; but ξ and K are real functions describing measurable physical quantities, so that the equation and its complex conjugate are identical. Eq. (2.23) will serve to illustrate the point. If the trial solution, $\xi = A\,e^{pt}$, is substituted the equation is reduced to an algebraic equation,

$$p^2 + kp + \omega_0^2 = 0,$$

yielding two possible values for p,

$$p = \tfrac{1}{2}[-k \pm (k^2 - 4\omega_0^2)^{\frac{1}{2}}]. \tag{4}$$

Let us keep ω_0 constant while allowing k to vary between $\pm\infty$. So long as $k^2 < 4\omega_0^2$ the two solutions are complex conjugates, but for larger values of k both are real. The solution of course matches (2.24) and (2.25) which were derived by the use of real functions alone, a somewhat more laborious process; it is one of the merits of complex exponentials in describing linear oscillations that they reduce differential equations to algebraic equations and minimize the effort of finding solutions.

The complex p-plane

As k changes, the solution of (2.23) represented by (4) describes a trajectory on the complex plane (fig. 1) whose essential features can be deduced directly from (2.23) before even proceeding to the solution. The product of the two values of p is ω_0^2, and so long as they are complex conjugates their product is pp^*, the square of the radius vector from the origin. Hence the complex solutions lie on a circle of radius ω_0, which is the natural frequency of the loss-free circuit. The overdamped solutions lie on the real axis, one inside and one outside the circle to keep the product equal to ω_0^2.

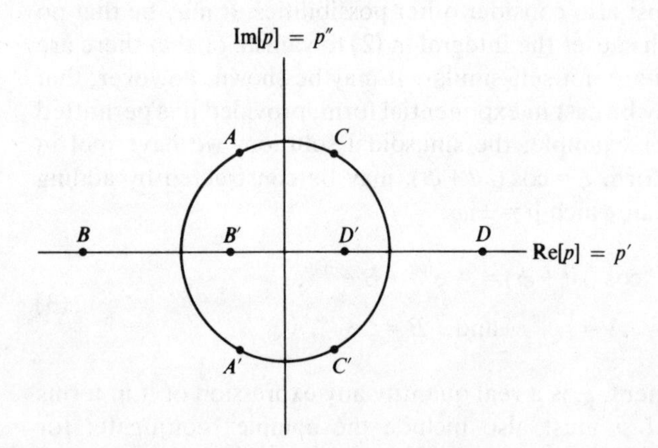

Fig. 1. The complex p-plane. Solutions of (2.23) lie in pairs, indicated by the same letters, on the circle of radius ω_0, or on the real axis; the left-hand side represents decaying modes, the right-hand side growing modes. A, A' are underdamped, B, B' overdamped, C, C' growing oscillatory modes, D, D' exponentially growing modes.

For a given value of k any linear combination of the two solutions is also a solution, and the most general solution takes the form

$$\xi = A\,e^{p_1 t} + B\,e^{p_2 t}. \tag{5}$$

Since ξ must be real, A and B must also be real, though otherwise arbitrary, in the overdamped case; in the underdamped case, where $p_2 = p_1^*$, $B = A^*$ but A may be complex. In both cases there are two arbitrary real constants in the solution, either A and B or the real and imaginary parts of A. These constants must be determined in any actual situation by the boundary conditions; if, for instance, at time $t = 0$ the displacement is ξ_0 and the velocity $\dot{\xi} = v_0$, we have from (5) that

$$A + B = \xi_0 \quad \text{and} \quad p_1 A + p_2 B = v_0,$$

which are enough to determine A and B. Physically this means that a simple oscillator set moving at a given speed from a given position is thenceforward fully determined in its motion. More complicated systems, described by higher order differential equations, will normally reduce, on substituting the trial solution e^{pt}, to algebraic equations of the same order, with as many characteristic values of p as the order of the highest differential coefficient; the number of arbitrary constants will be the same and so also will be the number of boundary conditions required to specify the behaviour completely. All complex values of p must occur in pairs, p and p^*, and obviously if the order is odd at least one solution must have a real value of p.

To return to the quadratic equation and its representation in fig. 1, it will be observed that the left-hand half of the circle, which arises when k is positive, describes decaying oscillations. Thus, two typical conjugate values of p may be written as $p' \pm ip''$, p' and p'' being real, and if the complex amplitude is written as $A\,e^{i\phi}$, A being real, the general solution has the form

$$\begin{aligned}\xi &= A\,e^{i\phi}\,e^{(p'+ip'')t} + A\,e^{-i\phi}\,e^{(p'-ip'')t}\\ &= 2A\,e^{p't}\cos(p''t+\phi).\end{aligned} \tag{6}$$

Negative values of p' clearly denote a decaying oscillation; positive values, on the right-hand side of the diagram, arise when k is negative and the oscillations build up exponentially. There is no mathematical reason to deny continuity to k between positive and negative values, and we have noted in chapter 2 systems in which external energy sources allow negative k to be realized; the whole of the diagram may be regarded therefore as accessible in principle. Its symmetry about the imaginary axis and its continuity lead to its crossing this axis normal to it. On the axis p is purely imaginary and describes a steady oscillation at frequency ω_0; as the lossy term k is added the decay represented by p' appears as a first-order effect in k, but the change in frequency is of second order. It is only when there is considerable decay during one cycle that the frequency suffers any significant change.

39

This is a rather general result, as may be seen by considering a differential equation of arbitrarily high order:

$$a + b\ddot{\xi} + c\ddddot{\xi} + \cdots = \alpha\dot{\xi} + \beta\dddot{\xi} + \cdots,$$

in which we have segregated even and odd terms. On substituting $\xi = e^{pt}$ the algebraic equation results:

$$a + bp^2 + cp^4 + \cdots = \alpha p + \beta p^3 + \cdots.$$

Pure imaginary values of p are only possible if either the left or right of the equation vanishes; a differential equation describing a non-dissipative system contains only even or only odd differential coefficients. Let us suppose that the system under consideration has even coefficients in its loss-free state, all the coefficients on the right vanishing, so that the introduction of weak losses is represented by small values of α, β etc. Then if a typical pair of loss-free solutions is $\pm p_0$, p_0 being imaginary,

$$a + bp_0^2 + cp_0^4 + \cdots = 0.$$

The first-order effect of the losses may be obtained by equating the first-order change in the left-hand side to the value of the right-hand side when $p=p_0$:

$$(2bp_0+4cp_0^3+\cdots)\,\delta p \approx \alpha p_0+\beta p_0^3+\cdots.$$

Since both sides now contain only odd orders, both brackets are pure imaginary and δp is therefore real. The trajectory of p must leave the imaginary axis in a normal direction.

In talking of a trajectory one implicitly assumes that the losses are controlled by a single physical parameter, i.e. that α, β etc. all vary in a well-defined way as the lossy mechanism (e.g. η, the viscosity of the oil in a number of dashpots) is systematically changed. The real and imaginary parts of p are then functions of η, and the trajectory is defined parametrically by $p'(\eta)$ and $p''(\eta)$. If, however, the coefficients α, β etc. can be varied independently the trajectory of p on the complex plane spreads out to cover whole areas (domains), any one point in which may be realized by appropriate adjustment of the lossy mechanisms. What the analysis just presented implies is that these domains narrow down to sharp points as they approach the imaginary axis so that one can only reach the axis, while remaining within a domain, in a normal direction.

This point may be illustrated by means of the simple pendulum shown in fig. 2, which is constrained to vibrate in one plane and has damping vanes attached to the bob and at a point B on the string. The mass of the vane at B can be made negligible, so that the equations of motion of B, whose displacement is x_1, and C, whose displacement is x_2, take the form

$$mg[x_1/l_1-(x_2-x_1)/l_2]+\lambda_1\dot{x}_1=0$$

and

$$m\ddot{x}_2+\lambda_2\dot{x}_2+mg(x_2-x_1)/l_2=0.$$

Here λ_1 and λ_2 define the viscous drag of the dashpots at B and C. When

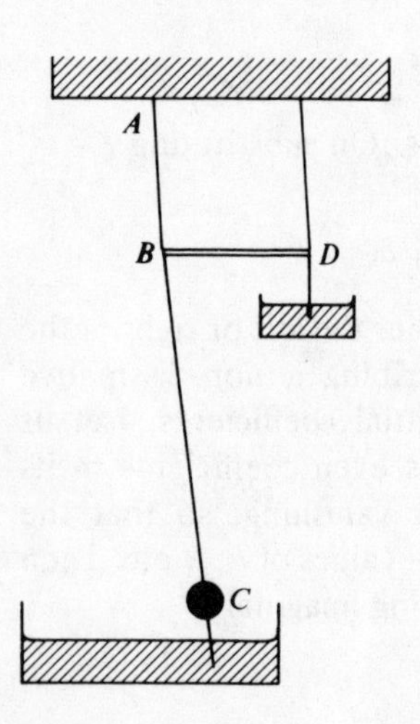

Fig. 2. The simple pendulum ABC has a light rod carrying a damping vane BD attached at B, and another damping vane attached to the bob. $AB=l_1$ and $BC=l_2$.

x_1 is eliminated from these equations the motion of the bob is found to be governed by the third-order equation:

$$\dddot{x}_2+(\omega_3+\omega_4)\ddot{x}_2+(\omega_2^2+\omega_3\omega_4)\dot{x}_2+\omega_1^2\omega_3 x_2=0,$$

in which

$$\omega_1^2=g/(l_1+l_2),\quad \omega_2^2=g/l_2,\quad \omega_3=mg(l_1+l_2)/\lambda_1 l_1 l_2$$

and

$$\omega_4=\lambda_2/m.$$

Correspondingly there are three values of p, obeying the equation

$$p^3+(\omega_3+\omega_4)p^2+(\omega_2^2+\omega_3\omega_4)p+\omega_1^2\omega_3=0. \tag{7}$$

There is always one real solution describing a rapidly decaying transient, which we ignore.

It is convenient to discuss the oscillatory solutions of this equation in two stages, first setting $\lambda_2=0$ so that only the damping vane at B is operative and p follows a trajectory in the complex plane:

$$p^3+\omega_3 p^2+\omega_2^2 p+\omega_1^2\omega_3=0. \tag{8}$$

The damping represented by ω_3 appears in two terms but they are not, of course, independently variable. The trajectory of p, like that for a simple harmonic vibrator, is still defined by a single parameter, though its form is markedly altered by ω_3 appearing twice in a third-order equation; there are now two situations, rather than one, where lossless vibration can occur. One is when ω_3 is very large (virtually no damping at B) and the pendulum swings as a simple pendulum from A ($p=\pm i\omega_1$); the other when $\omega_3=0$ (very heavy damping) and only the second and fourth terms in (8) matter; $p=\pm i\omega_2$, the pendulum swinging from B, again without decrement. The trajectory between these points may be found by computing solutions of (8) for a range of values of ω_3, or by a graphical construction based on rearranging (8) in the form

$$\frac{p^2+\omega_1^2}{p^2+\omega_2^2}=-\frac{p}{\omega_3}. \tag{9}$$

At every point on the trajectory, i.e. for every value of ω_3, the two sides must have the same argument (the value of θ when a complex number is represented as $a\,e^{i\theta}$, a being real). On the complex plane of fig. 3, with origin at O, M represents $(-\omega_1^2)$ and N represents $(-\omega_2^2)$. Choose a point X on the perpendicular bisector of MN and draw a circle C_1 centred on X to pass through M and N; draw C_2 centred on the origin and with the same radius, to cut the positive real axis in Q. Construct QR equal to MN and extend OR to cut C_1 in S_1 and S_2. Then S_1 and S_2 are points on the trajectory of p^2 between M and N;† other points are found by moving the

† *Proof.* By construction, the triangles NMX and ROQ are congruent. Therefore $\widehat{QOR}=\widehat{NXM}=2\,\widehat{NSM}$, where S is any point on C_1. The argument of S_1 is $\widehat{QOR}$, so that if S_1 represents p^2, the argument of p is $\frac{1}{2}\widehat{QOR}$, i.e. $\widehat{NS_1M}$. But if S_1 represents p^2, $MS_1=p^2+\omega_1^2$ and $NS_1=p^2+\omega_2^2$. The argument of $(p^2+\omega_1^2)/(p^2+\omega_2^2)$ is therefore $\widehat{NS_1M}$, the same as the argument of p; consequently it is possible to find a real value of ω_3 such that (9) is satisfied by this value of p.

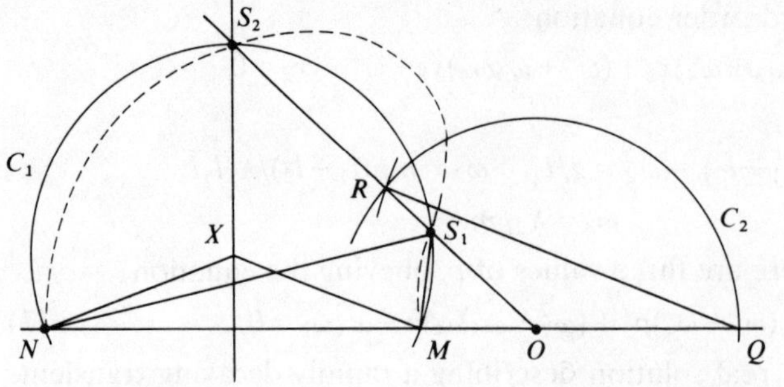

Fig. 3. Graphical construction of the solution of (8).

centre X to different positions, and the whole trajectory of p^2 above the real axis is shown as a broken line. Its mirror image below the axis completes the diagram. Every value of p^2 generates two values of p, and the single loop for p^2 becomes a double loop for p, having symmetry about both real and imaginary axes, as in fig. 4.

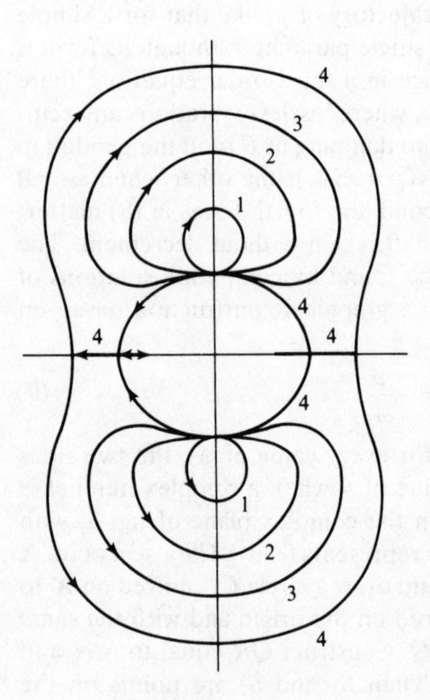

Fig. 4. Solutions of (8) for four different positions of B in fig. 2: (1) vane close to the top; (2) and (3) vane progressively lower; (4) vane low enough to allow critical damping – the trajectory consists of two loops joined by part of the real axis. The arrows show the direction of motion along the trajectory as the damping is increased.

A number of trajectories, corresponding to different positions of the damping vane on the pendulum, are shown in fig. 4. Note how they are all normally incident on the imaginary axis. The total length, l_1+l_2, and therefore ω_1 have been kept constant while l_2 and ω_2 are varied. In contrast to a pendulum whose bob alone is damped and whose trajectory, like that in fig. 1, exhibits one critical point where it meets the negative real axis, this system exhibits either none (if $l_1 < 8l_2$) or two. The difference clearly lies in the ability of the pendulum to swing without damping when B is locked by a very viscous dashpot, unless B actually coincides with the bob.

When both damping mechanisms are present, with λ_1 and λ_2 independently variable, p can occupy a domain whose bounds are fixed by the following considerations. Rearrange (7) in two alternative ways:

$$\omega_3(p^2+\omega_4 p+\omega_1^2)+p(p^2+\omega_4 p+\omega_2^2)=0, \qquad (7')$$

$$(p^3+\omega_3 p^2+\omega_2^2 p+\omega_1^2\omega_3)+\omega_4 p(p+\omega_3)=0. \qquad (7'')$$

The vanishing of each bracketed expression defines a trajectory, those in (7′) being parametrized by ω_4 and those in (7″) by ω_3. The trajectories are drawn in fig. 5. When $\omega_3=0$ and ω_4 is variable, p follows the trajectory C_2; while when ω_3 is made very large it is the bracket defining C_1 that dominates (7′) and C_1 is therefore the trajectory of p. It may now be noted that since C_1 and C_2 do not intersect there is no intermediate value of ω_3 which can allow any point on either C_1 or C_2 to satisfy (7′). As ω_3 is varied from 0 to ∞, ω_4 being constant, the trajectory of p moves from C_2 to C_1 without passing through either. By the same argument, as ω_4 is varied from 0 to ∞, ω_3 being constant, p moves from C_3 to C_4 without intersecting either. The domain available to p is thus the shaded region bounded by these curves. It may be noted in particular that as λ_1 and λ_2 are raised from zero in any way, the trajectory of p must leave the axis normally, as implied by the general argument given previously.†

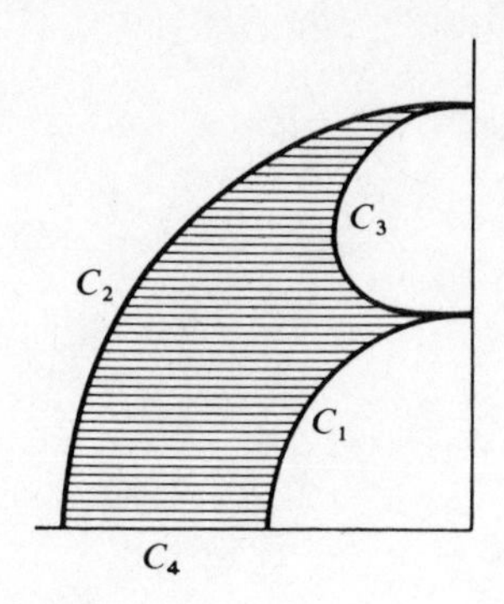

Fig. 5. Domain in which the solutions of (8) fall when ω_3 and ω_4 are positive; the complex conjugate solutions, mirrored below the real axis, are omitted. The circular arcs C_1 and C_2, of radius ω_1 and ω_2, are described parametrically by the equations:

$$p^2+\omega_4 p+\omega_{1,2}^2=0$$

$$(0<\omega_4<\infty).$$

C_3 is described by the equation, $p^3+\omega_3 p^2+\omega_2^2 p+\omega_1^2\omega_3=0$ $(0<\omega_3<\infty)$, and is the same as one of the trajectories in fig. 3. C_4 is described by the equation $p+\omega_3=0$ $(0<\omega_3<\infty)$.

† If λ_1 and λ_2 can be negative, other domains become accessible to p. Thus it is possible to choose λ_1 and λ_2 with opposite signs so that the pendulum vibrates at constant amplitude, p being pure imaginary but not $i\omega_1$ or $i\omega_2$. Further analysis of this system is left to the interested reader.

Complex frequency; conventions

If one is more interested in lightly damped vibrators than in heavily or overdamped systems, it is convenient to take as the standard form of solution $e^{i\omega t}$ or $e^{-i\omega t}$ rather than e^{pt}; in other words to introduce a new variable, the complex angular frequency, defined by the relation $\pm i\omega \equiv p$. Real values of ω describe undamped vibrations, complex values decaying or growing vibrations. There is no guide but convention to the choice of sign in the definition, since p and p^* always occur together in the solution of equations in real variables. If $p_1 = p' + ip''$ and $p_2 = p_1^* = p' - ip''$, the alternative forms of ω are as follows:

(i) if $i\omega \equiv p$, $\quad \omega_1 = p'' - ip'$, $\quad \omega_2 = -p'' - ip' = -\omega_1^*$.
(ii) if $-i\omega \equiv p$, $\quad \omega_1 = -p'' + ip'$, $\quad \omega_2 = p'' + ip' = -\omega_1^*$.

For a decaying vibration p' is negative, so that the imaginary part of ω is positive in (i), negative in (ii). The real part of ω takes either sign in both (i) and (ii); when numerical values of ω are quoted it is conventional to state the real part as a positive number.

Corresponding to these definitions there are differences in the representations of trajectories on the complex plane. Thus if convention (i) is adopted, a trajectory on the p-plane is converted to one on the ω-plane by multiplying p by $-i$, i.e. rotating the diagram through $\pi/2$ in a clockwise sense; but if convention (ii) is adopted the diagram must be rotated anticlockwise. In addition the sign of the real part of ω may be altered, i.e. the diagram may be reflected in the imaginary axis. The three representations are illustrated in fig. 6.

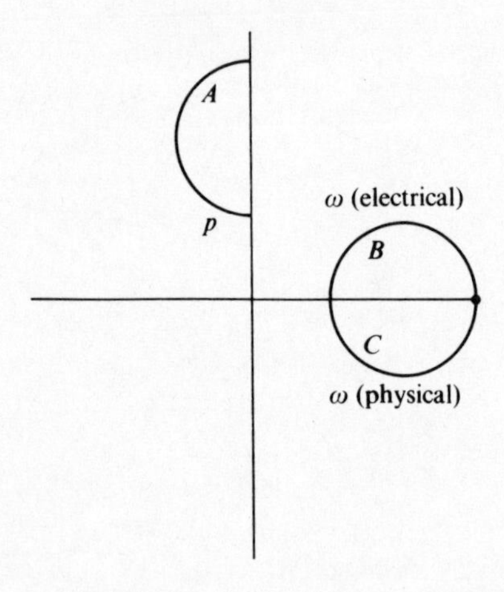

Fig. 6. Relation between p and ω; the p-trajectory is represented on the ω-plane by either B or C, according to the convention adopted.

The labelling of convention (i) as electrical and (ii) as physical springs from customary usage in electrical engineering on the one hand, and quantum mechanics on the other. In addition the electrical engineer, to avoid confusion with the symbol i as representing current, designates $\sqrt{(-1)}$ as j. The engineer, inserting a trial oscillatory solution into, for example,

(2.23), uses $e^{j\omega t}$, while the physicist uses $e^{-i\omega t}$. We shall be concerned with both fields and shall keep to the appropriate custom as far as possible, while adhering to the physical convention as the norm. The reader may like to think of j as being the same as $-i$, both being square roots of -1. It must be remembered, however, that when j is used decaying oscillations are represented by points in the upper half of the complex ω-plane, while when i is used they fall in the lower half.

The complex frequency ω may be broken up into its real and imaginary parts by writing it as either $\omega' - i\omega''$ or $\omega' + j\omega''$, the signs being chosen so that positive values of ω'' represent decaying oscillations. It has already been noted in (2.2) that, since $\tau_a = 1/\omega''$ and $\tau_e = \frac{1}{2}/\omega''$, the quality factor Q is $\frac{1}{2}\omega'/\omega''$. Slowly decaying vibrations, with high Q, are described by complex ω with very small arguments ($\tan^{-1}(\omega''/\omega') \ll 1$); but if $\omega''/\omega' = 1$, with the vector representing ω lying at $\pi/4$ to the real axis, the amplitude decays by a factor $e^{2\pi}$, i.e. 535, during one cycle. A considerable sector of the complex plane therefore corresponds to such heavy damping that the vibratory character of the system would hardly be discernible.

The vibration diagram

The typical solution (5) for an underdamped vibration, in which the two terms are complex conjugates, may be rewritten in terms of frequency as

either $$\xi = A\,e^{-i\omega t} + A^*\,e^{i\omega^* t},$$

or $$\xi = A\,e^{j\omega t} + A^*\,e^{-j\omega^* t}.$$

Because the second term in each case is the complex conjugate of the first, it is normal to take it for granted. Thus when it is stated that the vibration of a physical system is described by $A\,e^{-i\omega t}$, it is to be assumed that only the real part is meaningful:

$$\xi = A\,e^{-i\omega t}$$

signifies $$\xi = \mathrm{Re}\,[A\,e^{-i\omega t}] = \tfrac{1}{2}[A\,e^{-i\omega t} + A^*\,e^{i\omega^* t}].$$

The amplitude and phase of the vibration are both contained in the complex amplitude, A, but it must be remembered that the physical and electrical conventions involve different interpretations. If $A = a\,e^{i\phi}$, a being real, according to the physical convention $\xi = \mathrm{Re}\,[a\,e^{-i(\omega t - \phi)}]$; this is a vibration having maximum positive displacement when $\omega t - \phi = 0$, i.e. at a time ϕ/ω, so that vibrations of greater ϕ *lag* behind those of lesser ϕ. Conversely, according to the electrical convention an oscillation $A\,e^{j\phi}$ implies that $\xi = \mathrm{Re}\,[a\,e^{i(\omega t + \phi)}]$, and ϕ is a phase *lead*.

The possibility of describing a vibration by means of a single complex exponential leads to a helpful diagrammatic representation, based on the complex plane but with a quite different interpretation. In the previous discussion a point on the complex plane represented the complex frequency – now we use it to represent the phase and amplitude of the vibration itself. If ξ has the form $A\,e^{-i\omega t}$, or $a\,e^{-i(\omega t-\phi)}$, ω being real (constant amplitude of vibration),† the corresponding point on the plane is found by drawing a vector of length $|A|$, i.e. a, from the origin at an angle $(\omega t-\phi)$ from the positive horizontal axis in a clockwise direction. As time proceeds, the vector rotates clockwise at an angular velocity ω, and its projection on the real axis describes harmonic motion of amplitude $|A|$. Different vibrators having the same frequency, ω, but different phase angles ϕ are drawn as vectors making different angles with the horizontal, but all sweeping around together at the same speed. In describing the behaviour of different parts of a complicated system (such as an electrical circuit) which have different amplitudes and phases but a common frequency, it is normally only the relative phases that matter, the angles between the different vectors rather than their absolute orientation. All that need be shown in the diagram, then, to represent any one vibration, is the complex amplitude A, which is drawn as a fixed vector of length a making an anticlockwise angle ϕ with the horizontal axis; it is to be understood that $e^{-i\omega t}$ has been suppressed. This is the *vibration diagram* (or *phasor diagram* – there is no generally accepted terminology). It is illustrated in fig. 7, which also shows in elementary form one of the

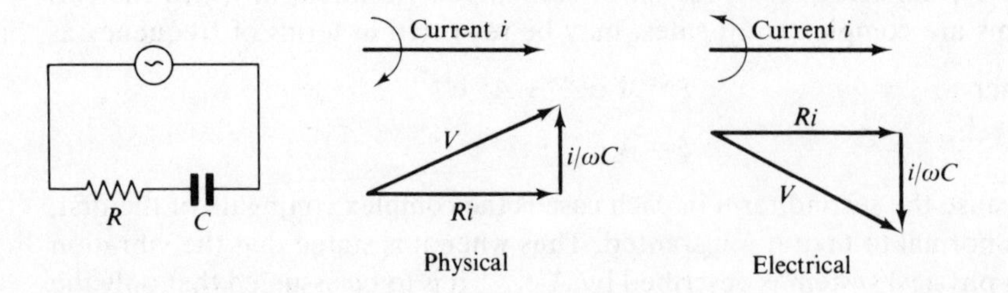

Fig. 7. Vibration diagram for the voltages in an electrical circuit, according to the physical and electrical conventions. The rotation direction is such that the current i, which has the same phase as Ri, the voltage across the resistor, leads the generator voltage V.

principal uses of the diagram in forming the resultant of a number of superimposed vibrations, by the simple process of vector addition. If the suppressed $e^{-i\omega t}$ is included just for the moment, each vector rotates at angular velocity ω, but while all components to be added have the same frequency, the vectors all rotate together, keeping the resultant constant and itself rotating at the common angular velocity to generate, as its real part, a vibration that is still harmonic. When the components differ in frequency the configuration changes as they turn relative to one another, and the resultant displacement oscillates in a non-sinusoidal manner. We shall discuss a variety of instances in the next chapter.

67
80

† If ω is complex, $\omega'-i\omega''$, $e^{-i\omega t}$ describes an equiangular spiral, the curve whose polar equation has the form $r=a\,e^{\mu\theta}$. The equiangular spiral is a self-similar curve in that bodily rotation though ϕ is equivalent to a uniform scale change, $r\rightarrow r\,e^{\mu\phi}$. The vibration diagram described in this section, though in principle applicable when $\omega''\neq 0$, is only really useful when $\omega''=0$.

Circuit analysis

The use of vibration diagrams for analysing electrical circuits, as in fig. 7, has largely been superseded by algebraic analysis, a technique too well known to need more than passing comment here. To write the impedance of a capacitor as $1/\mathrm{j}\omega C$, or of an inductor as $\mathrm{j}\omega L$, is to express the same relationships between e.m.f. and current as do the vibration diagrams. Both depend on the differential equations describing the behaviour of the circuit elements:

$$\text{for a capacitor,}\quad C\dot{V} = i,$$

$$\text{for an inductor,}\quad L\dot{i} = V.$$

If V and i are oscillatory and represented by $V\,\mathrm{e}^{\mathrm{j}\omega t}$ and $i\,\mathrm{e}^{\mathrm{j}\omega t}$, it is only necessary to replace each time-differentiation by the multiplier $\mathrm{j}\omega$ to obtain the phase and amplitude relationships for the circuit elements:

$$\mathrm{j}\omega CV = i, \quad \text{or} \quad Z \equiv V/i = 1/\mathrm{j}\omega C$$

and $$\mathrm{j}\omega Li = V, \quad \text{or} \quad Z \equiv V/i = \mathrm{j}\omega L.$$

This substitution is just as valid when ω is complex as when it is real, but it should be noted that while V and i remain in phase for a resistor, the quadrature relation for a capacitor or inductor is broken; the tangent to an equiangular spiral (see footnote on p. 54) is not normal to the radius vector.

To see how circuit analysis may replace the manipulation of differential equations, consider the series resonant circuit of fig. 2.1(c), reproduced in fig. 8(a). If the circuit is broken, at XX for instance, and the impedance Z between these points calculated, a complex frequency, ω, may be found at which $Z = 0$; at such a frequency current can flow in the circuit without establishing a potential difference across XX, which may therefore be

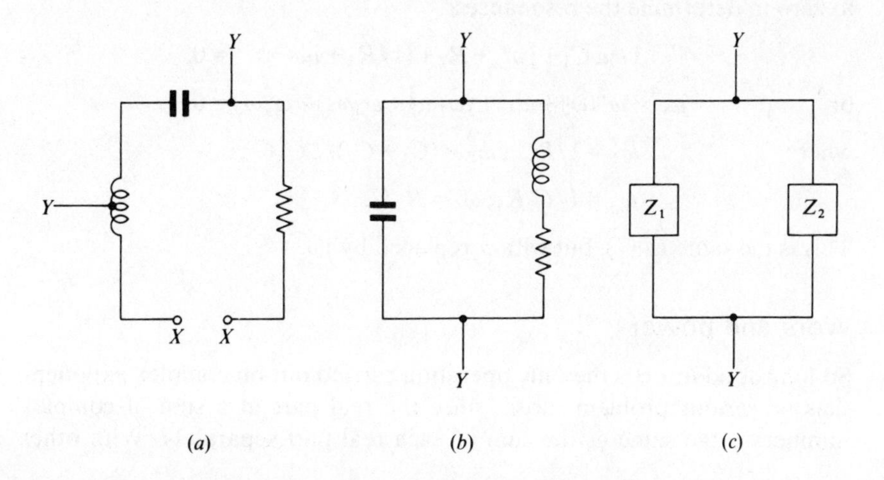

Fig. 8. Resonant circuits: (a) series resonant, when fed at XX, (b) parallel resonant, (c) generalized form of (b).

connected together. The circuit is thus proved capable of free oscillation at any frequency which causes its impedance to vanish. In this case

$$\mathrm{j}\omega L+R+1/\mathrm{j}\omega C=0,$$

or

$$L\omega^2-\mathrm{j}R\omega+1/C=0,$$

which is the same as is obtained from (2.22) when $\mathrm{d}/\mathrm{d}t$ is replaced by $\mathrm{j}\omega$. It is worth noting that since the total impedance round the circuit vanishes at resonance, the impedances, and therefore the admittances, along the two paths between any two points, YY say, must be equal and opposite. When an e.m.f. is applied to YY the resultant admittance, which determines the current, vanishes at resonance. The resonance condition may therefore be expressed in terms of a parallel resonance circuit, as in fig. 8(*b*) or, more generally, 8(*c*). If YY are not connected to an external source the only oscillations of e.m.f. that can occur must be such that the net current leaving or entering at Y is zero, which is achieved when $Z_1=-Z_2$.

30

The method may be applied with advantage to the less trivial problem presented by the pendulum in fig. 2, whose properties are modelled by the circuit of fig. 9. When R_1, which is analogous to λ_1, is small the circuit

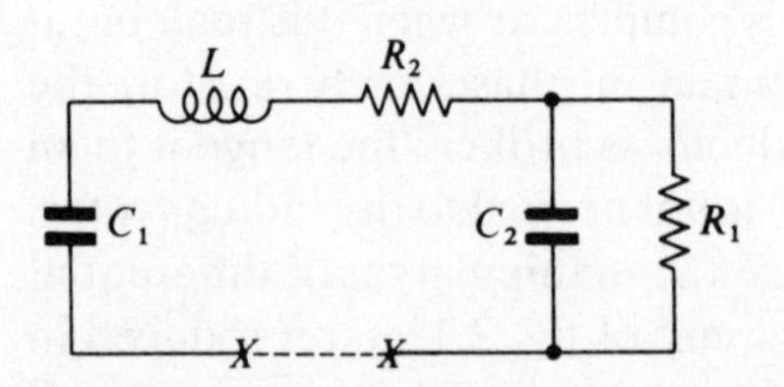

Fig. 9. Equivalent circuit for the pendulum of fig. 2.

behaves like L, C_1 and R_2 in series; when it is large, like L, C_1, C_2 and R_2 in series. As R_1 is varied from 0 to ∞, the resonance frequency goes from a lower to a higher value, and the damping passes through a maximum. We now break the circuit at XX and calculate its impedance, which we equate to zero to determine the resonances:

$$1/\mathrm{j}\omega C_1+\mathrm{j}\omega L+R_2+(1/R_1+\mathrm{j}\omega C_2)^{-1}=0,$$

or

$$-\mathrm{j}\omega^3-\omega^2(\omega_3+\omega_4)+\mathrm{j}\omega(\omega_2^2+\omega_3\omega_4)+\omega_1^2\omega_3=0,$$

where

$$\omega_1^2=1/LC_1,\ \omega_2^2=(C_1+C_2)/LC_1C_2,$$

$$\omega_3=1/C_2R_1,\ \omega_4=R_2/L.$$

This is the same as (7), but with p replaced by $\mathrm{j}\omega$.

Work and power

So long as addition is the only operation carried out on complex exponentials no serious problems arise, since the real part of a sum of complex numbers is the same as the sum of each real part separately. With other operations, however, it is essential to remember that only the real part has meaning, and in general the safe course is to take the real part at the outset, before any other operation is performed. There are certain manipulations, however, which can be accomplished on the complex numbers by knowing the appropriate rules, and of these multiplication is the most important. Suppose an oscillatory force $F_0 \cos(\omega t+\phi_1)$ acts on a body whose velocity component parallel to the force is $v_0 \cos(\omega t+\phi_2)$ – both having the same frequency but with a phase difference. The rate, P, at which work is done by the force is the product of force and velocity, correctly given by

$$\begin{aligned} P &= F_0 v_0 \cos(\omega t+\phi_1)\cos(\omega t+\phi_2) \\ &= \tfrac{1}{2}F_0 v_0[\cos(2\omega t+\phi_1+\phi_2)+\cos(\phi_1-\phi_2)]. \end{aligned} \tag{10}$$

The first term in the square brackets oscillates at frequency 2ω about a mean value zero, and represents energy being alternately transmitted to and taken from the body; the second term is a constant power transfer which may have either sign, depending on the phase difference, $\phi_1-\phi_2$. For the mean power transfer only the second term matters,

$$\bar{P} = \tfrac{1}{2}F_0 v_0 \cos(\phi_1-\phi_2). \tag{11}$$

In terms of the complex representation, F is expressed as $(F_0 \mathrm{e}^{-\mathrm{i}\phi_1})\mathrm{e}^{-\mathrm{i}\omega t}$ and v as $(v_0 \mathrm{e}^{-\mathrm{i}\phi_2})\mathrm{e}^{-\mathrm{i}\omega t}$, the amplitudes being placed in brackets. The real force is $\frac{1}{2}(F+F^*)$ and the real velocity $\frac{1}{2}(v+v^*)$, so that

$$\begin{aligned} P &= \tfrac{1}{4}(F+F^*)(v+v^*) \\ &= \tfrac{1}{2}\mathrm{Re}\,[Fv+Fv^*] \quad \text{or} \quad \tfrac{1}{2}\mathrm{Re}\,[Fv+F^*v]. \end{aligned} \tag{12}$$

Each term in the brackets is the same as the corresponding term in (10) so that, in particular, the mean power transfer is to be found by forming

$$\bar{P} = \tfrac{1}{2}\mathrm{Re}\,[Fv^*] \quad \text{or} \quad \tfrac{1}{2}\mathrm{Re}\,[F^*v]. \tag{13}$$

Waves

In many oscillatory systems the vibrations can best be described in terms of standing waves formed by the reflection and superposition of travelling waves. The complex representation derived for simple vibrators is readily extended to wave-motion. If the medium is linear in its properties, uniform so that any one part is similar to all others, and unchanging, so that translational invariance applies to time and space coordinates alike, all waveforms can be derived from exponential functions, though now functions of both time and space coordinates. To a certain degree this fact is obscured by the common habit in elementary texts of discussing almost exclusively the so-called Wave Equation, as if there were not an infinite number of differential and integral equations with wavelike solutions. This particular equation, with the one-dimensional form

$$\frac{\partial^2 \xi}{\partial x^2}-\frac{1}{v_0^2}\frac{\partial^2 \xi}{\partial t^2}=0, \tag{14}$$

describes the variation with x and t of a physical quantity ξ, which may be the transverse displacement of a stretched string, the longitudinal displacement of elements of air in a trombone, or the electric field between two parallel wires, to list but a few. It happens to have a general solution of particularly simple form

$$\xi = f_1(x - v_0 t) + f_2(x + v_0 t), \tag{15}$$

in which f_1 and f_2 are arbitrary functions, requiring only such properties of continuity and differentiability as will make (14) meaningful when they are substituted. If only f_1 is present the equation describes a displacement imposed on the system which runs along at velocity v_0 in the direction of increasing x, without changing its shape; this is because the simultaneous increase of t by t_0 and of x by $v_0 t_0$ leave the argument unchanged; if at time t the displacement at x was ξ, the same displacement will reach $x + v_0 t_0$ at time $t + t_0$. The second function, f_2, in (15) represents a disturbance travelling in the opposite direction at the same speed, and the two are independent, being able to pass one another as though each alone were present.

This is a non-dispersive wave, and indeed a very important example since, among others, the electromagnetic wave in free space is of this type. But waves in real materials tend to be dispersive, the shape of a disturbance does not propagate unchanged, and the simple wave equation is inadequate. A much more general linear equation may be constructed by analogy with (1). If the behaviour at any point is determined by what is going on actually at that point or in the neighbourhood, at that or some related instant of time, the linear superposition of all influences may be conflated into an integral equation of motion:

$$\int_{-\infty}^{\infty}\int_{-\infty}^{\infty} \xi(x', t') K(x' - x, t' - t)\, \mathrm{d}x'\, \mathrm{d}t' = 0,$$

of which a special type may be a partial differential equation involving derivatives other than those appearing in (14), and possibly including mixed derivatives like $\partial^{n+m}\xi/\partial x^n\, \partial t^m$. Such equations, like (14), have solutions in exponential form and the most general solution can be expressed as a sum of exponentials. A typical harmonic component may be written (adopting the physical convention)

$$\xi = A\, \mathrm{e}^{\mathrm{i}(kx - \omega t)}, \tag{16}$$

in which the *wave-number* k and the frequency ω may in general be complex, and A is a constant which may also be complex. When this is substituted in a partial differential equation the same transformation to an algebraic equation occurs as with the equation of an oscillator. For

$$\partial\xi/\partial x = \mathrm{i}kA\, \mathrm{e}^{\mathrm{i}(kx-\omega t)} = \mathrm{i}k\xi,$$

$$\partial\xi/\partial t = -\mathrm{i}\omega A\, \mathrm{e}^{\mathrm{i}(kx-\omega t)} = -\mathrm{i}\omega\xi;$$

in general

$$\partial^{n+m}\xi/\partial x^n\, \partial t^m = (\mathrm{i}k)^n(-\mathrm{i}\omega)^m\xi. \tag{17}$$

Thus (14) when treated in this way becomes, quite simply,

$$-k^2\xi+\omega^2\xi/v_0^2=0,$$
$$\text{i.e.}\qquad k=\pm\omega/v_0. \tag{18}$$

For each choice of frequency (achieved with a stretched string, for example, if ω is to be real, by waggling any one point on the string with regular sinusoidal motion), a wave propagates with velocity $v=\omega/k$, since this is the relation that must exist between changes of x and t if the argument $(kx-\omega t)$ is to stay constant. In this case $v=v_0$ and is constant – every sinusoidal wave, whatever its angular frequency ω or wavelength $2\pi/k$, travels at the same speed, a special case of the non-dispersive property that all disturbances travel unchanged.

In general, however, substitution of (16) in an equation containing several differential coefficients leads to a more complicated algebraic relationship between k and ω; an equation of the nth degree generates a polynomial of the same degree in ω and k whose vanishing gives the dispersion curve – the dependence of k on ω. A simple example is provided by waves of bending on an elastic rod, the lateral displacement ξ being governed by the differential equation (there is no occasion to derive it here):[2]

$$\partial^4\xi/\partial x^4+a^4\,\partial^2\xi/\partial t^2=0,\text{ where } a \text{ is a constant.} \tag{19}$$

On substituting (16) the dispersion equation is seen to take the form:

$$k^4=a^4\omega^2,$$
$$\text{i.e.}\qquad k=\pm a\omega^{\frac{1}{2}}\quad\text{or}\quad\pm ia\omega^{\frac{1}{2}}. \tag{20}$$

The two real solutions, $\pm a\omega^{\frac{1}{2}}$, show that sinusoidal waves can propagate in either direction with the same speed at a given frequency, but that higher frequency waves go faster ($v=\omega/k\propto\omega^{\frac{1}{2}}$) the wavelength varying in proportion to $\omega^{-\frac{1}{2}}$ rather than $1/\omega$ as in non-dispersive waves. In addition there are non-propagating disturbances, for the imaginary forms of k produce solutions of the form $e^{-i\omega t}\,e^{\pm a\sqrt{\omega}x}$, in which the amplitude of the oscillations grows or decays exponentially along the rod while every part vibrates in phase. In this system an arbitrary disturbance soon changes its shape, and only a sinusoidal wave continues with the same form.

As with simple oscillations, the complex wavelike solutions need to be combined if they are to describe real disturbances, but there are now more possibilities. Since ξ in (16) is the solution of an equation with real coefficients, ξ^* is also a solution; thus if (ω, k) defines a solution, so also does $(-\omega^*, -k^*)$. Moreover, if waves can propagate in the same form in both directions, $(\omega, -k)$ and $(-\omega^*, k^*)$ are also solutions. The equivalence of both directions (which does not necessarily hold for waves travelling in a rotating body or in the presence of a magnetic field) exhibits itself in the differential equation by the presence only of even differential coefficients, $\partial^2/\partial x^2$, $\partial^4/\partial x^4$ etc., which are unchanged when x is replaced by $-x$; in

212
240

consequence only even powers of k appear in the dispersion equations. If the medium is loss-free ω and k can be real, and real disturbances can be described in terms of the four solutions $(\pm\omega, \pm k)$. Thus we can construct:

travelling waves

$$e^{i(kx\pm\omega t)}+e^{-i(kx\pm\omega t)}=2\cos(kx\pm\omega t), \tag{21}$$

standing waves

$$e^{i(kx+\omega t)}+e^{i(kx-\omega t)}+e^{-i(kx+\omega t)}+e^{-i(kx-\omega t)}=4\cos kx\cos\omega t, \tag{22}$$

or, more generally, by an arbitrary combination of travelling waves in both directions, but of different amplitudes, partial standing waves.

If the medium is lossy, ω and k cannot both be real; in different situations either may be taken to be real and the other complex – more exceptionally both may be complex but there is no need to discuss examples of this last. It should be appreciated that the dispersion relation defines either ω or k in terms of the other, and it is therefore open to choose which, if any, shall be real. If a vibrator of unvarying amplitude generates a wave on a long string, it is obvious that one must take ω as real so that the time variation in $e^{i(kx-\omega t)}$ is constant in amplitude. Consequently k is complex, $k'+ik''$ say, and the variation along the string shows exponential attenuation if k'' is positive, e^{ikx} becoming $e^{ik'x}\,e^{-k''x}$. A wave travelling in the opposite direction with ω unchanged has both parts of k reversed in sign, so that the amplitude along the string is still attenuated in the direction of propagation.

This prescription will not work, however, for the free vibration of a standing wave. The half-wave pattern of the lowest mode of a string, as shown in fig. (2.15) can be constructed according to (22) if there are no losses, and the boundary condition that $\xi=0$ at both ends is satisfied if $kx=\pm\pi/2$ at the ends. Physically, one can imagine a travelling wave running to and fro, and being reflected with phase reversal at the ends, so that the incident and reflected waves always add to give zero displacement there. Provided the double length is an exact multiple of a wavelength there is no discontinuity anywhere and the wave equation is satisfied. But as soon as there is attenuation the travelling wave after traversing the string once in each direction has lost amplitude and can no longer join up without discontinuity. The boundary conditions can in fact only be satisfied by taking k to be real, and ω complex. This is not surprising, for we expect such a standing wave, unsustained by any source, to decay. If we write ω as $\omega'-i\omega''$, and take k as real, (22) may be rewritten as the sum of four complex components:

26

$$e^{i(kx-\omega t)}+e^{i(-kx-\omega t)}+e^{i(-kx+\omega^* t)}+e^{i(kx+\omega^* t)}$$

$$=2\cos kx(e^{-i\omega t}+e^{i\omega^* t})$$

$$=4\,e^{-\omega'' t}\cos kx\cos\omega' t.$$

The vector diagram is also useful here in visualizing different combinations, and here again there is no need to keep the conjugate complex functions in the diagram. By convention waves moving in opposite directions are ascribed the same ω and opposite signs of k. Travelling waves of amplitude A are thus expressed as $A\,e^{i(\pm kx-\omega t)}$ and represented by a single vector $A\,e^{ikx}$ or $A\,e^{-ikx}$, according to the direction of motion; this vector shows the amplitude and phase (at $t=0$) of the oscillation at the point x. In a lossless medium the oscillations at different points are represented by vectors of the same length but pointing in different directions, giving a graphic description of the effect of a travelling wave in causing every point to vibrate equally, but with a progressive phase shift. A partial standing wave is expressed as $A\,e^{i(kx-\omega t)}+B\,e^{i(-kx-\omega t)}$, where A and B may be complex, and represented as the sum of two vectors $A\,e^{ikx}+B\,e^{-ikx}$. As one moves along the line, increasing x, the vector of length A turns clockwise and that of length B anticlockwise, so that at some point the resultant takes its maximum value, $|A|+|B|$, and half a wavelength away its minimum $|A|-|B|$. Every point oscillates sinusoidally, but the amplitude varies with position. With the complete standing wave, when $|A|=|B|$, there are nodes, half a wavelength apart, where no oscillation occurs. The variation of amplitude with position in a partial standing wave can be written down immediately. If x is measured from a point where the amplitude is greatest, with the vectors in line, anywhere else they are inclined to one another at an angle $2kx$, and solution of the triangle gives for the resultant amplitude R:

$$|R|^2=|A|^2+|B|^2+2|A|\,|B|\cos 2kx. \tag{23}$$

It is the *intensity*, the square of the amplitude, that varies with position as a sinusoid superimposed on a steady mean level. When intensity is plotted the maxima and minima have the same shape, but in a plot of amplitude the minima are sharper than the maxima, becoming cusps when $|A|=|B|$ and the standing wave is complete. This is illustrated in fig. 10.

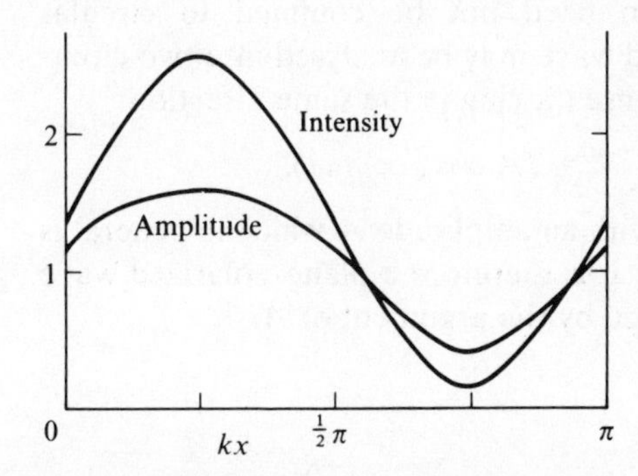

Fig. 10. Variation over half a wavelength of the amplitude and intensity in a partial standing wave. The amplitude is given by $|e^{ikx}+0.6\,e^{-ikx}|$, the intensity by the square of this, $1.36+1.20\cos 2kx$. Note that the minimum of the amplitude curve is sharper than the maximum.

Circular polarization

Finally, let us note yet another application of complex algebra in the description of a circularly polarized wave or vibration. If the end of a stretched string is moved in a circle at constant angular velocity ω, a circularly polarized disturbance runs down it. At any instant the string is displaced into a helical form, and as time proceeds the helix moves bodily down the string at the wave velocity. An observer watching the motion of a single point sees it moving in a circle. Since the disturbance is purely transverse, at any point it may be represented by a two-dimensional vector, just the thing that a complex number is most convenient to describe. To establish the coordinate system, we choose for a *real plane* any plane containing the direction of propagation (e.g. with a horizontal string we might choose the horizontal plane containing it); at each point on the string, x, we draw the *transverse plane* normal to the string, whose intersection with the real plane defines the real axis. Then e^{ikx} represents a unit vector everywhere, but the direction of the vector rotates progressively along the direction of propagation, so that the ends of all the vectors lie on a stationary helix. Looking along the x-axis we see the vector turning anticlockwise as x increases; this is a left-handed screw. By the same token e^{-ikx} describes a right-handed screw. Wave propagation is described by $e^{i(kx-\omega t)}$, which can be viewed either as the bodily rotation of the helix in a clockwise sense without translation (the effect of multiplying by $e^{-i\omega t}$ throughout) or as bodily translation at velocity ω/k without rotation (since changing x and t in the ratio ω/k leaves the argument unaltered). The fact that both descriptions are equivalent is only a way of saying that the helix is yet another self-similar curve, as it must be described by an exponential function.

This notation is particularly valuable for circularly polarized transverse electromagnetic waves, where the electric and magnetic field vectors can both be expressed as complex numbers. It should be noted that the complex number applied to transverse waves in this way is to be seen as a physically meaningful symbol in its entirety – the displacement it describes is a vector of two components. The original use, on the other hand, was as a convenient representation of a scalar displacement, for which the complex number contains too much information; only the real part therefore was needed. The vector interpretation need not be confined to circular polarization, since a plane-polarized wave may be analysed into two circularly polarized waves of opposite sense moving in the same direction:

$$A\,e^{i(kx-\omega t)}+A\,e^{-i(kx-\omega t)}=2A\cos(kx-\omega t).$$

This describes a progressive wave with an amplitude A which in general is complex and the same everywhere; it is therefore a plane-polarized wave with the plane of polarization defined by the argument of A.

4 Fourier series and integral

Synthesis of waveforms; Fourier series

The theory of Fourier series and the Fourier integral pervades mathematics and physics, and there is no lack of texts to which the student can refer for as much detail as he needs.[1] None of the applications in this book demands deep knowledge of mathematical technique, but the idea of analysing arbitrary functions into a series of sinusoids is so helpful that it is worth spending some time assimilating the concepts, so that they become a natural mode of thought, qualitative as much as quantitative. We shall concentrate on illustrating the application of Fourier methods to problems in vibration, with no more formal mathematics than is needed to make the arguments plausible. For the most part the discussion will be confined to functions of a single variable, such as the displacement of a vibrating system considered as a function of time. Once this is well assimilated there is no essential difficulty in extending the analysis to functions of many variables.

Periodic functions with a waveform other than sinusoidal may be built up by adding to a fundamental sinusoid others forming a harmonic series, that is, with frequencies that are integral multiples of the fundamental. Fig. 1 shows the progressive addition of terms of the Fourier series

$$y = 0.363 \sin x + \tfrac{1}{2} \sin 2x + 0.404 \sin 3x + \tfrac{1}{4} \sin 4x + \cdots$$

$$= \sum_1^\infty \frac{1}{n}\left[1 - \frac{\sin(\frac{1}{2}n\pi)}{\frac{1}{2}n\pi}\right] \sin nx. \qquad (1)$$

As terms are added the waveform is seen to approach the zig-zag represented by the complete series (1). Certain properties of the function are evident from (1); since $\sin nx$ is an odd function, y also must be an odd function, $y(-x) = -y(x)$, and the period is obviously 2π, since 2π is the smallest quantity that can be added to x while leaving $\sin nx$ unchanged, for all integral values of n. The vanishing of y for all x between $\pi/2$ and $3\pi/2$ is not obvious, but the sharp discontinuity at $x = 0$ may be inferred without difficulty. It arises from the first term in (1) and to discuss its origin we need only consider the simpler series:

$$y = \sum_1^\infty \frac{1}{n} \sin nx. \qquad (2)$$

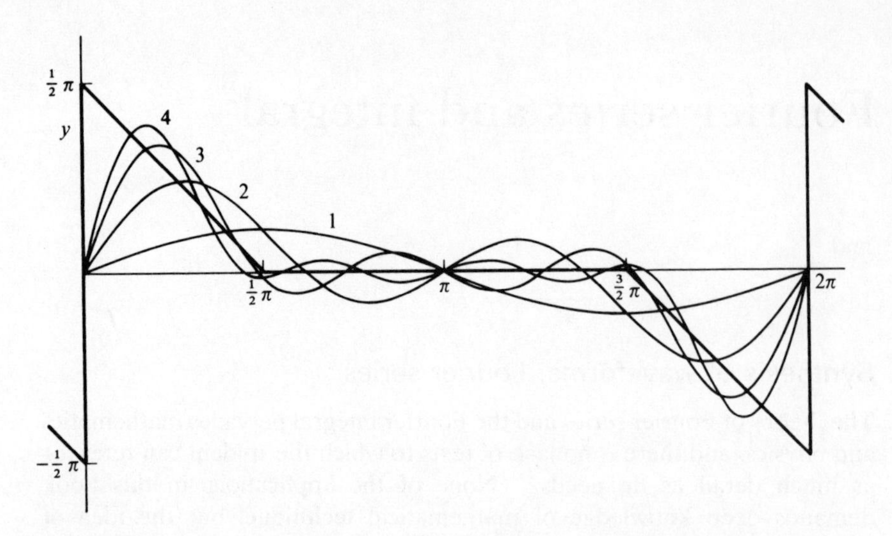

Fig. 1. Progressive stages in the Fourier synthesis of the saw-tooth function shown as a heavy line. The four curves are the first term, the first two, the first three and the first four, according to (1).

When x is very small, $\sin nx \sim nx$, at least for values of n up to something like $1/x$, and each term in (2) contributes an amount x to the value of y. If the series is terminated at a certain N, it is always possible to choose x so small that the approximation holds, and therefore the truncated series gives rise to a gradient equal to N at the origin; as N goes to infinity so does the gradient.

It may be observed in fig. 1 that the curve tends to overshoot its mark after the steep rise from the origin. This is an inevitable concomitant of the representation by a Fourier series of a discontinuous function, and it persists however many terms are employed, though confined to a steadily narrower region near the discontinuity. It can be demonstrated by use of (2). The truncated series,

$$y_N = \sum_1^N \frac{1}{n} \sin nx, \tag{3}$$

has a maximum when $\mathrm{d}y_N/\mathrm{d}x = 0$, i.e. when $\Sigma_1^N \cos nx = 0$. Now $\Sigma_1^N \cos nx$ is the real part of the geometric series $\Sigma_1^N \mathrm{e}^{inx}$, whose sum is $\mathrm{e}^{ix}(1-\mathrm{e}^{iNx})/(1-\mathrm{e}^{ix})$. The smallest value of x that causes the real part to vanish is π/N, which value, inserted into (3), shows that y_N rises to a maximum of $\Sigma_1^N 1/n \sin(\pi n/N)$. As N tends to infinity it becomes permissible to replace the sum by an integral; writing $\pi n/N$ as z, we have that

$$y_{\max} \sim \int_0^{\pi} \frac{\sin z}{z}\,\mathrm{d}z = 1.179 \times \pi/2. \tag{4}$$

Beyond the maximum y oscillates with diminishing amplitude about the value $\pi/2$. Since the oscillations are confined to a range of x that tends to zero as $1/N$, and since their amplitude remains finite, they have no practical significance and this mathematical curiosity, the Gibbs phenomenon, will be referred to no more.

The synthesis in fig. 1 is an explicit example of the representation of a function by a Fourier series, the function in this case being

$$\left.\begin{aligned} y(x) &= \frac{\pi}{2} - x \quad \text{when } 0 < x < \pi/2, \\ &= 0 \qquad\quad \text{when } \pi/2 < x < \pi, \\ \text{and} \qquad y(-x) &= -y(x). \end{aligned}\right\} \tag{5}$$

In general, when the function is periodic, with $y(x+2\pi) = y(x)$, but with no other symmetry properties, both sines and cosines must be used in the series:

$$y(x) = A_0 + \sum_1^\infty A_n \cos nx + \sum_1^\infty B_n \sin nx, \tag{6}$$

A_0 being the mean value of y. Fig. 2(a) shows an example involving both:

$$y = \sin x + \tfrac{1}{5} \cos 2x + \tfrac{1}{5} \sin 4x.$$

The function is, of course, periodic but has no point about which it is symmetric or antisymmetric.

If y is a complex variable, the A_n and B_n must also be complex. By introducing new coefficients,

$$a_n = \tfrac{1}{2}(A_n + \mathrm{i}B_n), \qquad a_{-n} = \tfrac{1}{2}(A_n + \mathrm{i}B_n), \tag{7}$$

(6) may be recast in the tidier and altogether more convenient form

$$y(x) = \sum_{-\infty}^{\infty} a_n \, \mathrm{e}^{\mathrm{i}nx}. \tag{8}$$

This is the Fourier expansion of a function whose period is 2π. Applied to a periodic vibration, $\xi(t)$, of period T_0, with the property that $\xi(t+T_0) = \xi(t)$, (8) takes the form

$$\xi(t) = \sum_{-\infty}^{\infty} a_n \, \mathrm{e}^{\mathrm{i}n\omega_0 t}, \tag{9}$$

where ω_0 is the fundamental frequency, $2\pi/T_0$.† If $\xi(t)$ is a real observable, A_n and B_n are real but a_n is in general complex, though (7) shows that $a_{-n} = a_n^*$. Only a_0, the mean value of ξ, and the values of a_n for positive n need then be specified to define $\xi(t)$ completely; a_1 is the amplitude of the *fundamental*, and a_n the amplitude of the nth *harmonic* (i.e. we call the first harmonic the fundamental – this is usual modern practice but there are exceptions in the literature).

† When it is clear from the context, as here, that the vibration is of constant amplitude, so that ω is real, we shall drop the prime from ω'.

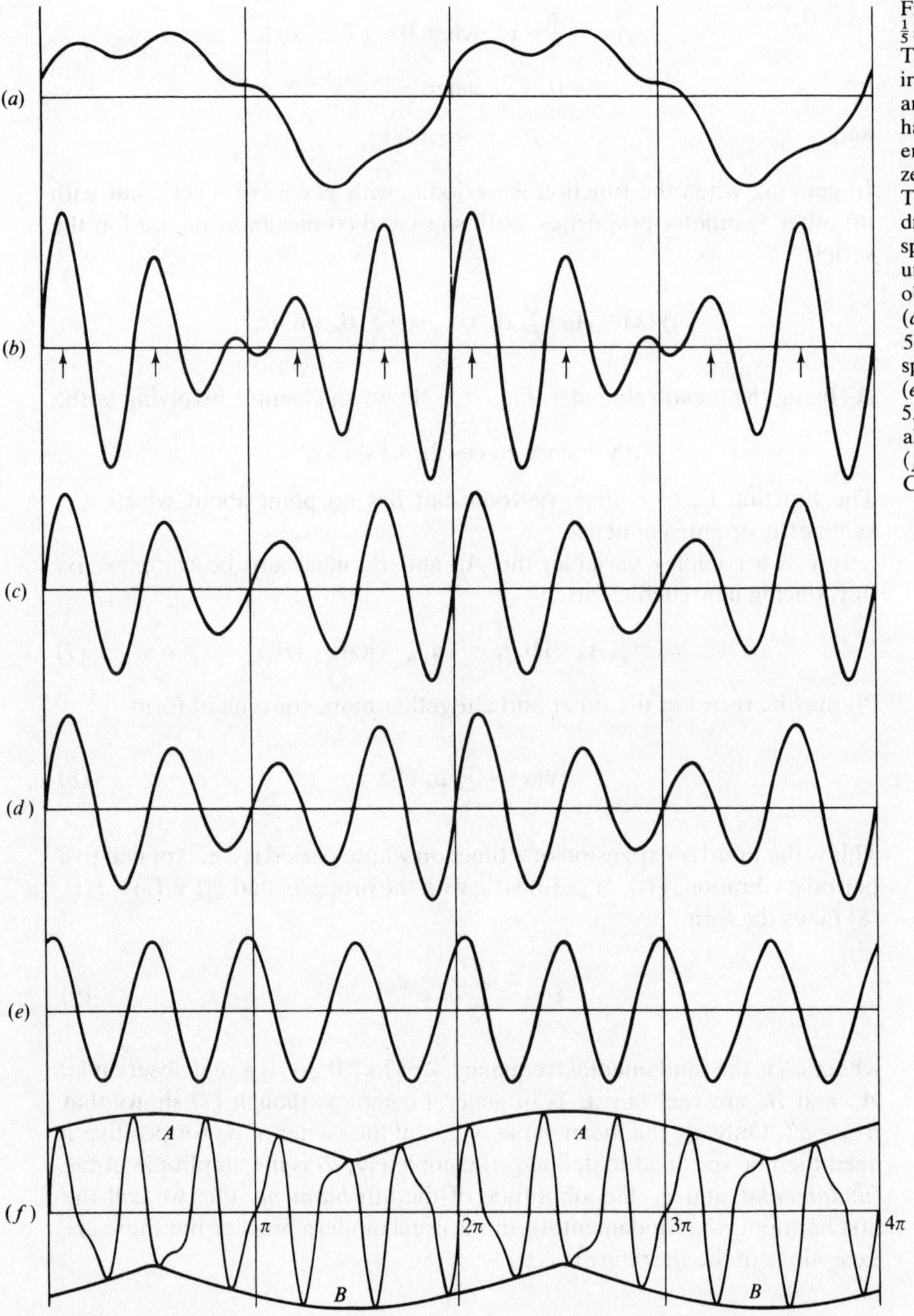

Fig. 2(*a*) $\sin x + \frac{1}{5}\cos 2x + \frac{1}{5}\sin 4x$. (*b*) $\sin 4x + \sin 5x$. The arrows show the irregular spacing of peaks, and especially the extra half-period when the envelope goes through zero. (*c*) $\sin 4x + \frac{2}{5}\sin 5x$. The envelope does not drop to zero, and the spacing of peaks is less uneven, though still obviously irregular. (*d*) $\sin 4x + \frac{1}{5}(\sin 3x + \sin 5x)$. Like (*c*) but with spacing evened out. (*e*) $\cos 4x + \frac{1}{5}(\sin 3x + \sin 5x)$. Frequency modulation at almost constant amplitude. (*f*) $\sin 6x + \frac{2}{5}\sin 11x$. Cusped envelope.

Once it is recognized that a function $y(x)$ may be represented by the Fourier series (8), it is easy to determine the values of a_n; for

$$\frac{1}{2\pi}\int_0^{2\pi} y(x)\,\mathrm{e}^{-imx}\,\mathrm{d}x = \frac{1}{2\pi}\sum_{n=-\infty}^{\infty}\int_0^{2\pi} a_n\,\mathrm{e}^{\mathrm{i}(n-m)x}\,\mathrm{d}x = a_m, \qquad (10)$$

the only term in the summation to survive integration over one period being that for which $n = m$. It requires a certain mathematical delicacy to prove the precise conditions for the validity of a Fourier expansion, but there is no need to discuss this here; it is sufficient to remark that periodic physical observables that are everywhere finite and either continuous or, if discontinuous, only at a finite number of points in the cycle, can always be Fourier-expanded, as in (8) or (9). It may be verified by substitution of (5) in (10) that the series (1) is what it purports to be.

Beats and related phenomena

Consider next the superposition of sinusoids that do not form a harmonic series but are rather close together in frequency. Two vibrations of the same amplitude produce the well-known beat pattern illustrated in fig. 2(*b*), where the frequencies have been chosen in the ratio 5 : 4. In general, for frequencies ω_1 and ω_2,

$$\cos\omega_1 t + \cos\omega_2 t = 2\cos\frac{\omega_1+\omega_2}{2}t\cos\frac{\omega_1-\omega_2}{2}t,$$

the first cosine representing a rapid oscillation at the mean frequency $\frac{1}{2}(\omega_1+\omega_2)$, the second a slow modulation of the amplitude with maxima occurring whenever $\frac{1}{2}(\omega_1-\omega_2)t = n\pi$, i.e. at a frequency $|\omega_1-\omega_2|$. If the two components have different amplitudes the depth of modulation is less but the beat frequency is the same. Fig. 2(*c*) shows the combination

$$y = \sin 4x + \tfrac{2}{5}\sin 5x.$$

It is helpful to visualize the combination as the imaginary part of a complex diagram, fig. 3, in which one vector of unit length rotates at an angular velocity of 4 units while another of length $\frac{2}{5}$, attached to it, rotates at 5 units. Since it turns, relative to the first vector, at 1 unit, this is the frequency of recurrence of maxima in the amplitude every time the vectors are momentarily in line.

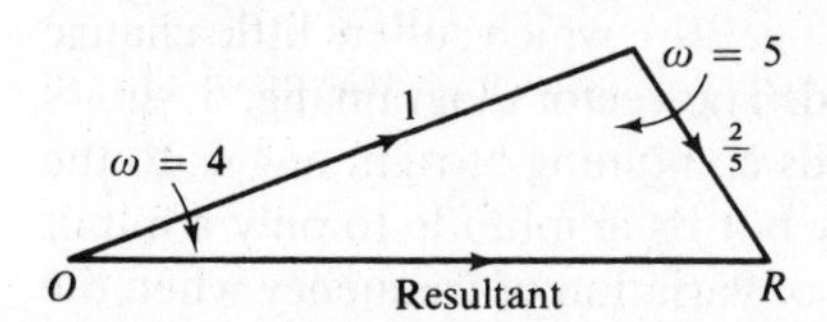

Fig. 3. Vector diagram for fig. 2(*c*).

When the two frequencies are very close together, the resultant vector changes its magnitude only a little in each revolution, and one may then attach some significance to the definition of the instantaneous frequency as the rate of change of phase angle of the resultant. If the stronger component, $a_1 \mathrm{e}^{-i\omega_1 t}$, has the lower frequency, then at the moment when they are in line, to give a maximum resultant, the two oscillations cooperate in giving the end point its maximum angular velocity. The linear velocity of the end point is $a_1\omega_1 + a_2\omega_2$ (a_1 and a_2 being taken as real and positive) and its angular velocity

$$\omega_{\max} = \frac{a_1\omega_1 + a_2\omega_2}{a_1 + a_2} = \omega_1 + \frac{a_2(\omega_2 - \omega_1)}{a_1 + a_2}.$$

By the same argument when the amplitude is at its minimum

$$\omega_{\min} = \frac{a_1\omega_1 - a_2\omega_2}{a_1 - a_2} = \omega_1 - \frac{a_2(\omega_2 - \omega_1)}{a_1 - a_2}.$$

The variation of instantaneous frequency can be discerned in fig. 2(*c*) from the uneven spacing of the zeros. The effect is most marked when $a_1 = a_2$, for then the zeros in the beat pattern are points of phase reversal, as can be clearly seen in fig. 2(*b*); $\omega_{\min}$ momentarily falls to $-\infty$.

The variation of instantaneous frequency is eliminated when the beat pattern results from the modulation of a dominant component (*carrier*) by symmetrically disposed sidebands, as, for example, in fig. 2(*d*) which illustrates the function

$$y = \sin 4x + \tfrac{1}{5}(\sin 3x + \sin 5x).$$

The vector diagram, fig. 4, shows how the two sidebands combine to alter the magnitude but not the phase angle of the carrier.

If now the phase of the carrier is advanced by $\pi/2$, to give the function

$$y = \cos 4x + \tfrac{1}{5}(\sin 3x + \sin 5x),$$

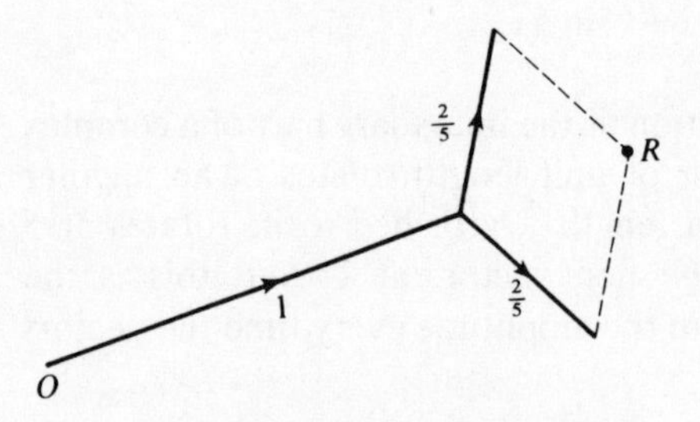

Fig. 4. Vector diagram for fig. 2(*d*).

the resulting oscillation is that shown in fig. 2(*e*), which suffers little change in amplitude but is frequency-modulated. The vector diagram, fig. 5, shows how this comes about, the two sidebands combining at right angles to the carrier to change its phase considerably but its amplitude to only a minor degree. It is easy to calculate the range of variation of frequency when the

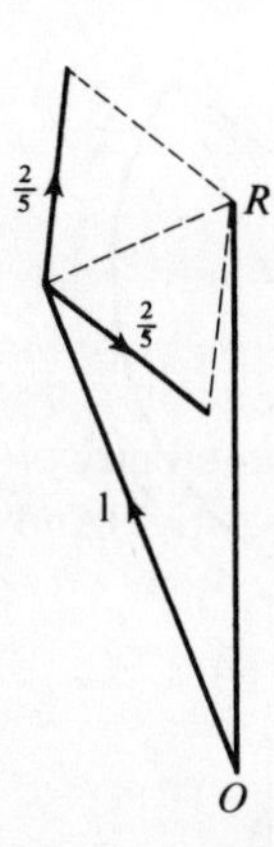

Fig. 5. Vector diagram for fig. 2(e).

sidebands are weak, but the more general case, of interest in f.m. radio transmission, involves a more thorough analysis which need not be embarked on here.

A final example of the superposition of a small number of sinusoids is worth showing as a curiosity which has been known to cause puzzlement. Two frequencies differing by a factor near 2 combine to give a cusped beat pattern as in fig. 2(f) which is the function

$$y = \sin 6x + \tfrac{2}{5} \sin 11x.$$

If the frequencies were in the ratio 2:1 and the oscillations were phased so that their positive peaks added, their negative peaks would have opposite sign. This would displace the envelope upwards, as at point A in the diagram. Because of the departure from the ratio 2:1, there comes a moment, as at B, where the relative phases are reversed and the envelope is displaced downwards. It should be noted that the 'local mean' value of y does not oscillate with the beat frequency – that can only happen if a component at the beat frequency is present in y. At A, where the envelope is displaced upwards, the positive peaks are spiky and the negative peaks blunt to the point of being doubled; this distortion compensates for the envelope shift and maintains the mean level of y at zero.

The Fourier integral

Let us now add together a larger number of sinusoids, equal in amplitude and in spacing of frequencies, and covering altogether a fairly small range of frequency. It is easiest to appreciate how the resultant arises by drawing a vibration diagram. In the simplest case the components are adjusted so that at some instant they are all in phase (fig. 6(a)). At other times there is a progressive phase variation from one to the next (fig. 6(b)). Thus if the nth component is $a\,\mathrm{e}^{-\mathrm{i}(\omega+n\varepsilon)t}$, ε being the frequency difference between

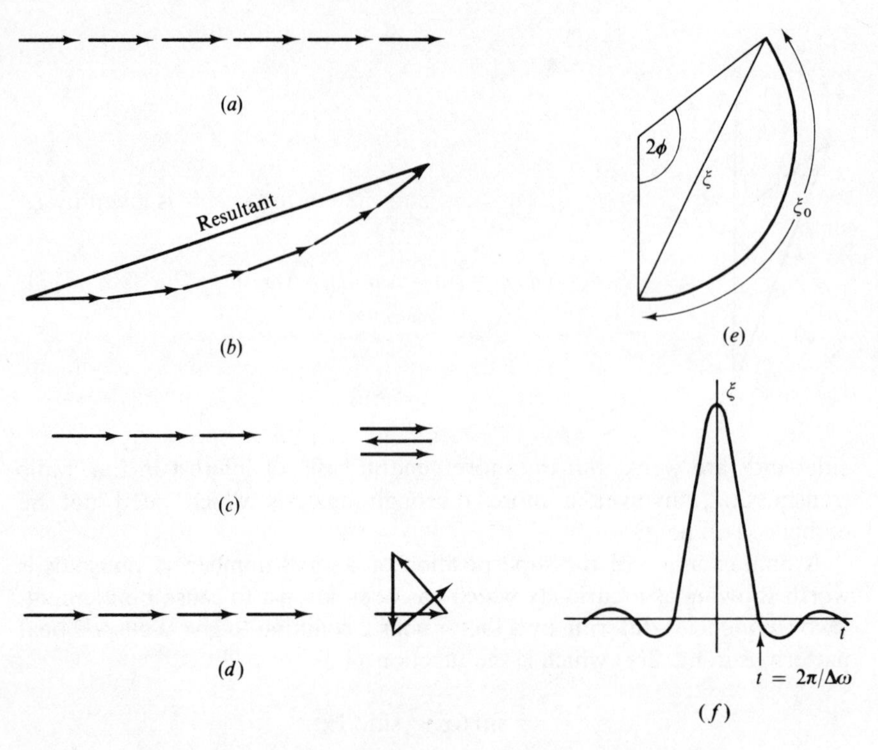

Fig. 6. Addition of several sinusoids of equal amplitude and evenly spaced frequencies. (*a*) All in phase; (*b*) a little later, with a constant phase difference between components; (*c*) subsidiary maximum with 3 components having phases 0, π and 2π; (*d*) subsidiary maximum for 4 components with phases in multiples of 133°; (*e*) construction for a continuous spectrum; (*f*) resultant obtained from (*e*), according to (12).

neighbouring components, the resultant amplitude is given by

$$\xi = \sum_{0}^{N-1} a\,\mathrm{e}^{-\mathrm{i}(\omega+n\varepsilon)t} = a\,\mathrm{e}^{-\mathrm{i}\omega t}\sum_{0}^{N-1}\mathrm{e}^{-\mathrm{i}n\varepsilon t} = a\,\mathrm{e}^{-\mathrm{i}\omega t}\frac{1-\mathrm{e}^{-\mathrm{i}N\varepsilon t}}{1-\mathrm{e}^{-\mathrm{i}\varepsilon t}}, \tag{11}$$

the same geometrical series as we have already met in a different context. **64** The presence of $\mathrm{e}^{-\mathrm{i}\omega t}$ makes the point that if we always draw the first vector in the same direction, horizontal in this case, the whole diagram must be thought of as spinning at angular velocity ω. The instantaneous resultant is the closing vector, and it is obvious from the diagram that it has the same phase as the central component, and is therefore spinning at a frequency $\omega+\frac{1}{2}N\varepsilon$. The resultant vibrates at this mean frequency, with an amplitude that is modulated according to the development with time of the length of the closing vector. When only two components are present this is a sinusoidal modulation, as shown in the simple beating pattern of fig. 2(*b*). With three components the condition that all are in phase recurs regularly at intervals T such that $\varepsilon T=2\pi$, but between each pair of principal maxima there is a subsidiary maximum of only one-third the amplitude, as illustrated in fig. 6(*c*). Similarly with four components there are two subsidiary maxima (fig. 6(*d*)). If the number of components is increased to N while their spacing is reduced to keep the total frequency spread, $N\varepsilon$, constant the principal maxima move further apart ($T=2\pi/\varepsilon \propto N$), but the number of subsidiary maxima increases to $N-2$, and the spacing of subsidiary maxima remains very nearly constant. As N is allowed to rise indefinitely, the pattern around each principal maximum quickly approaches a limiting form which is found by replacing the polygons of fig. 6 by the arc of a circle. If the frequencies of the components in this 'top-hat' spectrum lie evenly in the range $\omega_0 \pm \frac{1}{2}\Delta\omega$ the length of the arc remains constant at ξ_0 while its angular length (2ϕ in fig. 6 (*e*)) changes as $\Delta\omega \cdot t$; the radius of the arc is $\xi_0/2\phi$ and the closing vector is given by ξ, where

$$\xi/\xi_0 = \sin\phi/\phi = \sin(\tfrac{1}{2}\Delta\omega \cdot t)/\tfrac{1}{2}\Delta\omega \cdot t. \qquad (12)$$

This result, illustrated in fig. 6(*f*), is well known in Fraunhofer diffraction theory, and indeed the theory of the interference pattern of N equally spaced slits, with $N-2$ subsidiary maxima between the principal maxima, is exactly the same as that developed here, leading in the limit to an expression like (12) for the diffraction pattern of a finite aperture.

An analytical derivation of the result for a continuous distribution of components is easily given. The components lying in an infinitesimal frequency range between ω and $\omega + \mathrm{d}\omega$ all have the same phase, ωt, and combine to give a contribution whose amplitude is a fraction $\mathrm{d}\omega/\Delta\omega$ of the maximum resultant, ξ_0, that arises when all components are in phase. Their contribution is therefore $\xi_0\,\mathrm{d}\omega\,\mathrm{e}^{-\mathrm{i}\omega t}/\Delta\omega$, and the summation over all components takes the form

$$\xi = \frac{\xi_0}{\Delta\omega}\int_{\omega_0-\frac{1}{2}\Delta\omega}^{\omega_0+\frac{1}{2}\Delta\omega} \mathrm{e}^{-\mathrm{i}\omega t}\,\mathrm{d}\omega = \xi_0\,\mathrm{e}^{-\mathrm{i}\omega_0 t}\sin(\tfrac{1}{2}\Delta\omega \cdot t)/\tfrac{1}{2}\Delta\omega \cdot t, \qquad (13)$$

the same as (12), except that the spinning of the diagram at the mean frequency ω_0 was omitted there. It is included in (13), which shows the mean oscillation $\xi_0\,\mathrm{e}^{-\mathrm{i}\omega_0 t}$ modulated by the function $\sin\phi/\phi$, as in fig. 6(*f*).

The synthesis expressed in (13) may immediately be generalized. Let the sinusoidal components present a continuous range of frequencies, $-\infty < \omega < \infty$, with amplitude and phase described by a complex variable $f(\omega)$, whose meaning is that the components in an infinitesimal range $\mathrm{d}\omega$ combine to give a resultant $f(\omega)\,\mathrm{e}^{-\mathrm{i}\omega t}\,\mathrm{d}\omega$. Then the resultant of all components varies with time according to $\xi(t)$, where

$$\xi(t) = \int_{-\infty}^{\infty} f(\omega)\,\mathrm{e}^{-\mathrm{i}\omega t}\,\mathrm{d}\omega. \qquad (14)$$

The function $\xi(t)$ is the *Fourier integral transform* of $f(\omega)$. Fourier transforms possess the property of reciprocity – if ξ is the Fourier transform of f, then (with certain small modifications) f is the Fourier transform of ξ. To make the matter precise, let us write down $f_1(-p)$, the Fourier transform of $\xi(t)$ in terms of a third variable, $-p$,

$$f_1(-p) = \int_{-\infty}^{\infty} \xi(t)\,\mathrm{e}^{\mathrm{i}pt}\,\mathrm{d}t = \int_{-\infty}^{\infty}\int_{-\infty}^{\infty} f(\omega)\,\mathrm{e}^{\mathrm{i}(p-\omega)t}\,\mathrm{d}\omega\,\mathrm{d}t, \quad \text{from (14).} \qquad (15)$$

Integrate first with respect to t, taking the limits as $\pm T$ which will later be allowed to tend to infinity. Then

$$f_1(-p)=2\int_{-\infty}^{\infty}\left\{f(\omega)\,\mathrm{d}\omega\lim_{T\to\infty}\left[\frac{\sin(p-\omega)T}{p-\omega}\right]\right\}.$$

Now the quantity in square brackets has the form $\sin\phi/\phi$ as in fig. 6(f), with a peak value T when $\omega=p$, and dropping to zero when $|p-\omega|=\pi/T$. As $T\to\infty$, the area under the curve remains constant at π, but the width decreases to zero. Provided that $f(\omega)$ is a continuous function (like almost all physical variables), its variation over the range $p\pm\pi/T$ becomes negligible, so that we write

$$f_1(-p)=2\pi f(p),$$

or $$f(p)=\frac{1}{2\pi}f_1(-p)=\frac{1}{2\pi}\int_{-\infty}^{\infty}\xi(t)\,\mathrm{e}^{\mathrm{i}pt}\,\mathrm{d}t, \text{ from (15).}$$

To express this result in more usual terminology, rename $\xi(t)$ as $g(t)$ and replace p by ω. Then,

$$\left.\begin{aligned}&\text{if} && g(t)=\int_{-\infty}^{\infty}f(\omega)\,\mathrm{e}^{-\mathrm{i}\omega t}\,\mathrm{d}\omega,\\ &\text{it follows that} && f(\omega)=\frac{1}{2\pi}\int_{-\infty}^{\infty}g(t)\,\mathrm{e}^{\mathrm{i}\omega t}\,\mathrm{d}t.\end{aligned}\right\}\tag{16}$$

This, out of several equivalent forms of the Fourier transform theorem, is the one we shall take as standard. Since, mathematically speaking, ω and t are only symbols for variables, they may be interchanged without altering the truth of the proposition,

i.e. $$f(t)=\frac{1}{2\pi}\int_{-\infty}^{\infty}g(\omega)\,\mathrm{e}^{\mathrm{i}\omega t}\,\mathrm{d}\omega,\tag{17}$$

or $$f^*(t)=\frac{1}{2\pi}\int_{-\infty}^{\infty}g^*(\omega)\,\mathrm{e}^{-\mathrm{i}\omega t}\,\mathrm{d}\omega,\tag{18}$$

the latter, (18), having the same form, apart from the factor $1/2\pi$, as the first equation of (16).

It follows, therefore, that if $f(\omega)$ is a top-hat spectrum of frequencies generating $g(t)$, a $\sin\phi/\phi$ envelope, then $g(\omega)$, a $\sin\phi/\phi$ spectrum, will generate $f(t)$, a top-hat envelope, i.e. a wave train of constant amplitude but finite length. Let us verify this by explicit calculation, starting with a spectrum centred on ω_0, all components being in phase at time $t=0$, and with an amplitude distribution according to $\sin[t_0(\omega-\omega_0)]/(\omega-\omega_0)$, in which t_0 is a constant defining the width of the spectral distribution. Then the resulting amplitude follows from (14):

$$g(t)=\int_{-\infty}^{\infty}\frac{\sin[t_0(\omega-\omega_0)]}{\omega-\omega_0}\,\mathrm{e}^{-\mathrm{i}\omega t}\,\mathrm{d}\omega=\mathrm{e}^{-\mathrm{i}\omega_0 t}\int_{-\infty}^{\infty}\frac{\sin\phi}{\phi}\,\mathrm{e}^{-\mathrm{i}\phi t/t_0}\,\mathrm{d}\phi,$$

on substituting ϕ for $t_0(\omega-\omega_0)$. To evaluate the integral, note that $\sin\phi/\phi$ is an even function, so that only the even part of $e^{-i\phi t/t_0}$, that is, $\cos(\phi t/t_0)$, need be considered. Further

$$\sin\phi\cos(\phi t/t_0)=\tfrac{1}{2}\{\sin[(1+t/t_0)\phi]+\sin[(1-t/t_0)\phi]\}, \tag{19}$$

so that
$$g(t)=\tfrac{1}{2}\,e^{-i\omega_0 t}\int_{-\infty}^{\infty}\left(\frac{\sin\alpha_1\phi}{\phi}+\frac{\sin\alpha_2\phi}{\phi}\right)d\phi,$$

where $\alpha_1=1+t/t_0$ and $\alpha_2=1-t/t_0$.

Now $\int_{-\infty}^{\infty}(\sin\alpha\phi/\phi)\,d\phi=\pi$ if $\alpha>0$, $-\pi$ if $\alpha<0$. For such t that confer opposite signs on α_1 and α_2 the two contributions to $g(t)$ will cancel; $g(t)$ therefore only takes a non-zero value when t lies between $\pm t_0$:

$$g(t)=\pi\,e^{-i\omega_0 t}\quad\text{if } -t_0<t<t_0, \tag{20}$$

a limited train of constant amplitude as predicted by Fourier's theorem.

The length of the train is determined by the frequency range of its components; the central peak of the frequency spectrum is bounded by zeros when $t_0(\omega-\omega_0)=\pm\pi$, and therefore is $2\pi/t_0$ wide, while the train of oscillations is $2t_0$ long. Denoting these measures by $\Delta\omega$ and Δt, we have in this case

$$\Delta\omega\,\Delta t=4\pi, \tag{21}$$

independent of the choice of t_0. The same result follows from the original integration of a top-hat spectrum $\Delta\omega$ wide to form a train of oscillations (12) with its central peak $4\pi/\Delta\omega$ long. The qualitative reason for this result is easily understood. If a set of sinusoids with frequencies in a limited range $\Delta\omega$ are added together, and all combine in phase to give a maximum at some instant, a time $1/\Delta\omega$ must elapse before the extreme components have become dephased by one radian, and somewhat longer before they can add to give a resultant substantially smaller than their maximum. Since, as follows immediately from a sketch of the phase-amplitude diagram, the amplitude is symmetrical about the in-phase point, the total length of the train must be something in excess of $2/\Delta\omega$, i.e. $\Delta\omega\,\Delta t>2$. Obviously, the precise value of $\Delta\omega\,\Delta t$ in any particular instance, where there may not be zeros to provide a convenient definition of $\Delta\omega$ and Δt, must depend on the definitions adopted, but the close link between $\Delta\omega$ and Δt is no less inevitable for possessing a small measure of numerical imprecision.

A particular case of the foregoing example involves taking the mean frequency ω_0 in (20) as zero. A pulse of constant magnitude in the time interval between $\pm t_0$ can be resolved into Fourier components centred on $\omega=0$, with an amplitude distribution given by $\sin(t_0\omega)/\omega$, having the central zeros at $\pm\pi/t_0$. As the length of the pulse is reduced, so the width of the spectrum is increased, until in the limit we may describe an instantaneous impulse as made up of a uniform spectrum; all frequencies are equally represented and are phased so as to add up at one, and only one, moment of time – at all other times they mutually annihilate one another.

It can be seen immediately from the second equation of (16) that if the *impulse*, defined as $\int_{-\infty}^{\infty} g(t)\,\mathrm{d}t$, remains constant at some value P as $g(t)$ is compressed into an infinitesimal time interval at $t=0$, the transform $f(\omega)$ takes the limiting constant value $P/2\pi$.

In these examples the components were phased so as to give the maximum possible resultant at one instant, and this is normally the best synthesis of a given set of components if the aim is to generate the shortest train of oscillations. Consider, for example, a Gaussian spectrum of components, with the amplitude varying as $\exp[-t_0^2(\omega-\omega_0)^2]$, and falling to $1/\mathrm{e}$ of the peak at frequencies $\omega_0 \pm 1/t_0$. When these are compounded so as to be in phase at $t=0$, the resultant takes the form

$$g(t)=\int_{-\infty}^{\infty} \mathrm{e}^{-t_0^2(\omega-\omega_0)^2}\,\mathrm{e}^{-\mathrm{i}\omega t}\,\mathrm{d}\omega. \tag{22}$$

Now the exponents in the integrand may be combined and rearranged in the form

$$t_0^2(\omega-\omega_0)^2+\mathrm{i}\omega t=[t_0(\omega-\omega_0)+\tfrac{1}{2}\mathrm{i}t/t_0]^2+\mathrm{i}\omega_0 t+\tfrac{1}{4}t^2/t_0^2,$$

and if z is defined as $t_0(\omega-\omega_0)+\frac{1}{2}\mathrm{i}t/t_0$, (22) takes the form

$$g(t)=\mathrm{e}^{-\frac{1}{4}t^2/t_0^2}\,\mathrm{e}^{-\mathrm{i}\omega_0 t}\int_{-\infty}^{\infty}\mathrm{e}^{-z^2}\,\mathrm{d}z/t_0$$

$$=[(2\pi)^{\frac{1}{2}}/t_0]\,\mathrm{e}^{-\frac{1}{4}t^2/t_0^2}\,\mathrm{e}^{-\mathrm{i}\omega_0 t}. \tag{23}$$

This is a train of oscillations having the mean frequency ω_0 and an envelope which, like the spectral distributions, is also Gaussian. If we take the width to $1/\mathrm{e}$ as defining $\Delta\omega$ and Δt, we have

$$\Delta\omega=2/t_0 \qquad \Delta t=4t_0$$

and

$$\Delta\omega\,\Delta t=8. \tag{24}$$

The numerical value of $\Delta\omega\,\Delta t$ depends, as already remarked, on the form of $f(\omega)$ and on the definitions of $\Delta\omega$ and Δt, which here are rather more generous than is customary. More usually $\Delta\omega$ is taken as the root mean square deviation of the intensity distribution in the frequency spectrum. If the amplitude distribution is $f(\omega)$, the mean frequency $\bar{\omega}$ is defined by the equation

$$\bar{\omega}\equiv\int_{-\infty}^{\infty}\omega ff^*\,\mathrm{d}\omega\Big/\int_{-\infty}^{\infty}ff^*\,\mathrm{d}\omega,$$

and the width $\Delta\omega$ by the equation

$$(\Delta\omega)^2\equiv\int_{-\infty}^{\infty}(\omega-\bar{\omega})^2 ff^*\,\mathrm{d}\omega\Big/\int_{-\infty}^{\infty}ff^*\,\mathrm{d}\omega.$$

With the Gaussian distribution, $f(\omega)=\exp[-t_0^2(\omega-\omega_0)^2]$, evaluation of the integrals shows that $\bar{\omega}=\omega_0$ and $\Delta\omega=1/2t_0$, only one-quarter of the width as defined in the text.

By the same argument Δt would be t_0 and (24) would take the form

$$\Delta\omega\,\Delta t = \tfrac{1}{2}. \qquad (25)$$

In Schrödinger's wave mechanics, a particle with energy E has associated with it a wave of frequency $\omega = E/\hbar$. To describe a particle that was known to pass a certain point during an interval Δt it is necessary to associate with it a wave train no longer than this, which is a combination of sinusoidal waves lying in a frequency range $\Delta\omega \sim 1/\Delta t$. The energy of the particle is correspondingly uncertain, in accordance with Heisenberg's Uncertainty Relation

$$\Delta E\,\Delta t \sim \hbar. \qquad (26)$$

The relation (25) between the width of the envelope and the spectral range represents the best possible. If there is no instant when the components are all in phase the envelope becomes wider, as may be illustrated by assigning at $t=0$ a phase which varies as $\beta^2(\omega-\omega_0)^2$. It should be noted that a phase shift proportional to the first power, $(\omega-\omega_0)$, only moves the peak in time without preventing all components from adding in phase at some instant; but the quadratic shift proposed stops this from happening. We now have

$$g'(t) = \int_{-\infty}^{\infty} \mathrm{e}^{-t_0^2(\omega-\omega_0)^2}\,\mathrm{e}^{-\mathrm{i}[\omega t+\beta^2(\omega-\omega_0)^2]}\,\mathrm{d}\omega,$$

which is susceptible to the same procedure as was used in evaluating (22), with t_0^2 replaced by $t_0^2+\mathrm{i}\beta^2$. Consequently

$$g'(t) = \left(\frac{2\pi}{t_0^2+\mathrm{i}\beta^2}\right)^{\frac{1}{2}} \mathrm{e}^{-\frac{1}{4}t^2/(t_0^2+\mathrm{i}\beta^2)}\,e^{-\mathrm{i}\omega_0 t}$$

$$= \left(\frac{2\pi}{t_0^2+\mathrm{i}\beta^2}\right)^{\frac{1}{2}} \mathrm{e}^{-\frac{1}{4}t_0^2t^2/(t_0^4+\beta^4)}\,\mathrm{e}^{-\mathrm{i}[\omega_0 t-\frac{1}{4}\beta^2t^2(t_0^4+\beta^4)]}.$$

The envelope is still Gaussian, as in (23), but the amplitude at the centre is reduced, the local frequency is not constant and equal to ω_0 throughout, and the envelope is wider. The new value of Δt is $(t_0^2+\beta^4/t_0^2)^{\frac{1}{2}}$, from which it is clear that for a given choice of frequency spread as determined by t_0, the narrowest envelope results from phasing all components together, making β vanish.

In this example the envelope is still symmetrical, but in general when the components are not arranged so that they can all combine in phase an unsymmetrical envelope results. An important case is an envelope which is zero up to some instant which we take as the origin of t, then rises sharply and afterwards decays exponentially:

$$\left.\begin{aligned} g(t) &= \mathrm{e}^{-\alpha t}\,\mathrm{e}^{-\mathrm{i}\omega_0 t} && \text{when } t>0,\\ &= 0 && \text{when } t<0. \end{aligned}\right\} \qquad (27)$$

When this form of $g(t)$ is substituted in the second equation of (16), the spectral distribution follows immediately:

$$f(\omega)=\frac{1}{2\pi}\int_0^\infty e^{-(\alpha-i\Delta)t}\,dt=1/2\pi(\alpha-i\Delta), \quad (28)$$

where $\Delta=\omega-\omega_0$. The amplitude $|f(\omega)|$ varies as $(\alpha^2+\Delta^2)^{-\frac{1}{2}}$ and the argument as $\tan^{-1}(\Delta/\alpha)$. The significance of the latter is that, since the components between ω and $\omega+d\omega$ have the form $f(\omega)\,e^{-i\omega t}\,d\omega$, the argument of $f(\omega)$ is the phase of these components at $t=0$. These functions are shown in fig. 7; the spectrum represented by $f(\omega)$ in (28) is called *Lorentzian.* We shall meet this particular transform on several occasions – it dominates the mathematical analysis of resonant systems.

107, 129

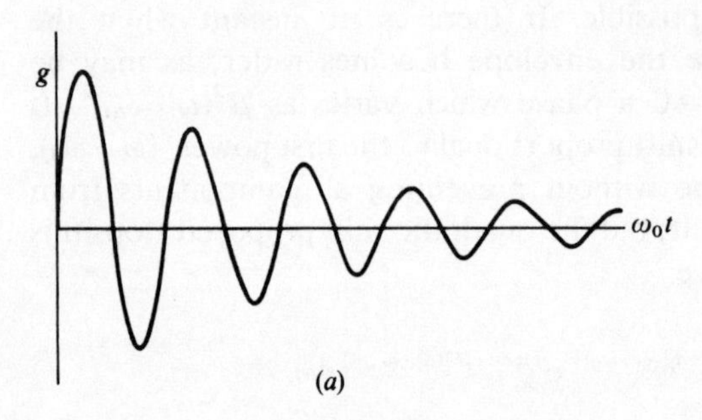

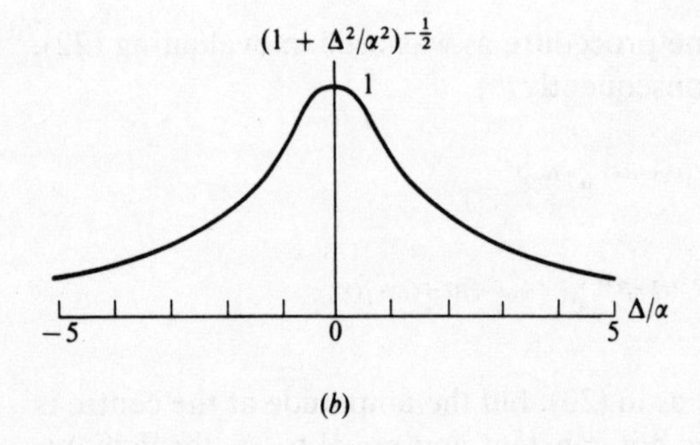

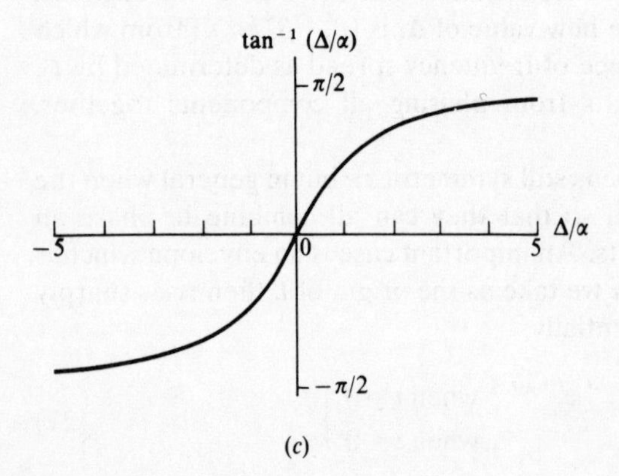

Fig. 7. The exponentially damped oscillation and its Fourier transform. (*a*) Illustrating (27), being actually $-\mathrm{Im}\,[g]$, i.e. $e^{-\alpha t}\sin\omega_0 t$; (*b*) the amplitude of $f(\omega)$ as given by (28); (*c*) the phase of $f(\omega)$.

One final example in this section applies the same ideas to wave propagation.† Suppose an impulse is applied at time $t = 0$ to the point $x = 0$ in an infinite one-dimensional medium, as, for example, by striking a piano string with a hammer. If, as in this case, the waves are non-dispersive, the kink created by the impulse runs away unchanged in both directions, but if they are dispersive the disturbance changes shape. One may now ask, in particular, what will be observed at a very great distance where the waves have had plenty of opportunity to alter their phase relationships. The formal solution may be written down immediately:

$$g(x, t) = a \int_{-\infty}^{\infty} e^{i(kx-\omega t)}\, dk, \quad \text{where } \omega = \omega(k). \tag{29}$$

This is the superposition of harmonic waves of all wave-numbers, the uniform spectrum of amplitudes and the absence of phase variation with frequency ensuring that at time $t = 0$ the displacement $g(x, 0)$ $(= a\int_{-\infty}^{\infty} e^{ikx}\, dx)$ vanishes everywhere except at the origin, where it is infinite; the integrated displacement $\int_{-\infty}^{\infty} g(x, 0)\, dx$ is finite and equal to $2\pi a$. In (29), therefore, we have an expression correctly describing the initial conditions and built up of wave-motions which are correct solutions of the equation of motion of the medium, provided the right dispersion law $\omega(k)$ is used.

74

To see what happens at a great distance X we may employ the vibration diagram to integrate (29), given that

53

$$g(X, t) = a \int_{-\infty}^{\infty} e^{i(kX-\omega t)}\, dk,$$

and that this can be represented as the vector addition, by head-to-tail joining, of innumerable vectors of which a typical one is $a\, e^{i(kX-\omega t)}\, dk$;

$$\left.\begin{aligned} &\text{length of elementary vector,} \quad ds = a\, dk, \\ &\text{phase of elementary vector,} \quad \phi = kX - \omega t. \end{aligned}\right\} \tag{30}$$

Note that since many different frequencies are involved, the rotation $e^{-i\omega t}$ is kept in the expression for the phase. The vibration diagram changes shape and spins incessantly, but we shall construct it at some chosen time t.

In fig. 8 is shown a typical variation of ω with k for a dispersive medium. A point (k_1, ω_1) is found at which the slope of the tangent is X/t; ϕ is constant on this tangent. We start drawing the vibration diagram with vectors representing points around (k_1, ω_1). If the dispersion curve followed the straight line, all vectors would have the same phase and the vibration diagram would be itself straight, with a very large resultant. This

† Since we do not intend to present a systematic account of the theory of wave-motion, no attempt is made to develop the elementary concepts of wave and group velocity. This section should be viewed simply as an illustration of how Fourier analysis may be applied to an interesting phenomenon. Moreover, since we shall need the theory of Cornu's spiral elsewhere, we have here a convenient excuse for presenting it.

149

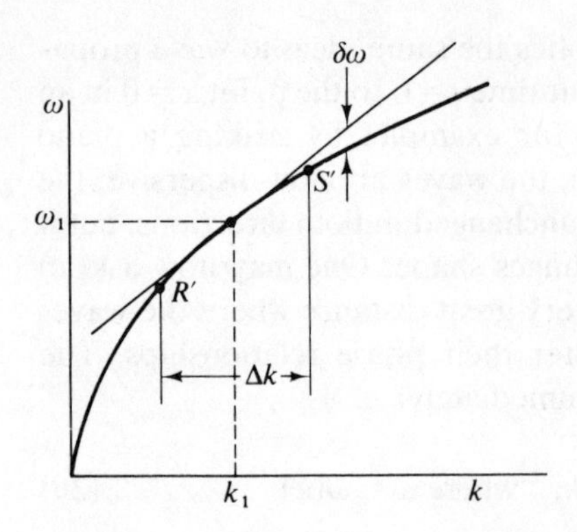

Fig. 8. Dispersion curve $\omega(k)$ for a wave-system.

is what happens with a non-dispersive medium, where $k=\omega/v$, v being constant; if X and t are such that $X/t=v$, the resultant is very large – otherwise it is zero, i.e. the pulse passes the point X at time X/v. Because of dispersion, however, the vibration diagram is rolled up into a form that is readily deduced when X is large. The significance of the intercept $\delta\omega$ shown in fig. 8 is that the vector contributed by this point on the curve has a phase differing by $\delta\omega \cdot t$ from that for a point on the line. When X is large t also is large, and a small intercept may imply a phase difference many times 2π. It is then adequate, as will become clear in a moment, to think only of the contributions from the immediate vicinity of k_1 and to suppose that $\delta\omega \propto (k-k_1)^2$. Hence, by use of (30), the vibration diagram has the properties:

(1) $\phi-\phi_1 \propto (k-k_1)^2$, where ϕ_1 is the phase of the vector due to k_1, or $\mathrm{d}\phi \propto (k-k_1)\,\mathrm{d}k$;

(2) $s=a(k-k_1)$, or $\mathrm{d}s \propto \mathrm{d}k$.

The curvature of the vibration diagram at any point is $\mathrm{d}\phi/\mathrm{d}s$, and is therefore proportional to s, which is distance measured along the curve from the point representing k_1. The curve with this property is Cornu's spiral, drawn in fig. 9, which is well known in the theory of Fresnel diffraction where it arises out of exactly parallel considerations.[(2)] Because,

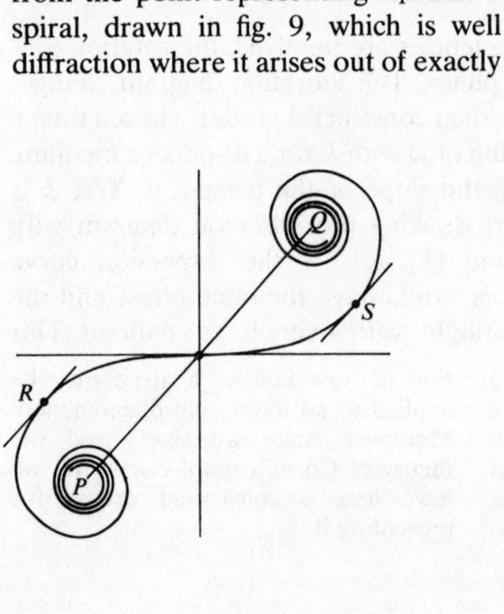

Fig. 9. Cornu's spiral.

when X is large, the spiral runs through many turns and comes close to its end-points while the quadratic approximation to $\delta\omega$ is still valid, it hardly matters precisely how $\delta\omega$ behaves further out.

Detailed analysis of the spiral can be found elsewhere, all that is needed here being the two facts that the closing vector PQ lies at an angle $\pi/4$ to the central vector, and that its length is the same as the length round the curve between R and S, where the tangents are parallel to the closing vector. From the constant angle between PQ and the central vector it follows that as the diagram spins the resultant spins at the same rate as the central vector, i.e. at ω_1. As time proceeds, an observer at X sees a resultant vibration whose frequency ω_1 itself changes gradually, being always that frequency at which $\mathrm{d}\omega/\mathrm{d}k = X/t$. The gradient, $\mathrm{d}\omega/\mathrm{d}k$, is the *group velocity*, *u*, of the wave; it is clear that the prescription for the gliding frequency experienced by the observer is that a frequency ω arrives at that time after the impulse that it would take a particle moving at velocity $u(\omega)$ to reach the observer. To determine the amplitude, find the points R' and S', on fig. 8, corresponding to R and S in fig. 9, where $t\,\delta\omega = \pi/4$, and let R' and S' be separated by Δk. Then the resultant amplitude is the same as if all contributions in the range Δk were added in phase, everything outside being ignored. From (30) it follows that $|g(X, t)| = a\,\Delta k$.

The propagation of magneto-ionic waves in the upper atmosphere illustrates the phenomenon.[3] Lightning strokes in, say, the Southern Hemisphere initiate a sharp electromagnetic pulse which through the action of the ionized upper layers is guided along the Earth's magnetic field, rising high above the equator before descending in the Northern Hemisphere. The magneto-ionic wave has a dispersion law $k^2 \propto \omega$, the constants of proportionality depending on ion density etc. Consequently the group velocity $u \propto \omega^{\frac{1}{2}}$, and over this long path the high frequencies travel faster and arrive first. A radio antenna connected straight to an audio-amplifier and loudspeaker signals the arrival of a dispersed pulse by emitting a descending whine (*whistler*) which traverses the audible range of frequencies in something like a second.

By contrast, waves in deep water obey a dispersion law $k \propto \omega^2$, so that $u \propto 1/\omega$, or $u \propto \lambda^{\frac{1}{2}}$ where λ is the wavelength, $2\pi/k$.[4] When a local storm far out in the ocean whips up the water suddenly, the watcher on a distant coast may first become aware of it by the arrival of a train of regular long-wavelength rollers, giving way to shorter and usually more irregular waves as the slower components reach shore.

Noise

Let us turn to an extreme situation in which many sinusoids are added with random phase relationships, such as our ears experience in heavy traffic or other situations described as noisy. The description of noise is important in the theory of communications systems, and, more generally, whenever organized signals are weak enough to need amplification to the point that they are confused by the irregular fluctuations inseparable from physical mechanisms.[5] There are many subtle difficulties attending the mathematical description of random noise which rule out a systematic treatment here. What follows is sufficient to give the background to later applications, but does not aspire to anything approaching rigour.

A vibration diagram representing all the separate components at any instant would consist of a large number of vectors of different length pointing at random, and their sum would be found by drawing them head to tail as in fig. 10. As time proceeds, each vector rotates at its characteristic frequency. If they all have the same frequency, the pattern spins unchanged and the resultant is a sinusoid of constant amplitude. If the components fall in a narrow range of frequencies, the resultant will at any instant spin at something close to the mean frequency, and the displacement will vary in an approximately sinusoidal fashion – approximate because of the slow variation of amplitude as the vibration diagram gradually changes in shape. With a wider range of frequencies the changes of amplitude will be more rapid. Given long enough, we may suppose that the relative phases run through all possible combinations, and we might ask what will the mean amplitude be, and how frequently will the resultant be found to have some other specified amplitude.

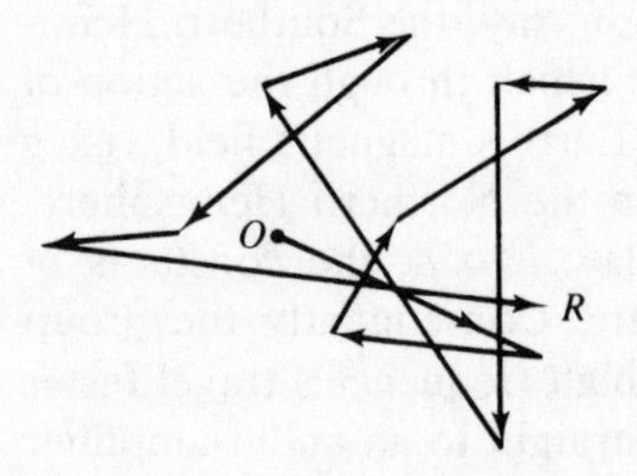

Fig. 10. Addition of vectors representing oscillations of random amplitude and phase.

It is more convenient, however, especially when the range of frequencies is so wide as to destroy the sinusoidal character of the resultant, to abandon the concept of amplitude of vibration, and consider instead the mean-squared value of the displacement, which will be termed the *mean intensity*, $\bar{I}$. Working in real numbers, a single component of the form $a\cos(\omega t+\phi)$ has an instantaneous intensity $I(t)$ given by $a^2\cos^2(\omega t+\phi)$, and since the mean value of $\cos^2\theta$ is $\frac{1}{2}$, $\bar{I}=\frac{1}{2}a^2$. With a number of superimposed sinusoids we similarly write for the instantaneous intensity

$$I(t)=\left[\sum_i a_i\cos(\omega_i t+\phi_i)\right]^2$$

$$=\sum_i a_i^2\cos^2(\omega_i t+\phi_i)+\sum_i\sum_{j\neq i} a_i a_j\cos(\omega_i t+\phi_i)\cos(\omega_j t+\phi_j). \qquad (31)$$

If there are a very large number of components, so that the phase angles ϕ_i

are randomly strewn throughout the range 0 to 2π, the mean value of $\cos^2(\omega_i t+\phi_i)$ is $\frac{1}{2}$, while the double summation vanishes; for if we choose a particular value of i and first carry out the summation over j, $a_i \cos(\omega_i t+\phi_i)$ remains constant while $a_j \cos(\omega_j t+\phi_j)$ runs through the gamut, with positive and negative values equally likely. Consequently

$$\bar{I}=\tfrac{1}{2}\sum_i a_i^2, \tag{32}$$

which may also be written as $\frac{1}{2}N\overline{a^2}$, or $\Sigma_i \bar{I}_i$ if we write $\bar{I}_i$ as the mean value of $a_i^2 \cos^2(\omega_i t+\phi_i)$. This is the well-known result that randomly phased vibrations are to be combined by summation of intensities. In particular, N equal vibrations combined in random phase lead to a root mean square displacement $N^{\frac{1}{2}}$ times that due to a single vibration.

If energy is associated with the vibrating quantity, as when the amplitude is the current in an inductor or the stretching of a spring, (32) may be interpreted as a statement that the total energy of the system is the sum of the energies of each component alone. This important result is not confined to randomly phased Fourier components. It depends on the fact that any two components of different frequency, however they are phased at any special instant, will in the course of time run evenly through all possible phase differences relative to one another. The time-average of I, as expressed by (31), therefore has no contribution from the double summation. With a continuous spectrum of frequencies present we may select those lying within a range $d\omega$ and define $I_\omega\, d\omega$ as $\frac{1}{2}\Sigma_i a_i^2$ for those components lying with $d\omega$. I_ω so defined is the *spectral intensity* or *power spectrum*, and represents the energy per unit frequency range.†

It must be emphasized that the result expressed in (32) represents the long-term average, with every opportunity given for the vibration diagram to take up all possible configurations. It does not imply that if we form a

† So long as we represent the noise signal as a sum of real oscillatory functions the above definition of I_ω is satisfactory. With the more convenient representation in terms of complex exponentials, such as the Fourier integral (14), a little care is needed, since the components at $\pm\omega$ are not randomly phased, but related by the reality condition. Thus $a\,e^{-i\omega t}$ is accompanied by $a^*\,e^{i\omega t}$, the two together describing a real oscillation of amplitude $2|a|$ and mean intensity $2aa^*$. It is therefore correct to write $I_\omega\, d\omega$ as $2\Sigma_i a_i a_i^*$, restricting the summation to those components whose frequencies are positive and lie within the range $d\omega$. This will be the normal usage in what follows; if we revert to representation by real functions we shall draw attention to the fact. There is, one must admit, scope for misunderstanding here on account of different conventions adopted by different users of complex notation. So long as only one frequency is present, as in many electrical applications and as was assumed in chapter 3, $A\,e^{-i\omega t}$ describes an oscillation of amplitude $|A|$. In Fourier theory, however, the presence of the conjugate $A^*\,e^{i\omega t}$ means that the amplitude at frequency ω is $2|A|$. No misunderstanding of physical principle is likely to arise from failure to recognize the convention in any particular case, but quantitative errors involving factors of 2 and 4 are all too easily incurred.

short-term average of I it will be more or less the same at all times. Take the case of a large number of components lying in a narrow frequency range, so that the resultant is quasi-sinusoidal but with varying amplitude. It is easy in this case to appreciate the meaning of the short-term average, which may be taken over a period several cycles long, yet not long enough to allow the pattern to change appreciably. Over an extended period this short-term average changes, but although it fluctuates about a mean value given by (32) it does not remain close to this value – on the contrary, we shall find that the fluctuations are comparable with the mean value itself, and do not become any less when more vibrations are compounded.

There is no need to confine oneself to this special case in determining the range of the fluctuations. When the frequency range is narrow the vibration diagram of fig. 10 changes shape slowly on the time scale of the spinning of each vector, while when it is wide the spinning and the shape changes proceed at comparable rates. In both cases, however, the range of configurations available to the diagram is the same and in the long term the probability of a given resultant displacement is not affected by whether the behaviour is quasi-sinusoidal or totally devoid of periodic character.

The determination of this probability is resolved into the question: given a vibration diagram consisting of N vectors, not necessarily equal in length, joined end-to-end with random phase, what is the chance that the resultant will have a certain length? To answer the question, we may imagine proceeding by drawing the diagram many times, using the same set of vector lengths but a different choice of phase angles, and marking the result by a dot in each case, so that in the end a pattern of dots is obtained, with a density $\rho(r)$ varying with distance from the origin. Then the chance that any one choice of phase angles will yield a resultant between r and $r+\mathrm{d}r$ may be written as $P_r\,\mathrm{d}r$, which expresses the fraction of the dots lying within the annulus between r and $r+\mathrm{d}r$:

$$P_r\,\mathrm{d}r = r\rho(r)\,\mathrm{d}r \Big/ \int_0^\infty r\rho(r)\,\mathrm{d}r. \tag{33}$$

To find the form of ρ and P_r, we ask what happens to ρ if one more vector, of length a, is added to the N already there, again with arbitrary phase angle; clearly every dot will be shifted to some point on a circle of radius a, centred on its original position, each one being displaced in a different direction. The radius of the circle is small on the scale of the distribution as a whole, if N is large, since the distribution has a mean radius about $N^{\frac{1}{2}}$ times the length of the average vector. In any region where ρ is constant, virtually nothing happens on adding the new vector, since for every dot that crosses a line drawn in the plane there is another that will cross in the opposite direction. Where, however, ρ varies with position in a manner described by the vector grad ρ, and a line is drawn normal to grad ρ, there will be an excess of dots crossing the line from the denser to the rarer region; we can therefore introduce a flux vector, **V**, describing the number crossing a line of unit length, and write

$$\mathbf{V} = -D \operatorname{grad} \rho. \tag{34}$$

The magnitude of the constant D need not be considered, since in the end its meaning becomes clear. This equation is the same as expresses Fick's law for diffusion, and in fact the movement of the dots as N increases is exactly analogous to Brownian motion or the diffusion of solute molecules resulting from random thermal agitation.

Variations of **V** from point to point are responsible for accumulation or depletion of the density of dots, according to the equation of continuity,

$$\operatorname{div} \mathbf{V} = -\partial\rho/\partial N. \tag{35}$$

Combining (34) and (35) results in the diffusion equation,

$$D \operatorname{div} \operatorname{grad} \rho \equiv D\nabla^2\rho = \partial\rho/\partial N, \tag{36}$$

whose solution $\rho(r, N)$ shows how the distribution of a given number of dots depends on N. Since when $N = 0$ all the dots necessarily lie at the origin, we seek a solution having this initial form, and the answer to this problem is well known from the theory of diffusion:

$$\rho = \frac{C}{N} \mathrm{e}^{-r^2/4DN}, \tag{37}$$

in which C is a constant depending on the number of dots in the pattern, i.e. the number of trials with different choices of the phase angles. A typical set is shown in fig. 11. As N is increased the radial scale increases as $N^{\frac{1}{2}}$ (since r^2/N is the only combination containing r) and the density scale decreases as $1/N$ to compensate for the area occupied by the dots increasing as N.

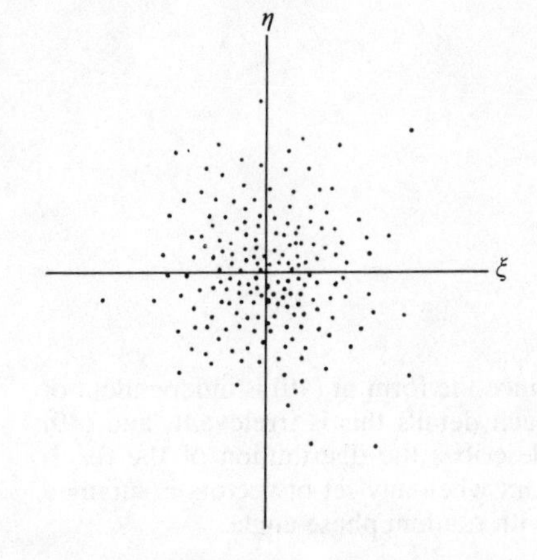

Fig. 11. Each point represents the resultant of adding the same large number of vectors with a different random choice of phases.

Since every point in fig. 11 represents the closing vector in a complex vibration diagram, one must project the whole distribution onto the real axis to give it physical significance. Thus the probability, $P_\xi\,\mathrm{d}\xi$, of observing a displacement between ξ and $\xi+\mathrm{d}\xi$ is proportional to the number of dots in a strip of width $\mathrm{d}\xi$ parallel to the η-axis. The integration of (37) with respect to η is easy, since $r^2=\xi^2+\eta^2$ and the variables separate; there is no need even to carry through the integration over the η-coordinate since it yields a constant; for a given value of N

$$P_\xi \propto \mathrm{e}^{-\xi^2/4DN}. \tag{38}$$

From this the mean intensity follows:

$$\bar{I}=\overline{\xi^2}=\int_{-\infty}^{\infty}\xi^2 P_\xi\,\mathrm{d}\xi\Big/\int_{-\infty}^{\infty}P_\xi\,\mathrm{d}\xi=2DN, \tag{39}$$

as may be seen by partial integration of the denominator; $\int_{-\infty}^{\infty}\mathrm{e}^{-z^2}\,\mathrm{d}z=[z\,\mathrm{e}^{-z^2}]_{-\infty}^{\infty}+2\int_{-\infty}^{\infty}z^2\,\mathrm{e}^{-z^2}\,\mathrm{d}z$, and the quantity in square brackets vanishes. Now we know from (32) that $\bar{I}=\frac{1}{2}Na^2$, so that we must assign to D the value $\frac{1}{4}a^2$. In the light of this (38) can be rewritten in the form

$$P_\xi=\mathrm{e}^{-\frac{1}{2}\xi^2/\bar{I}}/(2\pi\bar{I})^{\frac{1}{2}}, \tag{40}$$

the normalizing denominator being assigned such a value as ensures that $\int_{-\infty}^{\infty}P_\xi\,\mathrm{d}\xi=1$. The (Gaussian) form of P_ξ is shown in fig. 12.†

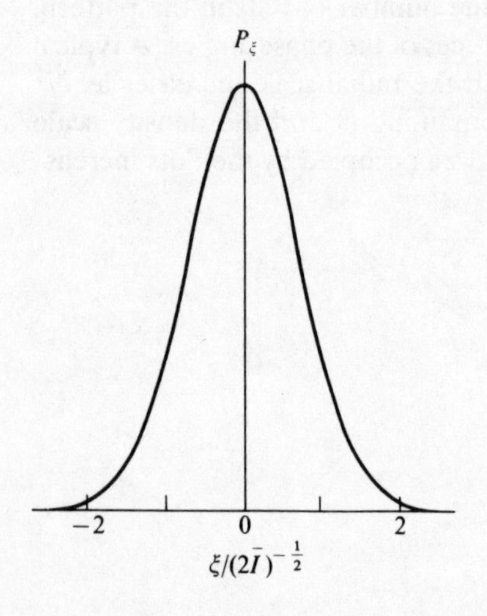

Fig. 12. The Gaussian function (40).

† In the derivation of (40) it may have seemed that each vector must have the same length a, but this is unnecessary. If there is a distribution of lengths the only change is that $D=\frac{1}{4}\overline{a^2}$, but since the form of (40) is independent of such details this is irrelevant, and (40) describes the distribution of the resultant when any set of vectors is summed with random phase angle.

As a measure of the extent of variation to which the resultant is subject we may use the standard deviation, σ, of I from its mean:

$$\sigma^2 \equiv \overline{(I-\bar{I})^2} = \int_{-\infty}^{\infty} (\xi^2 - \bar{I})^2 P_\xi \, d\xi = 2\bar{I}^2.$$

No matter how many oscillations are compounded, the variations in intensity are comparable to the mean.

It is easy to demonstrate the general character of random noise by displaying the output of a high-gain amplifier on an oscilloscope, for noise is always present in electrical circuits and can be amplified to any desired level. If the amplifier has a very wide band the noise output is entirely devoid of sinusoidal character and the characteristic time in which the amplitude can change appreciably is determined by the highest frequency that the amplifier can handle. The typical record of such noise ('grass') in fig. 13(*a*) was taken with a low-pass filter limiting the frequency range to virtually everthing below 10 kHz. This may be compared with fig. 13(*b*), taken with a narrow-band filter, so that only frequencies in the range 13.6–13.9 kHz were accepted. The quasi-sinusoidal character of the latter is evident, as are the wide variations of amplitude which we have seen are inescapable when random vibrations are compounded. It appears from the record that 1 ms is a fair measure of the time taken for the amplitude to change significantly. In this time the extreme components, 300 Hz apart in frequency, change relative phase by about 2 radians, slipping by 1 radian each from the mean. This is a very plausible figure for the sort of changes in direction of the vectors in the random vibration diagram that might alter its resultant appreciably, and in view of the difficulty of providing a rigorous argument we shall adopt it on empirical grounds.

The same process underlies an entirely different phenomenon – the speckle pattern observed when a laser beam is scattered off a rough surface (e.g. white card).[6] It is the surface, whose roughness has a scale many times the wavelength, that provides the random element, but now in the spatial rather than the temporal dimension. To enlarge on this point, the scattered light has the same character as would be produced by a myriad of point sources in the illuminated area, each emitting exactly the same frequency and vibrating with a perfectly definite, but random, phase with respect to all the rest.† When the observer's eye is focussed on the surface,

† Such a source of light is *random* but *coherent.* A better way of producing random coherent illumination of an area is by internal scattering in a translucent material. Take a cylinder of polyethylene, about 2 cm in diameter and 2 cm long, wrap it in aluminium foil (leaving the flat ends clear) and let a laser beam fall on one end; the other end glows brilliantly and uniformly, apart from the speckle pattern. The light intensity at any point is the resultant of wavelets that have reached it by multiple scattering, and one has only to move a wavelength or two along the end surface to alter the phases of the wavelets significantly. Thus the phase of the light emitted from different points varies randomly over a very short distance, yet this pattern of phases is coherent in the sense that it retains its form so long as the system is not disturbed.

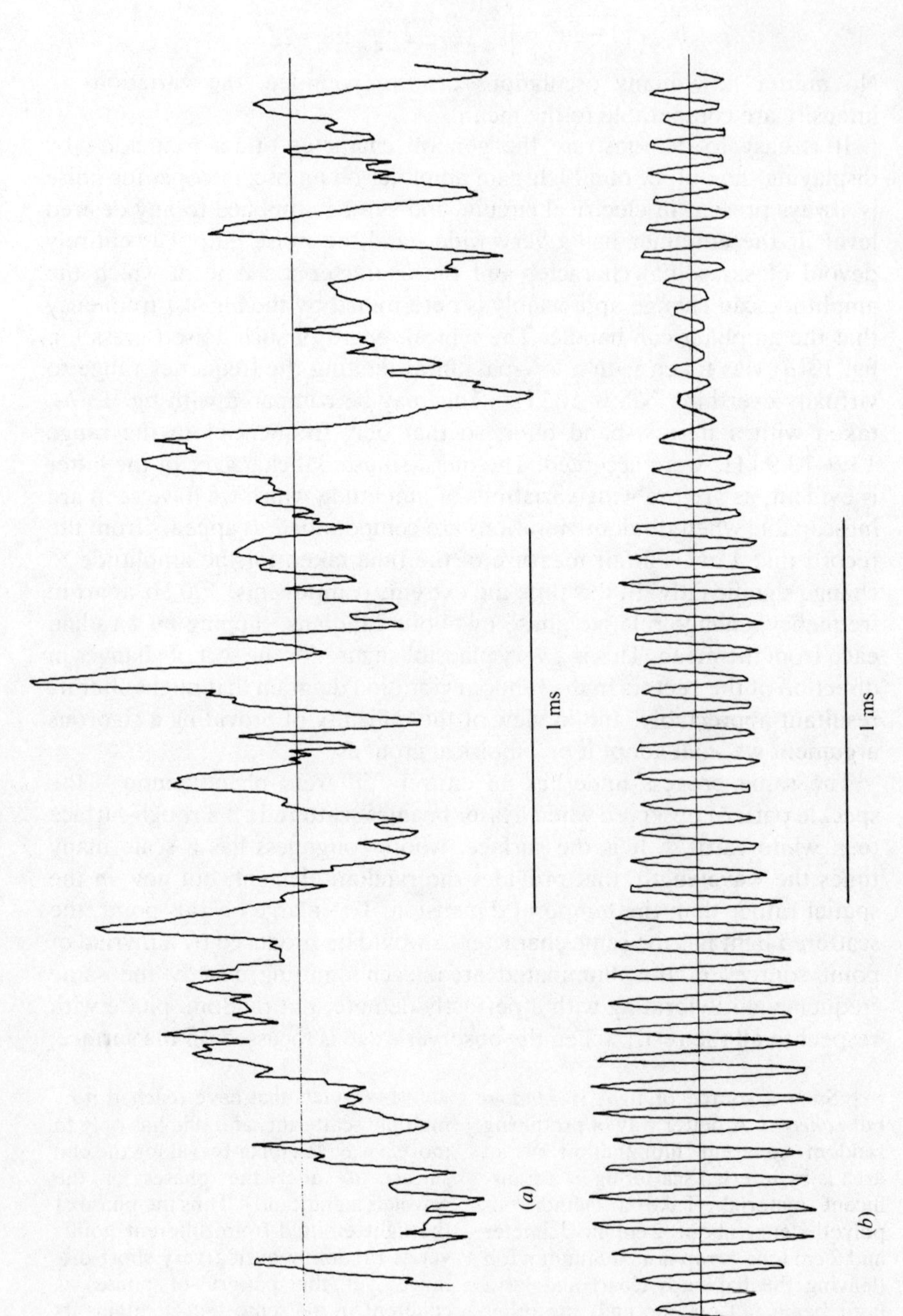

Fig. 13. Random noise (*a*) after low-pass filter, so that only frequency components below 10 kHz appear, (*b*) after band-pass filter, so that only components in the range 13.6–13.9 kHz appear. (Recording by Mr P. Mole.)

the light from each point source focusses to a point (or at any rate a very small area) on the retina and any one receptor receives light from a great many sources that lie close together on the rough surface. The response of the receptor is determined by the vector sum of all the contributions, and some receptors will happen to receive a collection of vectors that add up to a sizeable resultant, while others, though receiving just as many contributions, will experience a much smaller resultant. The distribution of resultants seen by the receptors will follow (40), and this accounts for the granular appearance of the illuminated surface. So long as the observer keeps his eye quite still, the pattern remains steady, for the phase-coherence of the laser beam is remarkably good and the random phase pattern imposed by a rigid rough surface does not alter. But as soon as the eye is moved, the phase relationships at the retina are altered and the pattern flickers. In explaining the phenomenon in terms of the response of individual receptors it has been assumed that each point source focusses to a point on the retina, but this is not necessary. In fact, if a foil, pierced with a very fine pinhole (say $\frac{1}{5}$ mm diameter), is held before the eye, diffraction at the hole causes each point source to illuminate a considerable area of the retina. Any one receptor receives weaker contributions from many more sources, but this does not alter the wide spread of the distribution of intensities at different parts of the retina. It does, however, enlarge the scale of the granular appearance. For neighbouring receptors, indeed all lying within the diffraction circle of the pinhole, receive very nearly the same collection of elementary contributions with the same phase relationships, and the resultant, correspondingly, is very nearly the same. This experiment is strongly recommended to any reader with access to a low-power continuous laser. For those who are without this facility the photographs in fig. 14 must suffice; the legend explains how they were obtained. In this case there was no lens and the pinhole may be taken as an array of randomly phased point sources, each illuminating the whole of the photographic plate. It is easy to appreciate from the previous example that the relative phases of the many contributions at any one point can only be significantly altered by moving far enough on the plate for the difference in path length from the two sides of the hole to approach a third or half a wavelength – anywhere within this radius the resultant illumination must be roughly constant. The spatial changes in phase relationships here are exactly analogous to the temporal changes when vibrations of slightly different frequency are compounded; making the pinhole smaller is equivalent to narrowing the frequency range and thus allowing only slower variations of the amplitude of quasi-sinusoidal vibration. To put the matter on a more quantitative basis, let the pinhole diameter be d, and the distance from pinhole to photographic plate D; then elementary geometry shows that moving $D\lambda/d$ on the plate imposes an extra λ between the rays from the extreme edges of the hole. Making allowance for the hole being circular, not square, and (in accordance with the known effect this has on the width of the diffraction pattern) treating it as a square of side $0.8d$, we

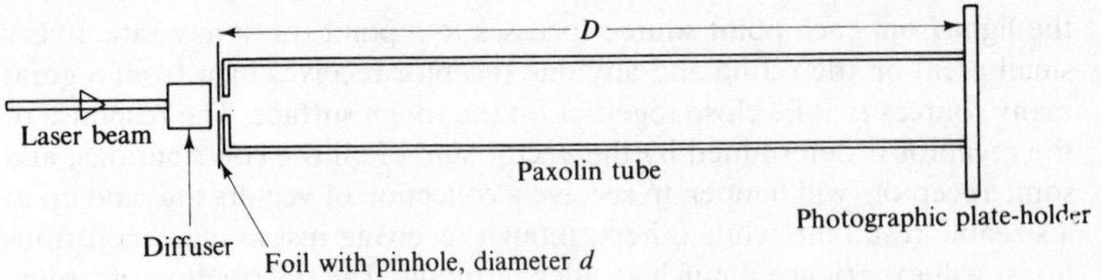

Fig. 14. Arrangement for photographing speckle pattern. The diffuser is a short rod of polyethylene wrapped in aluminium foil. The three photographs, reduced here by a factor 3/2, were taken with the following pinhole diameters: (*a*) 2.5 mm, (*b*) 0.50 mm, (*c*) 0.13 mm.

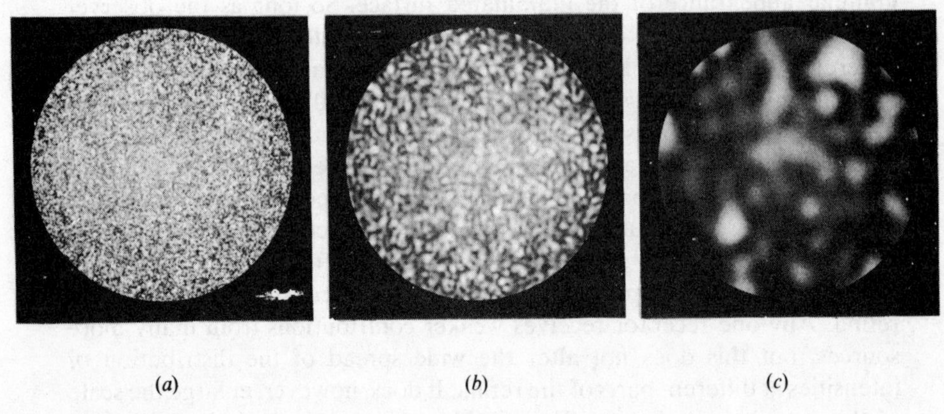

arrive at figures of 0.1 mm, 0.5 mm and 2 mm as the distance to move on the plate in order to impose variations in phase of 1 radian on the extreme rays with respect to the central ray. The observed granularity is on a scale perhaps three times as great, but this may be because the eye is caught by major peaks and troughs of intensity, between which there is quite a lot of unnoticed structure, such as shows up more plainly in a record like fig. 13(*b*).

The distribution (40) is quite different from that resulting from a constant-amplitude vibration; for if $\xi = X \cos \omega t$, the fraction of time spent between ξ and $\xi + \mathrm{d}\xi$ is given by

$$P'_\xi \, \mathrm{d}\xi = \mathrm{d}\xi / \pi (X^2 - \xi^2)^{\frac{1}{2}}. \tag{41}$$

This rises to infinite peaks at the two limits $\pm X$, markedly contrasting with the smooth fall to zero of P_ξ. Even when the random components are selected from a narrow frequency range, so that the resultant is a quasi-sinusoidal oscillation with slowly varying amplitude, $X(t)$, the wide variation of X ensures that the peaks in (41) are completely smoothed out. The actual distribution of amplitudes in this case can be derived from fig. 11. Any one choice of phase angles between different components, leading to a single dot in the pattern, remains generally constant over a number of cycles, so that the dot rotates very nearly in a circle, whose radius is the instantaneous value of X. The diagram as a whole is then to be interpreted, after dissection into circular annuli, as showing the relative probability of different values of X:

$$P_X\,\mathrm{d}X \propto X\rho(X)\,\mathrm{d}X,$$

i.e. in normalized form, and with ρ substituted from (37),

$$P_X\,\mathrm{d}X = X\,\mathrm{e}^{-\frac{1}{2}X^2/\bar{I}}\,\mathrm{d}X/\bar{I}. \qquad (42)$$

$P_X = 0$ when $X = 0$ and rises to a maximum when $X = (\bar{I})^{\frac{1}{2}}$. It can now be shown that the distribution (42) does indeed, as asserted above, smooth out the peaks of (41) into Gaussian form. For the probability of ξ lying in a given range, when all possible values of X are weighted according to (42), is given by the expression

$$P_\xi\,\mathrm{d}\xi = \mathrm{d}\xi \int_{|\xi|}^{\infty} P'_\xi P_X\,\mathrm{d}X = \frac{\mathrm{d}\xi}{\pi\bar{I}} \int_{|\xi|}^{\infty} \frac{X\,\mathrm{e}^{-\frac{1}{2}X^2/\bar{I}}\,\mathrm{d}X}{(X^2-\xi^2)^{\frac{1}{2}}},$$

which is readily integrated when z^2 is substituted for $X^2 - \xi^2$, and yields the expected result (40).

An alternative form of (42) is in terms of $P_I\,\mathrm{d}I$, the probability that the quasi-sinusoidal vibration will have an intensity between I and $I + \mathrm{d}I$, I being $\frac{1}{2}X^2$. If the intensity is in this range, X must lie between $(2I)^{\frac{1}{2}}$ and $(2I)^{\frac{1}{2}} + \mathrm{d}I/(2I)^{\frac{1}{2}}$; substituting $(2I)^{\frac{1}{2}}$ for X and $\mathrm{d}I/(2I)^{\frac{1}{2}}$ for $\mathrm{d}X$ in (42), we have that P_I is a simple exponential function:

$$P_I\,\mathrm{d}I = \mathrm{e}^{-I/\bar{I}}\,\mathrm{d}I/\bar{I}. \qquad (43)$$

This result is peculiar to quasi-sinusoidal vibrations, for which an instantaneous amplitude may be defined.†

The full extent of the variations of intensity of quasi-sinusoidal vibrations expressed by (43) will only be observed by instruments that have adequate time-resolution to react in times small compared with the characteristic time for changes in intensity, which we have empirically determined as $2/\Delta\omega$ when all Fourier components lie in the range $\Delta\omega$. It is not uncommon, however, for a recording instrument to respond only to

† It may be noted in passing that statistical mechanics leads one to expect the distribution (43) to apply to a harmonic vibrator subject to random disturbances by, for example, the impact of gas molecules.[7] If the temperature of the gas is T, the fraction of time during which the vibrator has energy between E and $E + \mathrm{d}E$ is proportional to $\exp(-E/\bar{E})\,\mathrm{d}E$, and $\bar{E} = k_\mathrm{B}T$, k_B being Boltzmann's constant. Each separate molecular collision imparts to the vibrator a little extra oscillatory motion, and the net effect of many collisions at random times is to add many vectors to its vibration diagram, just as in fig. 10. The form of the Boltzmann distribution which, as conventionally derived, possesses something of the mysterious quality of a rabbit emerging from the conjuror's hat, has an immediate molecular interpretation in this simple case. Other examples of this nature will be discussed in a later chapter.

stimulus over longer periods; thus a spectrometer may be able to select a very narrow band $\Delta\omega$, yet the intensity of light in the band may necessitate an exposure time much larger than $2/\Delta\omega$. In this case a resolving power of 10^6 (which means an extremely good spectrometer) implies that $\Delta\omega$ is 10^{-6} times the frequency of the light being studied ($\omega \sim 5\times10^{15}\,\mathrm{s}^{-1}$). With $\Delta\omega \sim 5\times10^9$ significant intensity fluctuations occur within a time of 10^{-9} s, and any normal photographic process records an average of the intensity over many such fluctuations. Consequently the recorded intensity lies very close to the mean $\bar{I}$. Just how close is by no means easy to derive rigorously, but a rough estimate can be made. For if the instrumental averaging time (e.g. photographic exposure) is T, we may imagine it divided into n intervals of $2/\Delta\omega$ each, n being $\frac{1}{2}T\,\Delta\omega$. Each interval is short enough for the average intensity to remain fairly constant throughout, but long enough for the mean intensities in successive intervals to be only weakly correlated. Thus what is recorded may be approximately described as n independent values of a quantity I that exhibits the full range of variation given by (43). The standard deviation of I is comparable with $\bar{I}$ itself, and in accordance with the theory of repeated measurements we expect n independent values to have an average which is subject to a standard deviation smaller by $n^{\frac{1}{2}}$. Thus the instrumental averaging should reduce the fluctuation of the recorded signal to about $(\frac{1}{2}T\,\Delta\omega)^{-\frac{1}{2}}$ of the signal itself. With a photographic exposure of 1 second, each grain on the plate will be subjected to the same mean intensity within less than one part in 10^4, so that the statistical fluctuations of intensity will be quite negligible compared with the other sources of unevenness, especially the graininess of the plate.

White noise

The examples just considered have involved the quasi-sinusoidal resultant of random vibrations chosen from a narrow frequency interval, $\Delta\omega$. Let us turn to the opposite extreme in which $\Delta\omega \to \infty$ and the components are spread evenly throughout the entire frequency range with their phases uncorrelated – this is the spectrum known as 'white noise', by analogy with white light which also in its ideal form includes all frequencies. There is no correlation between the displacements at any two instants, however close they may be, since for any choice Δt of time interval the range of frequencies greater than $1/\Delta t$ is infinitely larger than the range less than $1/\Delta t$, and the overwhelming majority of components suffer substantial relative phase changes during Δt. The displacement fluctuates with the distribution (40) but runs through the range of possibilities infinitely rapidly.

It should not be supposed that this is a description of an observable phenomenon; in every physical situation the spectral range is finite, sometimes because there are absorption processes or instrumental deficiencies that eliminate higher frequencies, but in the last resort because of quantization. To give a specific example, an electron accelerated in an X-ray tube, and brought to rest on hitting the target, emits a short pulse of electromagnetic radiation. However hard one tries to shorten the pulse by making the deceleration as sudden as possible, one never finds in the continuous spectrum of emitted X-rays any component with a frequency greater than $E/\hbar$, E being the kinetic energy of the electron and $\hbar$ Planck's constant; the quantum of radiation, $\hbar\omega$, can never carry away more energy than the whole of what the electron has available.

This limit to the frequency reveals itself with so sharp a definition as to have earned a place, for quite a long time, among the precision measurements which, taken together, allow the values of the fundamental constants to be determined. When an X-ray spectrometer is set to detect Bragg-reflected X-rays of a certain wavelength λ, it is found, as illustrated in fig. 15, that none are produced until the accelerating voltage reaches a certain critical value V_c. The electrons then have energy eV_c and emit X-rays whose maximum frequency is $eV_c/\hbar$; correspondingly the minimum wavelength, λ in this case, is $2\pi c\hbar/eV_c$, where c is the velocity of light.

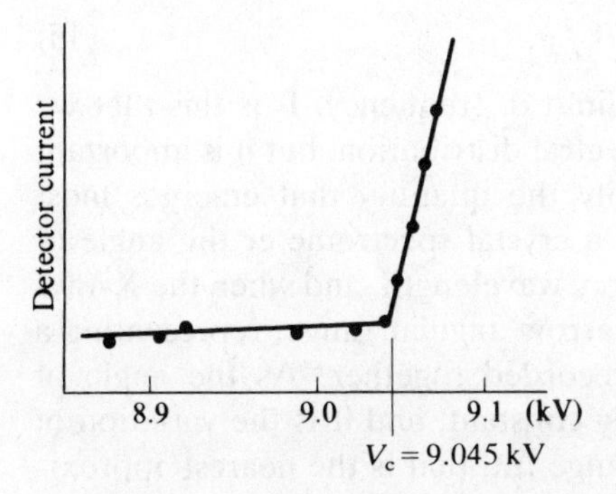

Fig. 15. Intensity of X-rays Bragg-reflected at 14° from NaCl crystal, as a function of the electron energy in the X-ray tube (H. Feder[8]).

Provided the unit cell dimensions of the crystal are well enough known for absolute wavelength measurements to be reliable, the experiment gives a value of $\hbar/e$, which can be combined with the results of other experiments, giving different combinations of fundamental constants, to determine each separately. In recent years the exploitation of the Josephson effect in superconductivity has superseded this particular determination of $\hbar/e$, but the observations are of no less interest for their clear demonstration of the overriding quantal control of the spectrum at the high-frequency end.

416

It is only very close to the cut-off, however, that this becomes apparent. To see this it is necessary to know what would be observed if a random sequence of pulses, due to electrons striking the target, were resolved by a spectrometer into its Fourier components. We start with a long sequence of NT_0 identical pulses randomly distributed in a time interval T_0, allowing this pattern to repeat indefinitely so that it generates a Fourier series. Later we allow T_0 to increase without limit while the mean frequency of pulses, N, remains constant. A single pulse in the sequence, having the form of a δ-function of strength P, generates a line spectrum with lines of equal amplitude P/T_0 (the pulse strength averaged over the interval) at a spacing

$\delta\omega = 2\pi/T_0$. If the pulse is not so sharp the amplitude of the lines will vary with frequency and we shall represent it by $f(\omega)/T_0$, whose magnitude we suppose to vary relatively slowly. From (16) it is readily seen that for a pulse of form $g(t)$, $f(\omega)$ takes the form $\int_{-\infty}^{\infty} g(t)\,\mathrm{e}^{\mathrm{i}\omega t}\,\mathrm{d}t$, independent of T_0. The lines in a narrow band $\Delta\omega$, $T_0\,\Delta\omega/2\pi$ in number, combine to form a quasi-periodic oscillation with mean intensity $(2ff^*/T_0^2)\times(T_0\,\Delta\omega/2\pi)$. And when the contributions to the intensity of all the other pulses are added we have for the spectral intensity

$$I_\omega\,\Delta\omega = (2ff^*/T_0^2)\times(T_0\,\Delta\omega/2\pi)\times NT_0 = Nff^*\,\Delta\omega/\pi. \tag{44}$$

Already T_0 has disappeared from the argument, though it is implicit in the sense that it must be large enough for $\Delta\omega$ to contain plenty of lines, i.e. $\Delta\omega \gg 1/T_0$. As $T_0 \to \infty$ this requirement is automatically satisfied and I_ω is the intensity of the continuous spectrum.

For the particular case of a sharp pulse, $g(t) = P\delta(t - t_0)$, $f(\omega)$ becomes $P\,\mathrm{e}^{\mathrm{i}\omega t_0}$ and

$$I_\omega = NP^2/\pi, \tag{45}$$

a constant spectral intensity without limit of frequency. It is this that we wish to compare with the observed spectral distribution, but it is important to recognize that I_ω is not necessarily the quantity that emerges most naturally from the measurements. In a crystal spectrometer the angle of Bragg reflection is proportional to X-ray wavelength, and when the X-rays are collected those within a certain narrow angular range, representing a certain wavelength range, $\Delta\lambda$, are recorded together. As the angle of reflection is changed $\Delta\lambda$ stays sensibly constant, and it is the variation of $I_\lambda\,\Delta\lambda$, the intensity in a wavelength range $\Delta\lambda$, that is the nearest approximation to the untreated output of the instrument, and is normally presented as the experimental observation. Although I_ω may be independent of ω, I_λ is not independent of λ. For $\omega = 2\pi c/\lambda$, and $\Delta\omega = -2\pi c\,\Delta\lambda/\lambda^2$; the energy in a narrow band $\Delta\omega$, amounting to $NP^2\,\Delta\omega/\pi$ according to (45), is spread over the wavelength range $\lambda^2\,\Delta\omega/2\pi c$, so that

$$I_\lambda = 2cNP^2/\lambda^2,$$

rising rapidly at short wavelengths. When the experiment is performed with very thin targets, so that most of the incident electrons either pass straight through, or radiate only one quantum, all radiation is from electrons of the same initial energy and conditions are optimal for comparison with theory. It is found that I_λ now varies as $1/\lambda^2$ right up to the quantum cut-off, and the classical model gives an extraordinarily accurate description of the mean spectral intensity at the lower frequencies which are not precluded by the necessity of emission in the form of quanta.

In this example the white noise spectrum results from a random sequence of more or less equal pulses, and it is often convenient in analysing the response of an instrument to a white noise input to assume that the input is such a sequence of pulses. Yet this is not strictly justifiable, for if we take the Fourier transform of a pulse sequence and arbitrarily alter the phases, it will certainly not transform back into a sequence of pulses but into a continuous random signal. There is, in fact, a certain residual order in the spectrum of the pulses and it may be that to ignore this is to falsify the analysis. To take a concrete example, consider the response of a lightly damped resonant system to a random pulse sequence (e.g. a pendulum struck at random intervals). At each pulse a damped oscillatory train is started and it is clear that the individual pulses reveal themselves as soon as their separation in time exceeds the decay time of the oscillations. On the other hand if the pulses arrive so frequently that the response at any moment is the combined effect of many, it will be difficult to discover in the fluctuations of the vibratory amplitude any feature that reveals the mean pulse rate. And, to go to extremes, if the pulse rate is faster even than the resonant frequency, so that many pulses arrive during one cycle, the precise character of the input noise certainly will not matter. To express the argument in terms of the Fourier transform of the input, let us replace a continuous noise signal, not consisting of separate pulses, by a pulse sequence constructed as follows: divide the signal into sections of random length, and integrate each section to yield the magnitude of the pulse to be placed at the centre of each section, replacing the original signal. Now take the Fourier transforms of the original signal and of its replacement. There will obviously be no difference for any component whose period takes in many pulses. It is clear, therefore, that in analysing the response of an instrument to white noise it is perfectly valid to imagine the noise as a pulse sequence with mean pulse rate higher than any frequency to which the instrument can respond. Nor will it matter exactly what detailed character is assigned to the pulse sequence. It might be thought that some difference would arise between identical pulses and pulses of random size and sign; at low frequencies, however, this is not so except, of course, at zero frequency where the identical pulses add to give a non-vanishing resultant. If one has any doubt about this, imagine the random pulses sorted into different sizes and treated as a large number of different random sequences of identical pulses. Since each sequence gives a low-frequency transform that is purely noisy, the sum of all will be no less, and can be no more, noisy.

Autocorrelation

Although we have defined I_ω in the context of noise spectra, for which it is particularly valuable, there is no objection to using it to describe spectral functions that are perfectly well phase-correlated, being Fourier transforms of smooth functions of time. Indeed, since many spectrometers and related instruments respond to intensity without regard to phase, the properties of I_ω deserve discussion in a wider context than noise. To start with a specific example, the Lorentzian spectrum described by (28) has a spectral intensity given by:

$$I_\omega = 2ff^* = 1/2\pi^2(\alpha^2+\Delta^2). \tag{46}$$

In using I_ω rather than f, all information about phases has been jettisoned, and (46) holds not only for the exponentially decaying oscillation of (27) but for innumerable other time sequences resulting from applying arbitrary phase changes to the components. This is irrelevant when the spectrum is uncorrelated in phase, being the output of a noise source or some other equivalent generator, so that I_ω contains all the information that one could wish to make use of. But one may still ask the question – what can one say about the time sequence of displacement when only I_ω is known? Clearly a complete specification of $g(t)$ is impossible, but there is still much of interest locked up in I_ω, notably the autocorrelation function which we now define and relate to I_ω.

If $g(t)$ is an arbitrary function, which may be complex, the autocorrelation function of $g(t)$ is defined as $\Gamma(\tau)$ where

$$\Gamma(\tau)=\overline{g(t)g^*(t-\tau)}/\overline{gg^*}$$
$$=\int g(t)g^*(t-\tau)\,\mathrm{d}t\bigg/\int g(t)g^*(t)\,\mathrm{d}t, \tag{47}$$

the integral being taken over the whole range of t, or a whole period if $g(t)$ is a periodic function. The general significance of $\Gamma(\tau)$ is clear from its definition and its name – it measures the way in which two values of $g(t)$, separated by τ, are related to one another on the average. At one extreme, if $g(t)$ is strictly periodic (not necessarily sinusoidal), so that $g(t-n\tau_1)=g(t)$ for all integral values of n, $\Gamma(n\tau_1)=1$; more than this, $\Gamma(\tau)$ is periodic in τ with period τ_1, since $g(t-\tau-n\tau_1)=g(t-\tau)$, and therefore $\Gamma(\tau+n\tau_1)=\Gamma(\tau)$. At the other extreme, if $g(t)$ is a white noise sequence, the values of g separated by an interval τ will be quite unrelated, however small τ may be, and $\Gamma(\tau)=0$ unless $\tau=0$. Other properties worth noting are the following:

(1) $\Gamma(0)=1$.

(2) $\Gamma(-\tau)=\Gamma^*(\tau)$. If $g(t)$ and therefore $\Gamma(\tau)$ are real, $\Gamma(-\tau)=\Gamma(\tau)$.

(3) If $g(t)=a\,\mathrm{e}^{-\mathrm{i}\omega t}$, a being a complex constant,

$$\Gamma(\tau)=\int_0^{2\pi/\omega}\mathrm{e}^{-\mathrm{i}\omega t}\,\mathrm{e}^{\mathrm{i}\omega(t-\tau)}\,\mathrm{d}t\bigg/\int_0^{2\pi/\omega}\mathrm{d}t=\mathrm{e}^{-\mathrm{i}\omega\tau}.$$

In this case the amplitude and phase of $g(t)$, as described by a, disappear from $\Gamma(\tau)$ which records only the periodicity. Already one can see why $\Gamma(\tau)$ may be useful when phase information is lacking.

(4) Following the notation of (16), define the Fourier transform

$$\gamma(\omega)=\frac{1}{2\pi}\int_{-\infty}^{\infty}\Gamma(\tau)\,\mathrm{e}^{\mathrm{i}\omega\tau}\,\mathrm{d}\tau=\frac{1}{\pi}\int_0^{\infty}\Gamma(\tau)\cos\omega\tau\,\mathrm{d}\omega\quad\text{if } g(t) \text{ is real.} \tag{48}$$

$$\text{Then}\quad\Gamma(\tau)=\int_{-\infty}^{\infty}\gamma(\omega)\,\mathrm{e}^{-\mathrm{i}\omega\tau}\,\mathrm{d}\omega=2\int_0^{\infty}\gamma(\omega)\cos\omega\tau\,\mathrm{d}\omega\quad\text{if } g(t) \text{ is real.} \tag{49}$$

(5) The *Wiener–Khinchin theorem.* Write $g(t)$ as a Fourier sum,

$$g(t)=\sum_i a_i\,\mathrm{e}^{-\mathrm{i}\omega_i t}.$$

Then, according to (47),

$$\Gamma(\tau)=\overline{\sum_i a_i\,\mathrm{e}^{-\mathrm{i}\omega_i t}\sum_j a_j^*\,\mathrm{e}^{\mathrm{i}\omega_j(t-\tau)}}\Big/\overline{\sum_i a_i\,\mathrm{e}^{-\mathrm{i}\omega_i t}\sum_j a_j^*\,\mathrm{e}^{\mathrm{i}\omega_j t}}.$$

When the sums in numerator and denominator are multiplied, all terms are periodic with frequency $\omega_i-\omega_j$ and vanish on averaging, unless $i=j$.

Hence $\quad \Gamma(\tau)=\sum_i a_i a_i^*\,\mathrm{e}^{-\mathrm{i}\omega_i\tau}\Big/\sum_i a_i a_i^*=\sum_i a_i a_i^*\cos\omega_i\tau\Big/\sum_i a_i a_i^*$

if $g(t)$ is real. As we proceed from a sum of discrete frequencies to a continuous spectrum, a given frequency range $\mathrm{d}\omega$ contributes to the sum a value of $\Sigma_i a_i a_i^*$ which is simply $\frac{1}{2}I_\omega\,\mathrm{d}\omega$, ω being restricted to positive values.

Hence $$\Gamma(\tau)=\int_0^\infty I_\omega\cos\omega\tau\,\mathrm{d}\omega\Big/\int_0^\infty I_\omega\,\mathrm{d}\omega, \tag{50}$$

and, from (49),

$$\gamma(\omega)=\tfrac{1}{2}I_\omega\Big/\int_0^\infty I_\omega\,\mathrm{d}\omega.$$

Apart from the normalizing coefficient, which ensures that $\Gamma(0)=1$, $\gamma(\omega)$, the Fourier transform of $\Gamma(\tau)$, is just the spectral intensity function; conversely, as expressed in (50), $\Gamma(\tau)$ is the Fourier transform of $I(\omega)$. This is the Wiener–Khinchin theorem, which gives precise expression to the information contained in the spectral density when all information about phase is thrown away.

With a white noise signal, I_ω is constant throughout the frequency range and (50) shows, as already noted, that $\Gamma(\tau)$ vanishes for all values of τ except zero. If, however, the constant intensity spectrum is cut off at a certain limiting frequency ω_0:

$$I_\omega=I_0\quad\text{when }\omega<\omega_0,$$
$$=0\quad\text{otherwise,}$$

then $\gamma(\omega)=1/2\omega_0$ so long as $\omega<\omega_0$ and, from (50),

$$\Gamma(\tau)=\frac{1}{\omega_0}\int_0^{\omega_0}\cos\omega\tau\,\mathrm{d}\omega=\frac{\sin\omega_0\tau}{\omega_0\tau}\qquad(\text{cf. }(13)). \tag{51}$$

The displacements are fairly strongly correlated over time intervals less than π/ω_0, but above that interval are hardly correlated at all – a result compatible, of course, with the Uncertainty Relation.

75

The top-hat form of $I(\omega)$ leading to the autocorrelation function (51) is perhaps met only rarely. More usual is a nearly random function of time whose autocorrelation decays exponentially,

$$\Gamma(\tau) = e^{-|\tau|/\tau_c}, \tag{52}$$

so that after an interval of τ_c (the *correlation time*) memory of earlier behaviour is largely lost. Then, from (48),

$$\gamma(\omega) = \frac{1}{\pi}\int_0^\infty e^{-\tau/\tau_c} \cos \omega\tau \, d\tau = \tau_c/\pi(1+\omega^2\tau_c^2).$$

The intensity spectrum I_ω is proportional to $\gamma(\omega)$, but its magnitude is left undetermined by $\Gamma(\tau)$. We have therefore that

$$I_\omega = I_\omega(0)/(1+\omega^2\tau_c^2), \quad I_\omega(0) \text{ being a constant.} \tag{53}$$

The intensity in the spectrum begins to fall off rapidly at frequencies above $1/\tau_c$, but the cut-off is not sharp as in the previous example. Both examples show that the absence of frequencies beyond ω_0 prevents the displacement changing arbitrarily quickly – not a surprising result, and indeed this analysis adds to our appreciation of what variations are possible with noise of a given spectral character very little that could not be guessed by taking thought. The autocorrelation function, however, has wide application as a quantitative tool in physical problems, and one or two examples will appear as we proceed.

103, 234, 364

5 Spectrum analysis

Computation of Fourier transforms[1]

Because of the importance of Fourier transforms in many branches of science, some far removed from vibrations as understood in this book, much effort has been devoted to the development of numerical procedures for converting an input time sequence into a spectrum of frequencies.† One in particular, the Fast Fourier Transform, is ingenious and subtle, but it is not our purpose to dwell on matters of mathematical or computational expertise. Apart from these the physical interest derives from the scale of the operations now possible, which were inconceivable to the manual computer. Very long sequences of numbers may be transformed so as to reveal periodicities, and instruments have been developed to use this facility to the full, with automatic data recording and analysis replacing the laborious personal efforts of earlier generations.

Certain points of interest arise from the character of the input which normally takes the form of a finite sequence of discrete numbers, for example the displacement at equal intervals of time, rather than a continuous function. The finite length of the input tape is of no consequence if the function to be transformed is periodic; one chooses a single period (or an integral number if there are small variations or experimental scatter and one desires to analyse the average waveform), and the Fourier series may then be computed by use of (4.10) directly. In terms of time and frequency, let the period be T_0, so that the fundamental frequency ω_0 is $2\pi/T_0$, and let the input data consist of N values of $F(t)$, a typical value being $F(nT_0/N)$ ($\equiv F_n$) with n running from 0 to $N-1$. If $F(t)$ were continuous, we should write the amplitude of the mth harmonic (neglecting constant factors) according to (4.10):

67

$$a_m = \int_0^{2\pi/\omega_0} F(t)\,e^{im\omega_0 t}\,dt.$$

† We shall continue to use time (t) and frequency (ω) as the variables in the transforms, but it will be appreciated that the applications are not restricted to these. Thus a function $f(x)$ of position on a line may be analysed into a spectrum of wave-numbers $g(k)$ – here x is analogous to t, wave-number k to frequency ω, and wavelength λ to period T_0.

Since, however, only discrete values of $F(t)$ are known, the integral is replaced by a summation: The nth input value is F_n, occurring at a time nT_0/N, i.e. $2\pi n/N\omega_0$, and

$$a_m \propto \sum_{n=0}^{N-1} F_n \, \mathrm{e}^{-\mathrm{i}m(2\pi n/N)}$$

$$= \sum_{n=0}^{N-1} \left[F_n \cos\left(mn\frac{2\pi}{N}\right) - \mathrm{i}F_n \sin\left(mn\frac{2\pi}{N}\right)\right]. \qquad (1)$$

This last form is appropriate to the computer's preference for real numbers.

Let us note what is implied by (1); $2\pi/N$ is the angular unit that results from dividing a circle into N equal parts, and any choice of the integers m and n denotes one of the set of N angles. A table of the cosines and sines of these angles, $2N$ numbers in all, forms the basic set of multipliers for the N data points F_n. The computation of the real part of a given a_m involves multiplying each value of F_n by the appropriate cosine multiplier and summing the lot. To compute another coefficient, m is changed to m' (say), and this means changing the order in which the multipliers are employed on successive F_n. Obviously the same process applies to the imaginary part of a_m, with the sine multipliers now brought into play. Each a_m therefore involves $2N$ processes of multiplication and a final addition, and given enough computing time one can build up a long sequence of a_m from a_0, the mean value of F, to very high harmonics.

There is a limit, however, set by the discrete nature of the input data. For if we write the form of (1) appropriate to a_{m+N}, the argument of the cosine and sine is $(m+N)2\pi n/N$, i.e. $2mn\pi/N+2n\pi$. Since $2n\pi$ is irrelevant in the argument, the multipliers in a_{m+N} occur in the same order as in a_m, and a_m is therefore a periodic function of m, repeating itself exactly after N values. Everything that can be said about the Fourier transform of F, as specified by N discrete data points, is contained in the amplitude of N harmonics. The reason is easily demonstrated by a simple example (fig. 1).

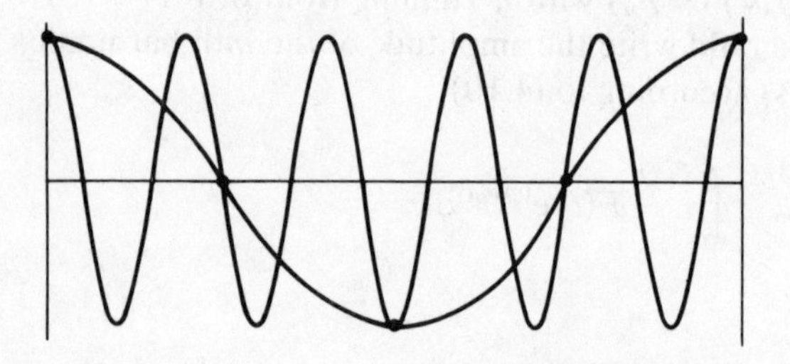

Fig. 1. The displacements represented by the black dots can be interpreted as a sinusoid in any number of ways, of which two are shown.

The points represent a possible set of input values for the case $N=4$, which are equally well fitted, if these points are all that is known, by the line of long wavelength, $\cos(n\pi/2)$, or by the line of short wavelength, cos $(5n\pi/2)$, the arguments of which differ by $2n\pi$. There is no way of telling, without intermediate data points, how irregular the true function F may be, and therefore no criterion for preferring one curve over the other (or for that matter cos $(9n\pi/2)$, cos $(13n\pi/2)$ etc.), except that the longest wavelength performs its limited task with the least fuss. Having recognized, then, that we should make a choice of not more than N Fourier components to describe the N data points, it is sensible to keep m as low as possible, letting it run between $\pm\frac{1}{2}N$. With real input, $a_{-m} = a_m^*$, so that $\frac{1}{2}N$ complex coefficients are all that can be usefully extracted from the data. The highest harmonic that can be determined in a waveform described by N values per cycle is the $\frac{1}{2}N$th, having a period equal to twice the interval between data points. Each harmonic needs both its amplitude and its phase to specify it completely, so that $\frac{1}{2}N$ harmonics need N numbers and use up the whole input material.

Another point of practical importance is the computer time involved. Each Fourier coefficient requires $2N$ multiplications and, with $\frac{1}{2}N$ coefficients accessible, the total number of multiplications needed to extract everything available from the data is N^2. A computer that takes less than a microsecond to perform each multiplication can processs thousands of data points in a minute, and tens of thousands are not ruled out, so that very considerable sequences of observations can be analysed into a correspondingly detailed spectral distribution, provided the computer can store and retrieve rapidly all the input data. With numbers as large as this, however, a more efficient program may bring great rewards, and the Fast Fourier Transform is such a one, requiring only about $N \ln N$ operations to process N data points. The saving in computer time, by something like $\ln N/N$, is a substantial gain – a factor of 100 when $N = 1000$.

These large numbers do not normally arise in the recording and analysis of periodic phenomena, where 1000 Fourier coefficients would provide a much better specification of the waveform than is generally required. It is another matter when a long sequence (e.g. a century of weather records) is to be examined for hidden periodicities, sometimes in very great numbers as in the output of the Michelson interferometer used as a spectrometer in the manner to be described presently. In such circumstances one must recognize that the practical considerations enforcing truncation of what should have been an endless sequence necessarily introduce distortions. A simple procedure is to treat the N data points as one cycle of a periodic variation, and extract N Fourier coefficients on this assumption. But any strong component with a period that does not fit into the cycle length will be altered from a continuous sinusoid to one which suffers a discontinuous phase change every N points. The Fourier transform of such a dislocated sinusoid has satellite components around its principal periodicity, and in interpreting the transform care must be taken not to attach significance to what may be a computational artefact. This is a problem that has received much attention, but beyond noting its existence there is no need here to enter into the matter more deeply.

103

The ear and the eye; optical spectrometers

There are many processes which allow the recording and analysis (other than by computer alone) of a time sequence – a gramophone disc, for example, or a magnetic tape, both of which convert the electrical signal from a microphone into permanent, retrievable forms; or the signal may be sampled at regular intervals and stored digitally.† We may, if we wish, transfer the recording of a symphony to punched tape; if we are satisfied to reject all frequencies above 10^4 Hz we need to record the amplitude of the signal at intervals of about 10^{-4} s, so that a Haydn symphony will demand over 10^7 entries in the form of numbers of, say, 10 binary digits each ($2^{10} = 1024$, and this number of steps in registering the amplitude at any instant is not excessive if very loud and very soft passages are to be adequately covered). Given this representation of the music we feed it into a computer and extract the Fourier transform, whose highest frequency components tell us more about the violins and flutes than about the double basses and bassoons, while the lowest frequencies (<20 Hz) are musically irrelevant since they are not perceived by the ear. There will be in the transform, one hopes, a marked periodicity in the amplitude as a function of frequency, with strong preference for those frequencies that correspond to the notes of the scale. Further, to select one note at random, in the vicinity of that frequency (440 Hz) that corresponds to *a*, the amplitude will vary in a complicated and highly significant fashion, for it is the beating of these close frequencies that determines the precise loudness of this note at all times. For example, if different movements are in different keys a given note may occur more frequently in one than in the others, a fact that is registered in the spectrum by the presence with roughly equal amplitude of two components perhaps only 10^{-3} Hz apart; it will be appreciated that a change in phase of one of these components will bring in this particular note during the wrong movement. Similarly, the most obvious feature of music, pulse or (in a subtler form) rhythm, shows in the Fourier transform as phase correlations between components differing by something around 1 Hz – very hard to pick out from the spectrum yet immediately apparent to even the tone-deaf.

Clearly the Fourier point of view is irrelevant to the music *qua* music. We do not take in half-an-hour's performance and then at leisure appreciate the complexity of its Fourier transform. The ear is a remarkable instrument in its ability to discriminate two notes of nearly the same frequency, but the degree of discrimination is, at its best, a few tenths of 1 Hz, not the 10^{-3} Hz mentioned above. In return for this we are able to appreciate sudden changes in the sound and thus listen to music (and everything else) as a sequence of events in time. The ear, in fact, occupies an intermediate position between the tape recording or disc which responds to the immediate sound pressure and a sound spectrograph

† This rather sophisticated process was used to record the sounds, and draw the diagrams, for figs. 7 and 8.

designed to sort out the sinusoidal components. The recorder ideally takes no account of the frequency spectrum but discriminates one instant from the next – the spectrograph discriminates one frequency from another but does not indicate the instant when a given frequency is present. Indeed, it cannot, for a sinusoidal train of finite length Δt is composed of frequencies in a range $1/\Delta t$ and to obtain fine discrimination in frequency involves analysing a correspondingly long sequence of oscillations. The ear, delighting in its sensitivity to sudden noises such as the percussive attack of each note played on a piano, must pay by a limited frequency discrimination.

Recognizing this, one must be careful not to over-emphasize the spectrum analysis aspect in treating particular cases. As already remarked, the most obvious frequency in a musical performance is the pulse rate – something that would not appear in a Fourier analysis of the time sequence of displacement (e.g. the voltage signal from a microphone) but would show if the instantaneous acoustic power level was recorded, with no attempt at frequency discrimination. Or, to take another example, when a nerve transmits information as a series of impulses, it is the frequency of arrival of impulses during a short interval that indicates the intensity of the signal. The concept of instantaneous frequency is of more value than that frequency we have labelled ω and derived from analysis of a long record. Instantaneous frequency may be defined in relation to the time interval between successive recognizable markers on the input signal – the short electrical impulse in a nerve-ending, or the momentary rise in acoustic power level every time the big drum is struck. Even with sinusoidal oscillations, or any others of well-defined waveform, instantaneous frequency may be defined through the interval between zeros of displacement; indeed many electronic frequency meters work on this principle. A frequency-modulated carrier wave is described in these terms by a graph of instantaneous frequency against time, and this graph is also, in normal f.m. communication, the variation of acoustic displacement with time that is to be extracted from the carrier and fed to the loudspeaker. It has therefore a much more direct significance than a Fourier analysis of the modulated carrier. But of course the concept of instantaneous frequency breaks down when the signal becomes complicated, without recognizable markers. With music or speech, therefore, neither Fourier analysis nor instantaneous frequency gives more than a faint inkling of what meaning is there to be appreciated, and to make progress it is necessary to devise descriptive techniques which are related to the peculiar abilities of the ear – rather good frequency discrimination, rather good time discrimination, and a pattern-recognition capacity in the brain that has so far eluded understanding.

68

Compared with the ear, the eye is a poor analyser of spectra. It accepts electromagnetic oscillations at frequencies between 4.5 and 8×10^{14} Hz, and only a very sensitive observer could distinguish 1% variations with assurance. Thus we might set $\Delta\omega\sim3\times10^{13}$, while the time discrimination of the eye is perhaps 10^{-2} s, as judged by the rather weak flicker due to mains lighting. The product $\Delta\omega\ \Delta t \sim 3 \times 10^{11}$, enormously larger than the theoretical lower limit of unity or thereabouts. There are compensations – the nervous system associated with the eyes has the far more important task of analysing directional variations and interpreting them as shapes, a faculty in which the ear is extremely rudimentary. **75**

90 In optical measurements, as was remarked in chapter 4, limitations to simultaneous discrimination in time and frequency by conventional instruments have until recently made consideration of the Uncertainty Principle irrelevant. The laser has changed this, but here we confine attention to the essential problem of normal spectroscopy – to analyse what is effectively a continuous source of phase-incoherent radiation, the result of many molecular systems emitting independently. The information extracted has been all in the frequency (or wavelength) domain – the frequencies of the various spectral lines and their profiles, i.e. the frequency spread around each discrete line, with hardly a thought for what goes on in any one emitter. The instruments have the property of accepting a continuous time sequence and converting it into its power spectrum, I_ω, or something from which I_ω may be derived without difficulty.

Two instruments deserve discussion as a start, because of the contrast they present in their principles of operation – the diffraction grating, and the Michelson interferometer used as a Fourier transform spectrometer.[(2)] The grating sorts into a multitude of channels the Fourier components of the input; the Michelson interferometer forms the autocorrelation coefficient which is then subjected to computer processing to yield the power spectrum. On the assumption that the general principles of design and use of these instruments is sufficiently well understood we may proceed to contrast the processing to which they subject an input signal in the form of a time sequence, $F(t)$, representing, for example, the variation of electric field strength in the electromagnetic wave. Each line of the grating scatters the incident wave into a wide fan of angles, and the telescope, set at a particular angle θ to the incident wave direction, collects and superimposes the scattered waves, but with a time lag τ between the waves from successive grating lines (when light is incident normally on a grating with spacing d, the extra distance travelled by successive waves is $d \sin\theta$, so that $\tau = d \sin\theta/c$). The output of the telescope is therefore $\sum_{n=0}^{N-1} F(t+n\tau)$ if the grating has N lines. This is a rapidly varying function if $F(t)$ is made up of components at optical frequencies, and the eye (or a bolometer, or some other detector) only records the mean square of the output, or possibly some other average. If N is large only those Fourier coefficients for which τ is an integral multiple of the period of vibration will survive the process of summation, so that in each direction of viewing, a particular frequency and its harmonics emerge in almost pure form. The degree of purity is determined by N, being proportionately better as N is made larger. This is readily seen by imagining a single pulse falling on the grating, to emerge in the direction θ as an evenly spaced succession of N pulses with an interval

τ between each. Now an endless succession of pulses has a spectrum consisting of perfectly sharp frequencies which are multiples of the fundamental $2\pi/\tau$; a finite succession, containing nN cycles of the nth harmonic, is represented by a slightly broadened band of frequencies, with a $\sin\phi/\phi$ type of amplitude distribution round the central maximum, and a width between zeros given by $\Delta\omega = 4\pi/N\tau$, as follows from (4.21). The relative spread, $\Delta\omega/\omega$, is thus $2/Nn$ and very small when $N \sim 10^4$–10^5.

The quantity Nn is taken as a measure of the resolving power of the grating, and structure in the power spectrum closer in frequency than ω/Nn is not revealed by the grating. This may be thought of as a result of the smearing process which turns every sharp line into a $\sin\phi/\phi$ profile or, what comes to the same thing, we may prefer to explain it in terms of the

94

autocorrelation coefficient of $F(t)$. If $F(t)$ has any tendency to repeat itself at time intervals greater than $N\tau$, nothing of this will show in the observed spectrum since the grating cannot superimpose two portions of $F(t)$ separated by so long an interval. But these long-term correlations arise from structure in the power spectrum involving frequency differences less than something like $2\pi/Nn$, and this leads to the same criterion for the resolution of fine structure in the spectrum.

Whereas the grating superimposes N replicas of $F(t)$ with progressive time delays, the Michelson interferometer takes only two replicas, by reflecting the input wave from two mirrors and combining the output in a detector (which we take to have a square-law characteristic). The time delay, τ, between the replicas is varied by moving one of the mirrors parallel to itself by a high-precision slide and lead-screw. If we designate the output of the detector by $S(\tau)$,

$$\begin{aligned} S(\tau) &= \overline{[F(t)+F(t-\tau)]^2} \\ &= 2\overline{F^2} + 2\overline{F(t)F(t-\tau)} \\ &= 2\overline{F^2}[1+\Gamma(\tau)]. \end{aligned}$$

This shows that if the mirror is steadily racked so as to vary τ at a uniform rate, the output of the detector is a record, apart from a constant, of the

95

autocorrelation function of the input. The Fourier transform, carried out automatically from the punched-tape record of $S(\tau)$, therefore displays the power spectrum of the source directly. The example shown in fig. 2 illustrates the power of the method in the sub-millimetre wavelength range where the extreme weakness of available sources, even the Sun, virtually precludes conventional spectroscopy. The resolution here is not especially high – only the first 2 cm of path difference produces significant variations of S; but in principle, especially at shorter wavelengths, very high resolution is attainable. If τ can be varied between 0 and some maximum value τ_0, $\Gamma(\tau)$ is known in this range and, by the same argument as in the last paragraph, fine structure involving frequency differences of the order of $2\pi/\tau_0$ is resolvable. The relative frequency difference resolvable is $2\pi/\omega\tau_0$, which is the reciprocal of the number of wavelengths in the path difference

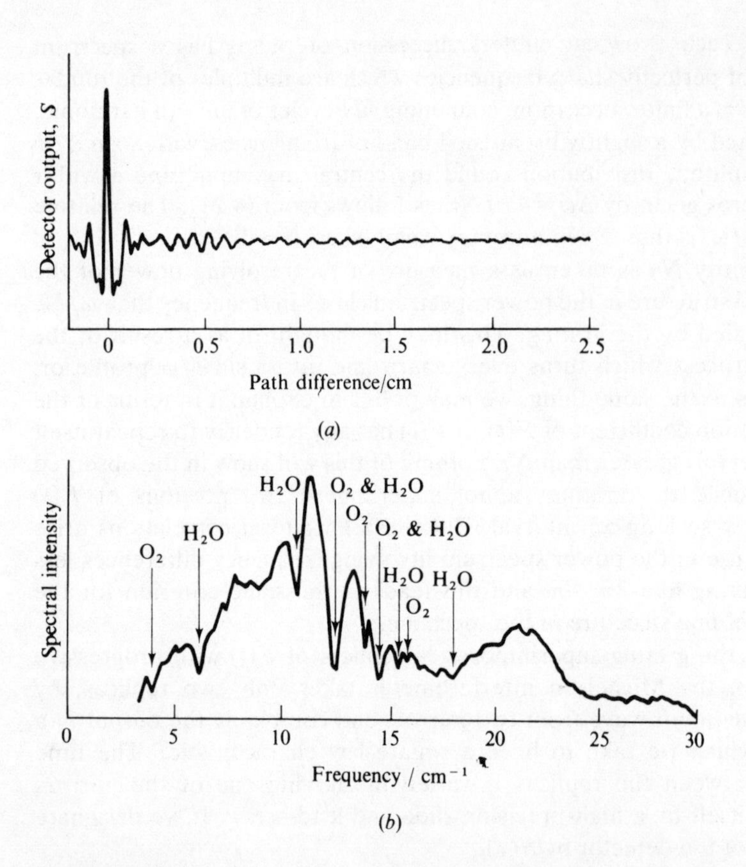

Fig. 2. Fourier transform spectroscopy in the extreme infrared (sub-millimetre wavelengths). The input to the spectrometer was sunlight as received on a high mountain, with everything except the very long wavelengths filtered out. The variation of detector output with path difference is shown in (*a*) and the spectrum derived by Fourier transformation in (*b*). The conventionally defined 'frequency' means the reciprocal of the wavelength, and the record covers the wavelength range from about 3 mm at the left to 0.3 mm at the right. Some of the absorption lines due to oxygen and water vapour in the atmosphere are labelled. Most of the other discernible features are real, but many are not yet identified. (Diagram provided by Prof. D. H. Martin and Mr E. Puplett.)

introduced by the mirror movement. We may take this number of wavelengths as a measure of the resolving power, and make immediate comparison with the corresponding quantity for the grating, which is the path difference in wavelengths between the wave trains from opposite ends; thus something like the width of the grating is to be compared to twice the mirror movement, and the latter may be tens of centimetres. It requires high skill to make a slide for the mirror that is optically true over tens of centimetres, but this is nothing to what would be involved in ruling and handling a high-quality grating as large as this.

A movement of 10 cm introduces a path difference of 20 cm, about 5×10^5 wavelengths of visible light. If N values of the detector output are recorded at equal intervals during the traverse, the Fourier transform of $\Gamma(\tau)$ has as its maximum frequency the $\frac{1}{2}N$th harmonic of an oscillation of period τ_0, i.e. $\omega_{\max}=\pi N/\tau_0$; in other words, to use the instrument up to a certain frequency, the number of data points must be twice the resolving power at that frequency, say 10^6 in the example quoted. If the computer available cannot handle this quantity of information, $\omega_{\max}\tau_0$ must be reduced, either by moving the mirror the same distance and recording the output at longer intervals, or by recording at the same intervals while reducing the movement. In the first case the high-frequency end of the spectrum is abandoned, in the second the resolution of fine structure is coarsened.

Linear transducers; response functions

In the acoustic field the classical instrument for spectrum analysis was the 22 Helmholtz resonator described earlier. The air in the cavity is set into resonance if the incident sound contains a Fourier component at or near its characteristic frequency, and the ear hears this note strongly amplified, while other notes are rejected. Nowadays it is easier to pick up the sound with a microphone, converting the pressure variations into voltage variations, and either to subject the record to computer analysis, or to feed the signal through a number of resonant circuits in parallel, recording the output of each separately so as to have a continuous record of the strength of a number of components in the incident sound. The Helmholtz resonator and the resonant circuit are examples of *linear transducers*, by which is meant a device which performs a linear transformation on an input signal 44 and feeds out the result. We have already discussed what constitutes a linear transformation with time-independent parameters. It may involve multiplication by a constant (linear amplifiers and attenuators have this property), differentiation (resistance–capacitance networks), time delay (by generating a wave at the input of a transmission line and picking it up when it reaches the output), and other transformations or mixtures of all these. The restriction to linearity in the transducer ensures that superposition holds – two signals added and fed into the transducer emerge as the sum of what each would have been converted into if it alone had been injected. Conversely, a complicated input may be imagined to be resolved into its Fourier components, each of which is transformed independently; and because of time-independence a sinusoidal input emerges as a sinusoid of the same frequency. It is therefore possible to specify the character of the transducer not by the processes or combinations of processes (multiplication, delay etc.) it performs, but by what it does to each Fourier coefficient. A typical sinusoid entering as $A\,\mathrm{e}^{-\mathrm{i}\omega t}$, ω being real, emerges with its amplitude changed by a multiplier characteristic of the frequency and having suffered a phase change. That is to say, $A\,\mathrm{e}^{-\mathrm{i}\omega t}$ is converted to $\chi(\omega)A\,\mathrm{e}^{-\mathrm{i}\omega t}$, where $\chi(\omega)$, the *transfer function*, is a complex function of frequency. Clearly since every input time sequence can be resolved into its Fourier coefficients, specification of $\chi(\omega)$ fully determines the action of the transducer:

If the input $\quad g_0(t)\;$ is $\displaystyle\int_{-\infty}^{\infty} f(\omega)\,\mathrm{e}^{-\mathrm{i}\omega t}\,\mathrm{d}\omega,$

the output $\quad g_1(t)\;$ is $\displaystyle\int_{-\infty}^{\infty} f(\omega)\chi(\omega)\,\mathrm{e}^{-\mathrm{i}\omega t}\,\mathrm{d}\omega.$

Just as the requirement that the input be real implies that $f(-\omega)=f^*(\omega)$, so for the output to be real $\chi(-\omega)f(-\omega)=\chi^*(\omega)f^*(\omega)$, and therefore

$$\chi(-\omega)=\chi^*(\omega). \tag{2}$$

An important particular case arises when the input $g_0(t)$ is a single impulse P at time $t=0$; then $f(\omega)$ is equal to $P/2\pi$, independent of frequency, and

74

$$g_1(t)=\frac{P}{2\pi}\int_{-\infty}^{\infty}\chi(\omega)\,\mathrm{e}^{-\mathrm{i}\omega t}\,\mathrm{d}\omega.$$

The *impulse response function* $h(t)$ of the transducer, which is the output signal $g_1(t)$ resulting from an input consisting of a unit impulse ($P=1$) at $t=0$, is $1/2\pi$ times the Fourier transform of the transfer function:

$$h(t)=\frac{1}{2\pi}\int_{-\infty}^{\infty}\chi(\omega)\,\mathrm{e}^{-\mathrm{i}\omega t}\,\mathrm{d}\omega. \tag{3}$$

Conversely, by (4.16),

72

$$\chi(\omega)=\int_{-\infty}^{\infty}h(t)\,\mathrm{e}^{\mathrm{i}\omega t}\,\mathrm{d}t. \tag{4}$$

This may be understood as a consequence of superposition if one calculates the response of the transducer to a sinusoidal input $\mathrm{e}^{-\mathrm{i}\omega t}$ by dissecting the input into a succession of impulses, each of which elicits the appropriate response. Thus in the time interval between t_0 and $t_0+\mathrm{d}t_0$ the input is equivalent to an impulse of magnitude $\mathrm{e}^{-\mathrm{i}\omega t_0}\,\mathrm{d}t_0$, and on integrating over all impulses comprising the sinusoidal input we have that the output is given by

$$g_1(t)=\int_{-\infty}^{\infty}h(t-t_0)\,\mathrm{e}^{-\mathrm{i}\omega t_0}\,\mathrm{d}t_0.$$

When z is written for $t-t_0$, this is converted into the form

$$g_1(t)=\mathrm{e}^{-\mathrm{i}\omega t}\int_{-\infty}^{\infty}h(z)\,\mathrm{e}^{\mathrm{i}\omega z}\,\mathrm{d}z.$$

But with a sinusoidal input, $g_1(t)=\chi(\omega)\,\mathrm{e}^{-\mathrm{i}\omega t}$, from which (4) follows directly.

It will be appreciated that since $\chi(\omega)$ and $h(t)$ are uniquely related, the impulse response function is just as good a specification of the properties of the transducer as is the transfer function, and indeed is often more convenient. If one wishes to know the output when the input, $g_0(t)$, is of arbitrary form, the argument just applied, involving breaking it down into a succession of impulses, yields the result

$$g_1(t)=\int_{-\infty}^{\infty}g_0(t_0)h(t-t_0)\,\mathrm{d}t_0, \tag{5}$$

and it may be easier to evaluate the integral directly than to find the spectrum of $g_0(t_0)$, multiply by $\chi(\omega)$ and take the Fourier transform of the resultant. The result (5) may be recognized as the convolution of g_0 with h, and application of the convolution theorem provides an alternative method for deriving the other results.

A specific example of an impulse response function is the response of a simple harmonic oscillator to an impulse – a pendulum, for instance, at rest and struck sharply at $t=0$. The resulting damped vibration is described by 32 (2.28) which, it must be remembered, applies only to positive values of t;

$$h(t)=h_0 \sin \omega' t \cdot \mathrm{e}^{-t/\tau_a} \qquad (t>0). \tag{6}$$

Hence, from (4),

$$\begin{aligned}\chi(\omega) &= \int_0^\infty h_0 \sin \omega' t \cdot \mathrm{e}^{-t/\tau_a + i\omega t}\, \mathrm{d}t \\ &= \frac{h_0}{2\mathrm{i}} \int_0^\infty [\mathrm{e}^{-(1/\tau_a - i\omega' - i\omega)t} - \mathrm{e}^{-(1/\tau_a + i\omega' - i\omega)t}]\, \mathrm{d}t \\ &= \omega' h_0/[(\omega'^2 - \omega^2 + 1/\tau_a^2) - 2i\omega/\tau_a].\end{aligned} \tag{7}$$

76 This gives the amplitude of steady oscillation set up by a periodic force, and has the characteristic resonant quality if $\omega'\tau_a \gg 1$. When ω is near ω' the denominator is small and the response correspondingly large, with a maximum $\sim \frac{1}{2}ih_0\tau_a$ at resonance, when $\omega^2 = \omega'^2 + 1/\tau_a^2$; as τ_a is increased the maximum becomes larger and the frequency range within which the response is comparable to this maximum becomes narrower. This 128 behaviour will be studied from a number of viewpoints in the next chapter, and at this stage it is enough to note that a reason for the ubiquity of resonant response obeying (7) is to be found in the simplicity of the impulse response in the form of a damped sinusoidal oscillation – so many systems behave in this way when struck, and all must of necessity show precisely the same resonant character. This is, of course, only another way of saying that simple linear systems with one natural frequency have very little choice of behaviour.

Causality

A point of considerable interest arises from the relation (3) between the transfer function $\chi(\omega)$ and the impulse response function $h(t)$. If we were to choose some arbitrary form of $\chi(\omega)$, satisfying only the reality condition (2), and evaluate $h(t)$, we should expect in general to arrive at a function occupying the whole range of positive and negative t. This, however, is not physically sensible – a passive transducer does not anticipate the arrival of an impulse, and any realistic $h(t)$ must vanish identically for all negative values of t; the pendulum just discussed is a typical example. There must therefore be certain restrictions on the form of $\chi(\omega)$ that any real transducer can generate. These restrictions are not very severe, for although we demand that $h(t)$ shall vanish in the negative range of t we have no physical arguments suggesting that its behaviour shall be in any way restricted in the positive range,† and a wide variety of functions must therefore still be allowable for $\chi(\omega)$.

Perhaps the easiest way to approach the matter is to ask how a transducer will respond to an oscillatory input that increases or decreases exponentially with time and is described by $A\,\mathrm{e}^{-i\omega t}$, with ω complex. There is no problem in extending the definition of $\chi(\omega)$ to include complex frequencies, and we therefore express the output as $\chi(\omega)A\,\mathrm{e}^{-i\omega t}$, having the same exponential variation of amplitude with time as the input. Now consider the case when the imaginary part of ω is positive and the input increases with time. At any instant the output may be regarded as the resultant of the impulse responses due to all previous values of the input. Since at earlier times the input was smaller, with an exponential decrease towards the past, and since the impulse response is finite, there are no circumstances in which the combined impulse responses can ever diverge; $\chi(\omega)$ is therefore finite at all values of ω lying in the upper half-plane in which $\mathrm{Im}\,[\omega]>0$. On the other hand, if $\mathrm{Im}\,[\omega]<0$ the response of the transducer at any instant is the resultant of a succession of impulse responses generated by an input which was exponentially larger the further back we look into the past. If the impulse response does not die away fast enough there is a very real likelihood that the largest contributions to the resultant originate in those impulses which came earliest; and when we come to form the resultant by integrating backwards towards $-\infty$ we can have no assurance that the integral will converge. In fact it is a very limited class of transducers that lack singularities of $\chi(\omega)$ in the lower half-plane, that is, localized values of ω at which χ becomes infinite. For there is a general theorem (Liouville's theorem) in complex variable theory which shows that the only function that is regular (i.e. continuous and differentiable) everywhere on the complex plane is a constant.[3] A transducer for which χ is constant produces an output that is exactly proportional to the input, so that an impulsive input yields an instantaneous impulsive output, $h(t)=\alpha\delta(t)$. Only the simplest and most idealized transducers such as networks of perfect resistors have this property, and one may therefore expect all real transducers to exhibit singularities in the lower half-plane, but none in the upper half-plane.

We are not, however, concerned at this stage to investigate the pathology of $\chi(\omega)$, since it is its regularity in the upper half-plane that is the characteristic expression of what is conventionally called *causality* – the fact that the effects never precede the impulse which causes them. The argument of the last paragraph showed that the regularity of χ in the upper half-plane was a necessary consequence of causality; and it is readily seen

† Except that $h(t)$ should be finite at all times; the analysis does not apply to an unstable transducer which, when excited by however small an impulse, goes out of control and develops an output that increases without limit.

that the converse is true, that if χ is thus regular causality is satisfied. For just as an impulse response function which is not a δ-function, vanishing for all $t>0$, must generate at least one singularity of χ in the lower half-plane, so one which does not vanish for all $t<0$ must generate at least one in the upper half-plane. The absence of singularities in the upper half-plane therefore guarantees the vanishing of the impulse response function at all negative times.

The restrictions placed by causality on the real-frequency response of a linear transducer may therefore be expressed in terms of the possible behaviour of $\chi(\omega)$ along the real axis that ensures regularity of χ in the upper half-plane. We shall assume that χ is finite for all real ω and that it tends to zero as $\omega \to \infty$. The first assumption is not necessarily true when $\omega = 0$ (in certain circuits incorporating superconductors, for example) or at a non-zero real value of ω if the transducer incorporates a strictly lossless resonant element. These are such special cases, however, as should not be allowed to lead us into more delicate analysis than is required by the vast majority of applications. As for the second assumption, concerning the behaviour as $\omega \to \infty$, it is something to be borne in mind when applying the theorem we are about to derive, for if it is not satisfied the integrals involved may diverge and make nonsense of the theorem; but again, the exceptional cases are rare and normally arise as artefacts when the real physical system is replaced by a simplified model whose limiting behaviour has not been considered with sufficient care.

If $\chi(\omega)$ is regular throughout the upper half-plane, so also is $\chi(\omega)/(\omega - \omega_0)$, ω_0 being a real constant, except for the simple pole at ω_0 where this function becomes infinite. An integral round the contour C shown in fig. 3, which incorporates an infinitesimal semicircle to avoid the pole, must therefore vanish, according to Cauchy's theorem:

$$\oint_C \frac{\chi(\omega)}{\omega - \omega_0}\,\mathrm{d}\omega = 0.$$

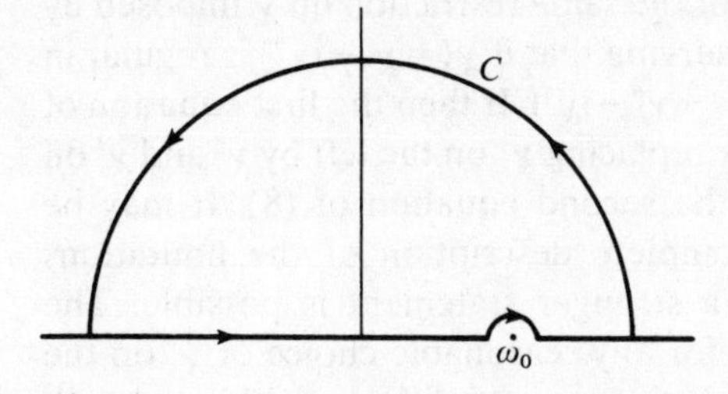

Fig. 3. Contour for integration in the complex plane.

If χ/ω falls to zero more sharply than ω^{-1}, as assumed, the large semicircle may be extended to infinite radius and there make no contribution to the contour integral. The contribution from the real axis may be dissected into that due to the semicircle at ω_0 and the rest, $\int\!\!\!\!-_{-\infty}^{\infty} [\chi(\omega)/(\omega - \omega_0)]\,\mathrm{d}\omega$, the stroke through the integral sign indicating that the principal part is to be taken (integrate from $-\infty$ to $\omega_0-\varepsilon$, and from $\omega_0+\varepsilon$ to ∞, and then let $\varepsilon\to 0$). Then

$$\int_{-\infty}^{\infty}\frac{\chi(\omega)}{\omega-\omega_0}\,\mathrm{d}\omega+\int_{\text{semicircle}}\frac{\chi(\omega)}{\omega-\omega_0}\,\mathrm{d}\omega=0.$$

The integral round the semicircle is carried out with $\chi(\omega)$ constant at its value $\chi(\omega_0)$; putting $\omega=\omega_0+r\,\mathrm{e}^{\mathrm{i}\theta}$, r being constant, and integrating θ from π to 0, the result is obtained:

$$\int_{-\infty}^{\infty}\frac{\chi(\omega)}{\omega-\omega_0}\,\mathrm{d}\omega=\mathrm{i}\pi\chi(\omega_0).$$

Finally, expressing $\chi(\omega)$ in terms of its real and imaginary parts, $\chi=\chi'+\mathrm{i}\chi''$, and separating real and imaginary terms in the equation, we find that

$$\left.\begin{aligned}\chi''(\omega_0)&=-\frac{1}{\pi}\int_{-\infty}^{\infty}\frac{\chi'(\omega)\,\mathrm{d}\omega}{\omega-\omega_0}\\ \text{and}\qquad \chi'(\omega_0)&=\frac{1}{\pi}\int_{-\infty}^{\infty}\frac{\chi''(\omega)\,\mathrm{d}\omega}{\omega-\omega_0}.\end{aligned}\right\}\qquad(8)$$

A perhaps more convenient form, involving integration only over the positive frequency range, results from applying the reality condition (2) in the form $\chi'(-\omega)=\chi'(\omega)$ and $\chi''(-\omega)=-\chi''(\omega)$. Then

$$\left.\begin{aligned}\chi''(\omega_0)&=-\frac{2\omega_0}{\pi}\int_{0}^{\infty}\frac{\chi'(\omega)\,\mathrm{d}\omega}{\omega^2-\omega_0^2}\\ \text{and}\qquad \chi'(\omega_0)&=\frac{2}{\pi}\int_{0}^{\infty}\frac{\omega\chi''(\omega)\,\mathrm{d}\omega}{\omega^2-\omega_0^2}.\end{aligned}\right\}\qquad(9)$$

These are the Kramers–Kronig relations, which show that knowledge of the real part of $\chi(\omega)$ for all real frequencies is sufficient to define the imaginary part completely; and vice versa.[4] The two relations are not independent, but alternative forms of the same restriction on χ imposed by causality. This is readily seen by observing that if $\chi(=\chi'+\mathrm{i}\chi'')$ is regular in the upper half-plane, so also is $\mathrm{i}\chi(=-\chi''+\mathrm{i}\chi')$. If then the first equation of (8) is true, so also is that obtained by replacing χ'' on the left by χ' and χ' on the right by $-\chi''$; and this yields the second equation of (8). It may be asked whether (8) represents a complete description of the limitations imposed by causality, or whether a stronger statement is possible. The answer is provided by the fact that for any reasonable choice of χ' on the real axis (by 'reasonable' is meant continuous and differentiable, and such as to allow the Kramers–Kronig integrals to converge) there exists one function and one only, extending over the whole complex plane, with the properties of having a real part on the real axis equal to χ' and of being regular in the upper half-plane. This means that causality places no restrictions on the form of χ', but allows only one form of χ'' for a given χ', i.e.

364

that given by (8), and this is as strong a statement as we can make. There is something rather satisfying about the way in which, having specified completely one half of the impulse response function (by making it vanish when $t<0$), we find we have halved our freedom of choice of $\chi(\omega)$, being allowed to specify either the real or the imaginary part† but not both.

As an illustration of the Kramers–Kronig relations consider the resonant response (7) whose essential form may be expressed, by redefining the variables,

$$\chi(\omega)=1/[(1-\omega^2)-\mathrm{i}\beta\omega], \text{ where } \beta \text{ is a positive constant; } \beta \ll 1. \quad (10)$$

Then

$$\chi'(\omega)=\frac{1-\omega^2}{(1-\omega^2)^2+\beta^2\omega^2}, \qquad \chi''(\omega)=\frac{\beta\omega}{(1-\omega^2)^2+\beta^2\omega^2}. \quad (11)$$

These two functions are shown in fig. 4. First we note that (10) satisfies the basic formulation of the causality principle, in that the poles of $\chi(\omega)$, at which the denominator vanishes, occur when $\omega=\frac{1}{2}[(4-\beta^2)^{\frac{1}{2}}-\mathrm{i}\beta]$, and are both in the lower half-plane. Unfortunately the formal verification of (8) involves a tedious integration, but the qualitative validity of the equations may be understood by drawing, which probably gives a better feel for what is going on. It is convenient in evaluating the second equation of (8) to note that $⨍_{-\infty}^{\infty} \mathrm{d}\omega/(\omega-\omega_0)=0$, so that a constant added to $\chi''(\omega)$ in the numerator makes no difference. We therefore write

$$\chi'(\omega_0)=\frac{1}{\pi}\int_{-\infty}^{\infty}\frac{\chi''(\omega)-\chi''(\omega_0)}{\omega-\omega_0}\,\mathrm{d}\omega;$$

the requirement of taking the principal value disappears, since the integrand remains finite even when the denominator vanishes. The shift of horizontal axis to pass through the curve of $\chi''(\omega)$ when $\omega=\omega_0$, which is all that this device amounts to, is shown in fig. 4(b). It is clear that just above ω_0 both numerator and denominator are positive, while just below both are negative; either way the integrand is positive throughout the range in which it is largest, and $\chi'(\omega)$ is correspondingly positive. On the other hand, if we were to choose ω_0 above the peak of χ'' the integrand would be predominantly negative. The general form of fig. 4(*a*) emerges clearly, and of course is substantiated precisely by carrying through the integration.

After this discussion of the Kramers–Kronig relations it must be confessed that they seem to have rather limited utility. There are circumstances in which measurements of either χ' or χ'' prove very troublesome and where information on one may be extended to the other by computer integration of (9). But the integrals are only very slowly convergent, and consequently the information on, say, χ' must extend over a very wide frequency range; moreover experimental errors in χ' can produce anomalies

† Or a suitable mixture of both, but we shall not try to delineate the latitude permitted in such circumstances.

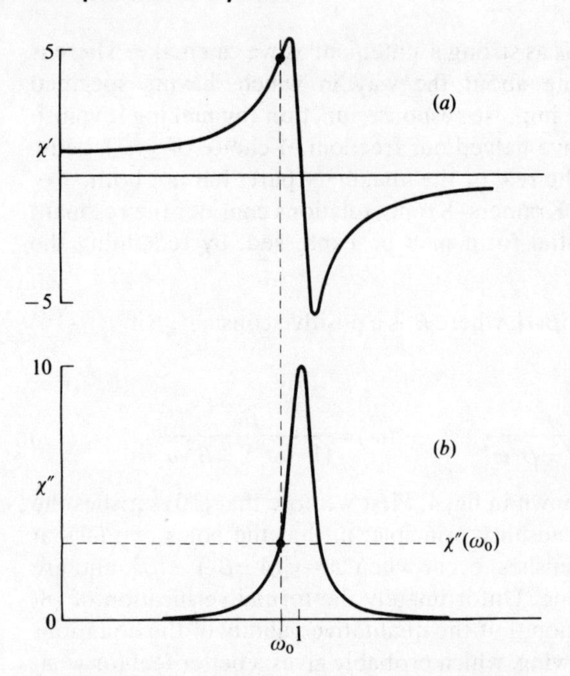

Fig. 4. Real (χ') and imaginary (χ'') parts of the response function of a resonant transducer, to illustrate Kramers–Kronig relations.

in χ'' which tempt the user to infer interesting physical behaviour where nothing special really exists. If both χ' and χ'' can be measured, but with limited accuracy (as for instance in certain studies of the optical properties of solids),[5] the relations may help in testing whether the results are compatible, and so provide a valuable critique of the experiment. But the case for presenting this theory here ultimately rests more on its very existence than on its utility; the qualitative idea of causality, involving a temporal ordering of causes and effects, is so fundamental to the physical universe that one should always try to discover its quantitative corollaries. In their way the Kramers–Kronig relations express a general truth that is independent of specific models, a distinction they share with thermodynamics, and deserve respect even if, unlike thermodynamics, they contribute little of practical value.

One special case is, however, worth noting briefly. If $\omega_0 = 0$ in (8), the second equation relates the real part of χ at zero frequency to the logarithmic integral of the imaginary part:

$$\chi'(0) = \frac{1}{\pi}\int_{-\infty}^{\infty} \chi''(\omega)\,\mathrm{d}\omega/\omega = \frac{1}{\pi}\int_{-\infty}^{\infty} \chi''(\omega)\,\mathrm{d}(\ln\omega).$$

In a resonant system $\chi''(\omega)$ is confined to a fairly narrow range of ω, and the integral can be computed from experimental measurements of loss. A complex system may have many loss peaks but, provided they are far enough apart in frequency to be resolved, each may be assigned its independent contribution to the zero frequency susceptibility. This idea has been put to good use in determining the contribution of electron spin paramagnetism (*Pauli paramagnetism*) to the magnetic susceptibility of a metal. There are many different contributions from various sources in a typical metal, which cannot be separated solely by measurement of the static susceptibility. This was the cause of much regret for many years since the Pauli contribution is potentially of great importance in testing the theory of electron–electron interaction in metals; if the electrons did not interact at all among themselves, the magnitude would be uniquely related to the thermal capacity of the electrons, a directly measurable quantity, and departures from this relation would give a quantitative estimate of the interaction strength. Now in a magnetic field of one or two tesla the electron spins should exhibit an absorption line in the microwave frequency range, and the logarithmic integral of this absorption gives the Pauli paramagnetism independently of any other mechanism. The line, though sharp in many insulators, is so broad in most metals as to disappear in the background contributed by other loss mechanisms, but in sodium it stands out strongly enough to be measured. It is fortunate, since the absorption line does not have the standard resonance profile, that the Kramers–Kronig theory is independent of any special model and can be applied even if the fine details of the line shape are imperfectly understood. The outcome of some years' devotion to this technically very demanding problem has been a firm value for the Pauli paramagnetic susceptibility in sodium, markedly different from what would be expected if the electrons were non-interacting, yet luckily not so different as make untenable theories of metals that start, as a first approximation, by neglecting this interaction.[6]

Resonant filters; sound spectrograph

Returning to experimental affairs, let us enquire what information can be discovered concerning the structure of a complicated input signal by passing it through a resonant filter, i.e. a transducer whose impulse response function is the decaying oscillation expressed by (6) and whose transfer function is the sharply peaked function (7). The formal solution has already been written down in (5), and applied to a pure sinusoidal input to show how the output is small unless the frequency is fairly close to the resonant frequency ω_0. If $\omega'\tau_a$ is large, the real and imaginary parts of the denominator in (7) are equal when the frequency is about $1/\tau_a$ away from resonance, and the power output, proportional to $\chi\chi^*$, is then reduced to half its maximum. We may therefore take $2/\tau_a$ as a convenient measure of the frequency bandwidth of the transducer, and this is also the reciprocal of the time-constant for decay of intensity in the impulse response.

When an arbitrary input $g_0(t)$ is fed into such a transducer, the input in the time interval t_0 to $t_0+\mathrm{d}t_0$ gives an impulse $g_0(t_0)\,\mathrm{d}t_0$ and starts a decaying oscillation which by time t is reduced by a factor $\mathrm{e}^{-(t-t_0)/\tau_a}$. The same output at time t would be produced from a lossless resonant transducer if the input were reduced by this factor. The two equivalent processes are illustrated in fig. 5 – the real time sequence fed into the real transducer is replaced by an exponentially attenuated time sequence fed into an ideal transducer with $1/\tau_a = 0$. Now the ideal transducer, stimulated by a pure sinusoid that has been applied to it ever since the start of time, responds infinitely more to an input that is exactly tuned than to any other, as is obvious from (7). The response at time t is therefore to be obtained by forming a Fourier transform of the input (fig. 5(b)) and picking out the amplitude of the component at ω_0, the transducer frequency. From this viewpoint the essential features of the transducer become clear. First, even if the input is a pure sinusoid, the exponential attenuation spreads its frequency spectrum over a range of the order $\omega_0 \pm 1/\tau_a$, as is clear from (4.27) or from the virtually equivalent analysis leading to (7). There is no possibility of making a spectral analysis of a non-sinusoidal input to a higher degree of resolution than this. Secondly, the output at any moment t does not refer to the spectral content at that exact moment (if such a statement has any meaning), but is related to the average over an interval of about τ_a before the time of observation. The process of analysis by resonant transducer is thus subject to exactly the same limitations as a expressed by the Uncertainty Principle $\Delta\omega\ \Delta t \gtrsim 1$.

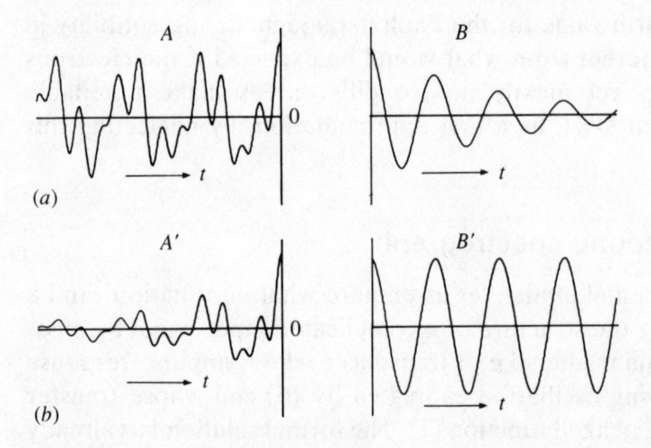

Fig. 5. A signal which, up to the instant $t = 0$, has had the form A enters a damped resonant transducer whose impulse response is B. The output at $t = 0$ is the same as if the damped signal A' had entered the ideal transducer B'.

Another way of looking at the same effect is to note that, since the real transducer only responds to frequency components of the input that lie within the range roughly described by $\omega_0 \pm 1/\tau_a$, we may eliminate from the input all but this narrow band of frequencies. Once this is done, what may have started as a jagged function has been reduced to something much smoother, for the resultant amplitude of the components in the range $1/\tau_a$ of frequency cannot significantly change in a time less than τ_a. The 'filtered' input is then approximately sinusoidal with mean frequency ω_0, but with amplitude and phase fluctuating slowly, at the sort of speed, in fact, that the transducer can respond to. It is these amplitude fluctuations that show up as output variations and are interpreted as fluctuations in the amplitude of components at frequency ω_0.

Clearly, then, if we are concerned to know how the spectral content of a signal changes with time we must be reconciled to limiting our frequency resolution to match any requirement to detect sudden changes. In the analysis of speech, for example, where an ordinary word may take $\frac{1}{3}$ second to say, and may contain several different noises as vowels and consonants alternate, we may wish a time resolution not worse than $\frac{1}{50}$ second. We must not then hope for a frequency resolution better than $\Delta\omega \sim 50\,\mathrm{s}^{-1}$, or $\Delta\nu \sim 8$ Hz. Even so we are in a position to obtain a very detailed analysis; neglecting all but the frequency range up to 3000 Hz, in which most of the interest of human speech lies, we may still divide it into over 300 separate frequency bands and record the variations of amplitude in each with the desired time resolution. Or if we are appalled at the thought of handling so many frequency channels we may cut them down to 30 and allows the luxury of a time resolution of 4 ms which is probably less than the ear can discern. One sound spectrograph, used for analysing speech, divides the range from 80 to 8000 Hz into as many as 500 channels, and records them one at a time. A short record of speech is speeded up and fed into a filter circuit whose bandwidth is adjustable down to about 45 Hz, and the time variation of the output recorded as a horizontal line of varying blackness; the same record is fed through again and again with the pass band slowly swept to higher frequencies, and the line displaced upwards as the frequency is increased. A typical result is reproduced in fig. 6, which shows how the title of this book appears when spoken in Southern English. The most prominent features are the wide band hissing noise, with the energy concentrated at higher frequencies, characteristic of the s's in 'physics', and the broader sh-sound towards the end of 'vibration'. The pure vowels are all short in duration, but the two diphthongs (the first i and the a in 'vibration') show clearly the variation of quality as the sound proceeds. Many other examples are to be found in the book of Potter, Kopp and Green.[7]

The pure vowels are sounded by concentrating the energy in certain bands characteristic of the vowel (*formants*). In normal speech the vocal chords emit a rather irregular sequence of impulses, at a mean frequency that is lower in men than in women, and fluctuates as the voice is inflected for emphasis, becoming higher in pitch at moments of rhetorical fervour or emotional stress. The cavities of nose, mouth and pharynx are tuned by manipulation of lips, tongue, etc. so as to respond selectively to certain frequencies in the rich spectrum of this output from the vocal chords, and it is these that form the vowel sound. In a well-sung note the irregular series of impulses is locked in frequency so that the output of the vocal chords is a fairly pure tone rich in harmonics; the appropriate harmonics are selected by the vocal cavities to give the note its vowel character. Sopranos and basses singing the same vowel will impose it on quite different fundamental frequencies by selecting different harmonics. For example, the first vowel in 'father' is distinguished by a strong formant around 900 Hz; a bass singing this word on the A♭ a tenth below middle C (about 100 Hz) will

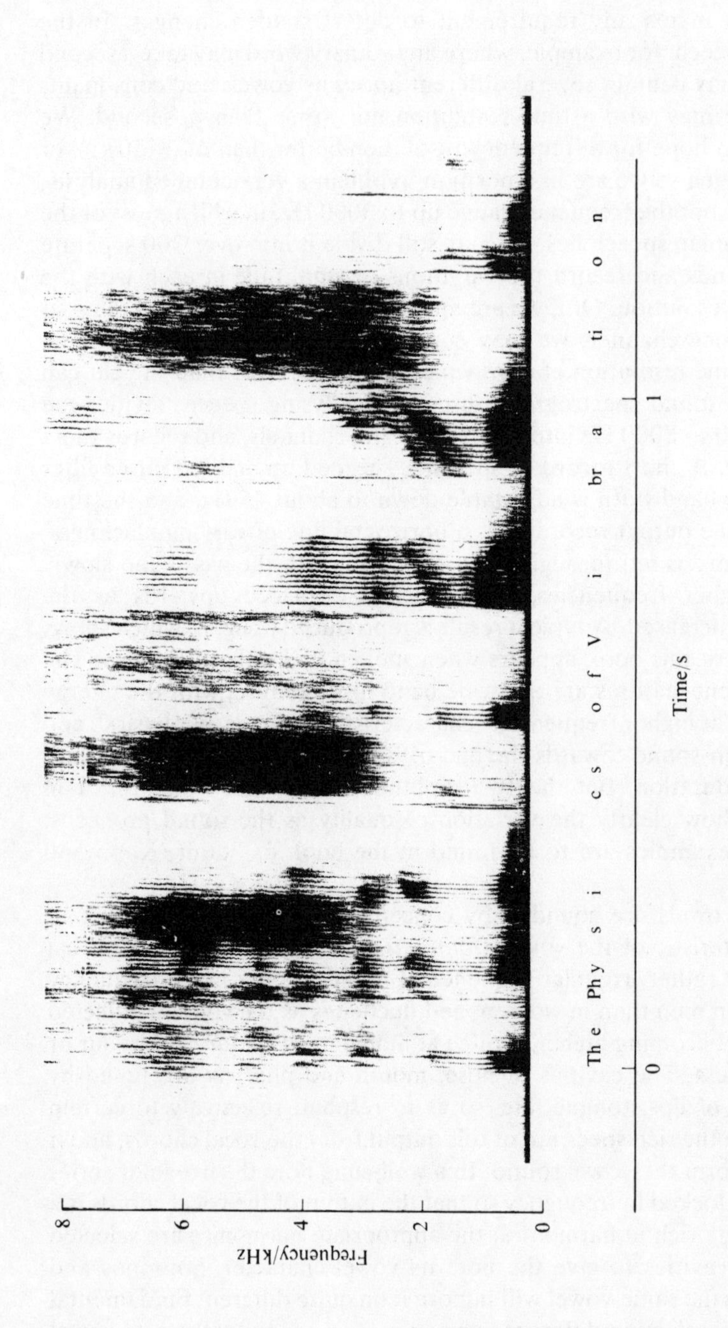

Fig. 6. Sound spectrograph, obtained with a Kay Sona-Graph, model 7029A, spectrum analyser. For explanation see text (Cambridge University Department of Linguistics.)

contrive to emphasize the ninth harmonic, a contralto singing an octave above will emphasize the fourth and/or fifth, while a soprano still another octave above, at 400 Hz, will have to make do with the second harmonic, though the higher components may emerge hardly shifted and the vowel will be perfectly recognizable. Fig. 7 illustrates this point. The fundamental periods differ by a factor of four, but the very strongly marked harmonics (the ninth for the baritone and the second for the falsetto) are not very different in frequency. The few cycles recorded here tell one little about such highly significant features as the attack (the first stages in producing the steady note) and variations of pitch (vibrato) and tone colour (harmonic content). These necessitate a sound spectrograph, and examples will be found, including analysis of a Caruso top note, in the book already cited.[7]

The analysis of speech by sound spectrograph is a fascinating adjunct to more general phonetic studies, but one that cannot be followed here. Let us turn to the rather simpler matter of the waveform produced by musical instruments, for which the sound spectrograph is usually unnecessarily elaborate. For we are primarily interested in a note of steady pitch with a periodic, nor normally sinusoidal, waveform. The oscillograph records in fig. 8 are typical of the many studies that have been made to relate the timbre of instruments to their waveform; in comparison with speech, their Fourier analysis presents no significant problem.

The mechanism of tone production differs considerably from one musical instrument to another, and for the present a very general account is sufficient to make the particular point that is relevant here, and which may be introduced by comparing the waveforms produced by a bassoon, a guitar and a bell. The first two show a periodicity that is absent from the last, and it is no accident that only the first two are used in Western music that depends on the harmony of combined notes.† For the basic intervals on which the diatonic scale is constructed are related by simple numerical ratios (2 for the octave, 3/2 for the fifth, 5/4 for the major third), and the overtones (harmonics) contained in a periodic waveform supply octaves, twelfths etc. above the fundamental that do not clash disturbingly with the notes of other instruments playing in a concord; it is possible for the ear to recognize without ambiguity the harmonic meaning of the sound. Most listeners in fact are unaware of the overtones when a wind-instrument plays a single note – their ears and minds interpret what they receive as a single note, but a note of a certain quality (timbre). Three oboes playing a major triad (C, E, G) are normally heard as these three notes, not as an extremely complex mixture of the fundamentals with ten or twelve audible overtones

† When bells, timpani and other percussion instruments, even xylophones, are used, it is either in a non-harmonic context or with such support from conventional instruments as will clarify the harmonic meaning beyond their powers of disruption. English ears, accustomed to the austere splendour of change-ringing, find the sound of a carillon playing harmonized music grotesque in the incompatibility of the bell timbre with the logic of diatonic harmony.

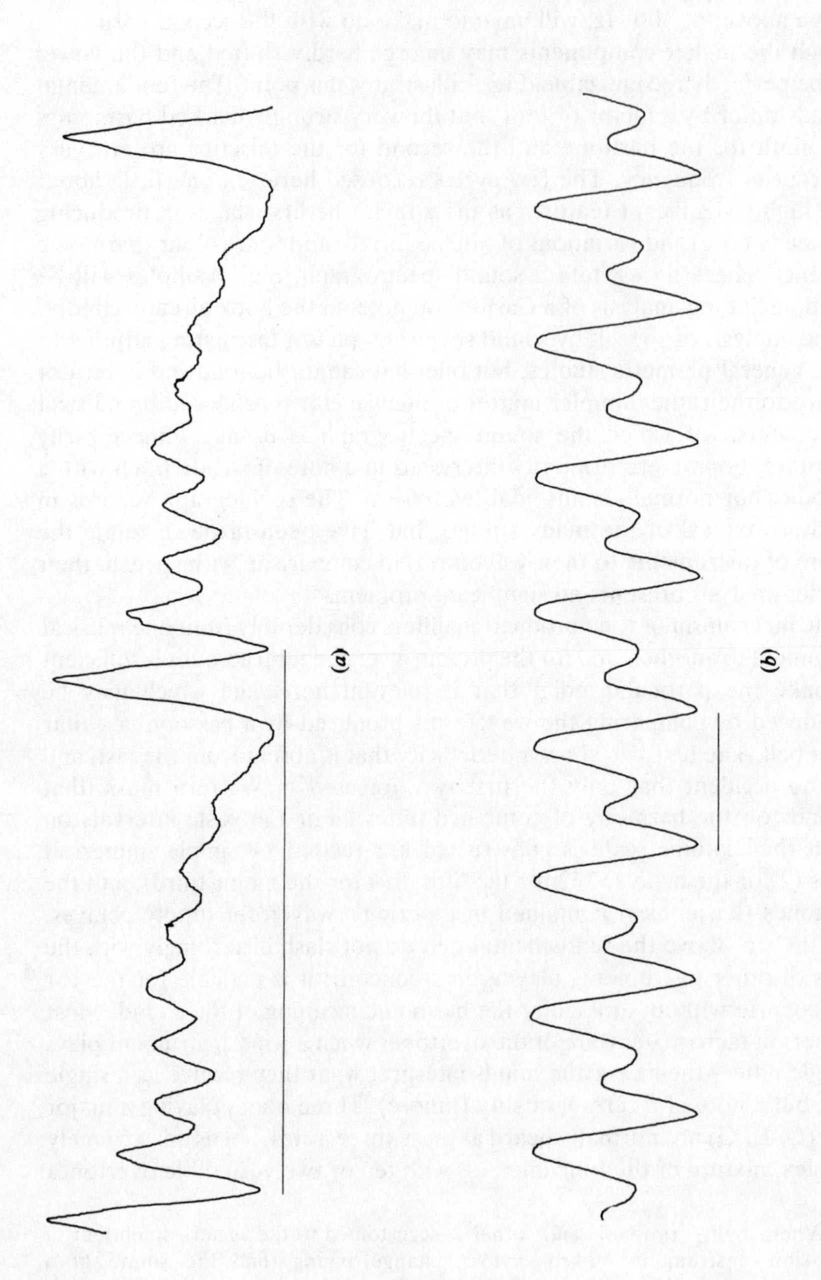

Fig. 7. Waveforms of the vowel *a* in *father* sung (*a*) on A♭ (104 Hz) in natural baritone and (*b*) two octaves higher (416 Hz) in falsetto. The strongly marked frequency giving the vowel its characteristic quality is the ninth harmonic in (*a*), i.e. 936 Hz, and the best approximation to this in (*b*), i.e. the second harmonic 832 Hz. Both traces are 20 ms long; here and in fig. 8 the fundamental period in each is indicated by a straight line under the trace.

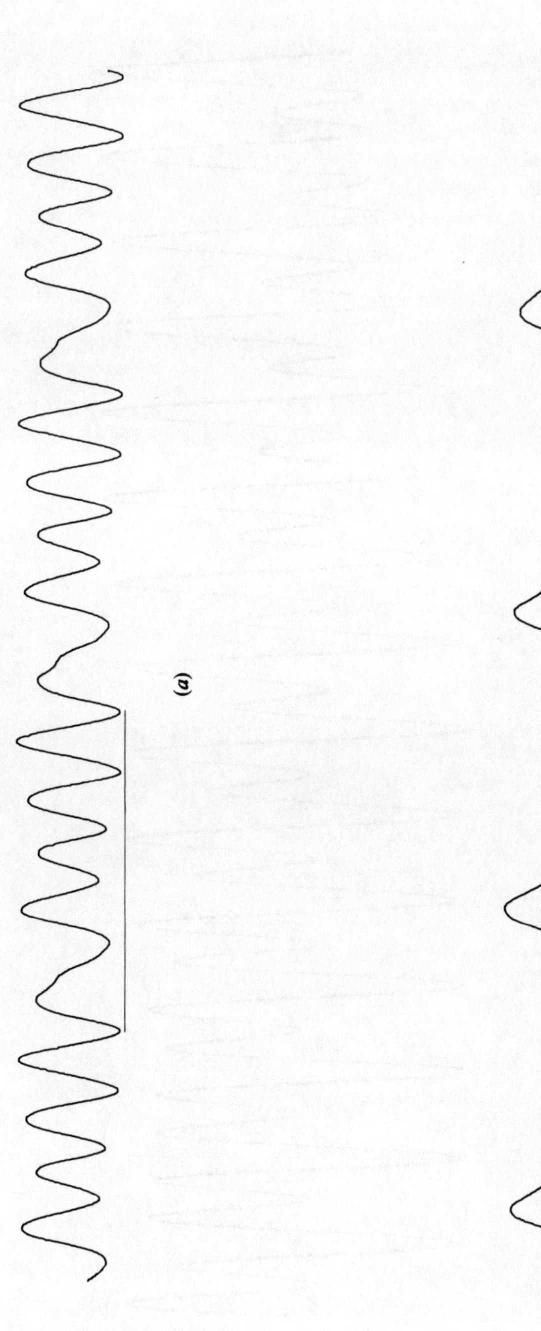

Fig. 8. Waveforms of (*a*) bassoon playing G (100 Hz), (*b*) bassoon playing g′ (400 Hz), (*c*) guitar about $\frac{1}{2}$ s after plucking e′ (338 Hz), (*d*) handbell about $\frac{1}{2}$ s after striking. The first two show highly reproducible waveforms from one cycle to the next, indicating a pure harmonic series; the guitar is slightly less perfect, indicating small departures from the harmonic series; the bell obviously suffers progressive changes of waveform – Fourier analysis gives the principal components the frequencies 2.120 and 4.355 kHz, in a ratio that is significantly different from 2 (cf. fig. 4.2(*f*)).

Fig. 8—*contd.*

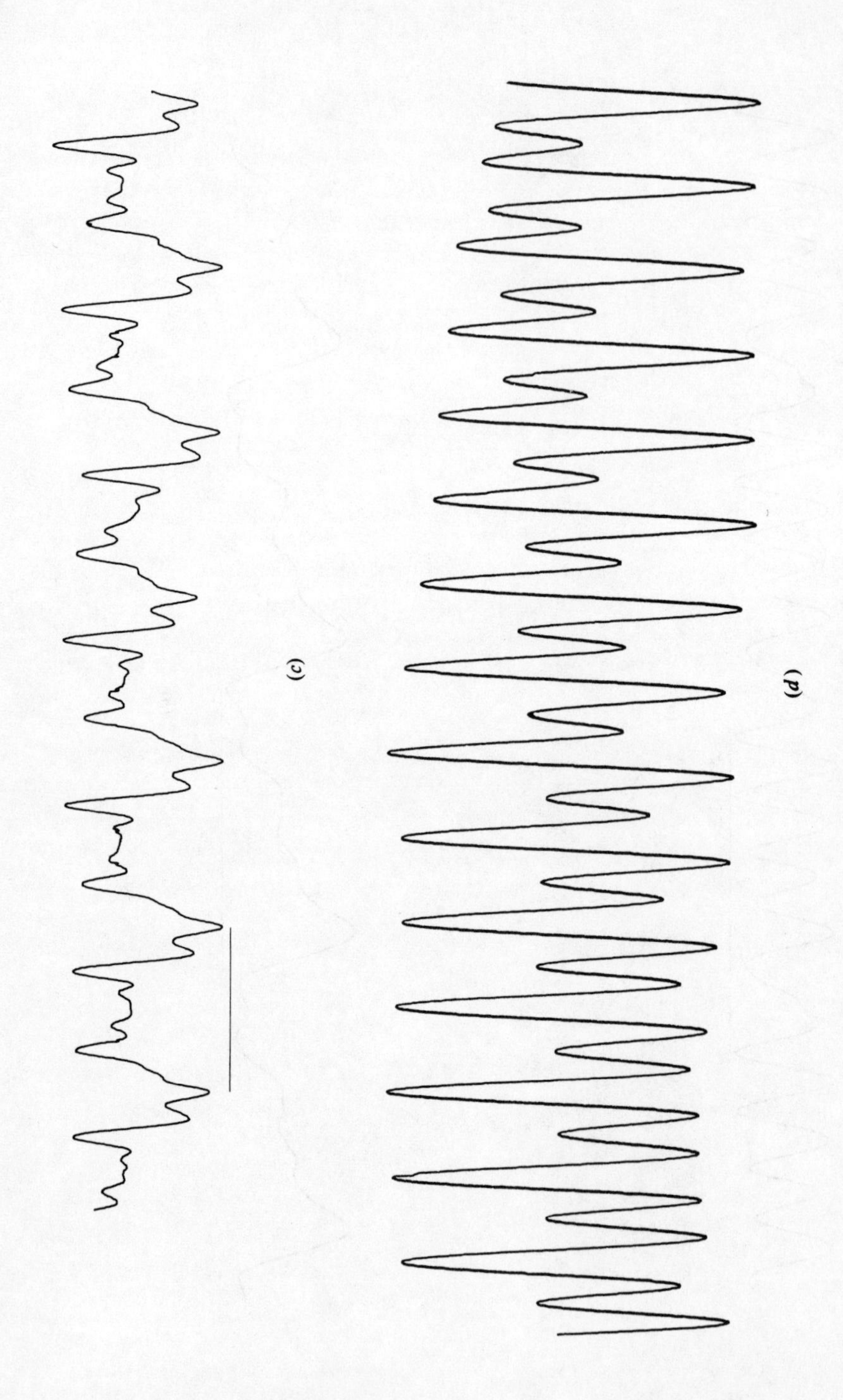

attached to each. A trained listener, however, can hear the separate constituents, and it is worth remembering that much progress in the analysis of tone quality, including the discovery of the nature of vowel sounds, was made before oscillographs were invented; Helmholtz used acoustic resonators, but Bell and Paget relied on their ears.

We must now note an important distinction between different types of musical instrument. Although both the guitar (and similarly the piano) and a wind-instrument produce waveforms that maintain their character over many cycles and therefore consist of a strict harmonic series of frequencies, the former has this property somewhat accidentally while in the latter it is intrinsic. The string, like the bell, is capable of executing independent harmonic vibrations in a large number of different modes, and when it is struck or plucked many of these are excited. As a consequence of the wave velocity on a string being independent of frequency, and the one-dimensional character of the wave, the modes have frequencies that are very nearly exact multiples of the fundamental; the bell, on the other hand, not having the simplicity of a uniform string, is far richer in non-integral multiples. The periodicity of the guitar or piano note arises therefore because the only vibrations it can maintain happen to be such as combine to give a periodic resultant. The wind-instrument, however, has the acoustic vibrations of the air in its tube excited by periodic puffs from the reed or lips; the back-reaction of the acoustic vibrations locks the reed frequency to a resonant frequency of the tube. If locking is complete the motion of the reed is now strictly periodic, though far from sinusoidal, and it may excite higher frequency acoustic vibrations if one or more of the higher resonant models of the tube lies close to a harmonic of the reed frequency. But if the instrument should be so badly designed that its higher modes are seriously out of tune with multiples of the reed frequency, they will not be appreciably excited and the emergent sound will still be periodic but with a lesser admixture of the higher harmonics. Nothing the untuned modes can do will break the periodicity which is controlled by the reed, though they may make it more difficult to lock the reed vibration to any acoustic mode; as an extreme example, the 'tacet horn' designed by Benade[(8)] is actually unplayable.

The reed is an example of a non-linear oscillator whose intrinsic waveform is not sinusoidal. This does not prevent one from carrying through a formal Fourier analysis of its motion into fundamental frequency and harmonics, for this is merely a mathematical procedure applicable to any periodic function. But the analysis is irrelevant to the physical processes at work; one could not excite any one of these sinusoidal vibrations alone, in the way that one can with a piano string by singing loudly when the dampers are raised.

Mixing and combination tones

With a wind-instrument the non-linear process occurs at the point where the sound is generated, and the resulting vibrations are modified by the tube acting as a linear transducer. The reverse process, in which independent vibrations in a linear system are passed through a non-linear transducer, distorts the waveform so as to generate new Fourier components whose frequencies are combinations of those originally present. The simplest example of practical importance is the rectifier which, in its ideal form, transmits unchanged all positive inputs and blocks everything negative. A sinusoidal voltage applied across a rectifier and resistance in series passes a current waveform consisting of sine loops interspersed with blank regions, as shown in fig. 9; the mean current is no longer zero and Fourier

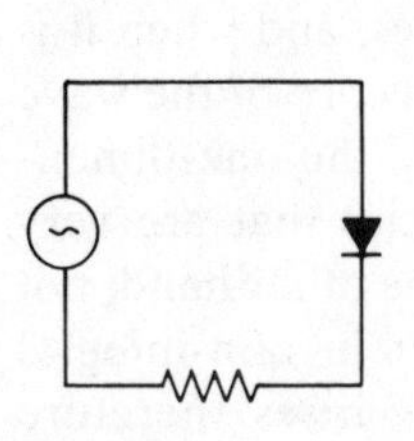

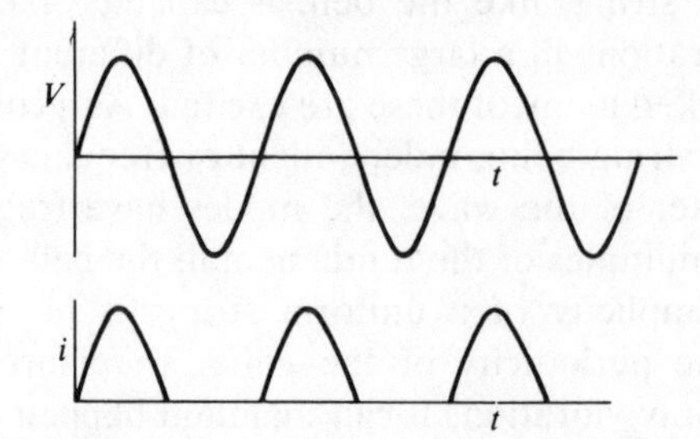

Fig. 9. Simple half-wave rectifier, without smoothing. The sinusoidal generator voltage V results in the intermittent current i.

analysis shows, as is obvious from the diagram, that a complete series of harmonics has been generated. If the output waveform is given by

$$y = \cos \omega t \quad \text{when} \quad -\pi/2 < y < \pi/2,$$
$$= 0 \text{ otherwise,}$$

then according to (4.10) the mth Fourier amplitude is given by

67

$$a_m = \frac{1}{2\pi} \int_{-\pi/2}^{\pi/2} \cos x \, e^{-imx} \, dx.$$

This leads to the following expansion:

$$y = \frac{1}{\pi}\Big[1 + \tfrac{1}{2}\pi \cos \omega t + \tfrac{2}{3} \cos 2\omega t - \tfrac{2}{15} \cos 4\omega t$$
$$- 2 \sum_{m\,\text{even}} (-1)^{\frac{1}{2}m} \cos m\omega t/(m^2 - 1)\Big]. \quad (12)$$

A more usual rectifier circuit incorporates a capacitor as in fig. 10, which also shows the current through the resistor. On the rising positive part of

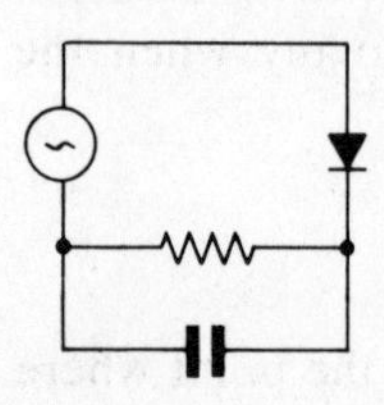

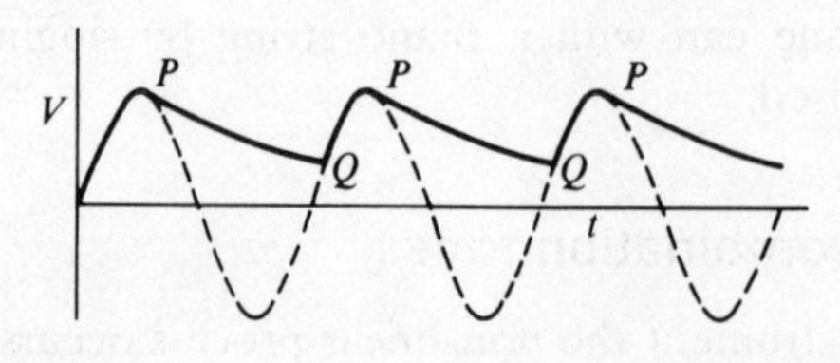

Fig. 10. Smoothed half-wave rectification.

the cycle the rectifier passes current through the resistor and simultaneously charges the capacitor to the point that the voltage across it equals the input voltage (this assumes zero resistance in the rectifier). A little time after the peak, at P, the applied voltage is falling at the same rate as the combination of resistor and capacitor can discharge itself if cut off from the input; the discharge curve PQ is of exponential form, $e^{-t/\tau}$, with $\tau = CR$, and is drawn to be tangential with the sine curve at P. Until P is reached, forward current through the rectifier is needed to hold the voltage to the sine curve. But beyond P the sine falls off more sharply than the exponential and the condition for isolation from the input is satisfied, the rectifier being back-biased. At Q forward biasing starts again, so that the heavy curve represents the variation of voltage across, and therefore current through, the resistor. If the time-constant τ is made much greater than the oscillation period, by choosing a large enough capacitor, the output is almost flat and the strong harmonic content expressed by (12) is greatly reduced. Still better is to use the full-wave rectifier circuit of fig. 11. When

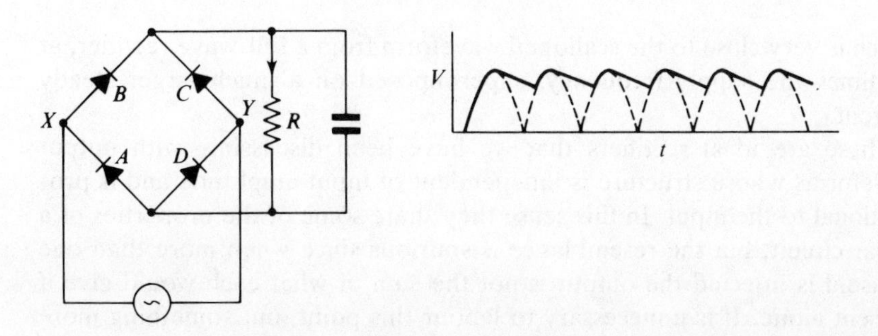

Fig. 11. Smoothed full-wave rectification.

X is positive, current cannot flow through A and C and passes through R in the direction of the arrow; when Y is positive, the current is blocked by B and D and passes through R in the same direction. The voltage input is thus two half-waves, and with the capacitor present the output is correspondingly flatter than with the half-wave rectifier of fig. 10.

When heavy current is required of a rectifier capacitive smoothing may be impracticable, and the use of a 3-phase a.c. supply is advantageous in reducing the ripple. The supply consists of a neutral line and three others with alternating voltages phased at 120° to each other. If each is rectified in a half-wave circuit and the outputs combined, the result is the superposition of three current waveforms like that in fig. 9, and gives the waveform shown in fig. 12(b). Still better is to use a full-wave rectifier circuit on each phase so that six outputs at 60° are added, and with suitable phase-changing networks the process may be multiplied further to give 12 or more outputs evenly spaced in phase. If a single-phase waveform is expressed as a Fourier series

$$y = \sum_{-\infty}^{\infty} a_n \, e^{-in\omega t},$$

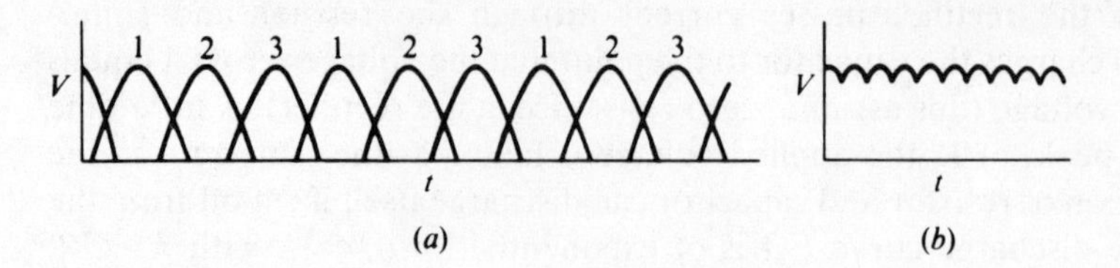

Fig. 12. Unsmoothed three-phase rectification. The intermittent outputs, labelled 1, 2 and 3, interlock to give a steady current with ripple superimposed.

the addition of N identical waveforms at phase spacing of $2\pi/N$ eliminates all those terms for which n is not a multiple of N. This is immediately obvious from the vibration diagram for any Fourier component, which is a closed polygon of N sides, having zero resultant if $n \neq mN$, but a straight line if $n = mN$. Thus the 12-phase rectifier just mentioned, feeding straight into a resistive load with no capacitive smoothing, has an output waveform whose Fourier components can be written down from (12),

$$y \propto 1 + \tfrac{2}{143}\cos(12\omega t) + \tfrac{2}{575}\cos(24\omega t) + \cdots,$$

which is very close to the scalloped waveform from a full-wave rectifier, at 12 times the supply frequency, superimposed on a much larger steady current.

These are ideal rectifiers that we have been discussing, with output waveforms whose structure is independent of input amplitude and is proportional to the input. In this sense they share some of the properties of a linear circuit, but the resemblance is spurious since when more than one sinusoid is injected the output is not the sum of what each would give if present alone. It is unnecessary to labour this point, but something more valuable can be learnt from an analysis of a less ideal rectifier, one in fact that is only slightly non-linear. A real rectifier (e.g. a p–n semiconductor junction) has a characteristic like that shown in fig. 13. When the voltage is

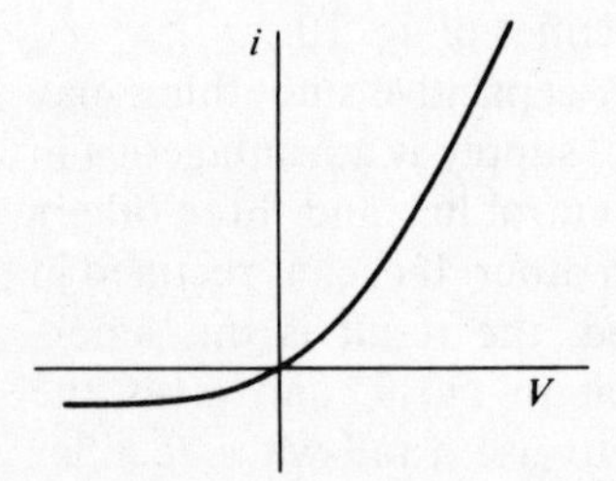

Fig. 13. Rectifier characteristic.

large and positive it passes current easily, when large and negative it passes so little as to be virtually blocked. For large voltages, then, it approximates to an ideal rectifier. For small voltages, however, the characteristic is only slightly curved, and now it is easily seen that the mean d.c. current is proportional to the square of the applied voltage. For this condition of square-law rectification to obtain, the voltage must be small enough for the characteristic to be adequately represented by the quadratic form

$$i = \alpha V + \beta V^2. \tag{13}$$

Substituting $V = V_0 \cos \omega t$ (note that since the system is non-linear we must beware of using the complex representation), we find that

$$\begin{aligned} i &= \alpha V_0 \cos \omega t + \beta V_0^2 \cos^2 \omega t \\ &= \tfrac{1}{2}\beta V_0^2 + \alpha V_0 \cos \omega t + \tfrac{1}{2}\beta V_0^2 \cos 2\omega t, \end{aligned} \tag{14}$$

The d.c. current being $\frac{1}{2}\beta V_0^2$ and the waveform being distorted by the production of a second harmonic whose relative magnitude depends on the strength of the applied voltage.

Rectification depends, of course, on the i–V characteristic having a different form for positive and negative voltages. A symmetrical characteristic, for example

$$i = \alpha V + \gamma V^3, \tag{15}$$

gives for an input $V_0 \cos \omega t$ an output

$$\begin{aligned} i &= \alpha V_0 \cos \omega t + \gamma V_0^3 \cos^3 \omega t \\ &= (\alpha V_0 + \tfrac{3}{4}\gamma V_0^3) \cos \omega t + \tfrac{1}{4}\gamma V_0^3 \cos 3\omega t, \end{aligned}$$

with odd harmonics only and no steady term.

When two different frequencies are fed into a non-linear circuit, mixing occurs and other frequencies are generated. Thus with $V = V_1 \cos \omega_1 t + V_2 \cos \omega_2 t$ substituted in (13) the output is given by

$$\begin{aligned} i = {} & \tfrac{1}{2}\beta(V_1^2 + V_2^2) + \alpha V_1 \cos \omega_1 t + \alpha V_2 \cos \omega_2 t + \beta V_1^2 \cos 2\omega_1 t \\ & + \beta V_2^2 \cos 2\omega_2 t + \tfrac{1}{2}\beta V_1 V_2 \cos(\omega_1 - \omega_2)t + \tfrac{1}{2}\beta V_1 V_2 \cos(\omega_1 + \omega_2)t. \end{aligned} \tag{16}$$

In addition to terms of the form (14) arising from each input component separately the output contains, in the last two terms, combination frequencies generated by the mixing of the inputs, with amplitudes determined by $V_1 V_2$ and frequencies equal to the sum and difference of the input frequencies. The same input applied to (15) yields no d.c. term, of course, but a rich collection of frequencies – ω_1, ω_2, $3\omega_1$ and $3\omega_2$ from each component alone, $\omega_1 \pm 2\omega_2$ and $2\omega_1 \pm \omega_2$ by mixing. There are no plain sum and differences tones, which require a rectifying characteristic for their production.

In general, if the characteristic can be expressed as a power series:

$$i = \alpha V + \beta V^2 + \gamma V^3 + \cdots \tag{17}$$

and the input takes the form $V_1 \cos \omega_1 t + V_2 \cos \omega_2 t + \cdots$, the output can contain only such combination frequencies as take the form $n_1\omega_1 + n_2\omega_2 + \cdots$, where the ns are integers and $|n_1 + n_2 + \cdots|$ is one of the exponents occurring in the power series (17), or is less than one of the exponents by an even number.

A fairly simple practical demonstration of mixing, which may also have useful applications, involves feeding three signals into a semiconductor diode whose output is amplified to excite a loudspeaker; two signals are the outputs of triode oscillators at about 10^9 Hz, which can be made to give very pure sinusoids, stable in frequency; and the third is the output of a square-wave oscillator at 10^5 Hz.[9] If one triode oscillator is kept at constant frequency while the other is slowly tuned, an audible note is heard when the difference in frequency, $\omega_1-\omega_2$, between the two is close to one of the harmonics of ω_3, the harmonic-rich square wave. What is heard is $\omega_1-\omega_2+n\omega_3$, and as ω_2 is varied the note is first heard very high and then descends below the audible range, rising again to a high and finally inaudible frequency. The successive tuning positions at which the note goes through zero frequency provide calibration points for the tunable oscillator at exactly equal spacings of ω_3. As many as 50 points can be obtained, corresponding to the difference $\omega_1-\omega_2$ running between $\pm 25\omega_3$.

The difference tone can be experienced directly if two audiofrequency signals at a high level of intensity are heard together. To demonstrate the effect convincingly as a property of the ear it is advisable to feed the amplified output of two signal generators into separate loudspeakers, so as to give no chance for non-linear effects in the loudspeakers to produce the difference tone. With the head between the loudspeakers two notes rather close together at a high audible frequency (several kHz) are accompanied by a low note which rises and falls rapidly as one of the signal generators is tuned through a relatively small range. The sound level required to hear the difference tone may prove unpleasantly loud unless one has an especially good (or unusual) ear. Not only the simple difference tone may be heard – other combinations like $2\omega_1-\omega_2$ are also produced and it is clear that the mechanism of the ear has rectifying powers and higher order terms in its response which are significant at a high intensity.

A somewhat obscure, and possibly related, phenomenon is made use of (though deprecated by purists) to produce the effect of very low notes from church organs that have not the space to accommodate 32 ft pipes. A 16 ft pipe at an octave above the desired note and a $10\frac{2}{3}$ ft pipe at an octave and a fifth are sounded together; that is, if ω is desired, 2ω and 3ω are sounded and the listener is persuaded that he has heard the difference frequency ω. This is called an *acoustic bass* and is sometimes explained as arising by the above mechanism of non-linearity in the ear. It is, however, hard to believe this explanation since the amplitude would depend, according to (16), on the product of the two component amplitudes, and listeners in different parts of the building would get a very different intensity of difference tone relative to the two generating tones. One should not rule out a psychological origin – possibly the brain hearing two low notes (the device only works for very low notes apparently) at 3ω and 2ω interprets them as arising from a still lower fundamental at ω. To carry the conjecture further, the audible illusion may be fostered by perfect tuning of the two constituents so that their waveform is strictly repetitive with a basic repetition frequency of ω, and this may provide a reason for the normal practice of strapping the two pipes together. For it is known that two organ pipes, like other maintained oscillators, in close proximity may influence one another (this is a non-linear effect which will be examined in some detail in chapter 12), and may pull their frequencies into a simple numerical ratio even if

391

they are initially not quite perfectly tuned. But there is yet another explanation possible, which is that the mechanical interactions between the two pipes clamped together and set into vibration by the air within may, through some unidentified rectifying effect, generate a real difference tone on the spot.

The non-linear generation of combination tones is the mechanism underlying the superheterodyne principle of radio detection. Audio-frequency information transmitted as amplitude modulation of a carrier wave having a high frequency, 1 MHz say, is amplified and detected in two stages; first the incoming signal, presumed to be very weak, is mixed in a non-linear device with a much stronger local oscillator signal of constant amplitude and differing in frequency from the carrier. With a square-law detector such as is described by the second term of (13) (the linear term being irrelevant to the mixing process), an input consisting of the incoming signal $f(t)\cos\omega_1 t$ and the local oscillator $A\cos\omega_2 t$ is transformed into $\beta[f(t)\cos\omega_1 t + A\cos\omega_2 t]^2$, which contains among others the term $\beta f(t)A\cos(\omega_1-\omega_2)t$. The information, expressed by $f(t)$, has been transferred to the lower difference frequency (above the audible limit) which can be chosen for convenience of amplification; at the same time its amplitude has been multiplied by βA, which can be made much greater than unity by suitable choice of local oscillator amplitude. In this way a very weak signal is enhanced before ever entering a conventional amplifier, and with a good local oscillator, producing an output almost free of random variations, the quality of the information is maintained in the process. After amplification the output is rectified, and the information appears as signal variations imposed on no carrier frequency but capable of actuating a loudspeaker or recorder directly.

Mixing of modulated radio waves sometimes occurs when they are reflected by the ionosphere.[10] Above the neighbourhood of a powerful transmitter the intensity may be so great as to impart extra kinetic energy to the electrons to an extent that is not negligible in comparison with their thermal energy. The reflecting properties of the ionosphere are thus slightly modified, and the radio waves from a second, distant, transmitter may be modulated, as it is reflected from this region, in synchronism with the rise and fall of the intensity of the nearby transmitter. Since this variation of intensity is the signal carried on an amplitude-modulated wave, a listener tuned to the second transmitter may be surprised to hear the program of the first breaking in on that of the second, even though the two transmitters have quite different carrier frequencies. This is the wave-interaction effect, sometimes called the Luxembourg effect after the first transmitting station powerful enough to produce a discernible non-linearity.

6 The driven harmonic vibrator

The simple harmonic oscillator, driven by a sinusoidally varying force, is central to the discussion of vibrating systems, being a model for so many real systems and therefore serving to unify the description of very diverse physical problems. In view of this it is worth spending some time examining it from several different aspects, even though it might be thought that a formal solution of the equation of motion contained everything useful to be said on the matter. Indeed, if one were concerned only with physical systems that could be modelled exactly in these terms a single comprehensive treatment would suffice for all. Real systems, however, normally only approximate to this idealization, and alternative approaches may then prove their worth in allowing the behaviour to be apprehended semi-intuitively, often enough with sufficient exactitude to make mathematical analysis unnecessary. The reader who has progressed to this point will be familiar enough with the most elementary analyses not to be worried that we approach the problem indirectly, picking up an argument that has already been partially developed. More familiar treatments will be introduced in due course.

Transfer function, compliance, susceptibility, admittance, impedance

The essential framework for this approach has already been laid down in chapter 5, where the concept of the transfer function $\chi(\omega)$ was introduced. We had in mind there a linear transducer into which a sinusoidal signal $A\,\mathrm{e}^{-\mathrm{i}\omega t}$ was fed, and from which emerged an output signal $\chi(\omega)A\,\mathrm{e}^{-\mathrm{i}\omega t}$. Precisely the same idea may be applied to a mechanical system or electrical circuit by reinterpreting the signal $A\,\mathrm{e}^{-\mathrm{i}\omega t}$ as the oscillating driving force and the output signal as the response of some part of the system, for example the displacement or velocity of one mechanical component, the charge on a capacitor, or the current in some part of the circuit. If $\chi(\omega)$ relates mechanical displacement to force it is called *compliance*, if it relates electrical or magnetic polarization to field strength it is *susceptibility*, if it relates current to e.m.f. it is *admittance*. Alternatively, the disturbing input signal may be a current and the output the resulting e.m.f. in which case the frequency-dependent relation between the two is the *impedance*, the electrical analogue of what in elastic systems is an *elastic modulus*. We do not attempt to define each of these quantities with high precision for it is not their exact meaning that matters – only the close analogy between them that allows the same discussion of $\chi(\omega)$ to be transferred to new situations with only minor modifications.

For a simple resonant system, with only one coordinate available to describe the response to an input force, the character of $\chi(\omega)$ is defined by the impulse response function – normally a damped sinusoid as in (5.6), leading to (5.7) for $\chi(\omega)$. Let us look at a number of illustrations, taking care to choose as input some form of disturbance that does not lead to an infinite response. Thus with a simple pendulum $\chi(\omega)$, as in (5.7), is appropriate to describe the relation between displacement as output and force as input, rather than the other way round. For a force $F(t)$ of finite magnitude and applied over a short interval δt results in a finite initial velocity and virtually zero displacement; on the other hand, an impulsive displacement $\xi(t)$ lasting for an interval δt would require infinite forces (an infinite forward impulse to displace the pendulum to ξ instantaneously, followed by an opposite impulse to bring it to rest there; after δt the same process in reverse). Immediately after the impulsive force assumed to be of unit strength, the pendulum bob, of mass m, is still at the origin but moving with velocity $1/m$. The subsequent motion is then described by (5.6), h_0 being assigned such a value as to ensure that the initial velocity is correct:

107

$$h(t) = (m\omega')^{-1} \sin \omega' t \cdot \mathrm{e}^{-\omega'' t} \qquad (t>0) \tag{1}$$

ω'' being written for $1/\tau_a$. Hence, from (5.7),

$$\chi(\omega) = -1/mY, \tag{2}$$

where Y is written for the resonant denominator that will recur continually:

31

$$Y = \omega^2 - \omega'^2 - \omega''^2 + 2\mathrm{i}\omega''\omega = \omega^2 - \omega_0^2 + 2\mathrm{i}\omega''\omega, \quad \text{from (2.24).} \tag{3}$$

Y may also be expressed in terms of the two complex frequencies that make up the damped oscillation (1). If $\Omega_1 = \omega' - \mathrm{i}\omega''$ and $\Omega_2 = -\Omega_1^* = -\omega' - \mathrm{i}\omega''$,

$$Y = (\omega - \Omega_1)(\omega - \Omega_2). \tag{4}$$

The response of the pendulum to an oscillatory force $F\,\mathrm{e}^{-\mathrm{i}\omega t}$ is given by the expression

$$\xi(t) = \chi(\omega) F\,\mathrm{e}^{-\mathrm{i}\omega t} = -F\,\mathrm{e}^{-\mathrm{i}\omega t}/mY. \tag{5}$$

If ω'' is small (light damping), $\chi(\omega)$ has a high peak of magnitude $\mathrm{i}/2\omega''\omega m$ when ω is very close to the natural frequency ω_0. We shall not discuss the form of $\chi(\omega)$ further at this point, but first present other illustrations.

The series resonant circuit of fig. 1 may be excited impulsively by attaching a cell across AB for an interval δt and straightway short-circuiting these terminals. The finite current excited by the cell can then oscillate in the closed circuit, dying out exponentially as the electrical

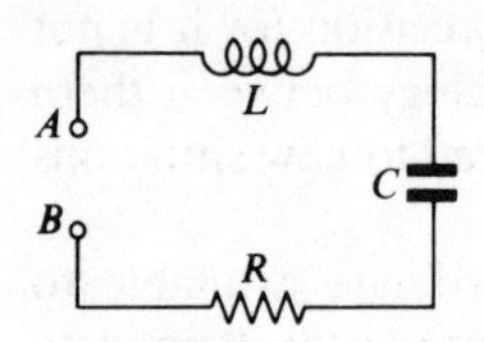

Fig. 1. Series resonant circuit.

energy is dissipated in the resistor. It is not appropriate to regard the current as the input driving force and the e.m.f. across AB as the resulting output, since it requires an infinite e.m.f. to establish a current suddenly through the inductor. Before jumping to the conclusion that $\chi(\omega)$ in (2) describes the relation of oscillatory current to oscillatory e.m.f., we must satisfy ourselves that the impulse response function has the form (1); but in fact it has not. In the initial stages after the cell is connected, the whole e.m.f., V, is taken up in forcing the current through the inductor to rise at a rate V/L. After time δt the current is $V\,\delta t/L$, and if $V\,\delta t$ is a unit impulse, the initial value of the current response, $h(0)$, is $1/L$. There is no significant potential drop across the capacitor since the charge that has so far flowed, $h(0)\,\delta t$ or $\delta t/L$, can be made infinitesimal by allowing δt to go to zero while increasing V to maintain $V\,\delta t$ at the value unity. Immediately after short-circuiting the terminals AB there is no e.m.f. round the circuit, so that if the resistance is small the e.m.f. across it, R/L, and the equal and opposite e.m.f. across the inductor, are both small, tending to zero as R is diminished. Initially, then, the current response $h(t)$ is nearly unchanging at its value $1/L$, consistent with a cosine rather than a sine form, as in (1), for $h(t)$. The precise form of $h(t)$ is most readily derived by noting that the charge $q(t)$ on the capacitor is completely analogous to the displacement of the pendulum. In particular, starting at zero it must be a sine function, and the initial rate of change $1/L$, determined by the initial value of the current, fixes the coefficient. Hence

$$q(t) = (\omega' L)^{-1}\,\mathrm{e}^{-\omega'' t} \sin \omega' t, \tag{6}$$

and for the current response we have that

$$h(t) = \dot{q}(t) = (\omega' L)^{-1}\,\mathrm{e}^{-\omega'' t}(\omega' \cos \omega' t - \omega'' \sin \omega' t), \tag{7}$$

a decaying sinusoidal oscillation, but one that starts at a different phase from (1).

Corresponding to (7) the admittance of the circuit follows by use of (5.4): **106**

$$\chi(\omega) = \mathrm{i}\omega/LY, \tag{8}$$

differing in phase from (2), through multiplication of the numerator by $-\mathrm{i}\omega$, but otherwise very similar in form and, when ω'' is small, with a sharp peak at the natural frequency ω_0. The reason for the appearance of $-\mathrm{i}\omega$ is clearly that this represents the operation of time-differentiation by which the response (7) is derived from that in (6).

The parallel resonant circuit of fig. 2 may be treated in the same way by reversing the significance of e.m.f. and current. It is now inappropriate to treat e.m.f. as the driving force, since applying a cell to the terminals immediately sends an infinite current through the capacitor. On the other hand, an impulsive current of unit strength deposits unit charge on the capacitor plates and establishes an e.m.f. $V = 1/C$ between the terminals.

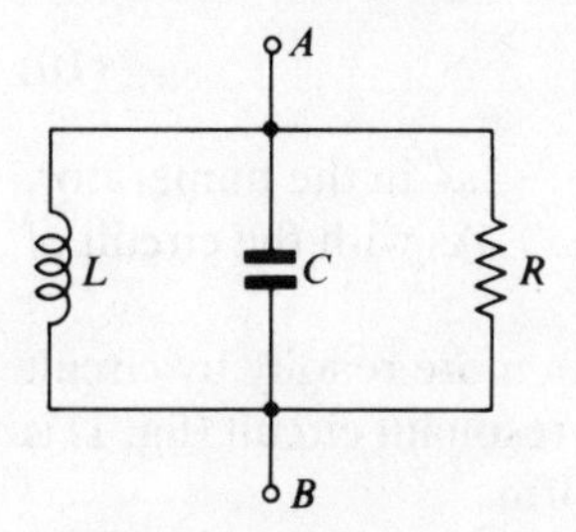

Fig. 2. Parallel resonant circuit.

When these are open-circuited after the charge has been injected, V oscillates as the capacitor discharges through the inductor, R being large enough to cause only slow damping, and the impulse response is roughly of cosine form. Once more, like (7), it is the derivative of a sine-like response function as may be seen from the fact that V is also given by $L\,\mathrm{d}i/\mathrm{d}T$, and that after the impulse there is no current through the inductor. In this case, then, with the required modifications made in (8),

$$\chi(\omega) = \mathrm{i}\omega/CY, \tag{9}$$

and $\chi(\omega)$ is now the impedance of the circuit. The close relationship between series and parallel resonant circuits is most obvious, as we shall see shortly, when they are analysed by the conventional circuit procedure.

Before proceeding to this, however, let us examine a slightly modified parallel resonant circuit, as drawn in fig. 3, the lossy element being provided by a small resistance in series with the inductor rather than a large resistance in parallel. The impulse response function, relating e.m.f. output

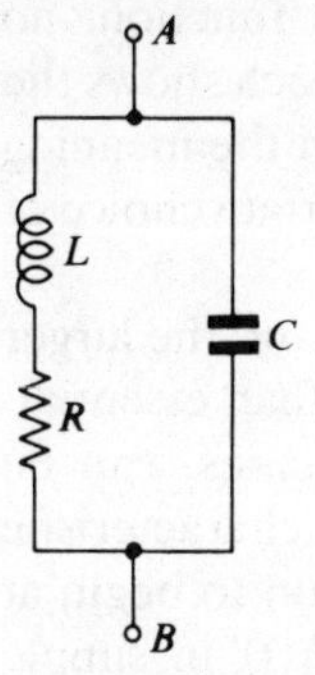

Fig. 3. Parallel resonant circuit (alternative arrangement to fig. 2).

to current impulse input, is almost the same as for fig. 2, but with a small phase shift. This is because V is not $L\,\mathrm{d}i/\mathrm{d}t$, being supplemented by a contribution Ri from the resistor. To see the exact form it is easier to note that after the impulse there is no current through L to discharge the capacitor, so that initially $\mathrm{d}V/\mathrm{d}t=0$. In this case, then, V is the integral, rather than the derivative, of a sine-like response function; the form of V is easily constructed to have this property and the correct initial value:

$$V(t)=(\omega' C)^{-1}\,\mathrm{e}^{-\omega'' t}(\omega' \cos \omega' t+\omega'' \sin \omega' t)$$

leading to

$$\chi(\omega)=(-2\omega''+\mathrm{i}\omega')/CY, \tag{10}$$

which differs from (9) only through the presence of $-2\omega''$ in the numerator, a small change when the resonance is sharp ($\omega'' \ll \omega'$). As with the circuit of fig. 2, $\chi(\omega)$ is the impedance.

The expressions (8) to (10) may be derived much more readily by circuit analysis. For example, the impedance of the series resonant circuit (fig. 1) is $R+\mathrm{j}\omega L-j/\omega C$, so that its admittance takes the form

$$\chi=(R+\mathrm{j}\omega L-\mathrm{j}/\omega C)^{-1} \tag{11}$$

To compare this with (8) it is necessary to identify ω_0 and ω'' with the circuit parameters. The free vibration of a circuit has been analysed in chapter 2, and reference to (2.24) shows that

31

$$\omega'^2=1/LC-\tfrac{1}{4}R^2/L^2,$$

and $$\omega''=\tfrac{1}{2}R/L.$$

When these values are substituted in (8), (11) follows immediately (remember $\mathrm{j}=-\mathrm{i}$). So also with the parallel resonant circuits. The close relationship between the formulae for χ for the circuits of figs. 1 and 2 is explained by the circuit analysis. For in the first case the impedance is that of three components added together – one real (R), one positive imaginary ($\mathrm{j}\omega L$) and one negative imaginary ($-\mathrm{j}/\omega C$); in the second case it is the admittance that is the sum of three analogous terms, $1/R$, $\mathrm{j}\omega C$ and $-\mathrm{j}/\omega L$. There is, from the point of view of the impulse response function, no essential difference between the series and parallel circuits. Each shows the characteristic resonance peak in $\chi(\omega)$, and they only differ in the meaning to be attached to χ, that is, in what we have called the appropriate choice of input and output parameters.

53

More generally, the small difference between (9) and (10) and the larger difference between these and (2) are again rather accidental than essential. A family of curves for $\chi(\omega)$, embracing these as special cases, can be generated from the exponentially damped oscillation that is characteristic of a simple linear vibratory system, by allowing the oscillation to begin at an arbitrary phase in its cycle. For the general form of $h(t)$ in simple systems we write

$$h(t)=A\,\mathrm{e}^{-\omega'' t}\cos(\omega' t-\theta), \tag{12}$$

yielding $$\chi(\omega)=A[-\omega' \sin\theta+\mathrm{i}(\omega+\mathrm{i}\omega'')\cos\theta]/Y. \tag{13}$$

The special cases we have analysed fall into this general form when A and θ are chosen correctly. For (2), $A=1/m\omega'$ and $\theta=\pi/2$; for (8), $A=(\omega'^2+\omega''^2)^{\frac{1}{2}}/\omega' L$ and $\theta=-\tan^{-1}(\omega''/\omega')$; for (10), $A=(\omega'^2+\omega''^2)^{\frac{1}{2}}/\omega' C$ and $\theta=\tan^{-1}(\omega''/\omega')$. There is no difficulty devising a circuit to fit any other value of θ. In fig. 4, for instance, the output, V_{out}, across the terminals CD is determined not solely by the current through r, nor by the rate of change of current through l, but by a mixture of both. By varying the ratio l/r the phase of the output may be changed relative to that of the oscillatory current established by the impulsive input across AB.

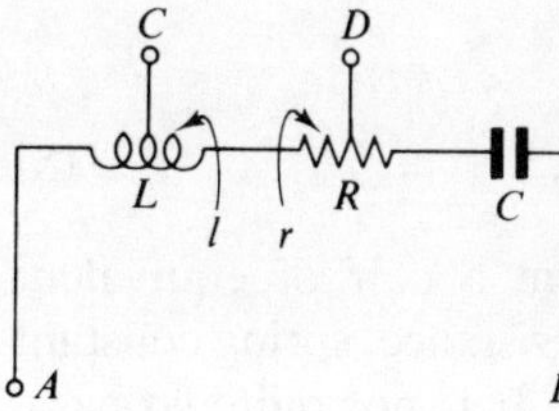

Fig. 4. Series resonant circuit tapped across inductance l and resistance r.

It is worth analysing this circuit in detail to bring out a point which has hitherto been irrelevant to the examples chosen. The most general response of a simple system to an impulse consists not only of the damped oscillatory term (12) but may contain an instantaneous response to the impulse itself. Thus when an impulsive e.m.f. is applied across AB, it appears in the circuit entirely across the inductor L, and a fraction l/L is incorporated in the output response, which therefore takes the form, for unit input,

$$h(t)=(l/L)\,\delta(t)+r\dot{q}+l\ddot{q},$$

where $\dot{q}$ is the current response of the circuit as given in (7). Hence

$$h(t)=(l/L)\delta(t)+(\omega' L)^{-1}\,\mathrm{e}^{-\omega'' t}[(r-2\omega'' l)\omega'\cos\omega' t$$
$$-(\omega'' r+\omega'^2 l-\omega''^2 l)\sin\omega' t].$$

The Fourier transform of $h(t)$ now includes a frequency-independent term, l/L, from the impulse, and gives after evaluation the transfer function

$$\chi(\omega)=\frac{\omega}{LY}(\omega l+\mathrm{i}r),$$

which is the same as the form that emerges immediately from circuit analysis

$$\chi(\omega)=\frac{r+\mathrm{j}\omega l}{R+\mathrm{j}\omega L-\mathrm{j}/\omega C}.$$

It will be appreciated that even with the additional complication of the direct response to the impulse, the most general form of response of a simple linear system is still determined by a very small number of parameters – the amplitudes of the direct and delayed (oscillatory) responses, and the phase, frequency and decay time of the latter – and these can be modelled by a simple circuit such as that of fig. 4. All examples in this general category, then, whether they be electrical, magnetic, mechanical, acoustic etc., may be replaced by their circuit analogue for the purpose of analysis. The advantage of doing this is that familiarity with the manipulation of impedances allows results to be derived economically, once the analogue has been set up, with the circuit parameters correctly related to those of the real system. It is easy enough to look at the equation of motion of a mass on a spring:

$$m\ddot{z}+\lambda\dot{z}+\mu z=F\cos\omega t, \tag{14}$$

and that of a series resonant circuit

$$L\ddot{q}+R\dot{q}+q/C=V\cos\omega t, \tag{15}$$

and to recognize that mass and inductance form a pair of equivalent parameters, as do the damping coefficient λ and resistance, spring constant μ and reciprocal of capacitance, force and e.m.f. It is not quite so easy, unless one is practised in the art, to do the same with, say, a Helmholtz resonator. For this reason we do not propose to lay any stress on the technique of modelling a non-electrical system by means of circuits, useful though this is in analysing complex systems. But it is essentially a practical tool, not a means of gaining insight into the physics of a system; as such it should be acquired when professional necessity drives, not elevated into an important principle. The far more important physical principle is that the similarities of behaviour of simple systems are not accidental, but spring from the necessary restrictions to the form of the impulse-response function.

22

The form of the resonance curve

Although some emphasis has been placed on the analogies between different systems, considerable variation is possible through the influence of the phase θ introduced in (12). This becomes clear when we examine the shape of the resonant response, $\chi(\omega)$; to begin with we shall assume the damping parameter ω'' to be small. This leads to a narrow, high-resonance peak and exhibits the essence of the phenomenon in its clearest form. If ω'' is small we are mainly concerned with the behaviour when ω is very near ω_0, so that $\omega-\Omega_1$ (see (4)) is small, and $\omega-\Omega_2$ large enough to be adequately represented by $2\omega_0$. To the same degree of approximation we may replace ω by ω_0 at all points except where the difference $\omega-\omega_0$ appears. Then the various forms of $\chi(\omega)$ that we have met may be summed up in the simplified form of (13):

$$\chi(\omega)\sim\tfrac{1}{2}A[\mathrm{i}\cos\theta-(\sin\theta+\tfrac{1}{2}Q^{-1}\cos\theta)]/(\omega-\Omega_1), \tag{16}$$

in which, following (2.2), Q replaces $\frac{1}{2}\omega'/\omega''$. When Q is large enough to make negligible the term in which it appears, the following result emerges:†

if $h(t)=\mathrm{Re}\,[a\,\mathrm{e}^{-\mathrm{i}\Omega_1 t}]$; $t>0$, a being in general complex,

$$\left.\begin{aligned} &\chi(\omega)\sim\tfrac{1}{2}\mathrm{i}a/(\omega-\Omega_1),\\ \text{or}\quad &\tfrac{1}{2}A\,\mathrm{e}^{\mathrm{i}(\theta+\pi/2)}/(\omega-\Omega_1),\end{aligned}\right\}\qquad(17)$$

if a is written as $A\,\mathrm{e}^{\mathrm{i}\theta}$. All the most important features of resonant response are contained in (17), which is the Lorentzian function already encountered in (4.28) as the Fourier transform of a decaying exponential. The extra phase factor in (17) obliges us to extend the discussion that was given there.

76

The form of χ is perhaps best appreciated graphically; $1/\chi$ varies as $(\omega-\Omega_1)\,\mathrm{e}^{-\mathrm{i}(\theta+\pi/2)}$, and if $\theta=\pi/2$ (as for the pendulum), $1/\chi\propto\Omega_1-\omega$ and is represented, as in fig. 5, by the vector LN between ω, a point on the real axis, and $\Omega_1(=\omega'-\mathrm{i}\omega'')$. As the applied frequency is changed L, representing ω, moves evenly along the real axis and the denominator

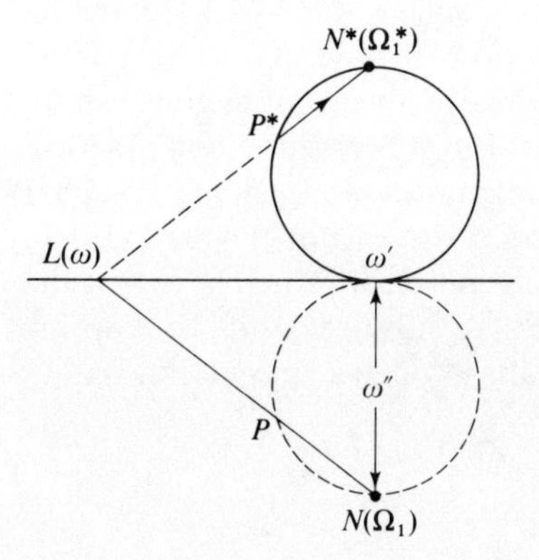

Fig. 5. Graphical construction for response of resonant system, for example displacement of a pendulum.

varies symmetrically about its minimum value when $\omega=\omega'$. The line NP has the right magnitude (apart from a constant) to represent $1/(\Omega_1-\omega)$, but the wrong phase, being in fact $1/(\Omega_1^*-\omega)$. If, however, one draws a circle on the diameter joining ω' and Ω_1^*, the vector P^*N^* correctly represents the variation of $1/(\Omega_1-\omega)$ in magnitude and phase angle. The variation of the real and imaginary parts of $\chi(\omega)$ are shown in fig. 6(*a*), which is virtually the same as fig. 5.4; χ' and χ'' are equal in magnitude when $|\omega-\omega'|=\omega''$, at which points the intensity (or power) of response, proportional to the square of the amplitude P^*N^*, has fallen to half its value at the peak. The width of the resonance, $\Delta\omega$, is defined as the

112

† A useful simplification of the impedance of a series resonant circuit involves the same degree of approximation. Instead of $R+\mathrm{j}(\omega L-1/\omega C)$ one writes ω as $\Omega_1+\delta$, and to first order in δ the impedance is $2\mathrm{j}L\delta$, i.e. $2\mathrm{j}L(\omega-\Omega_1)$. Similar linear approximations can of course be applied to other resonant systems.

366

difference in frequency between these points, i.e. $2\omega''$ or ω'/Q. The quality factor, Q, is just as useful in describing the forced vibration as in defining the decay of free vibration. Moreover this particular result does not depend on the value of the phase angle, θ, since the power response $P \propto \chi\chi^*$; i.e. from (17),

$$P(\omega) = P_0/(\omega - \Omega_1)(\omega - \Omega_1^*) = P_0/[(\omega - \omega')^2 + \omega''^2]. \tag{18}$$

The real and imaginary parts of $\chi(\omega)$ are, however, much affected by the phase angle θ; with $\theta = 0$, as for the current response (8) of a series resonant circuit, the real and imaginary parts in fig. 6 are interchanged

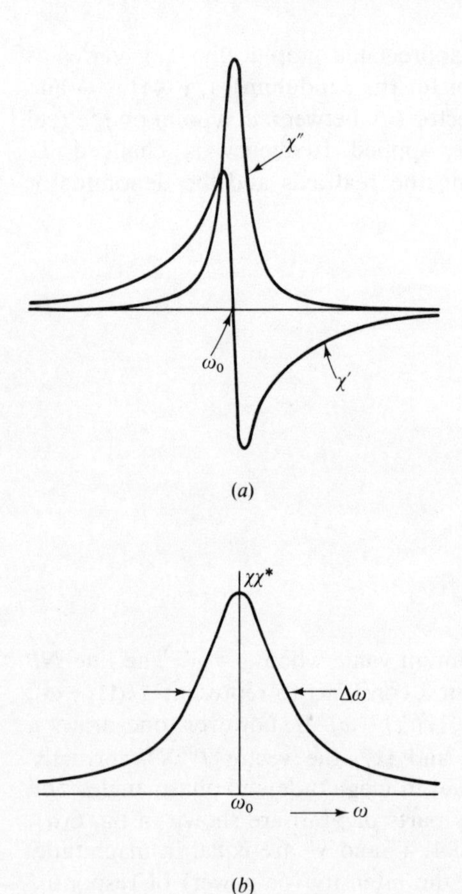

Fig. 6. (*a*) Real and imaginary parts of χ derived from fig. 5; (*b*) $\chi\chi^*$ for a resonant peak.

(apart from sign changes). In general, the horizontal line in fig. 5 should be tilted up at an angle $\theta - \pi/2$, as illustrated in fig. 7, but $\mathrm{Re}[\chi]$ and $\mathrm{Im}[\chi]$ are determined with respect to a horizontal real axis. The curves are now markedly asymmetric, as in fig. 8; $\mathrm{Re}[\chi]$ goes through zero at ω_1 and $\mathrm{Im}[\chi]$ has its maximum at ω_2, both below ω'. It is obvious from the diagram that the phase angle between input and output varies with

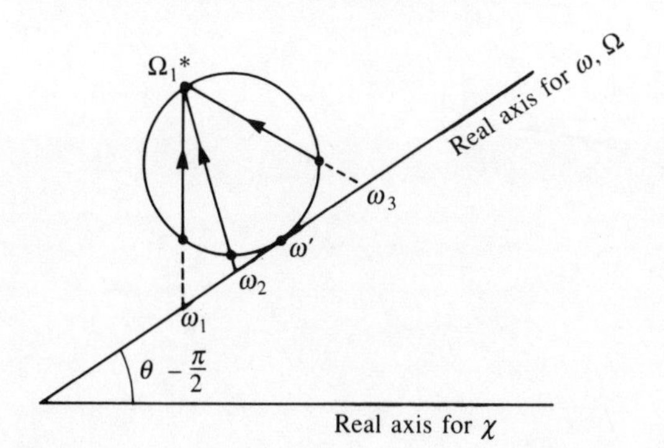

Fig. 7. The form of fig. 5 when the phase of resonant response is changed; $\chi' = 0$ at ω_1, χ' is maximal at ω_2, χ' is maximal at ω_3.

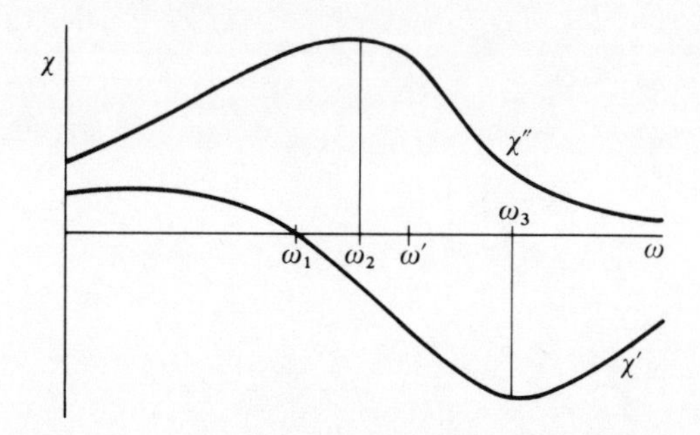

Fig. 8. Real and imaginary parts of χ derived from fig. 7; the horizontal scale is widened compared to fig. 6, to show asymmetries.

frequency in the same way for all θ, except that the overall range, which is always π, runs between $\theta \mp \pi/2$ according to the equation derived from (17),

$$\arg(\chi) = \theta + \pi/2 - \tan^{-1}\left(\frac{\omega''}{\omega - \omega'}\right). \tag{19}$$

This function, shown in fig. 9, together with (18) completely specify the form of $\chi(\omega)$.

The diagram of fig. 5 may be interpreted in physical terms so as to illuminate the phase and amplitude variations across the resonant peak. For the sake of clarity let us consider the specific case of a mass on a spring, damped by a viscous liquid. By turning the problem round and asking what force is needed to maintain a certain amplitude of motion, rather than what amplitude results from applying a certain force, we can divide the driving force into constituent parts, each performing a separate function. First consider the undamped vibrator which, left to itself, would oscillate at constant amplitude and at its natural frequency, $\omega_0 = (\mu/m)^{\frac{1}{2}}$. The vibration diagram in fig. 10(a) illustrates this situation; the vector A, rotating at angular velocity ω_0, represents the motion of the mass, the force required for its acceleration being supplied by the spring, and represented by the oppositely directed vector B. If, however, we wish the vector A to rotate at a different speed, corresponding to the mass being forced to vibrate at a frequency other than its natural resonant frequency, the spring force must

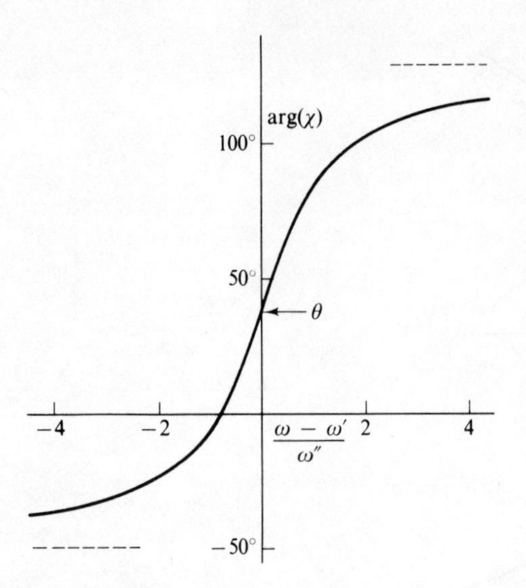

Fig. 9. Variation of the phase χ through resonance; the broken lines are the asymptotes, 180° apart.

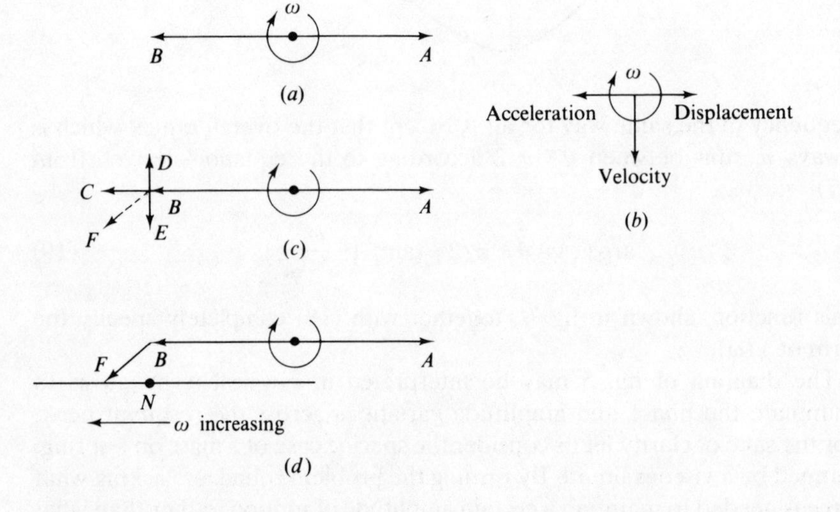

Fig. 10. Vector diagrams for forces needed to maintain vibration. The instantaneous displacement A and acceleration B for a free-running lossless vibrator are shown in (a); (b) shows the directions of displacement, velocity and acceleration in (a). For (c) and (d) see text.

be supplemented by an additional force in the same phase, increasing the magnitude of A if the frequency is to be raised. The required force is shown as C in fig. 10(c). If the viscous damping is now introduced, the mass experiences another force, directed opposite to the velocity and represented by D. For the vibration to be maintained at constant amplitude this force must be countered by the driving force, which must therefore contain a component E ($=-D$). The required driving force is then the vector sum of C, which changes the frequency and alters in magnitude as the resonance is traversed, and E which is sensibly independent of frequency across the narrow resonance. The final result is shown in fig.

10(d), with the end of the resultant force vector F moving along the horizontal line, and the force required to maintain a given amplitude being least at resonance, represented by N. At the low-frequency end F is parallel to the displacement A, being needed to stretch the spring, while at the high-frequency end it is parallel to the acceleration, being needed to supplement the spring which alone could not provide for so rapid a vibration. It will be recognized that this discussion has only constructed by physical reasoning the form of the denominator in (17), i.e. the line LN in fig. 5.

It is also instructive to examine the energy balance. The force C, supplementing the spring, can be considered to make a contribution to the potential energy of the vibrator, which is exchanged with kinetic energy twice per cycle; thus C is responsible for the oscillatory power described by the first term in (3.10). The force E, on the other hand, being in phase with the velocity, steadily transfers energy to the vibrator, to be dissipated as heat, the process described by the second term in (3.10) or by (3.13). From this point of view it is easy to see how a lightly damped resonant system will behave under non-linear damping, by frictional forces, for example. In order to maintain strictly sinusoidal motion the force C is needed as before; but the damping force is no longer represented by the real part of a rotating vector of constant length. Such a driving force E, supplying enough power to provide for the mean dissipation, will compensate the mean drag force but will be too large at some phases of the vibration and too small at others. This implies that if we drive the vibrator with a strictly sinusoidal force it cannot execute strictly sinusoidal vibrations, for this uncompensated part of the force will superimpose vibratory motion at a harmonic of the fundamental frequency. In a lightly damped system, however, the drag forces are small compared to the spring force, so that the uncompensated part cannot excite a large amplitude, especially at a frequency so far removed from the natural resonance. To a good approximation, then, we may ignore the harmonic generation and assume that sinusoidal vibrations require sinusoidal forces determined by the frequency shift from resonance and by the mean dissipation.

57

34

To treat in more detail the specific example of a constant frictional force, f, we note that when the amplitude is ξ_0 the work done per cycle is $4f\xi_0$, which is to be matched by a viscous force 'constant' λ giving $\pi\lambda\omega_0\xi_0^2$. The appropriate amplitude-dependent form of λ is therefore $4f/\pi\omega_0\xi_0$, inversely proportional to ξ_0. As ξ_0 is changed, the force C in phase with the acceleration changes in proportion, but the force E remains constant. The magnitude of F, instead of being given as for viscous damping by the equation

$$F^2 = 4m^2\omega_0^2\xi_0^2[(\omega_0-\omega)^2+\omega''^2], \quad \text{where } \omega'' = \tfrac{1}{2}\lambda/m,$$

is now determined by the equation

$$F^2 = 4m^2\omega_0^2\xi_0^2[(\omega_0-\omega)^2+4f^2/\pi^2m^2\omega_0^2\xi_0^2].$$

Hence for frictional forces the response to a force of constant magnitude but varying frequency is like that of an undamped vibrator:

$$\xi_0 = (F^2 - 16f^2/\pi^2)^{\frac{1}{2}}/2m\omega_0(\omega_0 - \omega).$$

The only essential difference is that F must be large enough to overcome the friction; also it should not be forgotten that if F is only a little larger than required for this purpose the assumption of light damping is hardly valid. It may be found instructive to compare this approximate result, obtained so simply, with the exact analysis, a considerably more troublesome calculation. The result of the exact treatment is

$$\xi_0 = \frac{1}{m(\omega_0^2 - \omega^2)}\left[F^2 - \frac{f^2(\omega_0^2 - \omega^2)^2}{\omega^4}\tan^2\left(\frac{\omega_0\pi}{2\omega}\right)\right]^{\frac{1}{2}},$$

which reduces to the approximate solution when ω is close to ω_0.

Frictionally damped vibrators may not be commonly encountered, but aerodynamic damping is certainly non-linear in many real systems, particularly vibrating strings. At low speeds the damping force is viscous, proportional to speed, but at higher speeds it varies as the square of the speed. In the latter case we have that

252

$$F^2 \propto \xi_0^2(\omega - \omega_0)^2 + \beta\xi_0^4, \quad \beta \text{ being a constant.}$$

The amplitude at resonance varies as $F^{\frac{1}{2}}$, but far from resonance, where it is controlled by spring constant or inertia rather than by losses, it varies as F. With increasing driving force the effective Q-value of the resonance falls.

Low-Q resonant peaks

If the damping term is not small, the approximation (17) is inadequate and the complete resonant denominator, Y, involving both natural frequencies, Ω_1 and Ω_2, must be used. To draw attention to but one point ignored in the high-Q approximation, the response at zero frequency is controlled equally by Ω_1 and Ω_2 which lie equidistant from the origin. The true response is twice as great as is suggested by the high-Q approximation. We shall consider only two special cases of low-Q resonances, the displacement response $\chi_d(\omega)$ as given by (2), and its 'time-derivative' $\chi_v(\omega)$ as given by (8), the velocity or current response. By means of (3) these two responses are expressed in terms of the displacement response at zero frequency, $\chi_d(0)$, which we write as χ_0:

$$\chi_d = -\omega_0^2\chi_0/(\omega^2 - \omega_0^2 + 2i\omega''\omega), \qquad (20)$$

$$\chi_v = -i\omega\chi_d = i\omega\omega_0^2\chi_0/(\omega^2 - \omega_0^2 + 2i\omega''\omega). \qquad (21)$$

The power responses take the form

$$P_d = \chi_d\chi_d^* = \omega_0^4\chi_0^2/[(\omega^2 - \omega_0^2)^2 + 4\omega''^2\omega^2], \qquad (22)$$

$$P_v = \chi_v\chi_v^* = \omega^2\omega_0^4\chi_0^2/[(\omega^2 - \omega_0^2)^2 + 4\omega''^2\omega^2]. \qquad (23)$$

These expressions for the power response, although derived for an input expressed in complex terms, are also correct when it is expressed in real terms; whether the input is expressed as $a_0\,e^{-i\omega t}$ or $a_0 \cos \omega t$, and the response as $a_1\,e^{-i(\omega t+\phi)}$ or $a_1 \cos(\omega t+\phi)$, P_d and P_v represent $(a_1/a_0)^2$.

The maximum of P_d occurs at a frequency ω_d (*displacement resonance*) given by $\omega_d^2=\omega_0^2-2\omega''^2$, which may be manipulated by use of (2.27) into the form

$$\omega_d=\omega_0\left(\frac{4Q^2-1}{4Q^2+1}\right)^{\frac{1}{2}}\sim\omega_0(1-1/Q^2+\cdots). \tag{24}$$

On the other hand, *velocity resonance*, the point at which P_v reaches its peak, is always at the undamped natural frequency, ω_0, no matter how low Q may be. The reason can be seen from the series resonant circuit of fig. 1; the amplitude of the impedance is smallest when L and C cancel each other, at ω_0, leaving only R. It is clear from (24) that the separation between the frequencies of displacement and velocity resonance is normally much less than the width of resonance, ω_0/Q, and it is not often that the distinction is worth making.

Substituting ω_d for ω in (22) yields the peak value of P_d:

$$(P_d)_{max}=\chi_0^2(Q+1/4Q)^2. \tag{25}$$

The displacement is enhanced by a factor very nearly equal to Q, compared with the response to a steady force. Similarly, substituting ω_0 for ω in (2) we find the peak velocity response to be nearly proportional to Q (not Q times the zero frequency response since $\chi_v(0)$ vanishes):

$$(P_v)_{max}\omega_0^2\chi_0^2(Q^2+\tfrac{1}{4}).$$

It is, of course, this property that makes tuned circuits useful in radio receivers, etc. They respond to a narrow band of frequencies as if the inductance and capacitance which normally limit the high-frequency current were absent. In themselves, they have no amplifying power – they simply prevent the signal being blocked by unfavourable impedances.

This point is brought out clearly when one considers the practical problem of generating an oscillatory magnetic field by passing an alternating current through a solenoid, represented by L and R in series. If the mains frequency is suitable there is available a source of comparatively low voltage capable of delivering an extremely large current, since the generator and feeder line resistance may be very small. It may then be convenient to put a capacitor in series with the coil to convert it into a series-resonant circuit (we ignore economic considerations – at mains frequency the capacitor may in fact be too large and expensive to be worth while). The impedance of the coil is reduced from $j\omega L+R$ to R, and the current increased by $(1+\omega^2L^2/R^2)^{\frac{1}{2}}$, which is approximately Q. The voltage across the coil is correspondingly large, about Q times as great as the supply voltage, and there is an equal, but oppositely phased, voltage across the capacitor which must be robust enough to stand this.

On the other hand, one may have as source a power oscillator delivering a high voltage but with a significant output resistance so that even short-circuited it can pass only a small current. The solenoid must then have many turns, with correspondingly high L and R, and the performance may be optimized by the use of a parallel resonant circuit as in fig. 3. At resonance the impedance is resistive and equal to L/CR, i.e. about Q^2 times greater than R; the oscillator delivers very little current when Q is high, and therefore its own resistance is less deleterious. But a current i entering the resonant circuit requires a voltage Q^2Ri across its terminals, and this produces a current $Q^2Ri/(R+\mathrm{j}\omega L)$ through the coil, approximately Q times as great as the input current from the oscillator. If the internal resistance of the oscillator is the controlling agent, limiting the current output, a parallel resonant circuit improves things by a factor Q, which may well be more than 100 for a copper coil at room temperature.

One last point is worth making here; when the losses in a resonant system are reduced so as to make the peak narrower, narrowing takes place through raising the peak to a greater extent than the wings, not by cutting down the wings. This is obvious from (22) and (23), whose denominators can only be decreased by reducing the value of ω''. Fig. 11 brings out this point by showing the resonant peaks for the same system but with Q varied by changing the losses. The diagram also shows how the difference between the displacement response at zero and infinite frequencies is χ_0, determined by the force constant (or its equivalent) and in no way by the losses. This is relevant to the theory of the optical properties of materials (especially anomalous dispersion) which we shall discuss in chapter 7.

204

Transient response

By building up the theory of the driven vibrator from the impulse response we have avoided formally solving the equation of motion, and indeed could continue in this vein since, as we have demonstrated fully, the impulse response contains the complete solution. Nevertheless it is desirable to give the formal solution also, for it often provides the most economical analysis of a problem. Typical equations of motion for resonant systems driven by sinusoidal forces are given in (14) and (15), and we may take as representative of all such systems the form

$$\ddot{\xi} + k\dot{\xi} + \omega_0^2\xi = f_0\,\mathrm{e}^{-\mathrm{i}\omega t}. \tag{26}$$

It is a standard theorem that the general solution of an equation of this type, in which the left-hand side is a linear function of ξ and the right-hand side a function of time alone, with ξ absent, is the sum of *any* solution of (26) plus the general solution of the equation formed by replacing the right-hand side by zero, i.e. (2.23). We know how to meet these requirements from what has gone before. To find some solution of (26), substitute $\xi = A\,\mathrm{e}^{-\mathrm{i}\omega t}$ which will generate the steady-state solution analogous to (5):

$$\xi = f_0\,\mathrm{e}^{-\mathrm{i}\omega t}/(-\omega^2 - \mathrm{i}\omega k + \omega_0^2) = -f_0\,\mathrm{e}^{-\mathrm{i}\omega t}/Y.$$

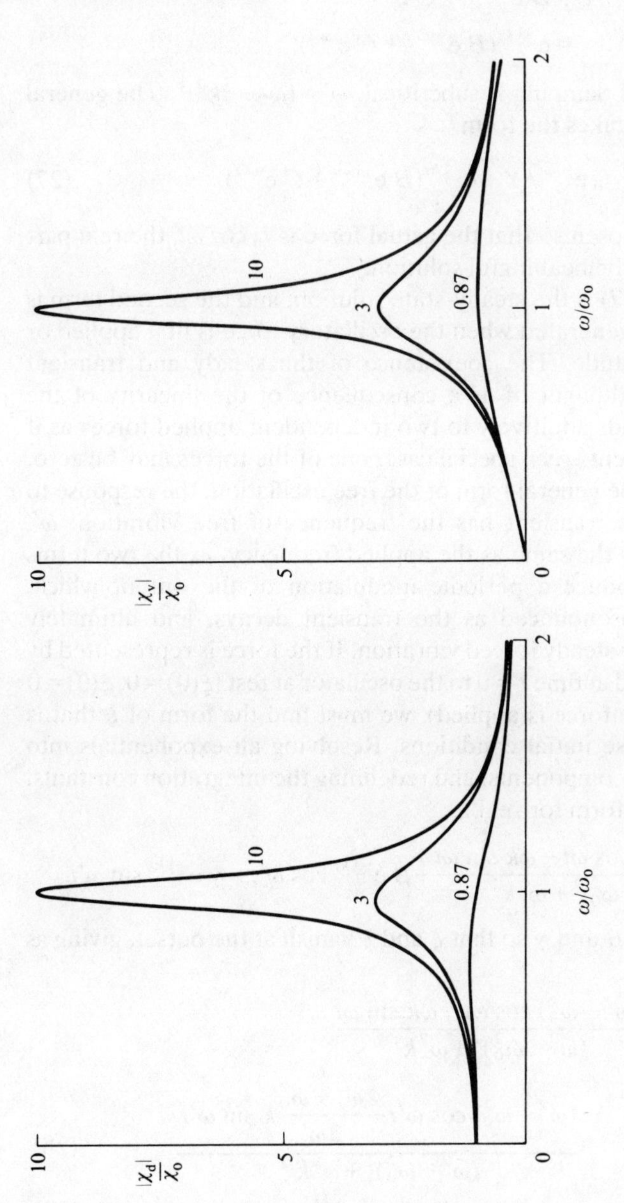

Fig. 11. Magnitudes of χ_d and χ_v for a given vibrator whose Q-value is varied by changing the damping term; the three values $Q = 0.87$, 3 and 10 are illustrated.

This must be supplemented by the general solution of (2.23) which can be written as the sum of two exponentials:

$$\xi = B\,\mathrm{e}^{-\mathrm{i}\Omega_1 t} + C\,\mathrm{e}^{-\mathrm{i}\Omega_2 t}$$
$$= \mathrm{e}^{-\frac{1}{2}kt}(B\,\mathrm{e}^{-\mathrm{i}\omega' t} + C\,\mathrm{e}^{\mathrm{i}\omega' t}),$$

where, as in (2.24), if damping is subcritical, $\omega' = (\omega_0^2 - \frac{1}{4}k^2)^{\frac{1}{2}}$. The general solution of (26) then takes the form

$$\xi = -f_0\,\mathrm{e}^{-\mathrm{i}\omega t}/Y + \mathrm{e}^{-\frac{1}{2}kt}(B\,\mathrm{e}^{-\mathrm{i}\omega' t} + C\,\mathrm{e}^{\mathrm{i}\omega' t}). \qquad (27)$$

If the origin of t is chosen so that the actual force is $f_0 \cos \omega t$, the real part of (27) is the physically meaningful solution.

The first term of (27) is the steady-state solution, and the second term is a decaying transient generated when the oscillatory force is first applied or is changed in magnitude. The coexistence of the steady and transient oscillations is to be thought of as a consequence of the linearity of the system, which responds additively to two independent applied forces as if each were alone present. As a special case, one of the forces may be zero, and the transient is the general form of the free oscillation, the response to zero force. Since the transient has the frequency of free vibration, ω', which is not normally the same as the applied frequency, ω, the two terms beat together to produce a periodic modulation of the output which, however, gets less pronounced as the transient decays, and ultimately resolves itself into the steady forced vibration. If the force is represented by $f_0 \cos \omega t$ and is applied at time $t = 0$ to the oscillator at rest ($\xi(0) = 0$, $\dot{\xi}(0) = 0$ immediately after the force is applied), we must find the form of ξ that is real and satisfies these initial conditions. Resolving all exponentials into their trigonometrical components, and redefining the integration constants, we have as a general form for real ξ,

$$\xi = -f_0 \frac{(\omega^2 - \omega_0^2)\cos \omega t - \omega k \sin \omega t}{(\omega^2 - \omega_0^2)^2 + \omega^2 k^2} + \beta\,\mathrm{e}^{-\frac{1}{2}kt}\cos \omega' t + \gamma\,\mathrm{e}^{-\frac{1}{2}kt}\sin \omega' t.$$

It is now easy to find β and γ so that ξ and $\dot{\xi}$ vanish at the outset, giving as the solution

$$\xi = -f_0 \frac{(\omega^2 - \omega_0^2)\cos \omega t - \omega k \sin \omega t}{(\omega^2 - \omega_0^2)^2 + \omega^2 k^2}$$
$$+ f_0\,\mathrm{e}^{-\frac{1}{2}kt} \frac{(\omega^2 - \omega_0^2)\cos \omega' t - \dfrac{\omega^2 + \omega_0^2}{2\omega'} k \sin \omega' t}{(\omega^2 - \omega_0^2)^2 + \omega^2 k^2} \qquad (28)$$
$$\doteqdot -\xi_0[\cos(\omega t + \varepsilon) - \mathrm{e}^{\frac{1}{2}kt}\cos(\omega' t + \varepsilon)], \qquad (29)$$

where $\tan \varepsilon = \omega k/(\omega^2 - \omega_0^2)$ and $\xi_0 = f_0/[(\omega^2 - \omega_0^2) + \omega^2 k^2]^{\frac{1}{2}}$, the approximation being valid for a reasonably sharp peak, excited at a frequency near resonance so that $\frac{1}{2}(\omega^2 + \omega_0^2)/\omega'$ can be replaced by ω. As ω increases through ω_0, ε varies from π to 0; ξ_0 is to be taken as positive.

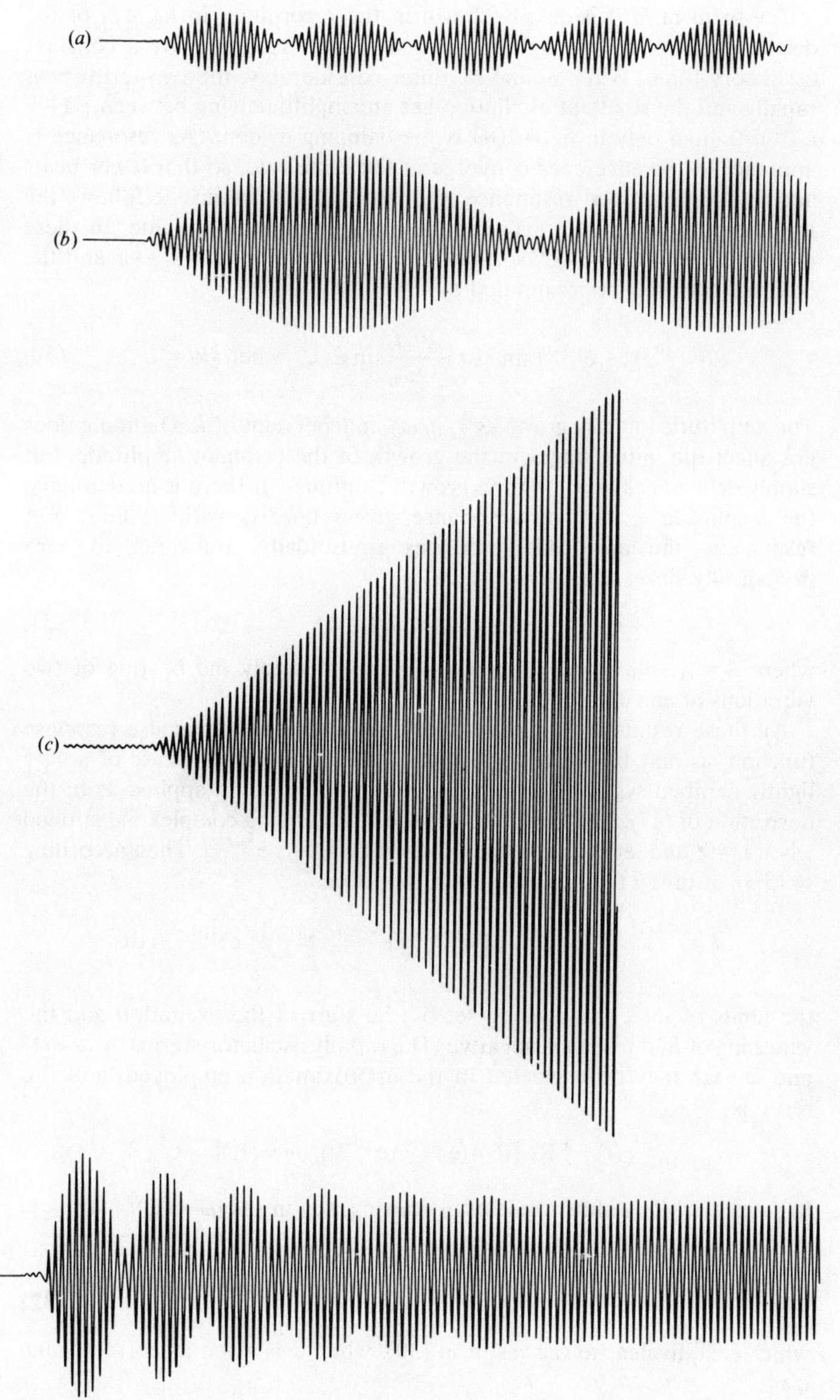

Fig. 12. Record of rigid pendulum displacement under the influence of an oscillatory force (*a*) some way off resonance, (*b*) near resonance, (*c*) tuned to resonance. The magnitude of the force is the same in all diagrams; note that the initial rate of growth is the same. For (*d*) the pendulum is more heavily damped, with an oil dashpot, to show the decay of the transient and the approach to the steady state. The scale in (*d*) is not the same as in the others.

The form of (29) is clearly shown in the recordings, in fig. 12, of the displacement of a very lightly damped pendulum driven by a constant oscillatory force. When ω and ω' differ considerably, the two terms beat rapidly and the resultant oscillation has an amplitude lying between $\xi_0(1 \pm e^{-\frac{1}{2}kt})$, though only in fig. 12(d) is the damping evident. As resonance is approached the envelopes converge at the same rate, so that fewer beats are visible. Exactly at resonance when $\omega = \omega'$, the amplitude follows the lower envelope, tending exponentially to its steady-state value. In these circumstances, $\tan \varepsilon = 4Q$, so that with a sharp resonance $\varepsilon \sim \frac{1}{2}\pi$ and the build-up is closely approximated by

$$\xi = \frac{f_0}{\omega_0 k}(1 - e^{-\frac{1}{2}kt}) \sin \omega_0 t \sim \frac{f_0 t}{2\omega_0} \sin \omega_0 t, \quad \text{when } \tfrac{1}{2}kt \ll 1. \tag{30}$$

The amplitude initially grows as $\frac{1}{2}f_0 t/\omega_0$, independent of k. Damping does not affect the initial stages in the growth of the resonant amplitude, but simply determines how long the growth continues. If there is no damping, the amplitude exactly on resonance grows linearly without limit. Off resonance, the amplitude oscillates sinusoidally, returning to zero periodically since, when $k = 0$ in (29),

$$\xi = 2\xi_0 \sin \bar{\omega} t \sin \Delta t, \tag{31}$$

where $\bar{\omega} = \frac{1}{2}(\omega + \omega')$ and $\Delta = \frac{1}{2}|\omega' - \omega|$. This is simply the beating of two vibrations of equal amplitude.

All these results are, of course, obtainable from the impulse response function, as may be illustrated for the specially interesting case of a very lightly damped system, the same approximations being applied as in the derivation of (17). Let the force $\mathrm{Re}\,[f_0\,e^{-i\omega t}]$, f_0 being complex, be applied when $t = 0$, and let the response function be $\mathrm{Re}\,[a\,e^{-i\Omega_1 t}]$. Then according to (5.5), at time t the response takes the form: **106**

$$\xi(t) = \tfrac{1}{4}\int_0^t (f_0\,e^{-i\omega t_0} + f_0^*\,e^{i\omega t_0})(a\,e^{-i\Omega_1(t-t_0)} + a^*\,e^{i\Omega_1^*(t-t_0)})\,dt_0,$$

the limits of integration being set by the start of the excitation and the vanishing of $h(t)$ when t is negative. The rapidly oscillatory terms in $\omega + \Omega_1$ and $\omega - \Omega_1^*$ may be neglected in the approximation employed, and the result is

$$\xi(t) \sim \tfrac{1}{2}\,\mathrm{Re}\,[\mathrm{i} f_0 a(e^{-i\omega t} - e^{-i\Omega_1 t})/(\omega - \Omega_1)].$$

In particular, for a short while after starting, so long as $(\omega - \Omega_1)t \ll 1$,

$$e^{i(\omega - \Omega_1)t} \sim 1 + i(\omega - \Omega_1)t,$$

and

$$\xi(t) \sim \tfrac{1}{2}\,\mathrm{Re}\,[f_0 a t\,e^{-i\omega t}], \tag{32}$$

which is equivalent to the result in (30), when a is given its correct value i/ω_0.

Let us look at the transient problem from the point of view of the vibration diagram, building up by vector summation the resultant of all the impulse responses initiated by successive increments of the applied force $f_0\,\mathrm{e}^{-\mathrm{i}\omega t}$. First we neglect the resonator losses, so that the solution of the equation,

53

$$\ddot{\xi}+\omega_0^2\xi=f_0\,\mathrm{e}^{-\mathrm{i}\omega t},$$

is built up out of increments, each being the vibration that would be excited in the system at rest if it were allowed to experience the applied force for a single element of time δt_0. We therefore look for a solution of the equation,

$$\ddot{\xi}+\omega_0^2\xi=f_0\,\mathrm{e}^{-\mathrm{i}\omega t_0}\,\delta t_0,$$

for $t>t_0$, given that $\xi=\dot{\xi}=0$ immediately before t_0. Integrating through the element of time δt_0 we see that ξ remains zero while $\dot{\xi}$ jumps to $f_0\,\mathrm{e}^{-\mathrm{i}\omega t_0}\,\delta t_0$. The general form of the subsequent oscillations is given by

$$\delta\xi=A\,\mathrm{e}^{-\mathrm{i}\omega_0 t}+B\,\mathrm{e}^{\mathrm{i}\omega_0 t},$$

and A and B must be chosen to satisfy the initial conditions. Hence

$$\delta\xi=\frac{\mathrm{i}f_0\,\delta t_0}{2\omega_0}[\mathrm{e}^{\mathrm{i}(\omega_0-\omega)t_0-\mathrm{i}\omega_0 t}-\mathrm{e}^{-\mathrm{i}(\omega_0+\omega)t_0+\mathrm{i}\omega_0 t}].$$

The first term, containing $\mathrm{e}^{\mathrm{i}(\omega_0-\omega)t_0}$ in its amplitude, changes phase only slowly with t_0 if ω and ω_0 are very close; subsequent increments can therefore build up to a substantial resultant. The second term, by contrast, is governed by $\mathrm{e}^{-\mathrm{i}(\omega_0+\omega)t_0}$ and is rapidly oscillatory. For excitation near resonance it never amounts to much and can be dropped in comparison with the first term. We therefore write

$$\delta\xi=\frac{f_0\,\delta t_0}{2\omega_0}\,\mathrm{e}^{\mathrm{i}(\omega_0-\omega)t_0+\frac{1}{2}\mathrm{i}\pi}\,\mathrm{e}^{-\mathrm{i}\omega_0 t},$$

which is represented on a vibration diagram by a short line element whose length is given by

$$\delta s=\tfrac{1}{2}f_0\,\delta t_0/\omega_0, \qquad (33)$$

and which lies at an angle ϕ to the real axis, where

$$\phi=(\omega_0-\omega)t_0+\tfrac{1}{2}\pi. \qquad (34)$$

As usual, the overall rotation of the curve represented by $\mathrm{e}^{-\mathrm{i}\omega_0 t}$ is set aside in constructing the vibration diagram. Different elements δt_0 at different times t_0 add extra elements to the curve, whose shape $\phi(s)$ is parametrically defined in terms of t_0 by (33) and (34). In this case the curve is a circle, since

$$\mathrm{d}\phi/\mathrm{d}s=2\omega_0(\omega_0-\omega)/f_0,$$

which is constant; $\mathrm{d}\phi/\mathrm{d}s$ is the curvature, whose reciprocal, $f_0/2\omega_0(\omega_0-\omega)$, is the radius of the circle. As time proceeds, the length of the arc grows steadily in an anticlockwise sense if $\omega < \omega_0$, starting from the point A in fig. 13(a) so that the first vector has phase $\pi/2$, in accordance with (34) when $t_0 = 0$. The resultant, AB, follows the behaviour of the amplitude expressed by (31), $2\xi_0 \sin \Delta t$, the maximum value being $2\xi_0$, i.e. $f_0/\omega_0(\omega_0 - \omega)$ if $\omega_0 + \omega$ is replaced by $2\omega_0$. This is just the diameter of the circle, and the time between successive zeros, π/Δ, is the period of revolution of B round the circle, $2\pi/\dot{\phi}$.

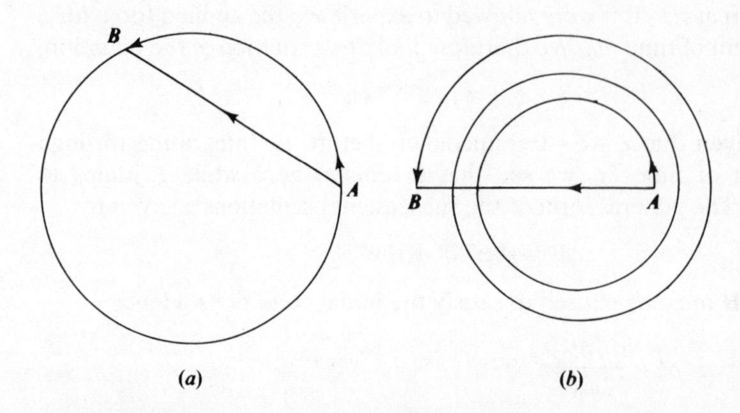

Fig. 13. Vibration diagrams for a driven resonant system; for explanation see text.

If there is damping we must imagine each vector element to start at the length given by (33) but to diminish exponentially as time proceeds. The resulting vibration diagram is then an equiangular spiral. For example, fig. 13(b) shows how the diagram might look after B has accomplished just over $2\frac{1}{2}$ revolutions, while after many revolutions the start of the curve has shrunk into the origin, and thenceforth the resultant amplitude AB stays constant as B goes round. Such diagrams represent the decay of the beat pattern due to the transient, and could be made with little extra difficulty to yield the correct form (28). It would be a real enthusiast for vibration diagrams, however, who would contend that this is an easier approach than the analytical solution of the equation. If, however, the problem is such as to make an analytical solution laborious, the vibration diagram can prove its worth by exhibiting in pictorial form the essential features of the solution.

To illustrate this point, let us examine the response of a resonant system to a gliding tone, for example an oscillatory force of constant amplitude but steadily varying frequency, such as might be expressed in the form

$$f = f_0 \, e^{-i(\frac{1}{2}\beta t^2 + \omega_0 t)}.$$

The phase of the applied force is $\frac{1}{2}\beta t^2 + \omega_0 t$, and if one defines instantaneous frequency as the rate of change of phase, in this case it is $\beta t + \omega_0$, increasing linearly with time and matching the natural frequency of the system when $t = 0$. When this force is applied to a lossless resonant system, the vibration diagram is no longer circular, for although (33) remains true (since the amplitude is constant) ωt_0 in (34) must be replaced by $\frac{1}{2}\beta t_0^2 + \omega_0 t_0$, so that

$$\phi = -\tfrac{1}{2}\beta t_0^2 + \tfrac{1}{2}\pi. \tag{35}$$

Hence $$\mathrm{d}\phi/\mathrm{d}s = \beta t_0 = 2\omega_0 \beta s/f_0. \tag{36}$$

The vibration diagram now has a curvature proportional to distance measured along it. Passing through the origin ($s=0$) the curvature becomes zero as it changes sign, so that there is an inflexion here; and as one proceeds along the curve the steady increase of curvature causes it to spiral in to a limiting point at each end. In fact (36) defines the Cornu spiral, which we have already met and seen illustrated in fig. 4.14. The following discussion refers to this diagram.

78

In this example of the use of the spiral, the origin is reached when $t=0$, all parts to the left being contributed by earlier responses to the gliding tone. The amplitude of the resonse is therefore the length of the vector from P to some point on the spiral, and this point moves along the spiral towards Q at a constant rate. At early times the amplitude grows smoothly, and the phase changes rather rapidly. This reflects the fact that the instantaneous frequency of response is not ω_0; at a point on the spiral near P, the tangent is always nearly normal to the vector from P and therefore the phase of the response changes at the same rate as that of the applied force. Before resonance is reached, then, the system follows the oscillations of the force, but with a small amplitude and a phase lag of $\pi/2$. As the gliding tone passes through resonance the response rises sharply and thereafter oscillates in amplitude and phase, but now remaining close in phase to the ultimate phase represented by PQ; i.e. the system is set ringing at its own natural frequency and subsequent excitation, off resonance, can only create small additional oscillations to beat with the principal vibration. The amplitude-variation with time is shown in fig. 14.

The ultimate amplitude is the same as if the force $f_0 \mathrm{e}^{-\mathrm{i}\omega_0 t}$, tuned to resonance, had been applied for as long as it takes the point to move from R to S on the spiral, i.e. for the gliding tone to slip in phase, relative to the free vibrator, from $\pi/4$ to zero and back to $\pi/4$. Since the phase error is $\frac{1}{2}\beta t^2$, the time needed is $(2\pi/\beta)^{\frac{1}{2}}$ and in this time, according to (30), the amplitude will rise from zero to $f_0(\frac{1}{2}\pi/\beta)^{\frac{1}{2}}/\omega_0$; this is the amplitude at which the lossless resonator is set ringing.

If the resonator is lossy, the left-hand end of the spiral must be imagined to shrink as B progresses along the curve, and the end-point A follows B at a suitable distance. The mathematical details involve Fresnel integrals with complex arguments, which are accessible in tabulated form but hardly enlightening. The physical character of the behaviour is fairly obvious without numerical work. If the gliding tone shifts frequency rapidly enough to pass through the middle region RS in much less than the decay time, the build-up follows fig. 14 very closely. After resonance the response oscillates in amplitude but gradually decays with the natural decay time of the

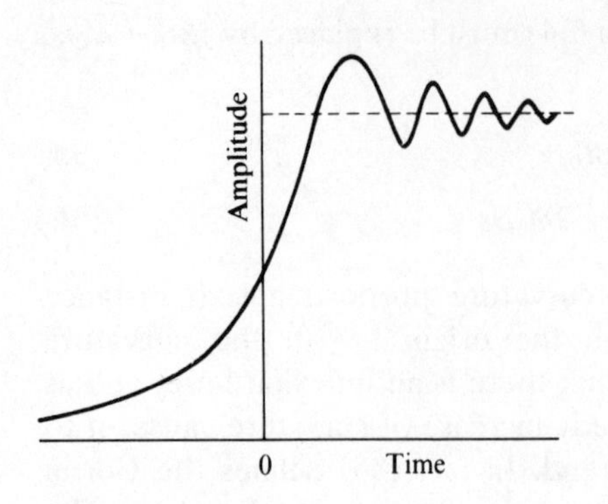

Fig. 14. Time variation of amplitude of response of a lossless resonator to a gliding force which passes through the resonant frequency when $t=0$.

resonator. On the other hand, if the tone glides so slowly that at each frequency there is time enough for transients to decay, the amplitude follows a resonance pattern as in fig. 11. It is only in between these limits that numerical computation may be helpful, but even then one can probably learn more, and more quickly, by sketching vibration diagrams than by searching through books of tables.

This discussion of transient phenomena may have conveyed the impression that there is more information locked up in the impulse response function than in, say, the steady-state response to an oscillatory disturbance. In fact, as is made clear by (5.3) and (5.4), they are equivalent descriptions; the steady-state response is in no sense an approximation which has to be corrected by the addition of the transient. To take a specific example, when a signal $f_0\,\mathrm{e}^{-\mathrm{i}\omega t}$ is switched on at time $t=0$ the fact that it has not existed for all time inevitably means it is not a pure sinusoid, but has a certain narrow Fourier spectrum. Each Fourier component, at frequency ω_1, say, is indeed pure and excites a response whose magnitude is determined by $\chi(\omega_1)$, the steady-state compliance at the same frequency. It is now a matter of routine calculation to add the responses over the whole range of the spectrum and to show that the resultant response is a gradual build-up to an ultimately steady oscillation. Because the Fourier transform of the applied signal is singular if it is allowed to continue indefinitely, at constant amplitude, we shall assume it to take the form:

106

$$g_0(t)=f_0\,\mathrm{e}^{-(\mathrm{i}\omega+\lambda)t}, \quad \text{when } t>0,$$

where λ is small and will eventually be allowed to vanish. Then the Fourier transform follows from (4.16):

$$f(\omega_1)=\frac{f_0}{2\pi}\int_0^\infty \mathrm{e}^{-[\mathrm{i}(\omega-\omega_1)+\lambda]t}\,\mathrm{d}t=\frac{f_0}{2\pi}\Big/[\mathrm{i}(\omega-\omega_1)+\lambda].$$

When this is applied to a transducer whose compliance is $\chi(\omega_1)$, the output is the Fourier transform of $\chi(\omega_1)f(\omega_1)$,

i.e. $$g_1(t)=\frac{f_0}{2\pi}\int_{-\infty}^{\infty}\chi(\omega_1)\,\mathrm{e}^{-\mathrm{i}\omega_1 t}\,\mathrm{d}\omega_1/[\mathrm{i}(\omega-\omega_1)+\lambda].$$

For a system governed by (26), $\chi(\omega_1)$ takes the form $-1/(\omega_1-\Omega_1)\times(\omega_1-\Omega_2)$, in accordance with (2) and (4) when $m=1$; $\Omega_1=\omega'-\mathrm{i}\omega''$, $\Omega_2=-\omega'-\mathrm{i}\omega''$. Hence

$$g_1(t)=\frac{-\mathrm{i}f_0}{2\pi}\int_{-\infty}^{\infty}\mathrm{e}^{-\mathrm{i}\omega_1 t}\,\mathrm{d}\omega_1/(\omega_1-\Omega)(\omega_1-\Omega_1)(\omega_1-\Omega_2),$$

where Ω is written for $\omega-\mathrm{i}\lambda$. This integral is most readily evaluated by contour integration, along a contour that runs the length of the real axis and returns via a semicircle at infinity. There are three simple poles at Ω, Ω_1 and Ω_2, all lying below the real axis. When $t<0$ the semicircle must lie in the positive half-plane to avoid divergence of $\mathrm{e}^{-\mathrm{i}\omega_1 t}$ at $-\mathrm{i}\infty$. There is no contribution to the integral from the semicircle and the integrand is regular within the contour; hence $g_1(t)=0$ for all negative t. For positive t, on the other hand, the contour must lie in the negative half-plane and enclose the three poles, whose contributions to the integral lead to a non-vanishing value:

$$g_1(t)=-f_0[\mathrm{e}^{-\mathrm{i}\Omega t}/(\Omega-\Omega_1)(\Omega-\Omega_2)+\mathrm{e}^{-\mathrm{i}\Omega_1 t}/(\Omega_1-\Omega)(\Omega_1-\Omega_2)$$
$$+\mathrm{e}^{-\mathrm{i}\Omega_2 t}/(\Omega_2-\Omega)(\Omega_2-\Omega_1)].$$

For a high-Q system excited near resonance $\Omega-\Omega_1$ is much smaller than $\Omega-\Omega_2$ or $\Omega_1-\Omega_2$, and the last term can be neglected without serious error. When λ is allowed to vanish, Ω becomes ω, the applied frequency and the first term is the steady-state response in the same form as (5); the second term is the transient, having the frequency and decrement of the free oscillation, and such amplitude as to cancel the first term (with the help of the neglected third term) at $t=0$. We have thus seen that the same result is obtained in three ways – by this approach, by the use of the impulse response function and by direct solution of the differential equation; it could not, of course, have been otherwise.

One last point is worth making before leaving the question of transient response. In discussing the idea of causality in the last chapter we noted that a resonant system excited by a decaying oscillatory force, whose frequency matched one of the natural (complex) frequencies Ω, would show an infinite amplitude of response. At any other frequency, including more rapidly decaying oscillations, the steady-state response, as expressed by $\chi(\omega)$, was finite. It might be thought, therefore, that except at the resonant frequencies $\chi(\omega)$ provides a satisfactory description of the ultimate steady-state response. In a sense this is true, but it is meaningless unless one can achieve the steady state. In particular $\chi(\omega)$ is a quite inadequate description of behaviour at all ω whose negative imaginary part is greater than that of Ω. For if we imagine a decaying oscillatory force switched on at some instant, exciting transients as well as the decaying 'steady-state' response expressed by $\chi(\omega)$, the transients decay less rapidly than the steady-state response when $\mathrm{Im}[-\omega]>\mathrm{Im}[-\Omega]$; the longer we wait, the further we find ourselves from the steady state – the behaviour is

108

ultimately controlled by the precise way in which the force was first switched on and the transients then excited. Neglect of the transients (to which physicists are prone, being so often concerned with steady-state response) can obviously lead to totally erroneous conclusions in this situation. This point, however, has no bearing on the validity of the Kramers–Kronig relations, which (apart from the fact that they are a mathematical theorem and valid as such even if they do not apply to any physical problem) are concerned with the behaviour of $\chi(\omega)$ only in that domain of ω in which transients can be neglected if one is prepared to wait long enough; the relations are not only mathematically sound but physically useful.

110

Response of a resonant system to noise

Let a series resonant circuit of high Q be excited by a randomly fluctuating e.m.f. from a source of negligible resistance and let the fluctuations occur at a much faster rate than the resonance frequency. Then, as discussed in chapter 4, we may replace the input by a closely spaced random succession of pulses which may be taken as equal, but have equal chance of being positive or negative if, as we suppose, the mean e.m.f. vanishes. The advantage of working with pulses is that the response of the circuit is already known. Each pulse excites a decaying oscillation, and the randomness of their times of arrival allows the mean effect of a sequence to be written by summing intensities. If each pulse has magnitude $\pm P(=\int V\,\mathrm{d}t$ integrated through the impulse), the intensity of the current response at a time t later is half the square of the amplitude of oscillation in (7):

92

$$\overline{i^2} = \tfrac{1}{2}P^2\,\mathrm{e}^{-2\omega'' t}/L^2, \tag{37}$$

if $\omega'' \ll \omega'$ and the second term is neglected. The mean intensity is found by adding all contributions like (37) from earlier pulses. With an average of N pulses arriving in unit time, the resultant is given by

$$\overline{i^2} = \frac{NP^2}{2L^2}\int_0^\infty \mathrm{e}^{-2\omega'' t}\,\mathrm{d}t = \tfrac{1}{2}NP^2\tau_\mathrm{e}/L^2, \tag{38}$$

where τ_e is $1/2\omega''$, the energy decay time; it is as if all pulses arriving in an interval τ_e are to be considered as contributing with their initial intensity $\frac{1}{2}P^2/L^2$, the rest being ignored.

One can, of course, derive the same result by determining the response of the resonant circuit to the Fourier components of a white noise source, making use of the power response function P_v, as expressed in (23). This is the appropriate current response, in power form, for a series resonant circuit whose zero frequency susceptibility $\chi_0 = C$, the capacitance. According to (4.45) the spectral intensity of the voltage noise due to a sequence of pulses is given by

92

$$I_\omega\,(\text{voltage}) = NP^2/\pi.$$

The circuit response is therefore given by

$$I_\omega\ (\text{current}) = NP^2/\pi \times P_v,$$

and when this is integrated over all positive frequencies the result is

$$\overline{i^2} = \frac{NP^2}{\pi}\int_0^\infty \frac{\omega^2\omega_0^4 C^2\,d\omega}{(\omega^2-\omega_0^2)^2+4\omega''^2\omega^2}$$

$$= \frac{NP^2\omega_0^2 C^2}{\pi}\int_0^\infty \frac{x^2\,dx}{(x^2-1)^2+\varepsilon^2 x^2}, \quad \text{where } \varepsilon = 2\omega''/\omega_0.$$

Since the integral takes the value $\frac{1}{2}\pi/\varepsilon$, as shown in the following square brackets, we have that

$$\overline{i^2} = \tfrac{1}{4}NP^2\omega_0^4 C^2/\omega'' = \tfrac{1}{2}NP^2\tau_e/L^2, \text{ as in (38)},$$

since $\tau_e = 1/2\omega''$ and $\omega_0^2 = 1/LC$.

[Note that $$\int_0^\infty \frac{x^2\,dx}{(x^2-1)^2+\varepsilon^2x^2} = \int_0^\infty \frac{dx}{(x-1/x)^2+\varepsilon^2}, \quad (39)$$

and $$\int_0^\infty \frac{dx}{(x^2-1)^2+\varepsilon^2x^2} = -\int_{x=0}^\infty \frac{d(1/x)}{(x-1/x)^2+\varepsilon^2} = \int_0^\infty \frac{dz}{(z-1/z)^2+\varepsilon^2}, \quad (40)$$

if z is written for $1/x$. Since the two integrals on the left of (39) and (40) are thus shown to be equal, each is equal to half their sum:

i.e. $$\int_0^\infty \frac{x^2\,dx}{(x^2-1)^2+\varepsilon^2x^2} = \int_0^\infty \frac{dx}{(x^2-1)^2+\varepsilon^2x^2} = \tfrac{1}{2}\int_{x=0}^\infty \frac{d(x-1/x)}{(x-1/x)^2+\varepsilon^2}$$

$$= \tfrac{1}{2}\int_{-\infty}^\infty \frac{dy}{y^2+\varepsilon^2} = \tfrac{1}{2}\pi/\varepsilon. \quad (41)]$$

In verifying the result of the simple derivation of (38) we have in fact gone beyond it, since there is no restriction in the second treatment to high-Q resonant circuits; such a restriction was apparently necessitated in the impulse treatment by the requirement that all phases of oscillatory response were equally likely to be superposed at any instant. This is not true if the damping is large during one cycle, yet the same mean square current is excited. Naturally the character of the response will differ. With a sharply tuned circuit only those noise components very close to ω_0 are effective, and the response will be quasi-sinusoidal; with a low-Q system, however, or one that is overdamped, the wide spectrum contributing to the response will ensure that all traces of periodicity are eliminated. Nevertheless $\overline{i^2}$ is the same, a result which seems very surprising at first sight, but one that is well verified by experiment.

The classic experiments in this field were carried out, not with circuits, but with light vanes and galvanometer coils suspended on torsion wires, and undergoing Brownian motion as a result of the random bombardment of gas molecules. We have here an explicit realization of a resonant system subject to noise in the form of a sequence of impulses, and it is worth working through a specific case in some detail. The example we choose is that in which the gas pressure is so low that each molecule behaves independently, arriving at the vane from deep within the gas and leaving again after striking it. There are then no collective motions of the gas contributing to the forces on the vane – only random individual bombardment at an easily calculated mean rate. Taking a flat vane for convenience, consider an element of area $\mathrm{d}A$ at a distance r from the axis of suspension (torsion wire). If there are n molecules per unit volume, of which a fraction $f(v)\,\mathrm{d}v$ have a velocity component, normal to the vane, lying between v and $v+\mathrm{d}v$, each side of the element will be struck by $nvf(v)\,\mathrm{d}v\,\mathrm{d}A$ molecules in unit time, and we suppose them to rebound elastically. Thus each collision communicates $2mvr$ of angular momentum, such as would set the vane, initially at rest, oscillating with energy $2m^2v^2r^2/I$, I being its moment of inertia. Different molecules striking the vane at different points communicate different increments of energy, and we may define the mean increment, $\overline{\Delta E}$, as the weighted average:

$$\overline{\Delta E}=2m^2\overline{v^3}\,\overline{r^2}/I\bar{v}, \tag{42}$$

in which $\overline{v^3}/\bar{v}=\int_0^\infty v^3f(v)\,\mathrm{d}v/\int_0^\infty vf(v)\,\mathrm{d}v$ and expresses the weighted mean of v^2 when allowance is made for the faster molecules striking the vane proportionately more frequently; $\overline{r^2}$ is a geometrical property of the vane, $\int r^2\,\mathrm{d}A/A$, A being its area (one side only).

Following the lines of the argument that led to (38) we now assert that the mean oscillatory energy $\bar{E}$ of the vane is $\overline{\Delta E}$ multiplied by the number of collisions occurring in τ_e, the natural decay time for the energy of oscillation of the vane. Since the number of impacts in unit time on both sides of a vane of area A is $n\bar{v}A$, we have that

$$\bar{E}=2nm^2\overline{v^3}A\overline{r^2}\tau_e/I. \tag{43}$$

It is now necessary to calculate τ_e, which we do on the assumption that the oscillations are only damped by the gas. When the vane is turning with angular velocity $\dot{\theta}=\omega$, the forward face receives more impacts at a higher relative velocity and a retarding force arises from this differential effect. The element $\mathrm{d}A$, now moving with velocity $r\omega$, receives $n(v+r\omega)f(v)\,\mathrm{d}v\,\mathrm{d}A$ impacts on its leading side, each of which transmits $2m(v+r\omega)r$ of angular momentum. The trailing side behaves similarly, but with r reversed in sign. The total angular momentum communicated by these molecules in unit time is therefore given by

$$\mathrm{d}G=2nmrf(v)\,\mathrm{d}v\,\mathrm{d}A[(v+r\omega)^2-(v-r\omega)^2],$$

$$=8nmr^2\omega vf(v)\,\mathrm{d}v\,\mathrm{d}A,$$

and this is the retarding torque due to collision. Integrating over the whole vane and over all collisions we must remember that only molecules moving towards the vane strike it, so that if $\bar{v}$ is defined as $\int_0^\infty vf(v)\,dv/\int_0^\infty f(v)\,dv$, then $\int_0^\infty vf(v)\,dv = \frac{1}{2}\bar{v}$. Hence

$$\Lambda \equiv G/\omega = 4nm\bar{v}A\overline{r^2}. \tag{44}$$

Now Λ is the damping parameter appearing in the equation of free motion when it is written in the form $I\ddot{\theta} + \Lambda\dot{\theta} + \mu\theta = 0$, and by analogy with similar equations we see that $\tau_e = I/\Lambda$. Hence, from (43) and (44),

$$\bar{E} = \tfrac{1}{2}m\overline{v^3}/\bar{v}. \tag{45}$$

Since $f(v)$ is the distribution function for one component of molecular velocity, $f(v) \propto e^{-mv^2/2k_BT}$, from which it follows that $m\overline{v^3}/\bar{v} = 2k_BT$ and

$$\bar{E} = k_BT. \tag{46}$$

89 This is a result which is required by Boltzmann's equipartition law, for the oscillating vane has two degrees of freedom with quadratic expressions for the energy: $E = \frac{1}{2}I\dot{\theta}^2 + \frac{1}{2}\mu\theta^2$, and the mean energy associated with each should be $\frac{1}{2}k_BT$. It is the overriding power of Boltzmann's law that confirms the necessity for the mean energy to be independent of the strength of damping, as we found for $\overline{i^2}$ in the series resonant circuit with which we began this section. It further follows that (46) is not a special result applicable only when molecular bombardments are independent of one another. Indeed the corollary is that the relationship between random bombardment and viscous drag discovered in this explicit calculation must in fact be more general, applying, for example, to denser gases and fluids even though interaction between the molecules precludes the elementary assumption that the random forces arise solely from uncorrelated impacts.

To see what the general relation is let us recast the preceding argument. If angular momentum δL is communicated to a vane at rest by an impulse, the kinetic energy δE communicated thereby is $\frac{1}{2}(\delta L)^2/I$. Hence we may use (42) and (44) to specify the mean strength of an impulsive torque:

$$\overline{(\delta L)^2} = 4m^2\overline{v^3r^2}/\bar{v} = 2k_BT\Lambda/N,$$

in which $N = n\bar{v}A$ and is the number of impacts on the vane in unit time.

92 But (4.44) provides the link between the spectral intensity of a white noise source and the strength and frequency of the random impulses that simulate it. We may therefore write for the random torque due to the medium:

$$I_\omega(\text{torque}) = N\overline{(\delta L)^2}/\pi = 2k_BT\Lambda/\pi. \tag{47}$$

In this form, as expected after the discussion of Boltzmann's law, all molecular details have vanished, only the damping parameter and the temperature remaining to define the noise intensity.

The expression (47) for the random torque can be applied to the suspended vane whatever its Q may be, and there is no need to go into the details which run exactly parallel to our analysis of the series resonant circuit. Naturally, for we have just made sure this will be so, the mean energy of the vane is always $k_B T$, but we find as well that this is divisible into potential and kinetic energy, each $\frac{1}{2}k_B T$ on the average. Thus,

$$I_\omega(\text{angular displacement}) = I_\omega(\text{torque}) \times P_d$$

and

$$I_\omega(\text{angular velocity}) = I_\omega(\text{torque}) \times P_v,$$

I_ω(torque) being given by (47), P_d and P_v by (22) and (23) with $1/\mu$ playing the role of χ_0. Integration over frequency yields the mean-square angular displacement and angular velocity, and for this the integrals evaluated in (41) are required. In the end the law of equipartition is verified in detail,

$$\tfrac{1}{2}\mu\overline{\theta^2} = \tfrac{1}{2}I\overline{\omega^2} = \tfrac{1}{2}k_B T.$$

Experimental study of these matters was first undertaken by Gerlach and Kappler,[1] using thin silvered mica vanes on very fine silica torsion suspensions, and making photographic records of the slow twisting Brownian motion. In spite of difficulties arising from wandering of the zero of the suspension, Kappler was able to measure the mean-square angular displacement, and hence the mean potential energy, over a period of 100 hours. The measurement yielded a value of Boltzmann's constant, k, within about 1% of the accepted value. Later, Jones and McCombie[2] used a moving-coil galvanometer with photocell amplification of the very small angular displacements. They could vary the damping by means of the resistance in the coil circuit (Kappler changed the air pressure), and could follow the changes in character of the motion as the Q-value was altered. Fig. 15 shows some of their traces, and the variation from quasi-periodic to aperiodic motion as Q is lowered is clear (cf. fig. 4.13, depicting entirely analogous behaviour). Like Kappler, they found the amplitude to be in accord with Boltzmann's law.

86

A point of some interest arising from this investigation is worth a brief digression. As the external resistance is changed, the damping of the galvanometer is shared in differing proportions by the air and by the resistor; but the Brownian motion retains the same mean energy. This means that the resistor must supply white noise related to its damping effect by the same formula (47) as applies to the air. To make the argument exact we need to introduce the parameters of the moving coil – if it has n turns, each of area A, and is hanging in a magnetic field $\mathcal{B}$, it experiences a couple αi when a current flows in it, α being $nA\mathcal{B}$. The damping effect arises from induction – when turning at a rate $\dot{\theta}$, the flux through the coil changes at a rate $\alpha\dot{\theta}$, inducing a current $\alpha\dot{\theta}/R$, if R is the resistance of the coil circuit, consisting of the coil itself and any external resistor connected across it. The induced current exerts a couple $\alpha^2\dot{\theta}/R$ on the coil, and this is the damping term already mentioned in chapter 2. Thus α^2/R acts as a damping parameter in the same way as Λ, and from (47) we can assert that the power spectrum of the injected random torque is given by

33

$$I_\omega(\text{torque}) = 2k_B T\alpha^2/\pi R.$$

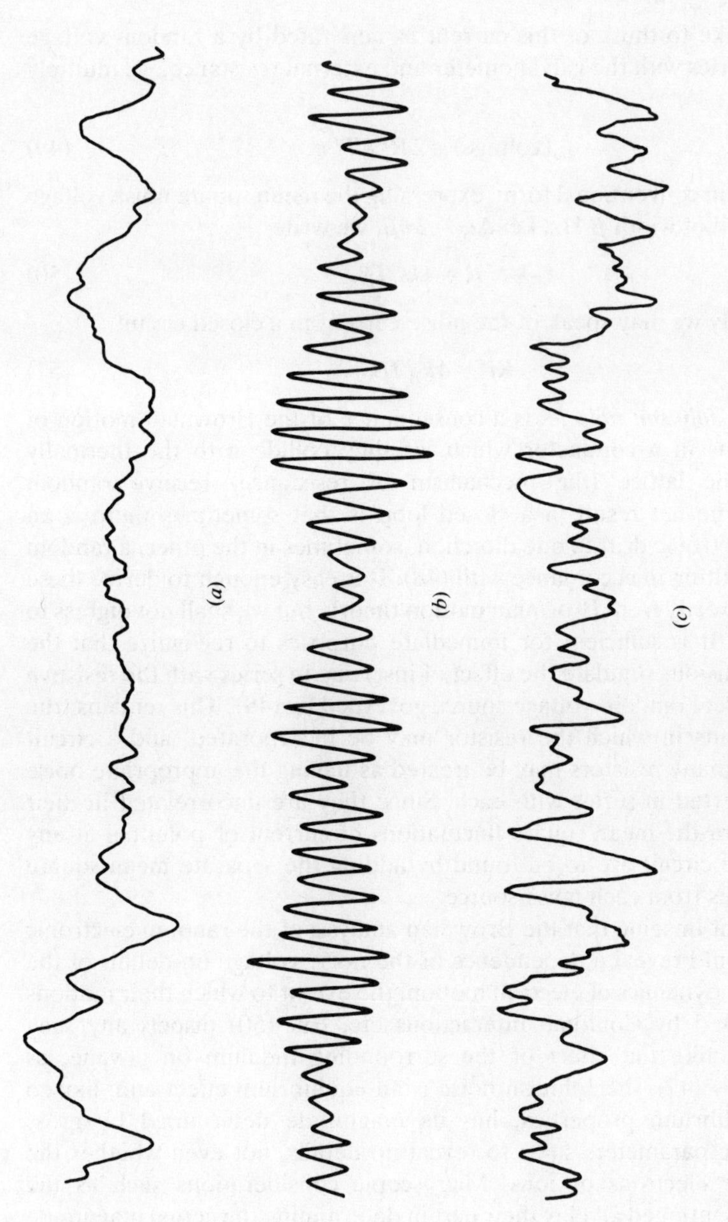

Fig. 15. Records of Brownian motion of a galvanometer coil (R. V. Jones and C. W. McCombie[2]; (*a*) coil short-circuited and heavily overdamped, (*b*) coil open-circuited and underdamped, (*c*) coil critically damped. These curves may be compared with fig. 4.13 showing how the appearance of noise depends on the bandwidth.

Since this torque results from a current in the circuit, we divide by α^2 to obtain the spectral intensity of the noise current,

$$I_\omega(\text{current}) = 2k_B T/\pi R; \tag{48}$$

and if we like to think of this current as generated by a random voltage source in series with the galvanometer and external resistance, we multiply by R^2:

$$I_\omega(\text{voltage}) = 2Rk_B T/\pi. \tag{49}$$

To cast this in conventional form, expressing the mean square noise voltage $\overline{V^2}$ in a band of width β Hz, i.e. $\Delta\omega = 2\pi\beta$, we write

$$\overline{V^2}/R = 4k_B T\beta. \tag{50}$$

Alternatively we may speak of the noise current in a closed circuit,

$$R\overline{i^2} = 4k_B T\beta. \tag{51}$$

This, the *Johnson noise*,[3] is a consequence of the Brownian motion of the electrons in a conductor which, as they collide with the thermally excited ionic lattice (the mechanism of resistance) receive random impulses. The net result in a closed loop is that sometimes there is an average electronic drift in one direction, sometimes in the other, a random current resulting in accordance with (48). It is easy enough to derive these formulae directly from Brownian motion theory, but we shall not digress to this extent. It is sufficient for immediate purposes to recognize that the random collisions simulate the effect of inserting in series with the resistive circuit an ideal random voltage source governed by (49). This remains true for all circuits in which the resistor may be incorporated, and a circuit containing many resistors may be treated as having the appropriate noise voltage inserted in series with each. Since they are uncorrelated in their noise spectra the mean square fluctuations of current or potential at any point in the circuit are to be found by adding the separate mean square contributions from each noise source.

One might imagine that the Brownian analysis of the random electronic motions would reveal a dependence of the noise voltage on details of the model – the dynamics of electron motion, the extent to which their motions are correlated by Coulomb interactions etc. But (50) dispels any such hopes; just like the effect of the surrounding medium on a vane, as expressed by (47), the Johnson noise is an equilibrium effect and, like so many equilibrium properties, has its magnitude determined by gross macroscopic parameters so as to reveal no details, not even whether the carriers are electrons or ions. Microscopic considerations such as the examples mentioned all play their part in determining the actual magnitude of the fluctuations and of the resistivity, but in such a way that the one is uniquely linked to the other.

There is, however, another situation – well away from thermodynamic equilibrium – where microscopic information is conveyed by the magnitude of the noise; this is the *Shot effect.* A thermionic emitter (e.g. a hot tungsten filament) in a simple vacuum diode can pass no more current than the filament emits. With a low voltage between collector (anode) and emitter (cathode), space charge builds up in the intervening vacuum, and the electrons in this gaseous cloud may re-enter the filament on striking it. A high enough voltage, however, prevents space charge accumulating, so that every electron as it leaves the cathode is inexorably drawn to the anode. The saturation current consists of a random stream of electrons, and it is of interest to see how these will excite a resonant circuit connected to the anode (fig. 16). One is tempted to ask whether the impulse due to the electron occurs at the moment it leaves the cathode or at the moment it arrives at the anode; analysis of the potential variations as a charged particle passes from one plate of a capacitor to the other shows that the impulse is not sharp but is spread over the transit time, which typically might be 10^{-8} s. If this is much shorter than the period of the resonant circuit the question becomes unimportant and each electron can be considered to generate a sharp impulse.

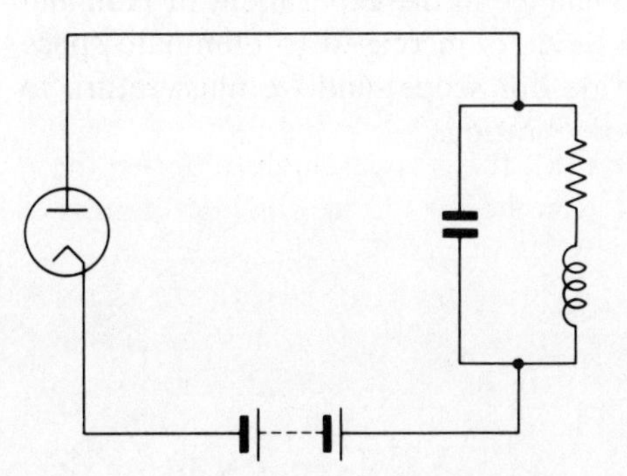

Fig. 16. Resonant circuit excited by shot noise of a saturated thermionic diode.

It is now straightforward to apply the impulse response method. An electron leaving the cathode and reaching the anode establishes virtually instantaneously a potential difference e/C across the capacitor, but no current in the circuit. If the Q-value is high the oscillations of voltage start with this amplitude and decay with energy decay time τ_e equal to L/R, during which time Li/Re electrons arrive if i is the mean current. Hence the mean square voltage across the capacitor may be written

$$\overline{V^2} = \frac{Li}{Re} \times \frac{1}{2}\frac{e^2}{C^2} = \frac{1}{2}\frac{Lei}{RC^2} = \frac{Qei}{2\omega_0 C^2}. \quad (52)$$

Hull and Williams,[4] who in 1925 made a careful study of the shot effect, used a resonant circuit of Q-value 100 at 725 kHz, the value of C being 600 pF. With a thermionic current of 20 mA the expected value of $\overline{V^2}$ was about $10^{-7}\,\mathrm{V}^2$, giving oscillatory voltages of about $\frac{1}{3}$ mV, quite large enough to measure accurately, even with the rather primitive electronics of the day. It is a tribute to their skill and patience over technical detail that their estimated value of the electronic charge was only in error by 1%.

Let us compare the shot noise (52) with that produced by Johnson noise in the circuit itself. A voltage source, whose spectral intensity is given by (49), in series with L, C and R produces a random voltage across the capacitor of magnitude

$$I_\omega(\text{capacitor voltage}) = \frac{2RkT/\pi}{(1-\omega^2 LC)^2 + \omega^2 C^2 R^2}.$$

On integrating over frequency, making use once more of (41), we find, not unexpectedly, that the mean square voltage is $k_B T/C$ – a further example of Boltzmann's law; the energy in the capacitor, $\frac{1}{2}CV^2$, is a quadratic function and takes a mean value $\frac{1}{2}k_B T$. With the capacitor used by Hull and Williams, the Johnson noise at room temperature had $\overline{V^2}$ equal to about 10^{-11} V^2, a power level 10^4 less than that of the shot noise.

Many points of physical interest arise from these considerations – why, for example, does shot noise, so large here, play no role in the Johnson noise, and how does the noise level change in the experiment of Hull and Williams as the voltage across the diode is increased to eliminate space charge? But these are matters outside our scope, and we must return to topics that are central to the study of vibrations.

7 Waves and resonators

The simple resonators that have been the principal theme up to this point possessed one resonant mode only and were, moreover, excited by external influences whose reaction to the effect they produced could be ignored. We turn now to extensions of these ideas in two directions. First, the resonator may consist of a string, a tube, a transmission line, a waveguide – all media for the propagation of waves, whether mechanical, acoustic or electromagnetic – and may therefore be capable of excitation in a number of different modes (see the discussion of the vibrations of a string in chapter

26

2). Secondly, the resonator, which may be one of the above or something simpler, may be embedded in a medium for wave-propagation, being excited by the incidence of a wave and re-radiating a wave as it responds to the excitation. For the most part we shall confine the discussion to one-dimensional systems, with only occasional excursions into the considerably

195

more versatile and complicated realm of three dimensions, such as the response of a small resonant system to a plane wave.

Preliminary remarks about one-dimensional waves; characteristic impedance and admittance

A well-developed calculus exists for treating the reflection and transmission of waves on one-dimensional transmission lines, and this is unquestionably the right approach to adopt in engineering design and in the analysis of all but the most elementary problems.[1] Since, however, this is fully explained in numerous textbooks and since we are concerned primarily with building up a physical picture of the behaviour of elementary systems, we shall go no more deeply into this than is necessary. It will soon become clear that very little formal technique is in fact needed at this stage. The concept of *characteristic wave impedance* is valuable in two ways; it greatly aids describing the response of the wave to an obstacle – how much of the energy is transmitted, how much reflected and how much absorbed – and it enables one to bring a variety of different systems into close analogy, so that the same procedures apply to all. It can be introduced most readily by means of the parallel-wire transmission line, carrying an electromagnetic wave of the form shown in fig. 1. From the point of

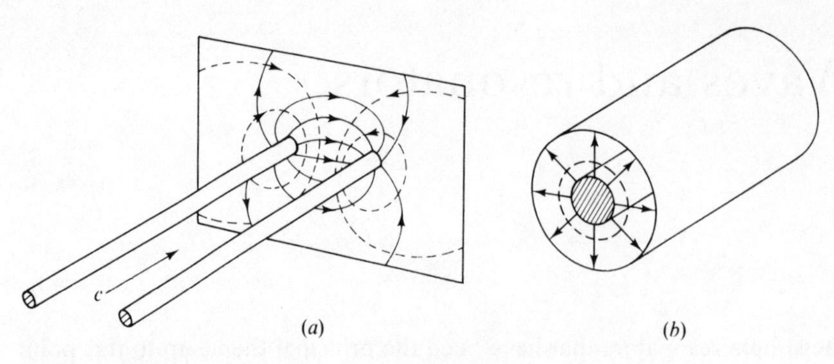

Fig. 1. Electric (full line) and magnetic (broken line) field configurations in the transverse plane of (*a*) twin-wire transmission line and (*b*) coaxial line. In both cases the wave is propagating into the paper.

view of Maxwell's equations we have here a wave travelling in the z-direction and described by the field patterns

$$\left.\begin{aligned}\mathscr{E} &= \mathscr{E}_0(x, y)\,e^{i(kz-\omega t)},\\ \mathscr{B} &= \mathscr{B}_0(x, y)\,e^{i(kz-\omega t)},\end{aligned}\right\} \tag{1}$$

for which the velocity of propagation, ω/k, is c, the velocity of light in free space, if the conductors are surrounded by empty space, not dielectrics. The conductors control the field pattern, since $\mathscr{E}$ must be normal and $\mathscr{B}$ tangential to them everywhere. In fact, solution of Maxwell's equations shows that for any system of lossless conductors of constant cross-section, carrying purely transverse waves, $\mathscr{E}$ and $\mathscr{B}$ are everywhere normal to and in phase with each other, with the constant ratio $\mathscr{E}/\mathscr{B} = c$. Thus the same holds for the coaxial line of fig. 1(*b*). Looking along the direction of propagation one must turn clockwise to go from the direction of $\mathscr{E}$ to that of $\mathscr{B}$ (i.e. $\mathscr{B} \wedge \mathbf{c} = \mathscr{E}$). This means that when there are waves travelling in both directions on the line, perhaps because of partial or complete reflection at an obstacle, at such points where the electric fields of each add in phase the magnetic fields are in antiphase, and vice versa (see fig. 2). Thus, to use real notation, if the strength of the electric field varies along some line, parallel to the conductors, according to the equation

$$\mathscr{E} = \mathscr{E}_0 \cos(\omega t - kz) + \mathscr{E}_1 \cos(\omega t + kz), \tag{2}$$

the magnetic field varies as

$$\mathscr{B} = \frac{\mathscr{E}_0}{c} \cos(\omega t - kz) - \frac{\mathscr{E}_1}{c} \cos(\omega t + kz). \tag{3}$$

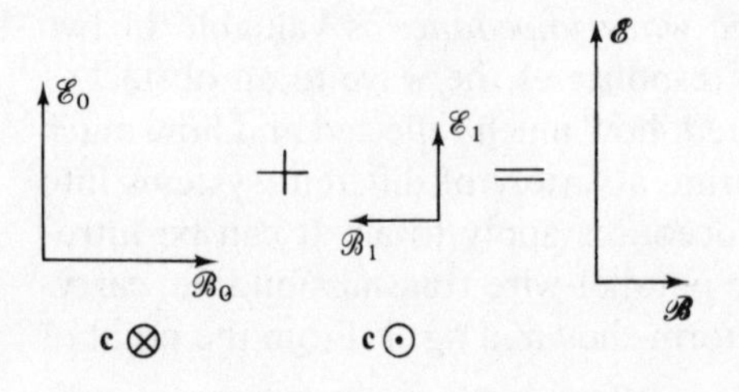

Fig. 2. Electric and magnetic fields in a partial standing wave, $\mathscr{E}_0$ and $\mathscr{B}_0$ referring to a wave propagating into the paper, and $\mathscr{E}_1$ and $\mathscr{B}_1$ to a wave coming out.

61 Both fields form a partial standing-wave pattern, as described in chapter 3, but with maxima and minima interchanged. The maxima of $\mathscr{E}$ are found where $kz = n\pi$, and at these points $\mathscr{E} = \pm(\mathscr{E}_0 + \mathscr{E}_1)\cos\omega t$ while $c\mathscr{B} = \pm(\mathscr{E}_0 - \mathscr{E}_1)\cos\omega t$; where $kz = (n+\frac{1}{2})\pi$, however, it is $c\mathscr{B}$ that has the amplitude $\mathscr{E}_0 + \mathscr{E}_1$ and $\mathscr{E}$ the amplitude $\mathscr{E}_0 - \mathscr{E}_1$. At the maxima and minima the oscillations of $\mathscr{E}$ and $\mathscr{B}$ are in phase, but elsewhere there is a phase difference. The ratio $\mathscr{E}/\mathscr{B}$ is in general complex and varies along the line; its value at any point is frequently referred to as the impedance.

An alternative point of view, which approaches more closely the idea of impedance as used in a.c. circuits, starts from the currents carried by the lines and the e.m.f. between them. The electric field lines terminating at the surface of the conductors demand a surface charge of density $\varepsilon_0\mathscr{E}$, and the magnetic field can only be prevented from penetrating the conductors by a surface current of density $\mathscr{B}/\mu_0$. If the conductors carry charge $\pm q$ per unit length and have an e.m.f. V between them ($V = \boldsymbol{\mathscr{E}} \cdot \mathrm{d}\boldsymbol{l}$ taken between the conductors along a line lying in a plane normal to the conductors), the capacitance per unit length, C, is defined by the equation $q = CV$; thus between the planes A and B in fig. 3(a) the lines carry charge $\pm CV\,\delta z$.

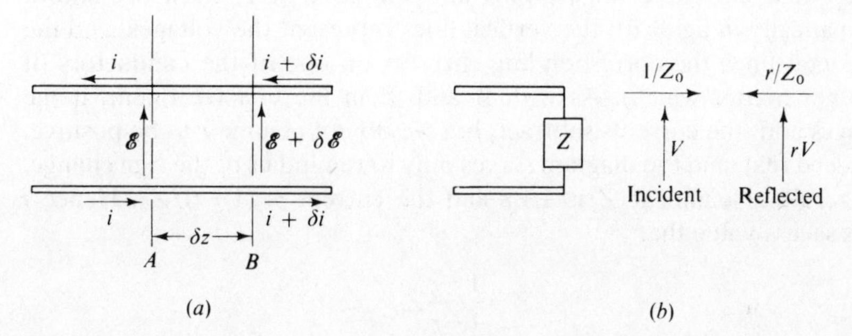

Fig. 3. Diagrammatic representation of current and electric field on a transmission line. In (a) the planes A and B are δz apart, and $\delta\boldsymbol{\mathscr{E}} = (\partial\boldsymbol{\mathscr{E}}/\partial z)\,\delta z$, $\delta i = (\partial i/\partial z)\,\delta z$. In ($b$) is shown the sign relationship between voltages and currents in the waves incident on, and reflected from, the terminating impedance Z.

With currents $\pm i$ flowing as shown, magnetic flux passes through the circuit defined by the conductors and the planes A and B. If this flux is written as $Li\,\delta z$, L is the inductance per unit length. It now follows from Faraday's law of induction, applied to this elementary circuit, that

$$L\,\partial i/\partial t = -\partial V/\partial z, \tag{4}$$

and from the conservation of charge that

$$C\,\partial V/\partial t = -\partial i/\partial z. \tag{5}$$

Hence

$$\partial^2 V/\partial z^2 - LC\,\partial^2 V/\partial t^2 = 0,$$

the equation of wave-motion with velocity $c = (LC)^{-\frac{1}{2}}$. Direct calculation of L and C shows that c is $(\varepsilon_0\mu_0)^{-\frac{1}{2}}$, the velocity of light.

We are, however, more interested at the moment in the relation between V and i, which is obvious from (4) and (5); if, as in a travelling wave, the ratio V/i is the same everywhere,

$$Z_0 \equiv V/i = (L/C)^{\frac{1}{2}} = Lc. \tag{6}$$

Because $\mathscr{E}$ and $\mathscr{B}$ in the field pattern are in phase, so too must V and i be, so that Z_0 is a pure resistance if, as assumed here, there is no dissipative process present. The interpretation of Z_0, the characteristic impedance of the line, is very straightforward in terms of the energy flux associated with the wave. When, for example, a wave travels from left to right in fig. 3(*a*), an observer at the plane A sees current i entering along one conductor and leaving along the other, while an e.m.f. V exists between them. If the line to the right of A were replaced by a resistor Z_0 between the conductors the product $\frac{1}{2}Vi$ would represent the mean rate of heat production therein. Instead, with the line extending to infinity there is no heat production but still a mean flux of energy $\frac{1}{2}Vi$ crossing the plane A and travelling down the line with the wave.

If the line is severed at A and the portion to the right replaced by a resistor Z_0, no reflection of the wave occurs, since voltage and current in the travelling wave are just right to supply the resistor. The line is then said to be correctly terminated. In general, however, an impedance Z terminating a line causes reflection. Let a wave of unit amplitude ($V=1$) be incident on Z, and let there be a reflected wave of amplitude r. The voltage and current configurations in each wave at Z itself are shown schematically in fig. 3(*b*); the vertical lines represent the voltages, and the horizontal lines the corresponding currents on one of the conductors (it does not matter which). As with $\mathscr{E}$ and $\mathscr{B}$ in the first treatment, if the voltages add, the currents subtract; but we do not assume r to be positive, or indeed real, and the diagram serves only to remind us of the sign change. The resultant e.m.f. at Z is $1+r$ and the current is $(1-r)/Z_0$. Hence r takes such a value that

$$Z=\frac{1+r}{1-r}Z_0$$

i.e.

$$r=\frac{Z-Z_0}{Z+Z_0}. \tag{7}$$

Unless Z is purely resistive, r is complex, indicating a phase change on reflection. Two special cases follow from (7): if $Z=0$ (short circuit) $r=-1$ and the e.m.f.s in the two waves cancel at the termination; if $Z=\infty$ (open circuit) $r=1$ and the currents cancel.

As one moves down the line away from the termination, the magnitudes and phases of V and i alter so that the apparent impedance between the conductors varies with a period of half a wavelength. We shall not be concerned to apply the general formula, but it is so easy to derive that we shall do so, and then draw attention to the more immediately relevant special cases. On moving from the termination towards the source through a distance z, one finds the current and e.m.f. of the incident wave advanced in phase by kz, while those of the reflected wave are equally retarded. Instead of an e.m.f. $1+r$ we now have $\mathrm{e}^{-ikz}+r\,\mathrm{e}^{ikz}$, and for current we have

$(e^{-ikz}-r\,e^{ikz})/Z_0$. Hence the length z of line terminated by Z_t behaves as an impedance Z', where

$$\frac{Z'}{Z_0}=\frac{e^{-ikz}+r\,e^{ikz}}{e^{-ikz}-r\,e^{ikz}}$$

$$=\frac{Z-iZ_0\tan kz}{Z_0-iZ\tan kz}, \quad \text{from (7),} \tag{8}$$

$$=\frac{Z+jZ_0\tan kz}{Z_0+jZ\tan kz}, \tag{9}$$

to express it in conventional electrical circuit terms. In particular, if the line is short-circuited ($Z=0$),

$$Z'=jZ_0\tan kz, \tag{10}$$

while if it is open-circuited ($Z=\infty$),

$$Z'=-jZ_0\cot kz. \tag{11}$$

In both cases the apparent impedance is imaginary, as it must be when no mechanisms for loss are present. A short length ($kz<\pi/2$) of short-circuited line is inductive, while a short length of open-circuited line is capacitive. When the line is one-quarter wavelength long ($kz=\pi/2$) the short-circuited line behaves as if open-circuited, and vice versa.

When an impedance is connected either in series with or across an otherwise unobstructed infinite line, as in fig. 4, an incident wave is partially reflected and partially transmitted. To take the series connection, Z

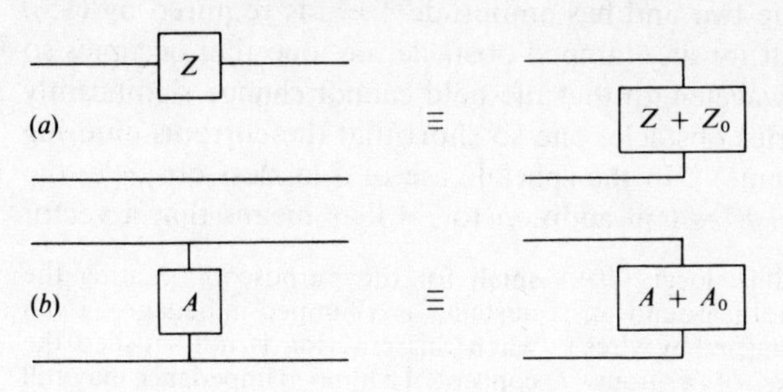

Fig. 4. (*a*) An impedance Z in series with an infinite line appears to an observer at the left as an impedance $Z+Z_0$; (*b*) an admittance A in parallel appears as an admittance $A+A_0$.

and the line to its right behave as Z_0 in series with Z, so that a wave incident from the left has a reflection coefficient given by substituting Z_0+Z for Z in (7):

$$r=Z/(Z+2Z_0). \tag{12}$$

For a wave of unit amplitude incident on the obstruction, the current through $Z+Z_0$ is $(1-r)/Z_0$, so that the e.m.f. across Z_0 is given by $1-r$:

$$t=1-r=2Z_0/(Z+2Z_0). \tag{13}$$

This is the amplitude of the transmitted wave. If the obstacle is non-dissipative, Z is purely imaginary, $\mathrm{j}X$ say, and the reflected and transmitted fractions of the incident power are given by

$$rr^* = X^2/(4Z_0^2+X^2), \qquad tt^* = 4Z_0^2/(4Z_0^2+X^2), \qquad rr^*+tt^* = 1. \tag{14}$$

Energy is of course conserved, and the reflected and transmitted waves are in phase quadrature at the obstacle, as follows from (12) and (13), r having an imaginary numerator while that of t is real.

For an obstacle of admittance A placed across the line very similar results obtain:

$$r = -A/(A+2A_0), \qquad t = 1+r = 2A_0/(A+2A_0), \text{ where } A_0 = 1/Z_0. \tag{15}$$

The relationships $t = 1 \pm r$ in (13) and (15) admit of a simple physical interpretation for which we take the case of the obstacle in parallel for the purpose of illustration. To the left of the obstacle there are both incident and reflected waves, and to the right a transmitted wave only, whose amplitudes and phases must be adjusted to give no discontinuity in e.m.f. at the obstacle. Now there are two progressive wave patterns that are continuous in the required sense, (1) a wave that passes the obstacle unchanged and unreflected; (2) equal waves radiating out in both directions from the obstacle and in phase at the obstacle. Neither satisfies all the boundary conditions separately, but a combination of the two does: a progressive wave moving from left to right, having amplitude unity at the obstacle, and a 're-radiated' wave of amplitude r at the obstacle. Then the incident and reflected waves are correctly described, while the transmitted wave is the sum of the two and has amplitude $1+r$, as required by (15). This is a general result for any lumped obstacle, i.e. one that occupies so small a fraction of a wavelength that the field cannot change significantly across it (or, for a series obstacle, one so short that the currents entering and leaving are the same).† In the special case of a lossless obstacle, the requirement that $rr^*+tt^* = 1$ in addition to $t = 1+r$ means that a vector

† There is considerable local distortion of the field pattern around an obstacle; a capacitor connected by wires between the conductors of a transmission line concentrates the electric field between its plates instead of having it distributed throughout the region between the conductors. The distortion is normally confined to a region round the obstacle of dimensions comparable with the lateral dimensions of the line. If this is much less than the wavelength it is possible to find planes on either side of the obstacle far enough apart for the fields to be undistorted, yet near enough for the space between to be regarded as small for the purpose of treating the obstacle as a lumped impedance. Even when this criterion is not satisfied the concept of a lumped impedance may still be applied to an obstacle whose physical dimensions are small, even though the field distortion is relatively extended. It ultimately breaks down when more than one mode of wave propagation is possible (as in waveguides), for the obstacle irradiated by one mode may generate and re-radiate another. Such a state of affairs obviously demands a more elaborate treatment than anything considered here.

diagram of r and t forms two sides of a right-angled triangle with unit hypotenuse.

185 Later in this chapter we shall refer to obstacles in the form of transducers that are not small in relation to the wavelength and which may connect transmission lines of different Z_0. If these are lossless we may derive all we need to know about them by energy conservation. Consider a non-dissipative transducer to which are connected two lines, with characteristic impedances Z_0 and Z_0', as shown in fig. 5 which also defines the two

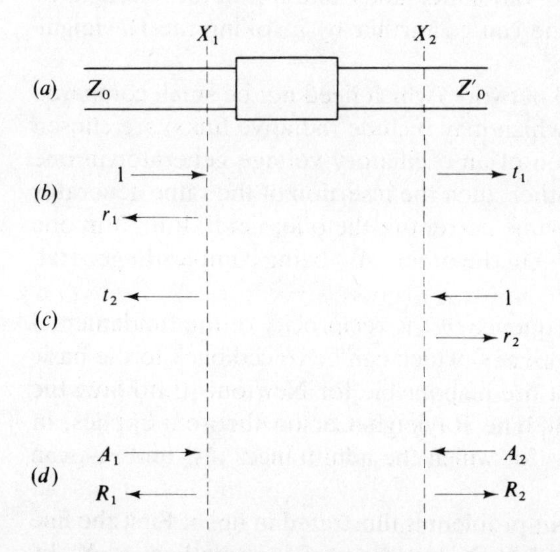

Fig. 5. Location of planes X_1 and X_2 so that all reflection and transmission coefficients are real. In (*b*) a wave of unit amplitude is incident from the left, in (*c*) from the right. In (*d*) the incident waves have amplitudes A_1 and A_2.

reflection coefficients and the two transmission coefficients in terms of the e.m.f. associated with the waves. The two planes X_1 and X_2 are chosen so that the incident and reflected waves are in phase on them; r_1 and r_2 are real and positive, but as yet we know nothing of the phases of t_1 and t_2. Since V^2/Z_0 represents the energy flux in a travelling wave, the absence of dissipation in the transducer implies that

$$\left.\begin{aligned} 1-r_1^2 &= at_1t_1^* \\ \text{and} \qquad 1-r_2^2 &= t_2t_2^*/a, \end{aligned}\right\} \tag{16}$$

where $a=Z_0/Z_0'$. Now let waves of amplitude A_1 and A_2 be incident as shown in fig. 5(*d*), and lead to outgoing waves R_1 and R_2. Then, since R_1 may be regarded as a reflected part of A_1 plus transmitted part of A_2, we have

$$R_1 = r_1A_1 + t_2A_2.$$

Similarly $$R_2 = r_2A_2 + t_1A_1.$$

But energy conservation demands that

$$R_1R_1^*/Z_0 + R_2R_2^*/Z_0' = A_1A_1^*/Z_0 + A_2A_2^*/Z_0', \quad \text{for all } A_1 \text{ and } A_2.$$

Hence, substituting for R_1 and R_2 in terms of A_1 and A_2, and equating the coefficients of $A_1A_1^*$, $A_2A_2^*$ and $A_1A_2^*$ to zero, we reproduce (16) and in addition have that

$$r_1t_2^* = -ar_2t_1.$$

These equations are only satisfied if

$$r_1 = r_2 \quad \text{and} \quad t_2^* = -at_1. \tag{17}$$

This is all that energy conservation demands, and it is in fact enough for most of our purposes. But one can go further by invoking the Rayleigh–Carson[(2)] reciprocity theorem:

> If two wires in an electrical network (which need not be small compared with the wavelength, and which may include radiative links) are chosen arbitrarily, and the insertion of an osciliatory voltage generator in one produces a current in the other, then the insertion of the same generator in the other produces the same current in the one; i.e. if $V\,\mathrm{e}^{-i\omega t}$ in one produces current $A_{12}V\,\mathrm{e}^{-i\omega t}$ in the other, A_{12} being complex in general, then $A_{12} = A_{21}$.

This is a very powerful consequence of the reciprocity of the fundamental laws (e.g. Coulomb's and Ampère's) which can be traced back to the basic properties of space–time that are responsible for Newton's third law, the earliest reciprocity law of all. The Rayleigh–Carson theorem applies, of course, only to linear systems for which the admittances A_{12} and A_{21} can be defined.

Its application to the present problem is illustrated in fig. 6. First the line on the right is short-circuited at X_2 and an e.m.f. is applied at X_1 by

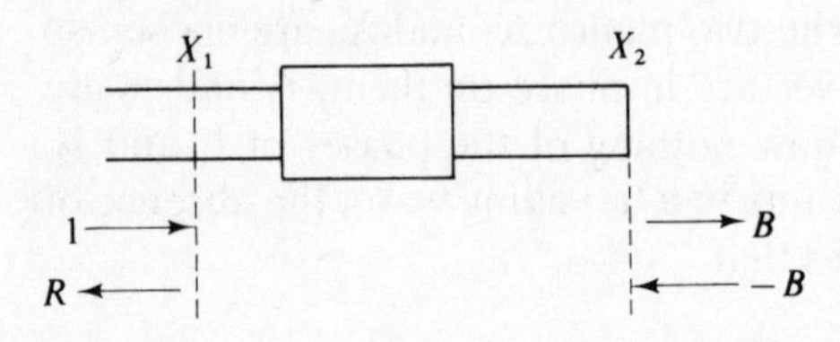

Fig. 6. As in fig. 5, but with the line short-circuited at X_2.

injecting a wave of unit amplitude. At X_2 let the amplitudes of incident and reflected waves be $\pm B$ as shown. Then R is made up of reflected and transmitted waves:

$$R = r_1 - t_2B$$

and

$$B = -r_2B + t_1.$$

Hence

$$B = t_1/(1+r_2) \quad \text{and} \quad R = r_1 - t_1t_2/(1+r_2).$$

The e.m.f. at X_1 is $1+R$ and the current at X_2 is $2B/Z_0'$.

Hence

$$A_{12} = \frac{2t_1/Z_0'}{(1+r_1)(1+r_2) - t_1t_2}.$$

We now repeat the calculation with the short circuit at X_1 and the input at X_2, which yields A_{21} simply by interchanging subscripts 1 and 2 and replacing Z_0' by Z_0. The denominator being unchanged, the Rayleigh–Carson theorem then demands that $t_1/Z_0' = t_2/Z_0$,

i.e.
$$t_2 = at_1. \tag{18}$$

It should be noted that this derivation does not make use of energy conservation, and (18) is therefore true for lossy as well as lossless transducers. In general, however, $r_1 \neq r_2$ – there is, for instance, no difficulty in arranging that a wave incident from the right shall be reflected, while one incident from the left shall be absorbed. But if there is no dissipation, (17) and (18) are both true, and t_1 and t_2 must be pure imaginaries, as has already been proved for the lumped obstacle. It follows that a lossless transducer may be replaced, for the purpose of calculation, by a lumped obstacle, a capacitor shunted across the line for example, and flanked by suitable lengths of line to produce the right phases of reflection and transmission. This simulation does not in general work over a range of frequencies, the form of the equivalent obstacle being a function of frequency.

Analogues of characteristic impedance

The characteristic impedance is valuable, as already remarked, as a unifying concept bringing many types of wave-motion into the same formalism. In order to calculate the behaviour of any given one-dimensional wave we need to know its dispersion law, i.e. how k and ω are related, and what reflection is caused by a given obstacle (there are further problems but they all yield to the same treatment and may be considered as examples arise). In each case we can find analogues of e.m.f. and current which enable results such as (9) to be taken over unchanged. In seeking these analogues we look for field variables (*a*) whose product, like Vi, represents power, (*b*) which are in phase for an unattenuated travelling wave and (*c*) which reverse their relative signs when the direction of propagation is reversed. We tabulate below a few examples.

(1) *String carrying transverse waves with velocity* $c_t = (T/\rho)^{\frac{1}{2}}$, *where* T = *tension and* ρ = *mass per unit length.*

If the transverse displacement is ξ, the transverse component of tension = $T\,\partial\xi/\partial z = \mathrm{i}kT\xi$. Taking this as the analogue of V, the analogue of i must be the transverse component of velocity, $-(\partial\xi/\partial t)$ or $\mathrm{i}\omega\xi$ (the sign is fixed by the requirement that Z_0 shall be positive and Vi represent the flux of energy in the direction of wave propagation).

Hence
$$Z_0 = \mathrm{i}kT\xi/\mathrm{i}\omega\xi = T/c_t = \rho c_t. \tag{19}$$

(2) *Sound waves with velocity* $c_l = (K/\rho)^{\frac{1}{2}}$, *where* K = *bulk modulus and* ρ = *density.*

First we consider waves in a cylindrical pipe of area S, and neglect drag at the walls. If ξ now represents the longitudinal displacement of the medium and p the local excess pressure, the force due to the wave acting across any cross-section is pS, and the velocity of the medium is $\partial\xi/\partial t$. The rate at which energy is transmitted in the positive direction is $pS\,\partial\xi/\partial t$, and this is the analogue of Vi. Clearly there are different ways of assigning the variables to V and i, but the convenient way is to regard p as equivalent to V and $S\,\partial\xi/\partial t$ as equivalent to i. Then, since the pressure gradient $\partial p/\partial z$ is responsible for acceleration of the medium according to the equation:

$$\partial p/\partial z = -\rho\,\partial^2\xi/\partial t^2, \quad \text{or} \quad \mathrm{i}kp = \mathrm{i}\omega\rho\,\partial\xi/\partial t,$$

we have that

$$Z_0 = p/(S\,\partial\xi/\partial t) = \rho c_1/S. \tag{20}$$

The cross-section of the pipe may be imagined to be made up of any number of smaller elements, as if one had a number of smaller pipes running along together and each carrying a wave of the same amplitude and phase. Then an element of unit area defines a system with characteristic impedance

$$Z_0 = \rho c_1. \tag{21}$$

(3) *Plane electromagnetic waves in free space.*
We can introduce here, as with the sound wave, a characteristic impedance or admittance per unit area, taking $\mathscr{E}$ as the analogue of V. Then since Poynting's vector, $\frac{1}{2}\boldsymbol{\mathscr{E}} \wedge \boldsymbol{\mathscr{B}}/\mu_0$, represents the mean energy flux per unit area, and since $\boldsymbol{\mathscr{E}}$ and $\boldsymbol{\mathscr{B}}$ are at right angles, $\mathscr{B}/\mu_0$ is the consequential choice for i.

Hence $$Z_0 = \mu_0\mathscr{E}/\mathscr{B} = \mu_0 c = 377\ \Omega. \tag{22}$$

When the wave is confined by parallel conductors, as in the transmission line of fig. 1, the various elements of plane wave systems must be regarded as being fed in series or in parallel according as they are stacked along $\boldsymbol{\mathscr{E}}$ or along $\boldsymbol{\mathscr{B}}$. Since, however, in such a system $\mu_0\mathscr{E}/\mathscr{B}$ takes the same value everywhere, we can avoid thinking in these terms by imagining electrodes, shaped exactly like the cross-section of the line, attached to a sheet of uniform resistive material of which any square has resistance between opposite edges of 377 Ω. The resistance between the electrodes is then the value of Z_0 for that line. It is clear that the line can be correctly terminated by attaching such a sheet across its end, for the current that flows is consistent, everywhere on the sheet, with the values of $\boldsymbol{\mathscr{E}}$ and $\boldsymbol{\mathscr{B}}$ in the travelling wave.

A marked similarity can be observed in the expressions for Z_0, in each case the wave velocity being multiplied by a density or its analogue. We have already seen how mass and inductance are analogous for vibrating systems, and the appearance of μ_0, being the specific inductance of free space, underlines this parallelism. There is no need to go into details of the use of Z_0 for other than electrical systems; one elementary example will suffice. Suppose a string hangs vertically under tension T provided by a

134

mass M immersed in a viscous liquid; we ask what fraction of the energy in a transverse wave on the string is reflected by the mass. It is necessary to determine the impedance of the mass in the same terms as were used to define Z_0, i.e. the ratio of applied force to velocity when the mass is vibrating with frequency ω. If its displacement has amplitude ξ_0, and we write v for the velocity $-\mathrm{i}\omega\xi_0$, the acceleration is $-\mathrm{i}\omega v$, so that

$$F = \lambda v - \mathrm{i}\omega M v,$$

where λ is the viscous drag constant. Hence

$$Z = F/v = \lambda - \mathrm{i}\omega M,$$

and from (7) and (19),

$$r = \frac{\lambda - \rho c_t - \mathrm{i}\omega M}{\lambda + \rho c_t - \mathrm{i}\omega M}.$$

If the tension is maintained by a light spring, and the mass M replaced by a light vane so that $\mathrm{i}\omega M$ can be neglected, it becomes possible by choosing $\lambda = \rho c_t$ to terminate the line correctly, all wave-motion being absorbed in the liquid without reflection.

Resonant lines

A transmission line short-circuited at both ends can resonate at any frequency that allows an integral number of half-wavelength standing-wave loops to fit in, with zeros (nodes of e.m.f.) at the ends. If there is no dispersion the resonant frequencies form a harmonic series. The same is true for an open-circuited line, but now the ends are antinodes. And for a line that is open-circuited at one end and short-circuited at the other there must be a half-integral number of loops, to allow a node at one end and an antinode at the other. If the length of the line is l, the resonant frequencies are thus determined:

Both ends open-circuited:	$kl = n\pi,$	$\omega = n\pi c/l.$
Both ends short-circuited:	$kl = n\pi,$	$\omega = n\pi c/l.$
One end short-circuited, One end open-circuited	$kl = (n-\frac{1}{2})\pi,$	$\omega = (n-\frac{1}{2})\pi c/l.$

Similar rules apply to acoustic resonators in the form of uniform pipes. If both ends are closed or open, $\omega = n\pi c/l$, with pressure antinodes at the ends in the first case, pressure nodes in the second; while with one end closed and the other open, as with the transmission line $\omega = (n-\frac{1}{2})\pi c/l$. The fundamental is at half the frequency of the fundamentals in the other cases, a fact made use of in designing deep-toned organ pipes for confined lofts. But only odd multiples of the fundamental now appear and the tone lacks richness.

These general remarks somewhat oversimplify the matter. It may not be easy to obtain a perfect short circuit or open circuit and as a result there are usually end-corrections, the resonator behaving as though it were a little longer than its actual length. With an acoustic resonator in the form of a uniform open-ended pipe, the air just outside the ends is in vibration with something like the same amplitude as just inside, and one has to go one or two radii beyond the end before the flow is spread out enough for the velocity to have fallen to a low value. The pressure must oscillate to some small degree at the end to provide the required acceleration of the air outside, and the apparent position of the pressure node therefore lies beyond the end. But so long as the end-correction is simply such as to give a constant extension, Δl, to l, the harmonic series is only displaced down in frequency without losing its simple numerical ratio of frequencies. This is commonly enough what happens when the lateral dimensions are much less than the length. Only when the wavelength is so small as to begin to compare with the lateral dimensions does the end-correction begin to change significantly. For many purposes, then, it can be ignored or treated as an empirical adjustment to l, and for the moment we shall adopt this attitude; but shall return to the matter in due course to discuss the radiation of power from the open end. 179

This type of resonant line may be thought of either as a series-resonant or parallel-resonant circuit according to the way it is excited. In fig. 7 the two sections of line to either side of the point of excitation can be replaced by reactances, whereupon the character becomes obvious. The circuit (*a*)

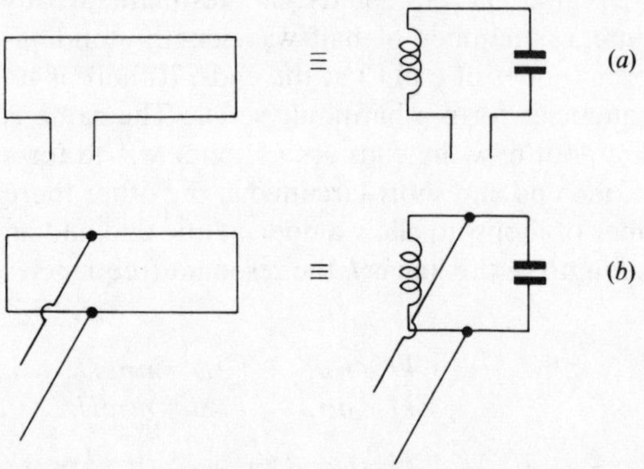

(*a*)

(*b*)

Fig. 7. A closed transmission line fed (*a*) in series, (*b*) in parallel, and the equivalent circuits.

fed by an oscillatory current source reveals a resonance at such frequencies as allow the e.m.f. to fall to zero; and in (*b*) an oscillatory e.m.f. can force no current through the circuit at resonance. The condition for resonance is that the reactance of each section must be equal and opposite. If their lengths are l_1 and l_2, according to (10)

$$\tan kl_1 = -\tan kl_2,$$

i.e.

$$k(l_1 + l_2) = n\pi,$$

the same resonance condition as before, of course.

The essential simplicity of the resonant response is revealed most clearly by the standing-wave pattern produced when a line resonator is excited by a current source at an open end, as in fig. 8. The e.m.f. variation is shown for frequencies around the fundamental resonant frequency, ω_0. The current pattern, with antinodes at the ends, remains virtually unchanged, but as the node of e.m.f. passes through the point of excitation, so the input impedance goes through zero as a linear function of frequency; and that is the essence of a high-Q resonance. If the wave suffers attenuation, so that when ω is real (steady excitation) k is complex, the standing wave may be thought of as a travelling wave totally reflected at the far end, but having lost some of its initial amplitude by the time it returns. As a result the input e.m.f. at resonance does not quite go through zero. In order to make it do so the input wave must have an amplitude that decreases with time so as to match the attenuation of the reflected wave. The true resonance frequency is thus seen to be complex to the extent required to describe the natural decrement of the resonator vibrating freely.

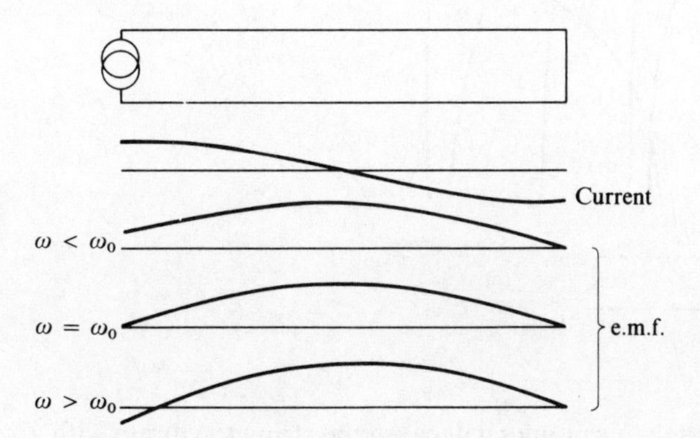

Fig. 8. A current source attached to a closed line, showing how the node of e.m.f. shifts as the resonance is traversed.

The regularity of the harmonic series is destroyed by dispersion, for an arithmetic series of wave-numbers, as demanded by the boundary conditions, does not imply an arithmetic series of frequencies unless the velocity is constant. To discuss this problem properly demands an investigation of boundary conditions as well as of the dispersion law – the law expressed by (3.20), for instance, arises from the fourth-order equation (3.19) and the resonances of a finite rod must be found by applying four boundary conditions, two at each end, rather than one at each end as with a string or transmission line. We shall not undertake this investigation here. An easier problem is the non-harmonic series that results when a resonant line is loaded at one point. The effect is easily demonstrated on a piano by threading a fine short nail in and out through one set of three strings such as are used for each of the treble notes. The fundamental is lowered (by perhaps a semitone), but quite as obvious is the strange timbre that results, and the way in which this itself alters as the nail is moved to different positions along the strings. Qualitatively it is easy to see why the harmonic

59

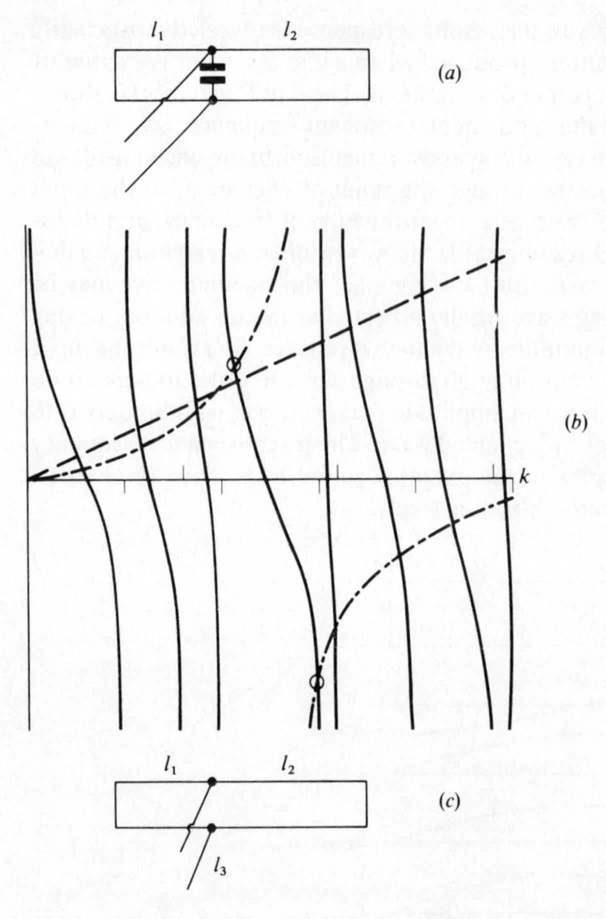

Fig. 9. (*a*) A closed transmission line with a lumped capacitor across it and excited at that point; (*b*) diagram for determining the resonances; the markers on the axis show the locations of the infinities of cot kl_1 and cot kl_2; (*c*) capacitor replaced by a length of transmission line (exciting line not shown).

series is destroyed. Certain harmonics will cause the string to vibrate with a node at or near the nail, and insertion of the nail hardly changes the pattern of vibration; but those harmonics with an antinode at the nail will be loaded thereby and caused to drop in pitch. For a more detailed analysis consider the transmission line resonator loaded by a capacitor, as in fig. 9, and imagine it excited at the point of loading. The input admittance follows from (10):

$$A = -\mathrm{j}A_0(\cot kl_1 + \cot kl_2) + \mathrm{j}\omega C,$$

and the resonance condition, $A = 0$, implies

$$\cot kl_1 + \cot kl_2 = \omega C Z_0. \tag{23}$$

The left-hand side is plotted in fig. 9, which shows the infinite discontinuities that occur whenever kl_1 or kl_2 is a multiple of π. Two discontinuities are close together if the two sides are almost in resonance together, in which case the capacitor lies near a node. The zeros are evenly spaced and mark the values of k at which the unloaded line is resonant. But the loaded line is resonant where the curves are intersected by the line

ωCZ_0, and immediately one sees how some resonances are shifted more than others, the least affected being those closely flanked by discontinuities.

Also shown on the diagram is the effect of loading the line not with a capacitor but with a short length of open-circuited line (fig. 9(c)), itself capable of resonances. The left-hand side of (23) must now equal $-Z_0$ times the admittance of the load, i.e. $-\cot kl_3$. The infinite discontinuities lead to extra resonances for, as shown by the ringed intersections in fig. 9(b), wherever $\cot kl_3$ has a discontinuity there will be two intersections with the same branch of the left-hand side. In this way the total number of resonant modes is conserved, those originally present in the unloaded line being shifted and supplemented by the resonances of the load. The complexity of even this simple example is enough to discourage a general analysis of lines loaded in more than one point, but fortunately the main characteristics of the behaviour are already plain.

494

Finger-holes in woodwind instruments

A special case of loading is the drilling of finger-holes in woodwind instruments, to change the pitch. Any real instrument, even the simplest recorder or tin-whistle, is to the physicist a complicated device by reason of the action of the mouthpiece in maintaining vibration. Moreover the bore is rarely uniform; in each instrument special variations of cross-section have been introduced empirically to realise a desirable tone-quality. To bring out the basic properties of finger-holes it is better to present measurements on a uniform pipe; what follows is intended to be merely illustrative and not a definitive contribution to acoustics – in fact all the experiments described in the next page or two were conducted, including the construction of the resonators, during a single day, and could equally readily be carried out by the reader for his own instruction, given access to normal laboratory facilities.

The resonant pipe was a brass tube 1 metre long and of diameter 35 mm with a wall-thickness 1.7 mm. A small loudspeaker, excited by a variable-frequency oscillator, was placed 20 mm from one end and the excitation observed by listening at the other end. It is easy to hear the resonances and to tune the oscillator to the peak of response within about ±2 Hz; with a few repetitions of each measurement a resonant frequency can be pin-pointed within 1 Hz. The first few resonances were at the following frequencies, the figures in brackets being successive multiples of 169: 167 (169), 338 (338), 507 (507), 676 (676), 843 (845), 1016 (1014). One would need to study the matter more carefully to be convinced of any departure from a strict harmonic series. The end-correction is therefore sensibly independent of frequency in this range. That such a correction is needed is demonstrated by repeating the experiment with a tube exactly half the length, the resonant frequencies then being 328, 660, 992, etc., markedly lower than the even harmonics of the longer tube. If one adds 25 mm to the length of each tube the ratio of lengths, 1.95, matches the mean frequency ratio of corresponding harmonics. This is not an accurate determination of the end-correction but is nevertheless quite good. Better investigations suggest that 0.6 times the radius should be added at each end, i.e. 21 mm altogether in this particular case. The coefficient 0.6 is empirical and applies to an unflanged pipe, for which the theory is difficult on account of the awkward flow back outside the pipe. A pipe with a large plane flange on its end, while still troublesome, is not intractable to analysis, and was investigated by Rayleigh,[3] who gave 0.82 times the radius as the correction. If one makes the simplifying assumption that the flow pattern outside is exactly like that outside a hole in a thin plane sheet (see the discussion of Helmholtz resonators in chapter 2) the coefficient comes out at $\pi/4$, which is very close to Rayleigh's result. This is relevant to the next part of the experiment, to which we now turn.

23

The effect of small holes is shown in fig. 10; the holes were drilled one-quarter of the length from one end, at a pressure antinode of the second harmonic where the effect is strongest. It is immediately clear that a

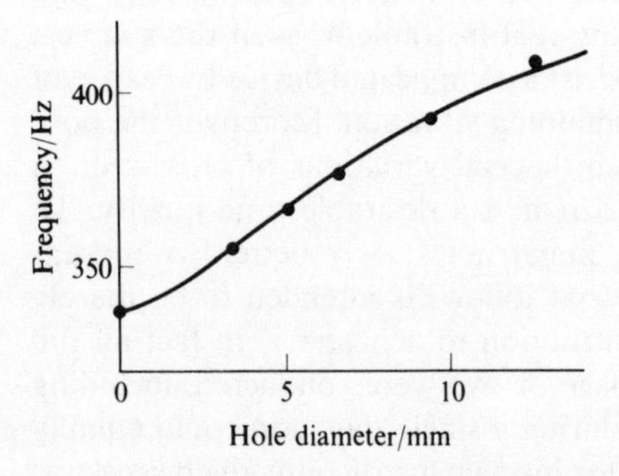

Fig. 10. Effect of hole size on the resonant frequency of an open-ended tube.

very small hole can cause a considerable change in resonant frequency, as anyone knows who has played a recorder and had difficulty in closing every hole completely. The full curve is the theoretical result, with no adjustable constants, derived in the following way. A hole of radius r would behave like an ideal tube of length $\frac{1}{2}\pi r$ if the wall-thickness were negligible, and we have just seen that a long flanged tube has very nearly the same coefficient for its end-correction; we may confidently take $\frac{1}{2}\pi r+t$, t being the wall-thickness, as the effective length. If the characteristic admittance of the main tube, of radius r_0, is A_0, that of the hole is $A_0 r^2/r_0^2$, and according to (10) the admittance it presents is given by

$$A_{\text{hole}} = -\mathrm{j}(A_0 r^2/r_0^2)\cot\left[k(\tfrac{1}{2}\pi r+t)\right] \doteqdot \frac{-2\mathrm{j}A_0}{\pi k r_0^2}\,\frac{r^2}{r+2t/\pi},$$

the approximation being justified since the effective length is very much less than a wavelength. The resonance condition (23) then takes the form

$$\cot kl_1 + \cot kl_2 = -\frac{2}{\pi k r_0^2} \frac{r^2}{r + 2t/\pi}. \tag{24}$$

When $r = 0$ (no hole), $k(l_1 + l_2) = 2\pi$; with the hole situated as it is, $kl_1 = \frac{3}{2}\pi$ and $kl_2 = \frac{1}{2}\pi$, so that $\cot kl_1 = \cot kl_2 = 0$. For small variations of k, then, (24) takes the form

$$(l_1 + l_2)\,\Delta k = \frac{2}{\pi k r_0^2} \frac{r^2}{r + 2t/\pi},$$

and, but for the wall thickness, Δk and hence $\Delta\omega$ would be proportional to the radius of the hole. With larger holes, this approximation fails, since the most effective hole cannot do more than shift the open end to the hole position, rendering the tube beyond ineffective and raising the frequency by a factor $(l_1 + l_2)/l_1$ which is $\frac{4}{3}$ here. It is at first sight somewhat surprising that a hole of diameter 7 mm in a tube 5 times as large should go halfway towards this limit. In woodwind instruments the holes are rather larger in proportion, but the wall thickness is also larger, so that they do not completely immobilize the tube beyond. This explains why finer tuning can be achieved by various combinations of open and closed holes, the open hole nearest the mouthpiece being always the principal factor in tuning.

To conclude the experiment, fig. 11 shows the changes in the resonant frequencies due to a side tube of variable length. The side tube, being closed by oil, is effectively open-circuited with admittance $jA_0(r^2/r_0^2)\tan kl_3$, l_3 being the actual length plus an end-correction at the junction with the main pipe. The broken lines on the graph show the unperturbed resonances of the pipe as horizontal lines, and the fundamental resonance of the side tube running through them. The coupling together of the resonators at the junction produces a characteristic effect which we shall meet several times – at each cross-over the lines are broken and rejoined with different connections. Thus the line ABC may be thought of as describing at the end A the resonance of the main tube in its third harmonic, slightly perturbed by the presence of the side tube, round B as the resonance of the side tube slightly perturbed by the main tube, and round C as the resonance of the main tube in its second harmonic. There is a continuous gradation as l_3 is changed. It will be observed that there are no experimental points over much of the section B; this is because the farther end of the main tube is so little excited that it is almost impossible to discern the resonance by listening there. This picture is confirmed, and can be made quantitative, by use of fig. 9(b). As l_3 is increased, the broken curve suffers a reduction in horizontal scale and the discontinuity moves steadily towards the origin, while the full curves stay fixed. Some resonances are strongly shifted and others less strongly. For those that have a pressure node at the point where the side tube enters, there is no coupling, and the lines in fig. 10 cross without influencing one another; the

368
371
401
529

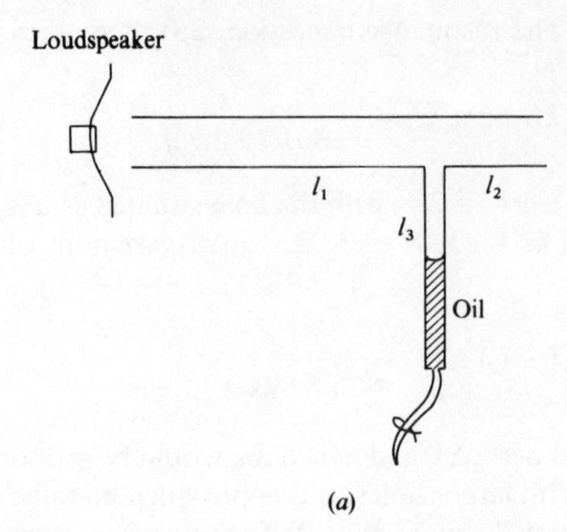

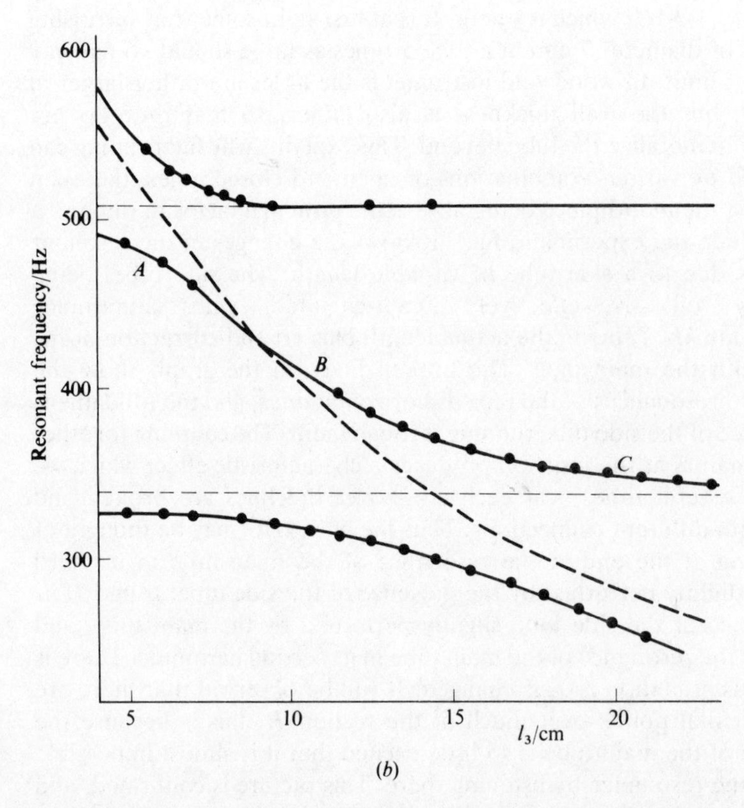

Fig. 11. The effect of a side arm on the resonant frequency of an open-ended tube; (*a*) schematic diagram of experiment, (*b*) shift of certain resonances with length of side arm. The curve *ABC* is theoretical; the horizontal broken curves are resonances of the main tube when $l_3 = 0$, and the third broken curve is the resonance of the side arm alone.

fourth harmonic in the present experiment is very nearly such a resonance, characterized in fig. 9(*b*) by a branch that plunges almost vertically between two very closely spaced discontinuities. In fact no shift in frequency could be discerned as l_3 was changed. Finally we note that the line *ABC* in fig. 11 is the theoretical curve, again with no adjustable parameters; the theory is verified in a most satisfactory manner.

Radiation from an open end

176 The treatment of the end-correction to an open-ended acoustic resonator, following Rayleigh's analysis, is incomplete in so far as it assumes no radiation of power in the form of a wave. His solution of the field equation was quite correct, but the equation chosen for solution was itself an approximation. The pressure of the sound field obeys the wave equation

$$\nabla^2 p = \frac{1}{c^2}\partial^2 p/\partial t^2 = -k^2 p$$

for a harmonic wave, and provided the region in which the disturbance exists (e.g. for a distance of a few radii of the tube beyond the end) is much smaller than the wavelength, the right-hand side of this equation is so much less than the cartesian components of the operator on the left that it is reasonable to replace the wave equation by Laplace's equation, $\nabla^2 p = 0$. This eliminates the radiation field, and causes all external regions to vibrate in phase, with a flow pattern of exactly the same form as for irrotational flow of an inviscid, incompressible fluid which also obeys Laplace's equation. A two-dimensional example of such flow is exhibited in fig. 12 which shows how short a path is needed for the fluid to spread evenly into the available space. In three dimensions the same happens, with the flow velocity soon becoming radial and isotropic, and falling as $1/r^2$ at greater distances. It is only at distances comparable to the wavelength that the inadequacies of Laplace's equation become obvious, and if the orifice is small we may at that distance assume perfect isotropy of flow. We can therefore confine our discussion of the wave equation to spherically symmetrical solutions. It is quite possible at this point to proceed purely

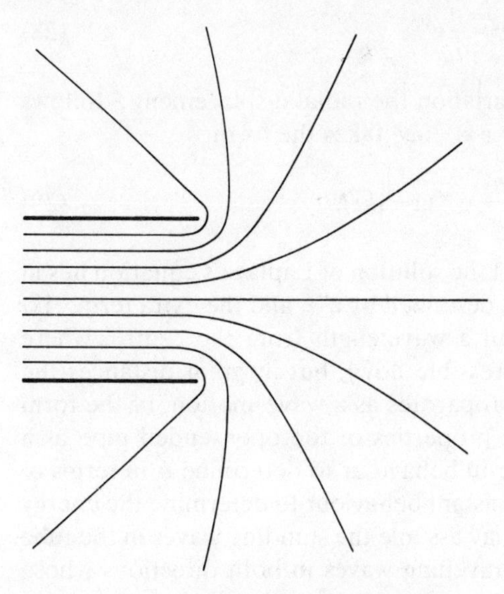

Fig. 12. Two-dimensional irrotational flow emerging from parallel plates. The stream-lines define equal fluxes between each.[4]

mathematically, by writing the wave equation in spherical coordinates, but we shall instead derive the required form from the basic physics, since we acquire in the process other useful relations.

Consider a spherically symmetrical radial displacement $\xi(r, t)$ applied to a volume element defined by the solid angle $\mathrm{d}\Omega$ and the spheres at r and $r+\mathrm{d}r$. Then the inner surface moves from r to $r+\xi$, and the outer surface to $r+\mathrm{d}r+\xi+(\partial\xi/\partial r)\,\mathrm{d}r$. The original volume $r^2\,\mathrm{d}\Omega\,\mathrm{d}r$ has become $(r+\xi)^2\,\mathrm{d}\Omega(1+\partial\xi/\partial r)\,\mathrm{d}r$, a relative dilation of $2\xi/r+\partial\xi/\partial r$, to first order in ξ. If the bulk modulus of the medium is K, the pressure at this point is given by

$$p=-K(2\xi/r+\partial\xi/\partial r)=-\frac{K}{r^2}\frac{\partial}{\partial r}(r^2\xi). \tag{25}$$

Now the force on unit volume of the medium is $-\mathrm{grad}\,p$, which in this case is a radial force of magnitude $-\partial p/\partial r$. Hence the equation of motion takes the form

$$\rho\,\partial^2\xi/\partial t^2=-\partial p/\partial r. \tag{26}$$

From (25) we have that

$$\frac{\partial^2 p}{\partial t^2}=-\frac{K}{r^2}\frac{\partial}{\partial r}\left(r^2\frac{\partial^2\xi}{\partial t^2}\right)=\frac{c_1^2}{r^2}\frac{\partial}{\partial r}\left(r^2\frac{\partial p}{\partial r}\right)$$

from (26), where $c_1^2=K/\rho$. For oscillatory motion

$$\frac{\partial}{\partial r}\left(r^2\frac{\partial p}{\partial r}\right)=-k^2r^2p, \tag{27}$$

whose solution describing an outgoing spherical wave has the form,

$$p=\frac{a}{r}\,\mathrm{e}^{\mathrm{i}kr}. \tag{28}$$

Corresponding to this pressure variation the radial displacement ξ follows from (26), and the radial velocity $v=-\mathrm{i}\omega\xi$ takes the form

$$v=\frac{\mathrm{i}a}{\omega\rho r^2}\,\mathrm{e}^{\mathrm{i}kr}(1-\mathrm{i}kr). \tag{29}$$

The difference between this and the solution of Laplace's equation lies in the phase variation with distance, described by $\mathrm{e}^{\mathrm{i}kr}$, and the extra term $-\mathrm{i}kr$ in the coefficient. Much less than a wavelength from the centre, where $kr\ll 1$, v varies as $1/r^2$ (incompressible flow), but at great distances the second term is dominant and v propagates as a wave-motion, of the form $\mathrm{e}^{\mathrm{i}kr}/r$. We can now calculate the properties of the open-ended pipe as a radiator of sound, using the close-in behaviour to determine a in terms of the motion at the orifice, and the distant behaviour to determine the energy flux. If the radiation is weak we may assume the standing waves in the tube to be perfect and composed of travelling waves in both directions whose pressure amplitude is p_0. The associated velocity oscillations have amplitude $p_0/\rho c_c$ and at the velocity antinode just beyond the end of the pipe they add to give an oscillatory flux Φ over the whole orifice, of area S, where

$$\Phi = \int v\,\mathrm{d}S = 2Sp_0/\rho c_1. \tag{30}$$

This must be matched to the flux in the close-in field described by (29), which we assume isotropic within a solid angle Ω (4π for all-round radiation, 2π for a flanged orifice). Hence

$$\Phi = 2Sp_0/\rho c_1 = \mathrm{i}a\Omega/\omega\rho;$$

i.e.
$$a = -2\mathrm{i}Sp_0k/\Omega. \tag{31}$$

At great distances the pressure, with amplitude $|a|/r$, and the velocity, with amplitude $k|a|/\omega\rho r$, vibrate in phase and are responsible for a mean flux of energy per unit area equal to $\frac{1}{2}k|a|^2/\omega\rho r^2$. Over the whole solid angle, then, the radiated power is P, where

$$P = \tfrac{1}{2}k|a|^2\Omega/\omega\rho = 2S^2k^2p_0^2/\rho c_1\Omega. \tag{32}$$

This result can be conveniently expressed in terms of the *radiation resistance* of the orifice. If a line of characteristic impedance Z_0 is terminated by a resistance R which is much less than Z_0, (7) shows that r^2, the fraction of power reflected, is approximately $1-4R/Z_0$, so that $4R/Z_0$ is absorbed in the resistor. We assign such a value to R as makes $4R/Z_0$ the fraction of the power arriving at the orifice which is radiated. Since the power carried by the travelling wave is $\frac{1}{2}p_0^2/Z_0$, we have that

$$\frac{R}{Z_0} = \frac{1}{4}\,\frac{2S^2k^2p_0^2}{\rho c_1\Omega}\,\frac{2Z_0}{p_0^2} = k^2S/\Omega, \tag{33}$$

since, according to (20), $Z_0 = \rho c_1/S$.

In the experiments described above, the cross-sectional area of the pipe was nearly 10^{-3} m^2, and at the frequency of the second harmonic the wavelength was close to 1 m, so that $k \sim 6$ m^{-1} and the fraction radiated into the full solid angle was about 1.3%. If this radiation loss is the major cause of dissipation we may estimate Q for the resonance by noting that a wave reflected to and fro at this frequency will suffer 1.3% loss of energy for every wavelength traversed, and will be reduced by a factor e in something like 80 wavelengths, or 500 radians. Radiation therefore limits Q to 500. As it happens there is a greater source of dissipation in the viscous drag near the walls of the tube. Though straightforward enough, it would take us too far afield to develop the theory of this effect, and we shall merely note the result of the calculation – if the viscosity of the gas is η and the radius of the tube r_0 the value of Q due to this source of loss alone is $(\frac{3}{2}\omega\rho/\eta)^{\frac{1}{2}}r_0$, about 250 in the present case. The two sources of loss acting together limit Q to 170, which is still high enough to explain how, even by the rather crude device of listening, it was possible to determine resonant frequencies within about 1 Hz.

Similar problems of end-correction and radiation arise with electromagnetic systems, but in detail the differences are considerable. Unlike sound waves, electromagnetic waves cannot be propagated in metal tubes whose cross-section is small compared with the wavelength; the boundary conditions to be satisfied by the electric and magnetic fields do not allow a wave of constant amplitude over the whole cross-section. Typical transmission systems for long waves have at least two conductors, like the parallel wire and coaxial lines of fig. 1. Near the end the lines are distorted, bulging out from the planes normal to the propagation direction, as illustrated in fig. 13. When the wavelength is much longer than the lateral dimensions, the lines take up the same shape as in a charged capacitor of this form. In fact the determination of the end-correction is entirely analogous to the process already described for the acoustic resonator. The boundary conditions are troublesome, however, and there does not seem to exist any systematic compilation of solutions for electromagnetic end-corrections. In practice, of course, it is not hard to determine the natural frequency of a resonator by experiment and thus measure the end-correction directly – its calculation would be an exercise in applied mathematics hardly warranted by the usefulness of the results.

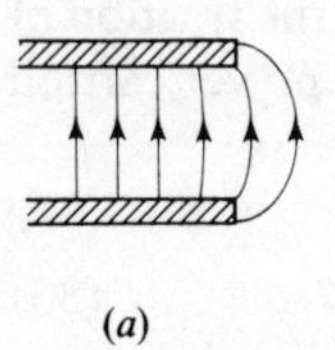

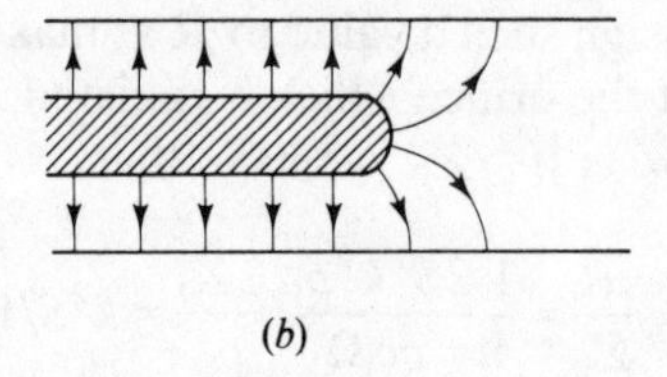

Fig. 13. Spread of field lines from the end of (*a*) a parallel wire transmission line and (*b*) the inner conductor of a coaxial line.

A long coaxial line carrying a standing wave alternates at quarter-cycle intervals between a pure electric field pattern and one that is purely magnetic; this is the analogue in electromagnetism of the exchange between potential and kinetic energy in a mechanical vibrator. The nodes of one field coincide with the antinodes of the other. If for one moment we ignore end-corrections, there can be no magnetic field at an open circuit, for such a field demands longitudinal currents in the conductors, but there is no objection to an electric field at this point. The field configurations are therefore as shown in fig. 14;

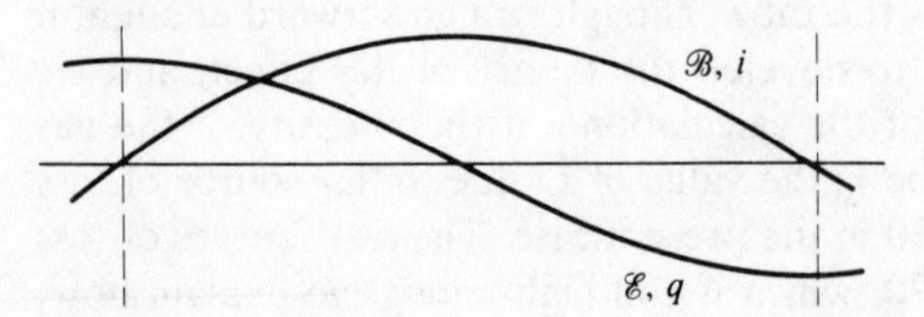

Fig. 14. Magnetic field and current, and electric field and surface charge density in one half-wavelength (marked by broken lines) of a standing wave on a long line.

the charge density on the surface of the conductors follows $\mathscr{E}$ and the current follows $\mathscr{B}$. The end-correction arises from the extra concentration of electric field lines, and therefore of surface charge, near the ends, as shown in fig. 15. In oscillation the extra charge must be alternately supplied and removed by the current, which does not fall to zero as a sinusoid but remains rather large until very near the end. The regions away from the end are thus fitted by a rather longer wavelength, and the resonant frequency is correspondingly reduced. For example, a resonator of this configuration, with outer and inner diameters of 14 and 1 mm, and an inner conductor 13 mm long, was found to have a fundamental resonance corresponding to a free space wavelength of 32 mm, rather than 26; the end-correction amounted to $1\frac{1}{2}$ mm at each end.

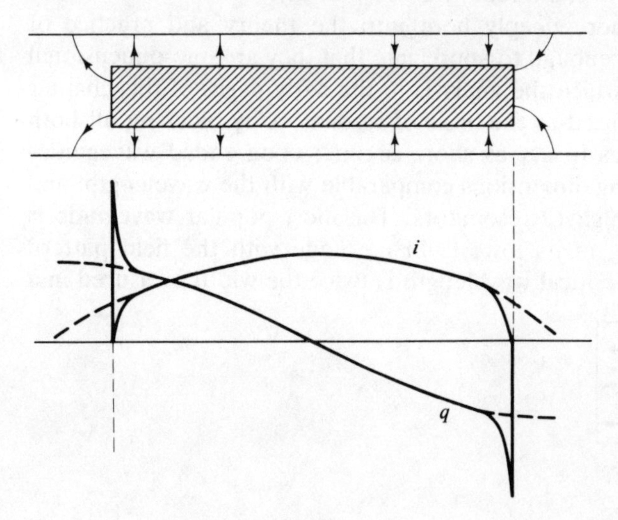

Fig. 15. Field lines, current and charge density on a short inner conductor of a coaxial line, to show origin of end-correction:

With the outer conductor extended in this way, no radiation is possible until the wavelength is reduced to the point where the outer conductor alone can carry propagating waves (waveguide modes).[1] In a tube much larger than the wavelength there are many different modes of waveguide propagation, each with its own critical free-space wavelength, λ_c. By λ_c we mean that the mode in question cannot be propagated at a frequency less than c/λ_c. The most readily propagated mode for a circular waveguide is the TE_{11} mode shown in fig. 16(*a*), for which λ_c is 3.41 times the radius of the tube. This is not one that can be excited by a perfectly centred coaxial line, for reasons of symmetry; in the coaxial mode $\boldsymbol{\mathscr{E}}(-\boldsymbol{r})=-\boldsymbol{\mathscr{E}}(\boldsymbol{r})$, while in the TE_{11} mode $\boldsymbol{\mathscr{E}}(-\boldsymbol{r})=\boldsymbol{\mathscr{E}}(\boldsymbol{r})$. The lowest mode that can be excited by the coaxial line is the TM_{01} of fig. 16(*b*), which has axial symmetry and a radial electric field to match that of the coaxial line. The critical wavelength for this mode is 2.61 times the radius; there was therefore no possibility of exciting it beyond the end of the inner conductor in the tube of 7 mm radius just mentioned, at a resonant frequency corresponding to a free

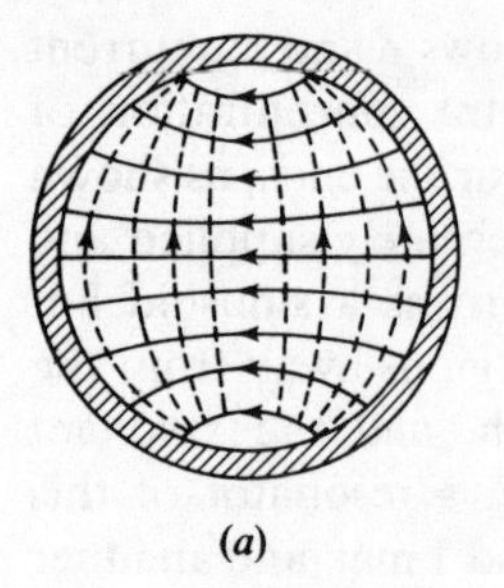

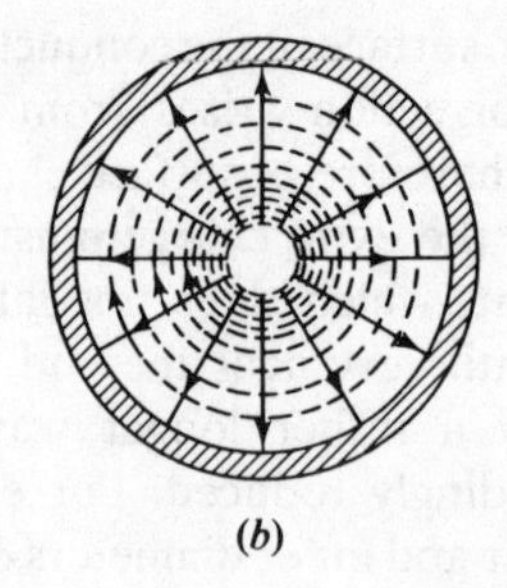

Fig. 16. Electric (full) and magnetic (broken) field lines in the transverse plane of a circular waveguide carrying (*a*) the TE_{11} mode and (*b*) the TM_{01} mode.

space wavelength of 32 mm; this resonator would be free of radiation losses until the free-space wavelength was reduced to 18 mm (or 25 mm if it was asymmetrical and could radiate a TE_{11} mode).

We need go no more deeply here into the theory and practice of waveguides, since it is enough to appreciate that they are one-dimensional wave propagators to which the analysis in the early pages of this chapter applies. They can be used as the basis of resonators by blocking off both ends with metal plates to act as short circuits; open-ended waveguides radiate strongly, having dimensions comparable with the wavelength, and are not suitable for high-Q resonators. The most popular waveguide is rectangular, operating in its lowest, TE_{10}, mode with the field pattern drawn in fig. 17. The critical wavelength is twice the width measured in a

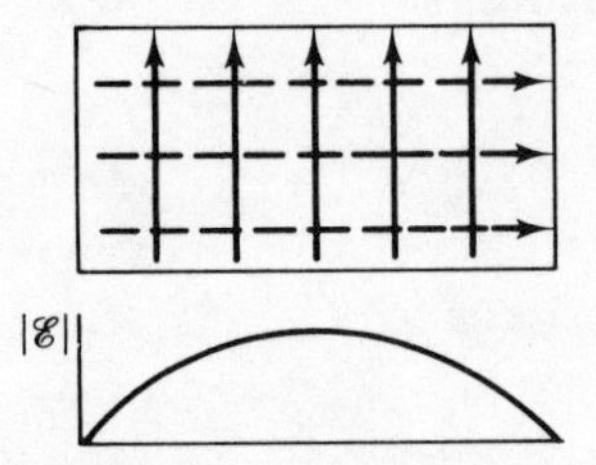

Fig. 17. Electric (full) and magnetic (broken) field lines in the transverse plane of a rectangular waveguide. The electrical field strength varies sinusoidally across the longer dimension as indicated below.

direction normal to $\mathscr{E}$. The mechanical simplicity of waveguides at microwave frequencies ($\lambda < 20$ cm, say) makes them the obvious choice for illustrating various arrangements in which resonators interact with waves.

Resonators attached to transmission lines

Typical arrangements are shown schematically in fig. 18. In each case the input transmission line, on the left, is imagined to contain a matched attenuating section, so that the wave reflected from the resonator disappears without trace. Similarly, in the cases (*c*)–(*f*) where there is an output line on the right, this is imagined to be matched. The three arrangements (*b*), (*d*) and (*f*) parallel (*a*), (*c*) and (*e*), the former being waveguide devices in which the resonant section is defined by diaphragms pierced with coupling holes, while the latter have lumped resonant circuits. There is no essential difference, apart from the lumped circuits having only one resonant frequency. It is the external features that concern us, and the

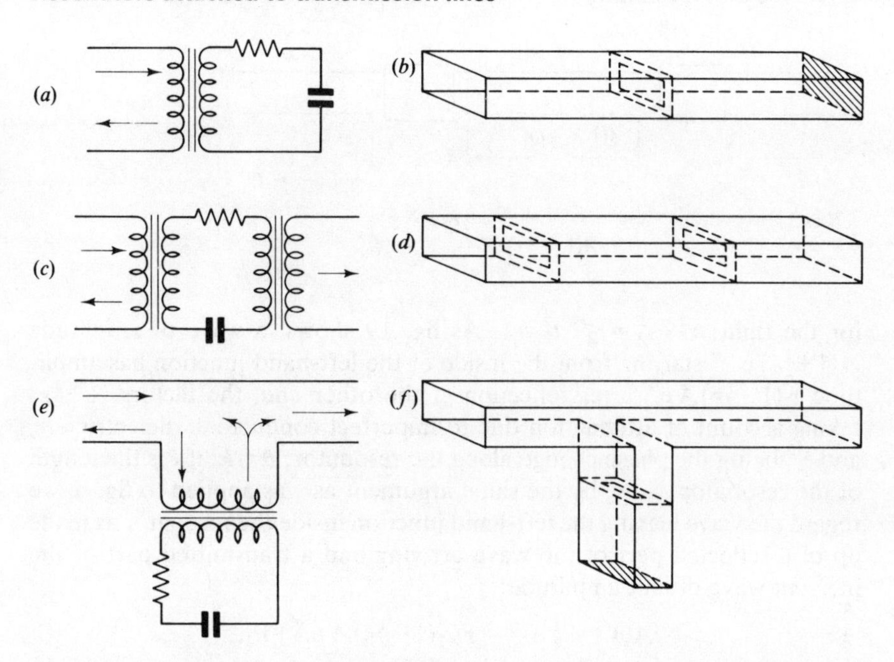

Fig. 18. Resonant circuits coupled to transmission lines, and their waveguide equivalents. A pierced diaphragm in a waveguide allows partial transmission and partial reflection, and serves to couple a resonant cavity to the waveguide; (*a*) and (*b*) line terminated by resonator; (*c*) and (*d*) feed-through resonator; (*e*) and (*f*) side arm terminated by resonator.

influence they have on the form of the resonance. The impulse response may be cited to indicate the differences due to the manner of excitation and detection. In (*c*), for instance, an incident impulse starts the resonator ringing, and what emerges on the output line is just a decaying oscillation; this arrangement has properties similar to those discussed in chapter 6. In (*a*), however, there is a reflected impulse followed by the oscillation, while in (*e*) there is a direct impulse as well as a later impulse and oscillation.

129

At this point we abandon the use of the impulse response in favour of the conventional approach by way of the response to a single-frequency input. The difficulty of handling the impulse response is particularly marked for the transmission line resonators, which have a large number of resonances at frequencies above the fundamental which we may suppose we are interested in. Instead of being excited to sinusoidal oscillation, the resonator may be thought of as reflecting the impulse to and fro, emitting a small fraction of its energy at each reflection. In general the impulse suffers progressive distortion on reflection, so that the emerging signal is a succession of pulses of gradually changing form. In principle, Fourier analysis of this output would tell one about all the resonances in one operation, but the game is not worth the candle.

Let us consider first the simplest case, (*c*) and (*d*), where the output signal is due entirely to the excitation of the resonator and is not mixed with any part of the input. We represent the resonator as a length of line connected by lossfree transducers to the input and output lines, and as in fig. 5 measure amplitudes at those points where the reflection coefficients are real and the transmission coefficients, as we now know, imaginary. We take these coefficients to be r_1 and $\mathrm{i}t_1$ for the left-hand junction, r_2 and $\mathrm{i}t_2$

169

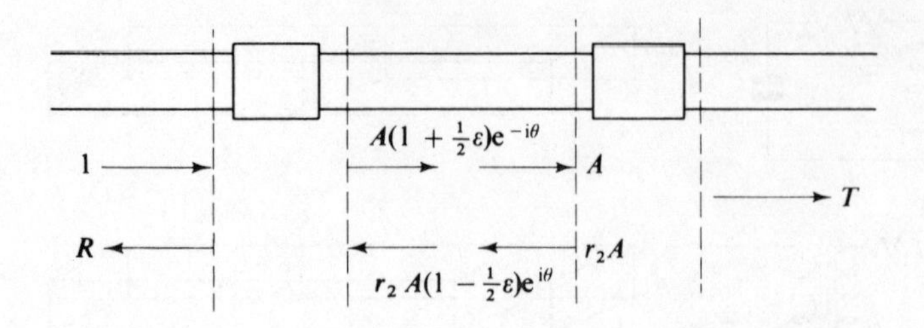

Fig. 19. Notation used in analysis of feed-through resonator.

for the right; $r_1^2+t_1^2=r_2^2+t_2^2=1$. As fig. 19 shows, a wave of amplitude $A(1+\frac{1}{2}\varepsilon)\,e^{-i\theta}$ starting from the inside of the left-hand junction has amplitude $r_2(1-\frac{1}{2}\varepsilon)A\,e^{i\theta}$ after reflection at the other end, the factors $(1\pm\frac{1}{2}\varepsilon)$ taking account of attenuation due to imperfect conductors, dielectric etc, and $e^{i\theta}$ being the phase change along the resonator; $\theta=kl$ if l is the length of the resonator. Now, by the same argument as was applied to fig. 6, we regard the wave leaving the left-hand junction inside the resonator as made up of a reflected part of the wave arriving and a transmitted part of the incident wave of unit amplitude:

$$A(1+\tfrac{1}{2}\varepsilon)\,e^{-i\theta}=r_1r_2(1-\tfrac{1}{2}\varepsilon)A\,e^{i\theta}+it_1.$$

Hence A is determined:

$$A=it_1/[(1+\tfrac{1}{2}\varepsilon)\,e^{-i\theta}-r_1r_2(1-\tfrac{1}{2}\varepsilon)\,e^{i\theta}],\tag{34}$$

and
$$T=it_2A.$$

If we confine attention to high-Q resonators, $\varepsilon\ll 1$ and $r_{1,2}\sim 1-\frac{1}{2}t_{1,2}^2$, so that

$$T\sim -t_1t_2\,e^{i\theta}/[1-(1-\alpha)\,e^{2i\theta}],$$

where $\alpha=\varepsilon+\frac{1}{2}(t_1^2+t_2^2)$. The nth resonant peak occurs when $\theta=n\pi$ and the denominator takes the very small value α. Only small changes in θ are needed to increase this substantially, so that to a good approximation

$$T\sim \pm t_1t_2/(\alpha-2i\,\Delta\theta),\tag{35}$$

which, as expected, is of the standard Lorentzian form. If a square-law detector is used to measure the power transmitted through the resonator, **76**

$$TT^*=t_1^2t_2^2/[\alpha^2+4(\Delta\theta)^2],\tag{36}$$

which falls to half its peak when $\Delta\theta=\pm\frac{1}{2}\alpha$. To traverse the peak at half power involves a relative change in θ, and hence in frequency, of $\alpha/n\pi$; therefore

$$Q=n\pi/\alpha.\tag{37}$$

136

This is what one would expect from the natural decrement of a standing wave excited in the resonator. In going once through the resonator in both directions a travelling wave loses a fraction ε of its amplitude and 2ε of its energy by attenuation, as well as a fraction $t_1^2+t_2^2$ into the coupling lines, making 2α in all. Since during this journey the phase of oscillation advances by $2n\pi$ radians, the loss per radian, which is $1/Q$, is $\alpha/n\pi$.

There is a relationship between the transmission through the resonator and the contribution of coupling losses to the width. According to (35), the amplitude transmission coefficient, ζ, the value of $|T|$ at resonance, is $t_1 t_2/\alpha$. We define Q_0 as the unloaded Q of the resonator, $n\pi/\varepsilon$, when coupling to the external circuits is negligible. Then it follows immediately that

$$Q/Q_0 + \tfrac{1}{2}(t_1/t_2 + t_2/t_1)\zeta = 1. \tag{38}$$

If, therefore, the coupling diaphragms are systematically changed in such a way that the ratio t_2/t_1 stays constant (it may be easiest to achieve this by taking pains that the holes in the diaphragms are identical), Q will vary linearly with ζ, and by extrapolating the line to $\zeta = 0$ one can determine the value of Q due to the resonator losses alone. The resonators shown in fig. 20 are designed to allow t_1 and t_2 to be varied systematically while keeping their ratio constant. In both cases the resonant elements are wires; the

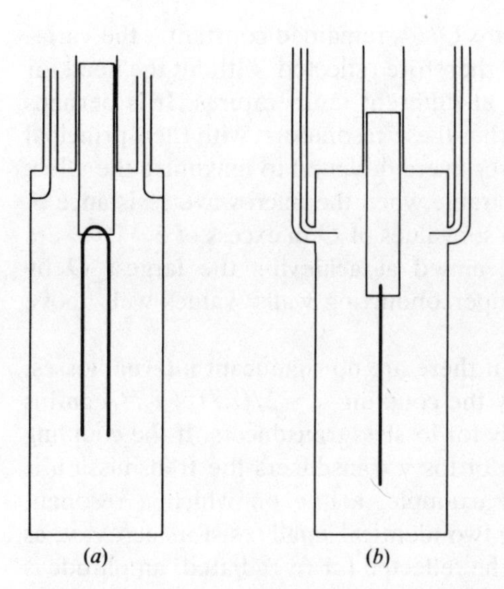

Fig. 20. Two examples of resonant wires coupled in feed-through fashion to coaxial lines. In both cases the total length of wire is a little less than half the free-space wavelength at the resonant frequency. Coupling is altered by raising or lowering the wire.

hairpin in (a) is a parallel wire transmission line open at the lower end and short-circuited at the upper, with magnetic coupling to the external coaxial lines which can be varied by moving the hairpin vertically. Similarly the coaxial resonator (b) is adjustable by vertical movement, but the coupling is electrical. An example of the verification of (38) is given in fig. 21. These resonators were used at very low temperatures to study the surface resistance of superconductors at microwave frequencies, and as the sample (the movable wire) was cooled below its superconducting transition temperature the resistance fell rapidly and Q rose to very high values.[5] By adjusting the coupling to keep the fraction of power transmitted at

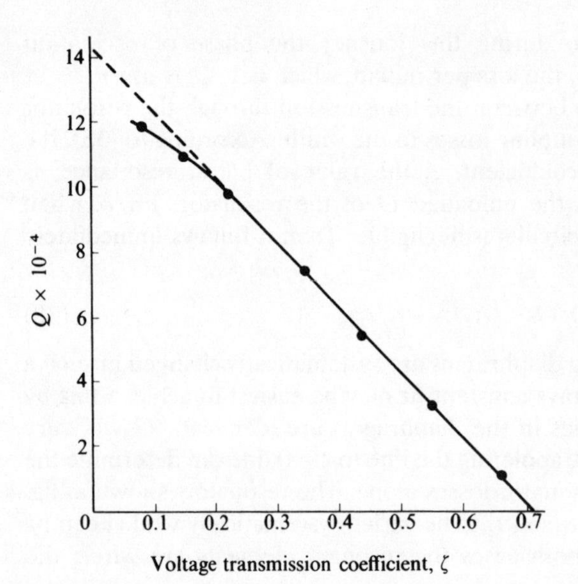

Fig. 21. Variation of Q with transmission coefficient ζ, when coupling is altered. This curve, obtained with the resonator of fig. 20(a), shows a slight departure from linearity at weak coupling, due to direct coupling between the loops. With better design the fault is easily reduced to insignificance.

resonance constant, ζ and therefore Q/Q_0 remained constant – the variations of the measured values of Q therefore reflected, without the need for correction, relative values of Q_0 at different temperatures. It is perhaps worth remarking, in parenthesis, that these resonators, with their principal conductors in the form of thin wires, were designed to maximize the effect of resistive losses, as is highly desirable when the microwave resistance of superconductors is so small. Even so, values of Q in excess of 5×10^5 were recorded. In other experiments, aimed at achieving the largest Q by building resonant cavities with superconducting walls, values well above 10^{10} have been reached.[(6)]

One implication of (38) is that if there are no significant internal losses, so that Q is determined solely by the coupling, $\zeta=2/(t_1/t_2+t_2/t_1)$ and is unity if $t_1=t_2$. But this is true only for lossless transducers. If the coupling to the external circuits is by way of lossy transducers the transmission is seriously reduced. Consider, for example, a line on which a resonant element is delineated by shunting two identical small resistors across it, as in fig. 22. From (7) we have that the reflected (or re-radiated) amplitude is $2\gamma-1$, if $\gamma\ll1$, and the transmitted amplitude must therefore be 2γ; r and t now are both real, instead of one being real and the other imaginary, as before. The calculation can be carried through in the same way as that leading to (35) and the answer is essentially the same, except that we can

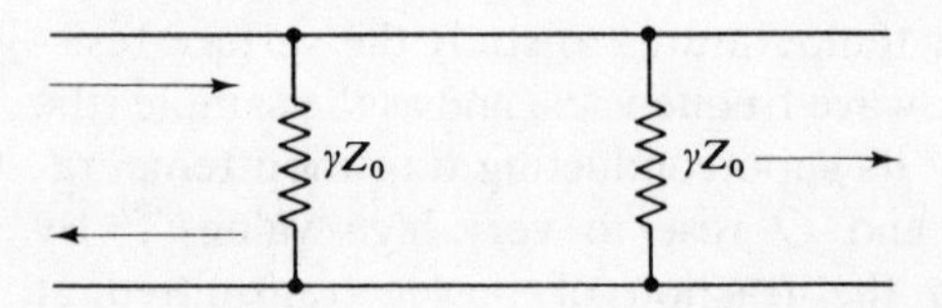

Fig. 22. Resonant length of line delineated by small shunt resistances, modelling the Fabry–Perot interferometer.

no longer take $r^2 + t^2 = 1$. The value of Q is still determined by the internal losses and by the loss on reflection, so that if we write $\alpha = \varepsilon + t'^2$ we must remember that t'^2 is the power not reflected, i.e. $1 - r^2$, but this includes both transducer loss and power into the output line. On the other hand, the factor $t_1^2 t_2^2$ in the numerator of (36) genuinely refers to the power entering the line and excludes what is lost in the resistors. For a given value of r, and therefore of α, the numerator of (36) is $(1 - r^2)^2$ for lossless transducers and $(1 - r)^4$ for resistors acting as transducers. The latter is smaller than the former by a factor $\frac{1}{4}(1 - r)^2$ if r is nearly unity, representing a very significant diminution in transmitted intensity, with the transducers absorbing almost all of what ideally they should transmit.

This presents a real problem in the design of reflecting coatings for optical interferometers. A Fabry–Perot interferometer, in which circular fringes are seen in transmission after multiple reflection between two optical flats, is an instrument governed by the same rules as we have been discussing.[7] A high value of Q is achieved in this case by using a spacing of many (perhaps 10^4–10^5) wavelengths between the optical flats (the transducers). With so large a value of n in (37), α need not be very small to obtain the resolution that a Q of, say, 10^6 confers. But if α is to be made $\frac{1}{10}$ by means of a resistive coating, such as a thin metal film may approximate to, the resulting transmission at resonance – the centres of the bright interference bands – is only 1/1600 of the incident power. Good approximations to lossless reflecting coatings can be achieved by superimposing layers, about one-quarter wavelength thick, of transparent materials with alternately high and low refractive index. By such means 99% reflection is possible with almost all the remaining 1% of power transmitted without absorption.

Because the high Q-value in a Fabry–Perot interferometer is commonly realized through a large value of n rather than almost perfect reflectivity, the resonant modes lies close together and their responses overlap. To return to (34) rather than the approximation (35), putting $\varepsilon = 0$ and $r_1 = r_2 = r$, we have

$$A = \mathrm{i}t/(\mathrm{e}^{-\mathrm{i}\theta} - r^2\,\mathrm{e}^{\mathrm{i}\theta}),$$

so that

$$AA^* = t^2/(1 + r^4 - 2r^2 \cos 2\theta). \tag{39}$$

The denominator is represented geometrically by the square of the distance D between a point P sweeping round a circle of radius r^2 and a fixed point O at unit distance from the centre of the circle, as in fig. 23. The difference between this and the Lorentzian resonance is that in the latter the circle is replaced by its tangent at the nearest point. If r^2 is close to unity the difference is negligible except far from the peak where the response is relatively very weak. The form of (39) when $1 - r^2$ is not so small is shown in fig. 24. For low reflectivity the curve approximates to a sinusoid. Resolution of the resonances is never totally lost, but the contrast between maxima and minima is weakened.

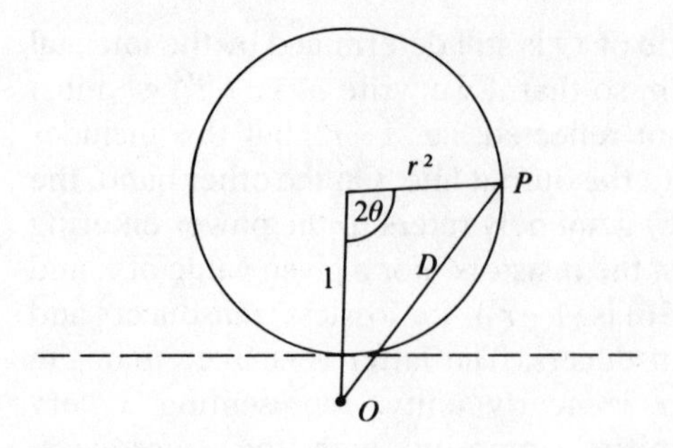

Fig. 23. Geometrical construction to find variation of transmitted intensity with frequency in the arrangement of fig. 22, according to (39).

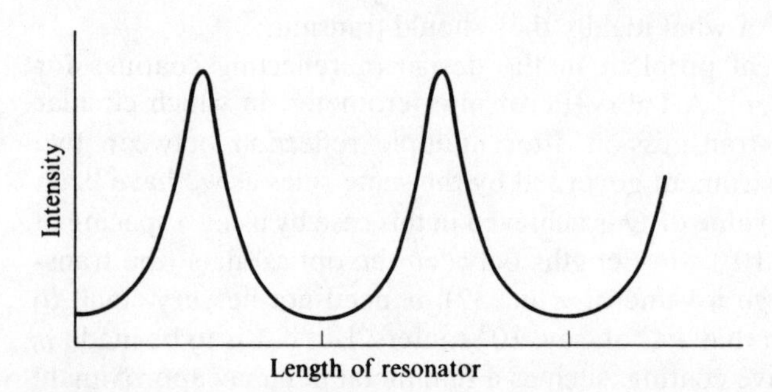

Fig. 24. Line-shape of transmitted intensity derived from (39) and fig. 23 with $r^2 = \frac{1}{2}$.

We have talked of resonances since this was the language with which the discussion started, but it will be realized that the conventional description of the variations of intensity is in terms of interference fringes. It is easy enough to make the logical connection between the two, by what may however appear an artificial device. In the optical interferometer θ is changed by altering not the frequency but the angle of incidence of a plane wave on the plates. The artificial device consists of regarding the wave incident at an angle to the axis as a wave travelling along the axis and having a variation of phase across its wavefront. The plane wave represented by $\xi = \mathrm{e}^{\mathrm{i}(\mathbf{k}\cdot\mathbf{r}-\omega t)}$ is to be thought of as moving along the z-axis:

$$\xi = \xi_0(x, y)\, \mathrm{e}^{\mathrm{i}(k_z z - \omega t)},$$

with its wavefront modulated according to the variations of $\xi_0(x, y)$,

where $$\xi_0(x, y) = \mathrm{e}^{\mathrm{i}(k_x x + k_y y)}. \tag{40}$$

The utility of this device depends on the reflecting surfaces being plane and lying in the (x, y)-plane, causing no change to the form of $\xi_0(x, y)$ – it is only the motion in the z-direction that is changed by reflection, so that in this way oblique waves can be brought into the ambit of our treatment of one-dimensional systems. The argument is illustrated in fig. 25. The true wavelength is λ, but measured along z the 'longitudinal wavelength' is greater, $\lambda_\ell = \lambda \sec\phi$, while the phase variation in the (x, y)-plane is governed by a 'transverse wavelength' $\lambda_t = \lambda \operatorname{cosec}\phi$. Changing ϕ for a wave of given frequency does not alter its real velocity, but its velocity along z is greater, being $c \operatorname{cosec}\phi$. Because of this (put otherwise, because

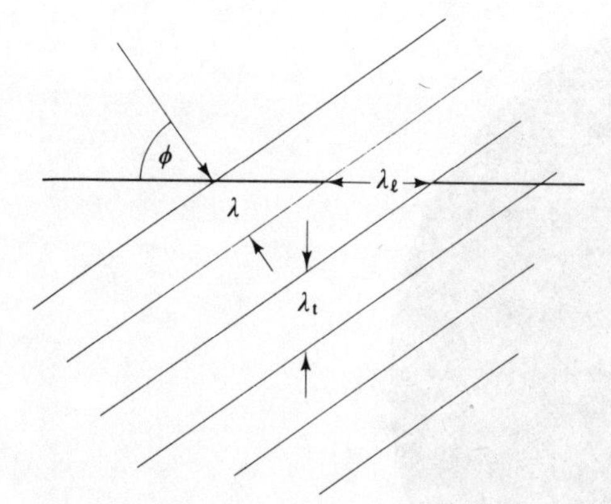

Fig. 25. Obliquely incident plane wave, to illustrate meaning of 'longitudinal wavelength', λ_ℓ, and 'transverse wavelength', λ_t.

the phase θ in (39) is $k_z l$, where l is the distance between the plates) θ can be changed by changing ϕ rather than ω. Hence fig. 24 is also a representation of the variation of intensity with angle when monochromatic light is allowed to fall on the interferometer from all directions.

To conclude this argument, let us note that one way of achieving omnidirectional illumination is to bring near the interferometer an extended monochromatic source (flame, discharge tube etc.). Since every point on the source is an independent emitter, the amplitude distribution in the (x, y)-plane is described by a function $\xi'(x, y)\, e^{-i\omega t}$ whose phase is random, however uniform the strength of the source may be. As a consequence of this randomness of phase, if we Fourier-analyse the spatial variation of amplitude we shall obtain a uniform spectrum. Now the functions ξ_0, as in (40), are nothing else but the basis functions of this Fourier transformation, and we therefore expect all k_x and k_y to appear with equal intensity. The eye or camera, focussed at infinity, channels each Fourier coefficient to a different point on the retina or photographic plate, and thus shows in one single image the response of fig. 24, as a set of interference rings. Fig. 26 is just such a photograph, taken with an exceptionally well-coated pair of plates, so that the bright fringes are very sharp. Even if the appearance of the rings left the matter in doubt, conclusive evidence for the enhanced sharpness of the fringes comes from the fact that alternate rings are formed by two different spectral lines. Sinusoidal fringes, as produced by the Michelson interferometer, are incapable of this resolution; two sinusoids added together produce a single resultant sinusoid.

Let us turn to the next-simplest arrangement, fig. 18(*a*) or (*b*), in which the response of the resonator is revealed by variations in the reflected wave amplitude or power. We need not delay to understand the technicalities of separating the reflected from the incident wave – it is enough to know that there are commercially available devices for so doing. The analysis of case (*b*) can be taken up directly from that of (*d*), as expressed in (34). After making the same approximations as led to (35) and which are adequate for

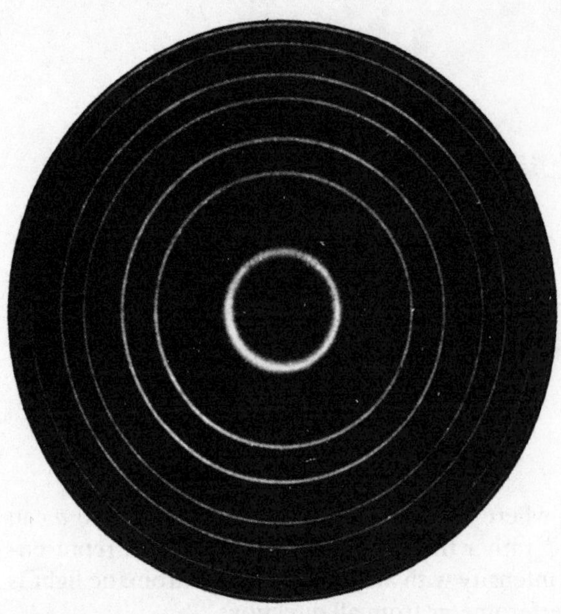

Fig. 26. Photograph of circular Fabry–Perot fringes obtained with highly reflecting surfaces, to show very narrow transmission peaks.[8]

high-Q resonators, we have that

$$A \sim \mathrm{i}t/(\alpha - 2\mathrm{i}\,\Delta\theta),$$

where $\alpha = \varepsilon + \frac{1}{2}t^2$, there being now only one external circuit to provide coupling loss. This value of A may also without extra error be taken as the amplitude of the wave falling on the transducer from inside, and hence

$$R = r + \mathrm{i}tA \sim 1 - t^2/(\alpha - 2\mathrm{i}\,\Delta\theta). \quad (41)$$

Well away from resonance there is virtually total reflection from the transducer, and resonance reveals itself as a drop in the reflected power. At the very centre of the resonance, when $\Delta\theta = 0$,

$$R_{\mathrm{min}} = 1 - t^2/\alpha.$$

If $\frac{1}{2}t^2$ is adjusted to equal ε, so that the resonant width is equally due to internal dissipation and to coupling losses, $R_{\mathrm{min}} = 0$. For stronger or weaker coupling the reflected power does not fall to zero, but otherwise the shape is always the same. With a square-law detector the measured output is given by

$$RR^* = 1 - 2\varepsilon t^2/[\alpha^2 + 4(\Delta\theta)^2],$$

being a Lorentzian resonance subtracted from a constant, unity, as in fig. 27.

It is with the arrangement of fig. 18(e) and (f) that the form of the resonance becomes less attractive. There is no need to analyse the general case, in which a wave arriving at the junction from any of the three arms is partially reflected and partitioned between the two other according to the impedances they present. Instead, we shall suppose that the side arm, terminated by the resonator, has a characteristic impedance twice that of

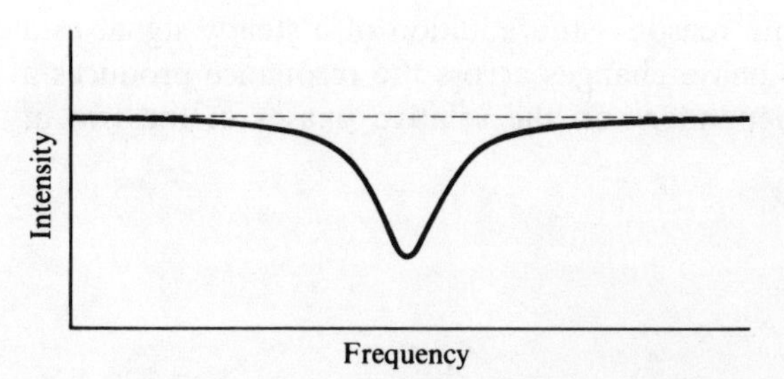

Fig. 27. Absorption resonance observed for reflected wave in arrangement of fig. 18(a) or (b).

the other two, which are equal. If the junction places all lines in series, in a ring, a wave incident from the side arm will see the sum of the two other characteristic impedances and will suffer no reflection, sharing its power equally between them. On the other hand a wave incident from the input line sees three times its characteristic impedance. According to (7), $r=\frac{1}{2}$, so that a quarter of the input power is reflected, while a further quarter carries straight on and the remaining half goes into the side arm, thus dividing the transmitted power in proportion to the impedances as is proper for resistances in series. If the incident wave has unit amplitude, that entering the side arm has amplitude $1/\sqrt{2}$, and is reflected as $R/\sqrt{2}$ according to (41). On dividing into two parts at the junction, an amplitude $\frac{1}{2}R\,\mathrm{e}^{\mathrm{i}\phi}$ enters the output line, the phase factor $\mathrm{e}^{\mathrm{i}\phi}$ being inserted to take account of the length of the side arm. The original wave that went straight past the junction has amplitude $\frac{1}{2}$, and combines with that from the side arm to give an output amplitude of the form:

$$T=\tfrac{1}{2}(1+R\,\mathrm{e}^{\mathrm{i}\phi}). \tag{42}$$

The behaviour of T is most readily followed diagrammatically. First we write R in the form, derived from (41),

$$R=1-C/(1-2\mathrm{i}Q\,\Delta\omega/\omega_0), \tag{43}$$

in which $C=t^2/\alpha$ and can take any value between 0 and 2. On an Argand diagram such as those in fig. 28 the second term in (43) is represented by AP, the chord of a circle of radius $\frac{1}{2}C$; since OA is equal to unity, OP represents R. As the point P passes quickly through B, at resonance, the characteristic dip in the reflected amplitude occurs. When the resonator terminates a side arm, (42) shows how R and unity are to be compounded with a phase difference depending on the length of the arm. Being concerned only with the amplitude, and not the phase, of T, we shall construct $2T\,\mathrm{e}^{-\mathrm{i}\phi}$ which involves adding the unit vector $\mathrm{e}^{-\mathrm{i}\phi}$ to R as given by (43). If Q is a point on the unit circle in fig. 28, representing $-\mathrm{e}^{-\mathrm{i}\phi}$, the output $|T|$ is proportional to the length of the vector QP. As drawn in fig. 28(a), $|T|$, which is QA well below the resonant frequency, rises with frequency to a peak when P is on the opposite side of the circle to Q, then falls and continues to fall past the resonant point B, reaching a minimum at a frequency above resonance and then reverting to the initial value. This distorted resonance is reminiscent of the case we examined in chapter 6 (fig. 6.8), and for the same reason – the addition of a steady signal to a resonant response whose phase changes across the resonance produces a resultant that is highly dependent on the relative phases of the two at resonance.

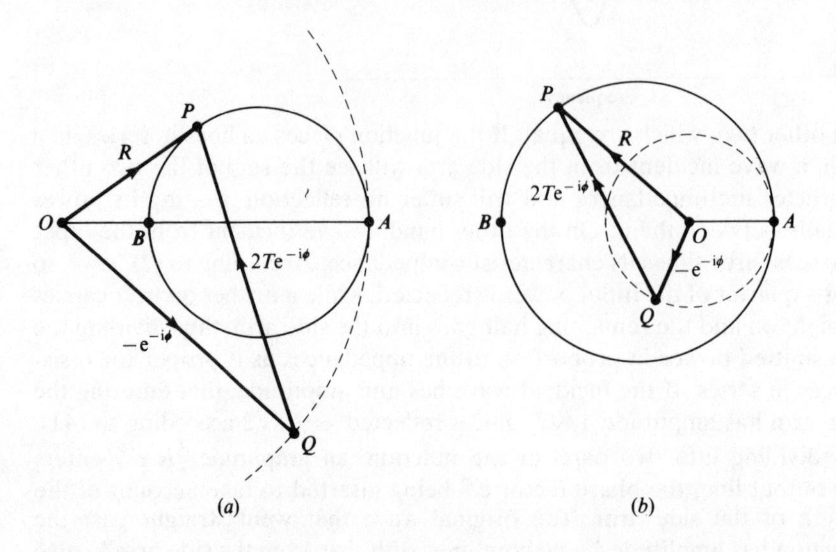

Fig. 28. Construction for transmitted amplitude in arrangement of fig. 18(e) or (f), according to (42) and (43). The full circles have radius $\frac{1}{2}C$, and the broken circles unit radius. In (a) $C < 1$ (weak coupling), and in (b) $C > 1$ (strong coupling).

Symmetrical responses with this arrangement may be obtained by ensuring that Q lies on the line AB. If Q coincides with A, there is virtually no output except near resonance; this is because it has been arranged so that when the resonator is perfectly reflecting, the length of the side arm is $\lambda/4$ (or an odd number of quarter-wavelengths) and acts as an open circuit at the junction, causing total reflection of the input wave. As the resonator is excited, near resonance, the barrier becomes less than perfect, and what is transmitted is proportional to AP, a Lorentzian resonance just as would be obtained with the arrangement of fig. 18(a) and (b). By making the side arm of zero length or an integral number of half-wavelengths, Q is caused to fall on the opposite side of its circle from A. Off-resonance the side arm now behaves like a continuous metal sheet closing the hole and allowing the wave to carry on unreflected. Near resonance the response is essentially the same as R in (41), with a resonant dip in the signal. This is the configuration normally used for absorption wavemeters; a calibrated adjustable length of waveguide attached to the side of a transmission section of a microwave circuit provides an easy way of measuring frequency and, since Q-values of 10^4 are fairly easy to reach, a precision of at least 1 in 10^6 can be reached if needed. The mechanism for tuning the wavemeter, usually by moving a reflecting termination at one end of the cavity, presents design problems, about which much detailed information is available.[9]

Excitation of a resonator by a plane wave

To illustrate the most elementary form of what has been developed into a powerful general formalism, especially important for describing in wave mechanics the scattering of particles by atoms and nuclei, consider a Helmholtz resonator held in mid-air and excited by a plane sound wave.[10] 22 The merit of this example lies in the sound wave being a scalar wave and much simpler to handle than the vector electromagnetic wave, and in the availability of real resonators much smaller than their natural wavelength; indeed the analysis closely follows the way in which such a resonator is (or was) used in practice.

The resonator interacts with the plane wave by virtue of the exaggerated motion of the air through its orifice near resonance. If the orifice is sealed the resonator is no more than a small passive obstacle, scattering a little of the sound wave but so little as to be negligible. The wave field can therefore be understood as composed of the undisturbed plane wave on which is superposed any wave radiated by the orifice – a picture of the process exactly parallel with our treatment earlier of an obstacle on a transmission line. 166 In the present example we know that the oscillatory air-stream from the orifice will have spread evenly, to become an isotropic radial pulsation, before it has travelled anything like a wavelength away from the resonator; thus the only sort of wave that is radiated is spherically symmetrical. This would not be so if the dimensions of the resonator were comparable with a wavelength, and it is in this situation that the more general treatment mentioned above is brought into play. But this belongs more to the physics of wave-motion than of vibrators and we shall go no further than the simplest case where the re-radiated wave is isotropic. The response of the resonator is then defined by a single complex parameter, giving the amplitude and phase of the re-radiated wave relative to that of the incident plane wave. We shall take the problem in two steps. First, writing down the pressure variation with position and time when spherical and plane waves are superimposed, we shall see how the relative phase and amplitude affects the flow of energy. Then we shall relate the re-radiated and incident waves to the variations of pressure and velocity in the vicinity of the Helmholtz resonator itself.

Fig. 29(*a*) shows plane and spherical wavefronts, and it is immediately obvious how important is the phase relationship. As shown, they are in antiphase in the forward direction, so that the amplitude of vibration near the axis is reduced – fig. 29(*b*) is a graph of amplitude on the plane marked in the diagram, on the assumption that the spherical wave has one-tenth of the amplitude of the plane wave at the centre. If x denotes distance from the axis, and ϕ the energy flux per unit area, the total flux across the plane is $\int_0^\infty \phi \cdot 2\pi x \, \mathrm{d}x$ or $\pi \int_0^\infty \phi \, \mathrm{d}(x^2)$. It is easier to see the effect of the scattered wave by plotting the square of the amplitude against x^2, as in fig. 29(*c*). It now appears that with this phase relationship the oscillations of flux around the mean give alternately equal positive and negative contributions to the

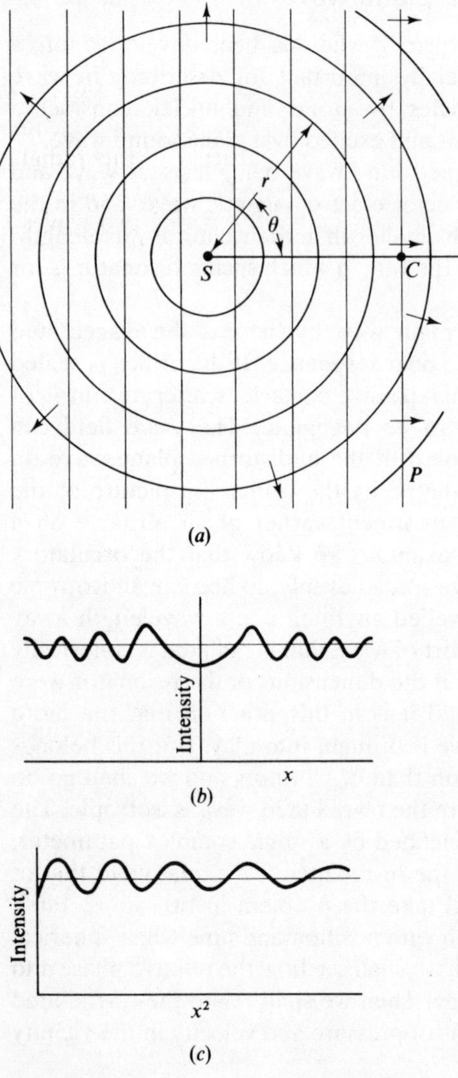

Fig. 29. (*a*) A spherical wave superposed on a plane wave, in antiphase in the forward direction; (*b*) variation of intensity across the plane *P* when the spherical wave has an amplitude at the centre *C* one-tenth that of the plane wave; (*c*) the same plotted against x^2, x being measured from *C*.

integral, with an odd half thrown in at the beginning, so that the resultant effect of the oscillations is zero, provided their amplitude decays steadily with increasing x. On the other hand, if the crests of the spherical wave lie a quarter-wavelength behind those of the plane wave, the resultant will be negative and equal to half the first loop. This is the sort of phase relationship that results when the scattering object is lossy. Once the amplitude and phase of the scattered wave have been chosen, the loss of energy flux in the ongoing travelling wave, and the energy flux in the back-scattered spherical wave, can both be calculated; thus the total dissipation by the scattering object is in principle determined as what is needed to satisfy the energy balance. This means that for a given obstacle the amplitude and phase of the scattered wave are not arbitrarily disposable. In particular, if the obstacle is loss-free a unique relationship must exist between amplitude and phase. To make the argument quantitative it is convenient to start afresh and to calculate the energy flux outwards through the surface of a sphere of radius r, centred on the obstacle. For this purpose we need expressions for the pressure and velocity fields, particularly the radial

57 velocity v_r since Φ, the flux per unit area of the sphere, is $\frac{1}{2}\,\mathrm{Re}\,[pv_r^*]$.

In terms of the polar coordinates defined by fig. 29(a), the two waves together produce pressure variations of the form

$$p(r, \theta, t) = \left(p_0\,\mathrm{e}^{\mathrm{i}kz} + \frac{a}{r}\,\mathrm{e}^{\mathrm{i}kr}\right)\mathrm{e}^{-\mathrm{i}\omega t}, \quad \text{where } z = r\cos\theta. \tag{44}$$

For the second term of (44), the spherical wave, integration of (26) gives ξ and hence v, which is $-\mathrm{i}\omega\xi$:

$$v_{\mathrm{sph}} = \frac{a}{Z_0 r}\left(1 + \frac{\mathrm{i}}{kr}\right)\mathrm{e}^{\mathrm{i}kr}, \quad \text{where } Z_0 = \rho c_1. \tag{45}$$

The corresponding expression for the plane wave takes the form:

$$v_{\mathrm{pl}} = (p_0/Z_0)\,\mathrm{e}^{\mathrm{i}kz}. \tag{46}$$

Note that the term $\mathrm{e}^{-\mathrm{i}\omega t}$ has been suppressed, according to convention, in (45) and (46). The radial component of velocity is $v_{\mathrm{sph}} + v_{\mathrm{pl}}\cos\theta$, so that

$$v_r = \frac{1}{Z_0}\left[\frac{a}{r}\left(1 + \frac{\mathrm{i}}{kr}\right)\mathrm{e}^{\mathrm{i}kr} + p_0\cos\theta\,\mathrm{e}^{\mathrm{i}kr\cos\theta}\right]. \tag{47}$$

For the total flux, P, through a sphere of radius r, we need to evaluate $\int_0^\pi 2\pi r^2\Phi\sin\theta\,\mathrm{d}\theta$:

$$P = \frac{\pi r^2}{Z_0}\int_{-1}^{1}\mathrm{Re}\left[p_0^2\zeta + \frac{aa^*}{r^2}\left(1 - \frac{\mathrm{i}}{kr}\right) + \frac{a^* p_0}{r}\left(1 - \frac{\mathrm{i}}{kr}\right)\mathrm{e}^{-\mathrm{i}kr(1-\zeta)} + \frac{ap_0}{r}\zeta\,\mathrm{e}^{\mathrm{i}kr(1-\zeta)}\right]\mathrm{d}\zeta,$$

in which ζ has been written for $\cos\theta$. The first two terms are the separate fluxes of the plane and spherical waves, the former vanishing on integration; the last two terms result from the simultaneous presence of both. The integrations are straightforward, yielding the result

$$P = \frac{2\pi}{Z_0}(aa^* - p_0\,\mathrm{Im}\,[a]/k).$$

It is seen from this expression that if the re-radiated wave is weak, the flux of energy is determined by the second term and requires a component of the radial wave in phase quadrature with the incident wave. Since P is the mean flux outwards, $-P$ represents P_{abs} the energy absorbed in the scatterer:

$$P_{abs} = \frac{2\pi p_0^2}{k^2 Z_0}(\text{Im}[s] - ss^*), \tag{48}$$

in which s is written for ka/p_0, being a dimensionless measure of the amplitude of the scattered wave. Since the energy flux in the incident plane wave is $\frac{1}{2}p_0^2/Z_0$, P_{abs} is as much power as is conveyed through an area A_{abs} of wavefront, where

$$A_{abs} = 2P_{abs}Z_0/p_0^2 = \frac{4\pi}{k^2}(\text{Im}[s] - ss^*) = \frac{\pi}{k^2}(1 - |2s - \text{i}|^2). \tag{49}$$

A_{abs} is the *absorption cross-section* of the scatterer, defined in a way that clearly allows it to be carried over to systems other than the particular acoustic resonator we have chosen as an example.

When the scatterer is lossless, but not necessarily weak, the vanishing of (49) forces a relationship between the phase and strength of scattering. For if s is written as $\alpha\, \text{e}^{\text{i}\phi}$, α being real, the condition $P_{abs} = 0$ implies that $\sin\phi = \alpha$. It is only necessary to know $\phi(\omega)$ to know all that matters of the scattering behaviour of a *small* loss-free scatterer. The word *small* must again be emphasized, since obstacles comparable with the wavelength can re-radiate waves that are not spherically symmetrical. In these circumstances each type of wave, described by a particular spherical harmonic, has its characteristic scattering phase $\phi(\omega)$, and the complete set of ϕs constitutes a description of the scattering process.

To confine attention to the Helmholtz resonator scattering only an isotropic wave, we need to fix s by reference to the properties of the actual scatterer. Let us idealize the resonator, replacing the single rather large aperture by a multitude of smaller apertures, spread evenly round the sphere, to give the same resonant frequency but with no awkward hiatus between R, the radius of the resonator, and a spherical shell at some rather larger R' where the flow can be considered to have evened itself out into a radial motion; with the idealized resonator R and R' coincide. The response of the resonator can be written immediately from (6.21) and the zero-frequency compliance. A pressure increment p immediately outside the sphere produces a volume change in the contained air, $\Delta V/V = -p/K$, where K is the bulk modulus, and a radial displacement $\xi/R = -\frac{1}{3}p/K$. Hence

140

$$\chi_0 \equiv \xi/p = -\tfrac{1}{3}R/\rho c_1^2, \quad \text{since } c_1^2 = K/\rho,$$

and the velocity response at frequency ω follows from (6.21):

$$\chi_v \equiv v/p = -\tfrac{1}{3}\text{i}\omega\omega_0^2 R/\rho c_1^2(\omega^2 + \text{i}\lambda\omega - \omega_0'^2), \tag{50}$$

where ω_0' is the resonant frequency found by this particular process of excitation (the reason for this remark will appear shortly). The damping term, $\text{i}\lambda\omega$, in which $\lambda \equiv 2\omega''$, is to be understood as describing dissipative effects within the resonator, not radiative damping, since χ_v describes the velocity response to an oscillatory pressure measured immediately outside and acting only on what lies within.

It is χ_v that enables the re-radiated wave to be matched to the local conditions outside the resonator. From (44) and (47) we write the values of v_r and p at radius R, expanding the exponentials as power series in R and retaining only the leading terms ($\propto 1/R^2$) and the first correction ($\propto 1/R$). After averaging round the sphere we have

$$\bar{v}_r(R) \sim \mathrm{i}a/\rho\omega R^2, \tag{51}$$

$$\bar{p}(R) \sim p_0 + a(1+\mathrm{i}kR)/R. \tag{52}$$

Then a takes such a value as makes $\bar{v}_r/\bar{p} = \chi_v$; after a little manipulation we find that

$$s \equiv ka/p_0 = -\lambda'\omega/[\omega^2 - \omega_0'^2(1-\tfrac{1}{3}k^2R^2) + \mathrm{i}(\lambda+\lambda')\omega], \tag{53}$$

where $$\lambda'\omega = \tfrac{1}{3}\omega_0'^2 k^3 R^3.$$

The correction term to $\omega_0'^2$ is easily understood. As used here, ω_0' defines the resonant frequency when excitation is by a pressure oscillation immediately outside. But the free resonance is at a slightly lower frequency, since the local velocity field outside, falling as $1/r^2$, also contributes to the kinetic energy, which is therefore not entirely attributable to the air within the sphere at R. What appears in (53) as $\omega_0'^2(1-\frac{1}{3}k^2R^2)$ is in fact the square of the true resonant frequency, ω_0. The difference between ω_0' and ω_0 is unimportant for the definition of λ'. What really matters is the extra term in the denominator, which depends on retaining the second term in (52); it is λ' which ensures that even a resonator without internal losses can never be excited beyond a certain limit, set by its ability to radiate. It is easily verified that if $\lambda = 0$, the denominator of (53) describes a resonator whose Q is governed by the same rate of dissipation as is expressed by (32), with Ω put equal to the total solid angle, 4π. Apart from the radiative loss, the resonator responds to the plane wave as it would to any other external pressure variation. It may be noted that λ' is not independent of frequency, like λ, but varies as ω^2. If we were to construct a model differential equation to simulate the behaviour, the radiation term would have to appear not in the first time-derivative of displacement but in the third.

It is interesting to look into this matter further, by setting up the differential equation for free vibrations of the resonator. For this we need to know the relation between p and ξ immediately outside, on the understanding that there are no external sources, so that all motion is to be attributed to spherical waves radiating outwards. Now an arbitrary time variation of $p(R)$ can be expressed as a Fourier series (or integral), each component of which describes the pressure at R due to a spherical wave of that frequency. We may therefore write $p(R, t)$ as the value at R of $p(r, t)$, where

$$p(r, t) = \sum_j \frac{a_j}{r} \mathrm{e}^{\mathrm{i}(k_j r - \omega_j t)}.$$

The point of this method of approach is that we can explicitly exclude converging waves of the type $(b/r)\,\mathrm{e}^{-\mathrm{i}(kr+\omega t)}$. Corresponding to $p(r, t)$ we have, from (26),

$$\rho\dot{\xi}(r, t)=\sum_j \frac{a_j}{r^2}(1-\mathrm{i}k_j r)\,\mathrm{e}^{\mathrm{i}(k_j r-\omega_j t)},$$

and in particular the displacement at R is given by

$$\rho\ddot{\xi}(R, t)=\sum_j \frac{a_j}{R^2}(1-\mathrm{i}k_j R)\,\mathrm{e}^{\mathrm{i}(k_j R-\omega_j t)}=p(R, t)/R+\dot{p}(R, t)/c_1. \qquad (54)$$

The pressure inside the resonator is $-3K\xi/R$, and the pressure differential, $p(R, t)+3K\xi/R$, is responsible for accelerating the air in the vicinity of the holes. If the total mass of this air is written as $4\pi R^2 M$, M is in effect the mass per unit area of the surface, and the equation of motion (local dissipation being neglected) takes the form

$$M\ddot{\xi}+3K\xi/R+p(R, t)=0.$$

Eliminating p from this and (54), we have

$$R\dddot{\xi}/c_1(1+\gamma)+\ddot{\xi}+R\omega_0^2\dot{\xi}/c_1+\omega_0^2\xi=0, \qquad (55)$$

where $\omega_0^2=3K/MR(1+\gamma)$, $\quad \gamma=\rho R/M \doteqdot \frac{1}{3}k_0^2R^2$ and $k_0=\omega_0/c_1$.

The significance of γ has already been noted in connection with (53); it corrects the resonant frequency to allow for the inertia of the gas outside. As expected, the equation of motion contains a third-order term, but paradoxically its sign is such as to cause regeneration rather than dissipation. It is, however, more than compensated at ω_0 by the dissipative first-order term, and it is the difference that describes the back-reaction of the radiated wave on its source and thus governs the radiative loss. The resonant frequency derived from (55) is approximately $\omega_0(1-\frac{1}{6}\mathrm{i}k_0^3R^3)$, which is the same as makes the denominator of (53) vanish.

A similar effect occurs when an accelerated charge radiates electromagnetic waves.[11] The field generated by the charge, when calculated to a sufficient order of approximation, includes an electric field acting on the charge itself with a force $\frac{2}{3}e^2\ddot{\mathbf{r}}/c^3$. The equation of motion of a charged mass bound by an elastic force to a fixed centre thus has the form

$$-\tfrac{2}{3}e^2\dddot{x}/c^3+m\ddot{x}+\mu x=0, \qquad (56)$$

with a natural frequency $\Omega \doteqdot \omega_0-\frac{1}{3}\mathrm{i}\,e^2\omega_0^2/mc^3$. The imaginary part describes the radiative power loss, which for an oscillatory dipole $p=p_0\,\mathrm{e}^{-\mathrm{i}\Omega t}$ amounts to $\frac{1}{3}\omega_0^4 p_0^2/c^3$, the classical result for the Hertzian dipole, normally obtained by evaluation of the energy flux in the distant radiation field. Unfortunately (56) also allows an exponentially growing solution and cannot therefore be a complete expression of the equation of motion. This is but one example of a ubiquitous problem – every naive attempt to extend physical concepts from mechanical to electrodynamical processes ultimately founders on a divergence. Quantum electrodynamics provides a new formalism for dealing with such difficulties.

505

After this digression we return to the acoustical resonator, for which (49) and (53) yield the variation with frequency of the absorption cross-section:

$$A_{abs} = 4\pi c_1^2 \lambda\lambda' / [(\omega^2 - \omega_0^2)^2 + (\lambda + \lambda')^2 \omega^2]. \tag{57}$$

At resonance the re-radiated wave has its greatest amplitude:

$$s_{res} = i\lambda'/(\lambda + \lambda'),$$

and

$$(A_{obs})_{res} = 4\pi\lambda\lambda'/k^2(\lambda + \lambda')^2.$$

The highest possible absorption results when the holes in the resonator are matched to the internal losses so that $\lambda = \lambda'$. Then

$$(A_{abs})_{max} = \pi/k^2. \tag{58}$$

This is the area of a circle of radius $1/k$, having a circumference equal to one wavelength, and it is the largest that can be attained by any isotropic radiator. The necessity for this maximum is inherent in (49), for $|2s - i|$ is essentially positive. It is, of course, quite a large cross-section for absorption; the resonator described in chapter 2, with a radius of 9 cm and a natural frequency of about 190 Hz (wavelength 1.77 m), is in principle capable of extracting all the energy in a plane wave passing through a circle of radius 28 cm.

23

The process just analysed has many analogues in different branches of physics. The de Broglie wave governing particle behaviour in quantum mechanics is a scalar wave whose frequency ω is $E/\hbar$ for a particle of total energy E. A typical process corresponding to resonance is the absorption of neutrons by nuclei.[12] The nucleus before the neutron enters is normally in its ground state, with a definite energy, and the compound nucleus after the neutron has entered has a wide spectrum of energy levels; if the incident neutron has energy ΔE that allows the compound nucleus to fall straight into a stationary state, as shown in fig. 30, the quantum mechanical equivalent of resonance occurs; the internal vibration frequency is determined by the energy difference between the initial and final states, and if this matches the neutron frequency the wave outside is depleted and the 'oscillation' within builds up; this reveals itself as a peak in the cross-section of the nucleus for capturing neutrons. The Q of the peak is determined by processes very closely analogous to those we have discussed. There is a natural breadth of resonance due to the possibility that the neutron may be spontaneously re-emitted – this is the analogue of λ', the radiation damping. It is also possible, however, that the compound nucleus will fall into a lower state by, for example, emitting energy in the form of a γ-ray – this is the analogue of λ, the internal dissipation, for the neutron is then permanently captured. The working out of this problem,

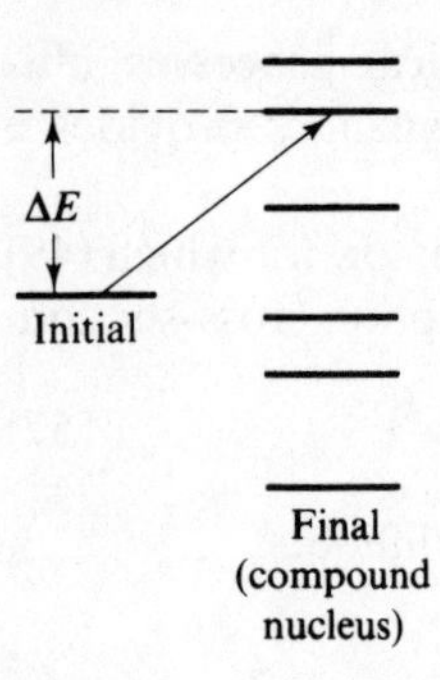

Fig. 30. Formation of an excited compound nucleus by the absorption of a neutron having the correct energy ΔE.

along strictly parallel lines to those used here, yields the Breit–Wigner formula for the variation with energy of absorption cross-section:

$$\sigma_{n\gamma} = \frac{\pi \Gamma_{nr} \Gamma_\gamma}{k k_r [(E - E_r)^2 + \frac{1}{4}(\Gamma_{nr} + \Gamma_\gamma)^2]}. \tag{59}$$

Here $\sigma_{n\gamma}$ parallels A_{abs} in (57); other parallels are as follows:

$$E \equiv \omega, \qquad E_r \equiv \omega_0, \qquad \Gamma_{nr} \equiv \lambda', \qquad \Gamma_\gamma \equiv \lambda.$$

The principal difference resides in the behaviour as ω or E goes to zero. In the acoustic case the cross-section tends to a constant, while with particles it increases as $1/k$, which for de Broglie waves is $1/E^{\frac{1}{2}}$. This difference results from the acoustic wave being non-dispersive ($\omega \propto k$) while the de Broglie wave is dispersive ($\omega \propto k^2$). It is profoundly important for the absorption of low energy (thermal) neutrons in reactors. These have a relatively large de Broglie wavelength, and there are a few nuclear species (e.g. Xe 135) with absorption cross-sections 10^6 times as great in area as the nucleus itself.[13] Only a small concentration of these is sufficient to reduce the efficiency of the reactor drastically, or even to prevent it from reaching criticality and becoming self-maintaining.

Electromagnetic scattering also behaves very similarly, if we overlook minor differences arising from the transverse character of the waves. A dipole feeding a resonant circuit (fig. 31) has an absorption cross-section

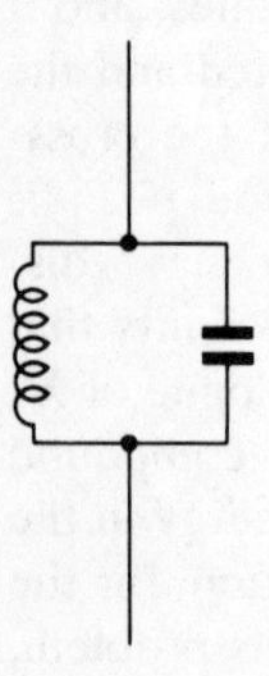

Fig. 31. Dipole feeding a resonant circuit.

which depends on the direction and polarization of the incident radiation, but the average of this cross-section over all directions is at most $\frac{1}{2}\pi/k^2$, one-half of the result expressed in (58). The factor $\frac{1}{2}$ may be considered a consequence of the polarization of transverse waves – an aerial cannot respond completely to all polarizations. If it has the maximum cross-section for one plane of polarization it is not responsive to the plane at right angles. The result (58) would apply if the radiation falling from every direction was optimally polarized, but for random polarization it must be halved. This average cross-section of $\frac{1}{2}\pi/k^2$ is achieved by every radio aerial that is perfectly matched to its feeder, as can be seen by a thermodynamic argument similar to that used by Kirchhoff in his theory of black-body radiation.[14] Imagine two cavities, each with walls maintained at the same temperature, and connected by a transmission line with an aerial at each end matched to the line. Any power picked up by aerial 1 from the black-body radiation in cavity 1 is transmitted to aerial 2 and radiated, without reflection, into cavity 2. Similarly cavity 1 is fed by whatever is picked up by aerial 2. Since there can be no resultant power transfer between cavities at the same temperature, both aerials must pick up the same amount; and the argument may be extended to any other aerial matched to its line. But the aerials are irradiated by isotropic radiation, and what they pick up is determined by the average absorption cross-section, which must therefore be the same for all matched aerials. Small aerials have a relatively small cross-section and pick up from a wide angular range; large aerials designed to receive only in a narrow pencil of directions have a correspondingly large cross-section for those directions from which anything is received.

Given that the aerial picks up as much power as would pass through a sphere of radius $1/\sqrt{2}k$ we may relate this to the energy density of radiation in the cavity. Consider a frequency range $\mathrm{d}\omega$, and let $u_\omega\,\mathrm{d}\omega$ be the energy per unit volume in this range. Draw a large spherical shell, lying between r and $r+\mathrm{d}r$, and centred on the aerial. Of all the energy, $4\pi r^2 u_\omega\,\mathrm{d}r\,\mathrm{d}\omega$, in the shell at any instant, only a fraction $1/8k^2r^2$ is moving in such a direction as to pass through the little sphere at the centre. Hence the energy absorbed from this shell is $\frac{1}{2}\pi u_\omega\,\mathrm{d}r\,\mathrm{d}\omega/k^2$, and as the different contributions from the shell arrive during an interval of $\mathrm{d}r/c$, the power absorption is $\frac{1}{2}\pi c u_\omega\,\mathrm{d}\omega/k^2$. Now according to Planck's law,

$$u_\omega = \hbar\omega^3/\pi c^3(\mathrm{e}^{\hbar\omega/k_\mathrm{B}T}-1),$$

and hence the power absorbed is given by

$$P=\hbar\omega\,\mathrm{d}\omega/2\pi(\mathrm{e}^{\hbar\omega/k_\mathrm{B}T}-1).$$

At low frequencies ($\hbar\omega \ll k_\mathrm{B}T$) the Planck law may be replaced by that of Rayleigh–Jeans, to give

$$P=\tfrac{1}{2}k_\mathrm{B}T\,\mathrm{d}\omega/\pi = k_\mathrm{B}T\beta, \tag{60}$$

where β is the bandwidth in Hz. This is reminiscent of the formula for

Johnson noise in a resistor (6.49), and not accidentally; for if we repeat the notional thermodynamical experiment with one of the aerials replaced by a resistor having R equal to Z_0, to terminate the line, the other aerial will feed power into it at the rate (60). This must be accompanied by an equal amount of power generated by the resistor, and indeed a Johnson noise generator for which $\overline{V^2}=4Rk_BT\beta$, feeding a line in series with itself, will generate just the right amount. This argument may be regarded as providing another derivation of Johnson noise or, alternatively, as confirming still further the idea of a maximum absorption cross-section, and throwing light on the close interrelation of different kinds of physical approach to a basic problem.

158

Anomalous dispersion

The interaction of an atom or molecule with visible light is an example of excitation by a wave whose wavelength greatly exceeds the size of the excited system. When the frequency is close to a characteristic absorption line (e.g. the Lyman series of ultraviolet lines in hydrogen or the two closely-spaced lines of the sodium doublet) the refractive index and absorption vary with frequency in a way that is fitted very closely by a resonance formula. Like neutron absorption, this is strictly a quantum phenomenon, yet one which has a precise parallel in classical physics. It is therefore appropriate at this stage, even though we make no attempt at developing the detailed argument, to state the quantum-mechanical result which validates the classical treatment, and then to examine briefly the consequences.

When a system occuping a quantum level with energy ε_k is irradiated with electromagnetic radiation of frequency ω, it is perturbed and acquires an oscillatory dipole moment. As is typical of quantum systems, the magnitude of this moment is not precisely definable for a single system; nevertheless the average taken over many identical systems is well-defined, and it is just this that matters when one makes an observation of the optical properties of a macroscopic sample containing many replicas of the atomic or molecular system under investigation. For such an assembly it is permissible to talk in classical deterministic terms when describing the evolution of the mean polarization. Moreover the solution of the quantum-mechanical problem shows that the polarization evolves exactly as if the system were a classical harmonic oscillator. Consider, then, a particle of mass m, carrying charge e, bound by an elastic spring to a fixed centre, so that under the influence of a varying electric field $\mathscr{E}_z(t)$ it obeys the equation of motion

439

$$m\ddot{z}+\mu z=e\mathscr{E}_z(t).$$

This is the same as the equation for a lossless pendulum, and the susceptibility therefore takes the form (6.2) with ω'' equated to zero:

129

$$\chi(\omega)=1/m(\omega_0^2-\omega^2),$$

where $\omega_0^2 = \mu/m$. Now χ describes the amplitude of oscillatory displacement under the influence of an oscillatory force of unit strength, but for present purposes the *atomic polarizability* is more useful – the amplitude of the oscillatory dipole moment due to an oscillatory electric field of unit strength. Since the force is $e\mathscr{E}_z$ and the dipole moment is ez, the atomic polarizability, α, is $e^2\chi$:

$$\alpha(\omega) = e^2/m(\omega_0^2 - \omega^2). \tag{61}$$

This is the expression for a classical oscillator. The average behaviour of a single quantized system can be simulated by a collection of such oscillators; thus with a system initially in the energy state ε_k the possibility of a transition to another state ε_j introduces a term of the form (61) into the polarizability, with ω_0 taking such a value that the quantum $\hbar\omega_0$ is equal to the energy difference $|\varepsilon_j - \varepsilon_k|$; resonance, the vanishing of the denominator in (61), corresponds to the quantum energy exactly matching what is required to cause the system to change energy levels from ε_k to ε_j. When such a transition is possible the system is highly sensitive to the irradiating electric field, while at a slightly different frequency it is less susceptible but nevertheless capable of being disturbed.† One modification of (61) is, however, called for – different energy levels are differently coupled to the initial state ε_k, in the sense that a given strength of $\mathscr{E}_z$ may readily stimulate a transition to some levels and only very weakly, or not at all, to others. For each transition, therefore, we introduce an *oscillator strength*, f_{jk}, writing for the total polarizability of the kth state,

$$\alpha_k(\omega) = \frac{e^2}{m}\sum_j \frac{f_{jk}}{(\omega_{jk}^2 - \omega^2)}. \tag{62}$$

It is worth noting, in passing, that when all the transitions involve only one electron changing its energy level, $\Sigma_j f_{jk} = 1$; as the reader will verify easily enough, this means that the system responds to a sharp impulse P as if the electron were free, moving off with initial velocity P/m. The oscillator strengths are positive when the system is being excited at resonance to absorb a quantum and make a transition to a higher level, $\varepsilon_j > \varepsilon_k$; but negative when $\varepsilon_j < \varepsilon_k$ and the effect of resonant irradiation is to extract a quantum from the system. This reversal of sign is highly relevant to the operation of masers and lasers (see chapters 11 and 20).

512

571 346

† We have ignored the transient effects arising from switching on the force, quite improperly in view of the assumption of no intrinsic loss in the system. In practice, however, there are various processes that cause or simulate loss, and in normal spectroscopic observations, which involve protracted irradiation of the sample, it is proper to consider only the steady-state response. When, on the other hand, a short wave-group travels in a dispersive medium, the transient response of the molecules to the leading edge, and the residual oscillations left behind the trailing edge, are significant factors in determining the speed of propagation and the evolution of the shape of the group.

Having stated the general result in (62), let us now concentrate on the special case of an isolated spectral line, where only one term in (62) need be considered. If, as in (61), the resonant frequency is ω_0, each individual system reacts to irradiation at a nearby frequency ω in a way that is exactly analogous to the Helmholtz resonator or the tuned dipole; being set into vibration it re-radiates a spherical wavelet. In a simple gas each molecule behaves similarly, and the superposition of all the spherical wavelets and the incident plane wave leads to two separate effects. In the forward direction there is a constant phase relationship between the wavelets and the plane wave, so that at any point all the wavelets add together into a resultant that is uniquely phase-linked to the plane wave – put otherwise, their superposition yields a resultant plane wave which combines with and modifies the propagation characteristics of the incident wave. This is, in rough pictorial terms, how the gas contrives to modify the wave velocity, and is the origin of the refractive index. At any other angle than the forward direction, however, the wavelets arrive at a given point with a phase that depends on precisely where they originated. The randomly distributed gas molecules contribute randomly phased wavelets, and the intensity of the wave scattered out of the forward direction is the sum of the intensities due to each molecule separately. This scattering is normally very weak, though the scattering of about 10% of sunlight in its passage through the atmosphere is obvious as the blue of the sky.†[15] Solids and liquids, in which the molecules are more ordered, normally scatter even less by this mechanism, but much more because of particulate impurities or structural defects.

To derive an expression for the refractive index there is no need to add the forward-scattered wavelets, though this is a possible approach to the problem. It is easier, however, to consider simply the volume polarizability of the medium, $N\alpha$ if N is the number of molecules per unit volume. The dielectric constant is $1+N\alpha/\varepsilon_0$ and the refractive index n is the square root of this. For a gas or any sparse distribution of scattering centres it is normally good enough to write

$$\delta n \equiv n - n_0 = \tfrac{1}{2}N\alpha/\varepsilon_0 = fNe^2/2m\varepsilon_0(\omega_0^2-\omega^2) \tag{63}$$

in the vicinity of an absorption line of oscillator strength f. In writing this we have introduced n_0 to represent other non-resonant contributions to the refractivity, which can be taken as sensibly constant around ω_0, where

† The characteristic absorption frequencies of nitrogen and oxygen are in the ultraviolet so that, in (62), $\omega_{jk} \gg \omega$ and α is nearly independent of frequency (i.e. air has rather low optical dispersion). An oscillating dipole of given magnitude scatters as the fourth power of its frequency, and blue light is much more heavily scattered than red. Hence the colour of the sky and the reddening of the setting sun, seen through a thick layer of atmosphere. The scattering of light by clouds, on the other hand, involves obstacles (water drops) larger than the wavelength, and all colours are more-or-less equally scattered, so that clouds tend to be white, especially when viewed from above.

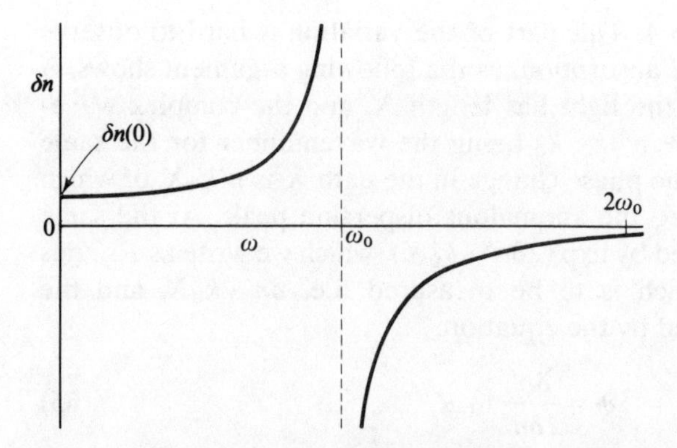

Fig. 32. Variation of refractive index around anomalous dispersion peak.

the absorption line results in a very strong dispersion. The characteristic anomalous dispersion curve, as given by (63), is shown in fig. 32.

It is will be observed that in traversing this curve δn drops from its low-frequency value $fNe^2/2m\varepsilon_0\omega_0^2$ – representing the static polarizability – to zero at frequencies well above ω_0. This can only happen if there is absorption somewhere in between, as follows immediately by applying the second of the Kramers–Kronig relations (5.9). The increment of refractive index, considered as a frequency-dependent complex number $\delta n' + \mathrm{i}\,\delta n''$, is a linear compliance to which the relation applies, and we see that

110

$$\delta n'(0) - \delta n'(\infty) = \frac{2}{\pi}\int_0^\infty \delta n''\,\mathrm{d}\omega/\omega \sim \frac{2}{\pi\omega_0}\int_0^\infty \delta n''\,\mathrm{d}\omega, \qquad (64)$$

the second, approximate, form being suitable for a narrow absorption line when $\delta n''$ can be taken as zero everywhere except very close to ω_0. It follows that the area under the absorption curve is $\frac{1}{2}\pi\omega_0$ times the refractive index loss in traversing the anomalous dispersion peak; and this result does not depend on any special assumption as to the origin of the loss, except that the system be linear, as is a very good assumption with light of ordinary intensity. If the intrinsic losses are very small, as assumed in setting up the model analysed here, the loss peak is narrow and correspondingly high. The anomalous dispersion curve (63) can therefore be followed very close to ω_0 and substantial variations of n observed. There are good accounts, with excellent photographs, in specialist optics texts of the ingenious experiments by which the phenomenon has been studied, and in the review by Korff and Breit.[16]

It is, however, worth noting that the ideal behaviour expressed by (63) must break down in the frequency range where the loss is appreciable. To take a specific example, the motion of gas molecules imposes a Doppler shift on the resonant frequency, and the loss peak, however narrow intrinsically, is smeared into a Gaussian form, reflecting the Gaussian distribution of molecular velocities parallel to the line of sight. Simultaneously the rise of $\delta n'$ towards $\pm\infty$ on either side of the resonance is halted, $\delta n'$ becoming a continuous function passing through zero at resonance, somewhat like χ' in fig. 5.4. This part of the variation is hard to observe experimentally because of absorption, as the following argument shows. If the sample traversed by the light has length X, and the complex wavenumber is written as $(n'+in'')k_0$, k_0 being the wavenumber for the same frequency in free space, the phase change in the path X is $n'k_0X$, of which $\delta n' \cdot k_0X$ is attributable to the anomalous dispersion peak. At the same time the intensity is divided by $\exp(2\delta n'' \cdot k_0X)$, which we write as R. Thus the phase change ϕ which is to be measured, i.e. $\delta n' \cdot k_0X$, and the intensity loss R are related by the equation:

112

$$\phi = \frac{\delta n'}{2\delta n''} \ln R, \tag{65}$$

and it will be noted that the length X has disappeared. Restrictions on the value of ϕ are intrinsic and cannot be eliminated by redesign of the experiment.

Let us now define the width, $\Delta\omega$, of the absorption peak by replacing the real peak by a rectangle of the same area, with the same height $\delta n''_{max}$ and width $\Delta\omega$ which must, according to (64), have the value

$$\Delta\omega = \frac{1}{2}\pi\omega_0[\delta n'(0) - \delta n'(\infty)]/\delta n''_{max}. \tag{66}$$

It is at a frequency about $\frac{1}{2}\Delta\omega$ from the peak that (63) begins to fail, and at that point $\delta n'$ can hardly be greater than the value given by (63), which may be rewritten to yield the inequality

$$\delta n' \lesssim \omega_0[\delta n'(0) - \delta n'(\infty)]/\Delta\omega = \frac{2}{\pi}\,\delta n''_{max} \quad \text{from (66)}.$$

Hence, from (65),

$$\phi \lesssim \frac{1}{\pi}\frac{\delta n''_{max}}{\delta n''} \ln R.$$

In the absorption region, then, where $\delta n'' \sim \delta n''_{max}$, ϕ can only reach 1 radian at the cost of something like twenty-fold attenuation. The best experimental procedures for studying the anomalous dispersion curve rely on interferometry, and it is no easy matter to measure fringe shifts of $1/2\pi$ with any accuracy in the presence of heavy attenuation. It is more satisfactory to measure the shape and strength of the absorption peak by passing the light through a cell, and to use the Kramers–Kronig transform to evaluate n' from this.

110

8 Velocity-dependent forces

Coriolis and Lorentz forces

In this chapter are collected together a somewhat disparate assembly of oscillatory (better, periodic) systems which have one thing in common – non-reversibility; not simply the lack of reversibility because of dissipation, but that which results from velocity-dependent forces not in themselves capable of causing dissipation. To classify such systems together in a single category may not be significant, and the reader who finds it pointless has probably missed nothing of value. The systems themselves, however, have interest and importance in their own right, and the whole chapter may therefore be regarded (like much else in this book) as a catalogue of phenomena that fit only loosely into any tidy pattern but must nevertheless form part of the mental furniture of anyone who wishes to appreciate the variety that oscillatory phenomena can exhibit. To a physicist the section on nuclear magnetic resonance may seem intrinsically more important than the rest. The reader should therefore take heed that it is also considerably more difficult, perhaps the hardest part of the book; it may, however, be omitted without losing the thread of any subsequent argument.

The velocity-dependent forces considered here, we repeat, are not the dissipative forces of viscosity or resistance responsible for the middle term, 30 $k\dot{\xi}$, in an equation of motion such as (2.23). They are, rather, the non-dissipative Coriolis force or Lorentz force – the former being involved when dynamical processes are observed from a rotating frame of reference, the latter being the force on a charged particle moving through a magnetic field.[1] Both forces act normal to the direction of motion and do no work, but nevertheless considerably affect the trajectory. They take similar forms:

Coriolis force – an observer in a frame rotating at constant angular velocity $\boldsymbol{\Omega}$ finds a particle of mass m, acted upon by a real force $\boldsymbol{F}$, to obey the equation of motion

$$m\ddot{\boldsymbol{r}} = \boldsymbol{F} - 2m\boldsymbol{\Omega} \wedge \dot{\boldsymbol{r}} - m\boldsymbol{\Omega} \wedge (\boldsymbol{\Omega} \wedge \boldsymbol{r}). \tag{1}$$

The second term is the Coriolis force and the third the centrifugal force.

Lorentz force – a particle of mass m and charge e moving in a magnetic field $\boldsymbol{B}$ obeys the equation of motion

$$m\ddot{\boldsymbol{r}} = \boldsymbol{F} - e\boldsymbol{B} \wedge \dot{\boldsymbol{r}}. \tag{2}$$

The force $\boldsymbol{F}$ includes the electrical force $e\boldsymbol{E}$; the second term is the Lorentz force.†

The similarity in form of these forces means that under suitable conditions the effect of a uniform field $\boldsymbol{B}$ on the trajectory of a particle can be eliminated by observing its motion from a frame rotating at angular velocity $\boldsymbol{\Omega}=-\frac{1}{2}e\boldsymbol{B}/m$. We need not discuss what these suitable conditions are (essentially that the trajectory be localized so that r does not become too large, and that $\boldsymbol{F}$ be a central force field). This matter is fully dealt with in standard textbook discussions of *Larmor's theorem.*[2]

When Coriolis or Lorentz forces are present a lossless system ceases to be reversible. To elaborate this point, we note that if $\boldsymbol{F}$ is determined by position alone the equation of motion,

$$m\ddot{\boldsymbol{r}}=\boldsymbol{F}(\boldsymbol{r}),$$

with solution $\boldsymbol{r}=\boldsymbol{r}(t)$, is equally satisfied by $\boldsymbol{r}=\boldsymbol{r}(-t)$, since only t^2 appears in the equation; if the particle can perform a certain trajectory it can equally well perform it backwards by being started at the end-point with its velocity reversed. This is not so when the equation of motion is (1) or (2), for reversing t reverses $\dot{\boldsymbol{r}}$ and the sign of the Coriolis and Lorentz forces, but not of $\boldsymbol{F}$; $\ddot{\boldsymbol{r}}$ is no longer uniquely determed by $\boldsymbol{r}$, and we expect a different trajectory in each direction. The simplest sort of vibrator, confined to a straight line or any other one-dimensional path, is of course unaffected by these transverse forces. To see their effect in the most elementary form we need to allow the vibrator two degrees of freedom. The Foucault pendulum provides an example. A conical pendulum can swing in a circle with the same angular velocity $\omega_0=(l/g)^{\frac{1}{2}}$ clockwise or anticlockwise if it is viewed from a non-rotating frame. Set up at the North Pole, however, and viewed from above by an earthbound observer, it will seem to have angular velocity of magnitude $\pm\omega_0-\Omega$ if Ω is the Earth's (negative) angular velocity, the positive sign applying to clockwise motion – the two frequencies are no longer equal and opposite. The trajectories in this simple case have the same shape, but this is exceptional.

13

When circular motion is to be described it is convenient to employ complex numbers, as discussed in chapter 3. The displacement of the bob of the pendulum is denoted by a complex number, r, the whole of which is significant; thus, $r=a\,\mathrm{e}^{-\mathrm{i}\omega t}$ represents clockwise motion (if $\omega>0$) round a circle of radius a, and the two stable modes of the pendulum are of the form $a_1\,\mathrm{e}^{-\mathrm{i}(\Omega+\omega_0)t}$ and $a_2\,\mathrm{e}^{-\mathrm{i}(\Omega-\omega_0)t}$. If they are equally excited, with $a_1=a_2=a$, superposition of the two yields the characteristic precessional linear motion of the Foucault pendulum,

62

$$r=a\,\mathrm{e}^{\mathrm{i}(\Omega+\omega_0)t}+a\,\mathrm{e}^{\mathrm{i}(\Omega-\omega_0)t}=2a\,\mathrm{e}^{\mathrm{i}\Omega t}\cos\omega_0 t.$$

† In this chapter and occasionally elsewhere, $\boldsymbol{B}$ and $\boldsymbol{E}$ refer to steady applied fields, which may be strong, $\boldsymbol{b}$, $\mathscr{B}$ and $\mathscr{E}$ to weak fields superimposed on them.

The bob swings very nearly in a plane with its characteristic frequency ω_0, but the plane rotates with angular velocity $-\Omega$, i.e. once a day, at the Pole, once in cosec θ days at latitude θ where the Coriolis force is reduced by a factor sin θ. Because of this precession the linear motion is not a normal mode of the vibrator – it is the circular motions that are. It may be noted, though we shall not dwell on it at this stage, that when the Coriolis force is absent the two circular motions have as much claim as linear vibrations in perpendicular planes to be called the normal modes. The two modes are *degenerate*, i.e. have frequencies of the same magnitude, and any linear combination gives rise to a stable pattern of vibration at the same frequency. Rotation, through the Coriolis effect, removes this degeneracy and with it all ambiguity in the definition of normal modes.

When a system whose normal modes are circularly polarized is driven by an oscillatory force directed along a fixed line, to calculate the response one may first represent the force as the sum of two rotating forces of constant magnitude. If F_0 is a complex number, $F_0 \cos(\omega t+\varepsilon)$ is an oscillatory force acting along a line whose direction is the argument of F_0. It is dissected into two components

$$F_0 \cos(\omega t+\varepsilon) = A_L \,\mathrm{e}^{\mathrm{i}\omega t} + A_R \,\mathrm{e}^{-\mathrm{i}\omega t},$$

where $A_L = \frac{1}{2}F_0 \,\mathrm{e}^{\mathrm{i}\varepsilon}$ and $A_R = \frac{1}{2}F_0 \,\mathrm{e}^{-\mathrm{i}\varepsilon}$; the left- and right-handed components, A_L and A_R, have the same magnitude, $\frac{1}{2}|F_0|$, but opposite arguments. If the system responds to a rotating force by itself rotating in synchronism, a compliance χ may be defined to describe the magnitude and phase of the response. Thus $\chi_L A_L \,\mathrm{e}^{\mathrm{i}\omega t}$ and $\chi_R A_R \,\mathrm{e}^{-\mathrm{i}\omega t}$ are the two rotating components of displacement. Real χ means that force and displacement rotate together and no work is done, but an imaginary part to χ implies a component of displacement normal to the force; if the displacement lags behind the force energy is steadily transferred to the system.

When the circular modes are degenerate the magnitudes of χ_L and χ_R are equal, but since displacement lags behind force by the same angle for both senses of rotation their arguments are opposite: $\chi_L = \chi_R^*$. An applied force $F_0 \cos \omega t$, with F_0 real, then produces a response

$$r = \tfrac{1}{2}\chi_2 F_0 \,\mathrm{e}^{-\mathrm{i}\omega t} + \tfrac{1}{2}\chi_R^* F_0 \,\mathrm{e}^{\mathrm{i}\omega t}.$$

If χ_R is written as $\chi \,\mathrm{e}^{\mathrm{i}\phi}$,

$$r = \chi F_0 \cos(\omega t - \phi),$$

a linear response lagging in phase behind the force by the phase angle of χ. In this case χ_R and χ_L have the same significance as the compliance defined for linear motion.

With non-degenerate circular modes, χ_L and χ_R are not complex conjugates and the general form of the response to a linear force is elliptical motion, being the composition of two circular motions of different amplitude at equal but opposite speeds. In most of the examples in this chapter the two modes are so different in character that only one matters at any time. The result of linear excitation is then circular motion. In cases like this we shall take it for granted that, although we may talk in terms of response to rotating forces, in practice it is usually easier to excite the system with an oscillating force, constant in direction, but the resulting response is the same.

Whirling[3]

The Foucault pendulum is an exceedingly gentle illustration of the Coriolis effect. A more dramatic example, and one with great nuisance value, arises when an oscillatory system is spun at an angular velocity exceeding its natural vibrational frequency. It may then spontaneously vibrate and possibly fragment, with great destruction to itself and objects in its vicinity. Consider a very simple case, illustrated in fig. 1, a small mass suspended by a length of piano wire from the spindle of a d.c. motor. The dimensions

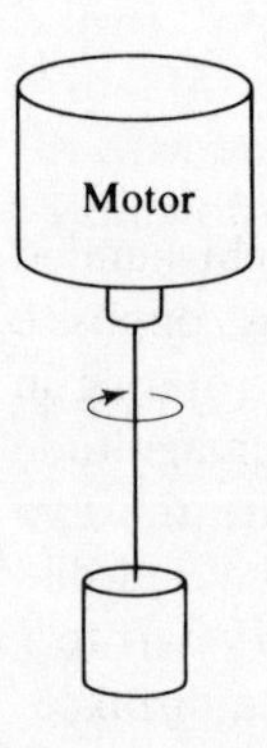

Fig. 1. Demonstration of whirling. A brass cylinder, $\frac{1}{2}$ in. diameter and $\frac{3}{4}$ in. long, soldered to a steel wire, 0.02 in. diameter and 4 in. long, is suitable. *Note warning in text.*

shown are suitable for private experimenting, though the observer should always interpose a cushion or screen between himself and the mass, which a moment's negligence may permit to fly off horizontally at great speed. This will happen if the motor is gradually accelerated through the critical angular velocity, equal to the natural vibration frequency of a few Hz. If, however, the wire is gently restrained with the fingers until this speed is exceeded, it will be found that the mass can continue to spin without significant horizontal deflection.

This arrangement is easily analysed, either in the laboratory frame of reference or in terms of a frame rotating with the motor spindle. The complex notation for horizontal displacement of the mass is convenient in both cases. To take the laboratory frame first, we suppose the wire to have a slight permanent set, so that the mass takes up a rest position r_0 from the axis when the spindle is not turning. When it is displaced to a neighbouring position r the linear restoring force due to the wire will have magnitude

$\mu(r_0 - r)$, directed back towards the rest position; the equation of motion therefore has the form

$$m\ddot{r} + \mu(r - r_0) = 0, \tag{3}$$

with solution $r - r_0 = A\,\mathrm{e}^{-\mathrm{i}\omega_0 t} + B\,\mathrm{e}^{\mathrm{i}\omega_0 t}$, the superposition of two circular motions in opposite senses at the frequency ω_0, equal to $(\mu/m)^{\frac{1}{2}}$.

Now let the spindle rotate at angular velocity Ω. Then the rest position varies with time as $r_0\,\mathrm{e}^{-\mathrm{i}\Omega t}$, and (3) takes the form

$$m\ddot{r} + \mu r = \mu r_0\,\mathrm{e}^{-\mathrm{i}\Omega t}, \tag{4}$$

the equation of undamped harmonic motion about the origin, not about r_0, with a rotating driving force arising from the initial imperfection as measured by r_0. The steady-state solution of (4) represents the mass rotating about the origin at angular velocity Ω,

$$r = r_0\,\mathrm{e}^{-\mathrm{i}\Omega t}/(1 - \Omega^2/\omega_0^2). \tag{5}$$

The resonant denominator is responsible for amplification of the initial permanent set. If Ω is sufficiently close to ω_0 any displacement r_0 can be amplified to the point where the linear approximation implicit in (5) fails. More probably, however, a new complication will set in when the wire is bent beyond its elastic limit, for the losses accompanying plastic flow may induce instability, r growing until the mass whirls around on a wire bent horizontal, or the wire breaks. We shall return shortly to the instability due to losses. For the present we consider only lossless systems, for which (5) shows that once the critical speed ω_0 is exceeded, the displacement falls again to manageable size. The process of acceleration through ω_0 is exactly analogous to the excitation of a resonant system by a gliding tone, treated in chapter 6; when Ω has reached a value much greater than ω_0, all that is left is a residual circular motion at ω_0, excited as Ω passed through the critical range.

148

The wire and mass in fig. 1 may be replaced by a length of wire (e.g. 30 cm of 24 SWG copper wire, straightened by pulling). In contrast to the former example, this has more than one resonant mode, as illustrated in fig. 2, each with its characteristic value of ω_0. These modes, though not

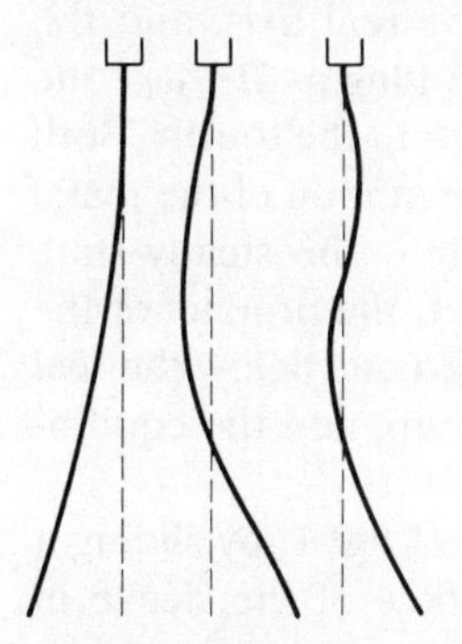

Fig. 2. The first three modes of a spinning wire, free at the lower end.

sinusoidal, constitute a suitable set of basis functions for a Fourier series describing the initial shape of the wire. A typical term in the series will have initial amplitude a_i, which will be changed to $a_i/(1-\Omega^2/\omega_{0i}^2)$ by spinning. Whirling can now occur near any one of the characteristic frequencies. It is easy to demonstrate that in between resonances the wire will spin stably; one has only to restrain the lower end with the hand as the fundamental frequency is traversed, and to stop somewhere between this point and the next frequency, which is very well separated from it.

Even easier is to let the lowest few centimetres of the wire dip into a beaker of water, which supplies a damping term to the equation of motion and prevents the amplitude building up without limit close to resonance. It is then possible to follow the development of the pattern as Ω is increased slowly, and to see how each mode in turn is amplified and then sinks into insignificance as its resonant frequency is passed, to be replaced by the next higher mode. With a very fine wire completely immersed one can run through several resonances, though the wide spacing of the frequencies makes it difficult to follow a long sequence.

If the damping mechanism, instead of being, like a beaker of water, fixed in the laboratory frame, rotates with the spindle the behaviour is quite different; at all speeds above the lowest resonance the motion is unstable and whirling is inevitable. It is now convenient to examine the behaviour from the point of view of a rotating reference frame, in which the equation of motion takes the form

$$m\ddot{r} = \mu(r_0 - r) + 2\mathrm{i}m\Omega\dot{r} + m\Omega^2 r - \lambda\dot{r}. \tag{6}$$

The first term on the right, taken from (3), is the elastic restoring force, the second and third the Coriolis and centrifugal forces from (1) expressed in complex notation, and the fourth a viscous damping term proportional to velocity relative to the rotating frame. The steady-state solution is obtained by putting $\dot{r}$ and $\ddot{r}$ equal to zero, and is the same as (5) without the exponential. To determine whether this solution is stable we must examine the oscillatory modes, and it is now permissible to set r_0 equal to zero. Substituting the trial solution $\mathrm{e}^{-\mathrm{i}\omega t}$ we see at once that

$$\omega^2 + (2\Omega + \mathrm{i}k)\omega + (\Omega^2 - \omega_0^2) = 0, \quad \text{where } k = \lambda/m. \tag{7}$$

The sum of the two roots, $-(2\Omega+\mathrm{i}k)$, lies below the real axis, and the product of the roots is $(\Omega^2-\omega_0^2)$, real and negative so long as $\Omega<\omega_0$; one of the roots must lie in the third quadrant and the other in the fourth. Both therefore have negative imaginary parts and cause the motion of the mass, displaced from equilibrium, to spiral in to the origin – the steady-state solution is stable when $\Omega<\omega_0$. When $\Omega>\omega_0$, however, the product of the roots is positive and real, so that one must lie above and one below the real axis. The former describes an outward spiralling motion, and the equilibrium is unstable.

This is readily demonstrated with the arrangement of fig. 1, by sliding a well-fitting PVC sleeve on to the wire. The plastic flow of the sleeve in bending is sufficient to induce instability; at speeds above ω_0 the mass, however well centred initially, gradually develops a displacement, spinning in an ever-increasing circle. The last stage is very sudden, probably aided by extra losses due to inelastic deformation of the wire. Since rotating machinery commonly carries its dissipative mechanisms around with it, this instability is a potential hazard that must always be borne in mind.

To return to the theoretical analysis, the transition from stable to unstable motion occurs at the angular velocity for which the elastic restoring force and the centrifugal force, both varying as r, balance. The two forces together define a parabolic potential energy function, which is bowl-like at low speeds, dome-like at high speeds. As the mass spins round, in the absence of losses, it is kept in its orbit, even on the dome, by the Coriolis force. When losses are introduced, the mass necessarily moves to orbits of lower energy, i.e. inwards when $\Omega < \omega_0$, outwards when $\Omega > \omega_0$. Looking at the phenomenon from the laboratory frame dispels any air of mystery which may attach to it. The effect is exhibited by a simple conical pendulum swinging in a vessel of fluid. If the fluid is caused to rotate more slowly than the pendulum bob in its circular orbit, the bob experiences a retarding force, and sinks towards the vertical. But if the fluid rotates faster than the bob, it urges the bob onwards and causes its orbit radius to increase. Exactly the same thing occurs with the whirling mass.

In the laboratory frame of reference, with the viscous damping fluid rotating, the equation of motion corresponding to (6) has the form, when $r_0 = 0$,

$$m\ddot{r} = -\mu r - \lambda(\dot{r} + \mathrm{i}\Omega r),$$

since $\dot{r} + \mathrm{i}\Omega r$ is the relative velocity of pendulum bob and fluid. Hence the natural frequencies are determined by the equation

$$\omega^2 + \mathrm{i}k(\omega - \Omega) - \omega_0^2 = 0,$$

which may be written in the form

$$(\omega + \tfrac{1}{2}\mathrm{i}k)^2 = (\omega_0^2 - \tfrac{1}{4}k^2) + \mathrm{i}\Omega k. \tag{8}$$

As Ω is varied, the trajectory of ω is easily found. For if $\omega + \frac{1}{2}\mathrm{i}k$ is written as $x + \mathrm{i}y$, equating the real parts on both sides of (8) shows that

$$x^2 - y^2 = \omega_0^2 - \tfrac{1}{4}k^2, \tag{9}$$

the equation of a rectangular hyperbola. The trajectory of ω is shown in fig. 3, the asymptotes of the hyperbola intersecting at $-\frac{1}{2}k$, the origin of x and y, shown as Q in the diagram. From the imaginary part of (8), $xy = \frac{1}{2}\Omega k$; this, with (9), shows that if one solution of (8) is (x', y'), the other is $(-x', -y')$, so that the two values of ω_0 for any given Ω lie on a line through Q, for example A and A' are a possible pair. Only when $\Omega = 0$ do the two solutions mirror each other in the imaginary axis, which is the condition that must be satisfied if the circularly polarized normal modes are to combine into a linear normal mode. So long as $\Omega < \omega_0$, both solutions lie

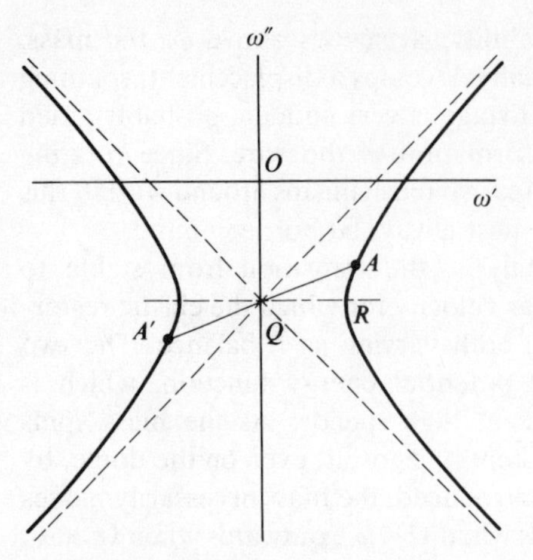

Fig. 3. Solution of (8).

below the real axis and represent decaying modes. But as Ω is increased above ω_0, A crosses over into the positive half-plane and the mode grows exponentially.

The gyro-pendulum

This example of a system with circular normal modes that are not time-reversed copies of each other is not important in itself, but it is easy to make and sufficiently amusing to be worth constructing; its discussion also serves as an introduction to the next section, dealing with a magnetized gyroscopic top, a matter of considerable physical significance. The gyro-pendulum illustrated in fig. 4 is essentially a rapidly spinning flywheel whose axis is mounted rigidly on a short pendulum which is free to swing in any direction. For small horizontal displacements described by the complex number r, the equation of motion is found by considering the horizontal components of angular momentum about the point of support, which changes because of the torque $\mathrm{i}mgr$, m being the total mass of flywheel and motor. The angular momentum is made up of two terms: (1) Lr/l, the horizontal component of the flywheel angular momentum, L, and (2) $-\mathrm{i}ml\dot{r}$, the angular momentum of the whole suspended mass as the pendulum swings. It is assumed that $r \ll l$ so that the normal linear approximation for pendulum motion is valid. It follows that, if losses are neglected,

$$\frac{\mathrm{d}}{\mathrm{d}t}(Lr/l - \mathrm{i}ml\dot{r}) = \mathrm{i}mgr,$$

i.e.

$$\ddot{r} + 2\mathrm{i}\omega_g\dot{r} + \omega_0^2 r = 0, \tag{10}$$

where $\omega_0 = (g/l)^{\frac{1}{2}}$ and $\omega_g = L/2ml^2$.

The characteristic frequencies resulting from (10) have the form

$$\omega_{1,2} = \omega_g \pm (\omega_0^2 + \omega_g^2)^{\frac{1}{2}}. \tag{11}$$

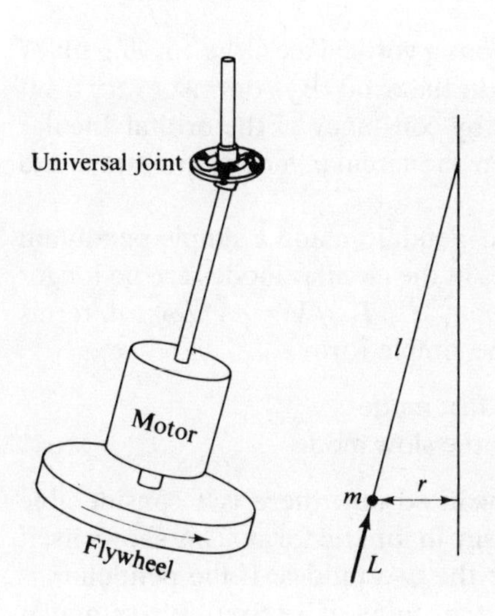

Fig. 4. The gyro-pendulum. A brass flywheel, 1 cm thick and 12 cm in diameter, spinning at about 3000 r.p.m. gives good results. The rigid rod holding the motor hangs from a standard flexible joint used in radio sets; a steel wire through the middle serves to support the weight of motor and flywheel. The flywheel is about 25 cm below the flexible support. On the right the notation used in the text is illustrated. *Note*: *this is a potentially dangerous experiment.* The flywheel must be unflawed, well-balanced and securely pinned to the motor shaft, which must itself be strong (say $\frac{1}{4}$ in. diameter) to resist the considerable gyroscopic torques.

If L is positive (clockwise rotation of the flywheel when viewed from above), the faster mode, ω_1, given by the positive sign in (11), is clockwise and the slower, ω_2, anticlockwise.

When the pendulum is deflected and released it executes an angular looping motion, the result of both modes being excited together. Fig. 5 shows the calculated epicyclic trajectory when the parameters take such

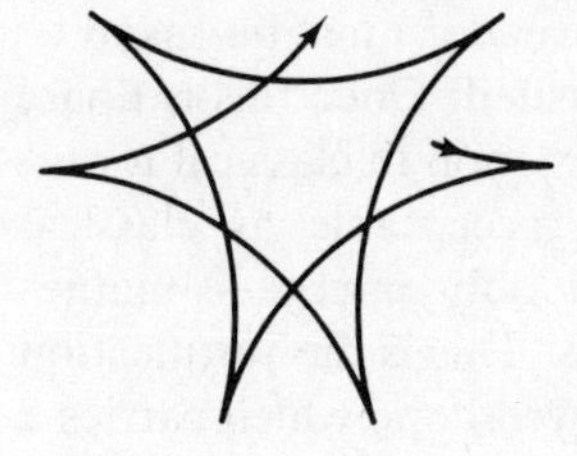

Fig. 5. Typical trajectory of gyro-pendulum released from rest.

values that $\omega_0 = 2\omega_g$; the two frequencies in (11) are then in the ratio $(\sqrt{5}+1)/(\sqrt{5}-1)$, about 2.6. The pendulum is supposed to have started at rest in its position of maximum deflection; this involves the amplitudes of the two modes being in inverse ratio to their frequencies, so that their linear velocities cancel when they are in phase. The orbit is then sharply cusped, with n cusps per revolution, n being easily shown to be $1-\omega_1/\omega_2$, 3.6 in this particular case. It may be observed that the pendulum has no orbital angular momentum at the cusps, but acquires angular momentum between cusps, in the reverse sense to the flywheel. This reflects the conservation of angular momentum about a vertical axis, the varying tilt of the flywheel demanding compensation in the orbit. By contrast every orbit of a simple pendulum is characterized by constancy of the orbital angular momentum. This is only possible when the circular modes are equal and opposite in frequency.

Another difference between the gyro-pendulum and a simple pendulum is that the potential and kinetic energies in the circular modes are no longer equal. Since $V=\frac{1}{2}mgr^2/l$ and $T_{1,2}=\frac{1}{2}m\omega_{1,2}^2r^2$, $T_{1,2}/V=\omega_{1,2}^2/\omega_0^2$; in terms of n, the number of cusps, this takes the simple form

$$T/V = n-1 \text{ in the fast mode}$$
$$1/(n-1) \text{ in the slow mode.}$$

With $n=3.6$, as in the example just worked out, there is a considerable difference in the proportion of the energy in kinetic form. This shows itself as a difference in the damping rate for the two modes. If the pendulum is supported on a spongy base so that the point of support sways and is subject to viscous damping the fast mode is far more susceptible and decays much sooner than the slow. As a particular example, readily demonstrated, if the pendulum is set in an epicyclic orbit and then the whole apparatus picked up in the hands the cusps vanish almost at once, leaving the pendulum executing a slow circular orbit.

The gyromagnetic top[4]

We turn to a system which is more difficult to demonstrate in the laboratory, but which is of great theoretical importance since it models in terms of Newtonian dynamics the processes involved in *nuclear magnetic resonance*. We have already, in anomalous dispersion, met one characteristically quantum-mechanical process that can be replaced by an equivalent classical model, and here is another. At this stage the reader must be asked to accept that the equivalence can be strictly established. Once this is done, there are advantages in thinking about the phenomenon in classical terms, many of the essential results being intuitively acceptable as classical mechanisms when from a quantal standpoint they only emerge as mathematical consequences of a rather involved analysis. This is the justification for discussing in some detail the mechanics of a gyroscope which carries a permanent magnetic moment $\boldsymbol{\mu}$, parallel to its angular momentum $\boldsymbol{L}$. For this discussion we abandon complex notation in favour of vector notation.†

204

516

In a steady magnetic field $\boldsymbol{B}$ the torque, and therefore the rate of change of angular momentum, is $\boldsymbol{\mu}\wedge\boldsymbol{B}$; this causes the axis of the gyroscope to precess around the direction of $\boldsymbol{B}$ with angular velocity $\boldsymbol{\Omega}$ such that the resulting change in angular momentum, $\boldsymbol{\Omega}\wedge\boldsymbol{L}$, is equal to $\boldsymbol{\mu}\wedge\boldsymbol{B}$. If we

† Although the system is magnetic, no Lorentz force is involved explicitly, the only magnetic force being the velocity-independent torque exerted by $\boldsymbol{B}$ on $\boldsymbol{\mu}$. The velocity-dependent processes are still purely mechanical, being those characteristic of gyroscopic motion.

write γ for the (scalar) gyromagnetic ratio μ/L, it follows immediately that

$$\boldsymbol{\Omega} = -\gamma \boldsymbol{B}. \quad (12)$$

This result is important in showing that the precessional rate is the same for all orientations of the axis. For an atomic nucleus in a given energy state (normally its ground state, of minimum energy) γ is uniquely defined by the state, so that Ω/B is a characteristic property of the nucleus. The value of γ may be determined experimentally by perturbing the precessional motion by means of a weak additional magnetic field rotating at something like $\boldsymbol{\Omega}$; at resonance the orientation of the axis is systematically altered, and the resulting magnetic effects may be detected. The geometry of the arrangement is shown in fig. 6; $\boldsymbol{L}$ precesses around $\boldsymbol{B}$ with angular velocity $\boldsymbol{\Omega}$, and the extra field $\boldsymbol{b}$ rotates with angular velocity $\boldsymbol{\omega}$, also around $\boldsymbol{B}$.

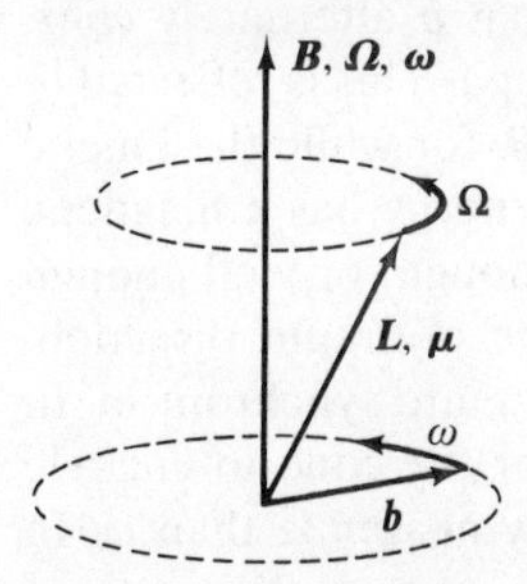

Fig. 6. Illustrating the notation for the gyromagnetic top.

It is convenient to picture the behaviour from the point of view of an observer rotating with $\boldsymbol{b}$, who will see $\boldsymbol{b}$ at rest in his frame of reference. Any vector $\boldsymbol{A}$ which is constant in the laboratory frame will appear to him to be suffering change as, from his viewpoint, it rotates at an angular velocity $-\boldsymbol{\omega}$; the perceived rate of change will be $-\boldsymbol{\omega} \wedge \boldsymbol{A}$. This is for constant $\boldsymbol{A}$, and in general it must be supplemented by any actual rate at which $\boldsymbol{A}$ changes in the laboratory frame. Thus the vector $\boldsymbol{L}$, precessing momentarily with angular velocity $-\gamma(\boldsymbol{B}+\boldsymbol{b})$ about the vector resultant of $\boldsymbol{B}$ and $\boldsymbol{b}$, will seem to the observer to be changing at a rate given by the equation

$$\begin{aligned}\dot{\boldsymbol{L}} &= -\gamma(\boldsymbol{B}+\boldsymbol{b}) \wedge \boldsymbol{L} - \boldsymbol{\omega} \wedge \boldsymbol{L} \\ &= -(\boldsymbol{\Delta} + \gamma \boldsymbol{b}) \wedge \boldsymbol{L}, \end{aligned} \quad (13)$$

where $\boldsymbol{\Delta} = \boldsymbol{\omega} - \boldsymbol{\Omega}$ and denotes the degree to which the rotation of $\boldsymbol{b}$ is mistuned to the precessional rate.

In the rotating frame $\boldsymbol{\Delta} + \gamma \boldsymbol{b}$ is fixed and the observer sees $\boldsymbol{L}$ precessing around the direction of the resultant, as shown in fig. 7. Since $\boldsymbol{\Delta}$ and $\boldsymbol{b}$ are perpendicular, the effective field has magnitude $(\Delta^2 + \gamma^2 b^2)^{\frac{1}{2}}/\gamma$, and the precessional angular velocity is $(\Delta^2 + \gamma^2 b^2)^{\frac{1}{2}}$, which we shall denote by χ. In the laboratory frame this pattern is seen as a slow perturbation of the principal precession. In case (a), where the cone does not include the

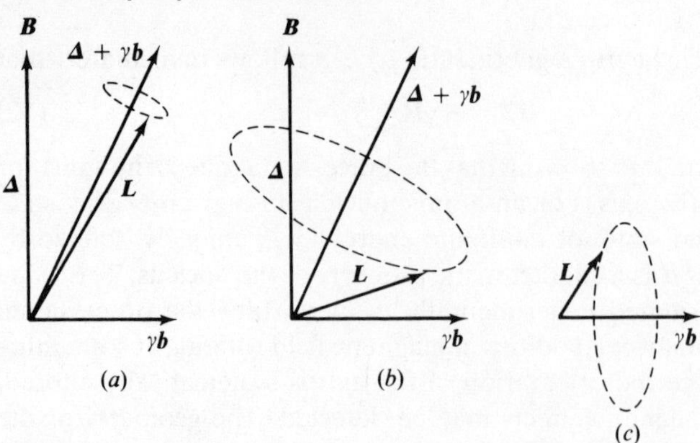

Fig. 7. Trajectories of $\boldsymbol{L}$ in the frame rotating with $\boldsymbol{b}$.

direction of $\boldsymbol{B}$, the vector $\boldsymbol{L}$ does not encircle the vertical axis, and in the laboratory frame the mean precession rate is ω; the moment does, however, alternately lag behind $\boldsymbol{b}$ and lead it, so that $\boldsymbol{b}$ alternately communicates energy to, and extracts it from the gyroscope. This is reflected in the periodic variations of the angle between $\boldsymbol{\mu}$ and $\boldsymbol{B}$, for while the kinetic energy is independent of orientation the potential energy has a magnetic term $-\boldsymbol{\mu}\cdot\boldsymbol{B}$. Case (a) is one in which $\boldsymbol{b}$ is strong enough, or well enough tuned in frequency, and $\boldsymbol{L}$ lies near enough to the optimum direction, parallel to $\Delta+\gamma\boldsymbol{b}$, for the precession rate to be drawn into synchronism. In case (b), however, $\boldsymbol{L}$ lies far enough from $\boldsymbol{\Delta}+\gamma\boldsymbol{b}$ for the cone to encircle the vertical axis. The mean precessional rate is now nearer Ω than ω. In case (c) the rotation of $\mathbf{b}$ is precisely tuned to the unperturbed rate of precession, so that $\boldsymbol{\Delta}=0$; the tilting motion of the axis is symmetrically disposed about the horizontal plane, and occurs at a rate γb.

The response of the system to $\boldsymbol{b}$ may be recorded either as variations in the component of $\boldsymbol{\mu}$ parallel to $\boldsymbol{b}$ or as variations in the transverse component. These measurements do not provide totally independent information because of energy conservation – the component of $\boldsymbol{\mu}$ normal to both $\boldsymbol{B}$ and $\boldsymbol{b}$ determines the rate of energy flow to or from the gyroscope, and hence the rate of change of the tilt angle and the parallel component of $\boldsymbol{\mu}$. Let us concentrate on the transverse component of $\boldsymbol{\mu}$ which, spinning at a mean rate of $\boldsymbol{\Omega}$ or something near, will induce an oscillatory signal in any coil with its axis normal to $\boldsymbol{B}$. It is of particular interest to take the special case when the moment $\boldsymbol{\mu}$ starts parallel to $\boldsymbol{B}$ at time $t=0$, so that the cone on which $\boldsymbol{\mu}$ precesses in the rotating frame has the appearance of fig. 8(a). The projection of $\boldsymbol{\mu}$ on the transverse plane follows round the ellipse of fig. 8(b) with frequency χ. The components parallel and perpendicular to $\boldsymbol{b}$ oscillate harmonically about a non-zero mean value represented by the centre of the ellipse. A little geometry shows that the components vary according to the expressions:

$$\mu_x=\mu\sin\theta\cos\theta(1-\cos\chi t),\qquad \mu_y=\mu\cos\theta\sin\chi t, \tag{14}$$

where θ is the tilt of the axis of the cone out of the transverse plane: $\tan\theta=\Delta/\gamma b$. This is in the rotating frame; in the laboratory frame (capital

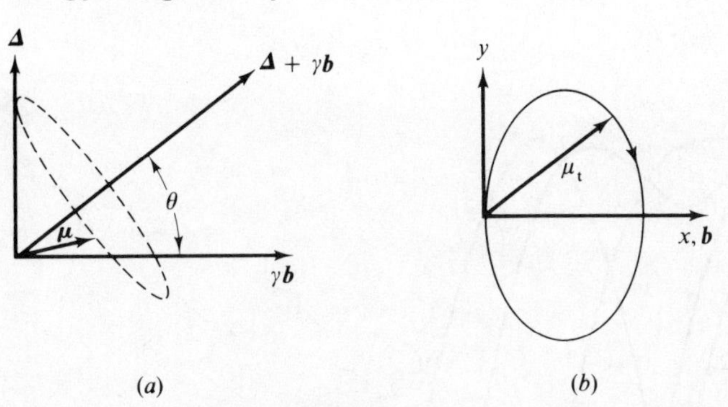

Fig. 8. Special case of fig. 7 when $\boldsymbol{L}$ and $\boldsymbol{\mu}$ pass through the direction $\boldsymbol{B}$. In (*b*) is shown the trajectory of μ_t the component of $\boldsymbol{\mu}$ in the plane normal to $\boldsymbol{B}$, as viewed in the rotating frame.

subscripts) $\mu_X = \mu_x \cos \omega t + \mu_y \sin \omega t$ and $\mu_Y = \mu_y \cos \omega t - \mu_x \sin \omega t$. Therefore

$$\left.\begin{aligned} \mu_X &= \mu \cos \theta[\sin \theta \cos \omega t(1-\cos \chi t) + \sin \omega t \sin \chi t] \\ \mu_Y &= \mu \cos \theta[\cos \omega t \sin \chi t - \sin \theta \sin \omega t(1-\cos \chi t)]. \end{aligned}\right. \tag{15}$$

It is easily seen from fig. 8(*b*) what these expressions mean. An observer in the laboratory frame sees μ_t spinning about the axis of $\boldsymbol{B}$ at the perturbing frequency ω, which is very nearly Ω, while its amplitude varies more slowly, with period $2\pi/\chi$, as μ_t traverses the elliptical locus. When $\gamma b \ll \Delta$, so that θ is nearly $\pi/2$, the ellipse shrinks to a small circle, and the transverse component μ_t oscillates harmonically:

$$\mu_t = 2\gamma b\mu \sin(\tfrac{1}{2}\chi t)/(\Delta^2 + \gamma^2 b^2)^{\frac{1}{2}} \sim \frac{2\gamma b\mu}{\Delta} \sin(\tfrac{1}{2}\chi t).$$

As this spins, the components in the laboratory frame vary according to the expressions

$$\left.\begin{aligned} \mu_X &= \frac{2\gamma b\mu}{\Delta} \sin(\omega + \tfrac{1}{2}\chi)t, \\ \text{and} \qquad \mu_Y &= \frac{2\gamma b\mu}{\Delta} \cos(\omega + \tfrac{1}{2}\chi)t. \end{aligned}\right\} \tag{16}$$

146 These should be compared with (6.31) and fig. 6.12(*b*). The behaviour is almost exactly the same as a lossless harmonic vibrator excited near resonance, and exhibiting a beat pattern due to the superposition of the steady-state response and the transient excitation.

In (16) the beat frequency is determined by the departure of the perturbing signal $\boldsymbol{b}$ from resonance, and depends little on the amplitude of $\boldsymbol{b}$. This is true only when $\gamma b \ll \Delta$, but as the perturbation approaches resonance and this inequality fails, the beat pattern changes its character. The circular locus of μ_t becomes markedly elliptical and ultimately, just on resonance, a straight line, when $\theta = 0$ and the circular orbit in fig. 8(*a*) lies in a vertical plane. The axis of the gyroscope, viewed in the laboratory frame, if its starts parallel to $\boldsymbol{B}$ begins to precess in a steadily widening cone, passes through the transverse plane and converges onto the opposite direction to $\boldsymbol{B}$; the process then repeats itself in reverse, and so on as long

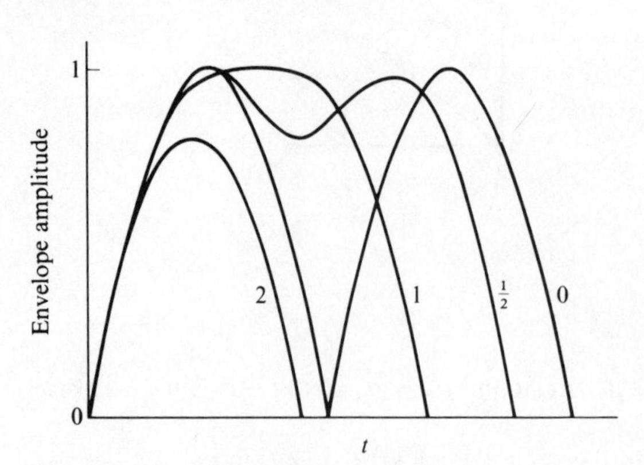

Fig. 9. Envelope curves showing the development of the component of $\boldsymbol{\mu}$ in the plane normal to $\boldsymbol{B}$ when the moment starts, at time $t=0$, parallel to $\boldsymbol{B}$. The amplitude of $\boldsymbol{b}$ is the same for all, but Δ is different; the numbers attached to the curves are values of $\Delta/\gamma b$. An envelope amplitude of 1 means that $\boldsymbol{\mu}$ lies in the transverse plane. In the absence of dissipative effects etc. the curves repeat themselves indefinitely.

as the resonant perturbation is maintained. A series of cases, with b constant and Δ varying through resonance, is shown in fig. 9, which gives only the envelope of the oscillations at frequency ω, i.e. the amplitude of μ_T. The saturation of the oscillations is clearly exhibited, and the reason for saturation is obvious: μ_T can never exceed μ. As saturation is approached the non-linearity of the system also reveals itself in the way the beat frequency does not fall to zero at resonance but reaches a minimum value γb. In contrast to fig. 6.12, the envelope becomes markedly non-sinusoidal as resonance is approached.

This is a quite different type of non-linear vibrator from anything we have met so far, in that it is isochronous but saturable. In this respect the behaviour is nothing like that of a particle in a non-parabolic potential, which may possibly be isochronous, but which is not saturable., The difference may be illustrated by the response to a gliding tone. In chapter 6 we showed how a lossfree harmonic vibrator, excited by a constant-amplitude signal whose frequency varied linearly with time ($\omega=\omega_0+\beta t$), ended up after passage through the resonance with an amplitude of vibration proportional to $\beta^{-\frac{1}{2}}$. With the gyromagnetic top the corresponding behaviour is a change in tilt of the axis relative to $\boldsymbol{B}$. When the frequency sweep is rapid the change is small and proportional to $\beta^{-\frac{1}{2}}$ – this is the linear approximation to the gyroscope equation which causes it to follow harmonic vibrator theory exactly. But if β is small enough for the predicted change to be well outside the linear approximation (for example, apparently conferring kinetic energy far in excess of the maximum permitted to the top) the matter obviously needs reconsideration.

149

In fact, in this limit the behaviour is very simply understood. Rather than sweeping the frequency, keeping $\boldsymbol{B}$ constant, let us keep ω constant and lower $\boldsymbol{B}$ slowly from a value well above resonance to one well below; this is an entirely equivalent process, and one that is carried out in certain procedures (*Adiabatic passage*) used in the study of nuclear magnetic resonance.† In the rotating frame the situation at some instant will resemble that illustrated in fig. 8(a). Let $\boldsymbol{B}$ now be suddenly changed by a small amount $\delta\boldsymbol{B}$ so as to depress the orientation of $\boldsymbol{\Delta}+\gamma\boldsymbol{b}$, without causing any immediate variation of the direction of $\boldsymbol{L}$. If, when this change is made, $\boldsymbol{L}$ is somewhere on the upper half of the cone whose axis is $\boldsymbol{\Delta}+\gamma\boldsymbol{b}$, the depression of the axis means that it must now sweep out a slightly wider cone; conversely if it happens to be on the lower half the cone is narrowed. When $\boldsymbol{B}$ is changed by a large number of small increments over a time that allows many cycles of precession, there is on the average no first-order change in the cone angle – its axis simply follows the direction of $\boldsymbol{\Delta}+\gamma\boldsymbol{b}$. Second-order change in the cone angle can be made as small as desired by changing $\boldsymbol{B}$ sufficiently slowly. This process of adiabatic passage results in the component of $\boldsymbol{M}$ parallel to $\boldsymbol{B}$ simply reversing in sign as $\boldsymbol{B}$ goes through the resonant value and $\boldsymbol{\Delta}+\gamma\boldsymbol{b}$ swings from vertically upwards to vertically downwards.

Nuclear magnetic resonance (NMR)

No attempt will be made to give a systematic account of the vast field of study opened up by the observation of magnetic resonance in solids, liquids and gases. We shall instead point out a few features that exemplify in practice the characteristics of this rather special oscillatory system. Adhering to the classical model we imagine a sample of water as a rather dense array (7×10^{28} to the cubic metre) of protons, the nuclei of the hydrogen atoms, and forget for the moment everything else. Each proton behaves like a magnetic top, with angular momentum L equal to $\frac{1}{2}\hbar$, or 0.527×10^{-34} Js, and magnetic moment μ equal to 1.41×10^{-26} J/T; the value of γ is 2.75×10^{8} T^{-1} s^{-1}, which means that in a field of 1 T the precession rate is $\frac{1}{2}\gamma/\pi$ or 42.6 MHz. If each precessed independently in a perfectly uniform field, all would have the same angular velocity Ω, so that in a frame rotating at this speed the whole distribution of moments could be represented by a stationary array of points on the surface of a sphere of unit radius, each point marking the end of a vector drawn from the centre. When this idealized system is perturbed by a field, $\boldsymbol{b}$, rotating with angular velocity $\boldsymbol{\omega}$, an observer rotating with $\boldsymbol{b}$ will see each point moving in a circle with the same angular velocity $\boldsymbol{\Delta}+\gamma\boldsymbol{b}$ about the same tilted axis, as in fig. 7. The whole distribution of points therefore turns bodily about this axis, and any resultant moment $\boldsymbol{M}$ of the whole assembly turns with it. This makes it a

† The term *adiabatic* is used in thermodynamics and in quantum mechanics with subtly different meanings. This particular usage is of the latter kind, and implies that the parameters of a physical system (in this case $\boldsymbol{B}$) are changed so gradually that the system evolves smoothly during the process. Thus far there is no conflict between thermodynamic and quantum mechanics. But it is further implied in quantum mechanics that the system is isolated from random processes such as collisions; thermodynamics does not require this restriction – indeed it would often make the process non-adiabatic.

very simple matter to calculate how the system responds to the perturbation.

Let us suppose that $\boldsymbol{B}$ alone is applied, and the system then allowed enough time (which may prove quite long) to settle into thermal equilibrium. Because the energy of a moment pointing along $\boldsymbol{B}$ is less, by $2\mu\boldsymbol{B}$, than that of one pointing against $\boldsymbol{B}$, there will be a slight excess of the former in thermal equilibrium. Making here our one concession to quantum mechanics (for the sake of simplicity rather than necessity) we recognize that only parallel and antiparallel states are permitted to the proton, and that these, according to the Boltzmann distribution, will be occupied so that out of N moments,

$N\,\mathrm{e}^{X}/(\mathrm{e}^{X}+\mathrm{e}^{-X})$ will point along $\boldsymbol{B}$

$N\,\mathrm{e}^{-X}/(\mathrm{e}^{X}+\mathrm{e}^{-X})$ will point against $\boldsymbol{B}$, where $X=\mu B/k_{\mathrm{B}}T$.

The resultant total moment M is due to the difference in these two occupations:

$$M = N\mu \tanh X \sim N\mu^2 B/k_{\mathrm{B}}T \quad \text{if } X \ll 1. \tag{17}$$

In a field of 1 T at room temperature, X is about 3.5×10^{-6}, so that the resultant moment is as if only this tiny fraction were aligned with the field, the rest being randomly oriented. Nevertheless it is quite enough to give a good signal when appropriately perturbed.

One method of detecting $\boldsymbol{M}$ is to apply the field $\boldsymbol{b}$ at the resonant frequency for just as long $(\pi/2\gamma b)$ as is needed to cause each moment, and therefore $\boldsymbol{M}$, to turn through $\pi/2$ about $\boldsymbol{b}$ in the rotating frame. Immediately afterwards $\boldsymbol{M}$, instead of pointing along $\boldsymbol{B}$, lies in the transverse plane and stays in it while it precesses (in the laboratory frame) with angular velocity $\boldsymbol{\Omega}$. Any pick-up coil surrounding the sample, with its axis in the transverse plane, will now have induced in it an oscillating e.m.f. at frequency Ω. If the applied field were perfectly uniform and the nuclei free from any perturbing interactions, the oscillation would persist indefinitely. In practice, however, it decays, either through inhomogeneity of the field which causes differently located nuclei to precess at different rates or, if this effect has been reduced to insignificance, through interactions of the nuclei among themselves and with their surroundings. This method is analogous to determining the impulse response function of a vibrator. Alternatively the spin assembly may be excited by a coil producing a continuous oscillatory magnetic field transverse to $\boldsymbol{B}$, and the response detected by means of a similar coil normal to the first. This arrangement, due to Bloch, avoids swamping the signal induced by the spinning moment, since the input and output coils are not magnetically coupled. The resonant response which can be traced out by varying the input frequency is related to the impulse response by the usual formula, provided the amplitude of excitation is low enough for linearity to be maintained.

106

The interactions between nuclei deserve some discussion here, since they serve to exemplify a number of points of general physical interest. For a fuller treatment the reader should seek specialized texts – the matter is complex and abounds with subleties. At first sight one might think that, isolated from everything else, the assembly of nuclei could not alter their total moment $\boldsymbol{M}$ by mutual interactions, for one would imagine the total angular momentum associated with their spin to be conserved. This is a misconception, however – the interaction between two magnetic moments is non-central and transfers angular momentum to the translational degrees of freedom. A typical example is shown in fig. 10; both moments are subjected to a clockwise torque, the sum of the two being compensated by an equal anticlockwise torque arising from translational forces experienced by each magnetic moment in the inhomogeneous field of the other. It is therefore possible for a molecule like water, containing two protons, to exchange spin angular momentum for rotation of the molecule as a whole. Since translational motions are readily exchanged during collisions, this process provides a mechanism by which the precessing moment of the assembly can be dissipated by conversion into bulk rotation.

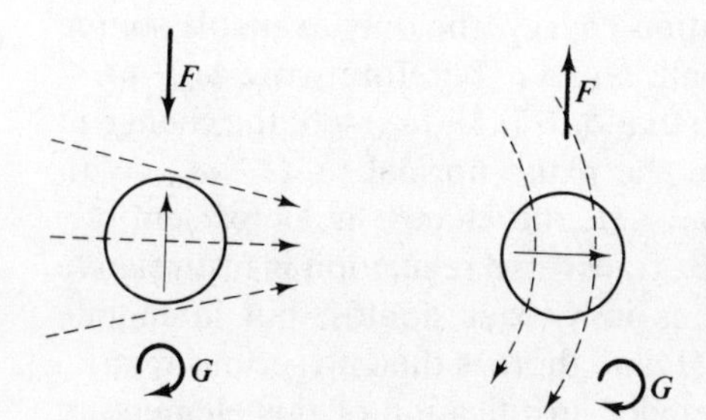

Fig. 10. Two protons with moments oriented as shown by the arrows in the circles; each produces a magnetic field at the other's location shown by the broken lines, leading to forces F and torques G.

Interaction between the protons in a molecule is the dominant dissipative process in water, but interaction between neighbouring molecules is also effective. Whatever the detailed mechanism, it is clear that the points representing individual moments can wander round the unit sphere in a diffusive manner, so that a lopsided distribution, established in the manner described, is gradually restored to the equilibrium configuration. In certain cases, of which we shall soon examine one in more detail, there is reason to expect that the decay of the transverse moment will be exponential, but this is not a general rule.

232

The process just outlined, whereby an ordered assembly of spins relaxes to thermal equilibrium, is slow compared to the precessional rate, for the magnetic field at one proton due to the other in a water molecule is only about 1 mT, an extremely weak perturbation of the main field $\boldsymbol{B}$. The response function is normally, therefore, a long train of oscillations and the resonance peak is correspondingly narrow; under the best conditions, indeed, Q-values of more than 10^9 are observed (see fig. 11). We shall now look into the matter in some detail, and it is important to recognize at the outset a distinction between relaxation processes that conserve the component of $\boldsymbol{M}$ parallel to $\boldsymbol{B}$ and those that do not. If the assembly has been polarized with $\boldsymbol{M}$ lying away from the direction of $\boldsymbol{B}$, it is possible for the individual moments to wander along lines of latitude (with respect to $\boldsymbol{B}$ as polar axis) and for the transverse component to decay while the longitudinal component stays constant; this involves no change in the polarization energy $\boldsymbol{M}.\boldsymbol{B}$, while relaxation of the longitudinal component requires energy exchange between the spins and the liquid as a whole. Let us estimate the energy exchange possible when spin angular momentum is converted into molecular angular momentum, a process strongly affected by the rate at which the molecule is turning at the time. We picture the molecule as a rigid body to which the protons are attached, without any constraint on their rotation except for their mutual interaction. As the molecule tumbles about under the impacts of its neighbours the proton spins stay pointing in the same direction, or precess in whatever magnetic field they experience, quite unaffected by the gross molecular motion.† Now suppose the interaction between the spins to cause $\delta\boldsymbol{L}$ of angular momentum to be transferred from them to the molecule at a time when its angular velocity is $\boldsymbol{\omega}_{\mathrm{m}}$. The accompanying energy exchange, $\boldsymbol{\omega}_{\mathrm{m}} \cdot \delta\boldsymbol{L}$, occurs at the expense of the polarization energy, the only available source since the interaction energy is so small, and we therefore write $\boldsymbol{\omega}_{\mathrm{m}} \cdot \delta\boldsymbol{L} = \boldsymbol{B} \cdot \delta\boldsymbol{\mu}$. But $\delta\boldsymbol{\mu} = \gamma\delta\boldsymbol{L}$ and $\boldsymbol{\Omega} = \gamma\boldsymbol{B}$, so that $\delta\boldsymbol{\mu} \cdot (\boldsymbol{\Omega} - \boldsymbol{\omega}_{\mathrm{m}}) = 0$; the change in spin moment can only take place in the plane normal to $\boldsymbol{\Omega} - \boldsymbol{\omega}_{\mathrm{m}}$. With slowly rotating molecules, having $\omega_{\mathrm{m}} \ll \Omega$, the change is almost entirely confined to the plane transverse to $\boldsymbol{B}$; transverse relaxation is fast relative to longitudinal. This is what happens in viscous liquids, but in mobile liquids ω_{m} may considerably exceed Ω, and there is then no serious restriction to the direction of $\delta\boldsymbol{\mu}$. More detailed justification of this elementary analysis will follow, in which it will be seen that increasing the tumbling speed does not make longitudinal relaxation easier – on the contrary both types of relaxation are inhibited, but the transverse more so than the longitudinal; it is in this way that they become comparable, as the Q of the resonance rises towards the enormous value quoted above.

Early in the history of nuclear magnetic resonance Bloch appreciated the possibility of transverse and longitudinal relaxation occurring at different rates, and introduced in his formulation of the equations of motion for the total moment $\boldsymbol{M}$ two characteristic relaxation times: T_1, the longitudinal relaxation time describing thermal equilibration in which the component of $\boldsymbol{M}$ parallel to $\boldsymbol{B}$ is changed, and T_2, the transverse relaxation time describing the diffusive decay of the transverse component. His introduction of these relaxation times was heuristic, expressing the essential idea in the simplest way, without any pretence that it described any real situation

534

† This is not so if the nucleus (not a proton) has an electric quadrupole moment, for this is acted on by the intramolecular electric field gradients which try to lock the spin direction relative to the molecule. We shall not consider the effects of this complicating process.

exactly; all the same, it does very well in a large number of applications.

We write Bloch's equations in complex form, using M_1 (real) to denote the parallel component of $\boldsymbol{M}$ and M_2 (complex) to denote the magnitude and direction of the transverse component. In addition a complex number b represents the rotating field in the transverse plane. The translation of the vector product $\boldsymbol{M} \wedge \boldsymbol{B}$ into this notation looks rather awkward but is quite straightforward. We quote the resulting equations and then expound them:

$$\dot{M}_1 + \gamma \,\mathrm{Im}\,[M_2 b^*] + (M_1 - M_0)/T_1 = 0, \qquad (18)$$

and

$$\dot{M}_2 + \mathrm{i}\gamma(M_2 B - M_1 b) + M_2/T_2 = 0. \qquad (19)$$

If $b=0$, the equations express the exponential relaxation of the magnetization to its equilibrium value M_0 in the steady field B. According to (18), M_1 returns to M_0 with relaxation time T_1; and according to (19), M_2 precesses with angular velocity Ω $(=\gamma B)$, while decaying to zero with relaxation time T_2. The perturbing terms containing b are the components of the vector product $\boldsymbol{M} \wedge \boldsymbol{b}$, and show how M_1 is changed by the interaction of M_2 with b, M_2 by the interaction of M_1 with b.

When b rotates at angular velocity ω we may write it as $b'\,\mathrm{e}^{-\mathrm{i}\omega t}$, b' being taken as real, and look for a steady-state solution in which M_1 is constant and M_2 varies as $(M_2' + \mathrm{i}M_2'')\,\mathrm{e}^{-\mathrm{i}\omega t}$. Substituting this trial solution we find that

$$M_1 = M_0 - \gamma b' T_1 M_2'' \quad \text{from (18)}, \qquad (20)$$

and

$$(1 - \mathrm{i}\,\Delta T_2)(M_2' + \mathrm{i}M_2'') = \mathrm{i}\gamma M_1 b' T_2 \quad \text{from (19)}, \qquad (21)$$

where, as before, Δ is written for $\omega - \Omega$. When M_1 from (20) is substituted in (21) and the real and imaginary parts of (21) separated, the solution emerges in the form

$$-M_2'/\Delta T_2 = M_2'' = \gamma b' T_2 M_0/(1 + \Delta^2 T_2^2 + \gamma^2 b'^2 T_1 T_2). \qquad (22)$$

The amplitude of the moment in the transverse plane is $(M_2'^2 + M_2''^2)^{\frac{1}{2}}$:

$$|M_2| = \gamma b' T_2 M_0 (1 + \Delta^2 T_2^2)^{\frac{1}{2}}/(1 + \Delta^2 T_2^2 + \gamma^2 b'^2 T_1 T_2). \qquad (23)$$

There is no need to describe the various techniques used to test and record the character of the response to a steady oscillatory field b, since these are fully dealt with in specialized texts. We shall consider only the salient physical processes that determine this character. So long as $\gamma b' \ll (T_1 T_2)^{-\frac{1}{2}}$ the last term in the denominator is negligible and the behaviour is a linear resonant response with a Q-value of $\frac{1}{2}\Omega T_2$; M_2 lags behind b by $\cot^{-1}(\Delta T_2)$ and causes the assembly of spins to receive energy continuously, to be dissipated as heat by the relaxation mechanism. At resonance the dissipation is greatest, with a phase lag of $\pi/2$ so that the power input W_{res} is $\Omega M_2'' b'$. From (22), without restricting b' to a small value,

$$W_{\mathrm{res}} = \gamma^2 b'^2 T_2 M_0 B/(1 + \gamma^2 b'^2 T_1 T_2), \qquad (24)$$

and in this state, according to (20), M_1 is reduced from its equilibrium value of M_0 to $M_0 - \delta M_1$, where

$$\delta M_1/M_0 = \gamma^2 b'^2 T_1 T_2/(1+\gamma^2 b'^2 T_1 T_2). \tag{25}$$

This change in M_1 implies that the energy of the assembly is increased by $B\,\delta M_1$. If this is constantly being dissipated to an external energy sink (e.g. molecular motions) at a rate $B\,\delta M_1/T_1$, comparison of (24) and (25) shows that, as must of course be true in the steady state, the assembly of nuclei is losing as much energy as it gains.

There is a limit to the power dissipation set by the saturation of the system, which shows itself in the term in b'^2 in the denominator of (22). However large b' may be, W_{res} cannot exceed $M_0 B/T_1$, when M_1 has fallen to zero. The amplitude of response, as measured by $|M_2|$ in (23), reaches a maximum at resonance when $\gamma^2 b'^2 T_1 T_2 = 1$; this maximum value is $\frac{1}{2}M_0(T_2/T_1)^{\frac{1}{2}}$, which is quite sizeable when T_1 and T_2 are comparable, as in many liquids. In solids, however, T_2 is commonly much shorter than T_1 and the signal is correspondingly weak.

As already remarked, T_1 and T_2 may be long in mobile liquids, making possible resonances of very high Q indeed. The experimental curve in fig. 11 has peaks as little as 0.04 Hz wide at a frequency of 200 MHz, so that $Q \sim 5\times 10^9$. With a decay time of seconds the systematic traversing of the spectrum would be a very slow process, but the acquisition of data is greatly accelerated by use of the impulse response, which is the superposition of decaying oscillations at each of the frequencies present. The information needed to obtain the spectrum shown here was provided by a single shot, with the impulse response recorded for 109 seconds only; from then on the spectrometer was available for other samples, while the response function was Fourier-analysed by computer, in accordance with the principle expressed by (5.4). The fine structure in the spectrum arises from the fact that the various protons (hydrogen nuclei) in the molecule of *ortho*-dichlorobenzene are coupled by their magnetic moments (it is essentially a quantum-mechanical feature that such coupling results in discrete lines) but the whole range of the spectral distribution, of which only a part is shown here, occupies less than 100 Hz; the frequencies differ by one part in 2×10^6 at most. The impulse response is therefore a virtually pure oscillation at a frequency of 200 MHz, with an envelope showing a slow but complicated beating pattern having a typical time-scale of 10 ms, and decaying with a time-constant of about 8 s. It is only necessary to record the variations of amplitude and phase of the basic oscillation at such intervals as will reveal the modulations of the envelope, and something like 10^4 data points suffice. To determine the phase a comparison oscillator is used (details of the electronic circuitry need not concern us) whose frequency stability must be such that no significant phase drift occurs in 109 s. This is no trivial matter, but well within the capability of modern techniques. It is perhaps more impressive to find that a superconducting solenoid can be designed and shimmed so that over the 5 mm extent of the

106

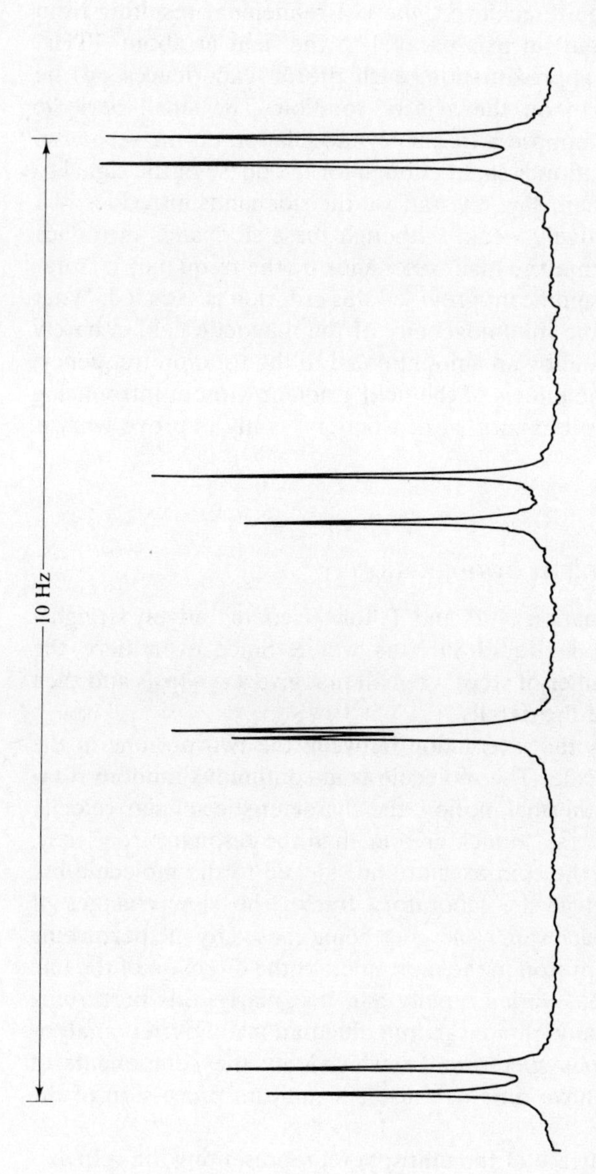

Fig. 11. Example of extremely sharp proton resonances at 200 MHz in a magnetic field of 4.7 T. The whole spectrum shown occupies a band of 10 Hz and the lines are typically 0.04 Hz wide at half-peak ($Q \sim 5 \times 10^9$). The spectrum was obtained with a WP 200 spectrometer (Bruker, Spectrospin Ltd.), the sample being a dilute solution of *ortho*-dichlorobenzene, $C_6H_4Cl_2$, in deuterated acetone, $(CD_3)_2CO$; having no hydrogen, the solvent contributes no resonances in this frequency range.

sample its field of 4.7 T is uniform to two parts in 10^{10}, as is necessary if the lines are not to be broadened as a result of protons in different regions of the sample precessing at different frequencies. In fact this degree of perfection is probably not quite achieved, the last refinement resulting from spinning the sample about an axis parallel to the field at about 30 Hz. Then, to first order of approximation, each proton experiences on the average the field strength on the axis of rotation. The small periodic fluctuations as it rotates impose a frequency modulation on the response, but so long as the modulation is slight enough for the phase of the signal to be shifted by no more than, say, one radian, the sidebands introduced by the modulation are relatively weak. Although these sidebands introduce 'ghost' lines separated from the main resonance by the frequency of rotation, they will remain insignificant provided this criterion is satisfied. What it amounts to is that if the inhomogeneity of the magnetic field is barely enough to broaden the line by an amount equal to the rotation frequency, rotation can eliminate the effects of the field gradient without introducing excessive sidebands; but it cannot work wonders – only improve what is already very good.

The physical mechanism of relaxation

We turn now to the estimation of T_1 and T_2 for the comparatively straightforward case of a mobile liquid such as water. Since even here the argument involves a number of steps we shall first give a synopsis and then proceed to fill in some of the details.

(i) We consider only the interaction between the two protons in the same water molecule. The molecule is in continuous random rotation as part of its thermal motion, the characteristic angular velocity being perhaps $10^{12}\,s^{-1}$, much greater than the resonant frequency. In this tumbling, the spin axes are not locked to the molecule but stay almost fixed in the laboratory frame, the slow changes of orientation that each spin undergoes being caused by the perturbing field of the other proton in the molecule. As the direction of the line joining the protons varies rapidly and irregularly, this perturbing field suffers equally rapid random fluctuations. Fourier analysis yields a continuous spectrum, of which only the components of lower frequency have time to cause a significant precession of the spin.

(ii) A point on the surface of the unit sphere, representing the spin axis of one proton, wanders randomly as a consequence of this perturbation, and the points representing all protons in the sample behave statistically as though they were governed by a diffusion constant D; if each point wanders independently the flux of points is described by a vector equal to $-D$ grad n, where n is the number of points per unit area, varying with position.

(iii) A polarized distribution of points on the sphere, such as describes a resultant magnetic moment $\boldsymbol{M}$, evens itself out by this diffusive process, and we shall show that, left to itself, $\boldsymbol{M}$ decays as $e^{-t/T}$, where $T = 1/2D$.

(iv) The calculation of D proceeds in three stages and makes use of ideas already presented in earlier chapters: (a) a particle moving on a line with nearly random velocity, $v(t)$, will be shown to be governed by a diffusion constant equal to $\frac{1}{2}\pi I_\omega(0)$, where $I_\omega(0)$ is the low-frequency limit of the intensity spectrum of v; (b) the intensity spectrum itself will be derived from the autocorrelation function of the tumbling motion of the molecules, supplemented by (c) an explicit calculation of the strength of the magnetic interaction between the two protons.

81
94

(v) The outcome of this analysis is that when the tumbling is much more rapid than the precessional motion

$$T_1 = T_2 = 160\pi^2 a^6 / 9\mu_0^2 \hbar^2 \gamma^4 \tau_c, \tag{26}$$

where a is the separation of the protons in the molecule and τ_c is the correlation time for the tumbling motion, the time it takes for memory of an earlier orientation of the molecule to be lost. For water, $a \doteqdot 0.15$ nm and $\tau_c \doteqdot 2 \times 10^{-12}$ s, while γ for the proton is $2.675 \times 10^8\ \mathrm{T^{-1}\,s^{-1}}$.† This yields a value of about 10 s for T_1 and T_2. The experimental value at room temperature is 3.6 s, and the difference is easily accounted for by the approximate value for τ_c and the neglect of interactions between protons on different molecules.

96

To put flesh on these bones, we start with (iii). Let the position of a point on the sphere be defined by polar angles (θ, ϕ), and let us consider M_1, the component of magnetization parallel to the polar axis. The ϕ-variable is irrelevant to M_1, all that matters being the mean concentration $n(\theta)$ around the line of latitude defined by θ. The diffusive flux across this line, whose length is $2\pi \sin\theta$, is $-2\pi \sin\theta \cdot D\, \partial n/\partial\theta$. Since the number of points is conserved, the variation of this quantity with θ determines the rate at which n changes:

$$\sin\theta\, \partial n/\partial t = D \frac{\partial}{\partial\theta}(\sin\theta\, \partial n/\partial\theta). \tag{27}$$

† The correlation time can be determined quite independently from the frequency-variation of the dielectric constant. In a slowly oscillating electric field, such that $\omega\tau_c \ll 1$, the electrical dipoles of the water molecule have time to follow the field changes, and ε is large; but when $\omega\tau_c \gg 1$ there is too little time available, and ε is much smaller. The similarity in magnitude of $1/\tau_c$ and the mean angular velocity of tumbling is coincidental, and means only that a water molecule can rotate through a substantial angle before collisions randomize its motion. In a viscous liquid this coincidence would not occur.

If sin θ is crossed out on both sides, this turns into the normal equation for diffusion on a line; the presence of sin θ makes allowance for the geometry of a spherical surface. The magnetization is given by the equation

$$M_1 = 2\pi\mu \int_0^\pi n \sin\theta \cos\theta \, d\theta, \tag{28}$$

so that $$dM_1/dt = 2\pi\mu \int_0^\pi \frac{\partial n}{\partial t} \sin\theta \cos\theta \, d\theta$$

$$= 2\pi D\mu \int_0^\pi \frac{\partial}{\partial\theta}\left(\sin\theta \frac{\partial n}{\partial\theta}\right) \cos\theta \, d\theta, \quad \text{from (27),}$$

$$= 2\pi D\mu \int_0^\pi \sin^2\theta \frac{\partial n}{\partial\theta} \, d\theta, \quad \text{on integrating by parts.}$$

A further integration by parts reduces the integral to $-\int_0^\pi 2n \sin\theta \cos\theta \, d\theta$, so that from (28) one sees that

$$dM_1/dt = -2DM_1.$$

Hence $$M_1 = M_0 e^{-t/T}, \quad \text{where } T = 1/2D. \tag{29}$$

This is the result quoted in (iii).

To relate D to the velocity fluctuations ((iv)a of the synopsis), consider a distribution of points whose density varies in one direction (which we label s) only. Then the diffusive motion obeys the equation:

$$\partial n/\partial t = D \, \partial^2 n/\partial s^2.$$

The spread of an initially compact group of points follows the solution

$$n \propto e^{-s^2/4Dt}/t^{\frac{1}{2}},$$

83

and the mean square displacement of a point in time t,

$$\overline{s^2}(t) = \int_0^\infty s^2 n(s,t) \, ds \Big/ \int_0^\infty n(s,t) \, ds = 2Dt. \tag{30}$$

This result is equally applicable to motion along a line drawn on a sphere; in any case we shall be dealing with small displacements only in calculating D, and the curvature is irrelevant.

Let us now calculate $\overline{s^2}(t)$ in terms of the velocity spectrum. If $\dot{s}(t)$ is a random real function with intensity spectrum I_ω, the components in a narrow range $\delta\omega$ combine to give a nearly periodic oscillation of $\dot{s}$ having the form $(2I_\omega\delta\omega)^{\frac{1}{2}} \cos(\omega t + \phi)$, ϕ being a random phase. By integration the oscillatory displacement due to the components in the narrow band is given by the expression

$$\delta s(t) = (2I_\omega\delta\omega)^{\frac{1}{2}}[\sin(\omega t + \phi) - \sin\phi]/\omega,$$

the integration constant being supplied to make $\delta s(0)$ vanish. This can be written in the form

$$\delta s(t) = 2(2I_\omega \delta\omega)^{\frac{1}{2}} \sin(\tfrac{1}{2}\omega t)\, \mathrm{Re}\,[\mathrm{e}^{\mathrm{i}(\frac{1}{2}\omega t+\phi)}]/\omega,$$

and interpreted as the real part of a randomly phased vector of magnitude $2(2I_\omega\delta\omega)^{\frac{1}{2}} \sin(\frac{1}{2}\omega t)/\omega$. Summation of all similar contributions from different ranges $\delta\omega$ to give an average resultant is achieved by adding the squares, so that the mean resultant vector has length R^2, where

81

$$R^2(t) = \int_0^\infty 8I_\omega \sin^2(\tfrac{1}{2}\omega t)\, \mathrm{d}\omega/\omega^2.$$

Since R is the length of a vector on a vibration diagram, of which only the real part is physically significant, we have that

$$\overline{s^2}(t) = \tfrac{1}{2}R^2(t) = 4\int_0^\infty I_\omega \sin^2(\tfrac{1}{2}\omega t)\, \mathrm{d}\omega/\omega^2. \tag{31}$$

The presence of $\sin^2(\frac{1}{2}\omega t)/\omega^2$ shows that the integral is dominated by values of $\omega \lesssim 2\pi/t$. If I_ω is flat at low frequencies, falling off at higher frequencies, the evaluation of (31) may be troublesome when t is short, but for longer times it is easy since I_ω may be given its low-frequency limit $I_\omega(0)$. Physically this means that memory of previous motion persists for a time (the reciprocal of the frequency at which I_ω begins to fall off) and during this interval the motion of the point on the sphere is not random; but if one is concerned with motion over a much longer interval, the short-term memory becomes insignificant compared with the random walk. In a viscous liquid the orientation of a molecule may stay sensibly constant for quite a while, during which the mutual interactions cause the spins to drift systematically through large angles; this invalidates our model of the process as essentially diffusive. In water, on the other hand, the combination of rapid tumbling and frequent collisions means that the systematic drifting is constantly disrupted and can be safely neglected. We are therefore justified in writing, from (31),

$$\overline{s^2}(t) = \pi I_\omega(0)t.$$

The variation of $\overline{s^2}$ as t, having the same form as (30), enables us to identify D immediately:

$$D = \tfrac{1}{2}\pi I_\omega(0), \quad \text{as quoted in (iv)}a. \tag{32}$$

We now have a picture of the points executing diffusive motion in the form of a nearly random velocity whose component in one direction, $\dot{s}$, is described by the intensity spectrum I_ω, which we now estimate (part (iv)b of the program). It is clear that during an interval of time shorter than is needed for the molecule to rotate very far the interaction between two protons is constant and v is also constant. There is therefore persistence of memory during a correlation time τ_c, which measures the time during which the molecule tumbles far enough and randomly enough to forget where it was before. We expect the autocorrelation function of $\dot{s}$ to decay (probably exponentially but the precise manner is unimportant) with some characteristic time τ_c. This is a case already discussed in chapter 4, where it is shown that $I_\omega = I_\omega(0)/(1+\omega^2\tau_c^2)$. By integrating over all positive frequencies we have that **96**

$$\overline{\dot{s}^2} = \int_0^\infty I_\omega \, d\omega = \tfrac{1}{2}\pi I_\omega(0)/\tau_c.$$

The value of $\overline{\dot{s}^2}$ depends only on the strength of the interaction between the dipoles. The effect of shortening τ_c is to spread I_ω over a wider frequency range and thus to lower its value at those low frequencies that are effective in the diffusive process. This is the reason for the especially sharp resonances in mobile fluids. As follows from (29) and (32)

$$T = 1/2\overline{\dot{s}^2}\tau_c. \tag{33}$$

This completes the program except for (iv)c, the evaluation of $\overline{\dot{s}^2}$ in terms of the magnetic field exerted by one proton on another. We shall carry out the calculation in cartesian coordinates, concentrating on v_y, the y-component of velocity on the unit sphere. Let one proton define the z-axis, as does A in fig. 12, and let the other, at B, have an arbitrary orientation.

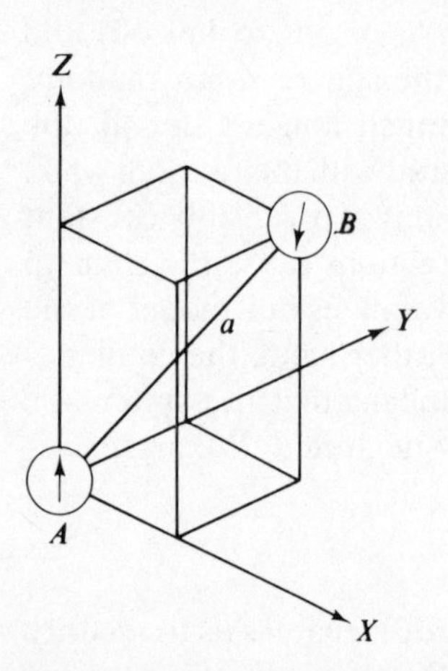

Fig. 12. Notation for calculating mutual torques of two magnetic dipoles.

By letting B occupy all positions on a sphere of radius a (the fixed separation of the protons in the molecule), and also allowing the spin at B to assume all orientations, we span the range of torques that one proton can exert on another, the latter having a definite orientation. The field at A due to the proton at B has an x-component (which is all that matters in calculating v_y) given by

$$B_x = \frac{\mu_0}{4\pi}\left[\frac{3x}{a^5}(\mu_x x + \mu_y y + \mu_z z) - \frac{\mu_x}{a^3}\right],$$

$\mu_{x,y,z}$ being the components of the dipole moment of the proton at B. The y-component of the torque on the dipole μ at A is then μB_x, and it is this that causes the point representing the dipole at A to move on the unit sphere with velocity $v_y = \mu B_x/L$:

$$v_{yA} = \frac{\mu_0 \mu}{4\pi L}\left[\frac{3x}{a^5}(\mu_x x + \mu_y y + \mu_z z) - \frac{\mu_x}{a^3}\right]. \tag{34}$$

At the same time the dipole at A exerts a torque on that at B causing it to move with velocity v_{yB}:

$$v_{yB} = \frac{\mu_0 \mu}{4\pi L}\left[\frac{3z}{a^5}(\mu_z x - \mu_x z) + \frac{\mu_x}{a^3}\right]. \tag{35}$$

Because the spins are coupled in this way their random motions are not independent, and the preceding analysis which assumed independence would be incorrect without allowance for this feature. To take an extreme example, if it were the case that the total spin angular momentum was conserved the two points would undergo such correlated random motion that no relaxation of the total moment occurred. As it happens, the correlation here has the reverse effect of enhancing the change in total moment over what would take place if each point wandered independently. The way to take account of correlation is to ascribe the whole movement to the proton at A, writing $v_y = v_{yA} + v_{yB}$; similarly when treating the motion of the proton at B we should add in the consequent motion of A. This looks like double counting but is not. What we are doing is to imagine in the first place that there is no correlation, in which case the two spins wander independently, each with a velocity spectrum of the form that would result from (34) alone, averaged over all arrangements of the second spin. We then recognize that the positive correlation represented by (35) adds to the magnitude of the wandering of each. By subsuming (35) in (34) we ensure that the variations of the total moment are correctly described, without having to abandon the model of independent motion.

This is the justification for writing v_y as $v_{yA} + v_{yB}$:

$$v_y = \frac{3\mu_0 \mu}{4\pi a^5 L}\left[\mu_x(x^2 - z^2) + \mu_y xy + 2\mu_z xz\right]. \tag{36}$$

Each dipole in the assembly oriented, like that at A, along the z-axis experiences a rapidly fluctuating motion v_y as the tumbling of the molecule takes B round the sphere of radius a centred on A. The mean value of v_y vanishes, since $\bar{x} = \bar{y} = \bar{z} = 0$ and $\overline{x^2} = \overline{y^2} = \overline{z^2} = \frac{1}{3}a^2$. To describe the magnitude of the fluctuating velocity we use $\overline{v_y^2}$, averaging not only over x, y and z but over all orientations of the dipole at B, by putting $\overline{\mu_x^2} = \overline{\mu_y^2} = \overline{\mu_z^2} = \frac{1}{3}\mu^2$. In this way we derive the mean square value of v_y for z-pointing dipoles:

$$\overline{v_y^2} = \frac{9\mu_0^2 \mu^2}{16\pi^2 a^{10} L^2}\left[\overline{\mu_x^2 (x^2 - z^2)^2} + \overline{\mu_y^2 x^2 y^2} + 4\overline{\mu_z^2 x^2 z^2}\right],$$

all cross-terms vanishing when the averages are taken; this reduces to

$$\overline{v_y^2} = \frac{9\mu_0^2\mu^4}{80\pi^2 a^6 L^2}, \tag{37}$$

when use is made of the averages $\overline{x^4} = \overline{z^4} = \frac{1}{5}a^4$ and $\overline{x^2y^2} = \overline{x^2z^2} = \frac{1}{15}a^4$. Hence, finally, by use of (33) we have

$$T = 40\pi^2 a^6 L^2 / 9\mu_0^2 \mu^4 \tau_c$$
$$= 160\pi^2 a^6 / 9\mu_0^2 \hbar^2 \gamma^4 \tau_c, \quad \text{as quoted in (26),}$$

since $\mu = \gamma L$ and $L = \frac{1}{2}\hbar$ for a proton.

There is no distinction in this calculation between T_1 and T_2. If τ_c is much smaller than the precession period a laboratory observer and one in a frame rotating at the precessional angular velocity could hardly detect any difference in the rapidly fluctuating interaction between protons, and would therefore see no distinction between diffusive motion parallel to, and transverse to, $\boldsymbol{B}$. If, on the other hand, the tumbling is slow, the rotating observer sees both protons maintaining the same spin orientation in his frame, while the line joining them rotates round $\boldsymbol{B}$ at $-\boldsymbol{\Omega}$, the angle to $\boldsymbol{B}$ only changing slowly. It is then appropriate to average the torque over one period of precession before considering the tumbling motion. When this is done (we omit the details, which are quite straightforward) and the changes in both spin directions added, it is found that there is no change in the component of spin parallel to $\boldsymbol{B}$ – as expected, the diffusive motion is confined to the transverse plane.† This means that as τ_c is increased, by an increase in viscosity, for example, T_1 and T_2 decrease together, according to (26); but when $\Omega\tau_c$ rises to unity and above, T_2 continues to decrease while T_1 increases again. In highly viscous liquids the resonance line is not very sharp, being determined by a short T_2, while the rate of equilibration is slow, being determined by a long T_1. These two effects combine to make detection of the resonance line difficult, since only very weak fields can be used without saturating the system.

With this we leave our discussion of nuclear magnetic resonance, which has been slanted towards the description of the phenomenon as a thing in itself, a fascinating example of a physical resonance process that is not time-reversible nor strictly linear, yet is still amenable to exact analysis. We

† This is one of these cases where the quantum-mechanical viewpoint leads to an easier intuitive understanding of the process. The energy levels of the proton are separated by $\hbar\Omega$, and transitions between them can only be excited if the perturbing field contains components with frequency around Ω. When $\Omega\tau_c \gg 1$, the spectrum of the perturbing interaction falls off at a frequency around $1/\tau_c$, well below Ω, and consequently the transitions, which are the quantal description of changes in the component of spin parallel to $\boldsymbol{B}$, cease to take place. On the other hand, motion round the transverse plane involves no energy changes, only disturbance to the phases of wave-functions, and is controlled by the zero-frequency strength of the perturbation, i.e. $I_\omega(0)$, as assumed in the derivation of (26).

return to our catalogue of velocity-dependent forces and their effects as revealed by resonant systems.

Cyclotron resonance

The simplest example of the Lorentz force is provided by a free charged particle moving in a uniform magnetic field. Motion along the direction of $\boldsymbol{B}$ is unaffected, but in the plane transverse to $\boldsymbol{B}$ the particle describes a circle. The complete motion is therefore a circular helix. We shall now disregard the motion along $\boldsymbol{B}$ and concentrate on the transverse plane. If the particle has a transverse component of velocity v and moves in a circle of radius r, the centripetal acceleration mv^2/r is caused by the Lorentz force eBv, so that

$$r = mv/eB = p/eB, \tag{38}$$

where p is the momentum of the particle, Written in terms of p, (38) holds even when v is large enough to produce relativistic changes of mass. So long as $v \ll c$, however, and m is constant, (38) shows that the angular velocity, v/r, is independent of v, taking the value.

$$\omega_c = eB/m. \tag{39}$$

This equation defines the *cyclotron frequency*, ω_c.

This is a system which, although two-dimensional, has only one natural frequency, for the equation of motion has the form, in complex notation,

$$m\ddot{r} + \mathrm{i}eB\dot{r} = 0, \tag{40}$$

with the general solution

$$r = r_0 + A\,\mathrm{e}^{-\mathrm{i}\omega_c t}, \; r_0 \text{ and } A \text{ being constants.}$$

The arbitrary positioning of the centre of the orbit, r_0, might be considered to account for the other degree of freedom that normally results in a second normal node; here the second mode has zero frequency.

In a cyclotron the particle being accelerated is not subjected to a uniform accelerating force, but instead the force is confined to one part of the large orbit.[5] This introduces extra complications which we shall not consider at this moment. Instead we investigate the response of electrons in a semi-conductor, where the orbits in a moderate field, say 1 T, are very small – less than 10^{-7} m in radius – so that an oscillatory electric field applied to the sample is virtually certain to be uniform over the trajectory of any one particle. If the semiconductor is pure, cold and shielded from light, the electron density can be made low enough for the charge displacements due to the electric field to have negligible disturbing influence on that field. Moreover, the electrons can traverse their orbits several times between collisions with impurities or other lattice irregularities, and this allows the resonance to manifest itself. In the absence of collisions, the equation of

275

motion takes the form of (40), supplemented by a rotating field (or one rotating component of a linear oscillatory field)

$$m\ddot{r}+\mathrm{i}eB\dot{r}=eE\,\mathrm{e}^{-\mathrm{i}\omega t}, \tag{41}$$

with the steady-state solution

$$\dot{r}=V\,\mathrm{e}^{-\mathrm{i}\omega t}, \quad \text{where } V=\mathrm{i}eE/m(\omega-\omega_c). \tag{42}$$

The velocity is always normal to E, as it must be in a lossless system as assumed here. The resonant denominator vanishes when $\omega=\omega_c$, the amplitude of response growing without limit, as in fig. 6.12(c).

145

To take account of collisions we make the usual assumption of randomness – an electron, chosen at any instant, has a probability $\mathrm{e}^{-t/\tau}$ of continuing to move without collision for more than a time t. This is the consequence of assuming that an electron always has a chance $\mathrm{d}t/\tau$ of colliding in the next interval $\mathrm{d}t$, irrespective of its past history. We assume further that collisions are catastrophic, in the sense that afterwards the electron is equally likely to move off in any direction. Let us now apply the impulse response method to a whole assembly of electrons, calculating their response to the rotating field $E\,\mathrm{e}^{-\mathrm{i}\omega t}$ from how they would respond if it was switched on only during the interval from t to $t+\mathrm{d}t$. If a typical electron has velocity $v_0\,\mathrm{e}^{\mathrm{i}\theta}$ before, it will have $v_0\,\mathrm{e}^{\mathrm{i}\theta}+(eE\,\mathrm{e}^{-\mathrm{i}\omega t}/m)\,\mathrm{d}t$ immediately after. It now travels in its cyclotron orbit so that at a later moment t_0, if it has not collided meanwhile, its velocity will be $v(t_0)$, where

$$v(t_0)=[v_0\,\mathrm{e}^{\mathrm{i}\theta}+(eE\,\mathrm{e}^{-\mathrm{i}\omega t}/m)\,\mathrm{d}t]\,\mathrm{e}^{-\mathrm{i}\omega_c(t_0-t)}.$$

When we apply this result to the whole assembly of electrons, moving randomly before the impulse, we must average evenly over all values of θ and multiply by $\mathrm{e}^{(t-t_0)/\tau}$ so as to count only those that have survived until t_0 without collision. The others, being randomly scattered, have zero mean velocity afterwards, and contribute nothing to the mean velocity of an electron at time t_0, which is now seen to be given by $\Delta\bar{v}$, where

$$\Delta\bar{v}(t_0)=(eE/m)\,\mathrm{e}^{-\mathrm{i}\omega t+(\mathrm{i}\omega_c+1/\tau)(t-t_0)}\,\mathrm{d}t.$$

To proceed from here to the mean velocity at t_0, when the rotating field has been applied at all times up to t_0, simply involves integrating $\Delta\bar{v}$ from $-\infty$ to t_0:

$$\bar{v}(t_0)=\frac{eE\tau}{m}\,\mathrm{e}^{-\mathrm{i}\omega t}/[1+\mathrm{i}(\omega-\omega_c)\tau]. \tag{43}$$

The collisions give rise to a standard Lorentzian resonant response, with $Q=\frac{1}{2}\omega_c\tau$. The phase difference between E and $\bar{v}$ shows that the collisions allow energy to be absorbed by the electrons. The resonance does not merely alter the dielectric constant of the semiconductor – this is the effect of the lossless part of the response – but also causes dielectric loss. This can be picked up experimentally by placing the sample in a microwave cavity excited at its own resonant frequency, and varying the applied field until ω_c

76

passes through the cavity frequency.[6] The additional loss lowers the response level of the cavity, and in this way the cyclotron resonance can be plotted out, not by keeping ω_c fixed and varying the exciting frequency ω but, as is common also in NMR, keeping ω fixed and varying ω_c. It is clear from (43) that this is only a technical difference. An example of what may be observed is shown in fig. 13. The appearance of several resonances indicates that the electrons in a semiconductor do not behave as free particles. Some

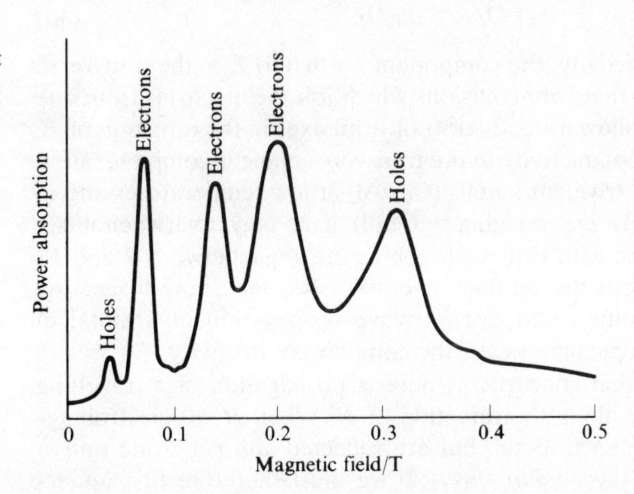

Fig. 13. Cyclotron resonance in germanium at 4 K and a frequency of 24 GHz (Dresselhaus, Kip and Kittel[7]).

indeed (*holes*) execute orbits in the reverse sense as if they were positively charged. Electron orbits and hole orbits can of course be distinguished by using circularly polarized rather than linearly polarized microwaves to probe them. Each resonance shows up only with one sense of polarization. This is technically more difficult, but valuable as a diagnostic tool, for each group of electrons or holes has a different cyclotron frequency, as if each had its own characteristic mass, different from that of a free electron. Knowledge of the effective mass to be associated with each group, and the way this mass changes with orientation of the magnetic field, provides numerical information that can be compared with quantum-mechanical calculations of the dynamical behaviour of electrons in the semiconductor, so that the approximations which are unavoidable in the latter may be tested and the theoretical tools thereby improved. Cyclotron resonance has indeed been a valuable aid to the development of semiconductor physics.

Helicons

The last exhibit in our menagerie of curious oscillatory systems also owes its character to the Lorentz force, revealed in this case by the Hall effect.[8] A free charged particle, acted upon by uniform perpendicular magnetic and electric fields, executes a helical orbit whose axis is parallel to $\boldsymbol{B}$ but drifts at constant speed E/B in a direction normal to both fields. Consequently the assembly of free electrons in a metal, n per unit volume, if they suffered no collisions would drift bodily at this speed, setting up a current density neE/B. Using a complex coordinate system to describe motion in the plane transverse to $\boldsymbol{B}$, we can write that if E is real, the current density J is purely imaginary, $J = \mathrm{i}neE/B$, and the conductivity in the transverse plane is also purely imaginary:

$$\sigma_\perp \equiv J/E = \mathrm{i}ne/B. \tag{44}$$

This is the Hall conductivity, the component normal to $\boldsymbol{E}$ in the transverse plane. In a real metal there are collisions which jolt the electrons from one orbit to another and allow a steady drift of orbit axes in the direction of $\boldsymbol{E}$. In general, then, the conductivity in the transverse plane is complex, but in pure monovalent and trivalent metals (Cu, Al) at low temperatures and in strong magnetic fields the imaginary (Hall) part may overwhelmingly dominate the real part, with Hall angle (being the angle between $\boldsymbol{J}$ and $\boldsymbol{E}$) less than 1° away from the perfection of 90°. We may now forget the microscopic interpretation and discuss wave propagation in a metal in terms of the macroscopic parameter, the complex conductivity.

The phenomenon that concerns us here is propagation of a travelling electromagnetic wave along the direction of $\boldsymbol{B}$. When $\boldsymbol{B} = 0$ electromagnetic waves cannot enter a metal, but are reflected and penetrate only a very shallow surface layer (*skin effect*);[9] we must therefore first understand how a magnetic field so changes the situation as to allow penetration. The essential condition for wave propagation to occur is that the Hall angle shall be large, so let us begin with the ideal case of no electron scattering and a Hall angle of 90°, $\sigma_\perp$ being purely imaginary, is, say. In view of the earlier examples in this chapter it should come as no surprise that the wave is circularly polarized; $\mathscr{E}$ and $\mathscr{B}$, the field vectors of the wave-motion, are constant in amplitude but spin around the z-axis, as in fig. 14, which shows travelling waves in free space moving away from and towards the observer. In both cases the sense of circular polarization has been chosen so that a fixed observer facing along the arrow of $\boldsymbol{B}$ would see $\mathscr{E}$ and $\mathscr{B}$ rotating clockwise. Then for both waves the displacement current, $\boldsymbol{J}_\mathrm{d} = \varepsilon_0 \partial\mathscr{E}/\partial t = -\mathrm{i}\omega\varepsilon_0\mathscr{E}$ in complex notation, and points at $\pi/2$ to $\mathscr{E}$ in a clockwise sense. Suppose now that the wave travels through a metal with perfect Hall conductivity; $\mathscr{E}$ will excite a current $\sigma_\perp\mathscr{E}$, i.e. $\mathrm{i}s\mathscr{E}$, parallel to $\boldsymbol{J}_\mathrm{d}$, and the total current will be $-\mathrm{i}(\omega\varepsilon_0 - s)\mathscr{E}$ or $-\mathrm{i}\omega\varepsilon\varepsilon_0\mathscr{E}$, the effective dielectric constant of the medium being given by

$$\varepsilon = 1 - s/\omega\varepsilon_0.$$

The imaginary character of $\sigma_\perp$ when the Hall effect is complete allows the electrons to contribute to the real dielectric constant. This is in marked contrast to the behaviour when no magnetic field is present and $\sigma_\perp$ is real; then s is imaginary and the effective dielectric constant is dominated by its imaginary component. The wave velocity, proportional to $\varepsilon^{-\frac{1}{2}}$, is complex

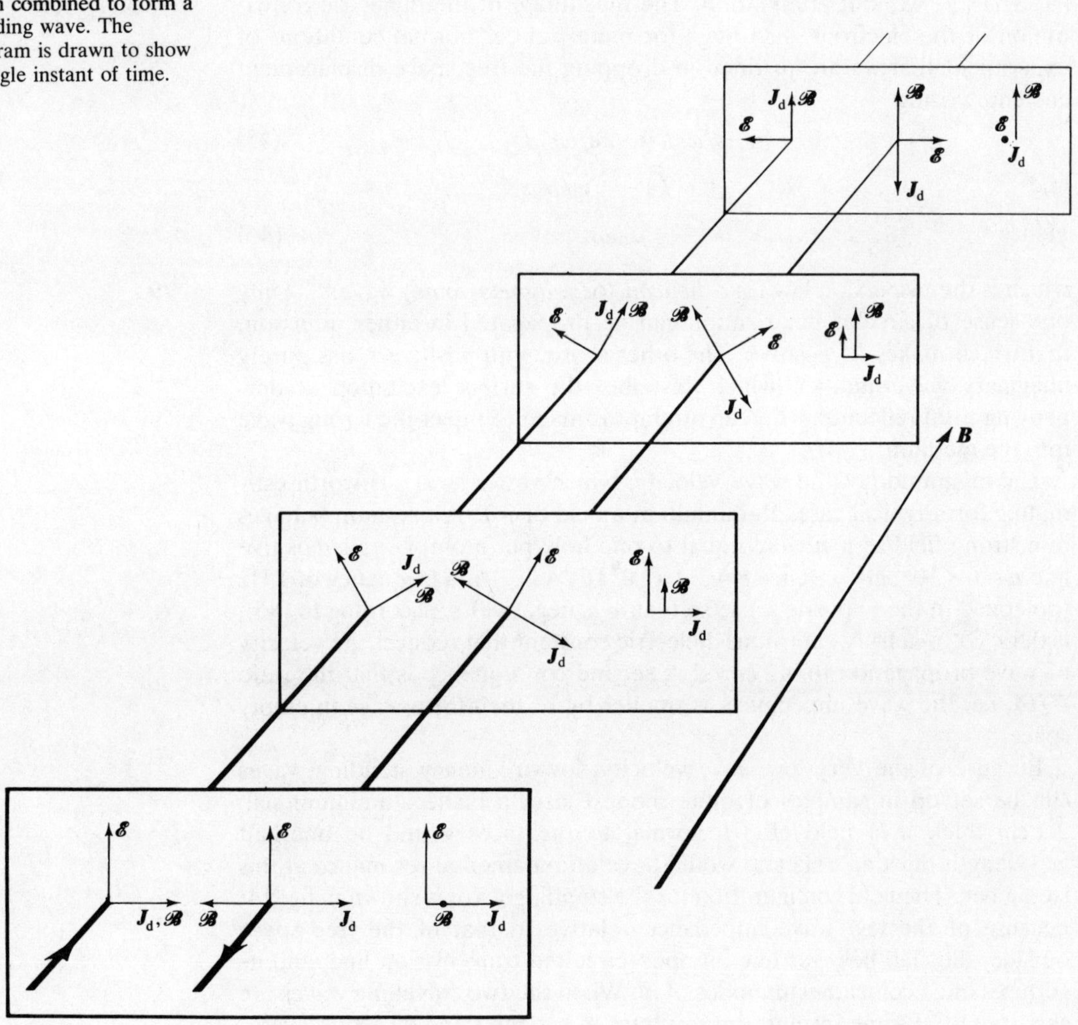

Fig. 14. Fields, $\mathscr{E}$ and $\mathscr{B}$ and displacement current, $\boldsymbol{J}_d$, in circularly polarized travelling waves and, on the right, their resultant when combined to form a standing wave. The diagram is drawn to show a single instant of time.

and describes a heavily damped wave – this is the skin effect. When s is real and negative, however, as for electrons when the Hall effect is complete, the additional term in ε raises the dielectric constant of the medium to a high positive real value, decreasing the velocity of propagation by a factor $(1-s/\omega\varepsilon_0)^{\frac{1}{2}}$, without attenuation. The magnitude of the dielectric contribution of the electrons is so huge for metals under normal conditions of experiment that we are justified in dropping the free space displacement current, writing

$$\varepsilon = -s/\omega\varepsilon_0 = -ne/\omega\varepsilon_0 B \tag{45}$$

and

$$\omega^2/k^2 = c^2/\varepsilon = -\omega/\mu_0 s.$$

Hence

$$k^2 = -\mu_0\omega s, \tag{46}$$

which is the dispersion law for a *helicon* (or *magneto-ionic*) wave.[10] Only one sense of circular polarization can be propagated in either direction, that which makes k^2 positive. The other sense, with positive s, has purely imaginary wave-number, which describes the surface excitation accompanying total reflection when an attempt is made to inject the wrong wave into the medium. **79**

The magnitude of the wave velocity, which varies as $\omega^{\frac{1}{2}}$, is worth estimating for a typical case, aluminium in a field of 3 T. Aluminium behaves in a strong field as if n were equal to one hole per atom, i.e. s is positive and $n = 6\times10^{28}\ \mathrm{m}^{-1}$. Hence $s = 3.2\times10^{9}\ \Omega^{-1}\ \mathrm{m}^{-1}$. At a frequency of 1 Hz (polarized in the opposite sense so that ω is negative) ε, according to (45), is 0.5×10^{20} – a truly enormous dielectric constant that reduces the velocity of wave propagation to 4.2 cm/s! A second consequence is that the ratio $\mathscr{E}/\mathscr{B}$, i.e. the wave impedance, is smaller by $\varepsilon^{\frac{1}{2}}$ than for a wave in empty space.

Because of the very low wave velocity, low-frequency standing waves can be set up in samples of quite modest size; a plane aluminium slab 2.1 cm thick in a field of 3 T normal to the faces would be one-half wavelength thick at 1 Hz and would have a fundamental resonance at this frequency. The field configuration for the standing wave is shown in fig. 14; because of the low wave impedance relative to that of the free space outside, the slab behaves like an open-circuited transmission line, and its surfaces must coincide with nodes of $\mathscr{B}$. When the two travelling waves are superposed at some instant, the resultant $\mathscr{E}$ has the same direction everywhere and its amplitude varies sinusoidally with position, just as in a plane polarized electromagnetic wave in free space; similarly $\mathscr{B}$ varies sinusoidally with its nodes and antinodes interchanged relative to those of $\mathscr{E}$. However, $\mathscr{B}$ and $\mathscr{E}$ are parallel rather than perpendicular, and as time proceeds they do not oscillate in amplitude like a plane standing wave, but spin around $\boldsymbol{B}$ keeping constant amplitude. The configuration of $\mathscr{B}$ confers on the sample as a whole a magnetic moment, transverse to $\boldsymbol{B}$, and spinning at the resonant frequency. By the inductive effect of this moment the helicon standing wave is easily detected. The experimental arrangement in

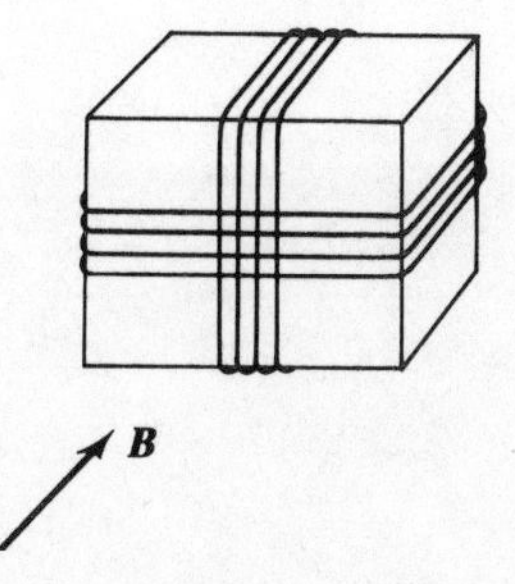

Fig. 15. Crossed coils and rectangular metal crystal, for generating and detecting helicons.

224

fig. 15, similar in principle to Bloch's nuclear induction arrangement, provides for one coil to excite helicons and another, normal to it, to detect them. By sweeping the frequency over a wide range a set of resonances can be detected at frequencies proportional to the squares of the integers, in accordance with the dispersion law (46). The odd harmonics are strong and the even weak (or absent altogether) with this arrangement, for the total magnetic moment is zero if there are an even number of loops, with alternating direction of magnetization, in the standing wave pattern. The resonances are not, of course, ideally sharp since electron scattering causes $\sigma_\perp$ to have a real component. The dispersion relation is now complex and, as discussed in chapter 3, in a freely oscillating standing wave k must be taken as real; (46) now shows that the phase angle of ω is equal and opposite to that of $\sigma_\perp$. Hence the natural decrement of the free helicon oscillations or, what is equivalent, the width of the resonance, provides a measure of ω''/ω' and hence of $\sigma''_\perp/\sigma'_\perp$, the tangent of the Hall angle. This makes the helicon a useful tool in studying the conductivity of metals in strong magnetic fields, with the advantage over conventional methods that no electrical contacts need to be made to the specimen.[11]

60

It is easy enough to solve the field equations for a standing wave in a parallel-sided slab of infinite lateral extent, but for a real sample, a finite slab or a body of any other shape, there are formidable complexities of applied mathematics. Even for the cylinder and the sphere the boundary conditions that must be satisfied are very troublesome indeed, but we need not attempt to discuss why, or how the difficulties have been overcome. It is enough for our purposes to note the general field pattern in the fundamental resonance of a sphere, as shown schematically in fig. 16. The transverse

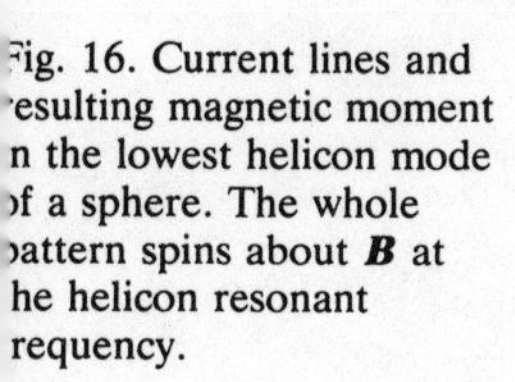
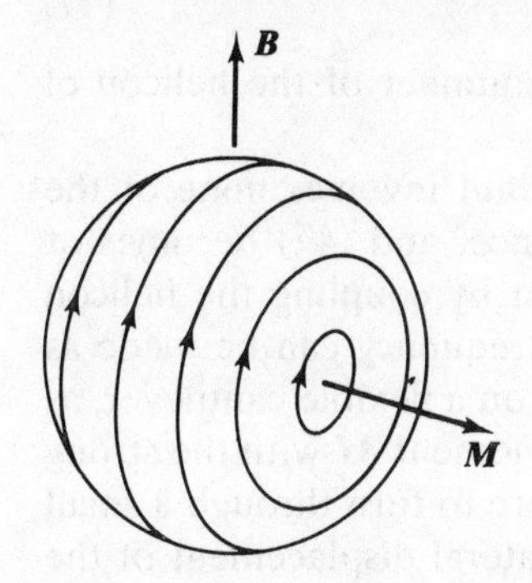

Fig. 16. Current lines and resulting magnetic moment in the lowest helicon mode of a sphere. The whole pattern spins about $\boldsymbol{B}$ at the helicon resonant frequency.

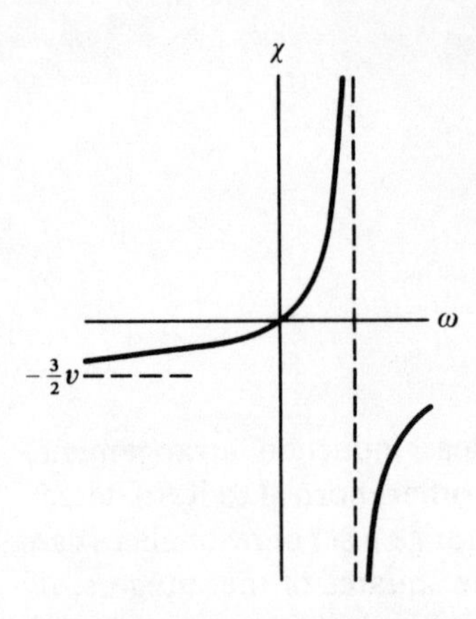

Fig. 17. Schematic diagram of susceptibility of a sphere, showing first helicon resonance.

magnetic moment is generated by circulating currents, and as with the slab the whole pattern of currents and moment spins around $\boldsymbol{B}$. If the sphere is subjected to a rotating field $\boldsymbol{b}$, lying in the plane transverse to $\boldsymbol{B}$, the resulting magnetic moment rotates with $\boldsymbol{b}$ but with a phase lag (so that there is a continuous energy flow into the sphere, dissipated as heat by electron scattering). The susceptibility χ, defined as $\mu_0 M/b$, is in general complex; its frequency variation shows a typical resonant peak at the fundamental mode frequency and, of course, at all resonances of higher order, of which the sphere exhibits a considerable and complicated variety.

The variation of χ to just beyond the first resonance is shown schematically in fig. 17; electron scattering has been assumed negligible, so that χ is wholly real,. It will be noted that when ω is in the opposite sense to the helicon rotation, χ tends to the limit $-\frac{3}{2}V$, V being the volume of the sphere. This is the value for a sphere that totally excludes magnetic field. At very low speeds, however, whether positive or negative, the induced currents are so weak that the applied field penetrates almost completely, and rotation of this uniform field within the sphere induces currents which are very simple in form, circulating around the equatorial axis parallel to $\boldsymbol{b}$ and with a strength proportional to distance from this axis. To quote the final result, if a is the radius of the sphere,

$$\chi \approx \tfrac{1}{5}\mu_0 V a^2 \omega \sigma_\perp = \tfrac{1}{5} V k^2 a^2, \tag{47}$$

in which k is taken from (46) and is the wavenumber of the helicon of frequency ω.

The calculation of χ in this low-frequency limit involves none of the complexities that are demanded by the resonance, and (47) becomes of interest and practical value when one finds that by coupling the helicon oscillation to mechanical motion the resonant frequency can be made as low as one wishes. Imagine the sphere mounted on a flexible cantilever, as in fig. 18, so that the interaction of the helicon moment $\boldsymbol{M}$ with the strong field $\boldsymbol{B}$ bends the cantilever and causes the sphere to turn through a small angle about an axis normal to $\boldsymbol{M}$ and $\boldsymbol{B}$ (the lateral displacement of the

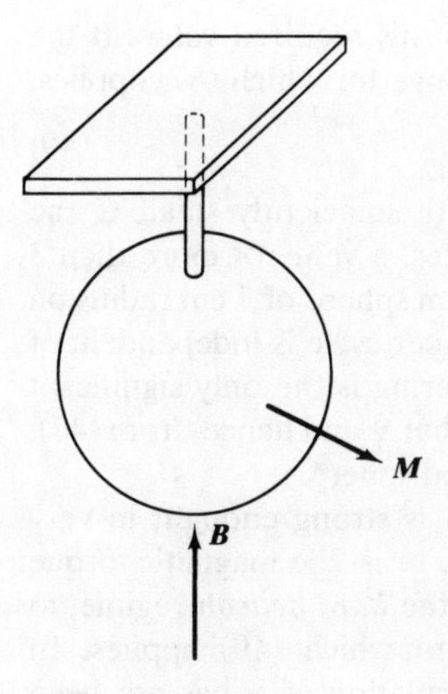

Fig. 18. Sphere supported on flexible cantilever to give 'soft helicon' resonance.

sphere is irrelevant – no currents are induced by the unrotating motion of a conductor through a uniform field). As $\boldsymbol{M}$ rotates about $\boldsymbol{B}$ with angular velocity ω, the axis of bending rotates with it, and the orientation of the sphere continually alters so that its originally vertical diameter describes a narrow-angled cone with vertical axis. If C is the torsion constant of the cantilever, the half-angle of the cone, θ, is equal to MB/C. In fig. 19(a) the diameter is shown as a broken line, displaced from the vertical by θ. From

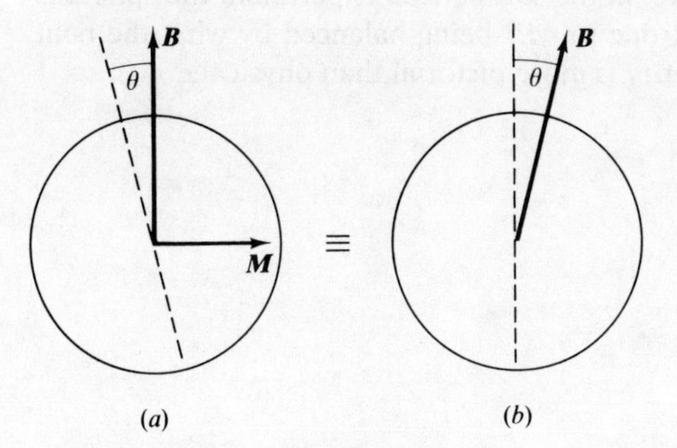

Fig. 19. (a) The interaction of $\boldsymbol{B}$ and $\boldsymbol{M}$ bends the cantilever of fig. 18 and tilts the originally vertical diameter of the sphere. This is equivalent, from the sphere's point of view, to (b) where $\boldsymbol{B}$ is shown tilted. As $\boldsymbol{M}$ spins round $\boldsymbol{B}$, $\boldsymbol{B}$ in (b) spins round the vertical axis.

the point of view of the sphere, it is as if $\boldsymbol{B}$ had been tilted through $-\theta$, and as $\boldsymbol{M}$ rotates the sphere sees $\boldsymbol{B}$ describing a cone, as if the steady vertical $\boldsymbol{B}$ had been supplemented by a rotating transverse $\boldsymbol{b}$ equal to $\boldsymbol{B}\theta$. Such a field establishes a moment $M = \chi B\theta$, and if the system is displaced and then allowed to oscillate freely, it must do so in such a way that the moment induced by its motion and the moment that is responsible for the bending of the cantilever are one and the same:

$$M = \chi B\theta \quad \text{and} \quad \theta = MB/C;$$

hence

$$\chi = C/B^2. \tag{48}$$

The natural frequency is that at which χ takes this required value. If the cantilever is weak enough for χ to lie in the range for which (47) applies,

$$\omega = 5C/\mu_0 B^2 V a^2 \sigma_\perp. \tag{49}$$

When B is large, ω varies as $1/B^2$ and with sufficiently small C the rotation period can be made as long as one likes, a value of more than 3 minutes being easily achieved with an aluminium sphere of 1 cm radius on a fairly flexible wire cantilever. The decrement per cycle is independent of the strength of the cantilever, if electron scattering is the only significant dissipative effect in the system, for (48) shows that χ and hence, from (47), $\omega\sigma_\perp$ must be real, just as for the rigidly mounted sphere.

The frequency only varies as $1/B^2$ when B is strong enough; in very weak fields the cantilever has enough rigidity to resist the magnetic torque and, according to (46) and (44), $\omega \propto B$. This is the *hard helicon* regime, to be contrasted with the *soft helicon* regime to which (49) applies. In between, ω has a maximum but the exact calculation of χ has not been carried to the point where the full variation with B can be computed and compared with experiment.

It is perhaps tempting to visualize the conical rocking of the sphere and the circulating current lines as analogous to a precessing gyroscope, but the analogy is unsound. There is no significant transverse angular momentum associated with the helicon. The precessing gyroscope is subjected to a resultant torque and precesses so that the rate of change of angular momentum equals the torque; in the soft helicon experiment the sphere is subjected to no torque, that due to MB being balanced by what the bent cantilever exerts. The similarity is more pictorial than physical.

9 The driven anharmonic vibrator; subharmonics; stability

Linear systems whose parameters are independent of time possess, as has been abundantly illustrated already, well-defined normal modes from which their motion can be synthesized by superposition; and the response to an applied force, varying with time, can be written in terms of the response to each separate Fourier component of the force. The same is not true of non-linear systems, since superposition is no longer a valid procedure for synthesizing the response. Every anharmonic system responds differently to a given form of time-dependent force, and even when the response has been found in any special case it will not scale up unchanged in response to an amplification of the force. Thus the response to a sinusoidal force is in general non-sinusoidal, the waveform changing with the amplitude of the force. There are very few general statements that can be made about the character of the response. One cannot even assert that the oscillations of the vibrator will have the same fundamental frequency as the applied force – it may respond at a subharmonic frequency, i.e. an integral submultiple of that of the force, or the response may be asynchronous to the point of randomness. Even when order prevails, with regular vibration at the fundamental or subharmonic frequency, changing the amplitude or frequency of the applied force to an infinitesimal degree may have the effect of throwing the response into an entirely new pattern.

The problems presented by asynchronous and unstable response have attracted much attention from mathematicians, and there is a very large body of analysis aimed at the systematic diagnosis and classification of pathological states.(1) It is difficult to reach a just evaluation of this work since the aims of a mathematician may be very different from those of a physicist or engineer. It is no denigration of pure mathematics to deny its usefulness, but when non-linear vibrations are under discussion one may feel that because it looks as if it deals with physical problems it must stand or fall on its practical utility. This is unfair to the mathematician, whose interest may be aroused primarily by the structural features underlying a great variety of systems, but so deeply that the topological and other techniques appropriate to their elucidation are inappropriate to working out the details in any particular case. The contrast with harmonic systems is striking; here the elementary mathematical tools that are useful in special cases also serve (if nothing more sophisticated comes to hand) to develop the general theory – consequently harmonic systems have long ceased to

See note on p. 284.

interest pure mathematicians. Anharmonic systems, on the other hand, are still full of mathematical interest, and this is enough to ensure that the average practising physicist or engineer will have the greatest difficulty in quarrying useful material from the professional literature. In time, it is to be hoped, the systematic assimilation of instabilities and related critical phenomena into a coherent pattern will lead to the sort of qualitative appreciation of their role that can be incorporated as a basic element in the scientist's world-picture.[(2)] The recent appearance of semi-popular articles expounding *catastrophe theory* points to an earnest desire on the part of mathematicians to break away from the traditional reductionist method that has given physical science its power. By contrast, these visionaries in the world of instabilities and critical phenomena appear to be seeking a new constructive approach to putting the pieces together again – the quest is for common patterns running through enormously diverse phenomena which will enable these phenomena to be described as complete entities and not as an ensemble of individually analysable fragments. This is an ambitious program which, even if only partially successful, will undoubtedly have far-reaching consequences. No revolution, however, is successful until its achievements have been consolidated, and it is too early to express confidence in the outcome of this particular high venture.

This preamble must be our excuse for shirking anything like a systematic account of what has been achieved by all the work to date. Instead we shall exhibit a few characteristic features of anharmonic systems by means of special examples which can be analysed in sufficient detail, and whose stability can be investigated, without excessive labour. The intent behind this procedure is to alert the reader's mind to the complexities that a real system may exhibit, at the same time warning him that he will usually find a complete solution a very taxing matter, involving long and tedious calculations. In practice it often suffices to be aware of the pitfalls, so that one ceases to rely on mathematics alone and develops a proper respect for experiment and observation.

Slightly anharmonic vibrators

The pendulum is a typical slightly anharmonic vibrator, with a restoring force proportional to $\sin\theta$ rather than directly proportional to the displacement θ. The resonant frequency decreases with amplitude and the response to a driving force of constant amplitude and variable frequency is correspondingly distorted from the Lorentzian form of resonance. Simple though it is, the pendulum is not a very convenient candidate for demonstrating anharmonicity in the laboratory, if only because its period is so long as to slow down the observations to an exasperating degree. A much more satisfactory example is a stretched wire executing transverse vibrations. It is almost inevitable that the tension will increase with deflection, and the frequency with amplitude as a result; by anchoring the ends of the

19

string in different ways the degree of anharmonicity can be adjusted. A suitable arrangement is shown in fig. 1; it is more elaborate than necessary, being designed for a dual purpose. In this chapter we shall consider only its response to a transverse exciting force at the end, but in the next chapter we shall discuss the response to a longitudinal force; the arrangement shown allows both forms of excitation to be applied without otherwise altering the system. The electromagnetic impeller acts on a comparatively massive bar, so that the amplitude of its vibrations is not seriously affected by the motion of the string, even at resonance. To a good approximation the amplitude of motion at the end is independent of frequency, and the argument illustrated in fig. 7.8 shows that the response would be of normal Lorentzian form were it not for the variation of tension and, as already noted, the non-linear damping at higher velocities.

299

173

31, 140

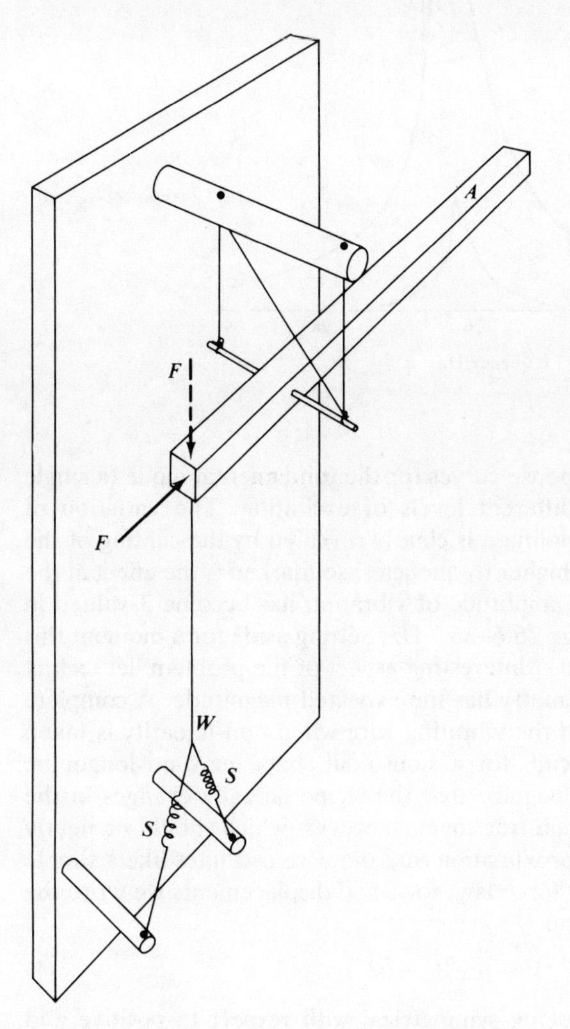

Fig. 1. Arrangement for studying vibrations of a stretched wire. The massive horizontal beam *A* is suspended by strings from a pivot one-third of the length from an end, and serves to tension the wire *W*, which at the lower end is attached to two springs *S* (this separates the two transverse modes and avoids parasitic excitation). An electromagnetic impeller applies an oscillatory force *F* either transversely (full arrow) or longitudinally (broken arrow; see chapter 10 for a discussion of this). Suspending *A* as shown makes the effective inertia of the beam the same for both directions of *F*. The diagram is not to scale – in the actual experiment *A* was about 25 cm long, while *W* was 150 cm.

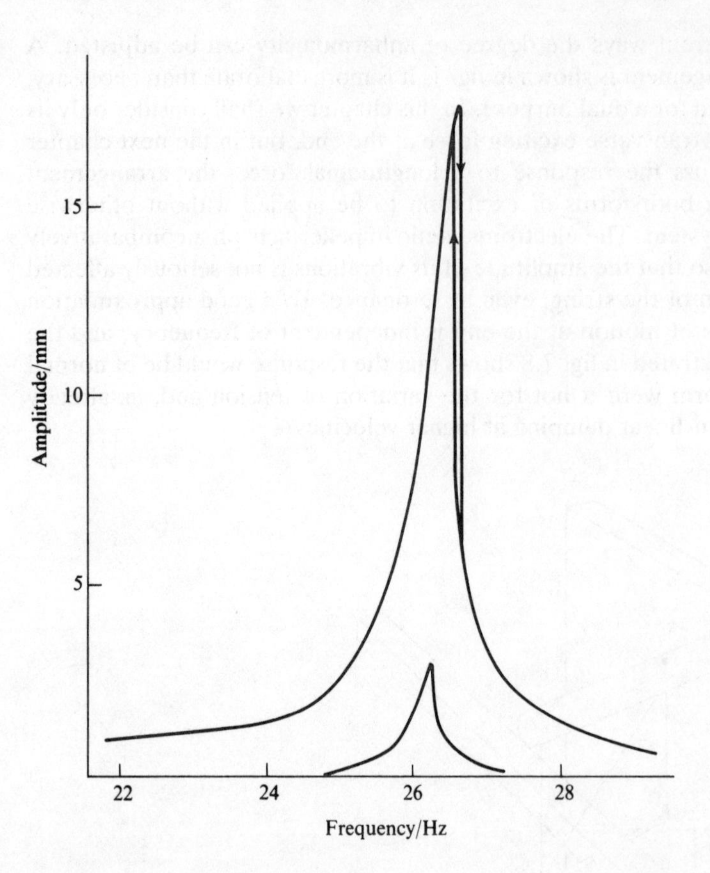

Fig. 2. Amplitude of centre of wire as a function of frequency. The upper curve was obtained with the force *F* ten times as strong as for the lower curve.

Fig. 2 shows typical response curves for the fundamental mode (a single half-wave loop) at two different levels of excitation. The variation of natural frequency with amplitude is clearly revealed by the canting of the top of the curves towards higher frequencies; so marked is the effect at the higher excitation that the amplitude of vibration has become 3-valued in the narrow frequency range 26.6–26.7 Hz. Setting aside for a moment this new effect, which is the most interesting aspect of the problem, let us first enquire whether the asymmetry has the expected magnitude. A complete solution of the problem of the vibrating wire when non-linearity is taken into account is very difficult, for a sinusoidal shape may no longer be assumed and one must recognize that the shape actually changes in the course of the cycle. A rough treatment, however, which should be nearly correct starts from the approximation that the wire oscillates like a simple vibrator with a non-linear force law; for small displacements we write the potential energy in the form

$$V = \tfrac{1}{2}\mu\xi^2(1+\beta\xi^2), \tag{1}$$

the correction term $\beta\xi^2$ being symmetrical with respect to positive and negative displacements. Corresponding to (1) the force depends on displacement according to the expression:

$$F = -\mu(\xi + 2\beta\xi^3). \tag{2}$$

14 The general result for the oscillation period, T, derived from (2.11), shows that when the amplitude of vibration is ξ_0 and the total energy E is $\frac{1}{2}\mu\xi_0^2(1+\beta\xi_0^2)$,

$$T = 2T_0/\pi \int_0^{\xi_0} \mathrm{d}\xi/[\xi_0^2(1+\beta\xi_0^2) - \xi^2(1+\beta\xi^2)]^{\frac{1}{2}}.$$

When $\beta = 0$, $T = T_0$ for all ξ_0, and we may find the first correction to T_0 by Taylor expansion, writing

$$T(\xi_0) \approx T_0 + \beta(\mathrm{d}T/\mathrm{d}\beta)_{\beta=0}.$$

Differentiating under the integral sign, keeping ξ_0 constant, and then putting β equal to zero, we find that

$$(\mathrm{d}T/\mathrm{d}\beta)_{\beta=0} = -T_0/\pi \int_0^{\xi_0} (\xi_0^4 - \xi^4)\,\mathrm{d}\xi/(\xi_0^2 - \xi^2)^{\frac{3}{2}} = -\tfrac{3}{4}\xi_0^2 T_0.$$

Hence $$T(\xi_0) = T_0(1 - \tfrac{3}{4}\beta\xi_0^2 + \cdots). \tag{3}$$

Since $\beta\xi_0^2$ is the only dimensionless parameter in the problem, there is no other term in ξ_0^2 that could have been overlooked by treating β rather than ξ_0^2 as the expansion parameter. According to (2) and (3), when the force constant at maximum amplitude ξ_0 has increased by a fractional amount f, the period has decreased, and the frequency increased, by $\frac{3}{8}f$.

In the experiment that produced the results in fig. 2 the wire was virtually inextensible and the mass at the top large enough to be little influenced by the wire; when the wire was displaced, therefore, the springs at the bottom were stretched. Direct measurement showed that when the vibration amplitude was 17.5 mm the spring tension was increased by 5%. We might therefore expect the frequency to increase by $\frac{3}{8}$ of this, i.e. 1.9%; perhaps a little more than is observed but near enough, in view of the approximate theory and rather inaccurate measurements. The effect can be magnified by removing the springs and giving the wire firmer anchorage at the bottom. The result is shown in fig. 3, where the instability and hysteresis are very clearly exhibited.

Even so, the frequency shift at the peak is still only 4%, and the form of the vibration must remain closely sinusoidal. We are therefore reasonably justified in supposing that if we were to sweep the frequency of the exciting force through resonance while adjusting its amplitude at every stage to maintain the vibration amplitude constant, the variation of force with frequency would be the same as for a linear resonant system (see fig. 138 6.10(d)). Translated into the terms of the experiment, if the force is constant the width of the resonant peak at any level of response is

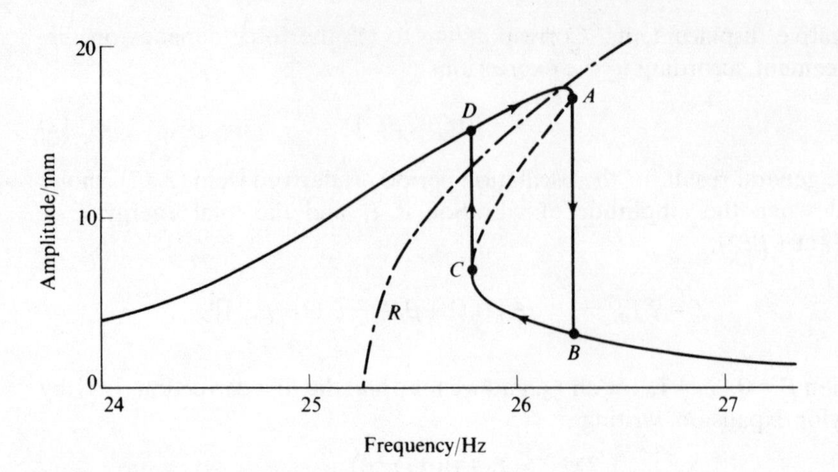

Fig. 3. As for figs. 1 and 2, but with springs S removed to increase asymmetry of resonance.

unaffected by the non-linearity. A Lorentzian peak, for example, still keeps the form (6.18),

$$P(\omega) = P_0/[(\omega - \omega')^2 + \omega''^2],$$

but now ω' is a function of P, for example, $\omega' = \omega'_0(1 + \frac{3}{4}\beta P)$ if $P = \xi^2$ and the period varies as in (3). In fact, the peaks in fig. 2 are probably not Lorentzian since the lower peak is only three-quarters as wide as the higher peak at half amplitude. The loss due to air resistance in the form of eddies increases with velocity at a considerably faster rate then viscous drag alone would predict, and the effective Q becomes correspondingly lower at large amplitudes. The extra losses are very noticeable in practice, the wire humming loudly when strongly excited.

140

We come at last to the many-valued character of the response. It is readily seen that the portion AC of the curve, shown broken in fig. 3, is unrealizable. At a point on this part of the line a small displacement upwards, i.e. an increase in amplitude at constant frequency, brings the point inside the resonance curve – the force needed to maintain the amplitude is less than that required for a point on the line. Consequently the amplitude increases further, and goes on doing so until the upper branch of the resonance curve is reached. By the same argument, if the point is displaced downwards it proceeds down until it reaches the lower branch. On the other hand, any point not on the broken line represents a stable response, small departures being followed by a natural reversion to the line. There is no difficulty, then, in understanding why the response curve takes a different form over part of its range according as the frequency is increased or decreased, or why there are jumps at the ends of the stable ranges.

If the curve of fig. 3 is traced out by increasing the frequency, the collapse of the amplitude from A to B takes a few seconds, being the natural decay time of any transient oscillation of the wire, and is accompanied by a marked beat pattern. By contrast on the return path the transition of the amplitude from C to D takes about the same time but is more nearly monotonic. Since the system is non-linear it is not quite permissible to discuss the behaviour in terms of the superposition of free transient excitations and the driven vibration at the excitation frequency; nevertheless the analogy will serve to explain the difference qualitatively. At most amplitudes along AB the driving frequency differs from the natural frequency (shown as the line R) by something comparable to, or greater than, the resonance width; this is the condition, illustrated in fig. 6.12, for beats to be clearly exhibited when transients are excited. On the other hand, CD cuts the line R and lies so close to it over most of its course that the long period of the beats prevents their showing themselves.

145

An example of subharmonic excitation

If a small loudspeaker is laid on its back, with a lightly loaded pin gently pressed against it by means of a spring, as in fig. 4, excitation at (say) 400 Hz with a signal generator can impart sufficient acceleration to cause the pin to lose contact over part of the cycle; as the critical amplitude is exceeded the emitted note suddenly becomes louder and harsher. With increased excitation there comes a point (possibly after a noisily unsettled interim) when a strong note sounds out an octave, a twelfth, a fifteenth or even more below the exciting frequency. The pin is now bouncing high enough to return only after 2, 3, 4 or more cycles. Each subharmonic is stable over a limited range of excitation, and it normally gives way to the next in the sequence, so that the note sounded drops by discrete steps, spelling out the harmonic series in inversion.

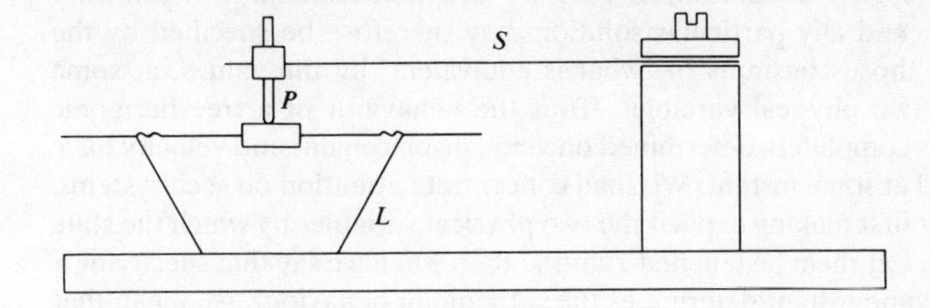

Fig. 4. A loudspeaker, L, lying on its back has a light pin, P, held by a spring, S, against the centre of its cone.

A simple, indeed oversimplified, picture of the process is shown in fig. 5, where the loudspeaker cone is assumed heavy enough to execute simple harmonic motion unperturbed by the bouncing pin. The pin is shown making contact with the cone every other cycle of the latter, and therefore responding in the octave subharmonic mode. The exact phase at which impact occurs is such that the upward movement of the cone compensates for the inelastic character of the bounce, the pin arriving and leaving at the same speed. The trajectory shown by a broken line is obviously unstable, for if the impact is slightly delayed it occurs when the cone is rising faster; the pin therefore rises next time to a greater height and arrives for the next

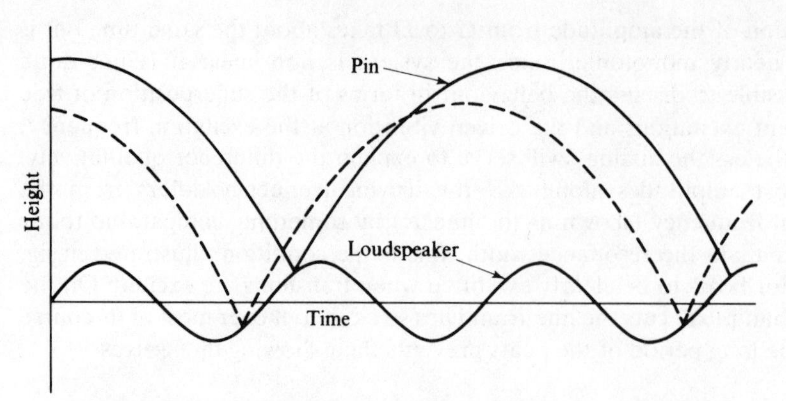

Fig. 5. Bouncing of the pin in the arrangement of fig. 4, generating the octave subharmonic.

bounce still further delayed. By contrast, the full curve represents a trajectory which, if delayed at one bounce, suffers correction and has the possibility of being restored to the correct phase. But one must not be rash in assuming stability, for overcorrection may lead to the phase of impact oscillating between advance and delay with ever-increasing amplitude. It is necessary therefore to examine the stability with greater care. Since the problem arises frequently we shall break off at this stage to develop rather general tests for stability, and return to this example later to illustrate their application.

Behaviour of simple systems near equilibrium

The simple systems we have encountered so far in this book have mostly been described by second-order differential equations, not necessarily linear. The general solution of such an equation contains two arbitrary constants, and any particular solution may therefore be specified by the values of those constants or, what is equivalent, by the values, at some point, of two physical variables. Thus the behaviour of a free harmonic vibrator is completely determined once the displacement and velocity have been fixed at some instant. We shall concentrate attention on such systems, without at first making explicit the two physical variables by which the state is defined. Let them be labelled x and y; then when we say that specifying x and y at some instant determines the subsequent behaviour, we mean that the motion can be represented by a well-defined trajectory on the (x, y)-plane, passing through the specified point (x, y). Normally different trajectories do not intersect, though there may be one or more singular points onto which trajectories converge; such a point is likely to be one at which the system is in stable equilibrium, while a point from which trajectories diverge is likely to be one of unstable equilibrium. Other types of singularities are possible, and our first task is to catalogue the permitted varieties, of which there are five only for a two-parameter system.

If (x_0, y_0) is an equilibrium point (which need not be stable), the system established exactly in this state can remain there indefinitely if undisturbed, but if shifted to a neighbouring point will in general find itself not in equilibrium, so that as time proceeds x and y will change. Since $\dot{x}=\dot{y}=0$ at (x_0, y_0), we expect a Taylor expansion of $\dot{x}$ and $\dot{y}$ in the vicinity of (x_0, y_0) to start with linear terms:

$$\dot{\xi}=\alpha_1\xi+\alpha_2\eta, \qquad \dot{\eta}=\alpha_3\xi+\alpha_4\eta, \tag{4}$$

where $\xi\equiv x-x_0$ and $\eta\equiv y-y_0$. Provided ξ and η are not too large it will usually be adequate to terminate the series at this point – the behaviour has been *linearized* for the purpose of describing small displacements around (x_0, y_0).† The possible trajectories are defined by (4), which can be manipulated into separate and identical differential equations for the time variation of ξ and η:

$$\ddot{\xi}-(\alpha_1+\alpha_4)\dot{\xi}+(\alpha_1\alpha_4-\alpha_2\alpha_3)\xi=0,$$

$$\ddot{\eta}-(\alpha_1+\alpha_4)\dot{\eta}+(\alpha_1\alpha_4-\alpha_2\alpha_3)\eta=0.$$

30 These are nothing but the harmonic oscillator equation (2.23), with k equal to $-(\alpha_1+\alpha_4)$ and ω_0^2 equal to $(\alpha_1\alpha_4-\alpha_2\alpha_3)$. There is here, however, no physical requirement for either k or ω_0^2 to be positive. We shall write $\alpha_1+\alpha_4$ as T, to indicate that it is the trace of the α-matrix relating $(\dot{\xi}, \dot{\eta})$ to (ξ, η), and also write $\alpha_1\alpha_4-\alpha_2\alpha_3$ as D to indicate that it is the determinant of this matrix:

$$\ddot{\xi}-T\dot{\xi}+D\xi=0 \quad \text{and similarly for } \eta. \tag{5}$$

The five regimes for the solutions of (5) are shown in fig. 6. Two stable regimes occupy the quadrant in which D is positive and T negative; two correspondingly unstable regimes occupy the quadrant in which D and T are both positive; and the last which is unstable, occupies the half-space in which D is negative. The criteria for the different regimes follow from the exponential solutions of (5). If $\xi=\xi_0\,e^{pt}$,

$$p^2-Tp+D=0,$$

so that

$$p=\tfrac{1}{2}[T\pm(T^2-4D)^{\frac{1}{2}}]. \tag{6}$$

Focal stability. $D>0$, $T<0$ and $T^2<4D$; the solutions are complex, causing ξ and η to execute decaying oscillations. Since the relative amplitude and phase of ξ and η are not determined by T and D alone, but

† In later applications we shall often compress (4) into the form:

$$d\eta/d\xi=(\alpha_3\xi+\alpha_4\eta)/(\alpha_1\xi+\alpha_2\eta),$$

from which t has been eliminated. This enables trajectories to be drawn on the (ξ, η)-plane to describe the solutions of (4), but there is no indication of the sense in which a given trajectory is followed. Accordingly, a spiral trajectory may imply either the convergence of the system onto a point of stable equilibrium or the divergence from a point of unstable equilibrium. To avoid this ambiguity we shall always ensure that the coefficients in the compressed form keep the signs they have in (4).

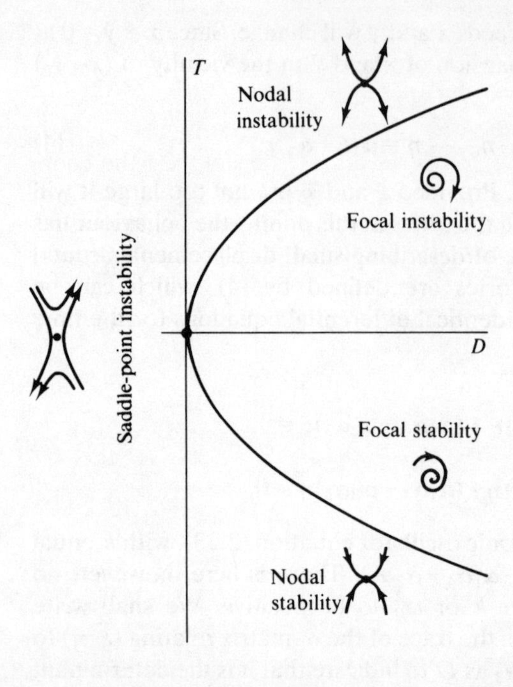

Fig. 6. Domains in which the solutions of (4) show different types of stability or instability.

depend on the magnitudes of the αs, the general form of the trajectory is an elliptical spiral converging onto the origin. The lightly damped harmonic vibrator is of this type. **30**

Focal instability. $D>0$, $T>0$ and $T^2<4D$; as above, but the oscillations grow and the elliptical spiral diverges from the origin. The harmonic vibrator with positive feedback, or some other gain mechanism overriding viscous losses, is of this type. **39** **311**

Nodal stability. $D>0$, $T<0$ and $T^2>4D$; the solutions are real so that the oscillations are replaced by exponential decay of ξ and η. The trajectories converge on the origin without spiralling round it. We shall examine the form of these trajectories in more detail soon. The overdamped harmonic vibrator is of this type. **32**

Nodal instability. $D>0$, $T>0$ and $T^2>4D$; as above but with trajectories diverging from the origin. A vibrator with very strong positive feedback is of this type.

Saddle-point instability. $D<0$; the solutions are real, one value of p being positive and the other negative. The positive solution guarantees instability, and the trajectories are of roughly hyperbolic form, avoiding the origin. A 'vibrator' with negative force constant, for example a ball sitting on top of a hill rather than in a valley, is of this type. No amount of damping or gain can stop it rolling off.

The general character of the trajectories just catalogued – spiral, non-spiral or hyperbolic; converging or diverging – is not altered by rotating or scaling the axes, and indeed the two parameters T and D are invariant under such coordinate transformations. It is helpful to use this fact to derive the form of the trajectories in more detail, transforming the coordinates so as to give (4) the simplest form. When the coordinates are rotated through an angle ϕ, followed by scaling up along one axis by a factor γ and down along the other by the same factor, the new variables X and Y are linearly related:

$$\dot{X} = \beta_1 X + \beta_2 Y, \qquad \dot{Y} = \beta_3 X + \beta_4 Y, \tag{7}$$

where

$$\left.\begin{aligned}
\beta_1 &= \alpha_1 \cos^2\phi - (\alpha_2 + \alpha_3)\cos\phi\sin\phi + \alpha_4 \sin^2\phi,\\
\beta_2 &= \frac{1}{\gamma}[\alpha_2 \cos^2\phi + (\alpha_1 - \alpha_4)\cos\phi\sin\phi - \alpha_3\sin^2\phi],\\
\beta_3 &= \gamma[\alpha_3\cos^2\phi + (\alpha_1 - \alpha_4)\cos\phi\sin\phi - \alpha_2\sin^2\phi],\\
\beta_4 &= \alpha_4\cos^2\phi + (\alpha_2 + \alpha_3)\cos\phi\sin\phi + \alpha_1\sin^2\phi.
\end{aligned}\right\} \tag{8}$$

It is clear that ϕ can always be chosen to that $\beta_1 = \beta_4$; the required choice is such as to make $\tan 2\phi$ equal to $(\alpha_1 - \alpha_4)/(\alpha_2 + \alpha_3)$. Further, unless either β_2 or β_3 then vanishes, γ can be chosen so that $\beta_3 = \pm\beta_2$ (since γ may have either sign the absolute signs of β_2 and β_3 are irrelevant). When these choices have been made, (7) takes either of two forms:

Case (a): $\beta_3 = -\beta_2$; $\quad T = 2\beta_1$ and $D = \beta_1^2 + \beta_2^2$.

$$\left.\begin{aligned}
\dot{X} &= \beta_1 X + \beta_2 Y,\\
\dot{Y} &= -\beta_2 X + \beta_1 Y.
\end{aligned}\right\} \tag{9}$$

Case (b): $\beta_3 = \beta_2$; $\quad T = 2\beta_1$ and $D = \beta_1^2 - \beta_2^2$.

$$\left.\begin{aligned}
\dot{X} &= \beta_1 X + \beta_2 Y,\\
\dot{Y} &= \beta_2 X + \beta_1 Y.
\end{aligned}\right\} \tag{10}$$

We discuss the two cases separately.

Case (a): T^2 is always less than $4D$, and the system exhibits either focal stability or focal instability. If X and Y are components of the radius vector, R, (9) shows that (since $R^2 = X^2 + Y^2$),

$$\frac{\mathrm{d}}{\mathrm{d}t}(R^2) = 2X\dot{X} + 2Y\dot{Y} = TR^2. \tag{11}$$

Further, if we write $\tan\theta = Y/X$,

$$\sec^2\theta\, \mathrm{d}\theta/\mathrm{d}t = (X\dot{Y} - Y\dot{X})/X^2 = -\beta_2 R^2/X^2,$$

therefore $\qquad \mathrm{d}\theta/\mathrm{d}t = -\beta_2.$

The representative point moves on an equiangular spiral, the radius vector increasing or decreasing exponentially, according to (11), while rotating at a constant speed, $-\beta_2$. For the spiral to converge, $T<0$. All this agrees with what we know already of decaying harmonic oscillations. The general form of trajectory in this case is an equiangular spiral distorted into elliptical form by scaling differently along any two orthogonal directions.

Case (*b*): it is convenient to carry out a further rotation of axes, through $\pi/4$, writing $X=(u+v)/\sqrt{2}$ and $Y=(u-v)/\sqrt{2}$. Then (10) takes the form

$$\dot{u}=(\beta_1-\beta_2)u, \qquad \dot{v}=(\beta_1+\beta_2)v. \tag{12}$$

These equations, showing that u and v develop exponentially and independently, define a non-spiralling trajectory which is stable only if $\beta_1-\beta_2$ and $\beta_1+\beta_2$ are both negative. Then $\beta_1^2-\beta_2^2$, i.e. D, must be positive and β_1, and hence T, must be negative. This is the case of nodal stability. In the coordinate system chosen, solution of (12) shows that

$$v=Au^{(\beta_1+\beta_2)/(\beta_1-\beta_2)}=Au^{\gamma}, \tag{13}$$

where A is an arbitrary constant, and $\gamma>0$. In the original coordinate system each value of γ defines a family of similar algebraic curves which may have any orientation, and describe the relaxation to equilibrium of an overdamped system. When $\beta_1-\beta_2$ and $\beta_1+\beta_2$ are both positive the trajectories have the same form, but the direction of motion is reversed; this is the case of nodal instability.

Finally we must consider the behaviour when $\beta_1-\beta_2$ and $\beta_1+\beta_2$ have opposite signs, i.e. when β_1 lies between $\pm\beta_2$, and $D<0$. Then γ in (13) is negative, and the trajectories avoid the origin. If $\beta_1=0$ they are rectangular hyperbolae; for non-vanishing β_1 they are not hyperbolae but have orthogonal asymptotes. Transforming back into the original coordinate system does not necessarily preserve this orthogonality.

The above analysis has assumed displacements from equilibrium small enough to be described by the linear equations (4), and this is permissible under all conditions save one – when the representative point lies exactly on the line dividing focal stability and instability, the horizontal axis in fig. 6. The example of an oscillating circuit with positive feedback will serve to illustrate the point. When the feedback is adjusted so that the total resistive loss is exactly cancelled ($Q=\infty$) the circuit may oscillate at any amplitude, poised between stability and instability, but only so long as everything is strictly linear. Commonly enough, however, the feedback is not so strong at higher amplitudes because the amplifier begins to saturate, and a strong oscillation experiences more resistive damping than the feedback can compensate. Consequently the system, which appears to be in neutral equilibrium if only linear terms are taken note of, is really in stable equilibrium; provided the higher order damping can be expressed by a power series expansion, so that a little damping is present at all amplitudes of oscillation, the current oscillations will decay to zero, though not exponentially, the effective time-constant steadily increasing as equilibrium is approached.

Now let us suppose the feedback to be increased slightly so that according to the linear theory the system is unstable. Oscillations will grow, but eventually the more rapidly increasing non-linear loss will compensate the linear gain, and the amplitude will saturate and thereafter remain constant. The representative point on the (x, y)-plane will execute a *limit cycle* about the point of unstable equilibrium. Further increase of positive feedback causes the limit cycle to expand and perhaps change its shape, so that the oscillations become non-sinusoidal. Eventually the limit cycle itself may become unstable so that the mode of oscillation switches to something entirely different. But this is well beyond the scope of what we are considering at this moment, which is confined to the almost-linear regime where the limit cycle will approximate to an ellipse and the waveform to a sinusoid.

348

An alternative form of behaviour along the line of infinite Q is for the total loss to become negative rather than positive as the amplitude grows. Then the system is unstable on this line, though the growth is not exponential, being slow at first and faster later. With slightly less feedback the linear lossy terms guarantee stability, but only while the amplitude is small enough for them to dominate the non-linear gain. In this case equilibrium is *metastable*, being stable for small displacements but unstable for larger.

Examples of instability

Since the problem, as formulated in the linearized first-order equations (4), has turned out to be reducible to the second-order differential equation of a simple vibrator, it is proper to indicate how the behaviour of the vibrator may be expressed as first-order equations. For this purpose we introduce displacement x and velocity v $(=\dot{x})$ as coordinates; then if μ is the force-constant and λ the viscous damping constant for a particle of mass m, as in (2.31),

37

$$\dot{x} = v \quad \text{and} \quad \dot{v} = -\mu x/m - \lambda v.$$

Interpreting ξ in (4) as x, η as v, we have that

$$\alpha_1 = 0, \qquad \alpha_2 = 1, \qquad \alpha_3 = -\mu/m, \qquad \alpha_4 = -\lambda/m.$$

Hence $$T = -\lambda/m, \qquad D = \mu/m.$$

In the notation of (2.23), $T = -k$ and $D = \omega_0^2$; the identity of (2.23) and (5) is immediately apparent. There is no need to discuss focal and nodal stability further – they are the underdamped and overdamped cases of harmonic vibrator behaviour. The vibration can be represented as a trajectory on the (x, v)-plane (*phase plane*), which turns out, as we shall discover, to be a very useful form of representation for non-linear vibrators. The general character of the trajectories is of course the same as

30

shown in fig. 6. To realize the unstable trajectories a source of energy is needed. Very commonly this is provided by some external agency, the power pack of an amplifier in an oscillatory circuit, for example; but the source may be internal, and the growth of instability accompanied by the conversion of stored energy into kinetic energy of vibration or heat.† If the energy source acts so as to make k negative, spontaneous vibration (focal instability) or nodal instability is the consequence. These will be discussed more fully in chapter 11. Saddle-point instability is a very common occurrence in the natural world, without the intervention of external energy sources. The collapse of a structure, involving the conversion of potential or elastic energy into, first, kinetic energy, is normally of this type, and it is no great extension of the concept to allow phase transitions such as spontaneous changes of crystal structure to be classified as saddle-point instabilities. These matters are not strictly within the compass of a book on vibration, but in view of their importance we shall devote a little space to some straightforward examples.

The Euler strut, shown in fig. 7, is a classic example of a system in which the force-constant μ may be varied at will from positive to negative, so that a critical point can be found at which μ vanishes and stability fails. The elastic strip, initially straight, remains vertical so long as the load on it is not too great. Above a certain critical load, however, it becomes unstable and, if displaced, continues to move away from the vertical, the lowering of potential energy as the mass sinks being more than enough to supply the stored elastic energy in the bent strut; the excess then provides kinetic energy for the weight to move further from equilibrium. Finding the equation of motion of the mass is a matter of determining the horizontal restoring force when it is displaced. With the notation of fig. 7(c), the

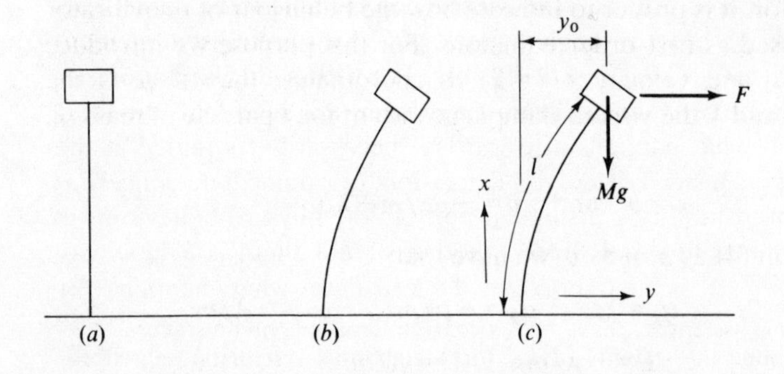

Fig. 7. The Euler strut: (a) subcritical load; (b) supercritical load; (c) notation used in analysis of displacement by a horizontal force.

curvature at a point on the strut is, for small displacements, $\mathrm{d}^2y/\mathrm{d}x^2$, and to create this curvature a bending moment $R\,\mathrm{d}^2y/\mathrm{d}x^2$ is needed, R being the flexural rigidity of the strut, constant for a strut of uniform cross-section.

† The 'one-shot' pulsed laser is an example. Irradiation of the active material stores energy by exciting the molecules to higher states, and this energy is then emitted in the form of a short burst of monochromatic light as the molecules undergo spontaneous and phase-coherent oscillation.

Hence the shape of the strut is fixed by the equation

$$R\,\mathrm{d}^2y/\mathrm{d}x^2 = Mg(y_0 - y) + F(l - x), \qquad (14)$$

with the general solution:

$$y = y_0 + F(l - x)/Mg + A\cos qx + B\sin qx,$$

where $q^2 = Mg/R$ and A and B are constants. When $x = 0$ both y and $\mathrm{d}y/\mathrm{d}x$ vanish. Hence $A = -Fl/Mg - y_0$ and $B = F/Mgq$. By putting $x = l$, $y = y_0$ we find y_0 and hence the force constant:

$$\mu \equiv F/y_0 = M\omega_p^2/(\tan ql/ql - 1), \qquad (15)$$

where ω_p is $(g/l)^{\frac{1}{2}}$, the frequency of a simple pendulum of length l.

The critical mass M_c is that which reduces μ to zero, by making ql equal to $\pi/2$. Hence

$$M_c = \tfrac{1}{4}\pi^2 R/gl^2.$$

The behaviour is easy to demonstrate with a straight length of steel strip, having a suitable mass attached to one end, and clamped in a vice. As the length of free strip is increased the natural frequency is seen to fall steadily, though the equilibrium state of the strip remains vertical. But after the point has been reached when $\mu = 0$ the vertical strip is unstable, and two new equilibrium positions now exist, one on each side. If the strip is perfectly elastic the bent equilibrium form may be worked out by rewriting the differential equation (14) with $F = 0$ and with the correct form for the curvature replacing the linear approximation $\mathrm{d}^2y/\mathrm{d}x^2$. There is something slightly academic about devoting great attention to this problem in applied mathematics, since most materials show brittleness or plasticity at small deformations; unless the strut is very thin in relation to its length the final configuration after the critical load is exceeded is liable to be some scattered fragments or a heap of tangled wreckage.[3] This is no disparagement of the elastic theory which, used sensibly, i.e. within the limitations of Hooke's law, has great value in predicting the critical load, but one must not suppose that what happens afterwards is necessarily contained in the linearized model. This is, however, a digression; our immediate concern is that the point of instability is a singular point in the physical behaviour, in the sense that continuity fails here, whether or not there is a Hookean regime beyond.† It is not possible to calculate what happens by

† The academic scientist may be satisfied to determine the stability of a system by means of linearized equations that correctly describe small perturbations from the stationary state. A practising engineer concerned with, for example, the flow of cooling gas in the core of a nuclear reactor, must worry about the possibility that the stable regime may not be proof against rather larger perturbations such as might occur in unusual circumstances. Unfortunately the systematic search for possible catastrophes over the whole range of variation of a large number of parameters is a formidable, often impracticable, undertaking. One of the characteristics of a great engineer is his capacity to foresee and guard against changes which routine analysis would overlook.

extrapolation of the properties through the critical point, even by very small steps. Thus a graph of the equilibrium position as a function of load shows a bifurcation at right angles (fig. 8). Symmetry about the vertical is maintained, but only in the sense that the positions available for equilibrium form a symmetrical pattern. Between what is allowed, and what actually occurs, symmetry is lost, for when two symmetrically disposed alternatives offer themselves a choice must be made between right and left. The outcome may be arbitrary or conditioned by external influences, but whichever choice is made the symmetry is inevitably broken. Obviously this critical point precludes extrapolation, since even though y_0 is a continuous function of load, its derivative with load is not.

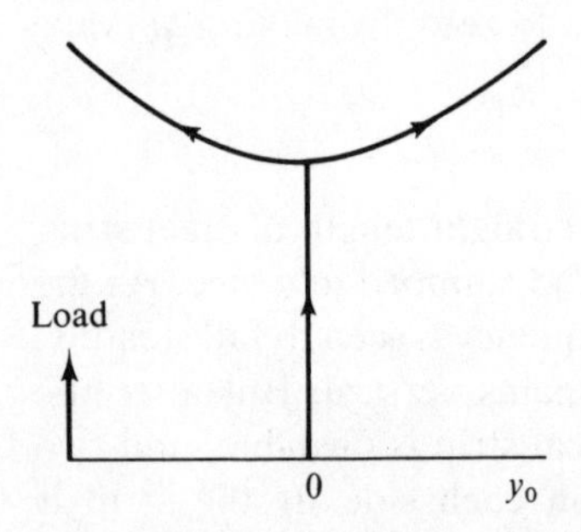

Fig. 8. Bifurcation of the solutions of the Euler strut problem at the critical load.

With other critical phenomena even y_0 need not be continuous; the hypothetical curve of fig. 9 is a perfectly possible form which would result in a sudden jump of y_0 at the critical load. The vibrating wire discussed at the beginning of this chapter shows a similar instability (see fig. 3), with hysteresis just like that indicated in fig. 9. This is hardly a structural instability, and perhaps a clearer illustration is provided by the sort of steel measuring tape that rolls into a small pocket container. The tape unrolls into a rather rigid straight strip by virtue of its slightly curved cross-section.

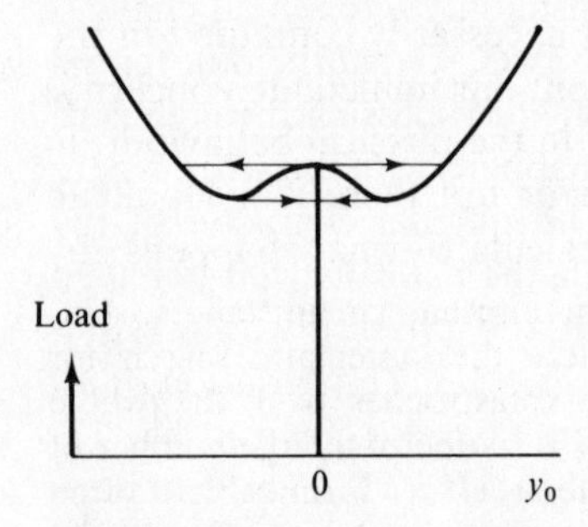

Fig. 9. An alternative form of bifurcation, with discontinuities and hysteresis.

Let the tape be placed flat on a table with its curvature concave upwards, and then the end gently pushed further and further beyond the edge. Quite suddenly, with a loud crack, the tape will kink and collapse to the floor. On pulling it back, it will remain kinked and hanging limp until much less projects over the edge, when suddenly it will jump back to its nearly horizontal, unkinked form. The transitions are sharp and the hysteresis very marked. This is not an easy problem to analyse in detail, but the short rope ladder shown in fig. 10 is quite simple; the diagram shows what torque is required to maintain a given angular displacement, and how it depends on past history. It is left to the reader to satisfy himself of the correctness of the graph. The inflected isotherms, below the critical temperature, in van der Waals' theory of imperfect gases, provide another example of an instability leading to a finite discontinuity. The fact that this is a simple model of a first-order phase transition may be taken as an indication of how the study of instability need not be confined to elementary dynamical systems but can be extended to include phenomena from all fields of science (and beyond).

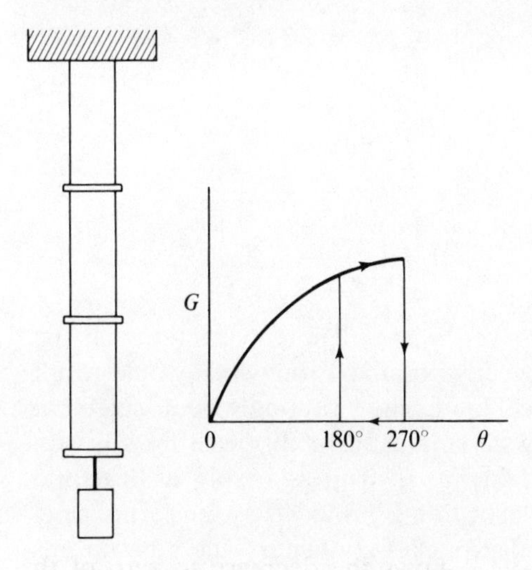

Fig. 10. Two parallel strings separated by equally spaced rods are held in tension, and the bottom rod is twisted through θ about a vertical axis. The graph shows how the torque G on the bottom rod varies with θ.

Returning to fig. 10, one observes that as G is increased the graph of G versus θ becomes horizontal at the point of instability – for small changes in θ near this point there is virtually no change in G, so that the differential torsion constant for small changes in displacement is passing through zero at the critical point. This is analogous to what happens with the strut, where μ passes through zero at the critical load. The singular character of the phenomenon results, of course, not from the mathematically well-behaved variation of μ itself, but from the fact that the physically interesting consequences are described by the frequency, $\omega_0 \propto \mu^{\frac{1}{2}}$. The variation of ω_0 with the load on the strut, as determined from (15), is shown in fig. 11. Since, near the critical load, μ is proportional to $M_c - M$, $\mu^{\frac{1}{2}}$ falls to zero with vertical tangent, another illustration of the impossibility of extrapolating further. The linear fall to zero of an elastic modulus or equivalent parameter is a clear diagnostic sign of approaching instability. Fig. 12 shows how the frequency of one of the lattice vibrational modes in strontium titanate (as determined by inelastic neutron scattering) falls with

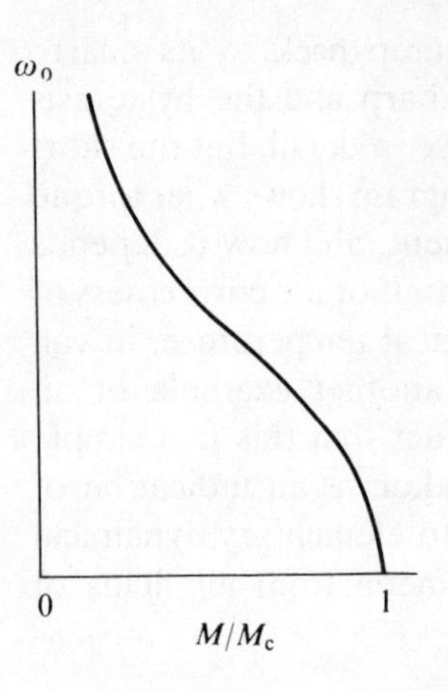

Fig. 11. Vibration frequency of the mass on an Euler strut.

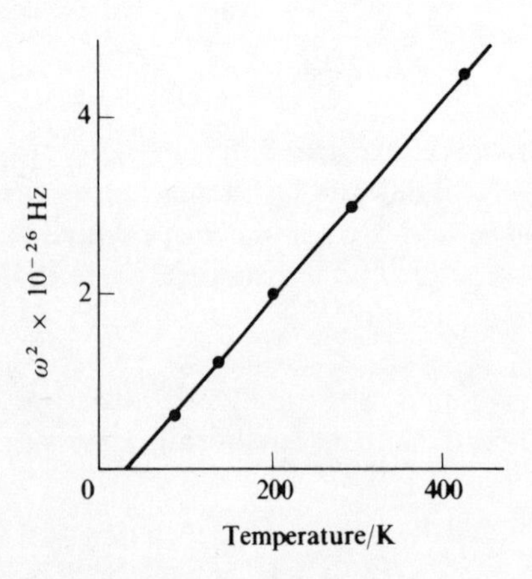

Fig. 12. Characteristic frequency of one of the vibrational modes in strontium titanate, showing how ω^2 falls linearly to zero as the temperature falls towards the structural transformation point at 35 K (R. A. Cowley[4]).

decrease of temperature, and fig. 13 shows the decrease to zero of the reciprocal of the magnetic susceptibility in nickel. Extrapolation to zero in each case yields a value, fairly close to what is determined by direct observation, of the temperature at which a phase transformation occurs, with changes in crystal symmetry and other, non-structural, properties; strontium titanate changes its crystal structure, barium titanate becomes ferroelectric, nickel ferromagnetic. The properties in the new phases, below the critical temperature, are not deducible from those above by extrapolation. It would take us too far afield to discuss these applications any further, but before leaving the topic a word of warning is in order, to avoid giving too strong an impression of a close analogy between phase transformations and static instabilities. In the vicinity of the critical temperature, for these and a host of other materials exhibiting phase instability, the linear approach of the modulus to zero breaks down, as shows up clearly in fig. 13. The *order parameter*, used to characterize a structural change in a solid, is a variable that might be expected to behave like the frequency in our simple examples, its square falling linearly to zero

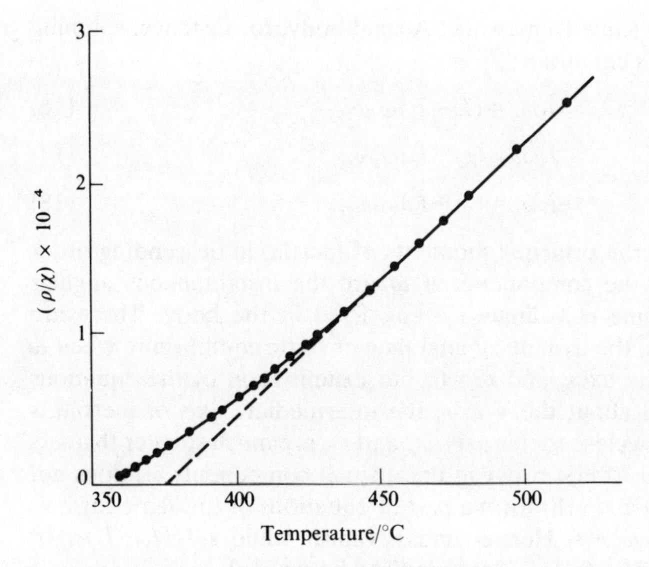

Fig. 13. Reciprocal of the magnetic susceptibility χ/ρ of nickel above its ferromagnetic Curie point, 358 °C (P. Weiss and R. Forrer[5]).

at the critical point. But, and fig. 14 shows this for two different materials, it is more usual for the cube to fall linearly in the very last stages before the transition. It is not surprising that a real material, built of a vast number of molecular units, should not be exactly modelled by a dynamical system of very few units. The study of real critical points is one of great difficulty and fascination.

284

To conclude this presentation of examples illustrating the theory of stability in simple systems, let us note that dynamical equilibrium may also

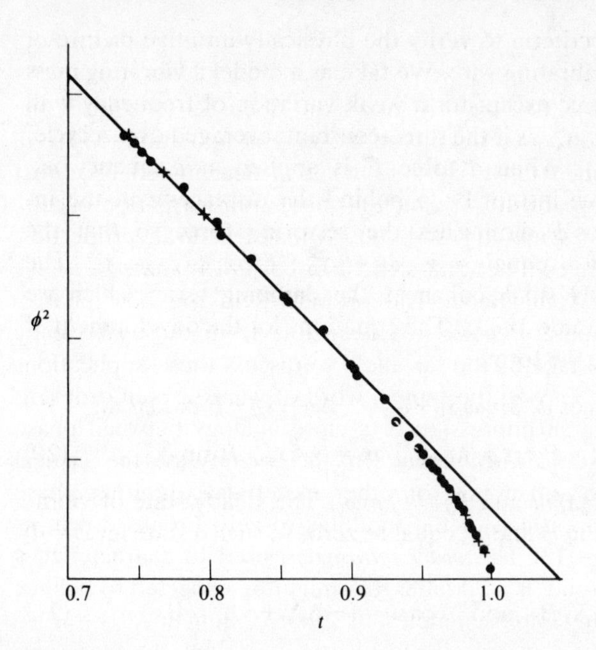

Fig. 14. The square of the order parameter, ϕ, in $LaAlO_3$ (crosses) and $SrTiO_3$ (points) as a function of reduced temperature, t, defined as T/T_c where T_c is the transition temperature. Just below T_c, $\phi \propto (1-t)^{\frac{1}{3}}$ (K. A. Müller and W. Berlinger[6]).

be brought into the same framework. A rigid body, for instance, spinning freely, obeys Euler's equations:[7]

$$I_x\dot{\omega}_x = (I_y - I_z)\omega_y\omega_z, \tag{16}$$

$$I_y\dot{\omega}_y = (I_z - I_x)\omega_z\omega_x, \tag{17}$$

$$I_z\dot{\omega}_z = (I_x - I_y)\omega_x\omega_y, \tag{18}$$

where I_x, I_y, I_z are the principal moments of inertia, in descending order of magnitude, and the components of $\boldsymbol{\omega}$ are the instantaneous angular velocities in the same coordinate system, fixed in the body. There are stationary solutions, the dynamical analogue of static equilibrium, when $\boldsymbol{\omega}$ lies along any of the axes, and $\dot{\boldsymbol{\omega}} = 0$, but examination of the equations shows that rotation about the y-axis, the intermediate axis of inertia, is unstable. For with $\boldsymbol{\omega}$ close to this axis, ω_x and ω_z are much smaller than ω_y, and (17) shows that to first order in these small components ω_y does not change. Hence (16) and (18) form a pair of equations of the same form as (4), with $\omega_x \equiv \xi$ and $\omega_z \equiv \eta$. Here α_1 and α_4 vanish, while $\alpha_2 = (I_y - I_z)\omega_y/I_x$ and $\alpha_3 = (I_x - I_y)\omega_y/I_z$. If I_y is intermediate between I_x and I_z, $\alpha_2\alpha_3$ is positive, D is negative and the motion shows saddle-point instability. On the other hand, the same analysis applied to motion about the axes where I is largest or smallest gives a positive value to D, while $T = 0$. The motion is oscillatory without decrement – in fact the axis of rotation wanders in an elliptical cone about the x- or z-axis, so that ω_y and ω_z (or ω_x) oscillate sinusoidally in phase quadrature.

The slightly anharmonic vibrator revisited

Let us use our stability criteria to verify the physically intuitive picture of mode-switching in the vibrating wire. We take as a model a vibrating mass whose motion is harmonic except for a weak variation of frequency with amplitude, $\omega(x_0) = \omega_0 + \gamma x_0^2$, as if the force-constant, averaged over a cycle, were $m\omega_0^2(1 + 2\gamma x_0^2/\omega_0)$. When a force F is applied at frequency ω_1, lagging in phase at some instant by ϕ behind the displacement, the in-phase component $F\cos\phi$ diminishes the restoring force so that the instantaneous vibration frequency is $\omega_0 + \gamma x_0^2 - F\cos\phi/2m\omega_0 x_0$. The quadrature component $F\sin\phi$ enhances the damping term, which we write as having an amplitude $\lambda\omega_0 x_0$. The equations for the development of ϕ and x_0 therefore take the form:

249

$$\dot{\phi} = \omega_0 + \gamma x_0^2 - F\cos\phi/2m\omega_0 x_0 - \omega_1 = -\Delta + \gamma x_0^2 - B\cos\phi/x_0; \tag{19}$$

and $$\dot{x}_0 = -\tfrac{1}{2}\lambda x_0(1 + F\sin\phi/\lambda\omega_0 x_0)/m = -Ax_0 - B\sin\phi, \tag{20}$$

where $\Delta = \omega_1 - \omega_0$, $A = \frac{1}{2}\lambda/m$ and $B = \frac{1}{2}F/m\omega_0$. The steady state of vibration is obtained by putting $\dot{\phi}$ and $\dot{x}_0$ equal to zero. In such a state let $\phi = \Phi$ and $x_0 = X$; then

$$\sin\Phi = -AX/B \quad \text{and} \quad \cos\Phi = (-\Delta X + \gamma X^3)/B. \tag{21}$$

By squaring and adding we derive a cubic equation in X^2, giving the possibility of three distinct amplitudes, as expected from fig. 3:

$$\gamma^2 X^6 - 2\Delta\gamma X^4 + (A^2+\Delta^2)X^2 - B^2 = 0. \tag{22}$$

To examine the stability of the state (Φ, X), put $\phi = \Phi + \xi$, $x_0 = X + \eta$. Then from (19) and (20), by differentiation, the linearized equations take the form:

$$\dot{\xi} = 2\gamma X\eta + B\eta \cos\Phi/X^2 + B\xi \sin\Phi/X = \alpha_1\xi + \alpha_2\eta,$$

and $$\dot{\eta} = -A\eta - B\xi\cos\Phi = \alpha_3\xi + \alpha_4\eta,$$

the coefficients α being expressed in terms of X alone by use of (21):

$$\alpha_1 = -A, \qquad \alpha_2 = 3\gamma X - \Delta/X, \qquad \alpha_3 = \Delta X - \gamma X^3, \qquad \alpha_4 = -A.$$

The trace $T = \alpha_1 + \alpha_4 = -2A$ and is always negative; it can only confer stability. The determinant, however, $D = \alpha_1\alpha_4 - \alpha_2\alpha_3 = 3\gamma^2 X^4 - 4\Delta\gamma X^2 + A^2 + \Delta^2$, and may be caused to go negative. The critical value of Δ, at which a given solution becomes unstable, is that which makes D equal to zero. It is easily verified, by differentiating (22), that this is also the value of Δ at which the response curve, $X(\Delta)$, has a vertical tangent. This is, of course, the same criterion as we took for granted in discussing fig. 3.

The instability of an anharmonic vibrator under the influence of too strong a force sets a practical limit to the validity of neglecting anharmonicity. So long as the slope of the resonance curve nowhere exceeds that of the line (e.g. the chain line in fig. 3) showing the variation of natural frequency with amplitude, the response is single-valued and the harmonic approximation may well prove adequate. The losses in the system, by broadening the response, help to save the harmonic approximation. For a given amplitude, however, the resonance curve can be made as steep as one wishes by reducing the losses, so that in principle it is always possible to achieve the regime of instability, however small the amplitude. It follows then that the ideal lossless vibrator has a single-valued response at all frequencies only if the force-constant is strictly constant – any departure, however small, introduces instabilities. The harmonic vibrator is indeed an idealization though in practice, as already remarked, the presence of loss mechanisms confers a limited validity on its use.

There is no need at this stage to give any more examples of the use of the stability criteria for systems governed by (4); they will turn up again on several occasions in chapter 12.

390, 395
400, 415

Stability of a system with more than two independent variables

More elaborate systems than the examples just mentioned may need more than two variables to specify their state, but in the vicinity of an equilibrium configuration the linearized equations of motion will be similar to (4):

$$\dot{\xi}_1 = \alpha_{11}\xi_1 + \alpha_{12}\xi_2 + \cdots,$$

or in general, for the ith coordinate:

$$\dot{\xi}_i = \sum_j \alpha_{ij}\xi_j.$$

If we substitute the trial solution, $\xi_i = \xi_{i0}\,\mathrm{e}^{pt}$, the equations reduce to linear algebraic equations:

$$(\alpha_{ii} - p)\xi_i + \sum_{j \neq i} \alpha_{ij}\xi_j = 0.$$

There are as many equations as there are independent variables, and since there are no constant terms the equations can only be mutually compatible if the determinant of the coefficients vanishes. The possible values of p are thus those which satisfy the equation:

$$\begin{vmatrix} \alpha_{11} - p & \alpha_{12} & \cdots \\ \alpha_{21} & \alpha_{22} - p & \cdots \\ \vdots & \vdots & \end{vmatrix} = 0,$$

which on multiplying out gives an equation of the same order in p as the number of variables, n. The system is capable of n independent modes of motion, vibratory if p is complex; only if the real part of every p is negative or zero can the system avoid instability, for any mode having a positive real part of p describes a growing disturbance. The criteria for stability are those rules which enable one to determine from the set of coefficients α_{ij} that there are no growing modes; the general solution to this algebraic problem was given by Routh, and in modified form by Hurwitz.[8] It is sufficient here to indicate the nature of the general problem and the existence of a solution; we do not propose to tackle any examples where more than two variables are present. There are full acounts available in standard textbooks. 284

Stability testing by discrete sampling (stroboscopic method)[9]

266

The rotating rigid body, like the simple vibrator, is a system whose coordinates can readily be followed continuously, and represented by a trajectory on the appropriate phase plane. It may be inconvenient, however, to keep note of so much detail if the behaviour is more complicated, and it may be sufficient to sample the coordinates at regular intervals to see how they are evolving, whether towards or away from steady values. This is particularly helpful for testing the stability of a dynamical system whose motion is periodic or approximately so, and especially if it is driven by a periodic force, for one can test at a standard moment in each cycle and cease to worry about what may be quite a complex pattern of behaviour between each sampling moment. This is what we did in deciding that the subharmonic generator in the arrangement of fig. 4 could not be stable in one of its possible configurations. Similarly if a maintained oscillator is found to be capable of executing a closed trajectory in the phase plane, the stability of this periodic motion can be examined by displacing it from its stationary cycle and noting each subsequent passage of a line intersecting the trajectory; in this case the state is determined by a single parameter, y say, and the stability depends on the relation between y_n and y_{n+1}. It will be realized that, merely by concentrating attention on discrete samples, one cannot in a detailed theoretical analysis avoid consideration of intermediate states of the system, since only by following the trajectory in detail between y_n and y_{n+1} can one discover how they are related; but the presentation of the results of such analysis is reduced to something manageable, and thereby it is made easier to catalogue the different types of behaviour that are possible in principle. A specific example of a fairly complicated trajectory reduced to simplicity by discrete sampling will be found in fig 10.6. **297**

Let us illustrate this by reference to a one-parameter system whose evolution is completely specified by the dependence of any sample y_{n+1} on the value of the previous sample y_n, $y_{n+1} = F(y_n)$. We shall not limit the discussion to the particular type of application mentioned above, for which the form of F is restricted to monotonic variation by the requirement that trajectories on the phase plane do not intersect. Much more variety is possible with biological applications, as for instance the annual variations in the population of an insect species which matures, breeds and dies in a single season. The population, y_{n+1}, in year $n+1$ is thus entirely the progeny of the previous population, y_n, and if (a very dubious assumption) we can neglect random disturbances such as weather we may hope to write $y_{n+1} = F(y_n)$. For small y_n one expects $y_{n+1} > y_n$, for otherwise the species can hardly survive to allow us to examine its ecology, but for denser populations y_n may saturate ($y_{n+1} \sim y_n$); alternatively, epidemic disease or other hazards of overgrowth may cause havoc and result in F decreasing at large values of y_n. Fig. 15 exhibits various cases of interest. Equilibrium

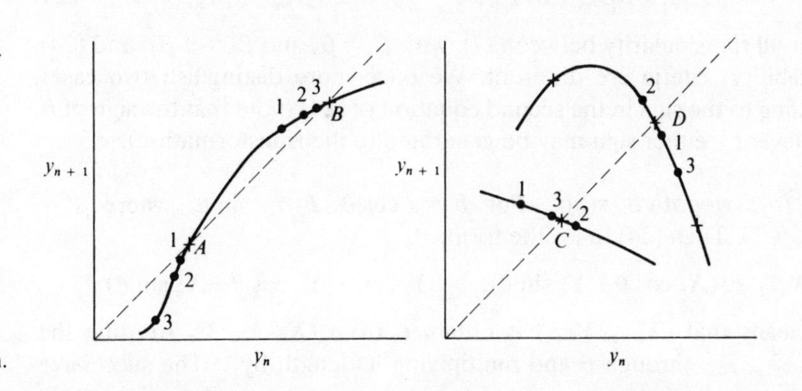

Fig. 15. The full curves are various hypothetical possibilities for the dependence of y_{n+1} on y_n. A stationary value is possible at an intersection with the line $y_{n+1} = y_n$, shown broken. Successive values shown here illustrate that A and D are unstable, while B and C are stable. The crosses on the line through D are mirror images in the 45° line and are a possible (not necessarily stable) alternating steady solution.

behaviour is possible only when $y_{n+1} = y_n$, at the intersection of the line representing F with a line of unit slope. The numbered points show successive values of y in the neighbourhood of an intersection. If the slope of F at the intersection is greater than unity, the point is unstable and divergence from the line is monotonic; if the slope is positive but less than unity it is stable, with monotonic convergence. These are the one-dimensional analogues of nodal stability and instability. If the slope is negative but not as steep as -1, there is oscillatory (analogous to focal) stability; if steeper than -1, oscillatory instability. The oscillation amplitude grows until two points are found on F which are mirror images in the line of unit slope; the solution may then become stationary with alternate values of y lying on each of these points. There is, however, no guarantee that this stationary solution is itself stable. If the negative slope at the intersection is too steep the stable behaviour may be a cycle of 4, 8 or more points; more remarkably there may be no periodic solution, and y may run almost at random through all points on finite segments of the curve. As we shall have no occasion, for the reason already explained, to treat cases in which the slope at intersection is negative, we shall go no further into this pathological behaviour, which has excited a certain amount of interest among mathematicians.(10)

284

Of more concern is the sampling of two-parameter systems. The device in fig. 4 can be characterized by the speed of the pin at the moment of contact with the base and the phase of oscillation of the base at this instant; given these two it is a matter of dynamics to calculate when the next contact will occur and what the pin speed will be. In general terms we may, as before, characterize a small variation from a stationary dynamical pattern by two parameters, ξ and η, which in the stationary configuration are found to vanish at every sampling. Close to such a configuration we expect linear behaviour:

$$\xi_{n+1} = a_1\xi_n + a_2\eta_n, \qquad \eta_{n+1} = a_3\xi_n + a_4\eta_n. \tag{23}$$

By the same transformations as before (23) may be expressed in terms of two coefficients only:

257

$$X_{n+1} = b_1X_n + b_2Y_n, \qquad Y_{n+1} = \pm b_2X_n + b_1Y_n, \tag{24}$$

but for all the similarity between (7), with $\beta_1 = \beta_4$ and $\beta_3 = \pm\beta_2$, and (24), the stability criteria are different. We once more distinguish two cases, according to the sign in the second equation of (24) (note that the sign of b_2 is irrelevant – either sign may be generated in the transformation).

Case (a), negative sign. Put $b_1 = s\cos\theta$, $b_2 = s\sin\theta$, where $s^2 = b_1^2 + b_2^2 = D$. Then (24) takes the form:

$$X_{n+1} = s(X_n\cos\theta + Y_n\sin\theta), \qquad Y_{n+1} = s(Y_n\cos\theta - X_n\sin\theta).$$

This means that (X_{n+1}, Y_{n+1}) is obtained from (X_n, Y_n) by rotating the vector (X_n, Y_n) through θ and multiplying its length by s. The successive points therefore lie at equal angular increments along an equiangular spiral which converges if $s<1$, but diverges if $s>1$. These are the discrete analogues of focal stability and instability.

Case (b), positive sign. As before, rotate the axes through $\pi/4$ to give the analogues of (12):

$$u_{n+1}=(b_1-b_2)u_n, \qquad v_{n+1}=(b_1+b_2)v_n.$$

Each coordinate evolves by geometrical progression and we distinguish three possible types of behaviour:

nodal stability – convergence when $|b_1-b_2|$ and $|b_1+b_2|<1$;
nodal instability – divergence along the same types of trajectories when $|b_1-b_2|$ and $|b_1+b_2|>1$;
saddle-point instability – when one of $|b_1-b_2|$ and $|b_1+b_2|$ is greater and the other less than unity.

Since the determinant and trace, D and T, of the a-matrix are unchanged by the transformation of coordinates, it is convenient to express these criteria in terms of them.

We have that

$$D=a_1a_4-a_2a_3=b_1^2\mp b_2^2, \tag{25}$$

and

$$T=a_1+a_4=2b_1; \tag{26}$$

therefore $b_2^2=\pm(b_1^2-D)=\pm\frac{1}{4}(T^2-4D)$, the upper sign referring to case (b). All points inside the parabola, $T^2=4D$, belong to case (a) and all points outside to case (b). On the straight line, $T=1+D$, which is tangent to the parabola and wholly outside it, (25) and (26) show that $b_2^2=(1-b_1)^2$; since the sign of b_2 is irrelevant, we choose the positive root, so that $b_1+b_2=1$ on this line. Similarly, on the line $T=-(1+D)$, $b_1-b_2=-1$. From here on it is easy to label the various regions of the (D, T)-plane according to their stability behaviour, as in fig. 16. The criterion for stability is simple enough: (D, T) must lie in the triangle defined by the points $(-1, 0)$ and $(1, \pm2)$.

284

We are now in a position to discuss the stability of the subharmonic generated by the bouncing pin.

The subharmonic generator revisited

To develop the model illustrated in fig. 4 quantitatively, let us assume the spring exerts constant pressure on the pin, so that it has constant acceleration f when free of the base, and let the coefficient of restitution at impact be ε. If the mass arrives with downward velocity V when the base is moving up with velocity v, the relative velocity after bouncing is $\varepsilon(V+v)$

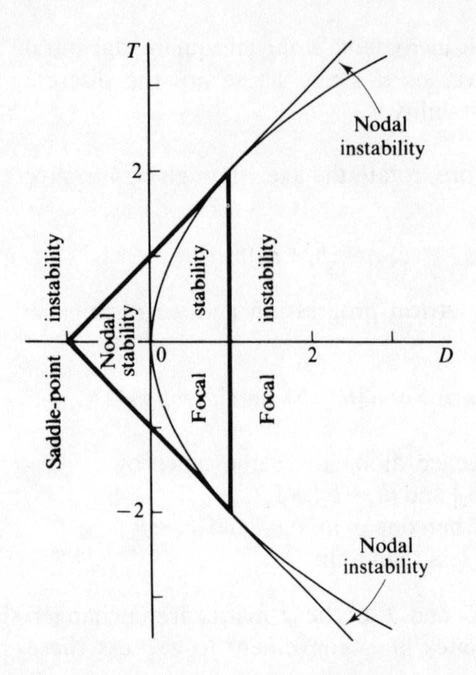

Fig. 16. Domains in which the solutions of (23) have various characters. The triangle defines the limits of stable solutions.

and the actual upward velocity of the mass is $\varepsilon V+(1+\varepsilon)v$. Since in a periodic solution this must equal V,

$$v/V=(1-\varepsilon)/(1+\varepsilon). \tag{27}$$

If the vertical position z of the base plate is $z_0 \sin \omega t$, and a bounce occurs at t_0,

$$v=\omega z_0 \cos \omega t_0. \tag{28}$$

Moreover, for the nth subharmonic the time between the bounces, which is $2V/f$, must be $2\pi n/\omega$. Hence the phase at the moment of contact is determined:

$$\cos \omega t_0=\frac{\pi n f}{\omega^2 z_0}\frac{1-\varepsilon}{1+\varepsilon}. \tag{29}$$

This equation has two alternative solutions, as shown in fig. 5, and although we know already that one is unstable we shall keep both solutions in mind and test both for stability.

The two parameters characterizing the periodic solution are conveniently chosen to be V and t_0, so that a slightly perturbed solution has the mass arriving with velocity $V+\delta V$ at $t_0+\delta t$. At the first bounce to be considered, when these parameters are δV_1 and δt_1, the velocity of the base is not v, as given by (28), but $v+\delta v_1$, where

$$\delta v_1=-\omega^2 z_0 \sin \omega t_0 \cdot \delta t_1.$$

Consequently, the mass rebounds with velocity $V+\delta V_1'$, where

$$\delta V_1' = \varepsilon\,\delta V_1 - (1+\varepsilon)\omega^2 z_0 \sin \omega t_0 \cdot \delta t_1. \tag{30}$$

If the base were at the same height when the mass returned at time $t_0+2\pi n/\omega+\delta t_2$, δV_2 would be the same as $\delta V_1'$. But the base has risen a further distance $\delta z = \omega z_0 \cos \omega t_0 \cdot (\delta t_2 - \delta t_1)$, and the velocity on arrival is thereby diminished by $f\,\delta z/V$. Hence

$$\delta V_2 = \varepsilon\,\delta V_1 - (1+\varepsilon)\omega^2 z_0 \sin \omega t_0 \cdot \delta t_1 - f\omega z_0 \cos \omega t_0 \cdot (\delta t_2 - \delta t_1)/V. \tag{31}$$

Further, the mass after rebounding with velocity $V+\delta V_1'$, as given by (30), arrives $2\pi n/\omega+\delta t_2-\delta t_1$ later with downward velocity $V+\delta V_2$, having suffered uniform acceleration in the interval. Hence.

$$2V+\delta V_1' + \delta V_2 = f(2\pi n/\omega + \delta t_2 - \delta t_1),$$

i.e.

$$\delta V_1' = f(\delta t_2 - \delta t_1) - \delta V_2.$$

From (30), by eliminating $\delta V_1'$, we have:

$$\varepsilon\,\delta V_1 + \delta V_2 = f\,\delta t_2 + [(1+\varepsilon)\omega^2 z_0 \sin \omega t_0 - f]\,\delta t_1. \tag{32}$$

Rearranging (31) and (32), and making use of (27) and (28), we have

$$(f\,\delta t_2) = a_1(f\,\delta t_1) + a_2 \delta V_1$$

and

$$\delta V_2 = a_3(f\,\delta t_1) + a_4\,\delta V_1,$$

where

$$\left.\begin{aligned} a_1 &= 1-(1+\varepsilon)^2 B,\\ a_2 &= \varepsilon(1+\varepsilon),\\ a_3 &= -\varepsilon(1+\varepsilon)B,\\ a_4 &= \varepsilon^2, \end{aligned}\right\} \tag{33}$$

and B is written for $\omega^2 z_0 \sin \omega t_0/f$. The problem is now formulated in the same way as (23); by choosing $f\,\delta t$ for ξ and δV for η we have arranged that the *a*s are dimensionless. From (29),

$$B^2 = \omega^4 z_0^2/f^2 - (1-\varepsilon)^2 \pi^2 n^2/(1+\varepsilon)^2, \tag{34}$$

and the positive and negative values of B define the two possible solutions shown in fig. 5.

As the amplitude of the base vibration is altered, B varies while ε remains constant; the *a*s are functions of one parameter, so that the point on the (D, T)-plane defining the behaviour traces out a line, in this case the vertical straight line $D=\varepsilon^2$, as follows from (33). Since $T = 1+\varepsilon^2-(1+\varepsilon)^2 B$, and B can take any value, we arrive at the following description:

(i) $\omega^2 z_0/f \geqslant (1-\varepsilon)\pi n/(1+\varepsilon)$ for the nth subharmonic to have any chance of being generated; this follows from (34).

(ii) At the critical value of z_0, $B=0$ and $T=1+\varepsilon^2=1+D$; the system is in neutral equilibrium at the upper boundary between nodal stability and saddle-point instability in fig. 16.

(iii) As z_0 increases, negative values of B increase T and lead to instability (the broken curve of fig. 5), while positive values of B diminish T at constant D to allow first nodal and then focal stability. Further increase of B results in a short stretch of nodal stability giving place to saddle-point instability. The overall range of B that can confer stability is thus determined:

$$0<B<2(1+\varepsilon^2)/(1+\varepsilon)^2,$$

or in alternative form, by use of (34), in terms of the range of amplitudes of base vibration that can excite the nth subharmonic:

$$\frac{1-\varepsilon}{1+\varepsilon}\pi n<f_0/f<\frac{1-\varepsilon}{1+\varepsilon}\pi n\left[1+\left(\frac{1+\varepsilon^2}{1-\varepsilon^2}\frac{2}{\pi n}\right)^2\right]^{\frac{1}{2}}, \tag{35}$$

where $f_0=\omega^2 z_0$ and is the maximum acceleration of the base. The ranges for different subharmonics may overlap but need not do so. For instance, if $\varepsilon=0.8$, the fundamental excitation ($n=1$) can occur for values of f_0/f between 0.35 and 1.07, the second subharmonic between 0.20 and 1.23, the third between 1.05 and 1.45, the fourth between 1.40 and 1.72, and the fifth between 1.75 and 2.01. The first four ranges overlap, but the rest do not. With ε rather smaller, none overlap, while with ε nearer unity several subharmonics may be stable with the same value of z_0. It will also be observed that stable bouncing motion is possible even when $f_0<f$ and, left to itself, the mass would move with the base and not be thrown up. Different modes of response may therefore be brought about by different initial conditions.

At once a number of questions raise themselves, to indicate how superficial our analysis has been even of this elementary example. When several subharmonics can coexist, what determines which, if any, actually occurs? As z_0 is varied so that a subharmonic ceases to be stable, how does the system respond, especially if there is no other stable subharmonic to which it can switch? Are there other forms of response, fractional harmonics possibly, or integral subharmonics with more than one impact during the cycle?

To answer the last question first, undoubtedly other patterns can occur. Examination of the system pictured in fig. 4 with a stroboscope is enough to show this, even if it also shows that the base is significantly disturbed by the impact, so that the model is not very close to reality. Nevertheless, double impacts can certainly occur, and even multiple impacts, especially if ε is low. Fig. 17 shows in schematic form such a possibility, the mass rebounding so as to be caught again and again by the rising base, but eventually being flung off as the downward acceleration of the base exceeds f. The waveform picked up by a microphone placed near the loudspeaker

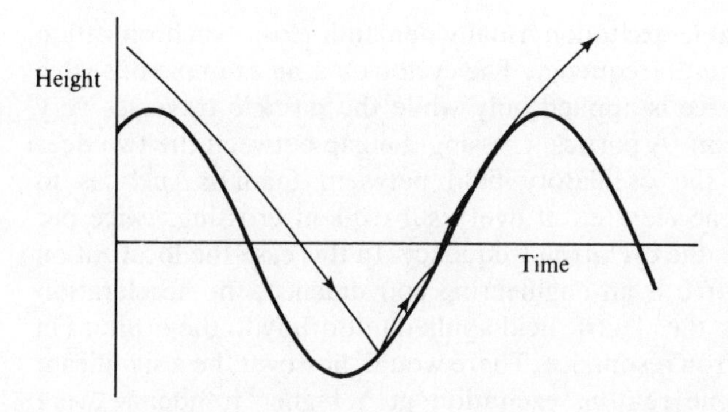

Fig. 17. Multiple bounces in the experiment of fig. 4.

confirms the complexity of the behaviour, with many spikes and discontinuities to indicate the multiple impacts. In spite of this, careful adjustment of the amplitude allows a given pattern to persist for a long time, and the possibility unquestionably exists of stable locking of the bouncing mass to a low subharmonic of the loudspeaker. At the same time, one is left in no doubt that most locking patterns are complex and that they are very numerous. Evidently an extended search of the possibilities would be needed to satisfy oneself that all stable modes had been located. This, however, would still be incomplete since there are almost certainly aperiodic trajectories, even at values of z_0 for which stable periodic behaviour is possible. The procedure for testing stability is now seen to be of limited utility if a full picture of what can happen is required. Probably nothing short of tracing out trajectories in detail, by computer for instance, would satisfy this need, should it arise; for the stability test is only the last stage – and the easiest – a critical examination of what has been discovered by hard work. As it happens, this particular system is not in itself of any great interest and there is no encouragement to undertake a full investigation. It is enough to recognize that as soon as one leaves harmonic or nearly harmonic problems even the most elementary cases may prove extremely formidable.

284

The cyclotron, and Azbel'–Kaner cyclotron resonance

A characteristic feature of the problem just discussed is that the applied force is not independent of the coordinate of the oscillatory system, as with all other cases of forced vibration examined so far. The equation of motion takes the form:

$$\ddot{\xi} + f(\xi) = F(\xi, t),$$

instead of a simple $F(t)$ on the right-hand side, the change being enough to generate almost unlimited possibilities for complication in the behaviour. If the restoring force $f(\xi)$ is proportional to ξ, so that the free motion is harmonic, the possible types of solution are severely restricted – the system is unable to adjust its frequency to that of the excitation by changing amplitude, so that stable excitation usually demands close synchronization of the force to the natural frequency. The cyclotron is an example of such a system, where the force is applied only while the particle traverses very short sectors of its orbit. A particle crossing the gap between the two dees when the phase of the oscillatory field between them is such as to accelerate it, is also accelerated at every subsequent crossing, twice per cycle, if the e.m.f. has the cyclotron frequency. In this case the localization of the accelerative force is an engineering convenience; the acceleration would take place with the electric field applied uniformly to the orbit, as in semiconductor cyclotron resonance. There would, however, be a significant difference if, for some reason, excitation at a higher frequency were desired. For with the dee arrangement an odd harmonic of the cyclotron frequency would also serve to accelerate particles; it would not matter that the e.m.f. ran through several cycles while the particles were inside the dees and isolated from electric fields. The point is the same as that illustrated in fig. 5.1 – if a sinusoid is sampled at equally spaced discrete points, π/ω_c apart in this case, the frequency remains indeterminate to the extent that a multiple of $2\omega_c$ may be added or subtracted.

237

Although excitation at a harmonic of the cyclotron frequency is not usual in particle accelerators, something very similar takes place in the arrangement used for the study of cyclotron resonance of conduction electrons in metals, originally proposed by Azbel' and Kaner.[(11)] In contrast to pure semiconductors, the skin effect in metals prevents a high-frequency field from penetrating deep into the interior. In a pure metal at low temperatures the free path between collisions is long and the conductivity proportionately high; a field at a microwave frequency of, say, 10^{10} Hz penetrates little more than 10^{-7} m into the metal, much less than the free path and also less than the radius of a typical cyclotron orbit of an electron. When a magnetic field is applied parallel to the surface some electrons are caught in helical orbits that just miss the surface (fig. 18), so that for a small fraction of each orbit they experience the electric field due to the microwave excitation. If successive entries into the surface layer catch the field in the same phase the electrons have their interaction with the field enhanced, while if there is phase reversal at successive entries the

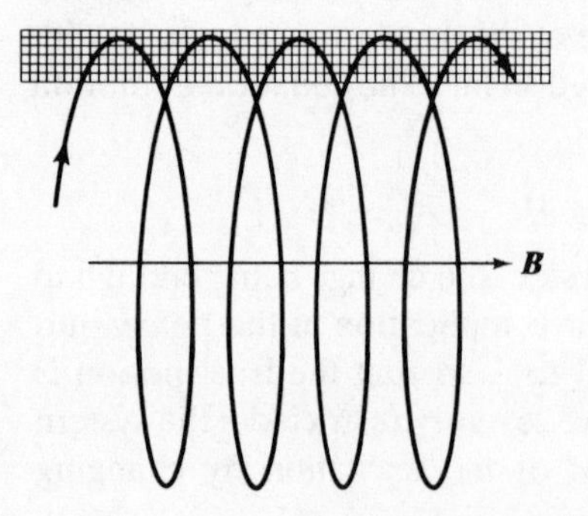

Fig. 18. Electron motion in Azbel'–Kaner cyclotron resonance. The shaded area represents the skin layer containing the high-frequency electric field.

interaction is diminished. With change of magnetic field, and therefore of ω_c, the effective conductivity of the metal alternates in magnitude, and this is detected as fluctuations in the surface resistance.

To make the argument more quantitative, let each entry into the surface layer result in an impulse $P\,e^{j\omega t}$ transmitted by the field to the electron. We define the response of the electron to this single impulse as the norm, the sum of all such individual responses being the surface current that would be set up by the field if the effects of resonance were negligible. In reality, however, each electron has a past history of similar impulses at previous entries, with the same phase difference θ between successive impulses. The average response is therefore as if to an impulse $P'\,e^{j\omega t}$, where

$$P'\,e^{j\omega t} = P\,e^{j\omega t} + \alpha P\,e^{j(\omega t-\theta)} + \alpha^2 P\,e^{j(\omega t-2\theta)} + \cdots,$$

θ being $2\pi(1-\omega/\omega_c)$; α is the probability that an electron will execute a cyclotron orbit without suffering a collision. If τ is the mean time between collisions, $\alpha = e^{-2\pi/\omega_c\tau}$. Summing the geometrical series, we have that

$$P'/P = (1-\alpha\,e^{-j\theta})^{-1}. \tag{36}$$

189

This resonant denominator is the same as that already met with in the discussion of transmission line resonators (chapter 7). When the free path is long ($\tau \gg 2\pi/\omega_c$), so that α is very nearly unity, P' has sharp peaks whenever θ passes through zero, at regular intervals of $1/\omega_c$, i.e. of $1/B$, as the magnetic field is changed. The electrons can be considered as responding in an integral subharmonic series to the applied oscillatory field. An experimental curve showing the observation of Azbel'–Kaner resonance in tungsten is given in fig. 19.

A number of points need to be made briefly in connection with this curve. In a real metal, different groups of electrons have different cyclotron

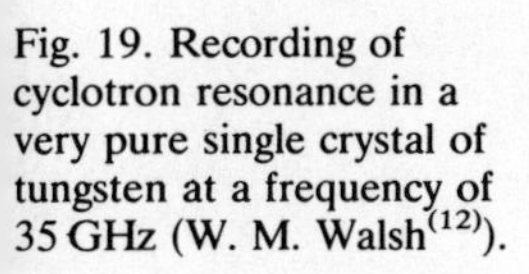
Fig. 19. Recording of cyclotron resonance in a very pure single crystal of tungsten at a frequency of 35 GHz (W. M. Walsh[12]).

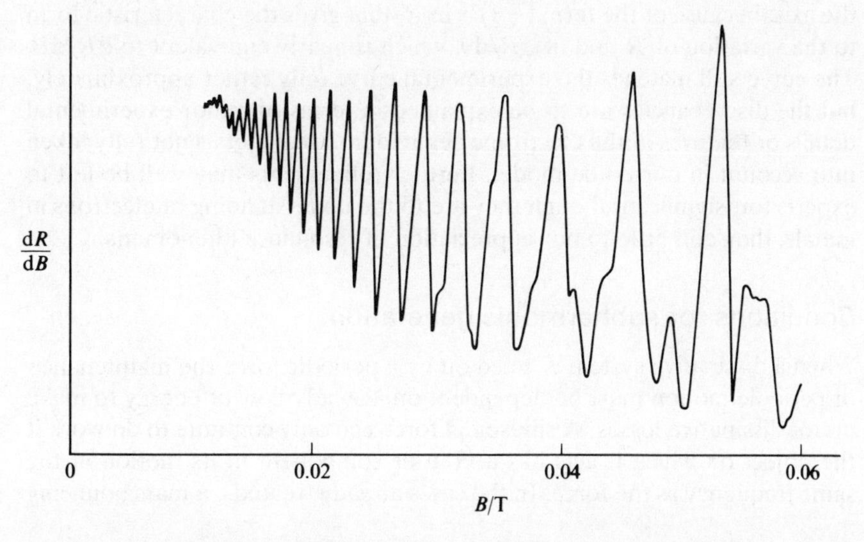

frequencies (more precisely, there is a continuous spread of ω_c) and their oscillations of conductivity mostly cancel each other out. Only dominant groups reveal themselves, and then as a comparatively minor oscillatory variation superimposed on a steady background. To observe the effect more readily it is common practice to record the derivative of the behaviour. For example, the surface resistance (the real part of the surface impedance Z, which is the value of $\mu_0\mathscr{E}/\mathscr{B}$ for the microwave fields at the surface of the metal) changes with B, and causes changes in Q and hence in the output of a resonant cavity incorporating the sample as one of its walls. A small oscillatory variation of B then causes oscillation of the output with an amplitude proportional to the derivative of output with respect to B. This is picked up and amplified, and presented as a measure of $\mathrm{d}R/\mathrm{d}B$, the derivative of surface resistance of the sample. Slowly varying changes of background are largely eliminated by this procedure, and the oscillation left correspondingly more visible.

It is obvious that the curve of fig. 19 bears a closer resemblance to a succession of low-Q resonances, as in fig. 7.24, than to the differential of such a function. To understand why this is so one must recognize that under the conditions of the experiment, the free path l of the electrons being very long, Z is proportional to $(-\omega^2 l/\sigma)^{\frac{1}{3}}$, where σ is the conductivity; this is the theoretical expression of the *anomalous skin effect*.[13] If the variations in σ expressed in (36) are superimposed on a larger steady background, the form of Z is given by the expression

190

$$Z \sim [-1 - a/(1-\alpha\,\mathrm{e}^{-\mathrm{j}\theta})]^{-\frac{1}{3}} \sim (-1)^{-\frac{1}{3}}[1 + \tfrac{1}{3}a/(1-\alpha\,\mathrm{e}^{-\mathrm{j}\theta})]$$

when a, measuring the amplitude of the σ-oscillations, is much less than unity. As θ changes, $a/(1-\alpha\,\mathrm{e}^{-\mathrm{j}\theta})$ runs round a circle, though at a non-uniform rate (see fig. 7.23 which illustrates a similar problem). In fig. 20 a typical trajectory of Z is plotted out, with points at equal intervals of θ marked. It is the orientation of the circle, lying along the 60° line instead of the axis, because of the term $(-1)^{-\frac{1}{3}}$ in Z, that gives the characteristic form to the variation of R and of $\mathrm{d}R/\mathrm{d}\theta$, which is nearly equivalent to $\mathrm{d}R/\mathrm{d}B$. The curve still matches the experimental curve only rather approximately, but the discrepancies are to be explained in terms of minor experimental details or features of the electronic behaviour in real metals not fully taken into account in our crude model. Further refinements may well be left to experts for, significant though they are to the understanding of electrons in metals, they add little to our appreciation of resonance phenomena.

Conditions for subharmonic generation

When a dissipative system is acted on by a periodic force the maintenance of periodic motion must be dependent on a steady flow of energy to make up for dissipative losses. A sinusoidal force can only continue to do work if the object on which it acts has a Fourier component in its motion at the same frequency as the force. In the cases already treated – a mass bouncing

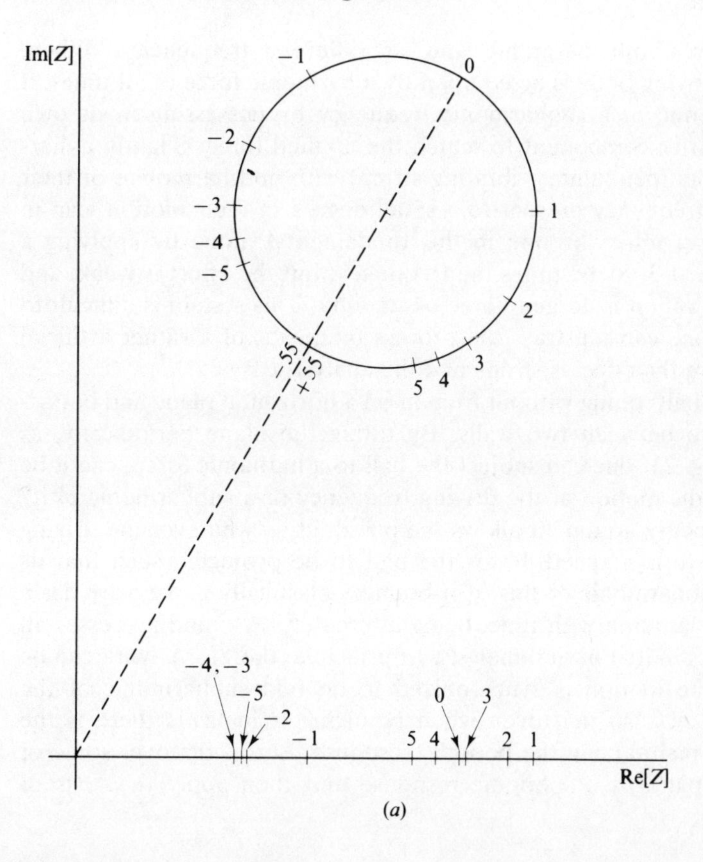

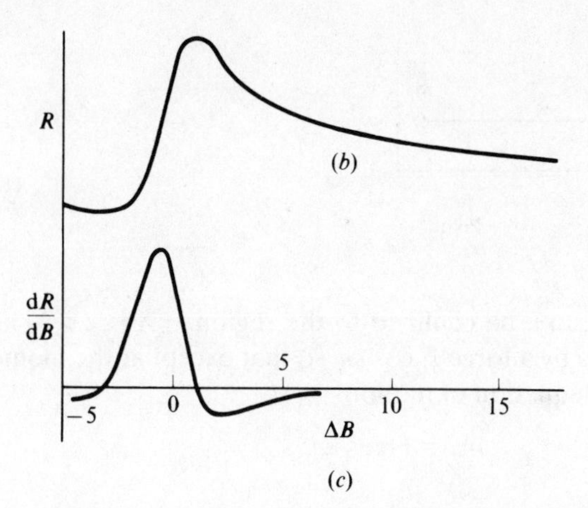

Fig. 20. The origin of the line shape in cyclotron resonance. As B is changed the point representing Z runs round a circle (shown here considerably enlarged). With α taken as 0.9 the marks show successive values at equal increments ΔB – there are 110 increments in a complete revolution, and most of the trajectory is described in a small fraction of a cycle. The real parts of Z are marked on the real axis in (a), and the variation of R and its derivative exhibited in (b) and (c).

on a vibrating base, and cyclotron resonance – the applied force was intermittent at a frequency determined by the motion of the object affected. It was this that generated a component of the applied force at the frequency of the resulting motion even though, as with cyclotron resonance, the electric field variations and the orbital motion of the electrons were both harmonic and at different frequencies. When, however, a moving body is acted upon by a harmonic force at all times, it can only respond at a subharmonic frequency by possessing in its own motion a Fourier component to match the applied force. Slightly anharmonic vibrators (pendulum, vibrating string) with odd harmonics of their fundamental frequency present to a small degree in their motion, can in principle be set into vibration in this fundamental mode by applying a periodic force at 3, 5 etc. times the frequency; but the effect is weak, and energy conservation is no guarantee of stability. This system is difficult to analyse, but one can illustrate the process by means of a rather artificial example where the effect is strong and the analysis easy.

Consider a ball rolling without friction on a horizontal plane and bouncing to and fro between two walls. By tilting the plane periodically, as sketched in fig. 21, one can subject the ball to a harmonic force; can it be set into periodic motion at the driving frequency or a subharmonic of it? Clearly the energy argument allows the possibility – whatever the driving frequency there is a speed V for the ball to be projected such that its period is a subharmonic of this; if it bounces elastically its velocity has a square-wave variation with time, being alternately $\pm V$, and possesses all odd harmonics in its Fourier analysis. In principle, therefore, work can be done when the motion is synchronized to an odd subharmonic of the driving frequency, so that even when bouncing is inelastic there is the possibility of maintaining the periodic response. Let us discover some of the possible patterns of periodic response and then apply the tests of stability.

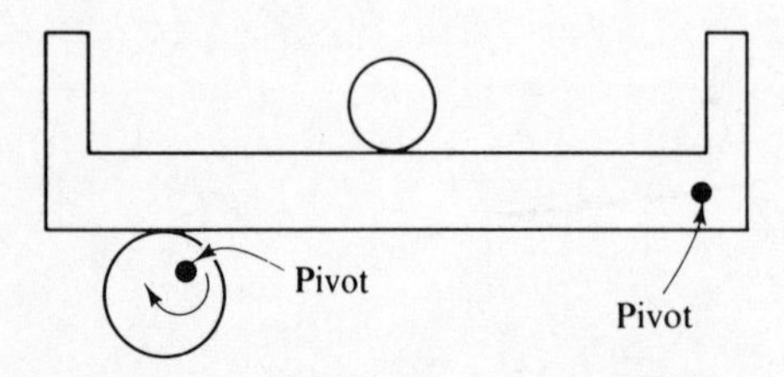

Fig. 21. A ball bouncing between end-plates is excited by periodic tilting.

Let the ball, of mass m, be confined to the region $-X<x<X$ and be acted upon at all times by a force $F \cos \omega t$, so that except at the moment of bouncing it obeys the equation of motion

$$m\ddot{x} = F \cos \omega t,$$

with general solution

$$x = -F \cos \omega t / m\omega^2 + vt + x_0,$$

where v and x_0 are constants during any one traverse of the range. In dimensionless form,

$$z = -\cos \theta + y\theta + c, \tag{37}$$

where $z = m\omega^2 x/F$, $\theta = \omega t$, $y = m\omega v/F$ and $c = n\omega^2 x_0/F$; $-Z<z<Z$, where $Z = m\omega^2 X/F$. The analysis is greatly simplified if only odd subharmonic solutions are considered. There are probably many other stable solutions, but we shall not search for them. The particular solutions chosen are those in which the range is traversed in time $n\pi/\omega$, i.e. in $\frac{1}{2}n$ periods, n being odd, so that the ball reaches Z when the phase is exactly opposite to that at which it left $-Z$, though one or more complete periods may have intervened. If the coefficient of restitution is such that at Z it rebounds with the same speed as it possessed on leaving $-Z$ it can repeat in the opposite direction the motion that took it from $-Z$ to Z, and this double traverse can then be repeated indefinitely, provided it is stable against small perturbations.

Let the phase be θ_0 at the moment the ball leaves $-Z$, and $\theta_0+n\pi$ on arrival at Z; then (37) shows that

$$Z = \cos\theta_0 + \tfrac{1}{2}n\pi y. \tag{38}$$

The (dimensionless) velocity is $\mathrm{d}z/\mathrm{d}\theta$, which is $\sin\theta_0 + y$ on leaving $-Z$ and $-\sin\theta_0 + y$ on arriving at Z. If the former is to be ε times the latter,

$$y = -\frac{1+\varepsilon}{1-\varepsilon}\sin\theta_0. \tag{39}$$

These two equations, (38) and (39), fix the values of θ_0 and y needed to give synchronous motion at the nth subharmonic. The physical significance is made clear in fig. 22. When $\theta = n\pi$, $\mathrm{d}z/\mathrm{d}\theta = y$; with perfectly elastic bouncing ($\varepsilon = 1$) the points A and B would be the end-points of the traverse for the third subharmonic, y being adjusted so that the vertical

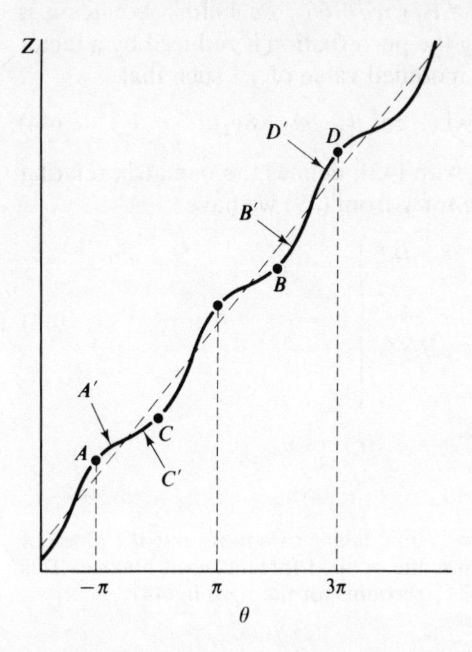

Fig. 22. Illustrating (38).

distance from A to B represented $2Z$. Equally well C and D are possible end-points for the same subharmonic when $\varepsilon = 1$, and both possibilities must be borne in mind. When $\varepsilon < 1$, the end-points must be shifted so that the initial velocity is ε times the final velocity. Thus A' and B' (at $-\pi+\Delta$ and $2\pi+\Delta$) or C' and D' (at $-\Delta'$ and $3\pi-\Delta'$) are possible end-points.

To investigate the stability of these two possible solutions, consider the ball leaving $-Z$ at a phase $\theta_0+\delta\theta$, with such a velocity that y takes the adjusted value $y+\delta y_1$; c also will be changed, but we shall eliminate it from the argument. The effect of these perturbations on the position of the ball follows by differentiating (37):

$$\delta z = (\sin\theta + y)\,\delta\theta_1 + \theta\,\delta y_1 + \delta c. \tag{40}$$

Since these perturbations refer to changes in the description of the motion at $-Z$, we have that $\delta z = 0$ when $\theta = \theta_0$,

$$\text{i.e. } 0 = (\sin\theta_0 + y)\,\delta\theta_1 + \theta_0\,\delta y + \delta c. \tag{41}$$

At the next bounce, when $z = Z$, let $\theta = \theta_0 + n\pi + \delta\theta_2$; the same argument shows that

$$0 = (-\sin\theta_0 + y)\,\delta\theta_2 + (\theta_0 + n\pi)\,\delta y_1 + \delta c. \tag{42}$$

Eliminating δc from (41) and (42), we have that

$$\delta\theta_2 = [-n\pi\,\delta y_1 + (y + \sin\theta_0)\,\delta\theta_1]/(y - \sin\theta_0). \tag{43}$$

This is one equation relating the perturbations at the start of the second traverse to those at the start of the first. For the other we note that the velocity, being $\sin\theta + y$, is perturbed by $\cos\theta\cdot\delta\theta + \delta y$, and that at the moment of arrival at Z, when $\theta = \theta_0 + n\pi + \delta\theta_2$, δv before bouncing is $-\cos\theta_0\cdot\delta\theta_2 + \delta y_1$. After bouncing the perturbation is reduced by a factor ε, and this must be described by a modified value of y,† such that

$$\cos\theta_0\cdot\delta\theta_2 + \delta y_2 = \varepsilon(-\cos\theta_0\cdot\delta\theta_2 + \delta y_1). \tag{44}$$

This is the second equation which, with (43), defines the a-matrix relating $(\delta y_2, \delta\theta_2)$ to $(\delta y_1, \delta\theta_1)$. Substituting for y from (39) we have

$$\left.\begin{aligned} a_1 &= \varepsilon - B, \\ a_2 &= -C, \\ a_3 &= \varepsilon B/C, \\ a_4 &= \varepsilon, \end{aligned}\right\} \tag{45}$$

where $B = \tfrac{1}{2}n\pi(1-\varepsilon^2)\cot\theta_0$ and $C = \varepsilon(1+\varepsilon)\cos\theta_0$.

† Since alternate traverses are in opposite senses, y changes sign at each bounce. To avoid this we describe each traverse as if it were in a positive direction, taking care to reverse the phase of the applied force at each bounce. This accounts for the signs in (44).

This a-matrix is rather similar to (33); in particular, $D=\varepsilon^2$, independent of θ_0, so that as the strength of the applied force, and hence Z and θ_0, are changed, the locus on the (D, T)-plane is once more a vertical line in fig. 16. Since $T=2\varepsilon-B$, the limits of stability, $T=\pm(1+D)$, are expressed by the equation:

$$2\varepsilon-\tfrac{1}{2}n\pi(1-\varepsilon^2)\cot\theta_0=\pm(1+\varepsilon^2),$$

i.e.
$$\cot\theta_0=-\frac{2}{n\pi}\frac{1-\varepsilon}{1+\varepsilon}\quad\text{or}\quad\frac{2}{n\pi}\frac{1+\varepsilon}{1-\varepsilon}. \tag{46}$$

Four values of θ_0 satisfy these equations, but only two are admissible since (39) shows that θ_0 must lie in the third or fourth quadrants.

To interpret these critical conditions in terms of the limits within which the exciting force F can lie, it is necessary to proceed from $\cot\theta_0$ through (39) to (38), evaluating y and then Z, the latter containing the required information. Numerical evaluation is easiest at this stage, and some typical results are shown in fig. 23. In contrast to the mass bouncing on a vibrating plate, the subharmonics appear at weaker, rather than stronger, excitation; this is because the natural frequency increases with energy here, while in the other case it decreases. The shape of the upper limit of fundamental excitation results from the fact that y, as computed from (46) and (39), becomes less than unity when ε is small. The velocity can now become

271

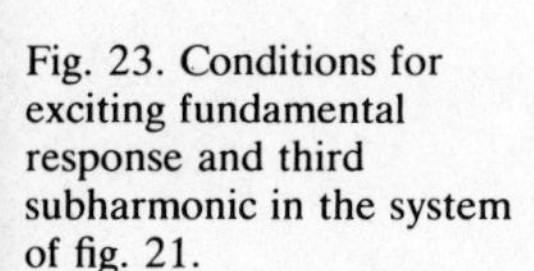

Fig. 23. Conditions for exciting fundamental response and third subharmonic in the system of fig. 21.

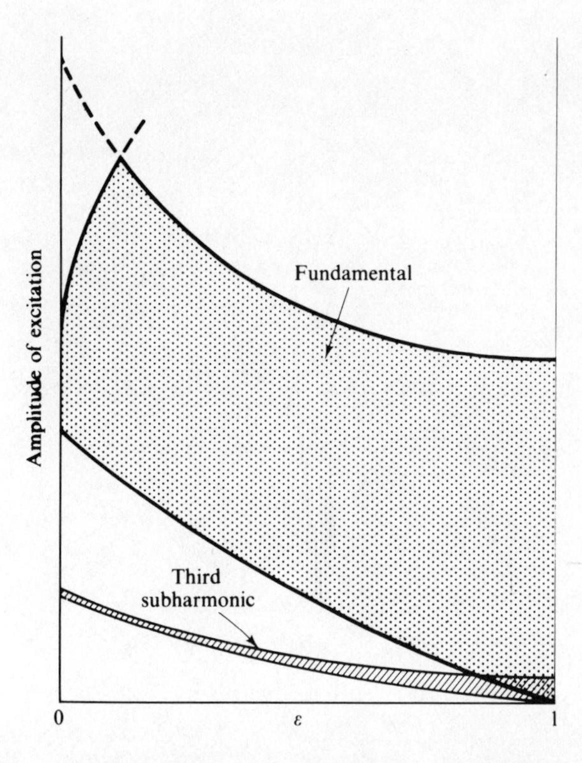

negative, and it is necessary to verify that it does not stray beyond the limits $\pm Z$. This additional feature precludes the whole of the stable range being realized.

It is seen from fig. 23 that the third subharmonic is stable only within a rather narrow range of force, especially if $\varepsilon \ll 1$. In particular, it should be noted that too strong a force precludes stability, though obviously not for energetic reasons. This has a consequence for the possibility of subharmonic excitation of systems whose anharmonicity is not so extreme as in this example. With correspondingly smaller terms in the Fourier expansion of the velocity–time behaviour, a given amplitude of response (and hence, roughly, of dissipative loss) requires a stronger force to supply the dissipated energy. If this force lies above the stability limit, no subharmonic excitation will occur, even though energetically it appears possible. We may therefore expect, for any given dissipation per cycle, that there will be a critical degree of anharmonicity needed to permit subharmonics, and that this critical anharmonicity will be greater in more lossy systems. So much we may aver fairly confidently; it is another matter to delineate the limits of the subharmonic regime and we shall take the matter no further.

Additional note (see p. 247 and elsewhere in the chapter)
This field has expanded greatly in recent years. A more up-to-date, though still elementary, account will be found in my *Response and Stability.*[14]

10 Parametric excitation

It has long been known that a resonant system can be set into oscillation by periodically varying the parameters. A very extensive analysis was given by Rayleigh,[1] who drew attention to analogies with rather different physical processes – the perturbation of a planetary orbit by another planet having a quite different period in some near-integral relationship, and the propagation of waves in periodically stratified media. This latter example relates to a phenomenon of much greater physical interest now than in Rayleigh's day since it is basic to Bragg reflection of X-rays and to the motion of electrons in solids. Nevertheless, these are analogies only in the sense that the same type of differential equation is involved in all of them, as well as in the theory of vibrating elliptical membranes. It can hardly be claimed that the solution of one of these problems adds anything to the intuitive understanding of another. A mathematical framework, however, is well worth having, for these are not easy problems. The examples presented here earn a place in their own right as showing yet another aspect of the variety of oscillatory phenomena in nature. At the same time, there are important technological applications in the form of *parametric amplifiers*, as will be discussed in outline at the end of the chapter.

Probably the easiest way to demonstrate parametric excitation is to set up a simple pendulum whose length can be slightly varied at twice the natural frequency. If the amplitude of this variation is too small, nothing happens except that the bob vibrates up and down on the end of the string, but with a larger excursion the pendulum spontaneously begins to swing and its amplitude grows exponentially. A suitable demonstration apparatus is shown in fig. 1. The pendulum has a brass bob of about 100 g hanging on a fine steel wire, which passes through a small hole in the plate A and is hung by a loop on an eccentric pin fixed to the wheel C. Rotation of the wheel at, say, 1 revolution a second causes the effective length of the pendulum below A to vary with this frequency, and conditions for parametric excitation are achieved by adjusting the length of the pendulum so that its period is twice that of the wheel C. In practice the experiment works rather better if the string is pulled sideways by the light cord attached at B. This serves to hold the string against the side of the hole in A, thus defining the length of the pendulum rather precisely, but it also ensures that oscillations in and perpendicular to the plane of the paper

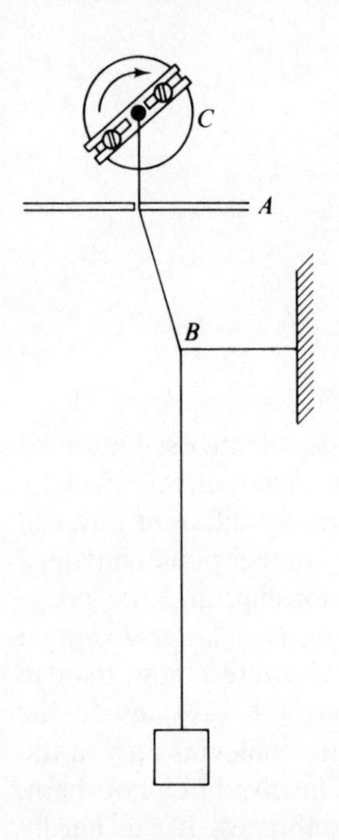

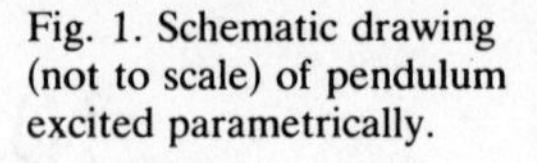
Fig. 1. Schematic drawing (not to scale) of pendulum excited parametrically.

have well-separated frequencies, so that only one mode is excited. Pendulum oscillations normal to the plane, with A as the effective point of support, are used to show the phenomena, and it is these whose frequency is carefully adjusted, by moving the bob on the threaded rod, to be exactly half that of the wheel. To get good results, it is worth being very patient in adjusting this frequency; the motor should run at a constant speed, for example synchronously with the a.c. mains. It is convenient to attach a side arm to the axle of C which can strike a metal spring once every turn; the ear and eye used together are very sensitive to slow departures of wheel and pendulum from synchronism. The pin on C is first adjusted to be central, so that C can turn without changing the pendulum length, and the frequencies are adjusted. Then the pin is moved to an eccentric position, say 5 mm off-centre, and the wheel started with the pendulum at rest. It will not be long before the pendulum begins to swing with ever-increasing amplitude, limited only by the breakdown of synchronism as its wide swing appreciably lengthens the period. With a smaller amplitude of excitation the growth will be found to be less rapid; if all is in good adjustment, the rate of growth (the reciprocal of the number of swings in which the amplitude is doubled) should be a linear function of the amplitude of excitation. We stress again the proviso about everything being well adjusted – it is in practice hard to obtain points on the straight line with weak excitation, when the rate of growth is slow. The reason will become clear in due course from the mathematical analysis. If the variation of pendulum length is insufficient to induce growth, the parametric excitation serves only to modify the rate of decay of any oscillation already present.

There is a certain critical excitation which allows the oscillation to persist at constant amplitude. It is virtually impossible to exhibit this in practice, but the reader is encouraged to try, since more can be learned about the physical processes by being frustrated than by accepting defeat at the bidding of the printed page.

An explanation of how the pendulum oscillation can be maintained follows from considering the energy supplied by the motor driving C. The tension, T, of the string is not constant but is slightly greater than mg as the pendulum swings through vertical, and slightly less as it pauses at the ends of its swing; T therefore runs through two cycles for each complete cycle of the pendulum. If the motion of C is such that the upward velocity of the pin, v, is in phase with the variations of T, energy is fed into the pendulum at all phases of the oscillation. To achieve this optimal condition, the vertical displacement, a, which determines the shortening of the effective length of the pendulum, must lag behind T, in phase quadrature.

To obtain a quantitative measure of the condition for spontaneous oscillation, let the pendulum have length l and angular displacement θ. The inward radial acceleration of the bob, $l\dot{\theta}^2$, results from the imbalance of T and the radial component, $mg\cos\theta$, of the weight of the bob.† For small amplitudes we approximate to $\cos\theta$, writing

$$T - mg(1-\tfrac{1}{2}\theta^2) = ml\dot{\theta}^2.$$

Substituting $\theta_0 \cos\omega t$ for θ, and $-\omega\theta_0 \sin\omega t$ for $\dot{\theta}$, yields the result

$$T = mg(1+\tfrac{1}{4}\theta_0^2 - \tfrac{3}{4}\theta_0^2 \cos 2\omega t),$$

of which only the oscillatory last term, $-\tfrac{3}{4}mg\theta_0^2 \cos 2\omega t$, concerns us here. Vertical displacement of the string according to $-a_0 \sin 2\omega t$ causes an upward velocity of $-2a_0\omega\cos 2\omega t$; the resulting instantaneous power input, Tv, is $\tfrac{3}{2}mga_0\omega\theta_0^2 \cos^2 2\omega t$, which is always positive and has mean value $\tfrac{3}{4}mga_0\omega\theta_0^2$. The energy E in the oscillating pendulum is $\tfrac{1}{2}ml^2\omega^2\theta_0^2$, and the equation for the increase in energy therefore takes the form

$$\dot{E} = E/\tau_e', \quad \text{where} \quad \tau_e' = \tfrac{2}{3}l/a_0\omega.$$

The build-up is exponential, with time-constant τ_e', and the system may be assigned a (negative) quality factor, $-Q'$, related to τ_e' in the usual way,

8

$$Q' = \omega\tau_e' = \tfrac{2}{3}l/a_0; \tag{1}$$

Q' is the number of radians of oscillator required for an increase in energy by a factor e.

Of course any real pendulum suffers internal damping which, if linear, is described by an intrinsic quality factor, Q. Spontaneous oscillation sets in if $Q > Q'$, the gain from the parametric process outweighing dissipative loss.

† No account need be taken of tension changes which result from variations of l, since these make no net contribution to the energy transfer.

The critical amplitude a_c for parametric excitation therefore follows by setting $Q' = Q$ in (1):

$$a_c = \tfrac{2}{3}l/Q. \tag{2}$$

The system described and illustrated in fig. 1 has typically $Q \sim 2500$ and $l \sim 1$ m, so that a_c is a little under $\frac{1}{3}$ mm. If a_0 is considerably greater than this the dissipation may be neglected without much harm, and the oscillations will build up exponentially at a rate governed essentially by Q'. Experimentally it was observed that with $l = 0.84$ m and $a_0 = 8.5$ mm the amplitude increased fourfold, and the energy sixteenfold, in 33 swings. The measured Q' is therefore $33 \times 2\pi/\ln 16$, i.e. 75. According to (1) it should have been 66, or 68 if the intrinsic Q is allowed for. The higher measured Q' means a slightly slower build-up, possibly explained as an error of observation but more likely due to imperfect tuning of the parametric excitation. We shall return to the question of tuning shortly.

There are many analogous examples of parametric excitation, since somewhere in most oscillators one can find a force (or its equivalent) oscillating at twice the frequency, which can be used to feed in energy:

(*a*) A trivial example is the toy illustrated in fig. 2, a disc suspended on a loop of string, which can be made to spin alternately clockwise and anticlockwise by pulling on the loops twice per complete cycle. The cycle involves many revolutions back and forth, and the system is highly anharmonic, but the parametric excitation does not depend on this; nor is it vitiated by variations in period, for the skilful operator adjusts his pulling frequency as required. The following two examples are more nearly harmonic, the second strictly so.

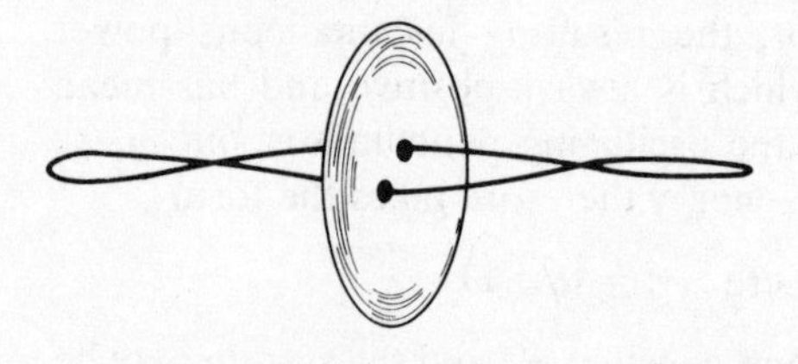

Fig. 2. Child's toy working by parametric excitation.

(*b*) A string vibrating transversely (fig. 3) exerts a different longitudinal component of force at the point of attachment S according as it is in the configuration A or in either of the equivalent configurations B and C. If the tension is constant the longitudinal force at S varies as $T \cos \theta$. More probably the tension increases with displacement and overcompensates the decrease due to $\cos \theta$, but this does not eliminate the essential property of the longitudinal component, that it oscillates at twice the natural frequency. Longitudinal motion of S at this double frequency may supply enough power to excite transverse oscillations parametrically. We shall return to this example in more detail.

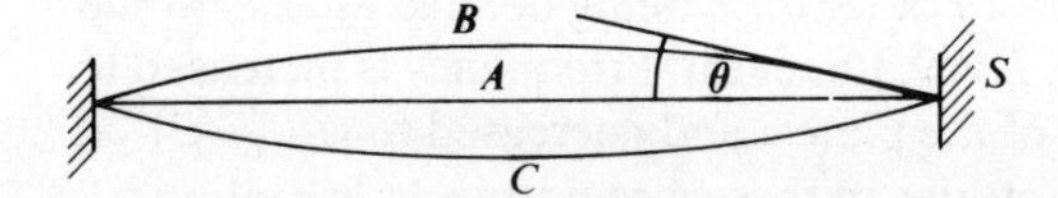

Fig. 3. Three configurations of a stretched string in its lowest mode.

(*c*) If a current is oscillating in the circuit of fig. 4 and the capacitance C is varied at twice the resonant frequency, for example by changing the plate separation, energy can be fed into the circuit. For the attractive force between the plates is $\frac{1}{2}\varepsilon_0\mathscr{E}^2$ per unit area when the electric field is $\mathscr{E}$, and takes a maximum value twice in each cycle as the current passes through zero and the charge on the plates reaches its peak positive or negative value. These are the moments when the separation should be increasing, the process being reversed when the charge is around zero.

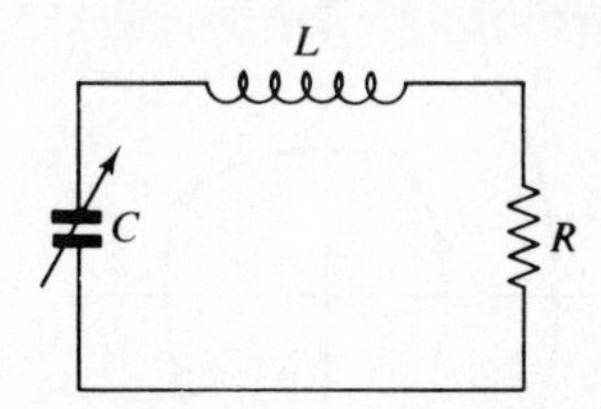

Fig. 4. Parametrically excited resonant circuit.

Detailed analysis of parametrically excited systems

The energy argument gives valuable insight into the conditions necessary for parametric excitation, but for a full discussion direct analysis is best. The equation of motion for the circuit of fig. 4 is readily written down, but not so readily solved. Let the spacing of the capacitor plates vary as $1+\beta\cos 2\Omega t$, β being much less than unity, so that the capacitance varies as $C_0(1+\beta\cos 2\Omega t)^{-1}$. Then the charge on the plates obeys the equation:

$$L\ddot{q}+R\dot{q}+q(1+\beta\cos 2\Omega t)/C_0=0, \tag{3}$$

which is linear but, in contrast to systems with fixed parameters, has one time-dependent coefficient. The principle of superposition applies, in the sense that two independent solutions of (3) can be combined by addition to yield all possible solutions. Translational invariance has, however, disappeared; a solution does not remain valid when shifted arbitrarily along the time axis, but only when the shift is a multiple of π/Ω.

It is not easy, except for a very practised performer, to follow the general behaviour of (3) by sketching solutions. Let us therefore start with a simplified version in which the spacing of the plates is changed by sudden steps, being alternately $1\pm\beta$ times the mean, and let the changes take place every quarter of a cycle at the resonant frequency. Further, let $R=0$, so that (3) can be written (in a shorthand notation):

$$\ddot{\xi}+\omega_0^2\xi(1\pm\beta)=0. \tag{4}$$

In a phase-plane plot the trajectory of (ξ, v), v being defined as $\omega_0\dot{\xi}$, would be a circle if $\beta = 0$. If, however, $\beta \neq 0$, the natural frequency is increased to $\omega_0(1+\beta)^{\frac{1}{2}}$ at the wider spacing of the plates, and decreased to $\omega_0(1-\beta)^{\frac{1}{2}}$ at the narrower spacing. The trajectories in these circumstances are elliptical, with the major axis lying along the v-axis in the former case, and along the ξ-axis in the latter. Fig. 5 shows how the trajectory is changed by switching from positive to negative sign (*a*) at the moments when $\xi = 0$ and (*b*) when $v = 0$. In (*a*) the response grows exponentially, in (*b*) it decays, and the general form of behaviour when the system is started at some arbitrary point on the plane is a mixture of both types, with the growing mode (*a*) ultimately dominating.

259

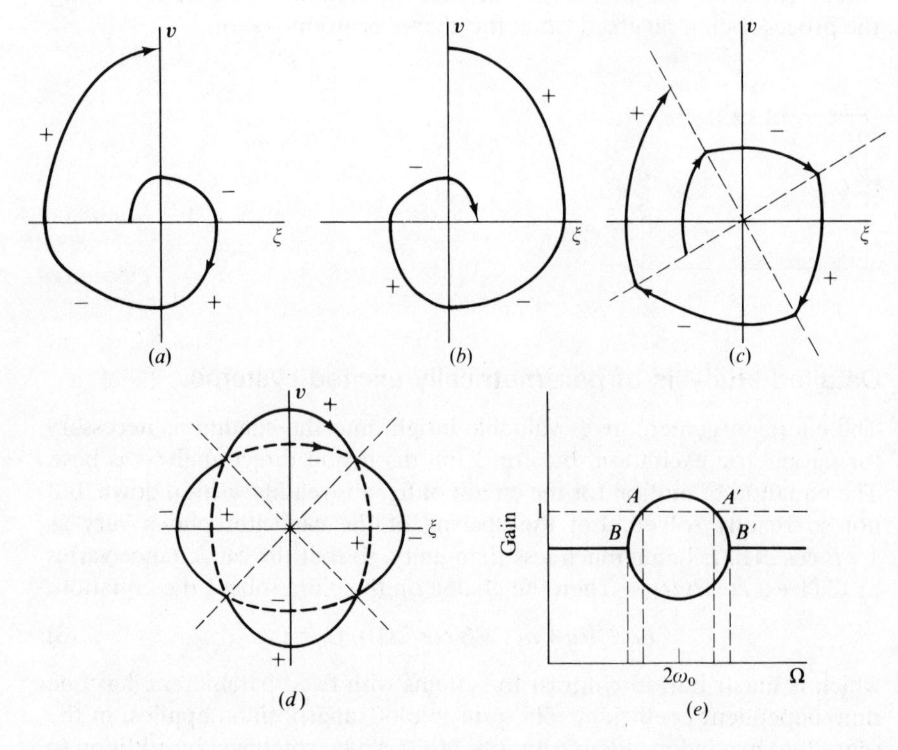

Fig. 5. Trajectories in phase-plane when capacitor in fig. 4 has its spacing increased and decreased suddenly twice per cycle (*a*) with oscillation so phased as to give maximum amplification, (*b*) phased to give maximum attenuation. The + and − signs indicate the quadrants in which the spacing is large and small. In (*c*) switching occurs at an intermediate phase, and the amplification is less than in (*a*). In (*d*) the oscillation has steady amplitude, and there are two modes – slow (full line) and fast (broken line). The graph (*e*) shows schematically how the gain (amplitude ratio of successive swings) changes with tuning. On the circle the modes are phase-locked to the excitation; outside the range *BB′* they are not and their mean gain, over many cycles, is independent of the excitation.

Considered as a stability problem like those treated in the last chapter, with the value of (ξ, v) noted only once each cycle of the excitation, the solution in terms of two exponentials is seen to have the characteristics of saddle-point instability. If a large enough resistance is inserted in the circuit, the parametric excitation will prove inadequate to sustain oscillation, and the behaviour will be described by two decaying exponentials, such as characterize nodal stability. Another way of stopping oscillation, even if the resistance is not large enough to do so in a perfectly tuned

270

system, is to detune the excitation. For small changes in excitation frequency the system will remain within the domain of saddle-point instability, its response being locked in phase with the excitation at exactly half the frequency, but as the detuning is increased a point will be reached where phase-locking cannot be maintained. At this stage, if the analogy with other systems is sound, one may expect a transition to nodal stability and thence to focal stability. This can be illustrated diagrammatically with the simplified model, though detailed analysis involves rather tedious algebra and will be omitted. When the switching frequency for the capacitor is not exactly $2\omega_0$ the oscillation may still be able to adjust itself to synchronism by suitable phasing, as in fig. 5(*c*), but the rate of growth or decay is reduced thereby. In the absence of losses, detuning may be carried to the point where steady oscillation can just persist in one of the modes shown in fig. 5(*d*). The switch-overs then occur at phase angles shifted by $\pi/4$ from those in (*a*) and (*b*); for the circuit to be made to follow the outer, rosette-shaped trajectory the switching speed must be lower than $2\omega_0$, and to follow the inner, cushion-shaped trajectory it must be higher. It is easy enough to calculate the time taken to execute one cycle of either, and to show that the tuning range for excitation is $2\omega_0(1\pm 2\beta/\pi)$. Schematically the amplification per cycle varies with the excitation frequency as shown in fig. 5(*e*), which has been drawn so as to include a small amount of dissipation. Between A and A' an amplified mode exists and the system is unstable. In the narrow ranges AB and $A'B'$ there is nodal stability – the oscillation is phase-locked to the excitation so that sampling once per cycle of the latter reveals nothing of the oscillation but only the two modes of exponential decay of the amplitude. Beyond B and B' phase-locking fails, and sampling shows up an oscillatory decay to rest, i.e. focal stability. It would be a mistake to make too much of the distinction between nodal and focal stability. Mathematically the two types of critical point, A and B, are clearly marked; to the observer the difference between stability and instability is more important than the precise category to which the stability should be assigned.

The general features revealed by this model are substantiated by detailed calculation of the behaviour when the capacitance is varied sinusoidally. Since we shall ultimately compare the calculation with measurements on a pendulum, we shall analyse this case rather than the circuit with variable capacitor. The same method applies to both and the general behaviour of the solutions is very similar. The equation of motion of the pendulum of variable length is readily set up. If the length l is $l_0(1+\beta\cos 2\Omega t)$, and the deflection θ is small, the tangential force on the bob is $-mg\theta-\lambda l\dot{\theta}$, λ being the viscous drag coefficient. Now a particle whose position is defined by l and θ has a tangential component of acceleration $l\ddot{\theta}+2\dot{l}\dot{\theta}$; the equation of motion therefore takes the form

$$ml_0(1+\beta\cos 2\Omega t)\ddot{\theta}-4m\beta l_0\sin 2\Omega t\cdot\dot{\theta}+\lambda l_0(1+\beta\cos 2\Omega t)\dot{\theta}+mg\theta=0. \tag{5}$$

We may restrict the analysis to systems of low loss, not too strongly excited, without losing any essential results; this allows second-order terms (e.g. $\lambda\beta$) to be neglected. If, further, we write θ as $y(t)\,e^{-\frac{1}{2}\lambda t/m}$, the intrinsic loss term disappears, and the simpler form results:

$$\ddot{y}(1+\beta\cos 2\Omega t)-4\beta\Omega\dot{y}\sin 2\Omega t+\omega_0^2 y=0, \tag{6}$$

where $\omega_0^2=g/l_0$. For comparison the same treatment converts (3) into the form

$$\ddot{y}+\omega_0^2 y(1+\beta\cos 2\Omega t)=0. \tag{7}$$

Both (6) and (7) are linear equations with coefficients that are periodic functions of t, having π/Ω as period. It is a general property of such equations (*Floquet's theorem*)[2] that their solutions can be cast in the form $e^{ipt}P(t)$, where p is a constant (possibly complex) and $P(t)$ is a periodic function with period π/Ω, i.e. having the property that $P(t+\pi/\Omega)=P(t)$. Since the equations are of second order, there will be two independent solutions of this form. We now express $P(t)$ as a Fourier series, writing y in the form

$$y=\sum_{n=-\infty}^{\infty} A_n\, e^{i(p+2n\Omega)t}. \tag{8}$$

On substituting in (6), and writing the trigonometrical terms in exponential form, we see that

$$\sum_{-\infty}^{\infty}\{-A_n(p+2n\Omega)^2(1+\tfrac{1}{2}\beta\, e^{2i\Omega t}+\tfrac{1}{2}\beta\, e^{-2i\Omega t})$$
$$-2\beta\Omega A_n(p+2n\Omega)\,(e^{2i\Omega t}-e^{-2i\Omega t})+\omega_0^2 A_n\}\,e^{2in\Omega t}=0.$$

This equation contains terms in $e^{2in\Omega t}$ with n taking all integral values, but by choosing a particular value of n, multiplying by $e^{-2in\Omega t}$ throughout and integrating over one period, π/Ω, everything can be made to disappear except the coefficient of the chosen $e^{2in\Omega t}$. It follows that the coefficient of each $e^{2in\Omega t}$ must vanish separately. Hence for all n,

$$[(p+2n\Omega)^2-\omega_0^2]A_n+\tfrac{1}{2}\beta[(p+2n\Omega)^2-4\Omega^2](A_{n-1}+A_{n+1})=0. \tag{9}$$

A recurrence relation involving three coefficients normally presents difficulties, and (9) is no exception. For a full discussion the reader should consult treatments of Mathieu's equation, of which (7) is an example.[3] It will be enough here to treat the case $\beta\ll 1$, for which simplifying approximations are available. If one arbitrarily fixes the ratio of two coefficients, A_1/A_0 say, the rest may be systematically calculated by repeated application of (9). As one proceeds to larger $|n|$ less error is occasioned by neglecting the difference between ω_0^2 and $4\Omega^2$ in the square brackets, and we can write the asymptotic form of (9):

$$A_n+\tfrac{1}{2}\beta(A_{n-1}+A_{n+1})\approx 0. \tag{10}$$

When the A_n follow a geometrical progression, $A_{n+1}=-cA_n$, (10) is satisfied, and when $\beta \ll 1$, c is either $\frac{1}{2}\beta$ or $2/\beta$. Although one may discover an initial value of A_1/A_0 that makes the values of A_n tend, as $n \to +\infty$, towards a series having $c=\frac{1}{2}\beta$ and converging to zero, in general one cannot expect the same choice to impose convergence as $n \to -\infty$. For convergence at both extremes the value of p must also be adjusted correctly. If this has somehow been achieved in the case where β is small, convergence is very rapid and only a few values of A_n will have any significant size; physically speaking, a very gently excited pendulum, with no damping, will execute nearly sinusoidal oscillations, possibly growing or decaying in amplitude, possibly showing beats between its free and its driven vibrational modes, but free from such irregularities of waveform that announce the presence of high Fourier terms.

With this in mind we concentrate attention on the behaviour of (9) when n is small. For example, when $n=-1$,

$$[(p-2\Omega)^2-\omega_0^2]A_{-1}+\tfrac{1}{2}\beta[(p-2\Omega)^2-4\Omega^2](A_{-2}+A_0)=0. \tag{11}$$

Now if the excitation is very nearly tuned to twice the natural frequency, $\omega_0 \sim \Omega$, and if $p \sim \omega_0 \sim \Omega$, the coefficient of A_{-1} is much smaller than that of $(A_{-2}+A_0)$; in spite, therefore, of the smallness of β we may expect A_{-1} to be sizeable. Putting $n=-2$ in (9) shows, on the other hand, that A_{-2} is much smaller than $\frac{1}{2}(A_{-1}+A_{-3})$ and can be neglected in (11) without much harm. Similarly by putting $n=1$ we can see that A_1 and terms above this are of minor importance, so that the solution is dominated by A_0 and A_{-1}. If all other terms are assumed to vanish, only two equations of the type (9) retain any interest, those that result from putting $n=-1$ and 0:

$$[(p-2\Omega)^2-\omega_0^2]A_{-1}+\tfrac{1}{2}\beta[(p-2\Omega)^2-4\Omega^2]A_0=0, \tag{12}$$

and

$$[p^2-\omega_0^2]A_0+\tfrac{1}{2}\beta[p^2-4\Omega^2]A_{-1}=0. \tag{13}$$

Here then is the explicit demonstration that arbitrary values of p are not permitted – (12) and (13) together demand that

$$[p^2-\omega_0^2][(p-2\Omega)^2-\omega_0^2]=\tfrac{1}{4}\beta^2[p^2-4\Omega^2][(p-2\Omega)^2-4\Omega^2]. \tag{14}$$

This quartic equation looks less formidable when we realize that the square brackets on the right are very close to $-3\Omega^2$ each, and that if p, ω_0 and Ω are nearly equal we may replace $(p^2-\omega_0^2)$ by $2\omega_0(p-\omega_0)$ and $[(p-2\Omega)^2-\omega_0^2]$ by $2\Omega(2\Omega-p-\omega_0)$. Let us take Δ, defined as $(\Omega-\omega_0)/\omega_0$, to measure the extent to which the parametric excitation is not tuned exactly to $2\omega_0$, and ε, defined as $(p-\omega_0)/\omega_0$ to measure the relative change in frequency caused by the parametric excitation. On making these and other equivalent approximations, after substituting the new variables, we arrive at a more transparent expression for the behaviour of the system:

$$\varepsilon(2\Delta-\varepsilon)=\Delta_c^2, \quad \text{where } \Delta_c=\tfrac{3}{4}\beta. \tag{15}$$

This has the solution

$$\varepsilon = \Delta \pm (\Delta^2 - \Delta_c^2)^{\frac{1}{2}},$$

i.e.

$$p = \Omega \pm \omega_0(\Delta^2 - \Delta_c^2)^{\frac{1}{2}}. \tag{16}$$

If the excitation is perfectly tuned ($\Omega = \omega_0$ and $\Delta = 0$)

$$p = \Omega \pm \mathrm{i}\omega_0 \Delta_c, \tag{17}$$

and the full solution is the superposition of the oscillations corresponding to both choices of p. From (13) it follows, with the same approximations as led to (15), that $A_{-1}/A_0 = \varepsilon/\Delta_c = \mathrm{i}$ if the positive sign is chosen in (17), $-\mathrm{i}$ if the negative sign is chosen. The character of the oscillation follows from (8):

$$y = A_0\, \mathrm{e}^{\mp\omega_0\Delta_c t}(\mathrm{e}^{\mathrm{i}\Omega t} \pm \mathrm{i}\, \mathrm{e}^{-\mathrm{i}\Omega t}),$$

and A_0 must be chosen so as to make y real, i.e. A_0 must be a real constant times $\mathrm{e}^{\mp \mathrm{i}\pi/4}$. Hence

$$y = C\, \mathrm{e}^{\mp\omega_0\Delta_c t} \cos(\Omega t \mp \tfrac{1}{4}\pi). \tag{18}$$

The physical meaning is immediately clear; the pendulum, whose length has been taken to vary as $l_0(1 + \beta \cos 2\Omega t)$, is being shortened most rapidly when $\Omega t = \pi/4$, $5\pi/4$ etc. If the lower signs are chosen in (18) these are the moments when y vanishes and the tension is greatest, while with the upper signs y is at its peak as the tension is least. The former choice leads, as expected, to growth of the oscillation, the latter to decay. All this, of course, is in addition to the natural decay $\mathrm{e}^{-\frac{1}{2}\lambda t/m}$ which was removed by the change of variables leading to (6). The full expression for the swing of the pendulum takes the form

287

$$\theta(t) = C_1\, \mathrm{e}^{-(\frac{1}{2}k + \omega_0\Delta_c)t} \cos(\Omega t - \tfrac{1}{4}\pi) + C_2\, \mathrm{e}^{-(\frac{1}{2}k - \omega_0\Delta_c)t} \cos(\Omega t + \tfrac{1}{4}\pi), \tag{19}$$

in which $k = \lambda/m$, and C_1 and C_2 are arbitrary real constants. The two independent solutions represent oscillations locked in different phases to the excitation, the second being locked so that energy is fed in and the natural decay counteracted, the first being locked so as to enhance the natural decay. No matter how the pendulum may have been swung initially, and this is what determines C_1 and C_2, after a long enough interval the second term will dominate the behaviour; and if β exceeds the critical value $\frac{2}{3}k/\omega_0$, i.e. $\frac{2}{3}/Q$, the second term represents a growing oscillation. This is the same result as obtained at the outset by energy considerations.

The word 'locked' in the last paragraph is perhaps not justified, since with $\Delta = 0$ there is no inherent tendency for the pendulum to change phase relative to the excitation. It is when we try to excite the pendulum with a frequency other than $2\omega_0$ that the question of locking becomes significant. As (16) shows, when the detuning is not excessive (to be precise, when $|\Delta| < \Delta_c$), the solution (17), for $\Delta = 0$, is modified in the imaginary part only, by replacing Δ_c by $(\Delta_c^2-\Delta^2)^{\frac{1}{2}}$. The full solution corresponding to (19) can be derived in the same way without difficulty:

$$\theta(t)=C_1\,e^{\alpha_1 t}\cos(\Omega t-\gamma)+C_2\,e^{\alpha_2 t}\cos(\Omega t+\gamma), \qquad (20)$$

where $\alpha_{1,2}=-\frac{1}{2}k\mp\omega_0(\Delta_c^2-\Delta^2)^{\frac{1}{2}}$ and $\tan 2\gamma=(\Delta_c^2/\Delta^2-1)^{\frac{1}{2}}$. The oscillations are still locked to the excitation, but at phases that are not optimal for transferring energy to or from the pendulum; the modifications of the natural decrement are correspondingly less. This agrees with the simple picture presented in fig. 5. Amplification sets in only when the excitation is rather larger than is required in the state of perfect tuning; the critical value of β is $\frac{2}{3}(1+4\Delta^2Q^2)^{\frac{1}{2}}/Q$ instead of $\frac{2}{3}/Q$. We now appreciate how carefully the system must be tuned to obtain quantitative agreement between theory and experiment. The pendulum mentioned earlier, having $Q\sim 2500$, needs its parametric excitation to be better tuned than 1 in 5000 (i.e. $|\Delta|<1/5000$) if the measured critical amplitude for spontaneous oscillation is to be within $\sqrt{2}$ of the theoretical minimum. This analysis also explains why it is hard to set the system up so that the rate of growth is a linear function of excitation; α_1 and α_2 are linear in Δ_c, and therefore in β, only when the excitation is perfectly tuned.

If the amplitude of excitation is held constant while the frequency is progressively detuned, two critical points are passed. The first marks the transition from saddle-point instability to nodal stability; when $\omega_0(\Delta_c^2-\Delta^2)^{\frac{1}{2}}$ is less than $\frac{1}{2}k$ (i.e. $\Delta^2>\Delta_c^2-\frac{1}{4}/Q^2$) the gain due to parametric excitation is too small to compensate the intrinsic loss. The second critical point occurs when Δ reaches the value Δ_c and, as (16) shows, p becomes real. Instead of the response being locked in frequency and suffering exponential gain or decay of amplitude, frequency locking now breaks down and, apart from the intrinsic loss, the amplitude is constant. This is the transition from nodal to focal stability. The behaviour when Δ exceeds Δ_c deserves elucidation, for it is not simply an arbitrary superposition of two oscillatory modes – the requirement that y shall be real ensures that the two modes coexist in a well-defined relationship.

Let us first define quantities $\delta\equiv\omega_0(\Delta^2-\Delta_c^2)^{\frac{1}{2}}$ and $X\equiv\Delta/\Delta_c+(\Delta^2/\Delta_c^2-1)^{\frac{1}{2}}$, and note that $\Delta/\Delta_c-(\Delta^2/\Delta_c^2-1)^{\frac{1}{2}}=1/X$. From (16) and (13) it follows that $p=\Omega\pm\delta$, with $A_{-1}/A_0=X$ for the positive sign, and $1/X$ for the negative sign. The general solution then takes the form

$$y=A_0'\,e^{i\delta t}(e^{i\Omega t}+X\,e^{-i\Omega t})+A_0''\,e^{-i\delta t}\left(e^{i\Omega t}+\frac{1}{X}\,e^{-i\Omega t}\right),$$

in which A_0' and A_0'' are arbitrary complex constants. It is convenient, for the purpose of writing y in its most general real form, to pair the first term in the first bracket with the second term in the second bracket, and take the other two terms as a second pair:

$$y=[A_0'\,e^{i(\Omega+\delta)t}+(A_0''/X)\,e^{-i(\Omega+\delta)t}]+[A_0''\,e^{i(\Omega-\delta)t}+XA_0'\,e^{-i(\Omega-\delta)t}]$$

If now we put $A_0'=C\,e^{i\zeta}$ and $A_0''=XC\,e^{-i\zeta}$, where C and ζ are arbitrary real constants, y assumes a real form with the required complement of two independent constants:

$$y = C\{\cos[(\Omega+\delta)t+\zeta] + X\cos[(\Omega-\delta)t-\zeta]\}, \tag{21}$$

and consequently:

$$\theta = C\,e^{-\frac{1}{2}kt}\{\cos[(\Omega+\delta)t+\zeta] + X\cos[(\Omega-\delta)t-\zeta]\}. \tag{22}$$

The requirement that y and θ be real prevents any single oscillatory mode occurring alone; the characteristic response (22) consists of two beating sinusoidal oscillations compounded with an exponential decay. The term in brackets has a beat period of $\frac{1}{2}\Omega/\delta$ swings between successive maxima, and the ratio of maxima to minima is $(X+1)/(X-1)$. In consequence, the amplitude may not decrease exponentially but may show a periodic rise and fall imposed on a steady decrease, a natural consequence of the inability of the excitation to lock the pendulum to exactly half its frequency; for, as time goes on, the phase of the pendulum drifts steadily with respect to the excitation, so that it passes through periods of being helped and being hindered by the parametric amplification, but with no mitigation in the long run of the intrinsic decay.

It may appear at first sight that this beating character, which may have a long period, is inconsistent with the relatively simple approach to equilibrium described as focal stability. It should be remembered, however, that the characteristic parameters, $(\theta, \dot{\theta})$ say, are not observed continuously but only at regular intervals dictated by the period of the excitation. All that happens in between reveals the beats, but the moments of observation pick up a steady progression of discrete values $(\theta, \dot{\theta})$ round a converging elliptical spiral. This is an immediate consequence of Floquet's theorem, as expressed in (8); successive values of y (and hence of θ), taken at intervals of π/Ω, reflect the geometric progression by a factor $e^{i\pi p/\Omega}$ at each step, but altogether miss the interesting structure described by the Fourier series. Of course there are two values of p, but direct substitution of $n\pi/\Omega$ for t in (21) shows that both terms in the braces oscillate as $(-1)^n\cos(n\pi\delta/\Omega\pm\zeta)$, having the same frequency and differing only in amplitude and phase, so that they compound to give a sequence of points lying evenly spaced on a simple sinusoid. The same holds for $\dot{\theta}$ as for θ, and the result is, as stated above, a sequence of points converging on a spiral path. This system, with its apparently more complicated behaviour, lies well within the stability formalism of the last chapter. An example is shown in fig. 6.

In an experiment designed to illustrate the foregoing analysis, the pendulum of fig. 1 was modified by replacing the brass bob by one of plastic which, being lighter, conferred a lower Q on the system. The pendulum was not of the correct length for exact tuning, and with a rather weak excitation amplitude a_0 of 3.9 mm ($\beta = 4.7\times10^{-3}$, or $\Delta_c = 3.5\times10^{-3}$) its swing, after being set going, varied in the way shown in fig. 7(a). The logarithmic plot 7(b) reveals a steady beat pattern imposed on an exponential decay which corresponds, as one would expect from (20), with

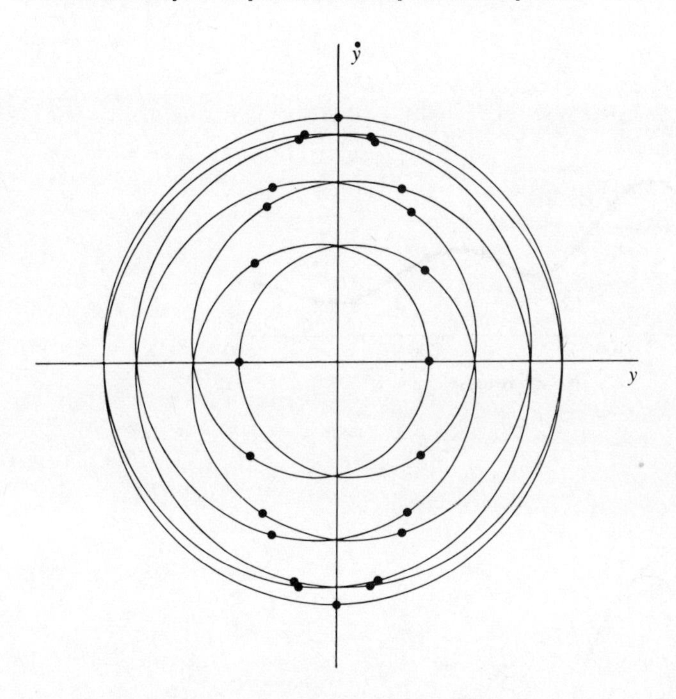

Fig. 6. Phase-plane trajectory for a typical solution of (21). Observations made at the same phase in successive cycles of the excitation lie on an ellipse, as indicated by the marked points.

the natural decay when there is no parametric excitation. To see how well this curve agrees with (22), we shall use it to evaluate the relevant parameters. Since the pendulum makes 103 swings during one beat period, δ/ω_0 is 1/206; as this is $(\Delta^2-\Delta_c^2)^{\frac{1}{2}}$ and $\Delta_c=3.5\times10^{-3}$, we find $(\Delta^2/\Delta_c^2-1)^{\frac{1}{2}}$ to be 1.39 and Δ/Δ_c to be 1.71. Hence $X=3.1$ and the ratio of maxima to minima in the beat pattern should be 4.1/2.1 or 1.95. The straight lines in fig. 7(*b*) are drawn to exhibit this ratio, and the comparison between theory and experiment leaves one in little doubt of the essential correctness of the theory.

By way of contrast, when the amplitude of excitation was increased above the critical value and the pendulum set swinging in an unfavourable phase, the amplitude initially decreased and then rose again, as in fig. 8, without any superimposed oscillations, since both modes were now locked to the excitation. The asymmetry of the curve is obvious; in the initial stages, dominated by the first term of (20), the excitation enhances the natural decrement and the fall is more rapid than the subsequent rise, dominated by the second term, where the natural decrement partially counteracts the parametric amplification.

To take a last look at the simple pendulum, let us remove the string tying it back at B (fig. 1), and so convert it into a conical pendulum, which is then set swinging in a circular orbit. The tension in the string now remains constant, so that no net energy is communicated to the pendulum by periodically changing its length. Moreover there are no tangential forces on the bob and in the absence of friction its angular momentum must stay

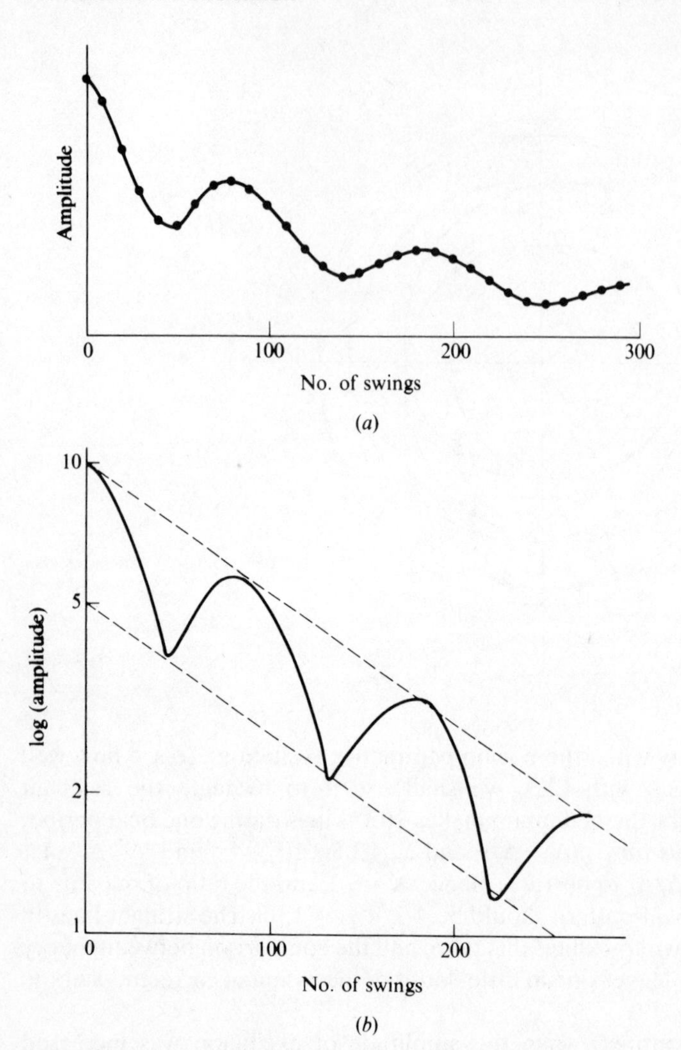

Fig. 7. Variation of amplitude of pendulum of fig. 1 when excitation is detuned so far as to prevent phase-locking; (*a*) direct observations, (*b*) logarithmic plot.

constant. It is tempting to think that no change could occur in the motion of the pendulum, but this is not so. As the bob is raised, energy is communicated to the pendulum, to be removed once more as the bob is lowered, and if this process is tuned to twice the orbital frequency, or near enough, a progressive deformation of the orbit occurs. The circular motion may be analysed into two linear harmonic oscillations, one phased so as to grow in purely exponential fashion and the other so as to decay exponentially, until in the end the motion is almost confined to the plane in which the phase is most favourable to growth. Since this means separating the two terms in (20), which are not necessarily in phase quadrature, the two linear components are not at right angles nor do they vibrate in phase. There is no need to work out precisely what combination is needed to synthesize circular motion since the conservation of angular momentum, which we

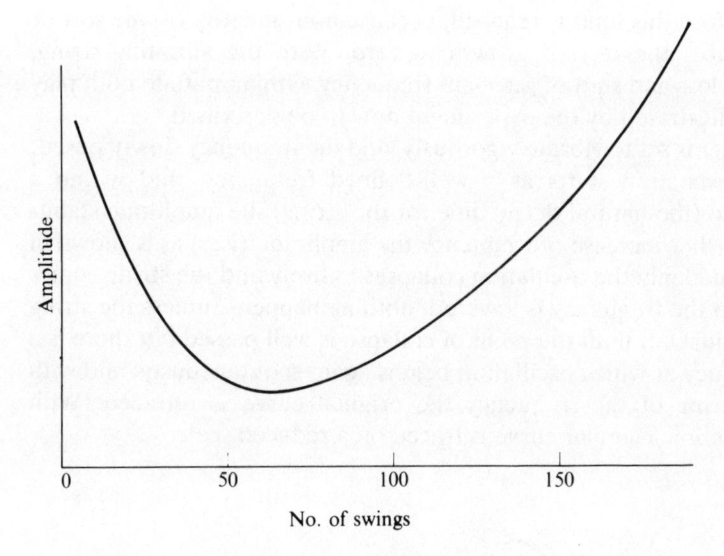

Fig. 8. As in fig. 7 but with excitation increased to allow phase-locking and parametric amplification.

desire to verify, does not depend on such detail. Let us start with two linear vibrations of amplitude A and B with an angle θ between them; treating the horizontal plane as a complex plane the displacement corresponding to the sum of the two motions is given by

$$r = A \cos \omega t + B \, \mathrm{e}^{\mathrm{i}\theta} \cos (\omega t + \phi), \tag{23}$$

ϕ being the phase difference between the vibrations. The angular momentum is the imaginary part of $mr\dot{r}^*$, which is equal to $ABm\omega \sin \theta \sin \phi$. Now let the first term in (23) grow by a factor $\mathrm{e}^{\alpha t}$ while the second decays by the same factor (note that if $k=0$, $\alpha_1 = -\alpha_2$ in (20)). Then $\mathrm{Im}\,[r\dot{r}^*]$ is unaffected, which shows that angular momentum is conserved. As for the total energy, which is proportional to $(rr^* + \dot{r}\dot{r}^*/\omega^2)$, this on evaluation is found to vary as $\cos \phi \cos \theta + \cosh 2\alpha t$, if t is taken as 0 at the moment when $A = B$. Since the derivative of $\cosh 2\alpha t$ vanishes when $t = 0$, the energy of a pendulum in circular orbit is indeed momentarily unchanged by the excitation, but as the orbit becomes elliptical the energy grows steadily. Observation confirms this analysis.

The pendulum, by virtue of its slow vibration, is easy to observe though tedious if much information is to be gathered. A more rapidly vibrating system which is readily excited parametrically is a string under tension, as in fig. 3. This can be realized by use of the arrangement shown in fig. 9.1, the impeller now being placed above the end to excite the string longitudinally. While it is not so easy with this as with the pendulum to verify the theory quantitatively, it is particularly well suited to illustrate how the process of parametric excitation is limited by non-linearity. The behaviour of a strictly linear system, as expressed in (19) and (20), for example, must continue as it started, growing exponentially without limit. A real system, however, will saturate either because the driver cannot deliver unlimited

249

power or, before this limit is reached, because non-linearity of one sort or another reduces the rate of growth to zero. With the vibrating string, increased air loss and shift of resonant frequency with amplitude both play a part, as is illustrated by the experiment now to be described.

The impeller is set to vibrate vigorously, and the frequency slowly raised; parametric excitation starts as a well-defined frequency, and within a second or two (the natural decay time for the string) the amplitude stabilizes. With further increase of frequency the amplitude rises, as is shown in fig. 9, until suddenly the oscillation collapses entirely and the string comes to rest. When the frequency is lowered, nothing happens (unless the string is violently plucked) until the point of collapse is well passed but there is a lower frequency at which oscillation begins again spontaneously, and with further lowering of the frequency the original curve is retraced. With smaller excitation a similar curve is traced on a reduced scale.

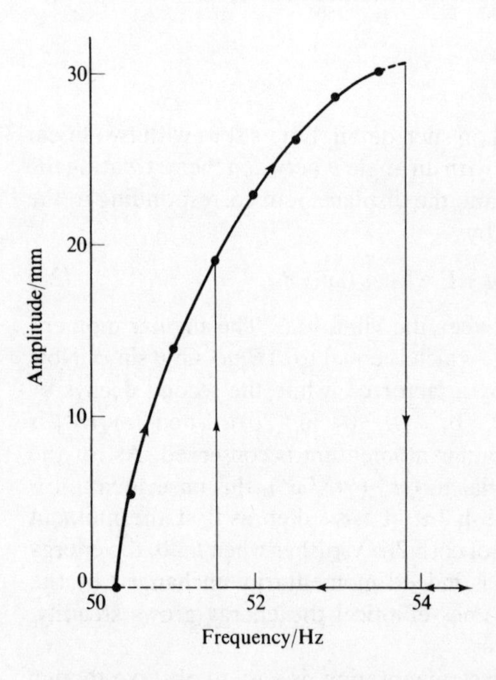

Fig. 9. The wire of fig. 9.1 excited parametrically.

The explanation of the shape of the curve follows naturally from the anharmonic behaviour exhibited and discussed in chapter 9 (see fig. 9.2). The increase of tension as the string is displaced is the dominant mechanism for parametric excitation, easily overcoming the cos θ reduction in the longitudinal component of tension at the point of excitation. Yet the motion is nearly enough harmonic for the assumption to be made that the linear theory need only be modified by allowing ω_0 to vary slightly with amplitude. Now the detuning Δ_1, at which the vibration need neither grow nor decay is given by $(\Delta_c^2 - \frac{1}{4}/Q^2)^{\frac{1}{2}}$, independent of amplitude if the excitation is kept constant (i.e. Δ_c fixed) and if Q does not change. We therefore

250

draw two lines on an amplitude–frequency graph, separated in frequency by $\pm\Delta_1$ from the line of the resonant frequency, to show the conditions in which vibration at a steady amplitude can occur (fig. 10). The lower line, however, is unstable since a small increase in amplitude brings the system to a point where the amplitude increases still further. If this were the whole story, as the frequency was raised nothing would happen until ω_1 was reached, when parametric excitation would begin to be effective. With further increase of frequency the amplitude would grow, following curve A. There would be no reason for the sudden collapse shown in fig. 9. This must arise from the lowering of Q at large amplitudes, already discussed in chapter 9, which diminishes the critical detuning and brings curves A and B together, as shown by the broken curve. The sequence of events shown in fig. 9 is now clear.

252

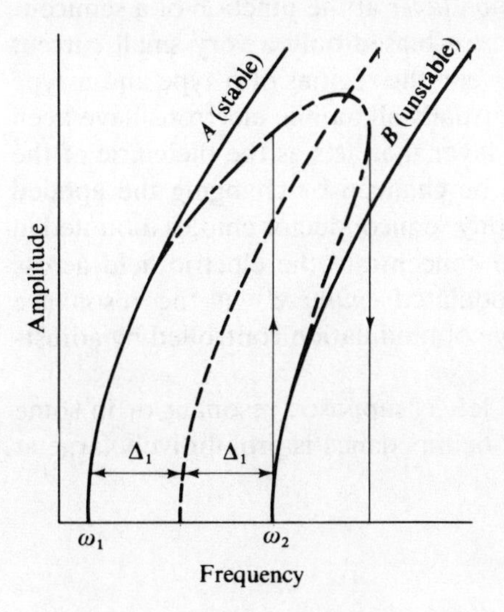

Fig. 10. Illustrating the explanation of fig. 9.

If this interpretation is correct, the observed frequency of 53.5 Hz just below the point of collapse must be very close to twice the natural frequency of the string when its amplitude is 30 mm, as against 50.9 Hz at low amplitudes. On the assumption that the change of natural frequency is proportional to the square of the amplitude, at 17.5 mm it should be 0.44 Hz, which is very close to what can be inferred from the results in fig. 9.2.

Parametric amplifiers[4]

The periodic variation of a circuit element, a capacitor for instance, produces distortion of any oscillatory current already present, generating sum and difference frequencies which may excite resonant response in a suitable tuned circuit. It is this effect which is employed in parametric amplifiers, particularly at the microwave frequencies used in satellite communication and radio astronomy, where incoming signals are very weak and easily swamped by extra noise generated at the first stage of amplification. At this stage what is needed above all is a modest degree of amplification with minimal noise; from then on the signal strength will be sufficiently far above the noise level of conventional amplifiers for the danger of losing information to be past. It is the merit of parametric amplifiers that their circuit elements can be cooled, so that Johnson noise is low, and that there are no elements generating shot noise. Precisely how little the noise can be, and how to minimize it, are matters that have been extensively studied. Here, however, we shall confine attention to the one basic question of how the device works. It is unnecessary to go into details about the physical processes underlying the operation of the *varactor*, the variable capacitor which lies at the heart of the amplifier.[5] Let it suffice to remark that it depends on the *depletion layer* at the junction of a semiconducting p–n rectifier; when it is reverse-biased only a very small current flows through the rectifier, since between the regions of p-type and n-type material is a thin layer from which virtually all mobile electrons have been swept by the applied field. It is this layer that acts as the dielectric of the capacitor, and whose thickness can be changed by changing the applied field. When the varactor, which is a tiny semiconductor chip, is mounted in a microwave cavity shaped so as to concentrate the electric field across it, its capacitance can then be modulated (*pumped*) at the resonance frequency of the cavity, and the range of modulation controlled by adjusting the pumping power.

158

In fig. 11 the input circuit on the left is supposed resonant or in some other way isolated by filters so that its impedance is prohibitively large at

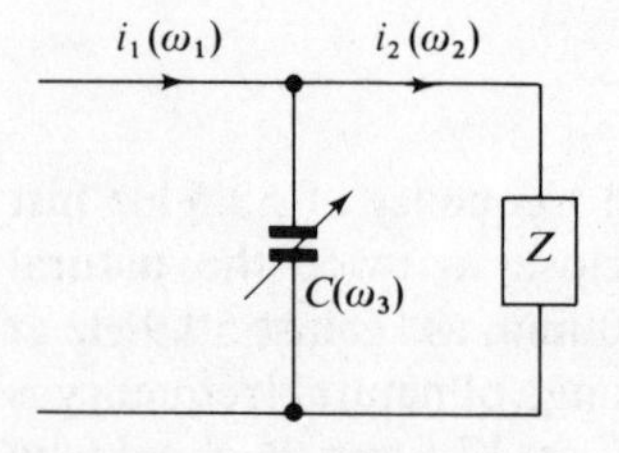

Fig. 11. Currents at different frequencies in a parametric amplifier circuit.

any other frequency than ω_1. In the same way the *idler* circuit on the right is resonant near ω_2 and is unaffected by anything happening elsewhere at ω_1 (input) or ω_3 (the pumping frequency). Now let the currents and capacitance vary as follows:

$$\left.\begin{aligned} i_1 &= a_1 \cos(\omega_1 t + \phi_1), \\ i_2 &= a_2 \cos(\omega_2 t + \phi_2), \\ 1/C &= [1 + \beta \cos(\omega_3 t + \phi_3)]/C_0. \end{aligned}\right\} \qquad (24)$$

The current flowing to the capacitor is $i_1 - i_2$, from which the charge follows by integration and hence the potential difference across it:

$$C_0 V = [1+\beta \cos(\omega_3 t+\phi_3)]\left[\frac{a_1}{\omega_1}\sin(\omega_1 t+\phi_1) - \frac{a_2}{\omega_2}\sin(\omega_2 t+\phi_2)\right]$$

$$= \frac{a_1}{\omega_1}\sin(\omega_1 t+\phi_1) - \frac{a_2}{\omega_2}\sin(\omega_2 t+\phi_2)$$

$$+\frac{\beta a_1}{2\omega_1}\{\sin[(\omega_1+\omega_3)t+\phi_1+\phi_3] - \sin[(\omega_1-\omega_3)t+\phi_1-\phi_3]\}$$

$$-\frac{\beta a_2}{2\omega_2}\{\sin[(\omega_2+\omega_3)t+\phi_2+\phi_3] - \sin[(\omega_2-\omega_3)t+\phi_2-\phi_3]\}. \quad (25)$$

This contains terms in ω_1, ω_2 and the combination frequencies $\omega_1 \pm \omega_3$ and $\omega_2 \pm \omega_3$. When the idler circuit is tuned so that $\omega_2 = \omega_3 - \omega_1$, four separate frequencies are present in V: ω_1, ω_2, $2\omega_1 + \omega_2$, $2\omega_2 + \omega_1$, of which the last two generate no current in the sharply tuned circuits.† In fact the only terms relevant to the individual circuits are the following:

$$\text{circuit 1:} \quad C_0 V_1 = \frac{a_1}{\omega_1}\sin(\omega_1 t+\phi_1) - \frac{\beta a_2}{2\omega_2}\sin(\omega_1 t+\phi_3-\phi_2). \quad (26)$$

$$\text{circuit 2:} \quad C_0 V_2 = -\frac{a_2}{\omega_2}\sin(\omega_2 t+\phi_2) + \frac{\beta a_1}{2\omega_1}\sin(\omega_2 t+\phi_3-\phi_1). \quad (27)$$

It is only by virtue of V_2 that any current i_2 flows in the idler circuit. If its resistance is R_2 and reactance X_2 the required form of V_2 is given by

$$V_2 = R_2 a_2 \cos(\omega_2 t+\phi_2) - X_2 a_2 \sin(\omega_2 t+\phi_2). \quad (28)$$

Since (27) and (28) must be identical the amplitude and phase of i_2 take such values that

$$\frac{\beta a_1}{2a_2}\sin\phi = \omega_1 C_0 R_2 \quad \text{and} \quad \frac{\beta a_1}{2a_2}\cos\phi = \frac{\omega_1}{\omega_2}(1-\omega_2 C_0 X_2), \quad (29)$$

where $\phi = \phi_3 - \phi_2 - \phi_1$. When the idler circuit, including the varactor in its average state, is exactly tuned to resonance, $X_2 - 1/\omega_2 C_0 = 0$, and therefore $\cos\phi = 0$; then

$$a_2/a_1 = \beta/2\omega_1 C_0 R_2. \quad (30)$$

Off resonance a_2/a_1 is less and (29) shows, not unexpectedly, that a significant response is only to be obtained if the idler circuit is tuned so that $\omega_3 - \omega_1$ (or $\omega_1 + \omega_3$) lies within its resonant peak.

† If the idler circuit is tuned so that $\omega_2 = \omega_1 + \omega_3$ the behaviour is very similar, and we shall not examine this case separately.

To evaluate the input impedance of the circuit at frequency ω_1, we note that with the idler on resonance (the only case we shall treat) (26) takes the form

$$C_0 V_1 = \frac{a_1}{\omega_1} \sin(\omega_1 t + \phi_1) - \frac{\beta a_2}{2\omega_2} \cos(\omega_1 t + \phi_1)$$

$$= \frac{a_1}{\omega_1} \sin(\omega_1 t + \phi_1) - \frac{\beta^2 a_1}{4\omega_1 \omega_2 C_0 R_2} \cos(\omega_1 t + \phi_1), \quad \text{from (30).}$$

The first term is in phase quadrature with the input current $a_1 \sin(\omega_1 t + \phi_1)$ and expresses the reactive part of the input impedance, which can be tuned out by adding an opposite reactance in series – this is just the operation of adjusting the input circuit to resonance. When this is done the circuit appears as a negative resistance

$$R_{\text{in}} = -\beta^2/4\omega_1\omega_2 C_0^2 R_2. \tag{31}$$

The negative resistance allows a larger voltage signal to be extracted than is present in the input. With the circuit of fig. 12 an input voltage V_{in} from a source of resistance R_1 (e.g. a transmission line of characteristic impedance R_1) produces a current $V_{\text{in}}/(R_1 + R_{\text{in}})$. Across the capacitor there is therefore a voltage V_1, at ω_1, of amplitude $V_{\text{in}} R_{\text{in}}/(R_1 + R_{\text{in}})$. The voltage amplification is $R_{\text{in}}/(R_1 + R_{\text{in}})$, which can be made equal to 10, say, giving a power gain of 20 dB, by adjusting $|R_{\text{in}}|$ to the value $\frac{10}{11} R_1$. If $|R_{\text{in}}|$ exceeds R_1 the circuit is unstable, and oscillates spontaneously even without an input signal. This is another form of parametric excitation of an oscillator, with two resonant systems excited simultaneously by an input at the sum (or difference) frequency, and is clearly to be avoided in an amplifier. The more precisely the pumping power can be controlled, the nearer can β be held to the critical value for instability, and the amplification benefits correspondingly.

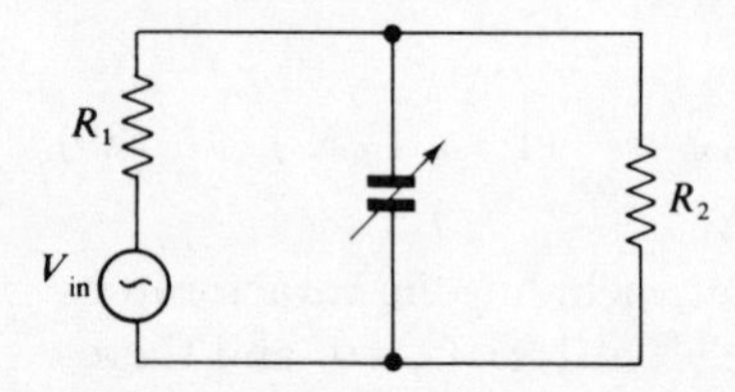

Fig. 12. Equivalent circuit of tuned parametric amplifier, fed by a transmission line of characteristic impedance R_1.

An amplified signal may also be extracted at frequency ω_2. The voltage across R_2 is $R_2 a_2$, i.e. $\beta a_1/2\omega_1 C_0$ from (30), when the idler circuit is well tuned. Hence

$$V_2/V_{\text{in}} = \beta/2\omega_1 C_0 (R_1 + R_{\text{in}}),$$

and this is the voltage amplification for this arrangement. To compare the two arrangements, note that

$$V_2/V_1 = -\beta/2\omega_1 C_0 R_{\text{in}},$$

and that β is adjusted so that $-R_{in}$ is very nearly R_1. Hence

$$V_2/V_1 \sim (\omega_2 R_2/\omega_1 R_1)^{\frac{1}{2}}.$$

Whether it is advantageous to pick off V_2 or V_1 for the next stage of amplification is a technical matter to be determined by the particular application. The analysis so far presented offers little guidance.

The design of real parametric amplifiers has progressed far beyond this point, and extended to wave–propagation devices in which parametric pumping causes progressive amplification as the wave proceeds. These are advanced matters which would be inappropriate here.

11 Maintained oscillators

This chapter is concerned only with simple systems in steady oscillation, such as were classified in chapter 2 under the heading of negative resistance devices, feedback oscillators and relaxation oscillators. Here we shall attempt to refine the description and classification of the different types though, as is common in such attempts, firm divisions are hard to find. At the same time we shall analyse a number of examples so as to understand what conditions must be satisfied for them to oscillate spontaneously, how the amplitude of oscillation is limited by non-linearity, and what determines the ultimate waveform. No attempt will be made to establish rigorously the general conditions for oscillation to occur. This is an important and well-studied problem, but one which deserves the fuller treatment that will be found in specialized texts.

39

It has already been remarked, in chapter 2 and elsewhere, that a resonant system governed by a second-order equation such as (2.23) will oscillate spontaneously if k is negative. A source of energy is required to overcome inevitable dissipative effects, and among the many examples of how the energy may be injected probably the commonest is by feedback. Let us start, then, with a general survey of the feedback principle, with particular reference to the influence feedback may have on the performance of a resonant system, and not solely in setting it in spontaneous oscillation. The argument will be conducted in terms of electrical circuits, which account for the overwhelming majority of applications.

30

The feedback principle

Baldly stated, feedback consists of taking a signal from a system, subjecting it to some processing (e.g. amplification, filtering) and injecting the result back into the system. Since the injected signal combines with what is already present, it has an influence on what is extracted for processing, and in this sense feedback establishes a cyclical flow pattern. This view of the matter is not, however, especially useful or, indeed, always valid. It is usually better to think of the electrical circuit, or mechanical system (or even the economic or social system to which the concept is also usefully applied) as having cross-links between different elements and as behaving in the only way that is consistent with the relationships that are dictated by the linkages. The idea of a sequence of events – extraction, processing, reinjection – derives from the non-reciprocal character of an amplifier, whose output is its response to its input, but not vice versa. It is thus natural to follow the process as if it were a time sequence, and it may be helpful to do so if the processing introduces a time delay. But the advantage is usually slight, and we shall have little occasion to discuss problems in these terms.

Consider a passive LCR circuit, modified so that the e.m.f. between any two points can be picked up and fed back, after amplification, across any two other points. As shown in fig. 1(a), pick-up and feedback are both across resistors, but the arrangement may be more general, as in fig. 1(b). Fig. 1(c) shows the idealized form of amplifier that is assumed, an input ε generating $A\varepsilon$ at the output stage. The input resistance R_{in} is supposed large enough to prevent any significant current being diverted from the circuit; the resistance R_{out} is similarly taken to be much larger than any impedance in the circuit, so that the current fed back, $A\varepsilon/R_{out}$ $(=g\varepsilon)$, is proportional to the input voltage. In addition, the input and output stages are considered to be insulated from one another, so that, among other things, there is freedom to reverse the sign of g by reversing the output connections; this is not impossible with modern devices, but is unusual and not to be taken for granted in any actual example – indeed, some of the difficulties experienced by electronic designers in the days of thermionic tubes sprang directly from the electrical connection between stages. So long as we are concerned with basic principle, however, we may disregard the practical problem, returning to reality later with a clearer idea of the distinction between essential features of the behaviour and those arising from practical limitations.

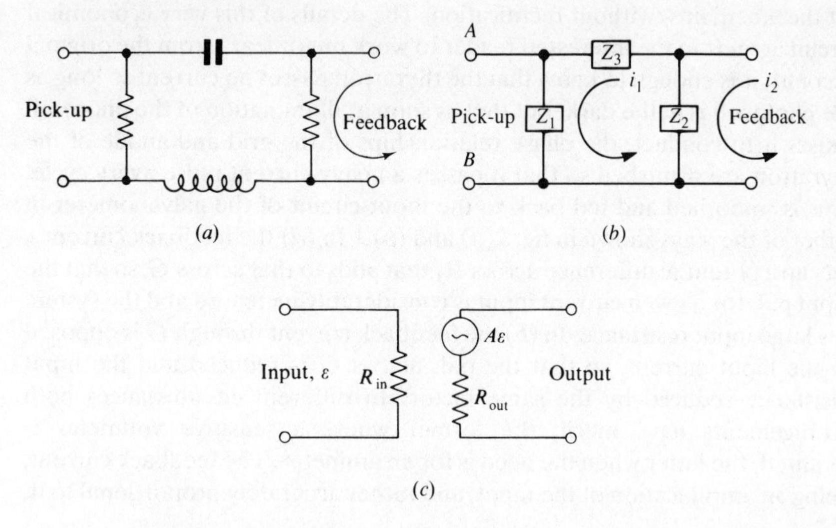

Fig. 1. (a) The e.m.f. across a resistor in a resonant circuit is picked up, amplified and fed back through another resistor; (b) generalized form of (a); (c) equivalent circuit of the amplifier.

To analyse the circuit of fig. 1(b) imagine the currents i_1 and i_2 to have complex frequency ω. If the impedance between the tapping points AB is $Z_1(\omega)$, $i_2=gZ_1i_1$ and the extra e.m.f. in the circuit, due to i_2, is $gZ_1Z_2i_1$. The effect of feedback is therefore to add another term to the equation for the circuit. Whereas without feedback the natural frequency is that at which Z_t, defined as $Z_1+Z_2+Z_3$, vanishes, with feedback the equation takes the form

$$Z_t - gZ_1Z_2 = 0. \tag{1}$$

For the circuit of fig. 1(a), the additional impedance is resistive, gR_1R_2. With the amplifier connected so that g is negative, the extra resistance is positive and the resonant properties of the circuit are damped (*negative feedback*). With the connections reversed, however, to make g positive (*positive feedback*) R_1 and R_2 may be chosen so that the total resistance in the circuit becomes negative, and the circuit goes into spontaneous oscillation, with an amplitude that rises exponentially until saturation of the amplifier lowers the effective value of g and prevents further growth. Systems of this sort will occupy much of the space in this chapter, and before becoming involved in them it is as well to appreciate that positive and negative feedback are not the only choices possible. One may, for example, choose a capacitor for Z_1 and feed back through a resistor, in which case $Z_1 = 1/j\omega C_1$ and the extra impedance is reactive, being equivalent to a capacitance C_1/gR_2. The resonant frequency, rather than Q, is now the property directly affected.

A rather early and ingenious realization of this sort of feedback, which shows how the restoring force constant in a resonant system can be manipulated to advantage, is the galvanometer amplifier invented and developed by Gall.[(1)] Light reflected from a galvanometer mirror is focussed on a photocell, which is part of a thyratron circuit (fig. 2(c)) running directly off the a.c. mains, without rectification. The details of this very economical circuit are left to the interested reader to work out or learn from the original account; it is enough to know that the thyratron passes no current so long as the photocell is in the dark, but that as soon as illumination of the photocell causes it to conduct, the phase relationships of the grid and anode of the thyratron are disturbed so that it passes a heavy current pulse every cycle. This is smoothed and fed back to the input circuit of the galvanometer in either of the ways shown in fig. 2(a) and (b).† In (a) the feedback current i_f sets up a potential difference across R_1 that adds to that across G, so that the input p.d. for a given current input is considerably increased and the system has large input resistance. In (b) the feedback current through G is opposed to the input current, so that the p.d. across G is reduced and the input resistance reduced by the same factor. In different circumstances both arrangements have merit, the former where a sensitive voltmeter is required, the latter when the need is for an ammeter. The feedback current, being an amplification of the input, and rather accurately proportional to it,

† It will be observed that the use of an optical lever enables the output and input stages of the amplifier to be electrically isolated. Much more recently optical stages have been incorporated into electronic circuits for the same purpose, a light-emitting diode serving as transmitter of information which is picked up by a phototransistor, all built into the same module.

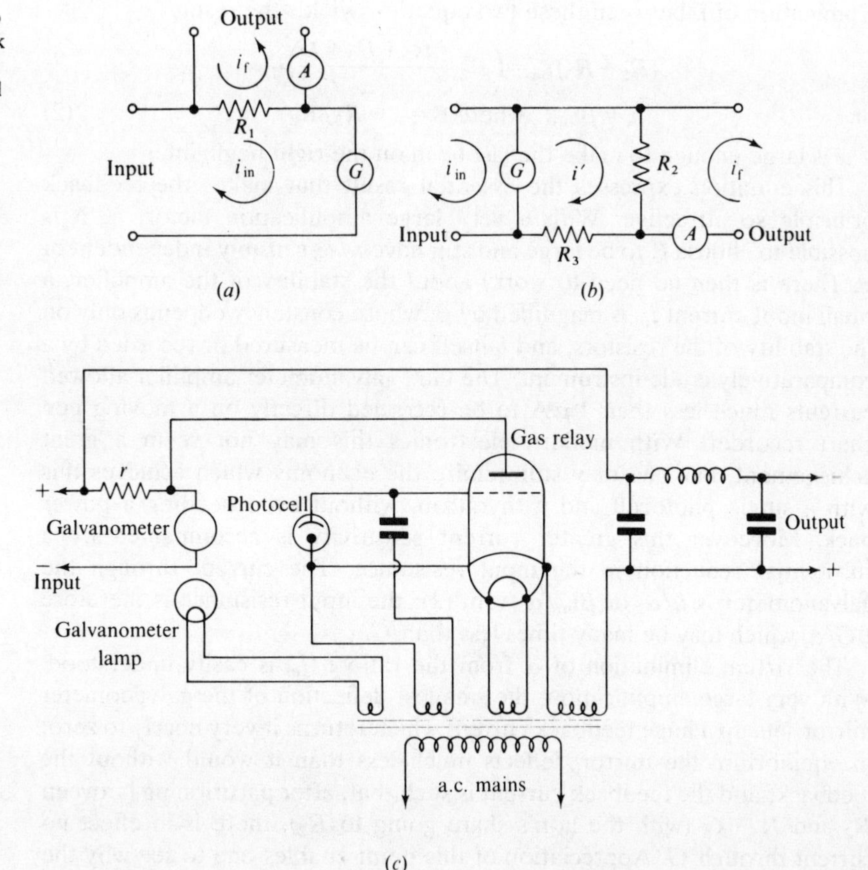

Fig. 2. (*a*) and (*b*) are two forms of feedback network for the galvanometer amplifier. Current injected through the input terminals deflects the galvanometer, and the amplifier injects the feedback current through the output terminals; the latter current is measured by the ammeter *A* and is proportional to the input current. (*c*) is the thyratron amplifier circuit (D. C. Gall[1]).

as we shall see when working through the circuit in detail, provides a very sensitive measure of the input. These matters, however, while explaining the utility of the device, are only incidental to the present discussion which is concerned with the effect of feedback on the dynamical behaviour of the suspended coil, a mechanical analogue of a resonant circuit. It will become clear that the changes due to feedback are wholly beneficial – the instrument is indeed a rare (if minor) example of an invention with virtually no disadvantages.

To develop a quantitative theory, we define the overall amplification α as the ratio of feedback current to current through the galvanometer; this may be 10^5. To consider only the arrangement 2(*b*), the input current i_{in} and feedback current i_f are accompanied by a current i' in the feedback loop; application of Kirchhoff's law to this loop shows that

$$(G+R_2+R_3)i'-Gi_{in}-R_2 i_f=0.$$

Moreover, the current through the galvanometer is $i_{in}-i'$, and therefore

$$i_f=\alpha(i_{in}-i').$$

Elimination of i' between these two equations yields the result:

$$(R_2+R_3)i_{\text{in}}=\left(R_2+\frac{R_2+R_3+G}{\alpha}\right)i_{\text{f}},$$

or

$$i_{\text{f}} \doteqdot \beta i_{\text{in}}, \quad \text{where } \beta=1+R_3/R_2, \tag{2}$$

if α is large enough to make the last term on the right negligible.

This equation expresses the essential result that makes the feedback principle so attractive. With a very large amplification factor, α, it is possible to choose β to be large and still have $i_{\text{f}}/i_{\text{in}}$ sensibly independent of α. There is then no need to worry about the stability of the amplifier; a small input current i_{in} is magnified by β, whose constancy depends only on the stability of the resistors, and i_{f} itself can be measured or recorded by a comparatively crude instrument. The Gall galvanometer amplifier allowed currents much less than 1 μA to be recorded directly on a moving pen chart recorder. With modern electronics this may not seem a great achievement, but one may still admire the economy which achieves this with a single photocell and a thyratron, without even needing a power pack. Moreover this greater current sensitivity is accompanied by a substantial reduction in the input resistance. The current through the galvanometer is i_{f}/α, or $\beta i_{\text{in}}/\alpha$ from (2); the input resistance is therefore $\beta G/\alpha$, which may be many times less than G.

The virtual elimination of α from the ratio $i_{\text{f}}/i_{\text{in}}$ is easily understood. With very large amplification, the smallest deflection of the galvanometer mirror sets up a huge feedback current which returns it very nearly to zero; in equilibrium the mirror deflects much less than it would without the feedback, and the feedback current is such that, after partitioning between R_2 and R_3+G (with the lion's share going to R_2), there is in effect no current through G. Appreciation of this point enables one to see why the dynamical properties are also greatly affected. The feedback current provides a strong restoring torque for any deflection from equilibrium, and therefore enhances the value of the torque constant of the suspension. The periodic time is thereby shortened, and so is the time required to reach equilibrium deflection after a current is applied – another significant gain in performance.

Let us analyse this in terms of the equation of motion of the galvanometer mirror. The free motion is governed by an equation of the form (2.29); when a current i flows through the suspended coil its contribution to the torque adds an extra term:

33

$$I\ddot{\theta}+(\lambda_0+A/R)\dot{\theta}+\mu\theta=S\mu i, \tag{3}$$

in which S, the current sensitivity, is the steady deflection produced by unit current.† From the definition of the amplification α it follows that a

† The damping term, $A\dot{\theta}/R$, arises from the e.m.f. $A\dot{\theta}/S\mu$ induced by movement of the coil. The value to be assigned to R is the resistance seen by a generator inserted in series with the galvanometer, i.e. the resistance of the network formed by G, R_2, R_3 and any resistance connected across the input terminals. It is unaffected by the feedback.

deflection Si of the galvanometer produces a feedback current αi, so that in general the relation between deflection and feedback current takes the form:

$$i_f = \alpha\theta/S.$$

Hence $$i = i_{in} - \alpha\theta/\beta S,$$

which, when substituted into (3) gives the modified equation of motion:

$$I\ddot{\theta} + (\lambda_0 + A/R)\dot{\theta} + \mu(1+\alpha/\beta)\theta = S\mu i_{in}.$$

This confirms the qualitative view of feedback as enhancing the torsion constant. The natural frequency is increased by $(1+\alpha/\beta)^{\frac{1}{2}}$, and the decrement per cycle reduced. The condition for critical damping is

$$\lambda_0 + A/R = 2[\mu I(1+\alpha/\beta)]^{\frac{1}{2}},$$

so that R can be much smaller without causing overdamping. This allows the instrument to be used to measure small currents in a low resistance circuit, and marks a considerable improvement over the unaided galvanometer for which current sensitivity is bought at the price of high resistance.

Negative resistance and spontaneous oscillation

Much of the rest of this chapter is concerned with the performance of maintained oscillators once they have attained a steady state. This, as already remarked, takes us out of the realm of linear processes, and it is therefore convenient to deal at the outset with that part of the discussion that belongs properly to linear theory; that is to say, the critical behaviour as the feedback parameters are changed infinitesimally so as to take the system from a stable equilibrium state to one in which oscillations grow spontaneously, but are still small in amplitude – the crossing of the line between focal stability and focal instability in fig. 9.6, for example. In the simplest cases, as with the circuit of fig. 1(a), g being independent of frequency, the behaviour can be seen at a glance; (1) takes the familiar form

256

$$\mathrm{j}\omega L + R(1 - gR_1R_2/R) + 1/\mathrm{j}\omega C = 0, \tag{4}$$

and the critical value of g is R/R_1R_2, at which value the second term vanishes. The matter is not so obvious when g is a (complex) function of frequency, as happens with amplifiers when internal capacitances, electron transit times etc. degrade the amplification at higher frequencies as well as introducing frequency-dependent phase shifts. Even if g is expressible as a polynomial in ω (4) may turn into an algebraic equation of high order, with a multitude of roots, any one of which having a positive imaginary part can be responsible for a growing spontaneous oscillation, making the circuit unstable. The same problem arises, even when g is constant, if the circuit containing the feedback loop is more complicated than fig. 1(a), having a number of linked networks and being described by a high-order differential equation. The resolution of this problem by application of the Routh–Hurwitz tests is one of the oldest successes in the theory of stability. It will be realized, however, that it is not a panacea, since in general the frequency-variation of g will not be representable by a polynomial. Fortunately an alternative procedure exists, and one which is far more closely related to experiment, in that it operates on $g(\omega)$ as a known function, whether measured or calculated, but does not assume any particular analytical form for g. This is the procedure developed by Nyquist[(2)], but while going just far enough into the argument to make it plausible, we shall not go to the considerable lengths of developing a rigorous proof, nor shall we multiply examples of its use.

268

The concept of *loop gain*, that plays a central role in Nyquist's theorem, may be understood from the circuit of fig. 1(b), in which ε, the e.m.f. across Z_1, is the input to the amplifier whose output is the source of i_2. Let us disconnect the input leads at AB, and with an external source supply an input ε to the amplifier. Then if R_{in} is large the output current $g\varepsilon$ is divided between Z_2 and Z_1+Z_3, so that the e.m.f. across Z_1 is $g\varepsilon Z_1Z_2/(Z_1+Z_2+Z_3)$. The loop gain, G, of the circuit is $gZ_1Z_2/(Z_1+Z_2+Z_3)$, being the ratio of output to input e.m.f. when the loop is open.† At one or more values of ω (in general complex) the loop gain is unity, and the loop can be closed without perturbing the currents and e.m.f.s anywhere in the circuit. These values of ω are the natural frequencies of the circuit with feedback; indeed, equating $gZ_1Z_2/(Z_1+Z_2+Z_3)$ to unity simply reproduces (1).

Our concern, however, is not with complex frequencies but with the trajectory of the loop gain on an Argand diagram as ω runs through all real values between $\pm\infty$. When, to take a specific example, Z_1 and Z_2 are resistances and Z_t $(=Z_1+Z_2+Z_3)$ is $\mathrm{j}\omega L+R+1/\mathrm{j}\omega C$, the impedance of the resonant circuit in fig. 1(a), the loop gain takes the form:

$$G(\omega)=gR_1R_2/(\mathrm{j}\omega L+R+1/\mathrm{j}\omega C).$$

If g is real and independent of ω, the trajectory of G is a circle with the points 0 and gR_1R_2/R at opposite ends of a diameter, as shown in fig. 3. At a certain value of g the trajectory passes through the point U at which G is real and equal to unity; it is then possible for the circuit to oscillate with real ω, that is to say it is neutrally poised between stability and instability. This condition is of course the same as that derived from (4), $g=R/R_1R_2$. For smaller g the circuit is stable, for larger it is unstable, and the onset of instability is marked by the trajectory expanding to enclose U. There are obviously many other feedback circuits with trajectories other than circular, especially if g is a complicated function of frequency, to

† Care must be taken in general that the open loop is terminated by the impedance which would lie ahead of it when it is closed. We have avoided the problem here by opening at a point of high impedance (the input to the amplifier).

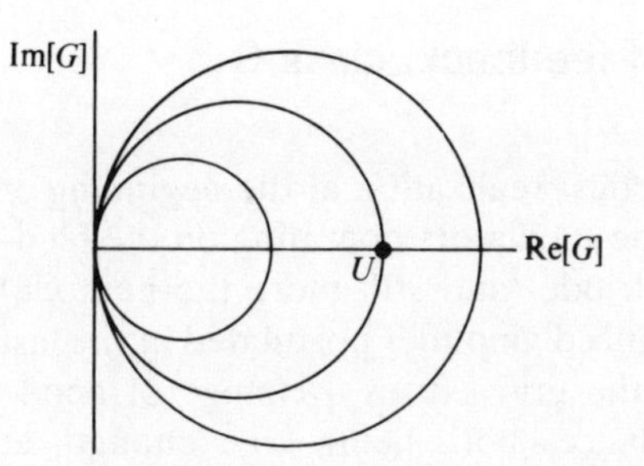

Fig. 3. Loop gain (Nyquist diagram) for circuit of fig. 1(a). In increasing order of g and radius of circle, the circuit is stable, neutral and unstable. U is the point, $G=1$.

which the same argument could be applied. For if the circuit is stable, as it should be, when g is very small and the trajectory of G confined to a region close to the origin, then if g is steadily increased so that the trajectory expands, a time will come when it passes through U and is in neutral equilibrium with respect to one mode of oscillation. A small increase in g will set it off into instability, and at the same time the point $G=1$ will have been enclosed within the trajectory. Although this does not constitute a proof of Nyquist's theorem, it does make it plausible that *a feedback circuit is stable if the trajectory of G does not encircle U.*

364

The argument also makes plausible the further statement that if the trajectory encircles U n times, there are n distinct modes of instability. For each increase in the value of n is achieved by the trajectory crossing U and allowing a new mode to pass through neutral equilibrium; though this can only be observed, of course, if the already existing unstable modes can be prevented from building up by the most careful choice of initial conditions. In the case of the circuit of fig. 1(a) the trajectory describes the circle twice as ω runs between $\pm\infty$, and this is consistent with there being two solutions of the equation of motion, Ω_1 and Ω_2; whether they are really different, seeing that $\Omega_2=-\Omega_1^*$, is a matter of definition, but if we allow ω to take negative as well as positive values, then in relating the number of encirclements of U to the number of unstable modes they must be counted separately.

In the practical application of Nyquist's theorem occasional difficulties arise, for example when the system is unstable in the absence of feedback but stabilized by the feedback itself. Again, it may happen that a system is stable when g is small, becomes unstable as g is increased, but then restabilizes at still higher g. In this case the first critical value of g marks the entry of U into the interior of the trajectory, and the second its re-emergence. Each special case serves to test, and ultimately to confirm, the great generality of the theorem.

At this point we leave the general and turn to particular examples of self-maintained oscillators, where more interest attaches to the steady oscillatory state than to diagnosis of the initial instability. It will be seen, indeed, that the latter is normally a trivial matter with the elementary examples we shall treat. The former may by contrast present challenging problems.

Negative resistance by means of feedback; class C oscillators

The idea summed up in (1) found practical realization at the beginning of electronic history, when the basic triode oscillators operating on the feedback principle were invented.[3] The triode (and still more the pentode) approximates rather closely to the idealized amplifier postulated in the last section, in that a voltage applied to the grid causes a change of anode current, the input and output impedances both being high enough, at moderate frequencies, for the analysis to give a satisfactory account of the observed behaviour. The quantity g is the *transconductance* of the triode, the change in output current due to unit change of grid voltage. Some typical primitive oscillator circuits are shown in fig. 4, from which various circuit components, inessential to understanding the mode of operation, have been eliminated. The capacitor C', resistor R' and choke L' in the Colpitts circuit are essential to the operation, in the sense that the grid and cathode cannot be allowed to float, but must have d.c. contact to the rest of the circuit to ensure that they are held at the correct mean potential. At the same time the oscillatory current must be channelled towards the right electrodes. Thus C' must be large enough to present no significant impedance at the oscillating frequency while R' and L' must be effective obstacles ($RC' \gg 1/\omega$ and $L'C \gg 1/\omega^2$). Then these components play little part in determining the character of the oscillation.

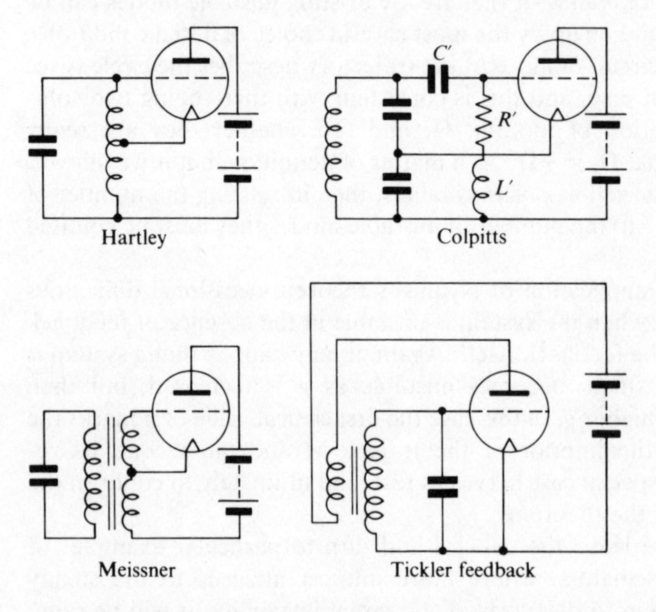

Fig. 4. Four early types of oscillator using triode amplifiers and feedback.

In the Hartley and Meissner circuits Z_1 and Z_2 are both inductive, so that gZ_1Z_2 is real and of order ω^2. In the Colpitts circuit gZ_1Z_2 is also real but of order ω^{-2}. In both cases the equations for ω are of third order; thus for the Meissner circuit the differential equation for a series resonant circuit must be supplemented by a term of the form $G\dddot{q}$, where $G = gM_1M_2$:

$$G\dddot{q} + L\ddot{q} + R\dot{q} + q/C = 0.$$

If we seek solutions of the form e^{pt}.

$$Gp^3 + Lp^2 + Rp + 1/C = 0. \tag{5}$$

A similar equation governs the Hartley circuit, but the Colpitts circuit has its new term at the other end,

$$Lp^3 + Rp^2 + p/C + G' = 0, \tag{6}$$

where $G' = g/C_1C_2$. In both (5) and (6), G and G' are positive. With all coefficients positive, and R small, the equations have one real and two complex roots as is easily verified by sketching the forms of (5) and (6). The real root is negative and therefore harmless, but if G or G' is large enough the other roots describe growing oscillations. The critical condition for spontaneous oscillation to start is that p shall be purely imaginary:

i.e. $$p = \pm \mathrm{j}\omega_0, \quad \text{where} \quad \omega_0^2 = 1/LC,$$

and $$G = R/\omega_0^2 \quad \text{or} \quad G' = R\omega_0^2.$$

For larger values of G or G' the effect of feedback is to introduce a negative resistance greater than R.

The Tickler feedback circuit differs from all these, in that the pick-up is from a capacitor and the feedback across an inductor; the pick-up e.m.f. is q/C and the feedback e.m.f. gMq/C. Either sign of feedback can be chosen (as also with the Meissner circuit), and for oscillation the feedback must counteract the resistor's action:

$$L\ddot{q} + (R - gM/C)\dot{q} + q/C = 0. \tag{7}$$

This is the simplest case of all those in fig. 4, feedback providing a frequency-independent resistive term of either sign and virtually unlimited magnitude. As with the other oscillators, at the critical feedback for initiating oscillation the frequency is exactly ω_0. With very strong feedback focal instability may give way to nodal.

256

In view of the simplicity of this oscillator we shall use it to illustrate the effect of saturation of the amplifier in limiting the amplitude of oscillation. When the amplitude is low the amplifier responds linearly, but the anode current, defined in terms of the rate of arrival of electrons, cannot be driven negative nor can it exceed the rate of emission from the cathode. A typical characteristic is shown in fig. 5, together with a convenient approximate form which is helpful in discussing saturation behaviour. The d.c. bias of the triode is irrelevant to the analysis; for treating oscillatory effects the origin is chosen to be the working point of the characteristic. We therefore write the anode current as a function of grid potential

$$i_a = i_a(v_g),$$

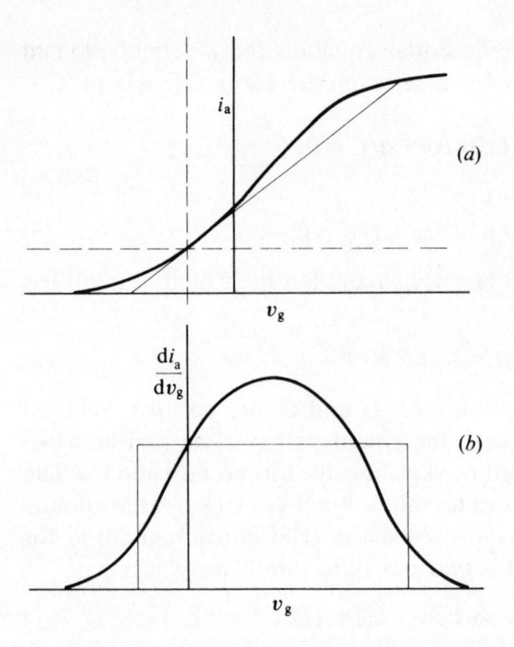

Fig. 5. (*a*) Schematic triode characteristic (heavy line) relating anode current, i_a, to grid voltage, v_g. The light line shows a simplified model characteristic, used in approximate analysis of the behaviour. The broken lines are the displaced axes with the operating point of the characteristic at the origin. (*b*) The derivative of the characteristics in (*a*), the top-hat being the approximation.

where $i_a(v_g) = gv_g$ when v_g is small. Instead of (7) the equation for q now has the form (since $v_g = q/C$)

$$LC\ddot{v}_g + RC\dot{v}_g + v_g = M\, di_a/dt = M\frac{di_a}{dv_g}\dot{v}_g. \qquad (8)$$

The form of di_a/dv_g is sketched in fig. 5(*b*), together with the 'top hat' which is the derivative of the approximate characteristic.

Assuming this approximate characteristic we can readily appreciate what happens. Once M has been increased above the critical value to initiate oscillation, there is nothing to prevent the oscillation growing until the amplitude of v_g extends to one limit of the amplifying range. It will continue to grow beyond this value, but the amplifier will not be operative over part of the cycle and the rate of growth will begin to diminish, until a steady state is reached in which the resistive dissipation and the power input from the amplifier are balanced. This is the *limit cycle* already introduced in general terms in chapter 9. If M is considerably greater than the critical value the final amplitude may reach far beyond the limits of the amplifying range. Then over much of the cycle $di_a/dv_g = 0$ and, as (8) shows, the circuit is in free oscillation. Provided its intrinsic Q is high, the crests and troughs of its waveform are almost pure sinusoids, very slightly modified by resistive loss. These sinusoidal sections are joined by sections lying within the amplifying range, and which may be parts of a growing sinusoid or even, if the gain is very high, exponential in form. But since these sections are only a minor part of the whole waveform, it hardly matters what their exact shape is – they are very nearly straight lines. Thus although the cycle is composed of sections which are mathematically distinct, to the eye of the observer there is little difference between the actual waveform and a pure sinusoid. During some of each cycle a small fraction of the energy is dissipated, and during the rest it is restored, and one may calculate the steady–state amplitude as that which allows the two effects to balance.

Alternatively one may calculate the effective amplification at the oscillating frequency and find at what amplitude it is just sufficient to neutralize the resistive effects. For this purpose we imagine a sinusoidal signal $v_{g0}\cos\omega t$ applied to the grid; the anode current is truncated by saturation and the induced e.m.f. therefore contains harmonics of ω which, however, we suppose to have a quite negligible influence on the current in the circuit, being so far removed from the resonant frequency. The effective transconductance, g_{eff}, on this assumption is determined by the amplitude of the fundamental component of the anode current.

If, for the purpose of illustration, the approximate characteristic is assumed symmetrical about the working point, the anode current is $gv_{g0}\cos\omega t$ so long as it is less than gv_{sat}, and gv_{sat} otherwise. The fundamental component in the Fourier transform is easily calculated and the details need not be given here:

$$g_{\text{eff}} = g\times\frac{2}{\pi}[\sin^{-1}(1/r)+(r^2-1)^{\frac{1}{2}}/r^2], \quad \text{if } r>1, \tag{9}$$

r being written for v_{g0}/v_{sat}. This function is shown in fig. 6; a more realistic characteristic would give a smooth curve but the limiting behaviour at large

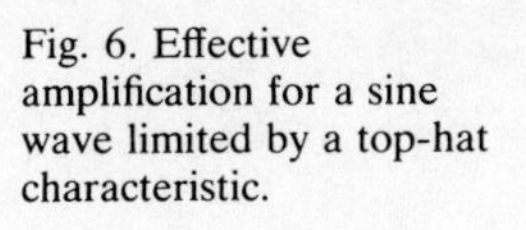

Fig. 6. Effective amplification for a sine wave limited by a top-hat characteristic.

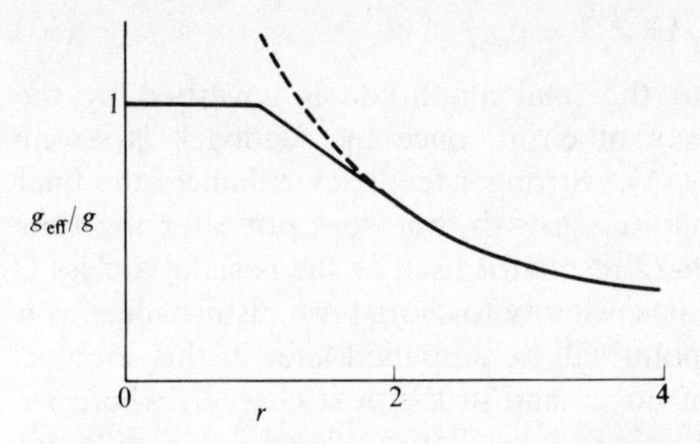

values of r would be the same, being the result for a square wave of amplitude $\pm gv_{\text{sat}}$, whose fundamental has amplitude $4gv_{\text{sat}}/\pi$;

$$g_{\text{eff}}\approx 4g/\pi r \quad \text{when } r\gg 1. \tag{10}$$

The oscillation builds up until, as follows from (7), $g_{\text{eff}} = RC/M$. Thus when there is strong enough feedback for (10) to apply,

$$v_{g0} = 4gMv_{\text{sat}}/\pi RC. \tag{11}$$

To reach the same conclusion by an energy argument, note that the feedback e.m.f. is $Mg\dot{v}_g$ while v_g runs between $\pm v_{\text{sat}}$, and that in this short interval the current in the resonant circuit is near its maximum i_0, which is equal to $\omega_0 v_{g0} C$. Each short burst of power introduces energy $Mgi_0 \int \dot{v}_g \,dt$, i.e. $2Mgi_0 v_{sat}$, and there are two bursts per cycle. The energy dissipated in the resistor during one cycle is $\pi R i_0/\omega_0$, equal to the energy fed in when (11) is satisfied. An oscillator excited in this way by short bursts of power is said to operate as a class C oscillator; the historical origin of this term, deriving from a classification of thermionic tube amplifiers, is irrelevant to our discussion. One merit of this type of oscillator is its intrinsic frequency stability, arising from minimal interference with the resonator circuit. The escapement of a pendulum clock makes use of the same principle, though the details are slightly different, as we shall discuss in the next section. 323

It is of interest to see how the oscillator approaches its final amplitude; when feedback is strong enough the energy argument may be extended without difficulty. It is enough to note that the dissipation by the resistor varies as the square of the amplitude while the power input varies as the first power. If y denotes the amplitude of oscillation of v_g at any instant, energy conservation determines the differential equation for y,

$$\frac{d}{dt}(y^2) = (v_{g0} y - y^2)/\tau_e. \tag{12}$$

The constants have been supplied by inspection; if there is no feedback the first term on the right vanishes, and τ_e is the time-constant for energy dissipation in the free vibrator, while with feedback the stable amplitude, v_{g0} in (11), must make the right-hand side vanish. The solution of (12) has the form

$$y = v_{g0} - A\,e^{-\frac{1}{2}t/\tau_e} = v_{g0} - A\,e^{-t/\tau_a} \tag{13}$$

showing that the approach to the final amplitude is governed by the time-constant of the free resonant ciruit, once the feedback is strong enough for v_{g0} to be given by (11).† Stronger feedback enhances the final amplitude and speeds up the initial growth, but does not alter the time scale of the last stages. A high-Q resonator used as the basis of a class C oscillator has the same sort of insensitivity to short-term disturbances as it has when running free. This point will be amplified later in this chapter, when we discuss the effects of noise, and in the next chapter, where we shall discuss how readily an injected oscillatory signal, at a different frequency, can entrain the oscillator, i.e. pull it into synchronism. 361 392

Although we have described the behaviour of a triode oscillator, it is more convenient for the purpose of checking the theory experimentally to use a transistor circuit or, better still, an integrated circuit designed as an

† An alternative approach is to note that twice per cycle the anode current switches rapidly between $\pm g v_{sat}$, producing in the resonant circuit an impulsive e.m.f. whose strength, $2M\dot{g}v_{sat}$, is independent of the oscillatory amplitude. The circuit responds like a free vibrator driven by a succession of constant impulses. This is only true once class C operation is fully developed and the impulses are much shorter than the period.

operational amplifier.[4] This can be treated as a 'black box' with certain characteristics, and for the low frequency application here the d.c. characteristics are sufficient. With the amplifier represented in fig. 7(*a*) the output voltage V_{out} is determined by the difference between the voltages of the two inputs, $V_{out}=f(V_+ - V_-)$. The input impedance is very high, and the output impedance very low, any current required to maintain V_{out} in the presence of a load being supplied through the earth lead (which is usually

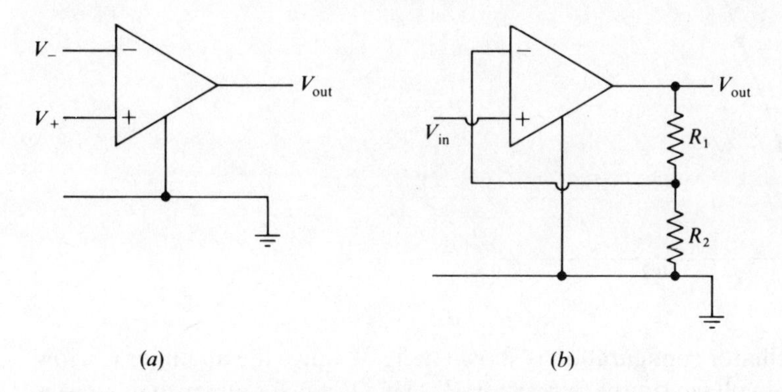

Fig. 7 Operational amplifier; (*a*) the basic device in schematic form; (*b*) with feedback to limit and stabilize the amplification.

omitted in circuit diagrams). So long as V_{out} is not too close to the supply voltage, between ± 10 V, say, the amplifier is linear, with extremely high gain, perhaps 10^5. Thus when $|V_+ - V_-| < 10^{-4}$ V it is linear, but for higher imput differentials it saturates. The high gain is normally not required, but it enables feedback to be introduced to lower the gain to a convenient level, which is then extremely well-defined, for the same reason as was explained in connection with the galvanometer amplifier. 310 In the present case it is clear from fig. 7(*b*) that the resistor chain fixes V_- in relation to V_{out}:

$$V_- = V_{out}/\beta, \quad \text{where } \beta = (R_1 + R_2)/R_2.$$

In the linear range of response, $V_{out} = A(V_+ - V_-)$, where $A \sim 10^5$, and these two equations show that

$$V_- = V_+/(\beta/A + 1)$$

and
$$V_{out} = \beta V_+/(\beta/A + 1) \doteqdot \beta V_+ \quad \text{if } \beta \ll A.$$

Provided the value chosen for β is much less than 10^5, the gain is β and is determined by the resistors, being extremely insensitive to the value of A; amplification is thus linear until saturation reduces the effective value of A to something approaching β. The range of V_+ is now of the order of $10/\beta$ V, much larger than the range before feedback is introduced, since feedback causes V_- to follow V_+ almost exactly. A measured characteristic is shown in fig. 8, with a slope of 50.5 instead of the expected 51, and a knee so sharp as to give confidence in the comparison of experimental behaviour with the theory developed above on the basis of the approximate characteristic in fig. 5.

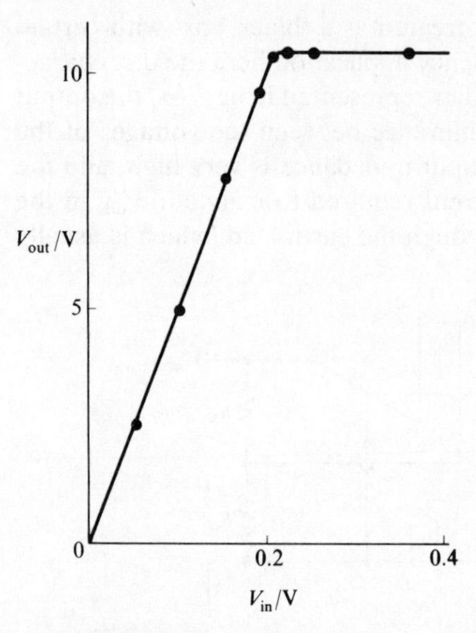

Fig. 8. Measured characteristic of the circuit of fig. 7(*b*), with $R_1 = 5000\ \Omega$, $R_2 = 100\ \Omega$.

The oscillator configuration is shown in fig. 9. Since the amplifier is a low impedance voltage source, a resistor R_3 ($10^4\ \Omega$, much greater than ωL_2) is inserted to ensure that the current in L_2 is proportional to the p.d. across the capacitor of the resonant circuit; β/R_3 plays the part of g in (7), and the negative resistance is $\beta M/R_3C$, i.e. $(R_1+R_2)M/R_2R_3C$, if $M/L_1 \ll \beta$. To eliminate non-linear complications in ferrite cores the inductors L_1 and L_2

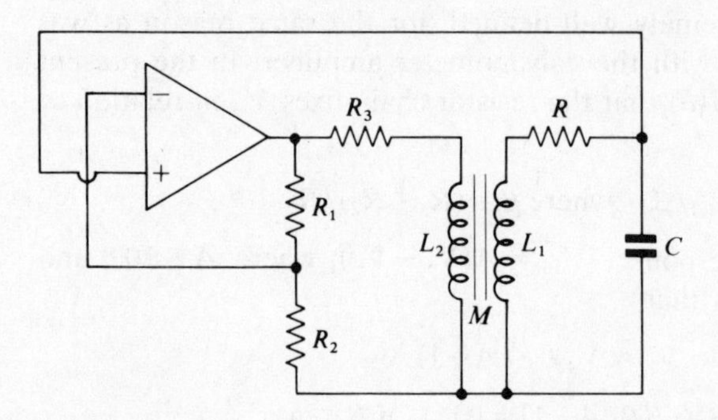

Fig. 9. Tickler feedback oscillator (see fig. 4) used to test theory.

were plain copper coils originally made as windings for an electromagnet. They were large and ungainly, but had the merit of high enough inductance ($L_1 = 69$ mH) combined with low resistance ($5.2\ \Omega$ at the operating frequency) to give a circuit of good Q (48) at so low a frequency (581 Hz) that the amplifier responded almost precisely in accordance with its d.c. characteristic. The two coils were spaced to give a value of M that permitted oscillations to begin when g was $3.78 \times 10^{-4}\ \Omega^{-1}$; if the theory is correct this means that $M = 15$ mH. The value of g was then increased to $50 \times 10^{-4}\ \Omega^{-1}$ and the spontaneous build-up of oscillations recorded, as shown in fig. 10(*a*), which is the p.d. across the capacitor. Only the first few cycles after switching on fall entirely on the linear amplifier characteristic and are therefore suitable for determining the rate of exponential growth. To compare experiment with theory note that unless Q is very small indeed (see (2.27)) the relative change in amplitude per cycle is a linear function of resistance, and hence of g. A change of g from 0 to 3.8×10^{-4} reduced the decrement from its free value to zero; the addition of 12 times as much should therefore give rise to as much increase in one cycle as the free oscillation loses in 12, a factor of 2.2 according to the trace of the decrement in fig. 10(b). The trace in fig. 10(a) is not very suitable for determining the increase in the first few cycles; it appears to be nearer 2.0 per cycle than 2.2, but a more careful study would be needed to see if this small discrepancy is real. Certainly there is no major disagreement here.

32

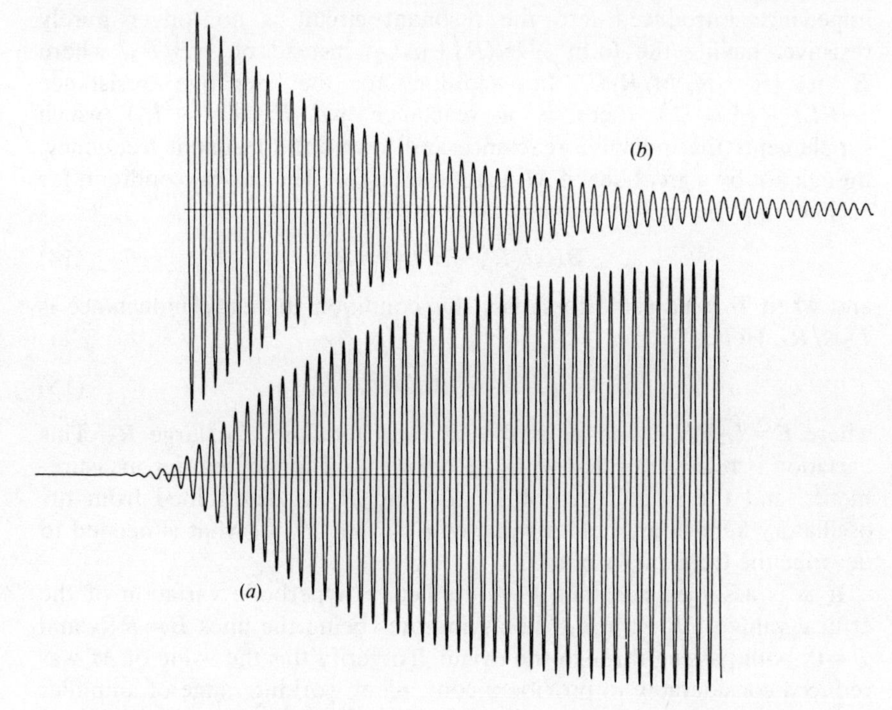

Fig. 10. (a) Record of the build-up of oscillation in the circuit of fig. 9 from the moment of connecting the amplifier in circuit; (b) free decrement of the resonant circuit with amplifier disconnected. The curves have been disposed so as to show the same time-constant governing the decrement in (b) and the approach to the steady state of oscillation in (a).

Turning to the behaviour near saturation, we see immediately from fig. 10 that the time-constant governing the approach to the maximum amplitude matches very closely that for free decay – the curves can be aligned so as to make their envelopes run parallel. This confirms (13), and a further simple check is provided by the amplitude in the steady state. The voltage amplitude across the capacitor is more than 3 V, at least 15 times what is needed to saturate the amplifier, whose output voltage therefore has an almost square waveform as it switches rapidly between the positive and negative limits. At the same time the oscillation in the resonant circuit is an almost perfect sinsusoid; this is true class C operation. The amplitude of the fundamental component of the output is $4/\pi$ times the amplitude of the square waves. Measurement on an oscilloscope showed the square wave to have an amplitude 3.1 times that of the sine-wave input, so that the effective value of β at the fundamental frequency was 3.95, i.e. $g_{\text{eff}} = 3.95\times10^{-4}\,\Omega^{-1}$. Since this maintains the oscillations at constant amplitude it should be the same as what is needed at a low level when the amplifier is linear, and this is the value $3.78\times10^{-4}\,\Omega^{-1}$ already quoted. Once more agreement is close enough to show that the theory is at least a very good approximation. We may be satisfied in this case that the oscillator is in essence a simple device.

Further confirmation is provided by the behaviour when the resistance R_3 is decreased to the point where the combination of R_3 and L_2 has a substantial reactive component. In the linear range of amplification the impedance introduced into the resonant circuit is no longer purely resistive, having the form $-B/(R_3+\mathrm{j}\omega L_2)$ instead of $-B/R_3$, where B is $(R_1+R_2)M/R_2C$. In addition to the negative resistance $-BR_3/(R_3^2+\omega^2L_2^2)$ there is a reactance $\mathrm{j}\omega L_2B/(R_3^2+\omega^2L_2^2)$ which supplements the inductive reactance and lowers the resonant frequency, though not by a great deal if M is not too large.† The critical condition for oscillation to be maintained now takes the form

$$BR_3/(R_3^2+\omega^2L_2^2)=R, \tag{14}$$

and when B is adjusted to satisfy this condition the extra inductance is L_2R/R_3. Hence

$$\omega_0^2/\omega^2=1+E/R_3, \tag{15}$$

where $E=L_2R/L_1$, and ω_0 is the limiting frequency for large R_3. This variation is readily verified with the circuit used for the earlier measurements, and the magnitudes of L_1, L_2 and R, as determined from the oscillatory behaviour, yield a value of E within 1% of what is needed to describe the frequency variation.

If ω is assumed constant, (14) predicts a hyperbolic variation of the critical value of B with R_3, the asymptotes being the lines $B=RR_3$ and $B=0$, both passing through the origin. To verify this the value of M was reduced considerably to provide a convenient working range of amplifier gain, and the curve shown in fig. 11 was obtained, having the expected form. In particular the diameter D, constructed graphically, passes very close to the origin, as predicted. The oscillatory regime lies within the hyperbola; as R_3 is reduced from a large value, the gain being kept constant, oscillation starts at one critical value and dies out again at another. The frequency varies with R_3, as is indicated by the values attached to the critical curve, which agree well with (15).

† By contrast, when R_3 is large and the impedance introduced by feedback is almost purely resistive, the frequency is (as expected) sensibly unaffected by the value of M, and does not alter by more than 0.1% even when the oscillator is in the extreme class C condition.

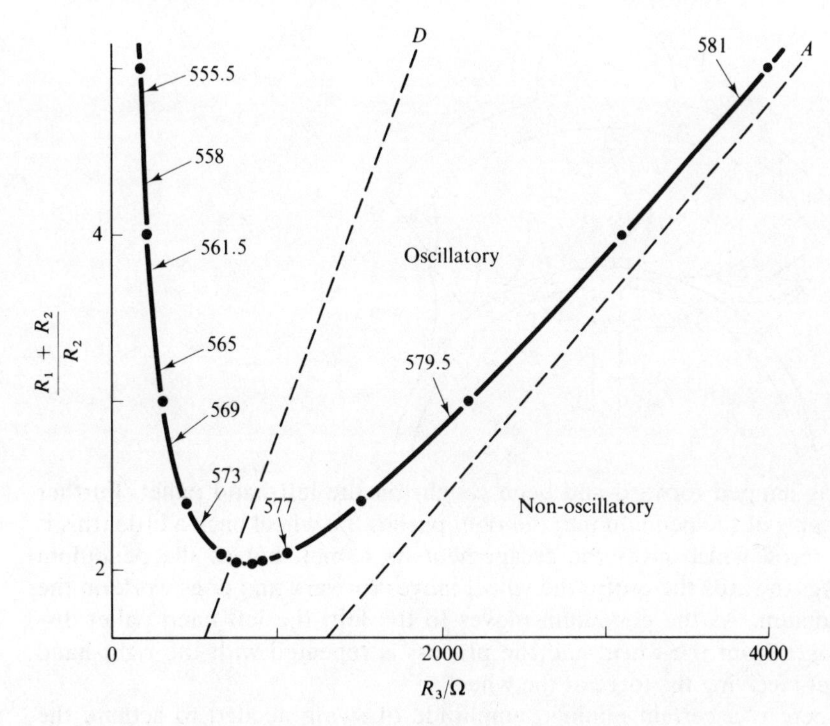

Fig. 11. The oscillator of fig. 9 with R_3 reduced to provide a significant inductive contribution to the feedback. The boundary between oscillatory and non-oscillatory conditions is hyperbolic, with the vertical axis and A as asymptotes, and the frequency varies along the critical curve as indicated by the arrowed values.

The pendulum clock, a non-linear class C oscillator

The traditional long-case pendulum clock has the typical characteristic of the class C oscillator in that its resonant element, the pendulum, is only stimulated during a small fraction of each vibration, more or less at the phases of maximum velocity so as to perturb the natural frequency as little as possible. In more recent designs (e.g. the Synchronome clock)[5] the stimulus is much more infrequent, perhaps twice a minute, and the time-keeping correspondingly dissociated from disturbing influences, apart from those affecting the pendulum's period directly. Here we shall discuss only the oldest type of escapement still in common use, the recoil escapement of Hooke, such as is illustrated in fig. 12. The eighteenth-century long-case clock in the author's possession has an escapement very much like this, and provides experimental data for comparison with the theory which we now present.

The swing of the pendulum rocks the pallets to and fro while the escapement wheel, which is gently impelled in a clockwise sense by the clock weights, through a gear train, keeps a light pressure on one or other of the pallets.† As the diagram is drawn, the pendulum has just swung far enough to the right to release the wheel from the right-hand pallet, so that

† The details of the suspension of the pendulum and the crutch which transmits the motion to the pallets are not essential for our discussion, though their correct design is important for the working of the clock.

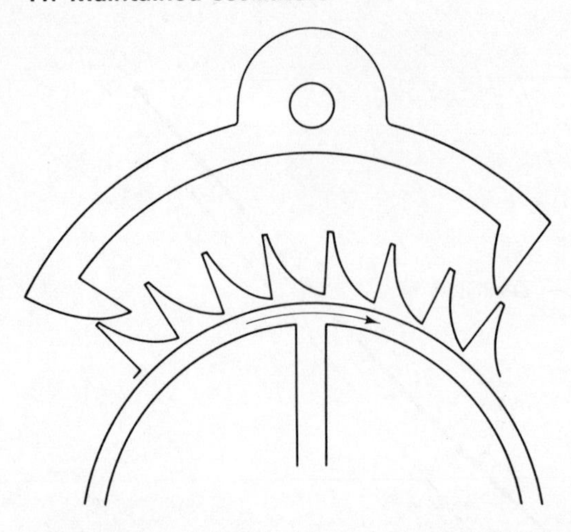

Fig. 12. The recoil escapement.[6]

it has jumped forward and been caught on the left-hand pallet. Further swinging of the pendulum to the right pushes the wheel back a little (this is the recoil which gives the escapement its name), but as the pendulum swings towards the centre the wheel moves forward and does work on the pendulum. As the pendulum moves to the left, the left-hand pallet disengages from the wheel and the process is repeated with the right-hand pallet receiving the force of the wheel.

There is a certain minimal amplitude of swing needed to actuate the escapement, but once this is exceeded the mechanism is such that, if friction could be neglected, the same amount of energy would be delivered during each swing, irrespective of the amplitude of the swing. For there is a certain angle at which one pallet disengages and the other is brought to bear, and another angle at which the process is reversed. Between these two instants in the swing the wheel has a well-defined net forward movement; with a larger swing, all that happens is that the wheel recoils further back, but the larger amount of work done on it by the pendulum is recovered on the return swing. The net work is just fixed by the forward movement of the wheel while in contact with the pallet. Indeed, if the escapement had no 'drop' – if it were so perfectly made that there was no movement of the wheel as the escapement switched over – the rate at which work was done by the wheel on the pendulum would be simply the product of the tangential force exerted by the wheel on the pallet and the mean tangential velocity of the wheel.

In detail, then, this maintained oscillator differs considerably in its mode of action from that in fig. 9, requiring a certain amplitude to get the process started, and thereafter being automatically self-limiting since the energy delivered per cycle does not increase with amplitude, while the dissipation does. A further difference lies in the dissipation being mostly due to friction at the escapement, so that effectively the motion of the pendulum is resisted by a constant force, not one that is proportional to its velocity. This reveals itself in the behaviour at an amplitude too small to actuate the escapement; the vibrations decay linearly, not exponentially (cf. fig. 2.22). **36**

It is reasonable, therefore, to consider the escapement-driven pendulum as an oscillatory system in which the energy communicated per swing is constant and the energy dissipated per swing is proportional to the amplitude, a. And since the stored energy is proportional to a^2, the differential equation for the variation of amplitude with the number of swings takes the form

$$a\,\mathrm{d}a/\mathrm{d}n = \varepsilon - \lambda a, \tag{16}$$

in which ε represents the energy delivered by the escapement and λa the frictional loss. The general solution has the form

$$n = -a/\lambda - (\varepsilon/\lambda^2)\ln(\varepsilon - \lambda a) + \text{const.} \tag{17}$$

By choosing the constant of integration to be $(\varepsilon/\lambda^2)\ln\varepsilon$, we elect to count n from the (hypothetical) starting point when $a = 0$. The limiting value of a, as n goes to infinity, is ε/λ to make the argument of the logarithm vanish. With f defined as $a\lambda/\varepsilon$, i.e. the amplitude measured as a fraction of its limiting amplitude, and N as ε/λ^2, (17) is converted to the form

$$n/N = -\ln(1-f) - f. \tag{18}$$

This function is plotted in fig. 13, but it should be remembered that the initial steep rise is not observable, since the escapement does not work at small amplitudes. The experimental points were taken with the clock weight increased to about half as much again in order to make the final amplitude considerably larger than was intended by the maker and so provide a larger range for observation. No great sophistication was attempted, the amplitude being read by means of a pointer on the pendulum swinging past a paper scale.

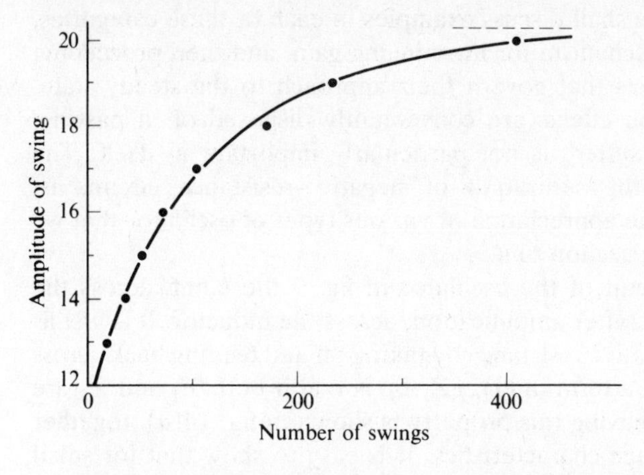

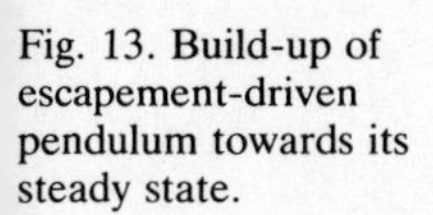
Fig. 13. Build-up of escapement-driven pendulum towards its steady state.

To fit the experimental points to the curve it was necessary to take N to be 150 and the limiting value of a to be 20.3 on the paper scale. From these values λ is found to be 0.136. If the same value applies also at low amplitudes when the escapement is inoperative and the equation of motion is like (16) but with ε put equal to zero, we expect the linear decay of amplitude to involve a reduction of 0.136 per swing. This is somewhat less than the observed value of 0.160, but near enough to indicate that the essence of the behaviour of the escapement is described by the simple model embodied in (16). One should not in such a rough experiment look for too close an agreement; apart from the crudities of observation the escapement is assumed in the theory to be free from eccentricity, something that is not very probable after a long working life. Any eccentricity in the pivot of the escapement wheel will cause the energy fed to the pendulum to vary periodically with a period of one minute, which is the rotation period of the wheel; this may well affect the details of the observations.

Further examples of negative-resistance oscillators

(a) A resistive feedback network. We have analysed a class C oscillator in some detail to show both the origin of its negative resistance and the final state of oscillation. The present section exhibits a varied selection of arrangements whose oscillatory instability may be attributed to negative resistance or an analogous property, but whose final states may differ considerably among themselves. It will be appreciated that these are only a few out of a vast range of choices, and that the selection is to a considerable degree a matter of personal whim. They have in common that they are based on resonant systems so that, as with the class C oscillator, the condition for oscillations to build up is that the loss mechanism must by some means be more than compensated; in a circuit the resistance must be made negative, in a mechanical vibrator the viscous drag must be reversed, in a electromagnetic cavity oscillator the dielectric loss must be replaced by dielectric gain, etc. We shall discuss examples in each of these categories, first elucidating the mechanism for introducing gain, and then proceeding to the non-linear effects that govern their approach to the steady state. Some of the saturation effects are conveniently disposed of in passing, where the oscillatory system is not particularly important in itself. The general question of the saturation of negative-resistance circuits is, however, so basic to the appreciation of various types of oscillator that we shall devote a separate section to it.

In the feedback circuit of the oscillator in fig. 9 the e.m.f. across the capacitor was fed back, after amplification, across the inductor. It is just as easy to construct a negative resistance by taking off and feeding back across resistors, since the extra term in (1), gZ_1Z_2, is real if both Z_1 and Z_2 are real. A simple circuit having this property is shown in fig. 14(a), together with two current–voltage characteristics. It is easy to show that for small currents, when the amplifier is operating in its linear range, the resistance between A and B is $-R_1R_3/R_2$. What happens after saturation depends on the values of the circuit components. If they have high resistance, so that the amplifier is not called upon to supply excessive current, the

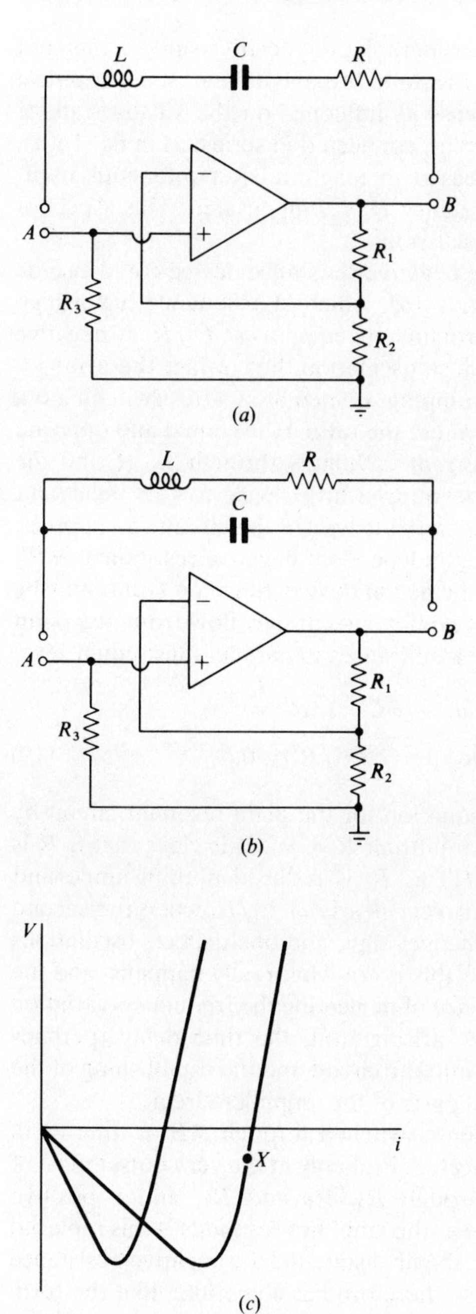

Fig. 14. Negative resistance device based on operational amplifier: (*a*) connected in series with resonant circuit; (*b*) connected in parallel; (*c*) two typical characteristics for the device.

potential at B saturates and thereafter any further current alters the potential of A relative to earth but leaves that of B unchanged. The gradient of the characteristic then reflects the value of R_3. With lower resistances in the circuit the current needed to saturate the amplifier is enough to modify its behaviour considerably; the positive slope is then not related solely to the values of the resistors. We may take it as an empirical parameter when we come to discuss its influence on the steady state of oscillation. With the oscillatory circuit connected in series, as in fig. 14(a), and the negative resistance increased in magnitude, spontaneous oscillation sets in when the circuit resistance R is compensated. This aspect of the behaviour needs no further discussion.

364

One might expect that the same negative resistance device could also be used in the parallel circuit of fig. 14(b), which at resonance has a high resistance between A and B, approximately equal to $\omega^2 L^2/R$. A negative resistance less than this should cause oscillation, but in fact the arrangement is always unstable, usually jumping, immediately after switching on, to a point such as X in fig. 14(c), where the ratio V/i is equal and opposite to the resistance R; a steady current circulates through L, R and the negative resistance. It is worth devoting a little space to this behaviour which is not predicted by an elementary analysis of the circuit. To appreciate this point, consider a frequency-independent negative resistance, $-R'$, connected across AB, instead of the actual device. Since an e.m.f. may be established between A and B, but no net current may flow from one point to the other, the sum of the three admittances in parallel must equal zero:

$$1/(R+\mathrm{j}\omega L)+\mathrm{j}\omega C-1/R'=0;$$

i.e.
$$LC\omega^2-\mathrm{j}(CR-L/R')\omega-(1-R/R')=0. \tag{19}$$

This equation is reduced to the equation for the plain resonant circuit by removing the two terms in R' (i.e. putting $R'=\infty$). It is clear that if R is small ($Q \gg 1$ in the absence of R') as R' is reduced nothing important happens until $R'<L/RC$, which is very nearly $\omega^2 L^2/R$, when the second term representing the damping changes sign, and one expects oscillations to grow, as already remarked. But this is not what really happens, and the mistaken prediction is a consequence of neglecting the frequency-variation of the negative resistance device, arising from the time delay (perhaps 10^{-4} s) between injecting current into the circuit and the establishing of the steady-state voltage at all internal parts of the amplifier circuit.

Imagine that a current is suddenly switched through AB at time $t=0$, the resonant circuit being disconnected. Probably at the very outset most of the current follows the path through R_3, R_2 and R_1, and a positive potential difference is set up. Only as the amplifier responds is this replaced by a negative potential difference, the final state for the negative-resistance circuit. The step-function response therefore has something like the form shown in fig. 15(a) – the exact shape is of no great consequence, only the crossing of the axis. Correspondingly the response to a short impulsive current injected at time $t=0$ is the derivative of this, having a positive instantaneous δ-function response, followed by a negative delayed response. Because the asymptote of fig. 15(a) is negative, the integral of the negative part of fig. 15(b) is greater than the δ-function. From the

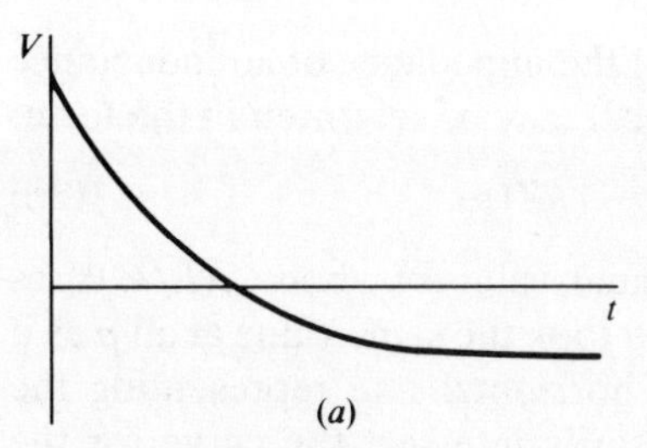

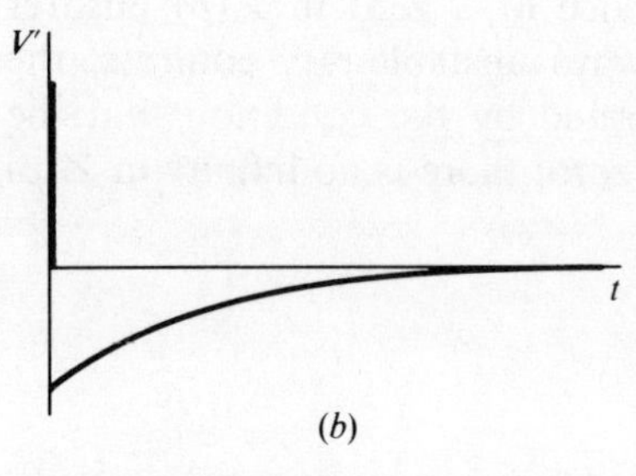

Fig. 15. (*a*) Schematic diagram of step-function voltage response of the device in fig. 14 when a steady current is switched into it at time $t = 0$; (*b*) the corresponding impulse response function, being the derivative of (*a*).

voltage response to a current impulse the impedance at frequency ω may be derived as the Fourier transform. It is more convenient here, however, since we are concerned to demonstrate exponential instability, to consider the response of the circuit to a growing exponential current of the form $i_0 \, e^{pt}$, p being real and positive. Since the voltage at any time t is the resultant of the responses to earlier current impulses, the impedance $Z(p)$ is seen to be derived from $V'(t)$ by forming the integral (*Laplace transform*):

$$Z(p) = \int_0^\infty V'(t) \, e^{-pt} \, dt.$$

Instead of being constant and negative for all p, as would be the consequence of neglecting the transient properties of the amplifier, $Z(p)$ is negative only when p is small enough for the exponentially reduced contribution of the negative tail of $V'(t)$ at large t to outweigh the positive δ-function. As p increases, $Z(p)$ goes through zero and is thenceforward positive, as in fig. 16.

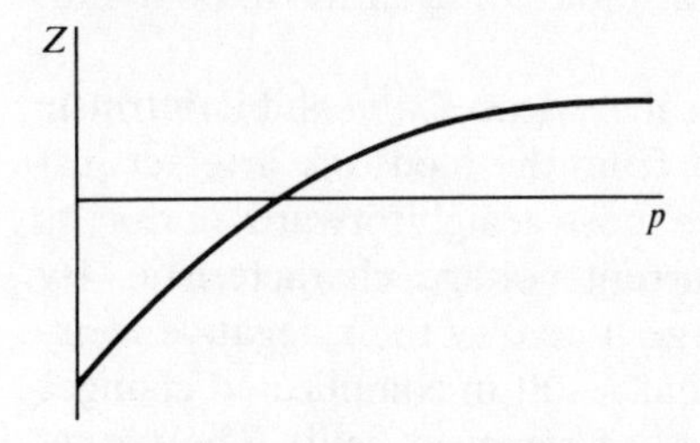

Fig. 16. Impedance of negative-resistance device as a function of *p*.

Let us apply this result to the parallel resonant circuit of fig. 14(*b*). If it is to be exponentially unstable there must be a positive value of p at which the admittances total zero. When p is real the impedance of an inductance is pL and of a capacitance $1/pC$. Hence (19) may be rewritten in the form:

$$1/(R+pL)+pC=-1/Z(p). \tag{20}$$

The left-hand side of (20) takes a minimal value of about $2(L/C)^{\frac{1}{2}}$, as shown in fig. 17. If $Z(p)$ were negative and took the same value at all p as it does when $p=0$, the resulting positive horizontal line representing the right-hand side of (20) would not necessarily intersect the curve for the left-hand side, but in practice the presence of a zero in $Z(p)$ ensures intersection. The circuit is therefore always unstable. By contrast, the series arrangement of fig. 14(a) is governed by the condition that the impedances, not the admittances, sum to zero; there is no infinity in $Z(p)$ to guarantee trouble.

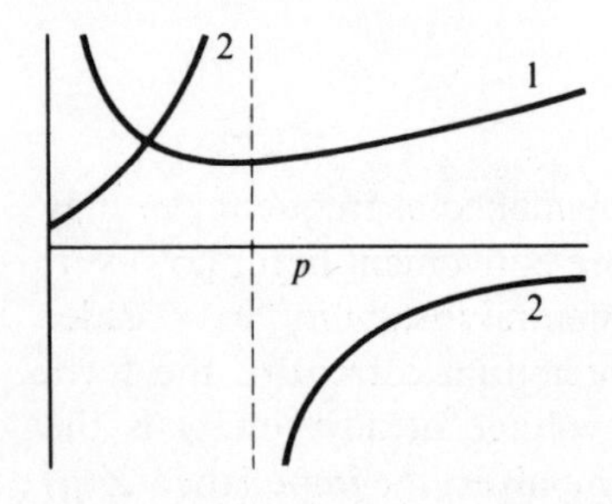

Fig. 17. Graphical solution of (20); curve 1 shows the variation with p of the left-hand side, and curve 2 the variation of the right-hand side.

It might be asked why the foregoing argument has been developed on the basis of the voltage-response to a step-function current, rather than vice versa – the current-response to a step-function voltage. The answer is simple – the circuit which enables the former to be determined is stable, but for the latter it is unstable, and the current-response function is not definable. To generate a constant current whose magnitude is unaffected by the behaviour of the load demands a high voltage source with a high resistance in series. So long as this resistance overwhelms the negative resistance of the load under test the current flowing is uniquely specified. But to generate a constant voltage across the load demands a very small series resistance, and the negative resistance of the load then causes the current to grow until non-linearity allows a stable configuation to be found.

(b) The tunnel diode[7]. Of the physical devices that exhibit intrinsic negative resistance behaviour, as distinct from the feedback artefact just discussed, the tunnel diode is probably the most straightforward, in that its behaviour is well represented by its current–voltage characteristic. By contrast, the Gunn diode and the discharge tube owe their negative resistance to spatially extended phenomena that result in complicated changes in internal structure (e.g. moving domains of reversed field). The tunnel diode shows none of these complexities. Unfortunately its behaviour is properly understandable only in the light of a basic course in solid-state

physics, and it would be out of place here to attempt an adequate explanation. The reader is referred to one of the texts cited in the references, which provide justification for the highly abbreviated and over-simplified description given in the square brackets that follow.

[In a semiconductor like germanium, electrons can only move freely if their energy lies within one of a series of bands which are separated by *energy gaps*, bands of forbidden energies. Moreover each band can only accommodate a definite number of electrons, and in pure germanium the number present is exactly sufficient to fill all bands up to what is called the *valence band*, leaving the next higher *conduction band* empty. In a filled band there are as many electrons moving in any one direction as in the opposite direction, and the band as a whole cannot be made to carry current. The only reason why pure germanium conducts at all is that a few electrons are thermally excited into the conduction band. If one dissolves in the germanium enough ($\sim 0.1\%$) *donor* impurity (Sb, As), of higher valency than Ge, the extra electrons occupy the lowest levels of the valence band, as shown schematically in fig. 18. If, on the other hand, an *acceptor* impurity (Ga, In) of lower valency is added, the valence band is depleted.

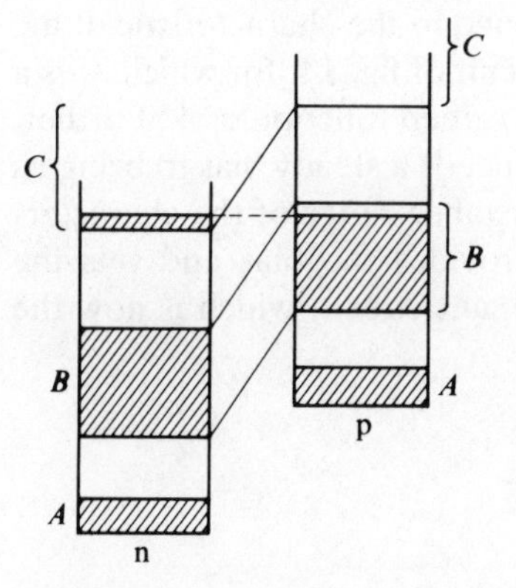

Fig. 18. Allowed and forbidden energy bands in heavily doped n- and p-type semiconductors. When electrons can be exchanged between them, through the tunnel barrier, the allowed bands fill to the same level, shown by a broken line; the higher electron concentration in n demands that some overflow into allowed band *C*, while in p some states in allowed band *B* remain empty.

When a thin wafer of Ge has donors diffused into it from one side, acceptors from the other, a very narrow junction between the two types of material can be formed, through which electrons may pass (though with difficulty) by the process of quantum-mechanical tunnelling. The junction region can support a moderate potential difference and, left to itself, the wafer settles down in equilibrium with such a potential difference as will bring the highest filled states to the same total energy, as shown in the diagram. If an external p.d. is now applied so as to change the field in the junction layer and raise the electron energy in the n-material relative to that in the p-material, electrons will tunnel through the layer from n to p, finding unoccupied states of the same energy into which they can go. A simple ohmic current results from this process. But as the external p.d. is increased further, a state of affairs is reached when the occupied part of the conduction band in n is opposite the forbidden band in p; when this

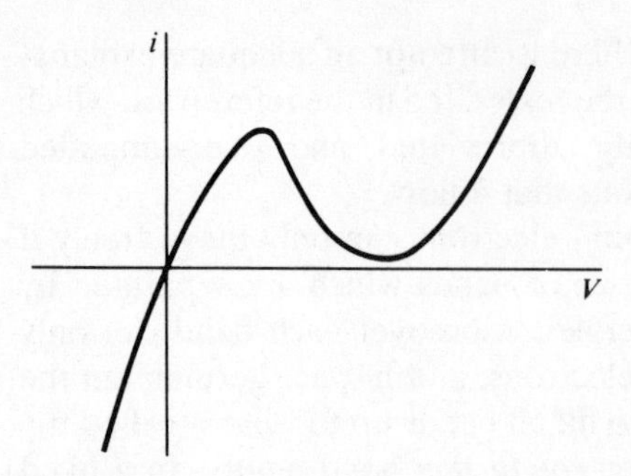

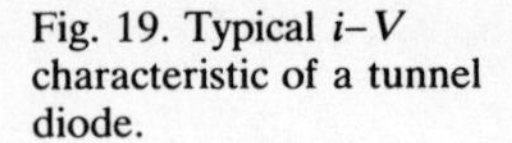
Fig. 19. Typical i–V characteristic of a tunnel diode.

happens tunnelling can no longer occur, for there is no adequate mechanism in the junction layer for electrons to lose energy as they tunnel and thus fall into the valence band of p. This is the reason for the drop of current as the p.d. is increased, as shown in the characteristic of fig. 19. The fact that it does not fall to zero, and increases again as the p.d. is further raised is due to the presence of thermally excited electrons which can pass between conduction bands on both sides.]

To understand the behaviour of the tunnel diode as a circuit element, almost all the required information is contained in the characteristic of fig. 19. In contrast to the negative-resistance circuit of fig. 14, for which V is a single-valued function of i, here i is a single-valued function of V. Further, though this is only a minor detail, the diode needs a steady bias to bring its mean operating point onto the negative resistance range of the characteristic. The arrangement shown in fig. 20 provides the bias and sets the negative resistance in parallel with the resonant circuit, which is now the

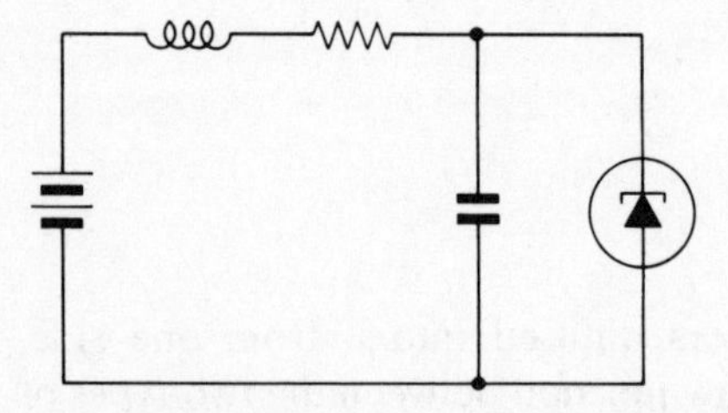

Fig. 20. Tunnel diode biased through a resistor to bring the operating point into the negative resistance range. The lead inductance and the intrinsic capacitance of the diode are shown explicitly.

appropriate configuration for exciting oscillation without incurring intrinsic instability of the circuit. In view of the preceding discussion this must not be taken as obvious, but demands an examination of the response-function of the diode. It is important to recognize that the diode has an intrinsic capacitance; the potential drop occurs across a very thin layer, perhaps 10^{-7} m thick, through which the electrons tunnel as indicated above, and this layer behaves as a parallel-plate capacitor with a negative leakage, i.e. a capacitor and negative resistance in parallel. In addition the leads connecting the diode to other parts of the circuit have a certain inductance, so that the equivalent circuit for small departures from the operating point is as shown in fig. 21. The cell is not shown, since it serves only to

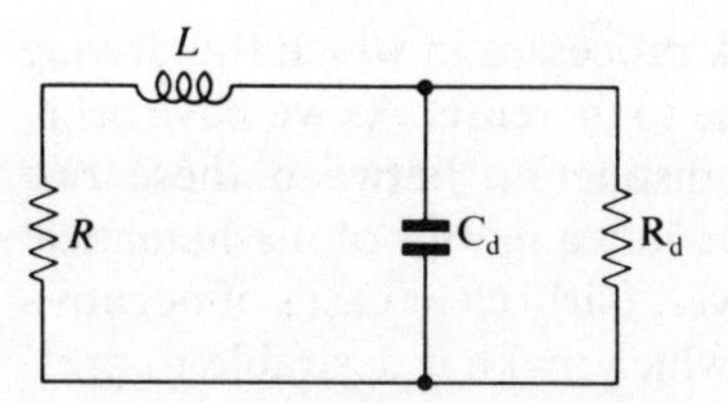

Fig. 21. Equivalent circuit of tunnel diode for small displacements from its operating point.

fix the operating point, where the value of dV/di is represented by R_d; the resistance R represents the resistance of the cell and other portions of the external circuit, while L is the lead inductance. This is the same circuit as is described by (19), and it is stable in the frequency-independent approximation provided that R_d, if negative, has a magnitude greater than R, and that $|L/R_d| < CR$. Keeping the lead inductance small obviates the spontaneous oscillations which will occur if the latter condition fails. The meaning of the former condition is seen by putting $L=0$; the circuit consists then of a capacitor across which are connected two resistors in parallel, one positive and one negative; the time-constant of the combination is dominated by whichever is smaller, and for stability the positive resistance must dominate.

There appears to be no bar, therefore, to setting up a low-resistance, low-inductance circuit to apply a step-function voltage to the diode and determine its current-response, which has no tendency to diverge as time goes on. Initially, when the biasing voltage is increased by a very small amount to provide the step-function excitation, a positive incremental current flows to change the charge on the capacitor, but this current decreases and eventually the negative resistance takes over so that the final incremental current is negative. The response is similar in form, therefore, to fig. 15(a) but for i rather than V. Correspondingly it is the admittance $A(p)$ that is well-behaved and remains finite for all real positive p, and there is no danger of exponential instability when the diode is connected across a parallel resonant circuit.

364

The tunnel diode can be used to maintain low-frequency oscillations, but it is a much more delicate device than a transistor amplifier and its real value lies in its ability to work up to much higher frequencies than the latter. In recent years, however, it has been superseded for this purpose by other devices, especially the Gunn diode, whose mode of operation is too complicated for it to be described as a simple negative resistance element, even if ultimately that is the significant feature of its behaviour.

(c) Mechanical analogues of negative and non-linear resistance; Liénard's construction. The phenomenon of whirling discussed in chapter 8 provided an example, quite different in kind from the circuits treated in the earlier sections, of a resonant system in which the damping term had had its sign reversed. Mechanical examples as simple as this are not encountered very frequently, most spontaneous mechanical vibrations

212

being more readily interpreted as feedback processes in which the driving force is so phased as to cause the amplitude to increase. As we have seen, particularly with class C oscillators, the distinction between these two interpretations is not easily drawn, being mainly a matter of mathematical formulation. The next examples are, however, fairly clear cases of negative resistance, complicated by non-linearities which make it desirable to preface their discussion with an account of a useful graphical technique, due to Liénard.[8]

Consider the standard form (2.23) for a simple vibrator, and let us choose such a unit of time that the natural frequency $\omega_0 = 1$. Then

30

$$\ddot{\xi} + k(\dot{\xi}) + \xi = 0, \tag{21}$$

and we have indicated that the dissipative force, $-k(\dot{\xi})$, is now to be regarded as a more general function of velocity than mere proportionality. In terms of the variables ξ and v, defined as $\dot{\xi}$, (21) may be rewritten:

$$\mathrm{d}v/\mathrm{d}\xi = -[\xi + k(v)]/v. \tag{22}$$

The slope of the trajectory on the (ξ, v)-plane (the phase plane) is defined at every point by (22), and may be determined graphically very easily, as illustrated in fig. 22(a). The curve K is $-k(v)$; to find the tangent to the trajectory at any point P, draw a horizontal line to meet K in Q, and drop a vertical line from Q to cut the ξ-axis at N. Then T, drawn normal to NP, is the tangent to the trajectory at P. The proof follows immediately from the fact that $PQ = \xi + k(v)$, while $QN = v$.

259

The argument may be extended to provide a construction, not just for the slope, but for the curvature as well, so that the trajectory may be synthesized as a succession of circular arcs. The construction of C, the centre of curvature of the trajectory S at P, is shown in fig. 22(b). Draw PQ, QN and NP as before. Construct the normal to PN at P (i.e. the tangent to S) and the tangent to K at Q, and find their intersection, G. Join GN and let it cut PQ, produced if necessary, in F. Then C lies on PN vertically below F.

Proof: Fig. 22(c) shows two neighbouring points on S, P and P', with their associated Q and N; C is the intersection of PN and $P'N'$ in the limit as P' is moved towards P. It is required to show that FC is a vertical line. GL is the tangent to K at Q, GM the tangent to S at P. Since the triangles $EQ'Q$ and $HP'P$ are respectively similar to NLQ and NMP, with the same scaling factor, it follows that $NN'/HP' = NL/NM = FQ/FP$. But in the limit, as $P' \to P$, $CN/CP = NN'/HP'$. Therefore $CN/CP = FQ/FP$ and FC is parallel to QN, i.e. is a vertical line.

Given a set square, and squared paper to speed up the determination of Q and N, one can find C in a very short time and hence draw a short arc of S through P. The trajectory that results from this process carries more conviction than one constructed solely from the tangents as found by Liénard's method.

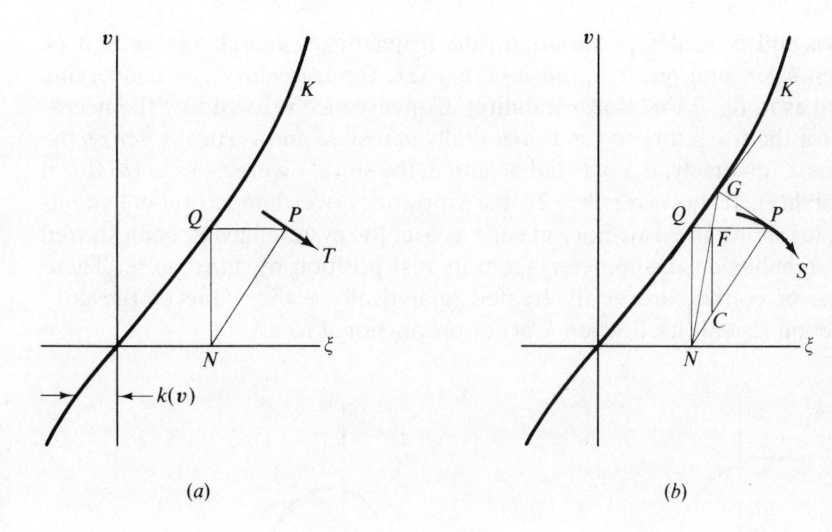

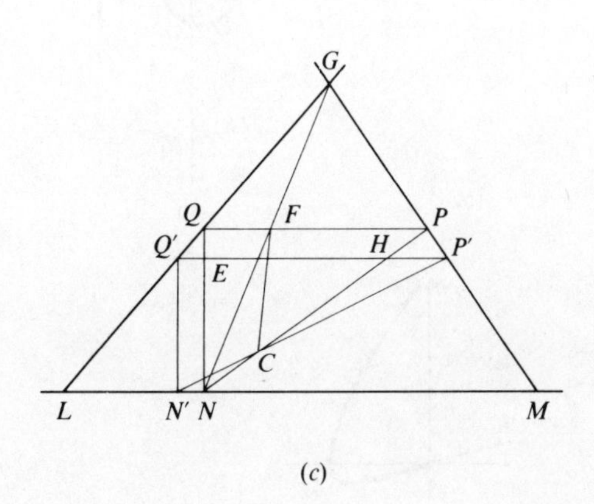

Fig. 22. (*a*) Liénard's construction for the tangent T at the point P on the phase-plane trajectory of a point obeying (22). (*b*) Construction for the centre of curvature, C, of the phase-plane trajectory at P. (*c*) Diagram used in proof of (*b*).

Apart from the graphical value of the construction, certain general points emerge, which the reader may readily prove, and which indicate why a trajectory has its particular shape:

(1) When S cuts the ξ-axis it does so vertically, and the centre of curvature C is the intersection of K with the ξ-axis.

(2) When S cuts K it does so horizontally, and the centre of curvature lies on the ξ-axis vertically below.

(3) C also lies on the ξ-axis when the corresponding point on K has a vertical tangent.

(4) If the tangents at corresponding points on K and S intersect on the ξ-axis, S has zero curvature. If they intersect at a vertical level between the ξ-axis and PQ, S is concave; otherwise it is convex.

Let us apply the graphical process to a few examples, starting with the simplest. For a lossless harmonic vibrator, $k = 0$, K coincides with the

v-axis and N is always the origin; the trajectory is a circle centred on O. When $k \propto v$ and positive, but less than $2v$, the trajectory is a converging spiral as in fig. 23(*a*) (focal stability). Convergence is forced by the necessity for the trajectory to run horizontally across K and vertically across the ξ-axis. Conversely, if $k \propto v$ and negative, the spiral diverges, as in (*b*) (focal instability). If, however, $k > 2v$ the vibrator is overdamped (nodal stability); (*c*) shows the trajectory in such a case, the system having been started (like a ballistic galvanometer) from its rest position by an impulse. These cases, of course, are easily treated analytically – the value of the construction reveals itself when k is not proportional to v.

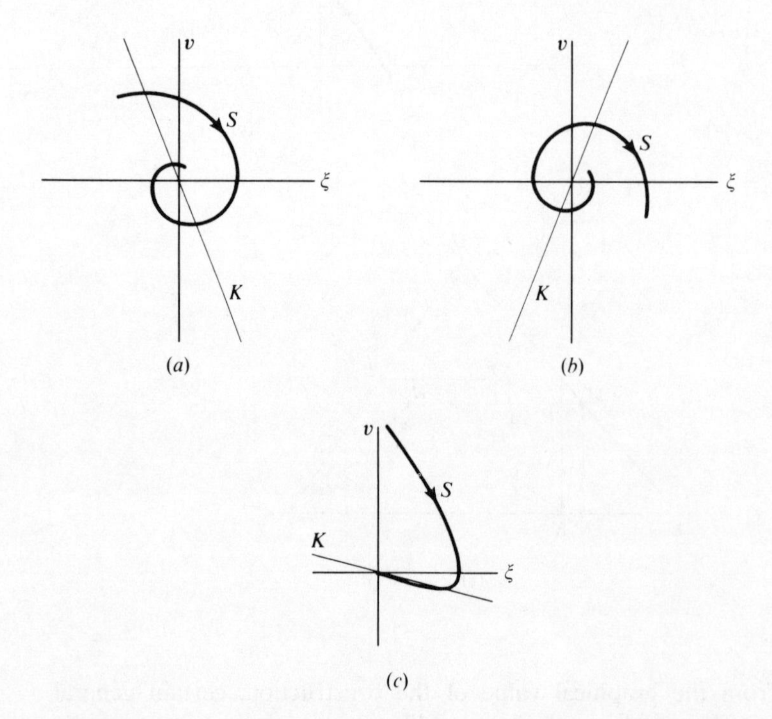

Fig. 23. Trajectories for the case $k \propto v$ in (21); (*a*) damped oscillation for $0 < k < 2v$, (*b*) growing oscillation for $-2v < k < 0$, (*c*) overdamped response for $k > 2v$.

The model system sketched in fig. 24 may oscillate spontaneously if the frictional force varies with slipping speed in the appropriate way, as the slipping speed rises. In fig. 24(*b*) the speed of the belt is represented by v_0. If $v < v_0$ the mass M is dragged forward, i.e. $k < 0$, while if $v > v_0$ it is dragged back; the frictional force is depicted as lessening with increase of $|v - v_0|$. Where K cuts the ξ-axis, at W, there is a point of unstable equilibrium, the spring being stretched so that its force just balances the frictional force at a slipping speed of v_0. But the slope of K, as in fig. 23(*b*), shows that oscillations will grow until M achieves a speed of v_0 at some moment in the cycle. The trajectory cannot cross the horizontal part of K, i.e. M cannot move faster than the belt. For if the trajectory were to continue above v_0, Liénard's construction gives a negative slope (shown as

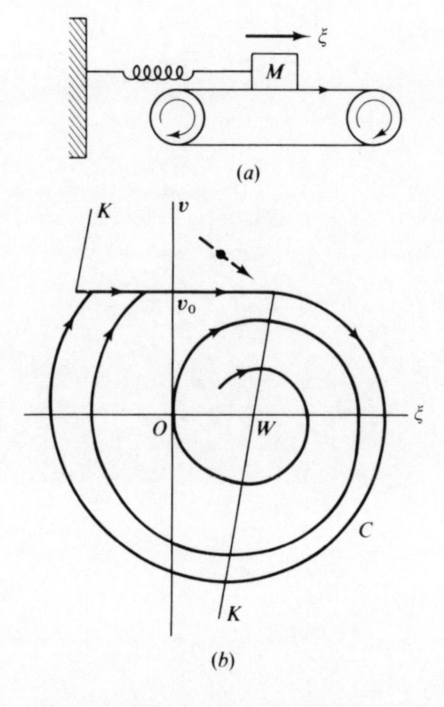

Fig. 24. (*a*) A mass M sliding on a moving frictional belt and restrained by a spring; (*b*) trajectory when the frictional force decreases with velocity.

a broken line); but v is positive and ξ must be increasing – motion on this part of the trajectory, as shown by the arrow, is towards the horizontal part of K, not away from it. Of course, what actually happens is that the trajectory follows K, the frictional force increasing as M moves at the same speed as the belt, until the maximum friction is being exerted. From then on the vibration repeats itself regularly, as shown by the limiting form C.

A commoner form of this type of process is the stick–slip vibration, arising when the frictional force is sensibly constant once slipping has begun, but less than the force needed to start slipping.[9] Real examples of this phenomenon were mentioned in chapter 2. The phase-plane trajectory is shown in fig. 25(*a*). For small vibrations about W there is no gain or loss in this idealized system, but if the vibration is started with enough amplitude to reach v_0 it is then stable against extra dissipative effects, if not too large. For example, if there is a viscous drag on M in addition to the friction of the belt, K takes the form shown in fig. 25(*b*). Small amplitude vibrations die away, but larger amplitude vibrations may survive thanks to the extra kick they get from the stickiness of the belt. The form of the vibration, $\xi(t)$, is easily constructed when viscous drag is neglected, since it consists of a pure sinusoidal curve interrupted by straight sections. In fig. 26 the sine curve is drawn about a base line at ξ_0, which is the displacement that causes the spring to balance the sliding friction; its amplitude is such that at ξ_1, which corresponds to the sticking friction, the slope $\dot{\xi}$ is just equal to the speed of the belt. Then the portion AB, where the sine curve has a greater slope, is excised and replaced by a straight line. Clearly the

42

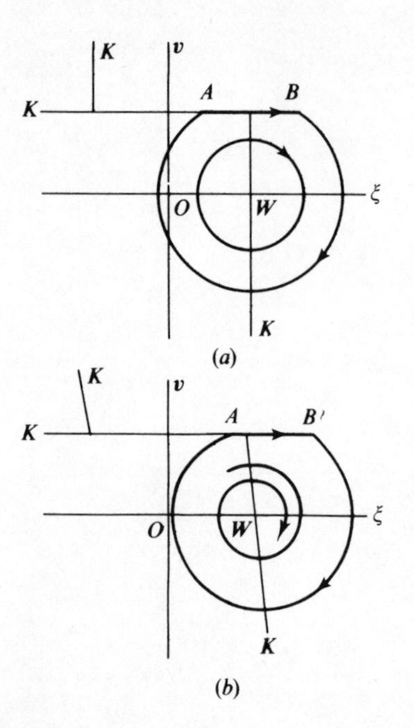

Fig. 25. Phase-plane trajectories for stick–slip oscillation (*a*) without other damping effects, (*b*) with additional viscous damping.

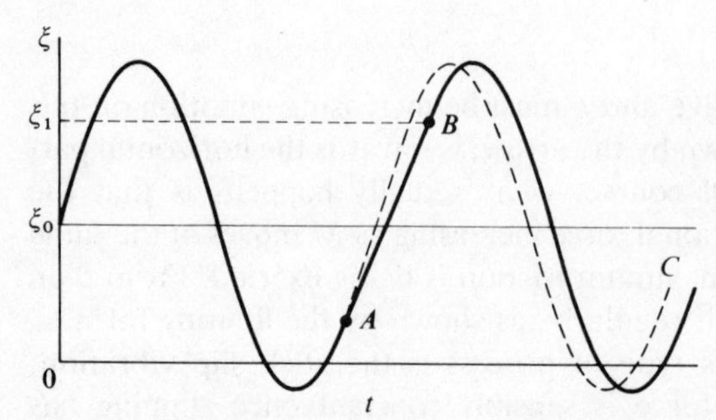

Fig. 26. waveform of stick–slip vibration, corresponding to fig. 25(*a*).

period of vibration is slightly increased over the natural period; this is equally clear from fig. 25(*a*), the stretch AB being traversed at a rather smaller value of v than along an arc of the circle connecting these points.

A quite different case of negative resistance may be observed when a metal block, suspended from a torsion wire, is rotated in a transverse magnetic field, as shown in fig. 27. When the block turns slowly, the motion of its parts in the field $\boldsymbol{B}$ induces eddy currents which create in the block a magnetic moment $\boldsymbol{M}$ at right angles to $\boldsymbol{B}$; the interaction of $\boldsymbol{M}$ with $\boldsymbol{B}$ results in a drag on the rotation, so that the torsion wire acquires a twist. As the angular velocity, Ω, of the torsion head is increased, $\boldsymbol{M}$ and the drag are at first proportional to Ω, so that the block behaves as though immersed in a viscous medium. The effect is surprisingly strong – a copper bar 5 cm in diameter held transversely in the field of a good electromagnet can only be turned very slowly by hand. At higher values of Ω the magnetic field of the eddy currents is strong enough to cause a marked diminution of the field

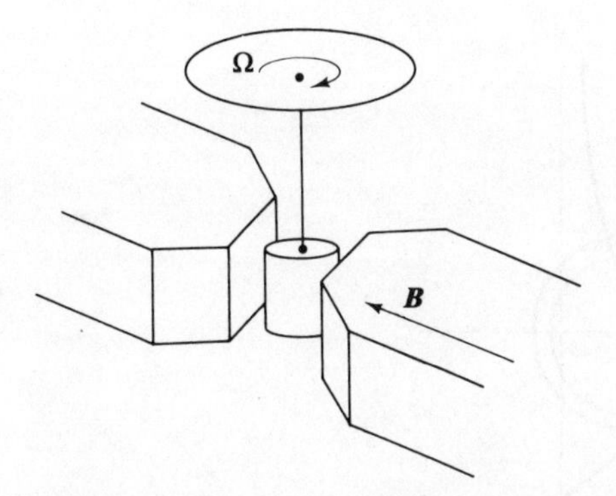

Fig. 27. A conducting cylinder supported in a horizontal magnetic field by a wire attached to a steadily rotating torsion head.

strength inside the block, and the value of M ceases to rise as fast as Ω. Indeed, the drag rises to a maximum at a certain Ω and at faster speeds decreases, in the way shown in fig. 28. At very high speeds the magnetic field is excluded from all but a thin layer at the surface of the block; this is the same as the well-known skin effect in metals, manifested here in a slightly different context.

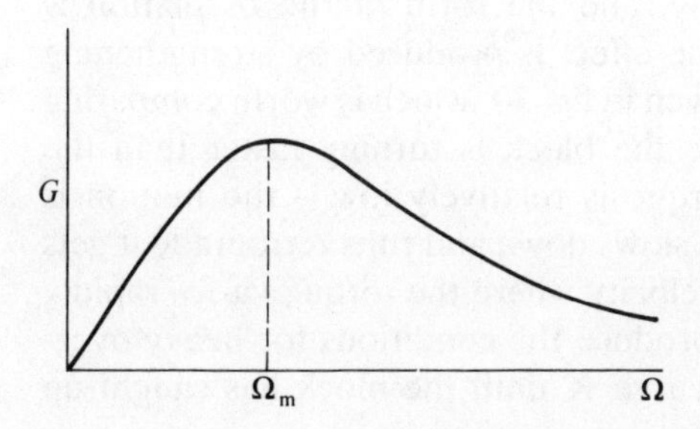

Fig. 28. Eddy-current torque G on cylinder as a function of rotational angular velocity, Ω.

So long as the drag $G(\Omega)$ is a function solely of Ω – and it is only fair to give warning that this proves an overoptimistic assumption – we can write the equation of rotary motion in a form analogous to (21). Let ξ be the angle of twist of the torsion wire, and let the torsion head have uniform angular velocity Ω_0. Then the angular velocity of the block is $\Omega_0+\dot{\xi}$, and

$$I\ddot{\xi}+G(\Omega_0+\dot{\xi})+\mu\xi=0,$$

where I is the moment of inertia of the block and μ the torsion constant of the wire. If $(I/\mu)^{\frac{1}{2}}$ be taken as the unit of time, and $\dot{\xi}$ written as v, this is reduced to the same form as (22), with

$$k(v)=G(\Omega_0+\dot{\xi})/\mu.$$

The phase-plane trajectory is now constructed as before, and fig. 29 shows how oscillations die out so long as Ω_0 is lower than the value Ω_m at the maximum of G, but grow to a limiting form if $\Omega_0>\Omega_m$. If the torsion wire is weak, the unit of time is long and Ω_m is correspondingly large. However,

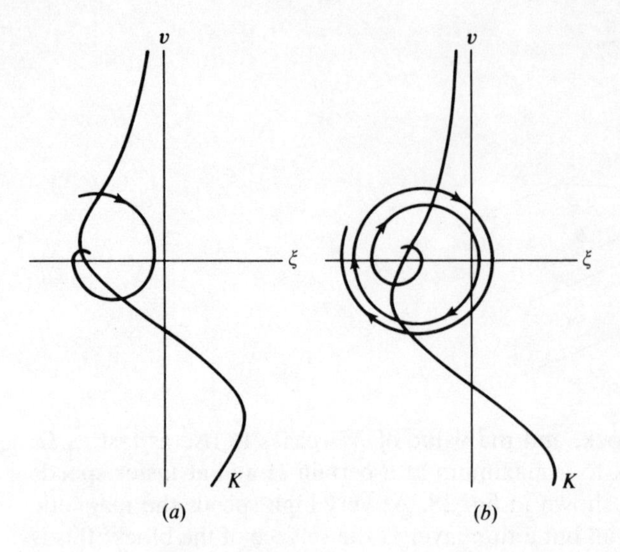

Fig. 29. Phase-plane trajectories for the arrangement in fig. 27: (*a*) damped oscillation when $\Omega < \Omega_m$; (*b*) maintained oscillation when $\Omega > \Omega_m$.

$\Omega_m \propto \mu^{-\frac{1}{2}}$, while $k(\Omega_m) \propto \mu^{-1}$, so that weakening the suspension causes the curve K to spread out horizontally, and the form of the oscillation is correspondingly distorted. The same effect is produced by strengthening the magnetic field. An example is given in fig. 30, which is worth comparing with fig. 23(*c*). When v is positive the block is turning faster than the torsion head, and the damping torque is relatively low – the motion is almost free. As soon, however, as it slows down and runs retrograde it gets caught up in the range of angular velocity where the torque varies rapidly with speed, in such a way as to reproduce the conditions for heavy over-damping. The trajectory hugs the curve K until the block has caught up

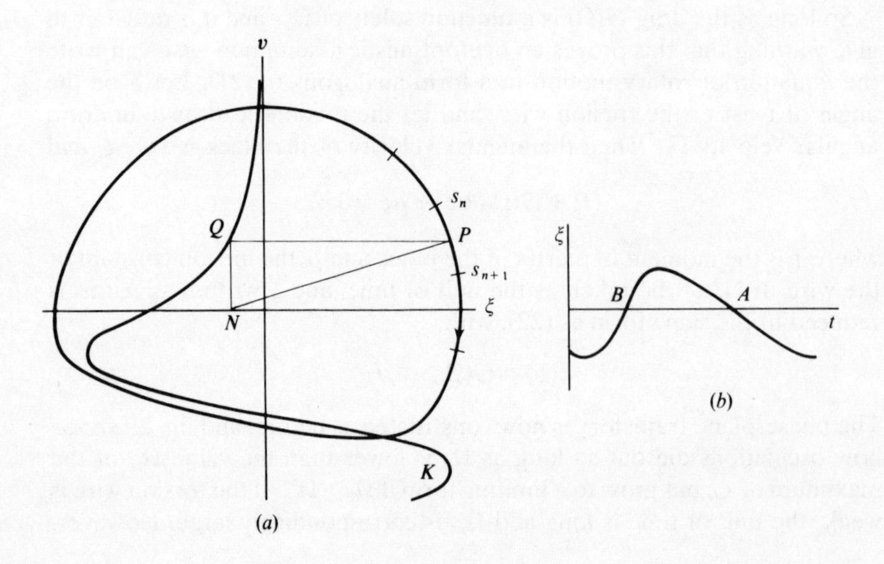

Fig. 30. Non-sinusoidal oscillation of conducting cylinder when damping is heavy; on the trajectory (*a*) is shown the construction by which the waveform (*b*) is derived.

enough speed to pass the maximum torque condition, when it begins to turn more freely again.

To translate the trajectory on the phase plane into a graph of ξ as a function of time, a simple graphical construction may be used, as illustrated in fig. 30. Divide the trajectory into equal arcs, each of length s, a typical arc being that between S_n and S_{n+1}. At the centre of each arc perform Liénard's construction. Then since QN is the ξ-component of the representative point's velocity along the trajectory, PN has the magnitude of the velocity itself. The time taken to traverse the arc s is therefore s/PN. In this way one may find the interval between each pair of points S_n and hence construct the waveform, as in fig. 30(b). It should be stressed that this construction, like the construction of the trajectory itself, is a qualitative graphical process that can only be made quantitatively accurate with very great care and skill. Its value lies in the insight it provides into the mechanisms at work and the possible forms the solution of the equation of motion may take. If one does not want insight but only a rather exact plot of the waveform, it is much better to integrate the equation of motion step-by-step with a computer.

To return to the physical system, experiment shows that the waveform is not quite like fig. 30(b). The observed forms in fig. 31 were recorded for a very pure copper cylinder cooled in liquid helium, having so high a conductivity that large eddy currents were induced by slow rotation and the maximum torque was generated at a speed of only $\frac{1}{10}$ revolution per second. So long as Ω was not much greater than Ω_m the oscillations, of

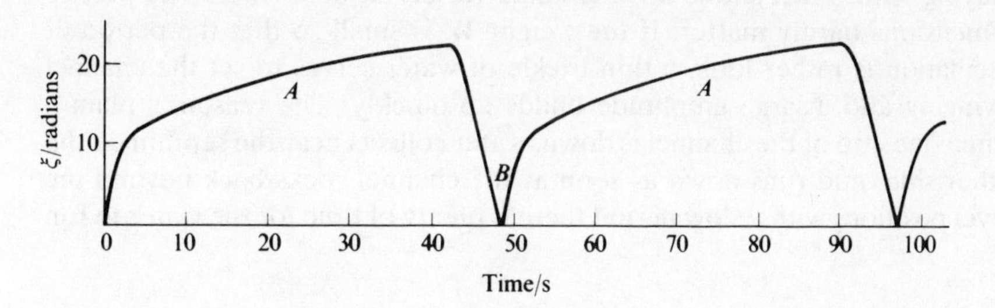

Fig. 31. Experimental record of oscillations of a conducting cylinder (unpublished observations by Dr. P. Martel).

rather small amplitude, were not far from sinusoidal, but when more violent oscillations occurred the 'overdamped' regions A between segments of fairly free motion B were elongated and opposite in slope to those of fig. 30(b). The assumption that G is determined by Ω alone is in fact not valid. The eddy currents take a long time ($\gg 1/\Omega_m$) to change their character, and when the block is turning through a large angle in such a time they and their magnetic moment are carried round with it, creating all manner of changes in the torque. The value of G at any instant, therefore, is determined by the entire past history and not by the instantaneous value of Ω. Instead of a non-linear differential equation, the motion is governed by a non-linear integral equation, so complicated indeed as to justify the effort of computation only if the solution were needed for some important purpose. Since this condition does not apply here we make no apology for shirking a fuller explanation of the observations. The phenomenon has been introduced here partly to show how to apply Liénard's construction, but partly also as an illustration of how difficult a mathematical problem may be posed by a very simple physical arrangement. Non-linear integro-differential equations are so troublesome that one tends to put out of mind physical problems whose mathematical formulation leads in this direction. The very next example to be discussed is just such a one, but fortunately, like the rotating block, of no importance. It is a measure of the task that so much remains to be accomplished in the theory of musical instruments, including the violin and all wood and brass wind instruments, in which the excitation of the sound involves a non-linear time-dependent interaction between (on the one hand) bow, reed or lips and (on the other) the strings or an acoustic resonator. The same could be said of the human voice, and it can hardly be contended that these are unimportant vibratory systems. It is well, then, to remember that the problems of classical mathematical physics are far from exhausted – if it sometimes appears that they are, the blame is more likely to lie with limitations of mathematical technique than with the problem-setting resources of the natural world.

Two other oscillatory systems are worth mentioning rather briefly, being rather easy to make and experiment with, though otherwise not especially notable. They will serve as an interlude before turning to the far more significant *maser*, which unfortunately is not suitable for do-it-yourself demonstrations. The water-driven rocker in fig. 32 is particularly worth playing with, for it shows a remarkable variety of behaviour. The precise dimensions hardly matter. If the weight W is small so that the period of oscillation is rather long, a thin trickle of water serves to set the channel swinging and a large amplitude builds up quickly. The reason is plain – when one end of the channel is down, water collects near the septum on the other side, and runs down as soon as the channel rocks back beyond the level position; with a slow period there is plenty of time for the water to run

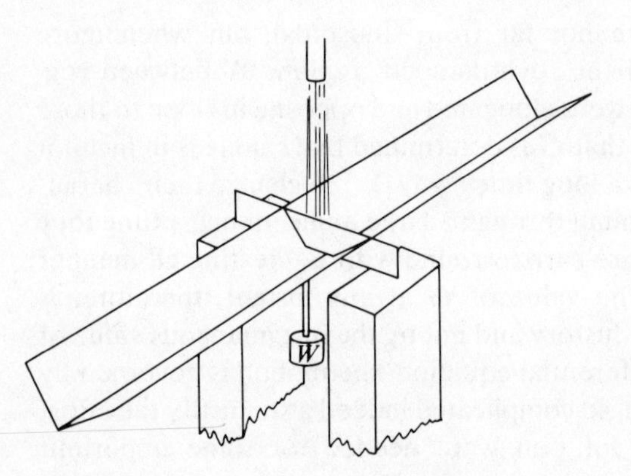

Fig. 32. Rocking channel driven by water stream. An aluminium channel, bent out of sheet, about 60 cm long and with sides of 4 cm, is suitable; a septum, cemented by epoxy resin into a saw-cut, serves also as pivot. The weight W can be used to adjust the rocking frequency. The behaviour can be studied with the gap below the septum blocked or unblocked.

off the end before the channel reaches its lowest point, and the weight of the water therefore exerts a couple in the same sense as the angular velocity, transmitting energy to the channel. If, however, W is heavy enough to make it rock fairly rapidly, it may start to rise before the water has reached the end, and the maximum torque is exerted against the angular velocity, causing damping of the vibration. This is particularly obvious when the amplitude is small and the flow of the water down the channel slow. At large amplitudes, however, there may still be time for the water to escape before the channel rises. Thus the vibration may be damped at small amplitudes and encouraged at large. This is, of course, a highly non-linear system, in which the time variation of the torque due to the water does not scale in amplitude alone, but suffers compression in time as the amplitude rises. It is worth constructing the rocker so that the septum can be removed, and also to provide for closing the ends of the channel. A number of quite surprising types of behaviour can be observed by making such minor changes, interpreting some of which offers a considerable challenge.

Thermally driven acoustic vibrations are not uncommon; almost every glass-blower, however inexperienced, has found a glass tube, after having one end sealed off, singing spontaneously. Rayleigh, in his *Theory of Sound*,[(10)] discusses several related instances, including one that is fairly easy to make and which was discovered by Rijke in 1859 – a large vertical pipe, open at both ends, and excited to emit a loud tone by a heated gauze placed about one-quarter of its length from the bottom. To understand what goes on we must first examine the dissipative effects present and then see how the heated gauze reverses their sign. The essential mechanism for loss arises from the oscillations of temperature of the vibrating gas column, which suffers alternately very nearly adiabatic compressions and rarefactions. In themselves these oscillations cause no dissipation, but if they lead to a transfer of heat to the walls of the resonator or other bodies in thermal contact with the air, there develops a phase difference between the pressure and volume oscillations of the air which has as its consequence the dissipation of acoustic energy as heat in the air.

This point is perhaps best appreciated in terms of the step response function of the pressure after a sudden small contraction in volume, as illustrated in fig. 33(a). At $t=0$ the gas is compressed adiabatically by ΔV and thereafter held at its new volume; the instantaneous response of the pressure is in accordance with the adiabatic equation $PV^\gamma=$ constant, so that $\Delta P_0/P=-\gamma\,\Delta V/V$. If there were no heat transfer to the walls or other bodies, the pressure would remain at ΔP_0, but if the gas can cool to the ambient temperature the subsequent change of ΔP is a (not necessarily exponential) relaxation to the value ΔP_∞ appropriate to isothermal compression, $\Delta P_0=\gamma\,\Delta P_\infty$. The time-derivative of this curve, shown as fig. 33(b), is the impulse response function, which in addition to its instantaneous component at $t=0$ shows a delayed response. It should be clear from the discussion of chapter 5 that the susceptibility (compressibility)

105

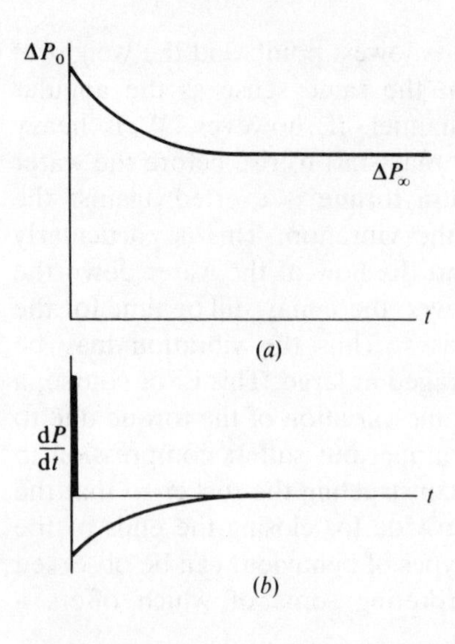

Fig. 33. (*a*) Excess pressure ΔP in a gas after a small step-function decrease in volume at time $t=0$; (*b*) the corresponding impulse response function.

$\chi(\omega)$ derived from this by a Fourier transform will in general be complex, with dissipation as an inevitable consequence.

For the effect to be reversed the gas must receive heat, not lose it, when it is hot, and/or lose heat when it is cold. Let us see how this is achieved by the heated gauze. When the air in the pipe is in vibration in its fundamental resonance all parts suffer oscillations of pressure in the same phase, the amplitude being greatest at the centre. Except for the effect of end-corrections, no fluctuations of pressure occur at the open ends; the air moving in and out, to provide for the fluctuations inside, has its greatest speed at the ends, and does not move at all at the very centre, except for a steady upward convective flow past the heated gauze. The gauze, about one-quarter of the way up, lies in a region where there are substantial oscillations both of pressure and of velocity. Let us now show diagrammatically, as in fig. 34, how various parameters change with time in the vicinity of the gauze. Curve (*a*) shows the fluctuations of pressure, ΔP being the increase above atmospheric pressure; the same curve serves for the temperature fluctuations since in adiabatic compression temperature and pressure march together. In curve (*b*) is shown the upward velocity, v, consisting of a sinusoidal fluctuation superposed on a steady convective movement; the phase of the fluctuating component is easily deduced from the fact that it must be greatest at the moment when the pressure is increasing most rapidly, as, for example, at A. By integrating curve (*b*) we are enabled to sketch the variation with time of the height, h, of an element of air, and this must clearly, as in (*c*), be a sinusoidal fluctuation around the uniform rise due to convection. Different elements of air will show the same pattern, but displaced vertically. It will be seen that there is a certain fraction of each cycle, drawn more heavily, during which the air reaches a height it has not previously attained. The first passage of the air through the heated gauze occurs during one or other of these periods, and if

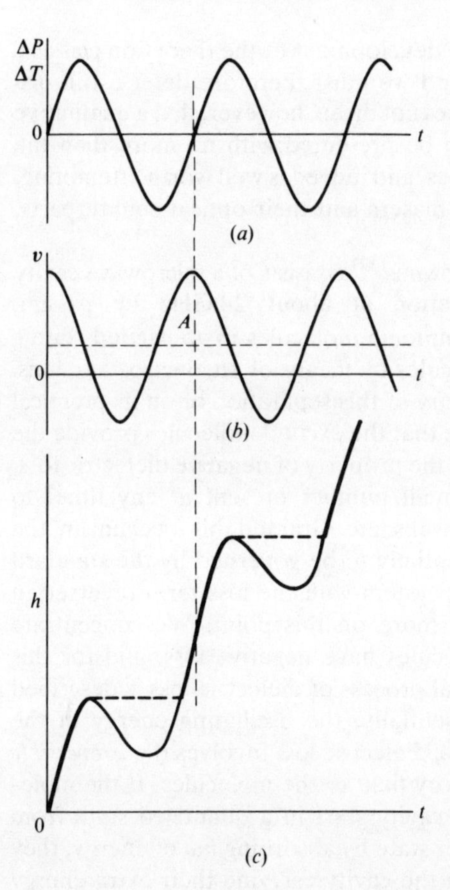

Fig. 34. Oscillations in Rijke tube of (*a*) ΔP and ΔT, (*b*) particle velocity, v, in lower half of tube, (*c*) particle height, h, being the integral of (*b*).

afterwards it falls back through the gauze and rises again, very little more happens since it does not have time to cool appreciably – it is the first passage that is important, and marks the moment when it receives a substantial input of heat. As the diagrams show, this moment is normally a little before the peak in the temperature fluctuation, and certainly not during that part of the cycle when the air is cooled by adiabatic expansion. The required conditions are present, therefore, for reversing the dissipative process, the heat being provided at the time the temperature is greatest, and we can understand in the light of this argument why the tube goes into spontaneous oscillation. As Rayleigh points out, if the gauze were to be put in the top half of the tube, it would communicate heat during the cold part of the cycle, and no spontaneous oscillation should occur – a prediction which is verified in practice.

(*d*) *The maser*[11]. Although it is often implied that maser action can only be understood as a quantum-mechanical process, this is an over-statement; there are processes which can be described equally well in classical or quantal terms, yet whose quantal description reveals the maser principle.

Nevertheless an attempt to base the development of the theory on classical ideas alone would be misguided, and we must therefore defer a full discussion until the second part. This does not mean, however, that a qualitative discussion is ruled out, for this can be presented with no more than the most general ideas of quantum physics, and indeed is well worth attempting, in view of the great importance of masers and their optical counterparts, lasers, in modern physics.

571

The original maser invented by Townes[12] consists of a microwave cavity maintained in spontaneous oscillation at about 24 GHz by passing through it, *in vacuo*, a stream of ammonia molecules in an excited state – separated from the unexcited molecules by means of an electrostatic lens. We need not spend time on the theory of this separation or on its practical realization; it is enough to recognize that the excited molecules provide the cavity with a dielectric filling having the property of negative dielectric loss, sufficient (in spite of the rather small number present at any time) to overcome the resistive losses of the walls, etc. Granted this mechanism, the oscillation in the cavity is seen essentially to be governed by the standard differential equation for a resonant system with the loss term reversed in sign, and there is no need to say more on this point. We concentrate therefore on why the excited molecules have negative loss, and for this purpose we consider how the normal process of dielectric loss is described in quantum-mechanical terms. Essentially, the oscillating energy in the cavity being quantized in units of $\hbar\omega$, dielectric loss involves the *irreversible* transfer of quanta from the oscillatory field to the molecules. If the molecules in their passage through the cavity start in a quantized state from which they can be excited to a higher state by absorbing $\hbar\omega$ of energy, they may make this transition and leave the cavity carrying their extra energy away with them.† Conversely, and this is the essence of the maser principle, molecules entering in the higher state may be stimulated by the oscillatory field to fall to the lower state, communicating one quantum to the field if they then leave the cavity before re-excitation occurs. It is not enough to pass an unseparated stream of ammonia molecules through the cavity, since in thermal equilibrium at room temperature there are very nearly equal numbers in the two states, the energy difference between the states, $\hbar\omega$, being much less than $k_B T$. In any case, the Boltzmann distribution ensures that there is an excess, amounting to $\hbar\omega/k_B T$ or something less than 1%, in the lower state, so that as a whole the stream of gas would be slightly lossy rather than the reverse. It is the separation of the

† This is a rather special process of dielectric loss. The more usual process, in dielectric materials as distinct from molecular beams, is that the quantum first excites internal motions of the molecule; subsequently this energy is transferred to other modes of motion which we characterize as heat, for example, translational kinetic energy or vibrations of the solid as a whole. But the end-result is the same from the point of view of the oscillatory field – the energy is irretrievably lost.

excited molecules that creates a stream of molecules capable of maintaining oscillation.

It should be noted that the negative-loss mechanism, unlike a negative resistance, is extremely frequency-sensitive; the natural frequency of the cavity must lie very close to the frequency $\Delta E/\hbar$, where ΔE is the level separation, if the quantum $\hbar\omega$ is to cause real transitions. How close may be estimated by a simple argument which may not carry conviction, standing alone, but in fact can be fully justified by a complete quantum-mechanical analysis. From the point of view of an ammonia molecule in its passage through the cavity, the radiation field is applied only for a time Δt equal to L/v, where L is the length of the cavity and v the molecular velocity. Correspondingly it experiences an oscillatory stimulation that is not perfectly sharp in frequency, but spread into a narrow range $\Delta\omega \sim 4\pi/\Delta t$ (cf. (4.21)). The cavity frequency must agree with the natural frequency of the ammonia excitation within this range for the process to work – if the natural frequency is not present in the field experienced by the molecule, no transition will occur. Since Δt is typically 10^{-4} s, $\Delta\omega$ is no more than $10^5\ \text{s}^{-1}$, a million times less than ω itself.

73

The foregoing description of maser action in terms of quantum jumps ignores the question of the dielectric response of the assembly of molecules – what we might call the classical description of the medium as a dielectric whose polarization is so phased relative to the exciting electric field as to transmit energy rather than absorbing it. A normal lossy dielectric has its oscillatory polarization lagging behind the field, but in the maser one must suppose it to lead. A quantum-mechanical calculation confirms this, and enables one to give a more precise account. An ammonia molecule in either of its two stationary states has no dipole moment, but after entering the cavity, and experiencing the electric field tuned exactly to synchronism with it, its moment grows steadily in magnitude, and is phased in quadrature with the field, lagging if it starts in the lower state, leading it in the upper state. With a weak field the rate of growth is constant and proportional to the field strength, and if the molecule leaves the cavity while in this linear regime the average dielectric response during its sojourn is linear and an effective dielectric constant ε can be assigned. But with stronger fields the dipole moment has time to reach its maximum and begin to fall; for the perfectly synchronized field causes it to shuttle continually between the two stationary states, and the dipole moment rises and falls in magnitude, sometimes leading and sometimes lagging. This is too pictorial a description to please a quantum physicist, but it is a good representation of the average behaviour of a large number of molecules, even if each separately cannot be described in such classical detail. With a very high field strength in the cavity the molecule makes so many transitions that it may emerge in either state. Each excited molecule then on the average communicates $\frac{1}{2}\hbar\omega$ to the cavity. If we define an effective (imaginary) dielectric constant in such a way that the rate of energy loss is $\frac{1}{2}\varepsilon''\mathscr{E}^2$ when the field amplitude is $\mathscr{E}$, the fact that the energy gain saturates means that

222

ε'' is negative and falls off as $1/\mathscr{E}^2$ at high field strengths. A schematic diagram showing how ε'' varies is given in fig. 35; the broken lines show the low and high field asymptotes, and the actual curve rises to something like twice the height of the latter when $\mathscr{E}$ is such as to cause each molecule to make a complete transition. In practice, of course, the amplitude grows until the effective value of ε'' is just right to supply as much energy as is dissipated elsewhere.

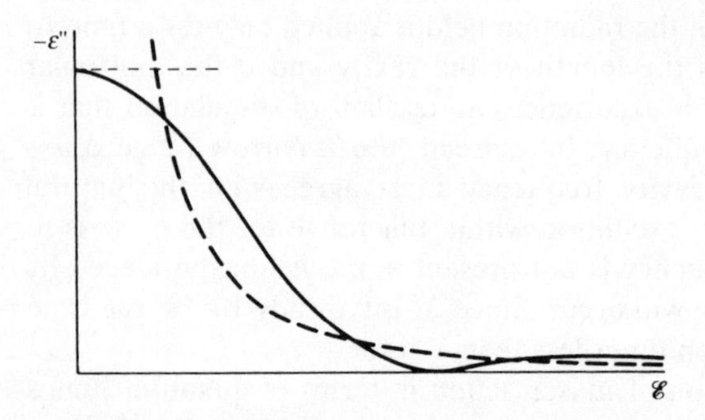

Fig. 35. The effective mean value of ε'' for ammonia molecules moving at a given speed through a microwave resonator, as a function of the oscillatory field-strength $\mathscr{E}$. The wide range of molecular velocities in practice smooths out the oscillatory tail so that it conforms to the asymptotic behaviour shown by the broken curve.

With this very sketchy account we must leave the maser, to return again when the time is ripe for a quantum-mechanical analysis. It may be remarked, however, that this analysis will establish an extraordinarily close correspondence between a two-level quantized system and the gyromagnetic top described in chapter 8 as a model of proton resonance (a proton in a magnetic field being itself a two-level system). Indeed, the account in the last paragraph of the growth of the oscillatory dipole to a maximum and its subsequent decay is simply an account of the way the transverse component of the moment of the top varies with time as its cone of precession expands and contracts under the influence of a perturbing field at the precessional frequency.

574
516
218

Saturation of negative-resistance oscillators

The maser reaches a steady level of oscillation when the source of energy just matches the losses, which are in any case very small. The molecules providing the compensating gain do so by being stimulated to develop dipole moments whose oscillations lead the oscillations of electric field. Both mechanisms of loss (resistance of walls or extraction of useful energy) and gain (dielectric) operate throughout the cycle of oscillation, cancelling each other's effects and leaving the cavity to a very good approximation in an ideally lossless state, executing pure sinusoidal vibrations. The class C oscillator obtained by inductive feedback also oscillates very nearly sinusoidally, but not quite so ideally as the maser, for it is a free-running oscillator, with decay, over most of its cycle and gets an occasional kick to keep each cycle up to the amplitude of its predecessors.

The approach to the steady state is different in these two cases. We have already considered the class C oscillator and now compare it with a high-Q resonant circuit just maintained by a slightly non-linear negative resistance; this is a convenient model of a maser. We assume the p.d. across the negative resistance to contain a positive cubic term, so that the current through this and the linear positive resistance of the circuit, taken together, generates a p.d. according to the equation:

$$V = -\alpha i + \beta i^3, \qquad (23)$$

α and β being small enough for the oscillation to build up very gently and equally gently approach its steady amplitude. It is then permissible to assume a pure sinusoidal waveform for the purpose of calculating the mean rate of dissipation. If the current is $i_0 \sin \omega t$, the dissipation in one cycle is ΔE, where

$$\Delta E = 4\int_0^{\frac{1}{2}\pi/\omega} (-\alpha i_0 \sin \omega t + \beta i_0^3 \sin^3 \omega t) i_0 \sin \omega t \, \mathrm{d}t$$

$$= (2\pi/\omega)(-\tfrac{1}{2}\alpha i_0^2 + \tfrac{3}{8}\beta i_0^4). \qquad (24)$$

The steady-state value of i_0 is $(\frac{4}{3}\alpha/\beta)^{\frac{1}{2}}$, such as to make ΔE vanish. If i_0 is increased by δi_0, the energy of oscillation is increased by $Li_0\,\delta i_0$, and the dissipation per cycle by $(\mathrm{d}/\mathrm{d}i_0)\,\Delta E \cdot \delta i_0$, i.e. $2\pi\alpha i_0 \delta i_0/\omega$. Hence the amplitude approaches its stationary value at a rate that may be expressed by a Q-value of $\omega L/\alpha$, which is the same as the Q for the initial build-up. This is in contrast to the class C oscillator, which has a Q for approaching the steady state of the same value as the resonant circuit without feedback, and a Q for initial build-up that can be very small indeed, if the amplifier is powerful.

Different again are the negative-resistance oscillators discussed earlier in which the current oscillations may, like the class C oscillator, have a rapid build-up, but as they approach the steady state may be subject to a resistance that alternates between large negative at low amplitudes and large positive at high. The amplification per cycle at low amplitude, and the decrement per cycle at high, are now such that it is quite unrealistic to pretend that the waveform approximates to a sinusoid. We shall not attempt to analyse the rate of approach to the steady state, but shall concentrate on the waveform in the steady state itself, and shall devote close attention to this because it will lead us to the important class of *relaxation oscillators.* There is a great similarity in the behaviour of a tunnel diode across a parallel resonant circuit and the series resonant circuit of fig. 14(a), and it is sufficient to treat the latter only. Let us consider, therefore, the general problem of a series resonant circuit consisting of an inductor, a capacitor, and a non-linear resistor across which the potential drop is an arbitrary function of current, $V(i)$, but with no time-dependence. The differential equation for the charge on the capacitor takes the form

$$L\ddot{q}+V(\dot{q})+q/C=0. \tag{25}$$

Now put $\xi=q(L/C)^{\frac{1}{2}}$, $\tau=t/(LC)^{\frac{1}{2}}$, $v=\mathrm{d}\xi/\mathrm{d}\tau=L\dot{q}$, and $k(v)=(LC)^{\frac{1}{2}}V(v/L)$. Then (25) takes the same form as (21) and can be analysed by the same methods. Fig. 36 shows an example of Lienard's construction applied to a non-linear characteristic corresponding in general terms to the negative-resistance feedback circuit whose characteristic is exhibited in fig. 14. The circuit oscillates with an amplitude that increases rapidly at first, and then reaches a steady state fairly quickly. The rapid expansion implies that the Q-value for the negative-resistance regime is low; in fact when $V=R\dot{q}$, the slope $\mathrm{d}k/\mathrm{d}v$ is just $R(C/L)^{\frac{1}{2}}$, or $1/Q$ if Q is not too small (cf. (2.27)). The Q-value chosen here is very close to unity, corresponding to an increase in amplitude by a factor $\mathrm{e}^{\pi/2}$, i.e. 4.8, each half-cycle.

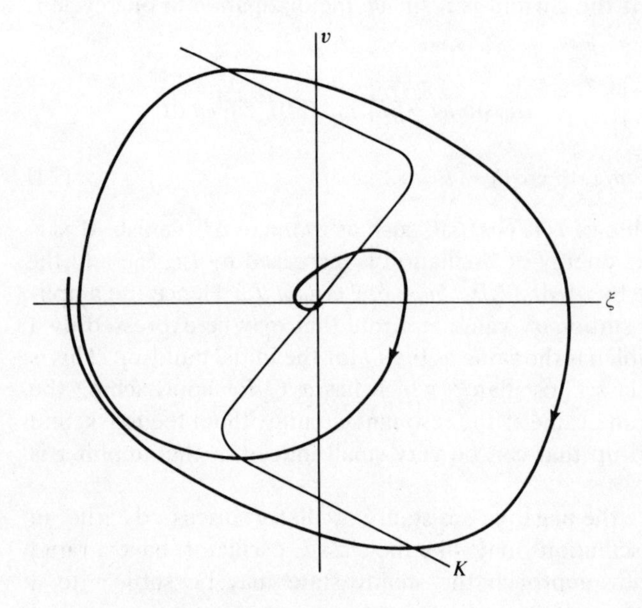

Fig. 36. Build-up to the steady state of the oscillations of the negative-resistance circuit in fig. 14(a), when feedback is not very strong and the oscillations are nearly sinusoidal.

The graphical procedure described earlier can be used to construct the waveform of the oscillation. In this particular case it is hardly worth exhibiting the final waveform, since it is only slightly distorted from a sinusoid. This is attributable to the choice of $k(v)$ with a small enough derivative everywhere for the circuit to remain oscillatory. Thus, as we have seen already, the value of Q at low amplitudes is about -1, while at high amplitudes, where the characteristic has also been drawn straight and with a slightly deeper gradient, Q is correspondingly slightly lower and positive. The fact that this linear stretch of the characteristic, at higher values of v, does not pass through the origin, does not prevent Q being defined, since by a change of variables the circuit equation may be reduced to the standard linear form. For if $V(\dot{q}) = V_0 + R\dot{q}$, (25) takes the form:

$$L\ddot{q}_1 + R\dot{q}_1 + q_1/C = 0, \quad \text{where } q_1 = q + CV_0. \tag{26}$$

So long as $|R| < 2(L/C)^{\frac{1}{2}}$, i.e. $|\mathrm{d}k/\mathrm{d}v| < 2$, the solutions of (26) are oscillatory about an offset mean value of $-CV_0$ (only q shows an offset; $\dot{q}$ oscillates about zero). But if, say, L is decreased so that the slope of $k(v)$ rises to high positive and negative values, the trajectory hugs the K-curve as close as possible and thus assumes a much more angular form, as illustrated in fig. 37 together with the wave form of the oscillation of current, v.

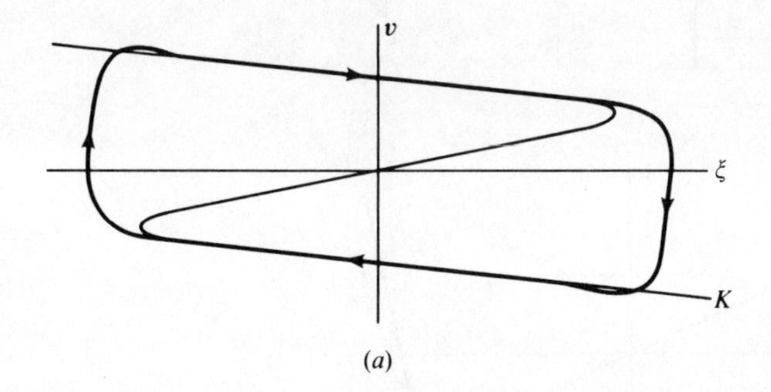

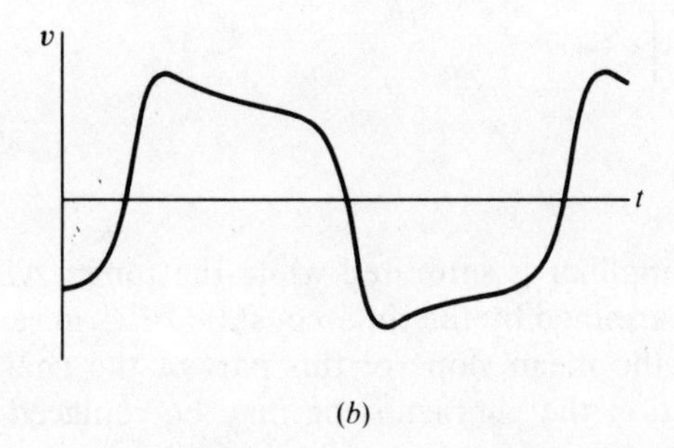

Fig. 37. The same as fig. 36 but with stronger feedback, to show (*b*) the squaring of the current waveform.

The general shape of the waveform follows from the solutions of (26). In the heavily overdamped case, the two time-constants for exponential growth or decay are L/R and RC. If R is negative, both exponentials are positive and that which grows faster dominates the growth; thus the low-current portions of the cycle are controlled by the short time-constant L/R. Conversely, if R is positive the rate of decay is limited by the slowest process, so that the high-current parts of the cycle are controlled by the long time-constant RC. With a sharply kinked characteristic $-V_0$, towards which q/C relaxes exponentially, may be well below the range of V traversed in the cycle; this part of the cycle will then be very nearly linear. When the inductor is removed altogether, the fast passage of the current through zero is limited only by the amplifier response or by stray inductances. The circuit of fig. 38 then shows a 'multivibrator' waveform,

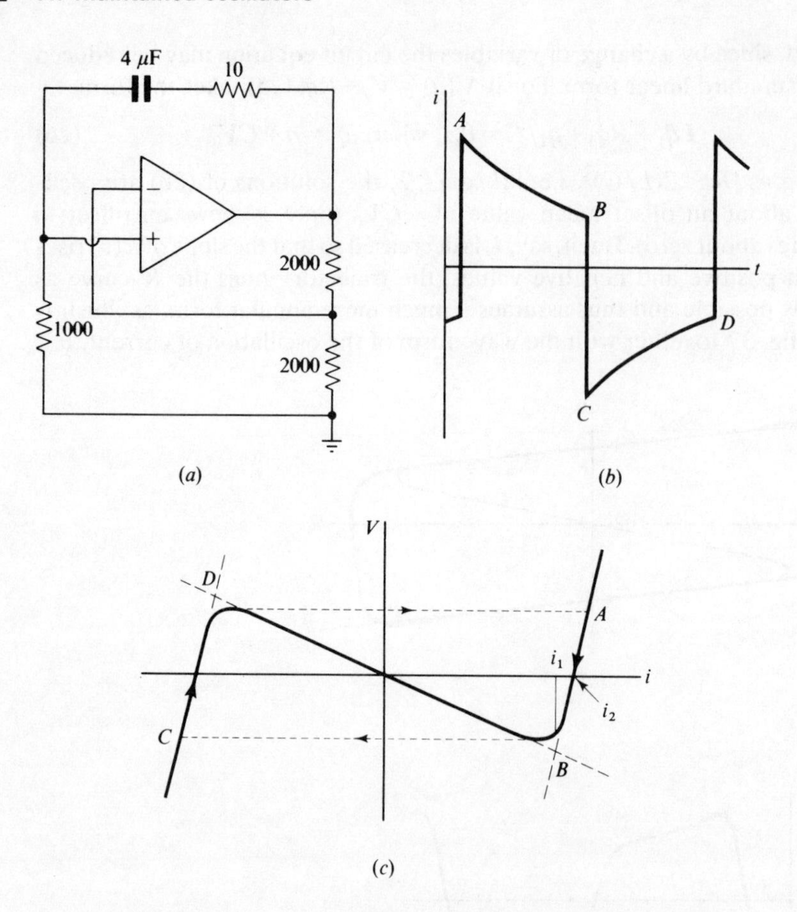

Fig. 38. (a) The circuit of fig. 14(a) with the inductor eliminated: (b) the 'multivibrator' waveform; (c) the trajectory on the i–V characteristic.

very rich in harmonics. The amplifier is saturated while the range AB is traversed, and the decay is determined by the time-constant of C in series with a resistor R_s defined by the mean slope of this part of the characteristic; to a good approximation the characteristic may be replaced by three straight lines, and the duration of AB, which is half the period, is $R_sC \ln (i_2/i_1)$. If i_2/i_1 is not much over unity the period may be substantially less than the time-constant, R_sC.

So long as L is present in the circuit, the phase trajectory is a continuous curve, as in fig. 37(a), and the i–V characteristic is traversed backwards and forwards during each cycle. As L is reduced the time taken for the negative-resistance range becomes steadily less, until when L is zero the transitions from D to A and from B to C are instantaneous, as represented by the broken lines. As AB is traversed the current discharges the capacitor and V decreases, but beyond B any attempt by the representative point to stray onto the negative-resistance range is foiled by the fact that the positive current must continue to reduce V. An inductor can allow this to occur, taking up the imbalance between the p.d. on the capacitor and that required to keep the point on the characteristic, though if L is small a very rapid change of current will be needed. Without L, the point has no option but to jump to the other side of the characteristic where the values of V and i are also consistent with the needs of the circuit elements. This argument has been spelt out in full to show that the jumping of the representative point is an idealized representation of what in practice is a continuous, but very rapid, process. We shall find similar behaviour in the next circuit to be analysed, the true multivibrator.

Multivibrator[13]

This is one of the classic circuits, invented early in the history of thermionic valves, and still used in updated versions as a generator of very angular waveforms. It also has the capability, as will be discussed in the following chapter, of being locked in subharmonic synchronism with an oscillator of much higher frequency, and therefore of acting as a frequency-divider. Apart from this, with small modifications it can be converted from a free-running oscillator into a bistable circuit with two equilibrium voltage configurations which can be switched over very rapidly; as such it has found application in logic circuits. The mode of operation is, rather naturally, treated in most textbooks in the semi-intuitive terms that electronic engineers come to adopt as their habitual descriptive language. In view of the complexity that would attend any direct mathematical analysis of the average practical circuit, especially if it depends for its operation on non-linear elements, one must respect the skill and economy with which a circuit designer picks out the relevant factors and sees how a given configuration will work. But the uninitiated reader often has difficulty in seeing how rigorous the argument is, and it is for his sake that we shall develop the theory of the multivibrator in some detail.

405

The multivibrator (fig. 39) is a two-stage RC-coupled amplifier connected back on itself so that if it were possible for the capacitors to be short-circuited without destroying the components there would be positive feedback at zero frequency. Two versions are shown in the diagram, the original with triode valves and a modern version with transistors. When the circuit is connected the resulting oscillations take the form of alternate stages switching fully on and off, the potentials at A, B, X and Y varying as shown in fig. 40. Almost all the time the amplifiers operate well outside their linear range of behaviour, and this means that their detailed characteristics are relatively unimportant, transistors and valves giving very similar results. We shall take the original circuit as the basis for discussion, simply because the connection of the resistors R to ground rather than to the positive line makes for less complexity while retaining the essential features. We shall also assume that the current passed by each valve is determined solely by the grid voltage, independent of the anode voltage. This is an oversimplification but not one that matters; we need know hardly more about the amplifier characteristics than that there is a range in which amplification occurs, bounded by cut-off when the input voltage is

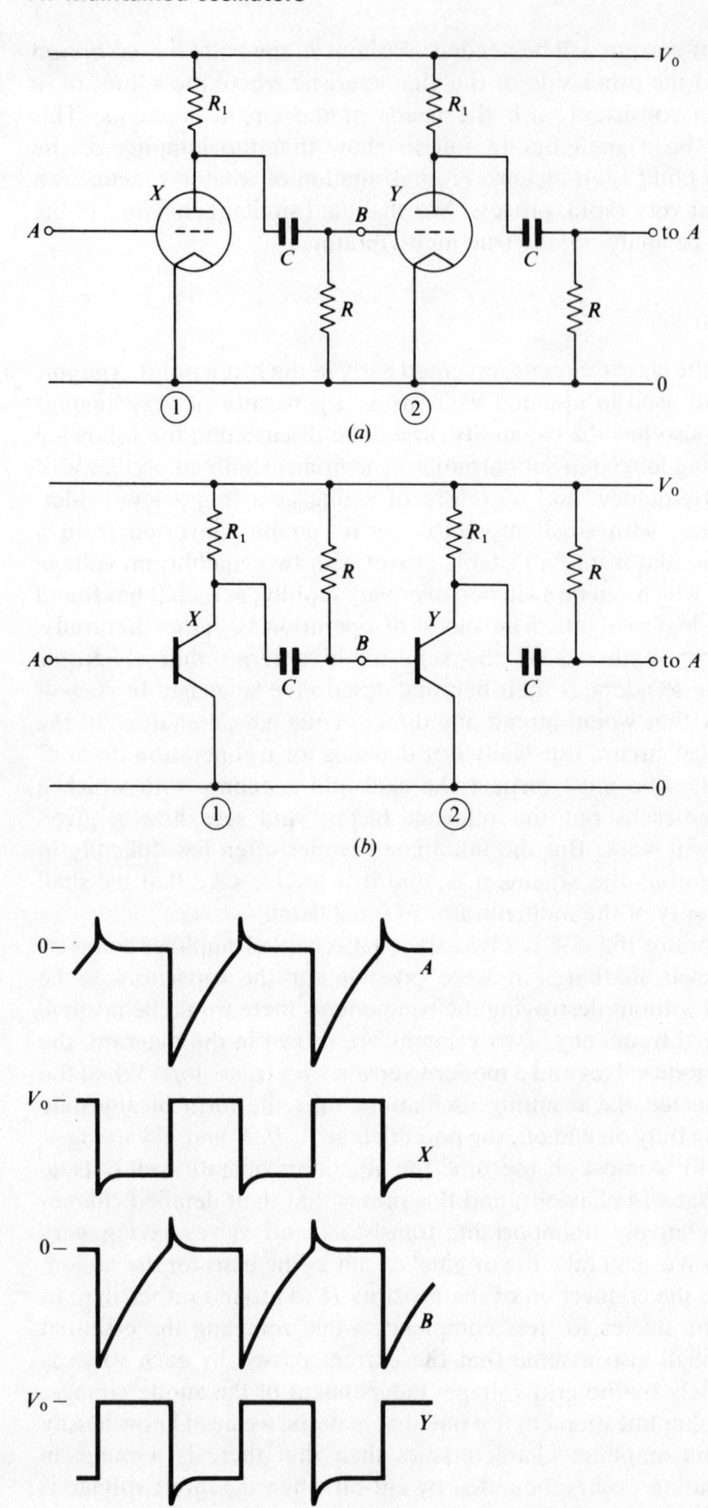

Fig. 39. (*a*) The original multivibrator and (*b*) a modern transistorized version. The resistors *R* which in (*a*) are connected from grid to earth can with advantage be connected instead to the HT line, as in (*b*), but the analysis is slightly simpler for (*a*).

Fig. 40. Waveforms of the voltages on the grids (*A* and *B*) and anodes (*X* and *Y*) of the multivibrator, fig. 39(*a*).

strongly negative and by saturation when it is strongly positive. In the latter circumstance, however, electrons readily move to the grid to eliminate the positive bias, but this will not concern us for the moment. We shall start by deriving the differential equations governing the circuit and, having seen that they are of uncomfortably high order, shall then discuss certain special cases to elucidate the processes taking place.

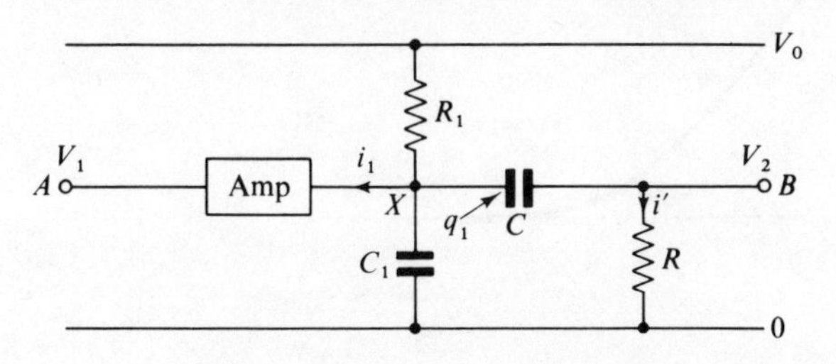

Fig. 41. Schematic diagram of one amplifier stage in a multivibrator, to show notation. It is assumed that no current flows out at *B*.

The left-hand stage of the circuit is represented diagrammatically in fig. 41, in which it will be seen that a capacitor C_1 has been introduced, to take account of the capacitance between the anode and other electrodes. It does not matter that the model is not exact – what is important, as will become clear later, is not to forget the existence of inter-electrode capacitance. The current i_1, as remarked above, is assumed to be determined solely by the grid voltage V_1, independent of the anode voltage, V_X:

364

$$i_1 = i_1(V_1). \tag{27}$$

The equation of continuity for the current at X takes the form:

$$i_1 + C_1 \dot{V}_X + (V_X - V_0)/R_1 + V_2/R = 0, \tag{28}$$

while continuity of the current through R and C gives the equation:

$$C(\dot{V}_2 - \dot{V}_X) + V_2/R = 0. \tag{29}$$

Eliminating V_X from (28) and (29), and using (27), we have:

$$\tau\tau_1 \ddot{V}_2 + \tau'[\dot{V}_2 + \mu(V_1)\dot{V}_1] + V_2 = 0, \tag{30}$$

where $\tau = RC$, $\tau_1 = R_1 C_1$, $\tau' = \tau + \tau_1 + R_1 C$ and $\tau'\mu(V_1) = \tau R_1 \mathrm{d}i_1/\mathrm{d}V_1$; μ is, as near as no matter, the differential amplification factor of the valve with R_1 and R in parallel in its anode circuit, i.e. the change in anode potential for unit change in grid potential. With pentode valves μ can be much larger than unity in the normal operating range, but falls to zero outside. A typical valve characteristic, and the derived form of μ, are shown in fig. 42. A similar equation applies to the right-hand stage:

$$\tau\tau_1 \ddot{V}_1 + \tau'[\dot{V}_1 + \mu(V_2)\dot{V}_2] + V_1 = 0. \tag{31}$$

Together, (30) and (31) yield a non-linear fourth-order equation for either V_1 or V_2, but fortunately one or two special cases tell us enough to make full solution unnecessary. First let us examine the behaviour near the

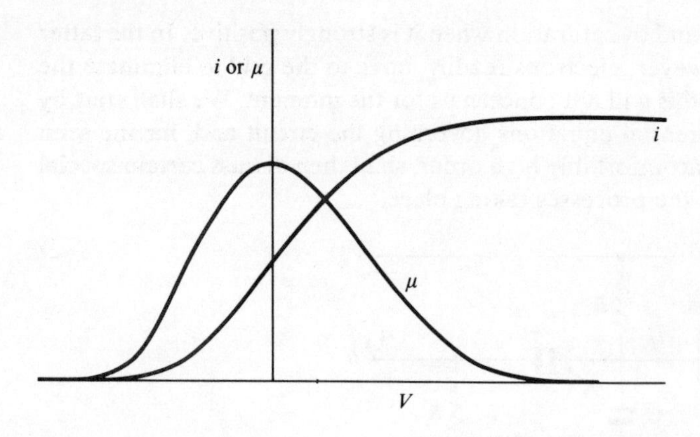

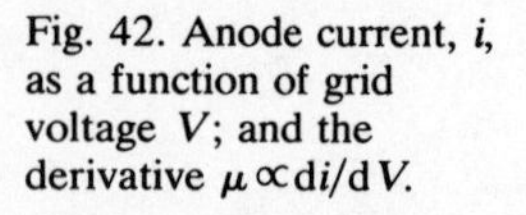
Fig. 42. Anode current, i, as a function of grid voltage V; and the derivative $\mu \propto \mathrm{d}i/\mathrm{d}V$.

equilbrium point, $V_1 = V_2 = 0$, on the assumption that μ is constant.† Then (30) and (31) are linear equations, inspection of which shows that solutions can be found in which $V_2 = \pm V_1$. Since four independent solutions are generated this way, the general solution is a sum of all four in arbitrary proportions. Each has the exponential form e^{pt}, where p obeys the equation

$$\tau_1 \tau p^2 + \tau'(1 \pm \mu)p + 1 = 0, \qquad (32)$$

the positive sign referring to the case $V_2 = V_1$ and the negative to the case $V_2 = -V_1$. The four values of p are therefore as follows:

$$p = \frac{1}{2\tau_1 \tau}\{-\tau' \mp \mu\tau' \pm [(1 \pm \mu)^2 \tau'^2 + 4\tau_1\tau_2]^{\frac{1}{2}}\}. \qquad (33)$$

So long as μ is large, there is always a positive real value of p which will guarantee the instability of the equilibrium state; any displacement grows exponentially with a time-constant approximately equal to τ_1/μ. This is obtained by taking the positive sign in (32), and corresponds to V_2 and V_1 diverging in opposite senses. The time-constant τ_1 is short enough in its own right, since C_1 is a very small capacitance, and the presence of μ makes the divergence even more rapid. If we had started by neglecting C_1 altogether we should not have deduced a discontinuous jump from equilibrium equivalent to taking τ_1 as zero, but should have missed this solution altogether, and hit on something entirely incorrect. For if $\tau_1 = 0$ in (32) the dominant positive value of p is approximately $1/\mu\tau'$, giving a characteristic time-constant μ times as long as the already substantial τ'. Having recognized this, however, we know enough to keep τ_1 in as long as it matters and thenceforth to neglect it if we are so minded.

The virtually abrupt breakdown of equilibrium involves a jump to quite different values of V_1 and V_2, well outside the linear regime of the

† Strictly we should have taken note that a small current flows to the grid when its potential is zero, and the equilibrium point has slightly negative values of V_1 and V_2. This is not important for the analysis.

amplifiers. To see what happens let us consider a special case, not too far from physical reality, in which $\mu(V)$ is a symmetrical function, $\mu(-V)=\mu(V)$. This allows V_1 and V_2 to stay equal and opposite at all times, and reduces the problem to easily manageable proportions. For (31) now takes the form, applying to either V_1 or V_2:

$$\tau_1\ddot{V}+\tau'[1-\mu(V)]\dot{V}/\tau+V/\tau=0, \qquad (34)$$

the equation of a harmonic vibrator for which the viscous damping term depends on position (i.e. V), being strongly negative for small displacements and not so strongly positive for large. One can immediately visualize how such a system will behave. When displaced from zero, it will be impelled by a force of the order of $\mu\dot{V}$ and will continue to accelerate away from the origin until μ has fallen nearly to unity. The motion will continue while the kinetic energy is either dissipated or stored as potential energy. On the return journey the speed will be limited by the longest time-constant in the solution of (34), i.e. $\tau'[1-\mu(V)]$ if τ_1 is very short. For much of the time this will be almost equal to τ', since $\mu\sim0$, but as the displacement approaches the value where μ begins to rise once more the motion will speed up until suddenly, as μ passes through unity, a very rapid reversal of displacement will occur, the motion being again dominated by the shortest time-constant. Something like the same process will now occur with reversed signs, and it will not be long before a steady state is reached, with V_1 and V_2 alternating in their positive and negative excursions. If we now neglect τ_1 the pattern is clear – everything happens outside the range in which $\mu>1$; every time the system reaches the edge of this range it jumps instantaneously to a far point on the opposite side, from which it relaxes back until it reaches a critical point once more. The essential feature of multivibrator action are displayed by this model, and we need take the argument only a little further.

In the light of this discussion, being assured that almost everything of importance takes place with one amplifier cut off and the other saturated, let us ignore τ_1 but let us no longer assume a symmetrical form for $\mu(V)$, so that V_1 and V_2 need not be equal and opposite. The fact that this model demands instantaneous jumps of V_1 and V_2 tells us that there must be different pairs of values of (V_1, V_2) consistent with the same charges on the capacitors C; for the charges cannot be altered abruptly. Let q_1 be the charge on the left-hand capacitor, as shown in fig. 41, and q_2 the charge on the right-hand. The condition for q_1 to take a given value is obtained by writing $V_X=V_2+Cq_1$ in (28), which then assumes the form:

$$V_2+\tau R_1 i_1(V_1)/\tau'=F(q_1). \qquad (35)$$

This equation defines a family of curves of constant q_1 on a (V_1, V_2)-plane, each identical in shape with the amplifier characteristic of fig. 42, except for a sign reversal. A similar set of curves of constant q_2 is obtained by interchanging q_1 and q_2. Two of each set are shown in fig. 43, the others being obtained by vertical displacement of any one constant-q_1 curve,

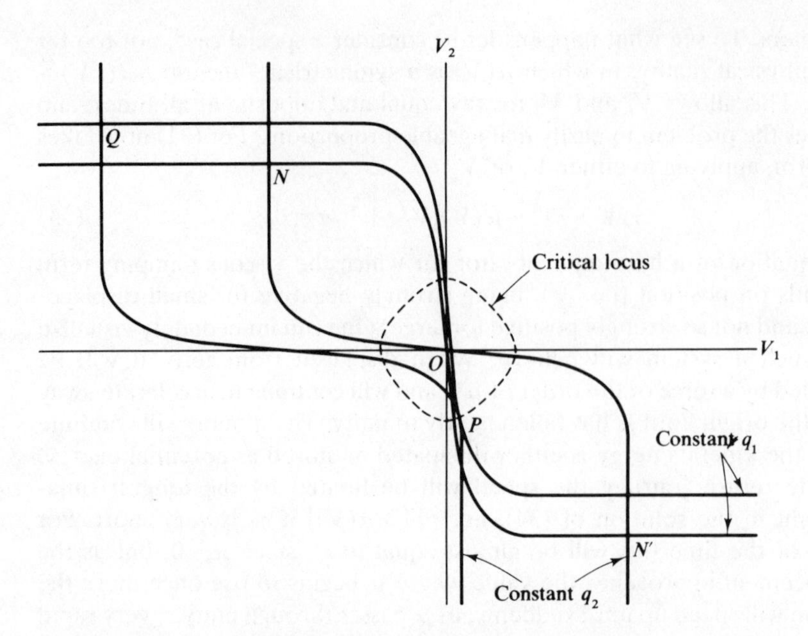

Fig. 43. Typical curves of constant q_1 and q_2; when the curves touch, as at P, P lies on the critical locus defined by the equation $\mu(V_1)\mu(V_2)=1$.

horizontal displacement of any one constant-q_2 curve. The two curves that intersect at the origin and also at N and N' illustrate the two possibilities immediately after switching on; the system is unstable at O and may switch to N or N', depending on chance. From then on the oscillation is under way, as already described; and the detailed process is made clear by considering the other two curves, drawn so as to touch at P and intersect at Q. The condition for touching follows from (35), which shows that the slope $\mathrm{d}V_2/\mathrm{d}V_1$ of the constant-q_1 curve is $-\mu(V_1)$; similarly $\mathrm{d}V_1/\mathrm{d}V_2$ for the constant-q_2 curve is $-\mu(V_2)$. For the curves to have the same slope at a given point (V_1, V_2),

$$\mu(V_1)\mu(V_2)=1. \tag{36}$$

This is the generalized form of the critical condition, $\mu=1$, which we derived for the symmetrical amplifier characteristic, and the points which satisfy (36) define the critical locus for the problem, in the following sense. It is easy to show from (30) and (31), with τ_1 neglected, that trajectories on the plane all tend towards this locus. Moreover V_1 and V_2 can move along the common tangent to the constant-q curves at P without incurring any changes in q_1 and q_2; hence the representative point arrives at the critical locus with infinite speed (finite if C_1 is included in the calculation), but can proceed no further. It therefore jumps immediately to Q. This is exactly analogous to the process illustrated in fig. 38.

At this point we must take account of a hitherto neglected factor. At Q, V_2 is highly positive, and the grid of amplifier 2 will quickly collect enough electrons to reduce V_2 (V_B in figs. 39 and 40) to a low value. As this happens the anode current falls slightly from its saturation value and the potential at Y rises a little. These effects can be traced in the waveforms of fig. 40, and are schematically represented in fig. 44. Starting at Q, the first stage QR is the rapid elimination of the positive grid potential V_2, while V_1 changes only slightly, being limited by the time-constant τ' which controls the discharge of the right-hand capacitor through R and R_1. From R to S this slow discharge continues and only as it comes very near to S, which lies on the critical locus, does it speed up. From S the point jumps to T, the mirror of Q in a 45° line. The elimination of the positive grid potential V_1 and the subsequent slow discharge of the other capacitor account for TU and UP, and finally the jump from P to Q completes the cycle. The period is essentially defined by the time taken on the paths RS and UP.

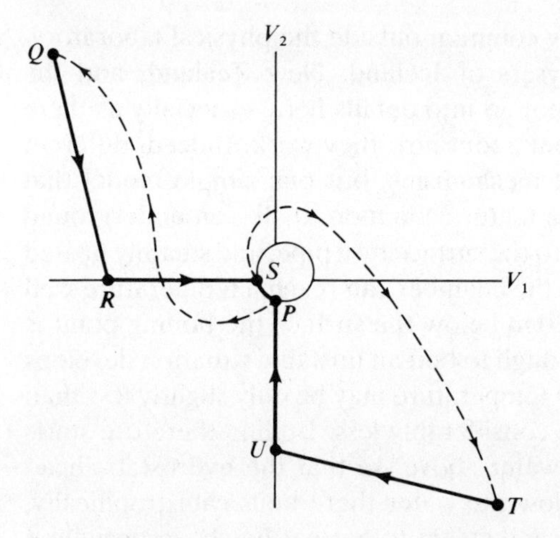

Fig. 44. Schematic representation of multivibrator cycle. The broken lines indicate virtually instantaneous jumps, from P to Q and from S to T.

It is now clear that the alternating saturation of each amplifier is automatic, depending in no way on fluctuations (as seems to be implied by some accounts), and that the whole process takes place outside the regime in which the amplifiers combine to give a gain greater than unity. There is a very close resemblance in both respects to the behaviour of the 'multivibrator' circuit of fig. 38, and we may feel justified in dropping the inverted commas and allowing the same name to be attached both to the pioneer circuit and to its simpler successor.

Relaxation oscillators

The multivibrator, whether in its original version or the simplified version of fig. 38, turns out to present a character remarkably similar to the linear time-base circuit of fig. 2.27, in which a gas-filled tube serves as an automatic switch, causing the RC combination, like Sisyphus, to strive ever vainly towards a state of equilibrium. The multivibrator is more symmetrical in its operation, having two states of opposite polarity from each

41

of which it relaxes exponentially (like an RC combination), only to be denied the attainment of equilibrium by a sudden switch to the alternate state. This behaviour may be taken as typical of the rather loosely defined *relaxation oscillator*, of which other examples were given in chapter 2, including the stick–slip phenomenon discussed earlier in the chapter. In this case the system has a natural frequency and perhaps cannot strictly be said to be a relaxation oscillator if by relaxation we mean exponential or quasi-exponential return to the equilibrium. More significant, however, is the fact that the driving force for the maintenance of oscillation comes from sudden changes in the physical parameters, and that the changes are triggered by the system itself, not by an external agency. This we take to be the fundamental feature, if any can be found, that defines this class of oscillators.

41
337

Relaxation oscillators are very common outside the physical laboratory, a very fine type being the geysers of Iceland, New Zealand, and the Yellowstone Park.[14] We shall not go into details here, especially as there seems to be no general agreement about how they work. Indeed, different geysers probably have different mechanisms, but one simple model that should apply to some builds on a feature common to all – an underground water-filled chamber connected to the surface by a pipe, and steadily heated by volcanic action. The water in the chamber can reach a temperature well above 100 °C without boiling (20 m below the surface the boiling point is 134 °C) and before it gets hot enough to boil an unstable situation develops some way up the pipe, where the temperature may be only slightly less than in the chamber but the pressure considerably less. Boiling therefore starts in the pipe, and blows out the water above, so that the hydrostatic head over the chamber is reduced. Now the water there boils catastrophically, blowing large volumes of water and steam to a great height (especially if the pipe enters the chamber well below its roof, and allows much of the water therein to be expelled before the pressure is released). At the end of the eruption, surface or underground water flows back into the chamber and cools it so that the steam condenses and the chamber is refilled almost completely. The heating process starts again and in due course another eruption takes place.

As described, this is clearly a relaxation process like the time-base circuit, but it is normally much more irregular. Among the large geysers, only Old Faithful in Yellowstone Park can be relied on to erupt at the predicted time, and this may be more legend than fact; most are irregular in their periods and violence. This need not be thought of as remarkable, for the process is far more complicated than a simple electrical circuit, and the exact sequence followed must depend on just where the first ebullition occurred, and how much water it managed to eject; after that there is scope for variation in every subsequent stage of the action. In really complex situations irregularity is the rule rather than the exception, for there is no resonant circuit to control the frequency, which is at the mercy of every stray influence. This is particularly true of oscillatory processes in biological systems, many of which (heart beat, menstrual cycle) can be seen to follow the general pattern of relaxation oscillations. Take, for example, the egg-laying of the domestic fowl; the alternating processes of build-up and discharge are clearly exhibited, and so is the variability in period between eggs and in their size, occasioned by too extensive a range of possible causes to be catalogued here. It is worth noting in this connection a study of the egg-weight in ducks which shows how a conflict between the natural period of egg laying and the daylight rhythm leads to systematic variations of weight and periodic non-laying days.[15] This is a process than can be 405 modelled by a relaxation oscillator under the influence of a periodic signal at a frequency other than the natural frequency of the oscillator. Not every such oscillator is susceptible to being forced out of its natural period, and even those that can be locked to the controlling frequency can only be persuaded to shift to a limited extent. The duck that continues to lay eggs of more or less the same weight day after day is one that has responded to the locking signal; the duck that lays successively lighter eggs for a few days, and then after a short gap begins the cycle again, is one whose natural period is significantly greater than a day and cannot be brought under control.

This has brought us to the threshhold of a major new topic, the mutual influences of oscillatory systems on one another, to which we shall turn in the next chapter. There is, however, one aspect of the disturbance of an oscillator by extraneous signals that is conveniently dealt with here, and that is the effect of noise voltages on the phase and amplitude of a class C oscillator, the simplest example to demonstrate the characteristic behaviour.

318

The effect of noise on a class C oscillator

158

When Johnson noise or other random disturbances are neglected, a class C oscillator settles down to regular oscillation at the equilibrium amplitude, measured, for example, by q_0, the peak charge on the capacitor. An impulsive e.m.f., of magnitude P, injected into the circuit, sets up an additional current P/L which, one-quarter of a cycle later, has been exchanged for an additional charge q, of magnitude $P/\omega_0 L$, on the capacitor. If this impulse is injected at a moment when the oscillatory current is at a peak, it alters the amplitude of oscillation without affecting its phase. Subsequently the amplitude relaxes back to its equilibrium value with a characteristic time-constant $2Q/\omega_0$, as illustrated in fig. 10(a). If, on the other hand, the impulse is injected at a moment when the oscillatory current is zero, it adds a small quadrature component to the oscillation, changing its phase but not its amplitude. There is no recovery from this disturbance, the oscillation proceeding thenceforward as though nothing had occurred. Noise, therefore, considered as a random sequence of impulses arriving at all phases of the oscillation, gives rise to fluctuations of the amplitude about its mean, and to an irregular drift of phase, so that the frequency spectrum is no longer perfectly sharp. As we shall see, under ideal conditions the effect is extremely small and cannot be held to blame for the frequency fluctuations exhibited by a typical oscillator in practice. These are caused by variations in the values of circuit parameters as a result of temperature changes or various aging processes. Nevertheless it is worth analysing the noise problem, if only to satisfy oneself how small the effect is.

A vibration diagram would show the vector representing q_0 rotating uniformly at angular velocity ω_0 in the absence of noise. A voltage impulse P would add a vector of length $P/\omega_0 L$, parallel to the imaginary axis. From the point of view of a frame rotating with q_0, the new vector may point in any direction according to the phase at the moment of the impulse. A random sequence of impulses therefore contributes a randomly oriented set of vectors, whose resultant has a Gaussian probability distribution, as expressed by (4.37). If for the moment we forget the tendency of the amplitude to relax to its equilibrium value, we picture the end-point of the vector q_0 as the source of a cloud of points which diffuse out in a spreading circular patch, depicting the way in which the probability distribution of the vibration amplitude evolves with time. Introducing formally a diffusion constant D for this process (similar to, but not identical with, the D in (4.37)), we have the normal diffusion law for ρ, the density of points on the plane,

83

$$D\nabla^2\rho = \dot{\rho}, \tag{37}$$

with solution after time t,

$$\rho \propto t^{-1}\,\mathrm{e}^{-r^2/4Dt}, \tag{38}$$

r being measured from the end of the initial vector q_0. The mean-square displacement follows from (38):

$$\overline{r^2} = \int_0^\infty r^3\rho(r)\,\mathrm{d}r \Big/ \int_0^\infty r\rho(r)\,\mathrm{d}r = 4Dt. \tag{39}$$

But we know that if there are N impulses per unit time, the resultant of Nt random vectors of length $P/\omega_0 L$ is given by

$$\overline{r^2} = NtP^2/(\omega_0 L)^2,$$

from which we identify the diffusion constant,

$$D = \tfrac{1}{4}NP^2/(\omega_0 L)^2 = \tfrac{1}{4}\pi I_\omega/(\omega_0 L)^2 \quad \text{from (4.45)}, \tag{40}$$

92

if I_ω is the spectral intensity of the white voltage noise.

Having determined D, we may now tackle the real problem, taking the amplitude fluctuations first. Because the amplitude tends to relax back to q_0, after a long time the cloud of points representing possible resultants will lie close to a circle of radius q_0, phase information having been lost. The radial distribution may be determined by the same argument as is applied to sedimentation in Brownian theory – a point at radius $q_0 + y$ is not merely wandering as a result of noise impulses, it also moves purposefully back to the radius q_0. For exponential relaxation the radial velocity must be proportional to y, so that $\dot{y} = -\frac{1}{2}\omega_0 y/Q$; this ensures a time-constant $2Q/\omega_0$ for the noise-free relaxation process shown in fig. 10. If, then, the cloud of points has a density $\rho(y)$, the flux across unit length from this cause is $-\frac{1}{2}\omega_0 y\rho/Q$. This, however, is opposed by the diffusive flux, $-D\,\mathrm{d}\rho/\mathrm{d}y$, and the equilibrium distribution results from the balance of the two fluxes:

$$D\,\mathrm{d}\rho/\mathrm{d}y = -\tfrac{1}{2}\omega_0 \rho y/Q.$$

Hence

$$\rho \propto \mathrm{e}^{-\omega_0 y^2/4QD}.$$

The distribution of amplitudes is Gaussian around the mean q_0, with

$$\overline{(\Delta q_0)^2} = \overline{y^2} = 2QD/\omega_0 = \tfrac{1}{2}\pi Q I_\omega/\omega_0^3 L^2 \quad \text{from (40).} \tag{41}$$

This may conveniently be expressed in terms of the energy, $\bar{\varepsilon}$, which would be possessed by an oscillation of this amplitude:

$$\bar{\varepsilon} = \tfrac{1}{2}\overline{(\Delta q_0)^2}/C = \tfrac{1}{4}\pi Q I_\omega/\omega_0^3 L^2 C = \tfrac{1}{4}\pi I_\omega/R, \tag{42}$$

since $\omega_0^2 = 1/LC$ and $Q = \omega_0 L/R$. To appreciate the magnitude of $\bar{\varepsilon}$, suppose the noise to be due entirely to the resistor R in the resonant circuit, giving a Johnson noise voltage whose intensity I_ω is $2Rk_\mathrm{B}T/\pi$, according to (6.49). Then

158

$$\bar{\varepsilon} = \tfrac{1}{2}k_\mathrm{B}T. \tag{43}$$

Johnson noise does no more than add to the existing oscillation a random element of about the same magnitude as would exist in the unexcited circuit anyhow. In reality there will be more than this, for the negative-resistance device will also contribute, probably several times as much but still not necessarily enough to raise $\bar{\varepsilon}$ to anything perceptible. For an oscillating circuit with a voltage amplitude of 5 V and a capacitor of 0.1 μF has oscillatory energy of about 1 μJ, more than 10^{14} times $\bar{\varepsilon}$ as given by (43).

In the light of this analysis it is clearly a good approximation to take the amplitude q_0 as constant when discussing the phase drift due to noise. The diffusion of the cloud of points round the circumference of a circle of radius q_0 is a one-dimensional process governed by the same diffusion constant D, and if $s(b)$ is the drift distance after time t, measured round the circle,

$$\overline{s^2} = 2Dt = \tfrac{1}{2}\pi I_\omega t/(\omega_0 L)^2.$$

The phase drift, θ, corresponding to s is s/q_0 and therefore

$$\overline{\theta^2} = \tfrac{1}{2}\pi I_\omega t/(\omega_0 L q_0)^2. \tag{44}$$

To put this result in perspective, let us use (41) to express I_ω in terms of $\overline{(\Delta q_0)^2}$:

$$\overline{\theta^2} = \omega_0 t\overline{(\Delta q_0/q_0)^2}/Q. \tag{45}$$

If we wait long enough for the mean-square phase drift to be 1 radian, the number of radians of oscillation that must elapse is the value of $\omega_0 t$ when $\overline{\theta^2}=1$, i.e. $Q/\overline{(\Delta q_0/q_0)^2}$, perhaps 10^{16} with the figures quoted above and with Q equal to 100. No real oscillator begins to approach this frequency stability – long-term stability of one part in 10^{12} is very good indeed – and it seems fair to conclude that, even with much less energetic oscillators, thermal noise is unlikely to be the limiting factor in precision timekeeping.

It may therefore be thought rather academic to pursue the matter further, but it is worth restating the last argument, if only as an illustration of the use of the autocorrelation function. The oscillator output takes the form $q_0 \exp[-\mathrm{i}(\omega_0 t-\theta)]$, θ being a phase angle that wanders randomly. In a time τ the change in θ, which we write as $\theta(\tau)$, has a Gaussian distribution about zero, with mean-square value $\overline{\theta^2(\tau)}$ proportional to τ, as in (44). Hence the autocorrelation function, $\Gamma(\tau)$, according to (4.47) takes the form $\overline{\exp \mathrm{i}\theta(\tau)}\exp(-\mathrm{i}\omega_0\tau)$, i.e. $\overline{\cos\theta(\tau)}\exp(-\mathrm{i}\omega_0\tau)$ since positive and negative values of $\theta(\tau)$ are equally likely. If the probability of $\theta(\tau)$ taking a certain value is proportional to $\exp[-\frac{1}{2}\theta^2/\overline{\theta^2(\tau)}]$ the mean value of $\cos\theta(\tau)$ is readily evaluated* as $\exp[-\frac{1}{2}\overline{\theta^2(\tau)}]$.†

94

Hence

$$\Gamma(\tau)=\mathrm{e}^{-|\tau|/\tau_c-\mathrm{i}\omega_0\tau}, \tag{46}$$

where, from (45),

$$\tau_c=2Q/\omega_0\overline{(\Delta q_0/q_0)^2}$$

The Fourier transform of $\Gamma(\tau)$ gives the spectral intensity of the oscillator output, being a Lorentzian line shape of the form

$$I_\omega \propto 1/[1+(\omega-\omega_0)^2\tau_c^2].$$

The effect of noise is to spread the output from a perfectly sharp frequency into a band of width about $2/\tau_c$, or $\omega_0\overline{(\Delta q_0/q_0)^2}/Q$. This is an alternative to (45) as a specification of the frequency stability, but of course it is equivalent.

Exercise. An alternative approach to some of the examples in this chapter is revealing. Apply the Nyquist diagram to the circuits of fig. 14, breaking the circuit at the positive amplifier input to determine the loop gain, $G(\omega)$; remember that $A(\omega)$ for the amplifier falls off as ω rises, and that the Kramers–Kronig relations demand that A be complex. The series and parallel arrangements, fig. 14(a) and (b), give significantly different Nyquist diagrams. Try this approach also on the tunnel-diode oscillator and the multivibrator. In the latter case, neglect C_1 in (28) and instead allow inter-electrode capacitance etc. to play their part in reducing μ at higher frequencies, so rendering it complex.

313, 327

110

333
355

† The reader may like to derive the form of $\Gamma(\tau)$ directly from the diffusion equation, using the same argument (but for diffusion on a circle rather than on a sphere) as was used to arrive at (8.29).

232

12 Coupled vibrators

A resonant system acted upon by an oscillatory force presents a straightforward enough problem if it is passive and linear, especially if the force is applied by some prime mover that is uninfluenced by the response it excites. Such problems are the subject matter of chapter 6, while non-linear passive systems, as discussed in chapter 9, are more complicated to handle. If the prime mover is influenced by the response, additional complexities enter, and this chapter treats of some of these. As is to be expected, linear systems present the least difficulty, and we shall begin with the behaviour of two coupled resonant systems, each of which may be thought of as driving the other and being perturbed by the reaction of the other back on it. Examples have already appeared earlier, as for instance **24** the coupled pendulums discussed in chapter 2, and the coupled resonant **175** lines in chapter 7. In both cases we noted a very general characteristic of such systems, that even if they are tuned to the same frequency before being coupled, they do not vibrate at this frequency when coupled, but have resonances which move progressively further from the original frequency as the coupling is strengthened. It would perhaps be logical, having considered this problem, to proceed to coupled, passive, non-linear vibrators; but these are so difficult that we shall leave them alone. It is not quite so hard to derive useful results for the behaviour of self-maintained oscillators when perturbed either by the injection of a steady signal or by being coupled to a **391** similar oscillator. These are significantly different from passive resonant systems in that they exhibit the Huygens phenomenon, locking in frequency to the input signal or to one another, with characteristic critical behaviour at that strength of coupling which initiates locking. There are also special **385** features worth investigating when a self-maintained oscillator is coupled to a passive resonant system, and these are particularly interesting when the **405** oscillator is a multivibrator or something similar with a highly non-sinusoidal waveform. Some aspects of the behaviour here are strongly reminiscent **416** of the way a Josephson junction between superconducting materials can be brought into oscillation and locked in frequency to a passive resonant circuit, and we shall conclude by giving a brief account of this phenomenon.

Coupled passive vibrators

It is convenient to develop a general picture of the mode-splitting phenomenon in coupled vibrators through electrical circuits, and to show the analogy with mechanical and other systems afterwards. In the circuit of fig. 1(a), the two series resonant circuits are coupled by means of a four-terminal element, of which a few realizations are given in figs. 1(b)–(f).† If it is a linear coupler, the e.m.f.s across its terminals in circuits 1 and 2 are expressible in the form:

$$V_1 = \mathfrak{z}_{11} i_1 + \mathfrak{z}_{12} i_2, \qquad V_2 = \mathfrak{z}_{21} i_1 + \mathfrak{z}_{22} i_2, \tag{1}$$

all $\mathfrak{z}$s being functions of frequency, though those shown in the diagram vary only slowly with ω. Except for the gyrator in fig. 1(f) the coupler is symmetrical, having $\mathfrak{z}_{21} = \mathfrak{z}_{12}$, but the gyrator has $\mathfrak{z}_{21} = -\mathfrak{z}_{12}$. We consider the gyrator separately and concentrate first on symmetrical couplers. The impedances $\mathfrak{z}_{11}$ and $\mathfrak{z}_{22}$ may be incorporated in the individual circuits along with the other elements, so that the natural resonant frequency of circuit 1, ω_1, is taken as that for which $\mathrm{j}\omega_1 L_1 + R_1 + 1/\mathrm{j}\omega_1 C_1 + \mathfrak{z}_{11}(\omega_1) = 0$; ω_1 in general is complex. If the circuit has a high Q, as we shall assume throughout this discussion, its impedance at frequency Ω, close to resonance, is very nearly $2\mathrm{j}L_1(\Omega - \omega_1)$. The equations for the coupled circuits then take the form

378

135

$$2\mathrm{j}L_1(\Omega - \omega_1) i_1 + \mathfrak{z}_{12} i_2 = 0 \tag{2}$$

and

$$2\mathrm{j}L_2(\Omega - \omega_2) i_2 + \mathfrak{z}_{21} i_1 = 0, \tag{3}$$

from which the possible resonant frequencies, $\Omega_{1,2}$, are seen to be solutions of the equation

$$(\Omega - \omega_1)(\Omega - \omega_2) = \kappa^2, \tag{4}$$

where $\kappa^2 \equiv -\mathfrak{z}_{12}\mathfrak{z}_{21}/4L_1L_2$. The coupling constant κ is real if $\mathfrak{z}_{12}$ and $\mathfrak{z}_{21}$ are equal and purely reactive (figs. 1(b)–(d)), imaginary if they are purely resistive (fig. 1(e)). There are two separate solutions (*normal modes*) to (4), and the system may, if properly excited, be set into free oscillation at either frequency. In general, however, both solutions coexist independently and the observed variations in frequency and amplitude of the currents in the two circuits are the result of the linear addition of the two modes.

† The circuit shown in fig. 1(g), in which two parallel resonant circuits are coupled by, for example, a capacitor C_3, is very similar in general behaviour to that in fig. 1(a) but a little more complicated. The difference can be pinpointed by observing that if $\mathfrak{z}_c$ is a capacitor, as in 1(b), the energy of oscillation in the circuit 1(a) can be expressed entirely in terms of q_1 and q_2, the charges on C_1 and C_2, and their first derivatives – the coupling introduces no more coordinates than were already needed for the uncoupled systems. The same is not true of the circuit 1(g), which has an additional degree of freedom in that there are three loops to the network instead of two.

Fig. 1. (*a*) Two resonant circuits coupled by $\mathfrak{z}_c$; (*b*)–(*e*) forms of $\mathfrak{z}_c$, having reciprocal properties; (*f*) non-reciprocal (gyratory) coupling; (*g*) an alternative form of coupling.

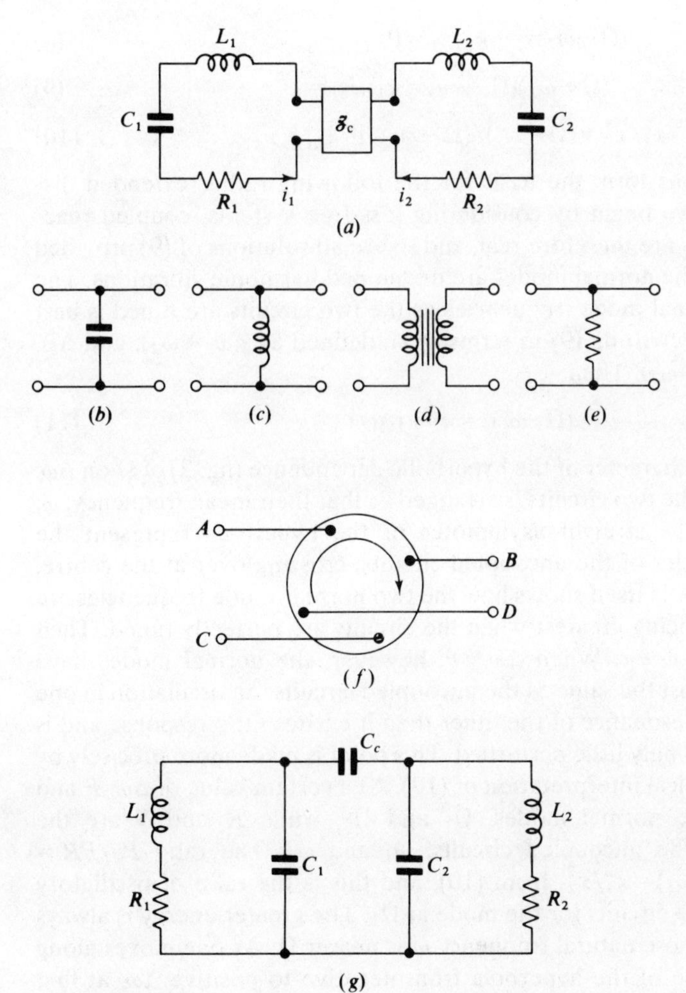

The currents in the two circuits are in general different when the system is excited in a pure normal mode. From (2) and (3) it is seen that

$$L_1^{\frac{1}{2}}i_1/L_2^{\frac{1}{2}}i_2=\kappa/(\Omega-\omega_1)=(\Omega-\omega_2)/\kappa=(\Omega-\omega_2)^{\frac{1}{2}}/(\Omega-\omega_1)^{\frac{1}{2}}. \qquad (5)$$

The form of (5) shows that it is natural to measure the response of a circuit not by the currents but by coordinates $x_{1,2}$, defined so that

$$L_1i_1^2=x_1^2, \qquad L_2i_2^2=x_2^2. \qquad (6)$$

In other words, something proportional to the square root of the energy of excitation is taken instead of current as a suitable coordinate, and then the different magnitudes of L_1 and L_2 play no further part in the equations, except in so far as they help determine ω_1 and ω_2. Rewriting (2), (3), (4) and (5), with κ_{12} defined as $\frac{1}{2}\mathrm{j}\mathfrak{z}_{12}/(L_1L_2)^{\frac{1}{2}}$, we have that

$$(\Omega-\omega_1)x_1-\kappa_{12}x_2=0, \qquad (7)$$

$$(\Omega-\omega_2)x_2-\kappa_{21}x_1=0, \tag{8}$$

$$(\Omega-\omega_1)(\Omega-\omega_2)=\kappa_{12}\kappa_{21}=\kappa^2, \tag{9}$$

and
$$x_1^2/x_2^2=(\Omega-\omega_2)/(\Omega-\omega_1),\ \text{if}\ \kappa_{21}=\kappa_{12}. \tag{10}$$

These equations form the basis for the following rather extended discussion, which we begin by considering loss-free systems, coupled reactively; ω_1 and ω_2 are therefore real, and so are all solutions of (9) provided κ^2 is positive – the normal modes are undamped harmonic vibrations. The variation of normal mode frequencies as the two circuits are tuned is best represented by rewriting (9) in terms of $\bar{\omega}$, defined as $\frac{1}{2}(\omega_1+\omega_2)$, and $\Delta\omega$, defined as $\frac{1}{2}(\omega_1-\omega_2)$. Then

$$(\Omega-\bar{\omega})^2=\kappa^2+(\Delta\omega)^2. \tag{11}$$

This reveals the character of the hyperbolic dependence (fig. 2) of Ω on $\Delta\omega$. If the tuning of the two circuits is arranged so that their mean frequency, $\bar{\omega}$, stays constant the straight asymptotes of the hyperbola represent the natural frequencies of the uncoupled circuits, crossing over at the centre, while the hyperbola itself shows how the two normal-mode frequencies are split, the effect being greatest when the circuits are perfectly tuned. Then $\Delta\omega=0$ and $\Omega=\bar{\omega}\pm\kappa$. When $\Delta\omega\gg\kappa$, however, the normal modes have frequencies almost the same as the uncoupled circuits; an oscillation in one is so far off the resonance of the other than it excites little response and is itself correspondingly little perturbed. This point is made more precisely by noting the graphical interpretation of (10). At a certain value of $\Delta\omega$, P and Q represent the normal modes, Ω_1 and Ω_2, while R and S are the frequencies of the uncoupled circuits, ω_1 and ω_2. The ratio $PS/PR = (\Omega_1-\omega_2)/(\Omega_1-\omega_1)=x_1^2/x_2^2$, from (10), and this is the ratio of oscillatory energy in the two circuits for the mode at Ω_1. The greater energy is always in that circuit whose natural frequency ω is nearer Ω. As one moves along the upper branch of the hyperbola from negative to positive $\Delta\omega$, at first $\omega_2>\omega_1$ and the upper mode has almost all its energy in circuit 2; when $\Delta\omega=0$ both circuits are equally excited, and ultimately as $\Delta\omega$ goes positive

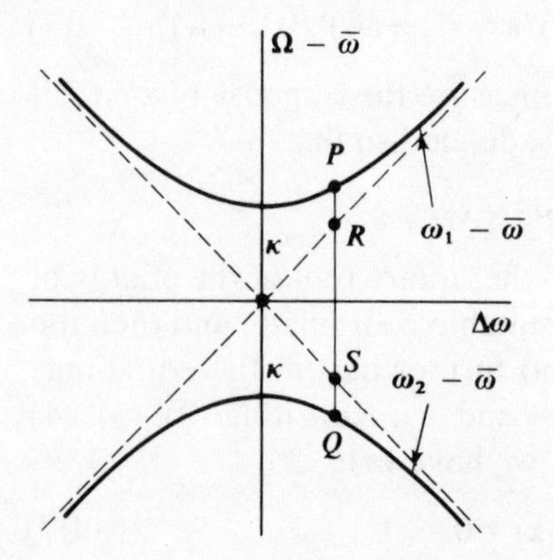

Fig. 2. Splitting of normal-mode frequencies.

the energy is concentrated in circuit 1, which now has the higher natural frequency. The opposite, of course, occurs with the lower mode, for which the energy partition is in the ratio QR/QS, the same as PS/PR.

The relative phase of the currents is given by (5); if κ is positive as with capacitive coupling, the currents are in phase in the upper mode ($\Omega > \omega_1$) and in antiphase in the lower mode. This is obvious from the circuit – if i_1 and i_2 oscillate in phase, the coupling capacitor, C_c of fig. 1(*b*), is charged and discharged periodically and introduces extra capacitance in series with C_1 in circuit 1, or C_2 in circuit 2. The combined capacitance being smaller, the natural frequency is increased; the effect is less marked, or even reversed, when the currents are in antiphase.† With inductive coupling ($\kappa < 0$), however, as with fig. 1(*c*), the effective inductance is increased and the frequency reduced when the currents are in phase. There is a sound energetic argument for the currents to be in phase or in antiphase in a pure normal mode – only thus can there be no mean power flowing from one circuit to the other, the current in one circuit being in phase quadrature with the e.m.f. due to the other. With capacitive coupling, for example, any mean power flow is the result of i_1 flowing through C_c in the presence of the e.m.f. $i_2/\mathrm{j}\Omega C_c$, and is zero.

The result just derived, that x_1/x_2 in one mode is the same as $-x_2/x_1$ in the other, makes possible an elegant and valuable graphical representation of the normal modes, for the lines $x_1 = cx_2$ and $x_2 = -cx_1$, c being a constant, are orthogonal when x_1 and x_2 are chosen as cartesian coordinates. On the (x_1, x_2)-plane of fig. 3, therefore, the normal modes appear

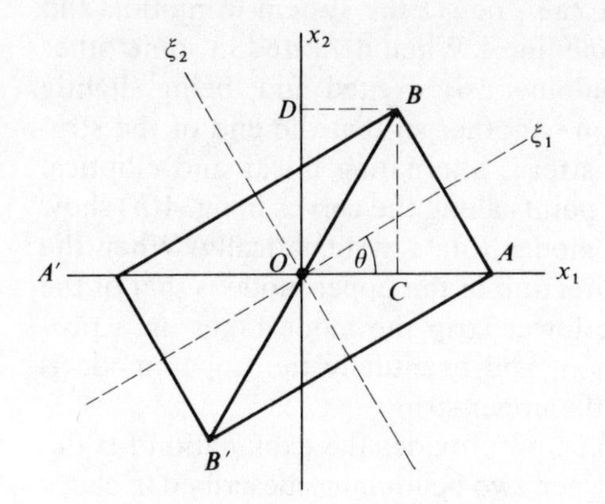

Fig. 3. Normal mode coordinates, ξ_1 and ξ_2, shown as a linear combination of x_1 and x_2 equivalent to a rotation of axes.

† If the currents are so phased that the effect of one circuit on the other is equivalent to introducing an extra capacitance, so also does the first circuit acquire extra capacitance. This reciprocity holds in general for reactive coupling when the currents are either in phase or in antiphase. Thus both natural frequencies are shifted in the same direction, and this explains why the normal modes lie outside the range between ω_1 and ω_2. The circuit whose frequency is less affected is obviously that which is more strongly excited, for the extra e.m.f. due to the other is relatively weak.

as harmonic vibrations along two orthogonal lines, ξ_1 and ξ_2. To find the orientation of the lines, put $\kappa/\Delta\omega = \tan 2\theta$; then from (11) the frequencies of the normal modes are given by

$$\Omega = \omega \pm \Delta\omega \sec 2\theta.$$

Taking the positive sign for the upper mode frequency, Ω_1, we evaluate $(\Omega_1 - \omega_2)/(\Omega_1 - \omega_1)$ and substitute in (10), to find that $x_1/x_2 = \cot\theta$, so that θ is the angle shown in fig. 3. This is for the case of capacitive coupling, when x_1 and x_2 have the same sign in the upper mode, represented by ξ_1; as $\Delta\omega$ runs from large negative to large positive values, θ swings round from 90° to 0, being 45° when $\Delta\omega = 0$. With inductive coupling the process is reversed in sign, θ swinging round from 0 to 90°; ξ_2 is now the upper mode, with the currents in antiphase.

A simple experiment to demonstrate these points is shown in fig. 4. When the two strips are joined at right angles their vibrations are independent; the natural frequency of the upper strip is unaffected by whatever length of lower strip may be protruding. The variations of normal mode frequencies are shown in fig. 4(*b*), and ideally even at the point where the curves cross there should be no coupling and no splitting of the modes. On the other hand, when the strips are not at right angles the bending of one exerts a bending moment on the other, and the vibrations are coupled. Consequently the modes are split in the characteristic 'hyperbolic' fashion, with the unperturbed modes as asymptotes far from the cross-over point. With a calibrated stroboscopic flash one can measure both the frequencies and the vibration directions of the normal modes. Only in two orthogonal directions can one set the system in motion and have it stay vibrating along the same line.† When deflected in some other direction the two modes are simultaneously excited and, being slightly different in frequency, beat with one another so that the end of the strip describes an evolving Lissajous pattern, alternating linear and elliptical motions. The indicators at various points along the curves in fig. 4(*b*) show how the directions of the normal modes rotate systematically. When the lower strip is short the vibration direction of the upper mode is that of the lower strip. As one lengthens the lower strip the mode takes on a proportion of the upper strip vibration, and eventually the upper mode is entirely composed of vibration of the upper strip.

The Lissajous figure just remarked on provides the explanation for the characteristic energy exchange between two pendulums, described in chapter 2. The effect is most marked when the two pendulums are exactly tuned, but something similar occurs even when they are not. With the parameters

24

† It is not strictly true that the normal modes have the end of the upper strip vibrating in orthogonal directions, but the discrepancy is hardly noticeable. The reader may care to work out for himself how the model does not quite fit the theoretical framework and is therefore enabled to violate, ever so slightly, this prediction of the theory.

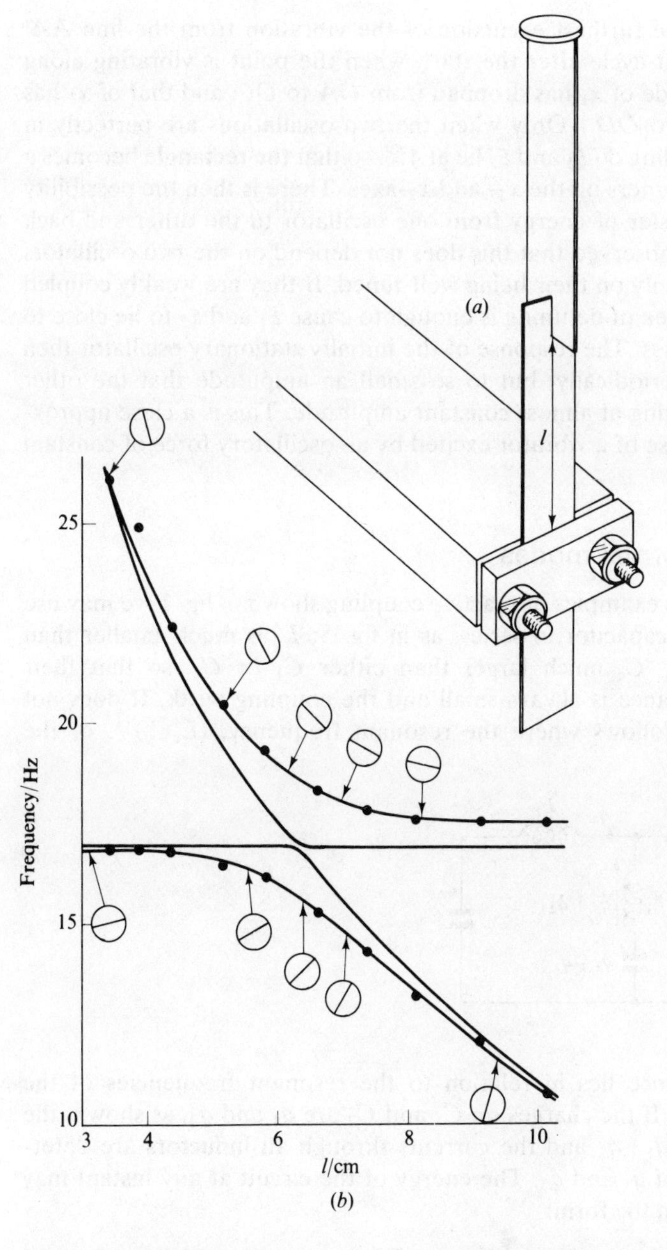

Fig. 4. (*a*) Simple experiment on coupled vibrations; (*b*) experimental results; the lines without points show the behaviour when the strips are nearly at right angles, those with points when the strips make an angle of 80°. The lines in circles indicate the vibration direction at various points along the curves.

such as to give normal mode orientations as shown in fig. 3, for example, displacing oscillator 1 to x_1 and releasing it, while $x_2 = 0$ at the moment of release, means starting the system at a point A, at which both normal modes are excited, ξ_1 with amplitude $x_1 \cos\theta$ and ξ_2 with amplitude $x_1 \sin\theta$. Subsequently, as each mode continues to vibrate independently of the other, the representative point executes a Lissajous figure contained within the rectangle $ABA'B'$, the time for a complete beat-cycle being

$2\pi/(\Omega_1-\Omega_2)$. The furthest excursion of the vibration from the line AA' occurs half a beat-cycle after the start, when the point is vibrating along BB'; the amplitude of x_1 has dropped from OA to OC, and that of x_2 has risen from zero to OD.† Only when the two oscillations are perfectly in tune before coupling do ξ_1 and ξ_2 lie at 45°, so that the rectangle becomes a square with its corners on the x_1- and x_2-axes. There is then the possibility of complete transfer of energy from one oscillator to the other and back again. It will be observed that this does not depend on the two oscillators being identical, only on their being well-tuned. If they are weakly coupled a very small degree of detuning is enough to cause ξ_1 and ξ_2 to lie close to the x_1- and x_2-axes. The response of the initially stationary oscillator then rises and falls periodically, but to so small an amplitude that the other continues oscillating at almost constant amplitude. This is a close approximation to the case of a vibrator excited by an oscillatory force of constant amplitude.

146

Energy and normal modes

In addition to the examples of reactive coupling shown in fig. 1 we may use an inductor and capacitor in series, as in fig. 5; L_c is much smaller than either L_1 or L_2, C_c much larger than either C_1 or C_2, so that their combined impedance is always small and the coupling weak. It does not matter in what follows where the resonant frequency, $(L_cC_c)^{-\frac{1}{2}}$, of the

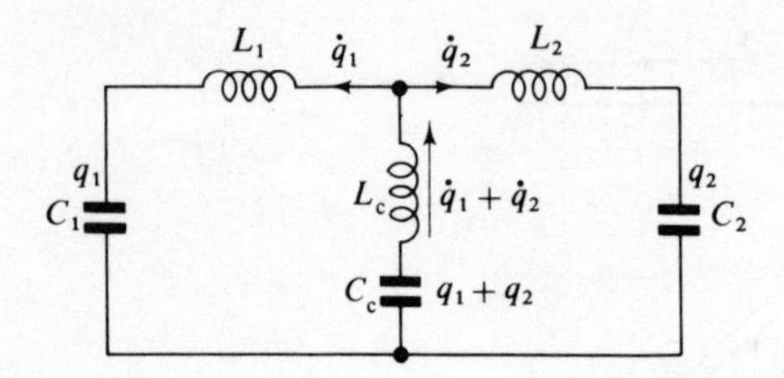

Fig. 5. As in fig. 1(a), with $\mathfrak{z}_c$ in the form of L_c and C_c in series.

coupling impedance lies in relation to the resonant frequencies of the separate circuits. If the charges on C_1 and C_2 are q_1 and q_2, as shown, the charge on C_c is q_1+q_2 and the currents through all inductors are determined in terms of $\dot{q}_1$ and $\dot{q}_2$. The energy of the circuit at any instant may then be written in the form:

$$E=\Phi+T,$$

† The reader may care to verify that when two normal modes are present together, the e.m.f. due to one circuit and the current in the other upon which it acts, at the coupling impedance, are not in phase quadrature. There is thus a steady power flow, as of course there must be when the beating phenomenon takes place. As the beat-cycle progresses, the phase relationship alters steadily, and energy flows to and fro through $\mathfrak{z}_c$.

where the potential energy,

$$\Phi = \tfrac{1}{2}q_1^2/C_1 + \tfrac{1}{2}q_2^2/C_2 + \tfrac{1}{2}(q_1+q_2)^2/C_c$$

27 $$= \tfrac{1}{2}q_1^2/\bar{C}_1 + \tfrac{1}{2}q_2^2/\bar{C}_2 + q_1q_2/C_c, \quad (12)$$

and the kinetic energy,

$$T = \tfrac{1}{2}L_1\dot{q}_1^2 + \tfrac{1}{2}L_2\dot{q}_2^2 + \tfrac{1}{2}L_c(\dot{q}_1+\dot{q}_2)^2$$
$$= \tfrac{1}{2}\bar{L}_1\dot{q}_1^2 + \tfrac{1}{2}\bar{L}_2\dot{q}_2^2 + L_c\dot{q}_1\dot{q}_2. \quad (13)$$

In these expressions $1/\bar{C}_{1,2}$ is written for $1/C_{1,2}+1/C_c$, $\bar{L}_{1,2}$ for $L_{1,2}+L_c$. The energy is a positive definite quadratic form in the variables q_1, q_2, $\dot{q}_1$ and $\dot{q}_2$. If contours of Φ are plotted on a (q_1, q_2)-plane they form a family of similar ellipses whose axial ratios are all the same and which are all oriented alike with respect to the coordinate axes; any one may therefore be taken as representative. Similarly for the contours of T on a $(\dot{q}_1, \dot{q}_2)$-plane. If the two planes are superposed, with $\dot{q}_1 \parallel q_1$, the orientations of the ellipses need bear no special relationship to one another. We now enquire what determines the normal modes – in which directions can one draw a line representing the vibrations of q_1 and q_2 in phase and unchanging in amplitude?

To answer this question, note first that

$$\partial\Phi/\partial q_1 = q_1/C_1 + (q_1+q_2)/C_c, \quad \text{from (12)}, \quad (14)$$

and $$\partial T/\partial \dot{q}_1 = L_1\dot{q}_1 + L_c(\dot{q}_1+\dot{q}_2), \quad \text{from (13)}.$$

Hence $$\frac{\mathrm{d}}{\mathrm{d}t}(\partial T/\partial \dot{q}_1) = L_1\ddot{q}_1 + L_c(\ddot{q}_1+\ddot{q}_2). \quad (15)$$

Now (14) is the sum of the potential differences across the capacitors in circuit 1, and (15) the sum of e.m.f.s across the inductors. Hence

$$\frac{\partial\Phi}{\partial q_1} + \frac{\mathrm{d}}{\mathrm{d}t}\left(\frac{\partial T}{\partial \dot{q}_1}\right) = 0, \quad (16)$$

and similarly, for circuit 2,

$$\frac{\partial\Phi}{\partial q_2} + \frac{\mathrm{d}}{\mathrm{d}t}\left(\frac{\partial T}{\partial \dot{q}_2}\right) = 0. \quad (17)$$

These equations could have been derived by Lagrangian methods, which provide the natural formalism in which to analyse more complicated problems in coupled oscillatory systems. The derivation from first principles is so easy here, however, that we have preferred to do this. Let us now introduce two vectors,

$$\text{grad}\,\Phi \equiv \left(\frac{\partial\Phi}{\partial q_1}, \frac{\partial\Phi}{\partial q_2}\right) \quad \text{and} \quad \text{grad}\,T \equiv \left(\frac{\partial T}{\partial \dot{q}_1}, \frac{\partial T}{\partial \dot{q}_2}\right).$$

Then (16) and (17) may be combined into the equation†

$$\text{grad}\,\Phi + \frac{\mathrm{d}}{\mathrm{d}t}(\text{grad}\,T) = 0. \tag{18}$$

Since Φ is constant on any Φ-contour, grad Φ at any point has no component along the direction of the tangent to the contour passing through that point; it is therefore directed normal to the contour. Similarly grad T is everywhere normal to the T-contour. If the representative point is to oscillate on a straight line through the origin, and still obey (18), the similarity of the T-ellipses implies that grad T remains constant in direction all along the line, and therefore that $(\mathrm{d}/\mathrm{d}t)(\text{grad}\,T)$ is parallel to grad T. It is only possible to satisfy (18) along such lines as cut the Φ and T ellipses at the same angle, so that grad Φ and grad T are parallel to one another, though not in general parallel to the line representing the trajectory. An example of the possible normal modes is shown in fig. 6.

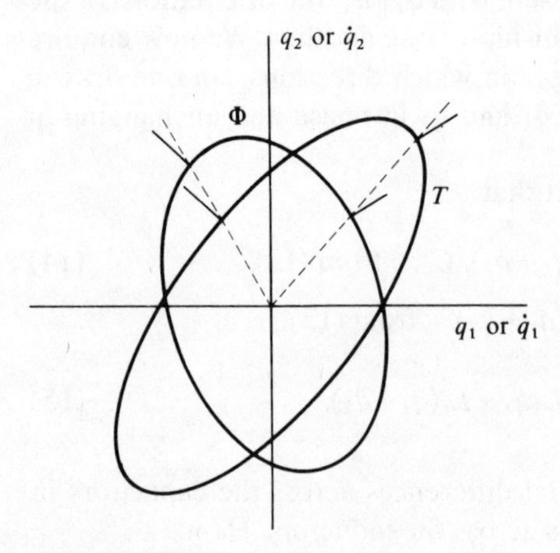

Fig. 6. Normal modes for the general case when Φ-contours are ellipses on the (q_1, q_2)-plane and T-contours different ellipses on the $(\dot{q}_1, \dot{q}_2)$-plane. The broken lines, on which the normals to the ellipses are parallel, define the ratio q_2/q_1 in each normal mode.

† The significance of this equation is perhaps more easily appreciated in the corresponding mechanical system. A mass moving on a line has kinetic energy $T = \frac{1}{2}m\dot{x}^2$, and $\partial T/\partial \dot{x} = m\dot{x}$, the momentum. Two masses, m_1 and m_2, with coordinates x_1 and x_2, have $T = \frac{1}{2}m_1\dot{x}_1^2 + \frac{1}{2}m_2\dot{x}_2^2$, represented by elliptical contours of constant T on an $(\dot{x}_1, \dot{x}_2)$-plane. The velocity vector $(\dot{x}_1, \dot{x}_2)$ and momentum vector $(m_1\dot{x}_1, m_2\dot{x}_2)$ do not point in the same direction; in fact grad T is the momentum vector, $\boldsymbol{P}$, normal to the contours of T and only parallel to the velocity on an axis of the ellipse. A body having tensorial mass, with principal values m_1 and m_2, is governed by the same equation. As for grad Φ, this is just $-\boldsymbol{F}$, the vector representing the forces on each mass. The general problem here is therefore to be thought of, if one wishes, as that of a sphere with tensorial mass rolling in an elliptical cup, and (18) is the equation of motion $\dot{\boldsymbol{P}} = \boldsymbol{F}$. The directions in which it can oscillate to and fro on a straight path are not in the principal planes of the cup, but oriented so that the restoring force, itself not radial, acts on the tensorial mass to accelerate it, not parallel to $\boldsymbol{F}$, but in a radial direction.

When q_1 and q_2 are chosen as variables the normal modes are not in general orthogonal. Unless one of them lies along an axis, however, it is always possible to scale the axes so as to make them orthogonal. The rule is simple when either Φ or T is free of cross-terms (q_1q_2 or $\dot{q}_1\dot{q}_2$). For one of the ellipses then has its axes along the coordinate axes and may be scaled into a perfect circle; the other ellipse, similarly scaled, remains elliptical and its axes are the orthogonal normal mode directions, for along these axes the normals to both ellipse and circle are radial. With capacitive coupling it is the T-ellipse which is to be scaled to a circle, and since the axis lying along $\dot{q}_1$ is $(L_2/L_1)^{\frac{1}{2}}$ times the length of the other axis, lying along $\dot{q}_2$, the q_2 and $\dot{q}_2$ values must be scaled up by $(L_2/L_1)^{\frac{1}{2}}$, leaving q_1 and $\dot{q}_1$ unchanged, to convert it into a circle. This is the same rule as yielded x_1 and x_2 as suitable coordinates. With inductive coupling the corresponding scaling factor is $(C_1/C_2)^{\frac{1}{2}}$, to convert the Φ-ellipse into a circle. In general, when both Φ and T have cross-terms, the factor is $[\bar{L}_2(\omega_2^2-\omega_c^2)/\bar{L}_1(\omega_1^2-\omega_c^2)]^{\frac{1}{2}}$, in which $1/\omega_{1,2}^2=\bar{L}_{1,2}\bar{C}_{1,2}$ and $1/\omega_c^2=L_cC_c$. The proof of this is left to the reader.

367

A very important property of normal modes is that they contribute independently to the energy of the vibrating system. To show this, we express an arbitrary state of oscillation as the superposition of the two normal modes, each vibrating at its own frequency, i.e. $\xi_1=A\cos(\Omega_1t+\phi_1)$ and $\xi_2=B\cos(\Omega_2t+\phi_2)$. Since the coordinates q_1, $\dot{q}_1$ etc. are linear combinations of ξ_1, $\dot{\xi}_1$ etc., and the total energy at any instant is a quadratic function of the qs and $\dot{q}$s, it is also a quadratic function of the ξs and $\dot{\xi}$s:

$$E=\tfrac{1}{2}a\xi_1^2+b\xi_1\xi_2+\tfrac{1}{2}c\xi_2^2+\tfrac{1}{2}d\dot{\xi}_1^2+e\dot{\xi}_1\dot{\xi}_2+\tfrac{1}{2}f\dot{\xi}_2^2, \quad (19)$$

As written, the cross-terms $b\xi_1\xi_2$ and $e\dot{\xi}_1\dot{\xi}_2$ imply that the energy is not a simple additive function of contributions due to each mode separately; but the requirement that E shall be independent of time allows one to show that b and e vanish. For the phase-constants ϕ_1 and ϕ_2 can be omitted, and when the oscillatory forms of ξ_1 and ξ_2 are substituted, (19) takes the form:

$$E=\tfrac{1}{2}A^2(a\cos^2\Omega_1t+d\Omega_1^2\sin^2\Omega_1t)+\tfrac{1}{2}B^2(c\cos^2\Omega_2t+f\Omega_2^2\sin^2\Omega_2t)$$
$$+\tfrac{1}{2}AB[(b-e)\cos(\Omega_1+\Omega_2)t+(b+e)\cos(\Omega_1-\Omega_2)t].$$

If E is to be independent of time the first two brackets, containing oscillatory terms at frequencies $2\Omega_1$ and $2\Omega_2$, must be constrained by choice of Ω_1 and Ω_2 to be constant; and the third bracket must vanish. Hence

$$a=d\Omega_1^2, \qquad c=f\Omega_2^2 \quad \text{and} \quad b=e=0.$$

The normal mode frequencies (which of course are the same as found by circuit analysis) are thus determined in terms of the energy coefficients, and the cross-terms eliminated. In the end (19) may be rewritten in terms of two coefficients only:

$$E=\tfrac{1}{2}d(\Omega_1^2\xi_1^2+\dot{\xi}_1^2)+\tfrac{1}{2}f(\Omega_2^2\xi_2^2+\dot{\xi}_2^2), \quad (20)$$

in which d and f play the roles of mass, inductance etc. associated with each independent normal mode, considered as a simple vibrator.

Examples of coupled vibrators

The general features of the behaviour of the following examples are very similar, and there is no need to analyse each in detail. The first three categories are distinguished by the expressions for their energies and the forms taken by Φ and T as defined in (12) and (13); the remaining two are outside this classification.

(1) *Potential coupling:* Φ contains a cross-term, but T does not. The circuit of fig. 1(a), with capacitive coupling as in fig. 1(b), is typical; so are the two pendulums on a slack string, shown in fig. 2.14, and the strip vibrator in fig. 4.

24

(2) *Kinetic coupling:* T contains a cross-term, but Φ does not. The inductive couplers, fig. 1(c) and (d), provide typical examples. So also does the circuit of fig. 7, a parallel resonant circuit with a series resonant circuit connected directly across it. If q_1 and q_2 are the charges on the capacitors,

$$\Phi = \tfrac{1}{2}q_1^2/C_1 + \tfrac{1}{2}q_2^2/C_2$$

and

$$T = \tfrac{1}{2}L_1(\dot{q}_1+\dot{q}_2)^2 + \tfrac{1}{2}L_2\dot{q}_2^2$$

$$= \tfrac{1}{2}L_1\dot{q}_1^2 + \tfrac{1}{2}(L_1+L_2)\dot{q}_2^2 + L_1\dot{q}_1\dot{q}_2.$$

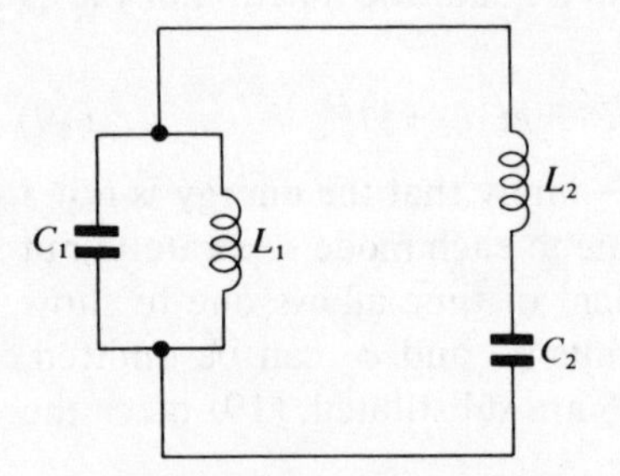

Fig. 7. Coupled series and parallel resonant circuits.

The circuit is no different from the standard form, fig. 1(a), with L_1 acting as coupling inductor, and C_1 having no inductor in its own arm. Then $\omega_1^2 = 1/\bar{L}_1C_1 = 1/L_1C_1$, $\omega_2^2 = 1/(L_1+L_2)C_2$ and $\kappa^2 = \tfrac{1}{4}L_1/L_2$. It is necessary that $L_2 \gg L_1$ for weak coupling. The acoustic resonator illustrated in fig. 7.11 is of this type, the side arm having the characteristics of a series resonant circuit, while the lengths of main pipe on either side are represented by L_1 and C_1 in parallel. As the coupling point is brought near a pressure node, L_1 and $1/C_1$ get smaller and the coupling is reduced. The splitting of the resonances has already been discussed in connection with this experiment, and is clearly typical of coupled resonant systems.

178

In a circular waveguide large enough to propagate both modes shown in fig. 7.16, it is possible to filter out the TE_{11} mode and let through the TM_{01}

184

mode by mounting a metal ring as shown in fig. 8. When the perimeter of the ring is about one wavelength it is excited by the TE_{11} mode, which is reflected. The other mode does not excite currents in the ring and is virtually unaffected by its presence. The ring is highly resonant and only works over a narrow frequency range. If one attempts to enhance the filtering by mounting a similar ring close behind it, the resonance is found to be split by magnetic (i.e. kinetic) coupling, and a measurable fraction of the microwave power is transmitted at the natural frequency of each ring alone, which now lies between the two peaks of reflection.[1]

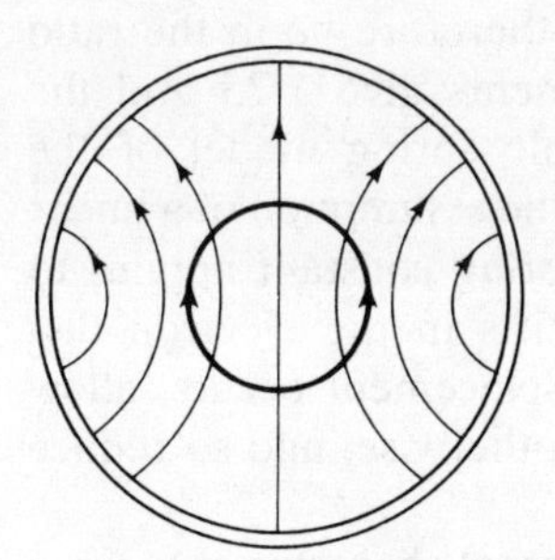

Fig. 8. Electric lines in a circular waveguide exciting resonant current oscillations in a metal ring (drawn heavily). The perimeter of the ring is about one free-space wavelength.

(3) *Potential and kinetic coupling:* both Φ and T contain cross-terms. A triatomic linear molecule modelled by the arrangement of masses and springs in fig. 9, and vibrating along its axis, falls into this category. If the displacements of the atoms are x_1, x_2 and x_3,

$$\Phi = \tfrac{1}{2}\mu_1(x_1 - x_2)^2 + \tfrac{1}{2}\mu_2(x_2 - x_3)^2,$$

and
$$T = \tfrac{1}{2}m_1\dot{x}_1^2 + \tfrac{1}{2}m_2\dot{x}_2^2 + \tfrac{1}{2}m_3\dot{x}_3^2.$$

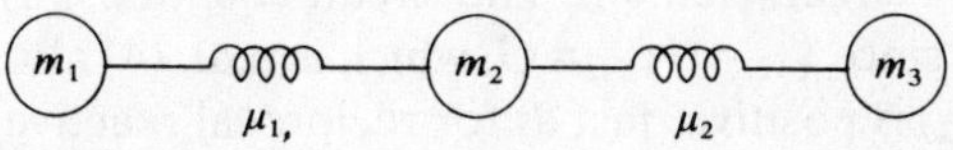

Fig. 9. Mechanical spring-and-mass model of a linear triatomic molecule.

Now one must remember that the centroid of the three atoms does not vibrate, so that $m_1x_1 + m_2x_2 + m_3x_3 = 0$. Using this to eliminate x_2, we have

$$\Phi = \tfrac{1}{2}\left[\mu_1\left(1+\frac{m_1}{m_2}\right)^2 + \mu_2\left(\frac{m_1}{m_2}\right)^2\right]x_1^2 + \tfrac{1}{2}\left[\mu_2\left(1+\frac{m_3}{m_2}\right)^2 + \mu_1\left(\frac{m_3}{m_2}\right)^2\right]x_3^2$$
$$+\left[\mu_1\left(1+\frac{m_1}{m_2}\right)\frac{m_3}{m_2} + \mu_2\left(1+\frac{m_3}{m_2}\right)\frac{m_1}{m_2}\right]x_1x_3,$$

and
$$T = \tfrac{1}{2}m_1\left(1+\frac{m_1}{m_2}\right)\dot{x}_1^2 + \tfrac{1}{2}m_3\left(1+\frac{m_3}{m_2}\right)\dot{x}_3^2 + \frac{m_1m_3}{m_2}\dot{x}_1\dot{x}_3.$$

These expressions provide the basis for a complete analysis, but we shall not pursue the matter in that generality. In the simpler case of a symmetrical molecule, like CO_2, with two oxygen atoms of mass M flanking a carbon of mass m, and the same spring constants for each bond, the two modes can be seen by inspection. The symmetric mode has both bonds stretching or contracting in phase, and the carbon atom at rest, while the antisymmetric has both oxygens moving in the same direction at the same speed while the carbon moves in the opposite direction at a speed $2M/m$ times as great. For a given displacement of an oxygen atom the restoring force in the latter case is greater than in the former by a factor $1+2M/m$, and the frequencies of these two modes should therefore be in the ratio $(1+2M/m)^{\frac{1}{2}}$, i.e. 1.91. Spectroscopic measurements give 1.75 and the discrepancy indicates the limitations of a simple spring model of the valence bond.[(2)] The fault does not lie so much in the assumption of a linear force law as in the assumption that the same spring constant applies to symmetrical and antisymmetrical vibration.† Shifts in the electron distribution, however, when an antisymmetrical displacement occurs, allow the potential energy to take a lower value than otherwise, and so reduce the spring constant and the resonance frequency.

(4) *Gyrator coupling:* because of the non-reciprocal character of a gyrator, for which $\mathfrak{z}_{21}=-\mathfrak{z}_{12}$, this example does not follow the pattern of those discussed previously. To consider only an idealized loss-free model, represented schematically in fig. 1(f), we assume that unit current flowing from A to C generates no p.d. across AC but does generate a p.d. of magnitude G across BD, B being positive; while unit current flowing from B to D generates no p.d. across BD, but generates G across AC, C being positive. The tendency of a gyrator to redirect current in a given sense (clockwise in this case) is only achievable at the expense of time-reversal symmetry, and requires a magnetic field or a rotating frame. The Hall effect in semiconductors offers the possibility of realizing something very near ideal gyratory properties in a circuit element. The circuit equations (2) and (3) keep the same form, and since $\mathfrak{z}_{12}=-\mathfrak{z}_{21}=G$, which is real, (4) also is unchanged and κ^2 $(=G^2/L_1L_2)$ is positive, just as for reciprocal reactive coupling. The splitting of the normal modes follows the same pattern, as in fig. 2, but their character is different; for (2) and (3) show that i_1 and i_2 are now in quadrature. In the representation used in fig. 3 the two modes are similar ellipses with major axes along x_1 and x_2 respectively, and with opposite senses of rotation.

239

† The force law does, of course, contain anharmonic terms which happen to play a significant role in determining some of the energy levels of CO_2. This arises because the bending vibration of the molecule accidentally has a natural frequency almost exactly half that of the symmetrical stretching vibration. Anharmonicity in the bending process causes the spacing of the atoms to vibrate at twice the frequency of the bending vibration, and to couple this vibration with the symmetrical stretching mode. The phenomenon of *Fermi Resonance* is clearly a quantum-mechanical analogue of parametric excitation.[(3)]

470

286

210

The Foucault pendulum illustrates this point, the Coriolis force providing gyrator coupling between the pendulum vibrations in two orthogonal planes. A perfectly constructed pendulum is isotropic, no direction of swing being different from any other; $\omega_1 = \omega_2$ and the normal modes are circular with slightly different frequencies for the two senses of rotation. It is the combination of these two in equal proportion that yields the usual manifestation of the effect, the slow precession of the plane of vibration. This is the analogue in a gyratory system of the energy exchange in a reciprocally coupled system. An imperfect Foucault pendulum, having different effective length for different directions of swing, behaves like two circuits with different ω_1 and ω_2; the normal modes are now elliptical, and when the pendulum is swung in a plane the beating of the modes causes the trajectory to open into an ellipse, whose axes may precess at a quite different rate from that expected with a perfect pendulum.[4] This is why it requires great care to construct a reliable Foucault pendulum.

(5) *Resistive coupling:* if $\mathfrak{z}_c$ is a resistor, as in fig. 1(*e*), $\mathfrak{z}_{12} = \mathfrak{z}_{21} = R_c$, and (4) holds but with κ^2 now negative. This implies that $\Omega_{1,2}$ lie between ω_1 and ω_2, the modes being attracted by the coupling rather than repelled. It must be remembered, however, that R_c is included in the circuits when ω_1 and ω_2 are derived, so that in general (not unexpectedly) ω_1, ω_2, Ω_1 and Ω_2 are complex. There is no point in taking separate note of resistively coupled loss-free circuits, which are a special case of the general problem of coupled lossy vibrators.

Coupled lossy vibrators

The determination of the normal modes of coupled vibrators whose Q-values are high, but not infinite, involves no extension of the foregoing theory beyond finding the solutions of (4) when ω_1 and ω_2 are complex. Proceeding as before to (11), we note that for reactive, resistive or gyratory coupling (though not for arbitrary mixtures) κ^2 is real; $(\Delta\omega)^2$ is in general complex but constant in an experiment in which only the coupling is changed, ω_1 and ω_2 remaining fixed. If then we write $\Omega - \bar{\omega}$ as $u + \mathrm{i}v$, we have that

$$u^2 - v^2 = \kappa^2 + \mathrm{Re}\,[(\Delta\omega)^2] \tag{21}$$

and

$$2uv = \mathrm{Im}\,[(\Delta\omega)^2]. \tag{22}$$

Since the right-hand side of (22) is constant, $\Omega - \bar{\omega}$ lies on a rectangular hyperbola, having the u- and v-axes as asymptotes, as shown in fig. 10, and passing through ω_1 and ω_2, which are the solutions when $\kappa^2 = 0$. Stronger coupling moves the solutions further from these points. For any real choice of κ^2, ω_1 and ω_2 together with the normal mode frequencies Ω_1 and Ω_2 form a parallelogram. The vectors $\Omega_1 - \omega_1$ and $\Omega_1 - \omega_2$, as shown in the diagram, have the real axis as bisector and their product is therefore real and positive; according to (4) Ω_1 and Ω_2 are solutions for reactive coupling

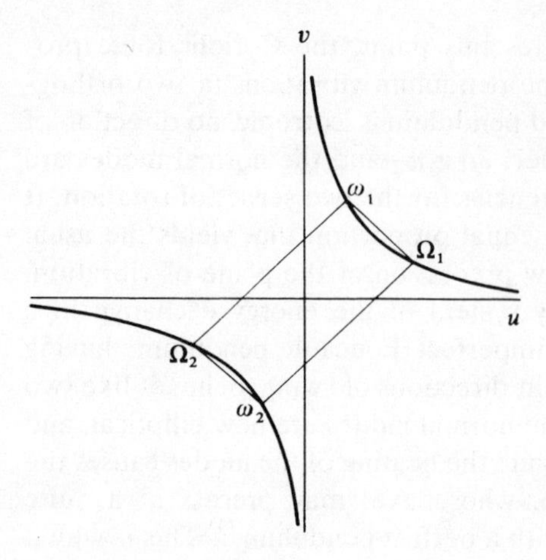

Fig. 10. Relationship of normal modes (Ω_1 and Ω_2) and natural frequencies of uncoupled lossy vibrators (ω_1 and ω_2), when represented on the complex frequency plane.

($\kappa^2>0$). Any point Ω on the hyperbola above ω_1 or below ω_2, on the other hand, has $\Omega-\omega_1$ and $\Omega-\omega_2$ with the imaginary axis as bisector, and corresponds to resistive coupling ($\kappa^2<0$). Reactive coupling repels the real parts of the normal mode frequencies, and attracts the imaginary parts; resistive coupling does the reverse. In the particular case when both circuits have the same Q, ω_1 and ω_2 lie at the same distance above the real axis and the hyperbola degenerates into two lines, vertical and horizontal, intersecting at $\bar{\omega}$. With reactive coupling the normal modes lie on the horizontal line and have the same Q at all strengths of coupling. The same degeneracy occurs when ω_1 and ω_2 have the same real part and lie on a vertical line. As κ^2 is increased the normal modes move towards one another on the vertical line and coalesce at $\bar{\omega}$, having the same frequency but different decrements until this point. Thereafter their frequencies diverge while their decrements remain fixed at the mean value for the uncoupled circuits.

If ω_1 and ω_2 do not have the same imaginary parts $\Omega-\omega_1$ does not approximate to a real quantity unless Ω is a great distance from ω_1 along the hyperbola. It follows from (5) that, even with reactive coupling, i_1 and i_2 may depart considerably from the in-phase or antiphase relationship characteristic of loss-free circuits. The requirement for different rates of dissipation in the two circuits makes a significant phase shift necessary, especially when the coupling is weak, to allow a steady transfer of energy from the less dissipative to the more dissipative circuit. Various expressions for the ratio of the currents in the two circuits follow from (5). If we write $x_1=cx_2$ in the upper mode, of frequency Ω_1,

$$c=\kappa/(\Omega_1-\omega_1)=-\kappa/(\Omega_2-\omega_2)=(\Omega_1-\omega_2)/\kappa=-(\Omega_2-\omega_1)/\kappa, \quad (23)$$

the complex numbers $(\Omega_1-\omega_1)$ etc. referring to the sides of the parallelogram in fig. 10; κ is positive for capacitive, negative for inductive, coupling. Further,

$$cc^*=|\Omega_1-\omega_2|/|\Omega_1-\omega_1|=|\Omega_2-\omega_1|/|\Omega_2-\omega_2|, \quad (24)$$

the ratio of the lengths of the sides. In the lower mode $x_1 = c'x_2$, where

$$c' = -1/c. \tag{25}$$

Only for loss-free circuits can one expect c to be real.

Impedances and impulse response functions for coupled circuits

Let us begin the discussion of the effect of an injected e.m.f. by calculating the input impedance of the coupled resonant circuits when the left-hand circuit in fig. 1(a) is opened and an e.m.f. is applied at frequency ω. The equations for the current are like (2) and (3), except that a source V is added to the right-hand side of (2). Since the input impedance of the circuit, $Z_{11} = V/i_1$, we may instead add $-Z_{11}$ to the coefficient of i_1 on the left-hand side of (2), and rewrite (4) in the form

$$[(\omega - \omega_1) - Z_{11}/2jL_1](\omega - \omega_2) = \kappa^2,$$

or

$$Z_{11} = \frac{2jL_1}{\omega - \omega_2}[(\omega - \omega_1)(\omega - \omega_2) - \kappa^2]$$

$$= \frac{2jL_1}{\omega - \omega_2}(\omega - \Omega_1)(\omega - \Omega_2), \tag{26}$$

since Ω_1 and Ω_2 are the solutions of (4). Similarly the input impedance of the right-hand circuit takes the form

$$Z_{22} = \frac{2jL_2}{\omega - \omega_1}(\omega - \Omega_1)(\omega - \Omega_2). \tag{27}$$

A third coefficient of the system is the transfer impedance, Z_{12} or Z_{21}, the e.m.f. needed in one circuit to set up unit current in the other. This can be derived from (26) by noting that Z_{11} is the e.m.f. needed to set up unit current in the same circuit, and that (3) gives the ratios of the currents. Hence

$$Z_{12} = -2jL_2(\omega - \omega_2)Z_{11}/\mathfrak{z}_{21} = -\frac{\mathfrak{z}_{12}}{\kappa^2}(\omega - \Omega_1)(\omega - \Omega_2). \tag{28}$$

This expression is symmetrical with respect to the two circuits if $\mathfrak{z}_{12} = \mathfrak{z}_{21}$, and provides another illustration of the Rayleigh–Carson reciprocity theorem.

168

The transfer impedance is low only in the vicinity of the resonances, and this makes the coupled circuit a useful feed-through arrangement, acting as a simple band-pass filter. For this purpose one is usually more interested in the magnitude of Z_{12} than in the phase relation between input e.m.f. and secondary current. Even so, (28) in its general form, with Ω_1 and Ω_2 different in both real and imaginary parts, develops into a cumbersome expression when one writes out $|Z_{12}|$ explicitly. A diagram is far more revealing, and fig. 11(a) shows $(\omega-\Omega_1)$ and $(\omega-\Omega_2)$ as vectors on a complex plane. Since Z_{12} is proportional to the product of the lengths of these vectors, one may easily sketch how this varies as ω moves along the real axis. Clearly if the imaginary parts of Ω_1 and Ω_2 are small compared with their separation, there will be minima of $|Z_{12}|$ at two points where ω lies almost as close as it can get to either Ω_1 or Ω_2. This is equivalent to the

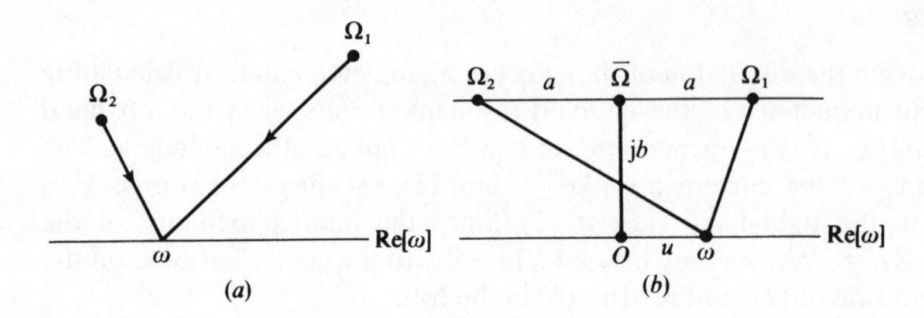

Fig. 11. To illustrate transfer impedance of coupled circuits.

statement that if the modes are separated by rather more than the widths of their resonances they will be resolved in the response function. On the other hand, if Ω_1 and Ω_2 are closer to each other than to the real axis, there is only a single minimum. The former case is referred to as over-coupled, the latter as under-coupled, and in between is the state of critical coupling when $|Z_{12}|$ has a flat minimum. For two circuits having the same Q, so that Ω_1 and Ω_2 are equidistant from the real axis, as in fig. 11(b), critical coupling occurs when $\text{Im}\,[\Omega_{1,2}]=\frac{1}{2}(\Omega_1-\Omega_2)$. The current response $|A_{12}|$ $(\equiv 1/|Z_{12}|)$ to an e.m.f. of constant magnitude but variable frequency, is shown in fig. 12 for this case and for typical under-coupled and over-coupled cases. To derive the form of these curves, take Re $[\bar{\Omega}]$ as the origin

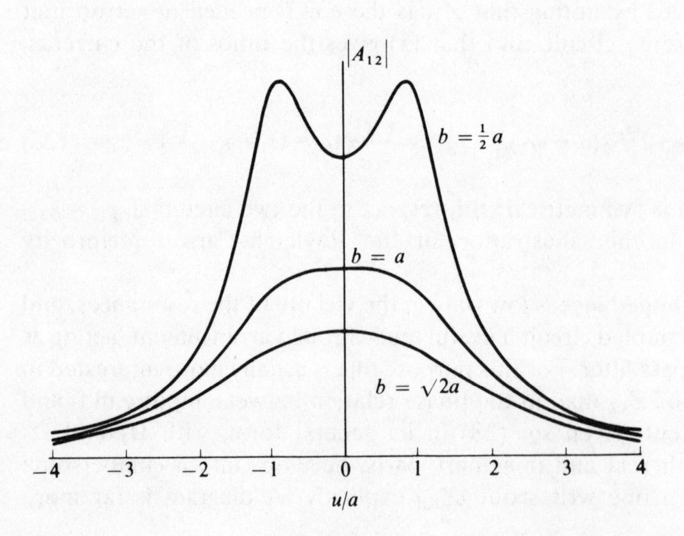

Fig. 12. Transfer admittance of coupled identical circuits, undercoupled ($b=\sqrt{2}a$), critically coupled ($b=a$) and overcoupled ($b=\frac{1}{2}a$).

on the complex plane, so that $\Omega_{1,2}=\pm a+\mathrm{j}b$, and write ω as $\mathrm{Re}[\bar{\Omega}]+u$. Then

$$|(\omega-\Omega_1)(\omega-\Omega_2)|^2=u^4+2(b^2-a^2)u^2+(a^2+b^2)^2$$

and
$$|A_{12}|=\frac{\kappa^2}{|\mathfrak{z}_{12}|}[u^4+2(b^2-a^2)u^2+(a^2+b^2)^2]^{-\frac{1}{2}}. \tag{29}$$

Critical coupling occurs when the coefficient of u^2 vanishes, i.e. when $b=a$. If the normal modes arise from the coupling of two identical resonant circuits, $\omega_2=\omega_1$ in (4) and $\Omega_1-\Omega_2=2\kappa$; the condition for critical coupling is then that $\kappa=\frac{1}{2}\bar{\Omega}/Q$.

The transfer admittance A_{12} may be written, from (28), in the form

$$A_{12}=-\frac{\kappa^2}{\mathfrak{z}_{12}(\Omega_1-\Omega_2)}[1/(\omega-\Omega_1)-1/(\omega-\Omega_2)]. \tag{30}$$

The current is made up of contributions due to each normal mode responding independently to the applied e.m.f. The character of each contribution is just the same as for a simple sharply resonant oscillator, as given in (6.17). In the secondary circuit the amplitudes are equal and opposite, for a reason which becomes obvious when one thinks of the impulse response function. An impulsive e.m.f. sets up an instantaneous current in the primary circuit, to which it is applied, but there is either no immediate response by the secondary circuit (when coupling is capacitive) or a much weaker response (when coupling is inductive). In the weak-coupling limit, which is implied in (30) by the approximate form of the resonant response, one may take the secondary current as initially zero, whatever the mode of coupling, and that means that the two damped normal-mode oscillations initiated by the impulse must be such that in the secondary circuit they start equal in amplitude and in antiphase.

135

To put the argument quantitatively, let an impulsive e.m.f. of unit strength be applied to circuit 1, so that immediately afterwards a current $1/L$ flows in circuit 1, but none in circuit 2. The impulse response function for x_1 has the form of decaying vibrations due to each normal mode, with as yet undetermined initial amplitudes a_1 and a_2:

$$h_{11}(t)=a_1\,\mathrm{e}^{\mathrm{j}\Omega_1 t}+a_2\,\mathrm{e}^{\mathrm{j}\Omega_2 t},$$

from which and from (25) it follows that in circuit 2, due to injection in circuit 1, the corresponding response has the form:

$$h_{12}(t)=\frac{a_1}{c}\,\mathrm{e}^{\mathrm{j}\Omega_1 t}-a_2 c\,\mathrm{e}^{\mathrm{j}\Omega_2 t}.$$

Now not only does h_{12} vanish at the outset, but its derivative does also, since until the coupling capacitor begins to be charged there is no e.m.f. to drive current through the inductor of the second circuit. Both the real and imaginary parts of $h_{12}(0)$ must vanish, so that

$$a_2 = a_1/c^2,$$

$$\left.\begin{aligned} h_{11}(t) &= a_1\,\mathrm{e}^{\mathrm{j}\Omega_1 t} + a_1/c^2\,\mathrm{e}^{\mathrm{j}\Omega_2 t} \\ \text{and}\qquad h_{12}(t) &= a_1(\mathrm{e}^{\mathrm{j}\Omega_1 t} - \mathrm{e}^{\mathrm{j}\Omega_2 t})/c. \end{aligned}\right\} \tag{31}$$

For a short while, until the difference between Ω_1 and Ω_2 asserts itself, $h_{11}(t)$ behaves like $a_1(1+1/c^2)\,\mathrm{e}^{\mathrm{j}\Omega_1 t}$. This must be matched to the response of circuit 1 which, just like a simple series resonant circuit of high Q, starts with the current taking its maximum value of $1/L$, and x its maximum value of $1/L^{\frac{1}{2}}$. Hence $a_1(1+1/c^2)$ must be equated to $1/L_1^{\frac{1}{2}}$; and

$$a_1 = c^2/L_1^{\frac{1}{2}}(1+c^2), \qquad a_2 = 1/L_1^{\frac{1}{2}}(1+c^2).$$

Translated into currents by writing $i_{11}(t) = h_{11}(t)/L_1^{\frac{1}{2}}$, $i_{12}(t) = h_{12}(t)/L_2^{\frac{1}{2}}$, (31) gives for the current impulse response functions:

$$\left.\begin{aligned} i_{11}(t) &= \frac{c}{L_1(1+c^2)}\left(c\,\mathrm{e}^{\mathrm{j}\Omega_1 t} + \frac{1}{c}\,\mathrm{e}^{\mathrm{j}\Omega_2 t}\right) \\ \text{and}\qquad i_{12}(t) &= \frac{c}{(L_1L_2)^{\frac{1}{2}}(1+c^2)}(\mathrm{e}^{\mathrm{j}\Omega_1 t} - \mathrm{e}^{\mathrm{j}\Omega_2 t}). \end{aligned}\right\} \tag{32}$$

Since the first and fourth forms of (23) show that $c + 1/c = (\Omega_1 - \Omega_2)/\kappa$, i_{12} may be recast in the form:

$$i_{12}(t) = \frac{\kappa}{(L_1L_2)^{\frac{1}{2}}(\Omega_1 - \Omega_2)}(\mathrm{e}^{\mathrm{j}\Omega_1 t} - \mathrm{e}^{\mathrm{j}\Omega_2 t}),$$

from which, by application of (5.4) to each term separately, $A_{12}(\omega)$ may be derived: **106**

$$A_{12} = \frac{-\frac{1}{2}\mathrm{j}\kappa}{(L_1L_2)^{\frac{1}{2}}(\Omega_1 - \Omega_2)}[1/(\omega - \Omega_1) - 1/(\omega - \Omega_2)],$$

which is the same as (30) since $\kappa = \frac{1}{2}\mathrm{j}\mathfrak{z}_{12}/(L_1L_2)^{\frac{1}{2}}$. Similarly A_{11} and A_{22} may be derived from i_{11} and i_{22}, and shown after a little manipulation to be reciprocals of Z_{11} and Z_{22} as given by (26) and (27).

This argument has been presented in some detail for two reasons: first, to show that one must exercise care in using the normal modes of lossy coupled resonators, even if the losses are quite small, since the complex nature of c introduces significant phase shifts into the impulse responses given by (32). Secondly, the ideas expressed and some of the results derived here will find immediate application to the next example of coupled systems, where the description in terms of normal modes in non-linear circuits proves to be both feasible and helpful, but where the physical principles must be kept firmly in mind if one is not to fall into plausible errors.

Coupled active and passive vibrators[5]

There is nothing in the foregoing analysis to prevent it being applied to coupled circuits one of which has negative resistance, so that ω_1 (say) has negative imaginary part. The parallelogram of fig. 10 then has ω_1 lying below the real axis, but if it is not too far below it may be possible to find a strong enough coupling constant for both normal modes to lie above the real axis; the circuit is then stable, behaving like any of the examples already discussed. If, however, a normal mode lies below the axis, it will grow exponentially until saturation of the amplifier leads to a steady state of maintained oscillation. With strong enough amplification, both normal modes will initially lie below the axis and be capable of growing, and it may be that the mode which happens to be more strongly excited when saturation sets in will survive in the steady state. On the other hand it may be that only one of the two is capable of survival. This is the problem which we shall resolve in this section, taking the model of a class C oscillator that was developed in chapter 11; a resonant circuit incorporates a resistor which is negative at low currents and positive at high, the steady state being characterized by such a current that the resistance is zero. It must be remembered that the circuit itself is of such high Q that it oscillates sinusoidally and does not respond appreciably to harmonics of its own frequency. We are interested therefore only in the average response of the amplifier and feedback over one cycle, and it is this that we express in the model by a resistance which differs from that of a passive resistor only in the fact that it is a function of the amplitude of the current oscillation. Once the oscillation has settled in its steady state the resistance in the active circuit stays constant, and one may therefore assign two frequencies, ω_1 and ω_2, to the uncoupled circuits, exactly as was done on setting up (2) and (3) (subscript 1 refers to the active circuit, 2 to the passive); The only difference is that we do not know *a priori* the imaginary part of ω_1; we only know that ω_1 moves on a vertical line to such a point that one or other of the normal modes, Ω_1 or Ω_2, lies on the real axis, representing steady oscillation. An example of the modified form of fig. 10 is given in fig. 13. For a given ω_2 and with ω_1 lying on a given vertical line there are, in this case, three separate solutions for the same real value of the coupling constant, κ^2. Each parallelogram is drawn so that the horizontal axis bisects the angle through which it passes; the product of the two sides, considered as complex numbers, is therefore real and is, moreover, the same for each. Since the negative resistance depends on the current, each of these solutions can only exist at a certain level of excitation. For example, steady oscillation at Ω_{2c} demands a large negative resistance, and therefore a small current, to make ω''_{1c} large enough. This arrangement, however, is unlikely to be stable since the same circuit parameters allow an alternative mode, Ω_{1c}, which can grow and swamp Ω_{2c}. If this were a passive circuit, with current-independent resistances, Ω_{2c} would certainly be unstable for this reason, but we shall find the matter less certain when the negative resistance varies with current.

318

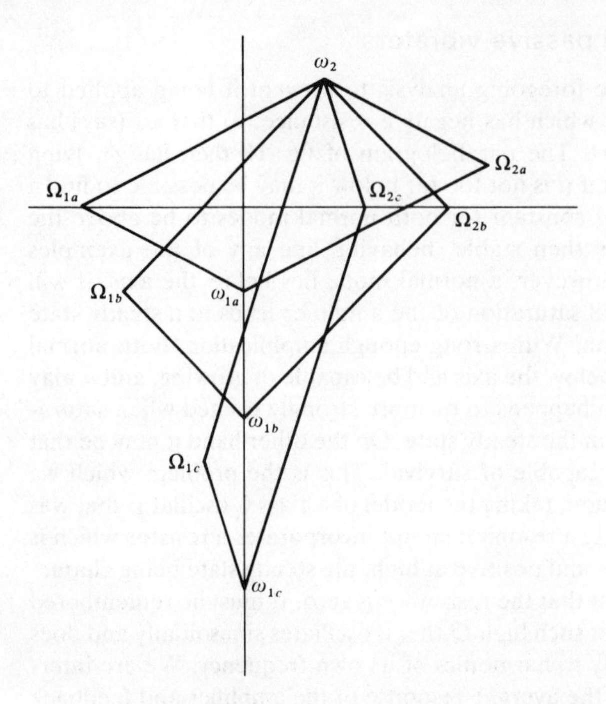

Fig. 13. Three possible normal-mode diagrams (after fig. 10) for a passive resonant circuit of natural frequency ω_2 coupled to an active circuit for which ω_1 lies on a given vertical line. In each parallelogram one corner lies on the real axis and is bisected by the axis; also the product of the sides is the same for each.

Nevertheless, if it is true in this case that the mode at Ω_{2c} is swamped by the growth of Ω_{1c}, the rise of current will bring ω_1 up towards the real axis, and the process may continue so long as any growing mode is available. If it continues to the point where the circuit oscillates at Ω_{1a} it is undoubtedly stable, for the alternative, Ω_{2a}, is a decaying mode. We shall find on proceeding with the analysis that in an active circuit stability may be possible even though an alternative mode lies below the real axis, and our task is to determine under what conditions this is so. The essence of the explanation lies in the fact that any attempt on the part of the alternative mode to grow in amplitude changes the negative resistance, and the ensuing changes in the properties of the circuit may suppress this growth.

The first stage in a quantitative discussion is to enumerate the possible steady-state solutions, whether stable or not, but being steady having the resistances constant. Writing (4) with the real and imaginary parts of ω_1 and ω_2 displayed, we have that a real solution Ω obeys the equation:

$$\Omega-\omega_1'-\mathrm{j}\omega_1''=\kappa^2/(\Omega-\omega_2'-\mathrm{j}\omega_2'')=\kappa^2(\Omega-\omega_2'+\mathrm{j}\omega_2'')/[(\Omega-\omega_2')^2+\omega_2''^2].$$

Since it is only ω_1'' that adjusts itself to circumstances, the real part of this equation is free from uncertainty:

$$(\Omega-\omega_1')/\kappa^2=(\Omega-\omega_2')/[(\Omega-\omega_2')^2+\omega_2''^2]. \tag{33}$$

The solutions, one or three in number, are exhibited in fig. 14 as the intersection(s) of the straight line L representing the left-hand side with the

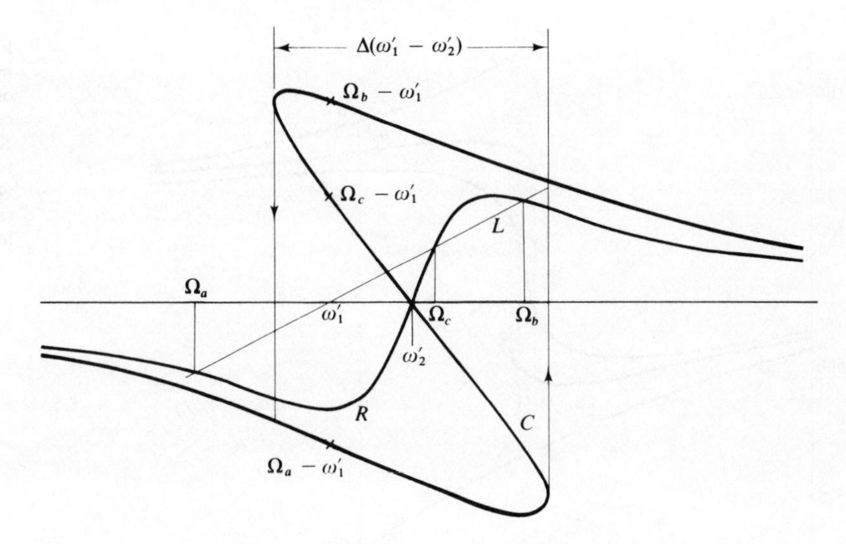

Fig. 14. The solutions $\Omega_{a,b,c}$ of (33) exhibited as the intersections of the resonance curve R with the straight line L. In curve C the solutions are replotted with $\Omega_{a,b,c} - \omega'_1$ as ordinates above ω'_1.

resonance curve R representing the right. For three solutions to be possible, as here, ω'_1 and ω'_2 must be close enough, or κ^2 large enough. The diagram can be replotted as curve C to show how the possible real values of Ω vary with $\omega'_2 - \omega'_1$, κ^2 being kept constant as one of the circuits is tuned. For any choice of ω'_1 relative to ω'_2 draw a line L through ω'_1 at a gradient $1/\kappa^2$; drop perpendiculars from the intersections to find the real solutions, here shown as Ω_a, Ω_b and Ω_c; replot the distances $\Omega_a - \omega'_1$ etc. as ordinates above ω'_1. The resulting curve is C, the required variation of $\Omega - \omega'_1$ with ω'_1. If all these solutions were stable one might expect, on tuning the passive circuit so that ω'_2 fell from a value well above ω'_1 to one well below, to find the system oscillating at a frequency initially slightly below ω'_1, as indicated by the left-hand side of the curve, falling steadily as the lower heavy curve was followed, and switching modes sharply at a certain point when $\omega'_2 < \omega'_1$, so that thenceforward Ω was greater than ω'_1 but by a steadily decreasing amount. On reversing the procedure the upper heavy curve would be followed, and there would be hysteresis. That something of this sort happens is clear from the experimental curves in fig. 15. The coupling was the same in each case, but extra resistances were inserted in the circuits to obtain curves b and c. The resistance in case c is high enough to preclude triple intersections, the maximum slope of the resonance curve R in fig. 14 being just less than $1/\kappa^2$. In curve b there is a pronounced mode switch with a narrow range of hysteresis. In curve a the hysteresis is much more marked, but clearly the whole range indicated in C of fig. 14 is not achieved, since the switch occurs before Ω turns over towards ω'_1. The very fact that any hysteresis occurs shows that linear theory is inadequate, since under no conditions is there more than one solution where the alternative mode is dissipative (e.g. Ω_{2a} in fig. 13).

Since graphical representations will prove convenient in what follows, we may start by recasting the above argument in geometrical form. Some

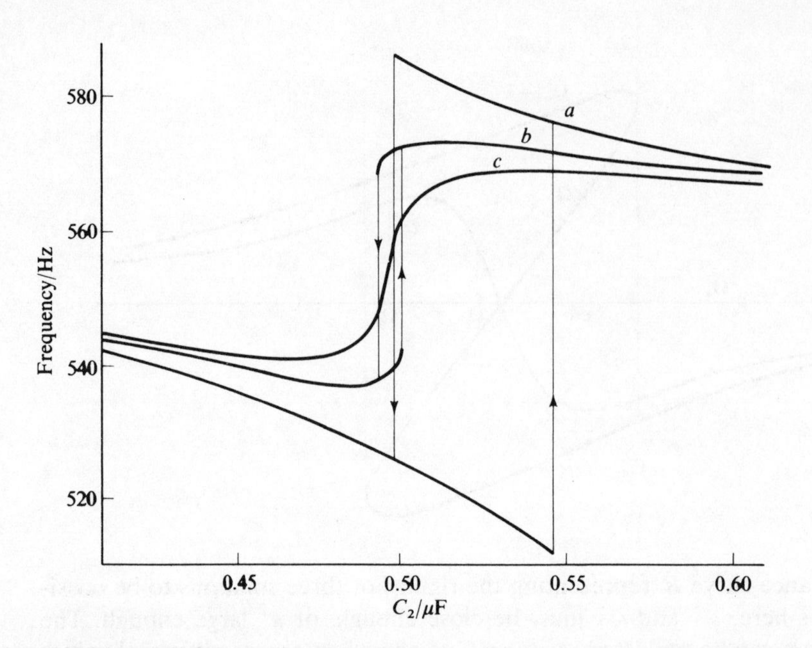

Fig. 15. Experimental measurements of oscillation frequency of coupled active and passive circuits; the natural frequency of the latter was varied by varying its capacitance C_2. See text for further details.

of the assertions made will be left unproved; in each case it is fairly easy, by coordinate geometry, to verify the statements. For convenience, the passive circuit is assumed to remain unchanged, with natural frequency ω_2, while the active circuit is tuned with respect to it. Thus ω_1' is variable, but κ^2 is taken to be constant. The possible solutions to (33), including now the imaginary parts, are represented as in fig. 16, by a family of parallelograms of which one corner (ω_2) is fixed, a second (Ω_1) lies on the real axis, a third (Ω_2) lies on the circle whose diameter is κ^2/ω_2'', and the fourth is ω_1. All these parallelograms have the required properties: $\omega_2\Omega_1$ and $\omega_1\Omega_1$ are bisected by the real axis, and the product of the lengths of the sides is constant ($=\kappa^2$). As Ω_2 runs round the circle, ω_1 describes a cubic trajectory;† the range labelled $\Delta(\omega_1'-\omega_2')$ in fig. 14 and fig. 16 is the tuning range which permits three separate solutions. The diagram is drawn for a value of ω_1' at the extreme edge of this range, which can be shown to possess the convenient property that the diagonal $\Omega_1\Omega_2$ makes a right angle with the side $\omega_1\Omega_2$.

Let us now suppose the system to be oscillating steadily in a pure normal mode at Ω_1, and enquire into the effect of adding a little of the normal mode at Ω_2. If the current due to the principal mode is $i^{(1)}$ in the active circuit, it is $i^{(1)}/c$ in the passive circuit,‡ and $i^{(1)}$ takes such a value that the

† Note that the line joining Ω_1 and ω_1 always goes through ω_2^*. This makes it easy to construct the trajectory of ω_1.

‡ From now on we take the two circuits to be virtually identical, with $L_1=L_2=L$.

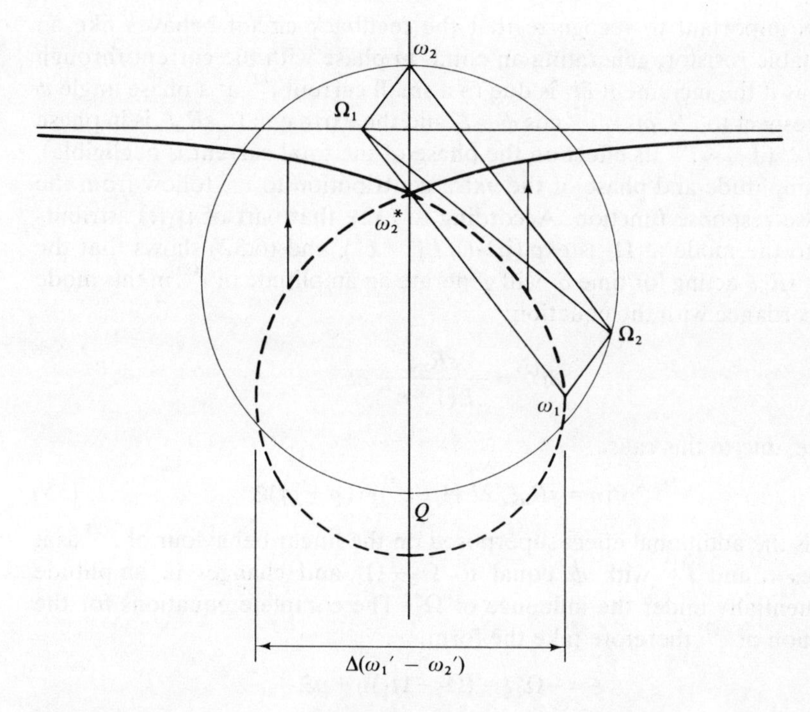

Fig. 16. Alternative to fig. 13, with ω_2 fixed, showing trajectory of ω_1 that is consistent with steady oscillation. Q is the point $\omega_2 - \mathrm{j}\kappa^2/\omega_2''$.

negative resistance R_a just balances dissipation by the positive resistances R in each circuit;

$$(R+R_a)|i^{(1)}|^2 + R|i^{(1)}|^2/cc^* = 0,$$

i.e.

$$R_a = -R(1+1/cc^*). \tag{34}$$

381 A small extra current $i^{(2)}$ in the active circuit, accompanied by $-ci^{(2)}$ in the passive circuit, and oscillating at the subsidiary mode frequency Ω_2, would develop independently of $i^{(1)}$ if R_a remained unchanged. But if the magnitude of the total current i_1 in the active circuit is changed, by δi_1 say, with the appearance of the subsidiary mode, so too is R_a changed, by an amount $-sR_a\,\delta i_1/i_1$ if s is defined as $-\mathrm{d}(\ln|R_a|)/\mathrm{d}(\ln i_1)$, having the value $+1$ for a

317 fully excited class C oscillator, as follows from (11.10). The effect is the same as if a source had been inserted into the active circuit, with an e.m.f. of $sR_a\delta i_1$. By this device we are enabled to continue treating the coupled circuits as a linear network, with the normal modes developing independently; but the new e.m.f. generates increments of the two modes, so that the changes of amplitude and phase they suffer are different from what they would be in a truly linear system. Since the e.m.f. generates each mode in roughly equal amplitude the relative effect on $i^{(2)}$ is much greater than on $i^{(1)}$ and we may neglect the latter while concentrating on the development of the alternative mode. We shall now take $i^{(1)}$ as real, and express $i^{(2)}$ at any instant by the complex number $\xi + \mathrm{j}\eta$.

It is important to recognize that the feedback circuit behaves like an adjustable resistor, generating an e.m.f. in phase with the current through it. Thus if the increment δi_1 is due to a small current $i^{(2)}$ at a phase angle ϕ with respect to $i^{(1)}$, $\delta i_1 = i^{(2)} \cos\phi = \xi$, and the extra e.m.f., $sR_a\xi$, is in phase with $i^{(1)}$ (if $\eta \ll i^{(1)}$ its effect on the phase of the total current is negligible). The amplitude and phase of the extra contribution to $i^{(2)}$ follow from the impulse response function. According to (32), that part of $i_{11}(t)$ attributable to the mode at Ω_2 is $\exp(\mathrm{j}\Omega_2 t)/L(1+c^2)$, and (6.32) shows that the e.m.f. $sR_a\xi$ acting for time δt will generate an amplitude of $i^{(2)}$ in this mode in accordance with the equation:

146

$$\delta i^{(2)} = \frac{sR_a\xi}{2L(1+c^2)}\,\delta t.$$

Hence, due to this cause,

$$\dot{\xi} + \mathrm{j}\dot{\eta} = sR_a\xi/2L(1+c^2) \equiv (p + \mathrm{j}q)\xi. \tag{35}$$

This is the additional effect superposed on the linear behaviour of $i^{(2)}$ as it rotates round $i^{(1)}$ with $\dot{\phi}$ equal to $\Omega_2' - \Omega_1$ and changes in amplitude exponentially under the influence of Ω_2''. The complete equations for the variation of $i^{(2)}$ therefore take the form:

$$\dot{\xi} = -\Omega_2''\xi - (\Omega_2' - \Omega_1)\eta + p\xi,$$

$$\dot{\eta} = -\Omega_2''\eta + (\Omega_2' - \Omega_1)\xi + q\xi;$$

that is to say, in the notation of chapter 9,

255

$$\dot{\xi} = \alpha_1\xi + \alpha_2\eta, \qquad \dot{\eta} = \alpha_3\xi + \alpha_4\eta,$$

where

$$\left.\begin{aligned} \alpha_1 &= -\Omega_2'' + p, \\ \alpha_2 &= -\Delta, \\ \alpha_3 &= \Delta + q, \\ \alpha_4 &= -\Omega_2''; \end{aligned}\right\} \tag{36}$$

$\Delta = \Omega_2' - \Omega_1$ and $p + \mathrm{j}q = sR_a/2L(1+c^2) = -s\omega_2''(1+1/cc^*)/(1+c^2)$, by use of (34) and the fact that $\omega_2'' = \frac{1}{2}R/L$. Then $T = \alpha_1 + \alpha_4 = -2\Omega_2'' + p$, and $D = \alpha_1\alpha_4 - \alpha_2\alpha_3 = \Omega_2''^2 + \Delta^2 - p\Omega'' + q\Delta$.

So long as $T < 0$ and $D > 0$ the alternative mode cannot grow spontaneously and the oscillation at Ω_1 is stable. In the absence of p and q instability would occur as Ω_2'' moved down through the real axis, and T became positive. In fig. 16, ω_1 would follow that part of its cubic trajectory drawn as a full line, the cusp marking the point at which $\omega_2' = \omega_1'$ and a reversible mode switch would occur. The presence of p (which is negative), as a consequence of non-linearity, delays the instability and allows ω_1 to proceed beyond the cusp in fig. 16 into the region indicated by the heavy broken curve; T may become positive before the extreme limit of ω_1' is reached, in which case ω_1 makes a vertical mode switch as shown in the diagram. This is what happens in curve *a* of fig. 15. In curve *b*, on the other hand, there is no instability along the heavy broken curve, and the mode switch occurs as ω_1' reaches the edge of the three-valued region. It will be appreciated that fig. 16 shows the mode-switching in terms of the behaviour of ω_1, which is not directly measured. The corresponding behaviour of $\Omega_1-\omega_1'$ is shown as C in fig. 14. This indicates how far the passive circuit is able to pull the frequency of the active circuit.

In the last paragraph it was taken for granted that the critical condition for instability was the vanishing of T rather than of D. The latter is intrinsically positive when p and q are absent, and usually, if not always, remains positive when they are present until T reaches zero. We shall not carry through the rather tedious reduction of these quantities in terms of the parameters of the system; the geometrical interpretation of ω_2'', κ^2, Ω_2'', Δ, p and q is given in fig. 17, and it is very quick in any given case to measure off the diagram and check the stability criteria that way. On the assumption that mode-switching occurs when $T=0$, i.e. $p=2\Omega_2''$, the information in the diagram leads by elementary geometry to the criterion:

$$RU/RS=2/s. \tag{37}$$

For a class C oscillator this means that mode-switching occurs when S bisects RU, unless this condition does not operate before ω_1 reaches the end of the heavy broken lines in fig. 16. So long as the coupling is weak enough for $\kappa^2<\frac{9}{2}\omega_2''^2$ the full range of hysteresis is observed (as in curve (*b*) of fig. 15), while for stronger coupling the criterion (37) applies (as in curve (*a*)). Finally we note that the transition is smooth (curve (*c*)) when $\kappa^2<\omega_2''^2$.

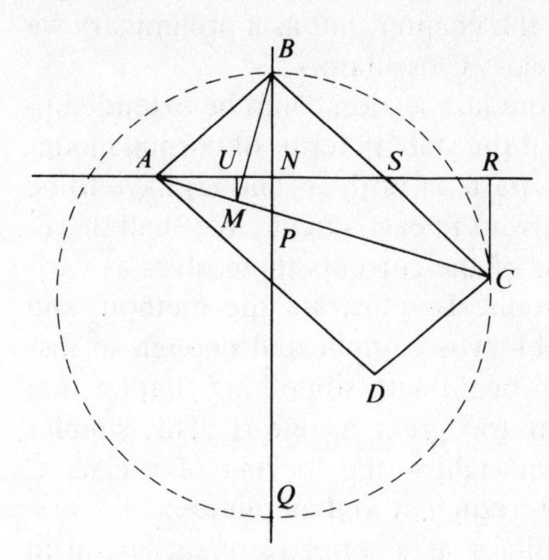

Fig. 17. Geometrical interpretation of the parameters in (36); $\omega_2''=BN$, $\kappa^2=BN\cdot BQ$, $\Omega_2''=-RC$, $\Delta=AR$, $p=-s\cdot BN\cdot MC/AP$, $q=s\cdot BN\cdot BM/AP$.

The Huygens phenomenon (entrainment)

Writing to his father in 1665, Christian Huygens, the inventor of the pendulum escapement, reported that two clocks mounted on the same yielding panel or bracket could pull each other into synchronism. The effect has frequently been commented on since in connection with many different self-excited oscillators, including electronic circuits. A not dissimilar biological example was remarked on in chapter 11. A critical degree of coupling is required; with less the two oscillators carry on at their own natural frequencies, slightly perturbed in frequency perhaps and exchanging a fraction of their energy at the beat frequency. The transition to the frequency-locked state is abrupt and is sometimes accompanied by hysteresis – before they can resume their natural oscillations the coupling must be reduced to a value significantly below that which initiated locking. Entrainment can be a nuisance in precise measurements. The ring laser, in which light is made to circulate by means of mirrors and is maintained by the amplifying action of the excited low-pressure gas through which it travels, is potentially a valuable navigating instrument because of its sensitivity to rotation.[6] So long as the frame supporting mirrors and gas tubes is at rest, the natural frequency is the same for both senses of circulation of the light; but rotation of the frame causes one mode to increase and the other to decrease in frequency, and the beat frequency which can be detected should, in ideal conditions, measure the angular velocity with very great precision and sensitivity. Unfortunately, however, scattering of light from one mode into the other, and other effects of minute imperfections, lock the modes together when their frequencies are not too different, and the instrument fails at low angular velocities, just where it is most needed as a substitute for gyroscopes. On the other hand, entrainment can be put to practical use in, for example, stabilizing the frequency of a multivibrator and using it to generate harmonics and subharmonics of the output from a primary clock, which might be a quartz or caesium frequency standard. We shall consider this application later in the chapter, but as a preliminary we analyse the Huygens phenomenon in class C oscillators.

361

405

In principle the method applied in the last section could be extended to two active oscillators, but description of the state in terms of normal modes is not convenient when the imaginary parts of both ω_1 and ω_2 have to be adjusted to be compatible with the current in each circuit. We shall therefore abandon normal modes in favour of the currents themselves as variables describing the state at any instant. To illustrate the method, and because it will turn out that the problem is complicated enough to discourage a complete solution, we shall begin with something simpler that can be fully worked through without too great an effort. This simpler problem is one of interest in its own right – the locking of a class C oscillator to an injected signal of fixed frequency and amplitude.

As usual, let us represent the oscillator as a series resonant circuit in which the resistance has a constant positive term, R, and a negative term whose magnitude is inversely proportional to the amplitude of current oscillations. The total resistance can be written as $R(1-i_0/i)$, i_0 being the amplitude at which the oscillation at the natural frequency ω_0 establishes itself if left unperturbed. Into the circuit we now inject an e.m.f. $v\,\mathrm{e}^{j\omega t}$, v being taken as real, which in accordance with (6.32) causes an oscillatory

389

146

current, in phase with v, to build up at a rate $\frac{1}{2}v/L$. If at any instant the total current is represented in magnitude and phase by $i\,e^{j(\omega t+\phi)}$, i and ϕ change in accordance with the equations†

$$\frac{di}{dt}=-\tfrac{1}{2}R(i-i_0)/L+\tfrac{1}{2}v\cos\phi/L, \tag{38}$$

and

$$\frac{d\phi}{dt}=(\omega_0-\omega)-\tfrac{1}{2}v\sin\phi/Li. \tag{39}$$

In these two equations the first terms describe the behaviour in the absence of injected signal, the amplitude of oscillation relaxing towards its equilibrium, i_0, with time-constant $2L/R$ just like a passive circuit, while the phase difference evolves at the difference frequency. The second terms describe the effect of the injected oscillation on the amplitude and phase of what is already present. These are the basic equations from which t may be eliminated by division:

321

$$\frac{dy}{d\phi}=\frac{-y(y-y_0-\cos\phi)}{\varepsilon y-\sin\phi} \tag{40}$$

in which $y\equiv Ri/v$, $y_0\equiv Ri_0/v$ and $\varepsilon\equiv 2L(\omega_0-\omega)/R$.

To sketch the solutions of (40) we construct on a y–ϕ diagram the loci on which the numerator and denominator vanish; on the former, labelled 0, the trajectories are horizontal while on the latter, labelled ∞, they are vertical. The two examples in fig. 18 differ in the choice of ε, only the ∞-loci being affected; they refer to poor tuning (ε large) and better tuning (ε small), the amplitude v of the injected signal being kept constant. Only if ε is small enough does the ∞-locus rise high enough to intersect the 0-locus and provide the possibility of a stable solution in which $\dot{\phi}=0$ and mode-locking has occurred. Where there is no intersection, as in fig. 18(a), there is a certain trajectory, labelled P, which enters and leaves the diagram at the same ordinate, y, and which is clearly periodic. As the oscillator slips in phase relative to the injected signal, the amplitude fluctuates periodically, and the appearance is of a beating between the free oscillation and the signal. When y_0 is small (strong perturbing signal) the beat modulation can be very pronounced, the oscillation amplitude dropping very nearly to zero at the beat minima. An expression for the beat period, T_b, can be derived from (39) which, rearranged and integrated, takes the form

$$T_b=\frac{2L}{R\varepsilon}\int_0^{2\pi} d\phi/(1-y_1/y), \tag{41}$$

† An alternative derivation, not involving the impulse response function, treats the injected e.m.f. as if it were due to an extra impedance, of magnitude $-v\,e^{-j\phi}/i$, inserted in the circuit. The real part, $-v\cos\phi/i$, gives rise to additional decrement according to the standard formula, $di/dt=-\frac{1}{2}Ri/L$, and this produces the second term in (38). The imaginary part, $v\sin\phi/i$, acts like an additional inductance and lowers the frequency, as expressed by the second term in (39).

Fig. 18. 0-loci, ∞-loci and sketched trajectories for solutions of (40); (*a*) $y_0 = 2$, $\varepsilon = 1$, (*b*) $y_0 = 2$, $\varepsilon = 0.287$, (*c*) illustrating form of trajectories when F and S are about to merge.

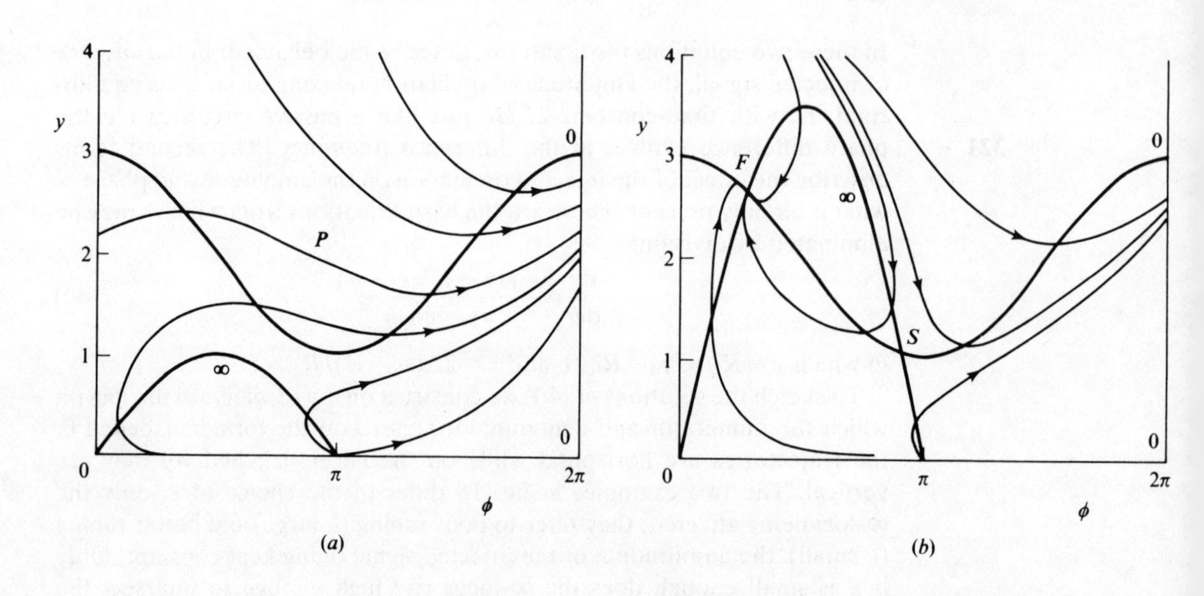

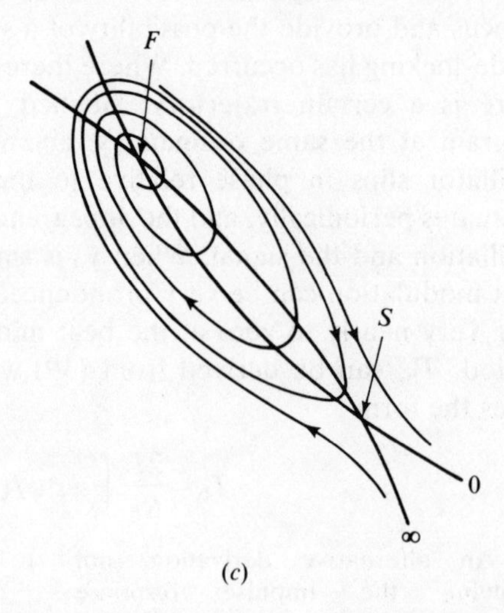

in which y_1 is written for $\sin\phi/\varepsilon$, the ordinate of the ∞-locus. The speed of the traverse, $\dot{\phi}$, is not uniform, but is much diminished when the trajectory runs close to the ∞-locus, and the denominator in (41) becomes small. This is particularly noticeable just before locking takes place, the oscillator feeling, as it were, a strong attraction towards the perturbing signal but breaking away only to repeat the process one beat period later.

When the loci intersect, as in fig. 18(b), curve-drawing shows immediately that S is a saddle-point but that F has possibilities for being a stable locked state. To investigate this further, let the coordinates of F be (Φ, Y); then, since both numerator and denominator of (40) vanish at F,

$$Y-y_0-\cos\Phi=0=\varepsilon Y-\sin\Phi, \tag{42}$$

from which it follows that

$$(Y-y_0)^2+\varepsilon^2Y^2=1, \tag{43}$$

i.e.

$$Y=\{y_0\pm[1+\varepsilon^2(1-y_0^2)]^{\frac{1}{2}}\}/(1+\varepsilon^2). \tag{44}$$

The solution with positive sign is F; that with negative sign is S. When $y_0<1$ and the 0-locus drops below the axis, intersection occurs at all values of ε, but at weaker signal inputs, when $y_0>1$, (44) has a solution only if $\varepsilon>\varepsilon_c$, where

$$\varepsilon_c=(y_0^2-1)^{-\frac{1}{2}}. \tag{45}$$

In the vicinity of F, write $\phi=\Phi+\xi$ and $y=Y+\eta$; then by differentiating numerator and denominator of (40) the linear approximation is derived in the form

$$\frac{d\eta}{d\xi}=-Y\frac{\eta+\xi\sin\Phi}{\varepsilon\eta-\xi\cos\Phi}=\frac{\alpha_3\xi+\alpha_4\eta}{\alpha_1\xi+\alpha_2\eta},$$

where the coefficients can be put in the following form by use of (42):

$$\alpha_1=y_0-Y,\quad \alpha_2=\varepsilon,\quad \alpha_3=-\varepsilon Y^2\quad\text{and}\quad \alpha_4=-Y.$$

255 For F to be a stable solution, $D>0$ and $T<0$. The first criterion demands that $Y(1+\varepsilon^2)>y_0$, which is seen from (44) to be satisfied by F, if it exists; it fails at the same moment as the loci cease to intersect. Two intersecting loci such as we find here cannot generate the same type of stationary point at both intersections, and the presence of one saddle-point guarantees that the other is not a saddle-point.† This argument is illustrated in fig. 18(c). The stability of F can therefore be tested by the second criterion alone; this demands that $Y>\frac{1}{2}y_0$, and is obviously satisfied in fig. 18(b) but is not necessarily satisfied when $y_0<1$. For ε can then be made as small as one likes without losing the intersection, and there is no problem in arranging for Y to be less than $\frac{1}{2}y_0$, whereupon F becomes an unstable focus.

† The same point is made by noting that at the moment the loci cease to intersect they run parallel at the point of contact. Hence $\alpha_4/\alpha_3=\alpha_2/\alpha_1$, or $\alpha_1\alpha_4-\alpha_2\alpha_3=0$.

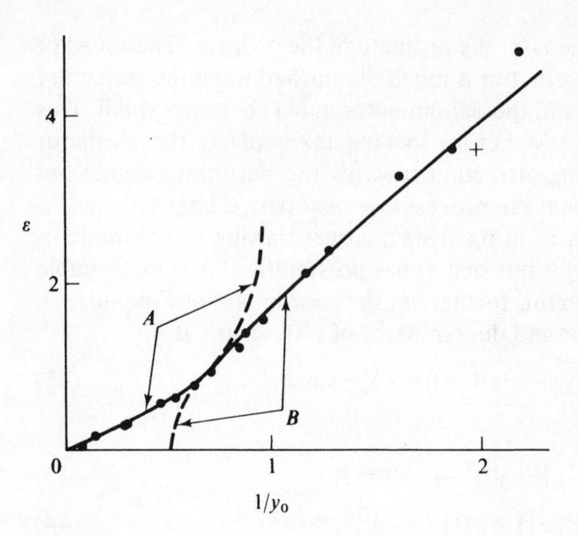

Fig. 19. Stability curve derived from (40); entrainment occurs everywhere below the line. The points are experimental values for the stability limit.

Fig. 19 illustrates the stability criteria as a plot of critical mistuning (ε) against strength of injected signal ($1/y_0$). Curve A is a plot of (45); when $1/y_0 < 1$ it is only for points below this curve that F exists and the possibility of locking need be considered. Curve B is the hyperbola $\varepsilon^2 = 4/y_0^2 - 1$ on which, as follows from (43), $Y = \frac{1}{2}y_0$. On the upper part of B, beyond the point $(1/\sqrt{2}, 1)$ at which it touches A, the saddle-point S either cannot have $Y = \frac{1}{2}y_0$ (in the range $1/\sqrt{2} < 1/y_0 < 1$) or does not exist ($1/y_0 > 1$); there is no question therefore but that this part of B refers to F alone, and that only points lying below B can show stable locking. The lower part of B, shown broken, refers to S, not F, and is irrelevant to this discussion. The full curve is therefore the limit of stability of the locked mode. The experimental points were taken by fixing the level of the injected signal and detuning its frequency away from ω_0 (both upwards and downwards) until locking failed – a very obvious moment since strong beats suddenly appear on the waveform of the oscillation. The value of ε has been taken as proportional to the difference between the upper and lower critical frequencies, the constant of proportionality being adjusted to get a good fit. No arbitrary adjustment was needed for $1/y_0$, for (44) shows that when the signal is tuned to ω_0, making ε vanish, the oscillation amplitude Y rises to $1 + y_0$; far from the locking condition, however, $y = y_0$ and from observation of the amplitude peak y_0 is unambiguously derived. The experimental points confirm the theory very adequately. Further confirmation is provided by the observation that when the injected signal is strong, but not when it is weak, the amplitude has dropped to half its unperturbed value, i.e. $Y = \frac{1}{2}y_0$ as demanded by the stability analysis, at the moment of unlocking.

Finally it is worth noting that when the injected signal is strong hysteresis appears, entrainment occurring, as ε is reduced, at a smaller value than that at which it fails when ε is increased again. This means that under certain conditions a periodic unlocked trajectory and a stable focus can be found on the same diagram. A set of computed trajectories confirms this; fig. 20 is drawn for the case $\varepsilon = 3.65$, $y_0 = \frac{1}{2}$ which lies, as shown in fig. 19, just below the critical curve B. A slowly converging trajectory surrounds the focus F which is clearly stable. The linear approximation begins to fail as one moves away from F; convergence becomes slower, and L is an unstable closed trajectory; trajectories within L converge, and those outside it diverge, breaking away after some circuits of F and gradually approaching the stationary unlocked trajectory, which lies about midway between T_1 and T_2. It is clear that between the limiting forms of behaviour – on the one hand a stable focus into which all trajectories converge, and on the other an unstable focus and a stable unlocked trajectory – one must normally expect a transition regime in which both stabilities coexist, and that hysteresis will accompany the switch from one to the other. That it is not observed at lower signal strengths is probably because the width of the hysteretic regime is narrow. There is no very obvious reason, even in the situation covered by curve A of fig. 19, when locking takes place as soon as the 0-loci and ∞-loci intersect, why the unlocked trajectory should simultaneously become unstable. We shall meet the same behaviour in the next example, and experimental data will be presented in fig. 23 to show the marked differences in detail between hysteretic and non-hysteretic locking. Observations of the system just analysed show it to have very similar characteristics, and there is no need to give an account of any further measurements in this case.

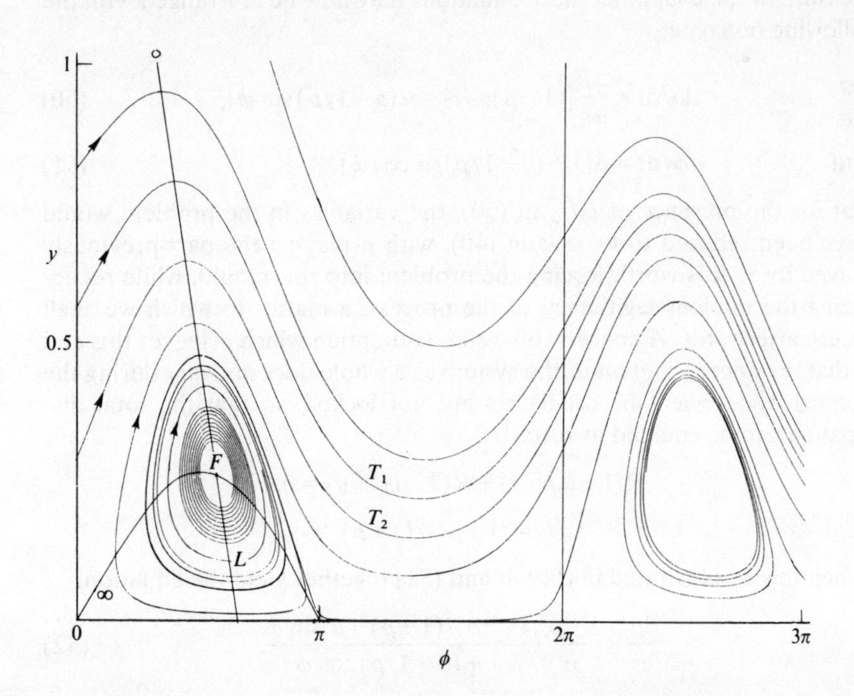

Fig. 20. Computed trajectories for (40) with $y_0 = \frac{1}{2}$, $\varepsilon = 3.65$, corresponding to cross in fig. 19.

Let us proceed, therefore, to the Huygens phenomenon proper, extending the argument to cover two class C oscillators coupled by a mutual inductance M, so that each injects a perturbing signal into the other, tending to lock it but inevitably being itself perturbed in the process. Let the oscillators be nearly identical, except for oscillator 2 having a slightly higher natural frequency than oscillator 1. At any instant the currents in the two circuits may be written as $i_1 \mathrm{e}^{\mathrm{j}\phi_1}$ and $i_2 \mathrm{e}^{\mathrm{j}\phi_2}$, i_1 and i_2 being real, and the time variation of ϕ_1 and ϕ_2 including the natural frequency as well as any perturbing effects. Then by the same argument as led to (38) and (39), we write the equations for the currents and phases:

$$\frac{\mathrm{d}i_1}{\mathrm{d}t} = -\tfrac{1}{2}R(i_1 - i_0)/L - \mathrm{Re}\,[\tfrac{1}{2}\mathrm{j}\omega M i_2\, \mathrm{e}^{\mathrm{j}(\phi_2-\phi_1)}/L], \tag{46}$$

$$\frac{\mathrm{d}i_2}{\mathrm{d}t} = -\tfrac{1}{2}R(i_2 - i_0)/L - \mathrm{Re}\,[\tfrac{1}{2}\mathrm{j}\omega M i_1\, \mathrm{e}^{\mathrm{j}(\phi_1-\phi_2)}/L], \tag{47}$$

$$\frac{\mathrm{d}\phi_1}{\mathrm{d}t} = \omega_1 - \tfrac{1}{2}\,\mathrm{Im}\,[\mathrm{j}\omega M i_2\, \mathrm{e}^{\mathrm{j}(\phi_2-\phi_1)}/Li_1], \tag{48}$$

$$\frac{\mathrm{d}\phi_2}{\mathrm{d}t} = \omega_2 - \tfrac{1}{2}\,\mathrm{Im}\,[\mathrm{j}\omega M i_1\, \mathrm{e}^{\mathrm{j}(\phi_1-\phi_2)}/Li_2], \tag{49}$$

in which ω_1 and ω_2 are the frequencies of the two oscillators when decoupled. Write $\omega_2 - \omega_1 = \Delta$, $i_2/i_1 = \rho$, $\phi_2 - \phi_1 = \phi$, $2L\Delta/\omega M = \varepsilon$, $\omega M/R = \mu$; as before, ε is a measure of the mistuning of the oscillators, while μ is a measure of the coupling. These equations may now be rearranged with the following outcome:

$$\mathrm{d}\rho/\mathrm{d}t = \frac{\Delta\rho}{\mu\varepsilon}[(1-\rho)i_0/i_2 - \mu(\rho + 1/\rho)\sin\phi], \tag{50}$$

and

$$\mathrm{d}\phi/\mathrm{d}t = \Delta[1 + (\rho - 1/\rho)/\varepsilon\,\cos\phi]. \tag{51}$$

But for the presence of i_0/i_2 in (50), the variables in the problem would have been reduced to two, as in (40), with ρ playing the part previously played by y. It is worth forcing the problem into this mould, while recognizing the dubious legitimacy of the process, a matter to which we shall return afterwards. A conceivably valid assumption which achieves this end is that the energy content of the system as a whole does not vary during the beating cycle when the oscillators are not locked, so that the total dissipation can be equated to zero:

$$R(1 - i_0/i_1)i_1^2 + R(1 - i_0/i_2)i_2^2 = 0,$$

i.e.

$$i_0/i_2 = (1+\rho^2)/(1+\rho).$$

When this is substituted in (50), it and (51) together yield the equation:

$$\frac{\mathrm{d}\rho}{\mathrm{d}\phi} = \frac{1+\rho^2}{\mu}\,\frac{(1-\rho)/(1+\rho) - \mu\sin\phi}{\varepsilon + (\rho - 1/\rho)\cos\phi}. \tag{52}$$

In like manner to fig. 18 we plot the loci of zero and infinite gradient, which intersect either four times or not at all; fig. 21 is an example where intersecting loci give the possibility of locking at either F_1 or F_2, but not at S_1 or S_2 which are saddle-points. The two modes of locking, F_1 and F_2, are closely related in that they occur at values of ϕ that are π apart and at values of ρ that are reciprocals; substituting $1/\rho$ for ρ and $\phi+\pi$ for ϕ in (52) leaves the equations unchanged.† It is only necessary, therefore, to consider the conditions for the occurrence of one intersection, F_1, and its stability, and we shall assume from now on that $\rho<1$. We proceed as before; if F_1 is (Φ, P), then

$$\mu \sin \Phi=(1-P)/(1+P) \quad \text{and} \quad \cos \Phi=-\varepsilon/(P-1/P). \tag{53}$$

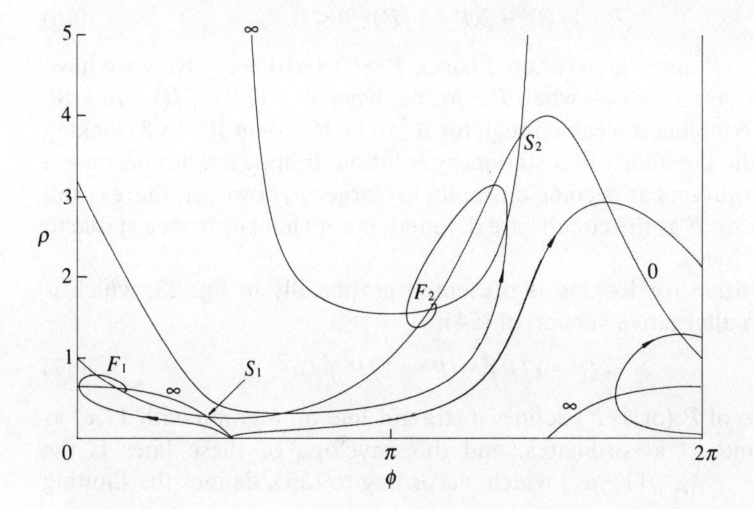

Fig. 21. 0-loci, ∞-loci and computed trajectories for solutions of (52).

By squaring and adding, to eliminate Φ, we find after rearrangement that

$$(P-\rho_1)^2(P-1/\rho_1)^2=\mu^2P^2[4\mu^2-\varepsilon^2(1-\mu^2)]/(1-\mu^2)^2, \tag{54}$$

where $\rho_1+1/\rho_1=2/(1-\mu^2)$. Now every term except that in square brackets on the right is essentially positive. The condition for P to exist is then seen to be that the term in square brackets must be positive. If $\mu \geqslant 1$ this is automatically satisfied, but if $\mu<1$ it is necessary that

$$\varepsilon^2 \leqslant 4\mu^2/(1-\mu^2). \quad 55)$$

At the moment this criterion fails to be satisfied, each focus $F_{1,2}$ coincides with its accompanying saddle-point $S_{1,2}$ at values of ρ equal to ρ_1 and $1/\rho_1$.

† At F_1, $i_2<i_1$ and the circuit of higher natural frequency, being less excited than the other, suffers the greater shift of frequency; the two are therefore locked at a frequency below the natural frequency of either. At F_2, $i_2>i_1$, and the frequency of locking is higher than either natural frequency.

To investigate the stability of F_1, put $\phi = \Phi + \xi$, $\rho = P + \eta$ to derive, with the help of (53), the linearized versions of the numerator and denominator of (52):

$$\frac{d\eta}{d\xi} = \frac{\alpha_3\xi + \alpha_4\eta}{\alpha_1\xi + \alpha_2\eta},$$

where $\alpha_1 = (1-P)^2/P$, $\alpha_2 = \mu\varepsilon(1+P^2)/P(1-P^2)$, $\alpha_3 = -\mu\varepsilon P(1+P^2)/(1-P^2)$ and $\alpha_4 = -2(1+P^2)/(1+P)^2$. As in the previous example, the merging of F and S at the moment when the loci cease to intersect is also the moment of failure of the stability criterion $D > 0$. This criterion is therefore the same as (55), and it is to the second criterion, $T < 0$, that attention must be directed. It may be written in the form

255

$$(P + 1/P)^2 - 2(P + 1/P) - 4 < 0, \tag{56}$$

so that $P + 1/P$ must lie between 2 (since $P > 0$) and $1 + \sqrt{5}$. Now we have seen that intersection fails when $P = \rho_1$, i.e. when $P + 1/P = 2/(1 - \mu^2)$. So long as the coupling is weak enough for μ^2 to be less than $\frac{1}{2}(3 - \sqrt{5})$ locking fails when the possibility of a stationary solution disappears, not because a stationary solution has become unstable; for larger μ, however, there is still an intersection F as the circuits are detuned, but it changes from a stable to an unstable focus.

The condition for locking is presented graphically in fig. 22, which is based on an alternative version of (54):

$$\varepsilon^2 = (P - 1/P)^2 - (P^{\frac{1}{2}} - 1/P^{\frac{1}{2}})^4/\mu^2. \tag{57}$$

Each choice of P (or $1/P$) defines a straight line on a graph with $1/\mu^2$ as abscissae and ε^2 as ordinates, and the envelope of these lines is the hyperbola $\varepsilon^2 = 4\mu^2/(1 - \mu^2)$ which, according to (55), defines the limiting

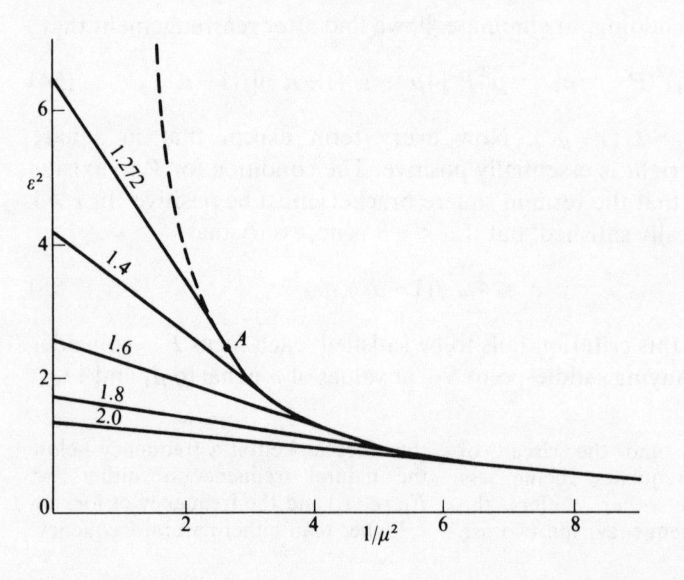

Fig. 22. Stability curve derived from (52), consisting of the straight line labelled 1.272 and the curve to the right of A. The numbers on the lines are values of δ/Δ, δ being the separation of the two frequencies at which the entrained system can oscillate, for a given separation, Δ, of the frequencies when the circuits are uncoupled.

condition for intersection of the loci. Not all points below the envelope allow stable locking, however, but only those that also satisfy (56) by lying below the line for $P+1/P$ equal to $1+\sqrt{5}$, i.e. $P=0.346$. Every point below the envelope lies on two straight lines defined by different values of P. This is because for a given choice of μ and ε, F is always accompanied by S. Since the value of P at the unstable intersection S is always less than at F, the diagram may be limited to F by drawing the lines only on the left-hand side of the point where they touch their envelope, and this has been done in fig. 22. The diagram serves also to indicate the frequency ω at which the oscillators lock, which is the value of $\mathrm{d}\phi_1/\mathrm{d}t$ in (48) or $\mathrm{d}\phi_2/\mathrm{d}t$ in (49), both being the same. Hence

$$\omega = \tfrac{1}{2}(\mathrm{d}\phi_1/\mathrm{d}t + \mathrm{d}\phi_2/\mathrm{d}t) = \bar{\omega} \pm \tfrac{1}{2}\delta,$$

where $\bar{\omega} = \frac{1}{2}(\omega_1 + \omega_2)$ and

$$\delta = \frac{1}{2}\frac{\omega M}{L}(P+1/P)\cos\Phi = \Delta(1+P^2)/(1-P^2) \quad \text{from (53).}$$

As with reactively coupled passive resonant circuits, $\delta > \Delta$ and the frequency at which steady oscillation can occur lies outside the range defined by the natural frequencies, and for the same reason, that the reactive coupling either depresses both or elevates both.

Some results of an experimental test of this analysis are shown in fig. 23. Two class C oscillators, with inductors of rather large radius (10 cm) were  arranged so that the spacing between the inductors, and thus the coupling,

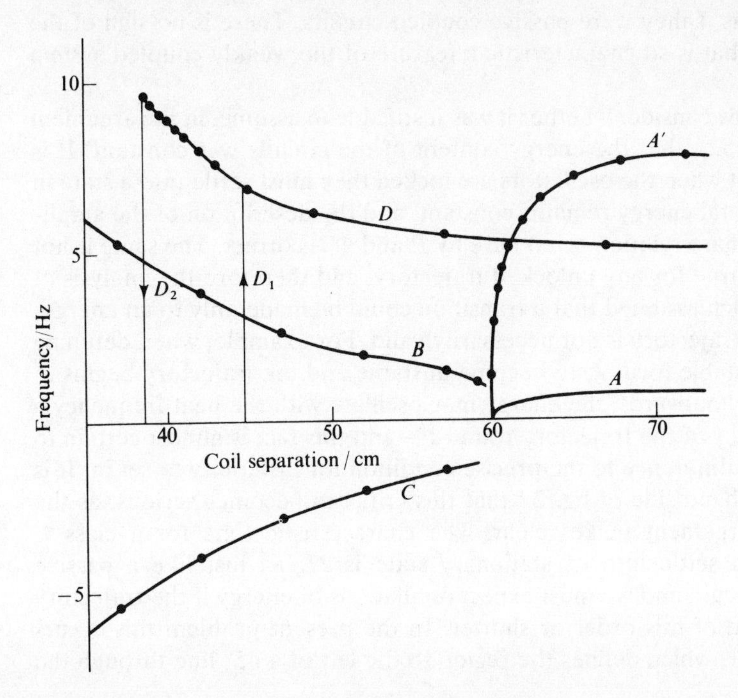

Fig. 23. The Huygens phenomenon with two class C oscillators. The circuit parameters were held constant while the coupling was changed by changing the coil separation. For curves A, B and C the frequencies of the uncoupled oscillators were 556.0 and 556.7 Hz. Curve A is the beat frequency (multiplied by 10 in A') when the coupling is too weak for entrainment; B and C are the two modes of entrainment, the frequencies here being measured from the mean, 556.4 Hz. For curve D the frequencies of the uncoupled oscillators were 555.9 and 560.8 Hz.

could be varied. When the frequency difference was small weak coupling sufficed to produce entrainment, and the approach to the critical point had the character shown in curve A; as the spacing is reduced the beat frequency drops, ultimately with a vertical tangent at the moment entrainment begins. From then on, further increase of coupling causes a progressive frequency shift away from the mean, along either curve B or curve C according as entrainment has taken place in the upper or lower mode. It is worth remarking that the parameters (e.g. amplifier gain) of the two oscillators must be carefully matched in order to obtain either B or C with equal likelihood. If one amplifier is stronger it will cause the system to lock in the mode whose frequency is nearer to its own. With carefully matched oscillators, however, it is possible to select either mode by pushing the two coils together at the right phase in the beat cycle. As with the previous example of entrainment, there is no sign of hysteresis when the frequency difference is small and the critical point is that at which 0-loci and ∞-loci first intersect. With a larger frequency difference stronger coupling is needed and hysteresis then appears, as in curve D. Failure of entrainment (D_1) on reducing the coupling takes place as T changes sign and the focus F becomes unstable; the onset of entrainment (D_2) on increasing the coupling takes place when the periodic unlocked trajectory becomes unstable. This is a problem which we have not analysed – it is distinctly more difficult than anything tackled here. Attention may be drawn to the steady increase in the beat frequency along D as the coupling is increased. A rough comparison with the difference between B and C is enough to show that the unlocked oscillators are beating at much the same frequency as if they were passive coupled circuits. There is no sign of the attraction that is so characteristic a feature of the weakly coupled system (curve A).

Let us now consider whether it was justifiable to assume, in the argument leading to (52), that the energy content of the circuits was constant. It is obvious that when the oscillators are locked they must settle into a state in which the total energy remains constant, and the description of the amplitudes and phase relation given here by P and Φ is correct. The same is not necessarily true for any unlocked trajectory, and therefore the analysis of stability which assumed that a transition could be made only to an energy-conserving trajectory is not necessarily valid. For example, when detuning causes the stable focus F to become unstable and the trajectory begins to spiral slowly outwards, the energy may oscillate with the beat frequency – the frequency of the trajectory round F – and this fact is almost certain to make some difference to the precise condition for instability to set in. It is on the left-hand side of fig. 22 that this criticism becomes serious, as the following argument makes clear. The characteristic time for a class C oscillator to settle into its stationary state is $2L/R$, just like a passive resonant circuit, and we must expect oscillations of energy if the trajectory has a period of this order or shorter. In the present problem this occurs when $\varepsilon\mu \gtrsim 1$, which defines the region to the left of a 45° line through the origin in fig. 22. On the other hand, at the breakdown of locking on the right-hand side, when D goes to zero as the loci cease to intersect, the trajectories around where F has just disappeared are traversed exceedingly slowly. There is every expectation, therefore, that the assumption of zero net energy production applies here both to the stable and to the just-unstable regime, and that the condition for stability (i.e. the existence of F) is correctly applied. We may hazard a guess that the boundary of the stable region in fig. 22 is the whole of the hyperbolic envelope up to A where it joins the limiting straight line, and that as we proceed along the straight line it becomes progressively less correct, the true limit falling below as other possibilities for instability make themselves felt. To go into the matter in more detail involves considerably more effort. Alternative approaches are (a) to define the system by means of three parameters, i_1, i_2 and ϕ, rather than two, and consider the conditions of stability of a linearized three-variable system, with three criteria to be satisfied; or (b) to extend the normal mode treatment used for the coupled active and passive oscillators. Both approaches should lead to precise criteria rather than the imprecise ones developed here but, as we have remarked before, there must be some doubt as to the value of the effort in terms of improved understanding of the mechanisms involved. We propose to remain satisfied with a treatment that reveals the underlying physics, even if the answer is not correct in all details.

An altogether simpler example of the Huygens phenomenon is provided by the *resistive* coupling of two active oscillators, such as the arrangement of fig. 1(e) but with both circuits maintained as class C oscillators. The equations have the same form as (46)–(49), except that $\mathrm{j}\omega M$ is replaced by R_c. Then if ε is redefined as $2L\Delta/R_\mathrm{c}$, and μ as R_c/R,

$$\mathrm{d}i_1/\mathrm{d}t=\frac{\Delta}{\varepsilon}[(i_0-i_1)/\mu-i_2\cos\phi], \tag{58}$$

$$\mathrm{d}i_2/\mathrm{d}t=\frac{\Delta}{\varepsilon}[(i_0-i_2)/\mu-i_1\cos\phi] \tag{59}$$

and

$$\mathrm{d}\phi/\mathrm{d}t=\frac{\Delta}{\varepsilon}[\varepsilon+(\rho+1/\rho)\sin\phi]. \tag{60}$$

From (58) and (59) we have that

$$\frac{\mathrm{d}}{\mathrm{d}t}(i_2-i_1)=\frac{\Delta}{\varepsilon}[-(i_2-i_1)/\mu+(i_2-i_1)\cos\phi]. \tag{61}$$

If entrainment occurs, $\mathrm{d}i_1/\mathrm{d}t=\mathrm{d}i_2/\mathrm{d}t=0$, and (58) and (59) show that $i_1=i_2$. As we shall see in a moment, only states in which $\cos\phi$ is negative can be stable; it is clear from (61) that in the neighbourhood of such a state any departure of i_2-i_1 from zero must eliminate itself automatically. In assuming $i_1=i_2$, then, when determining the conditions for locking we are on much safer ground than we were in the previous calculation. Writing $i_1 = i_2 = i$, we have from (58) and (60)

$$\frac{\mathrm{d}i}{\mathrm{d}\phi} = \frac{1}{\mu} \frac{i_0 - (1 + \mu \cos \phi)i}{\varepsilon + 2 \sin \phi}. \tag{62}$$

Typical loci of zero and infinite gradient are shown in fig. 24 for $\mu < 1$, and it needs no further analysis to see that S is a saddle-point and N a stable node, which always exists provided $\varepsilon < 2$ so that the denominator of (62) can vanish. Thus the oscillators lock when $\Delta < R_c/L$, and there is no possibility of hysteretic behaviour since non-locking trajectories crossing the diagram can only be present when the ∞-loci are absent.

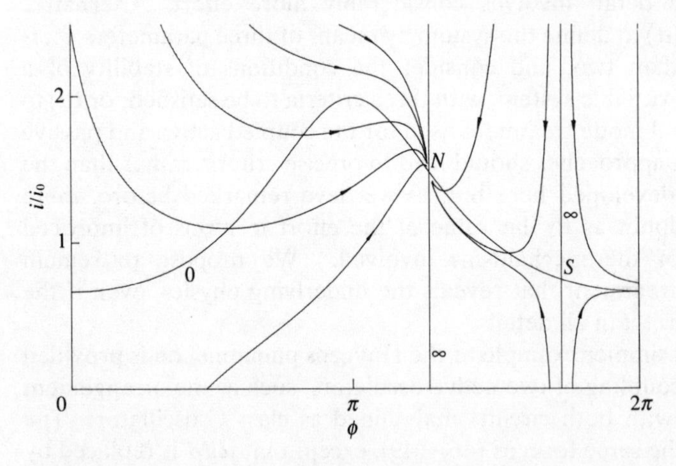

Fig. 24. 0-loci, ∞-loci and computed trajectories for solutions of (62).

This analysis is verified by experiment. When the coupling is weak the currents oscillate in amplitude with a beat-like pattern, and the amplitude oscillations in the two circuits are in phase. This contrasts with the reactively-coupled system where they oscillate in antiphase. As the coupling is increased the beat period gets longer without limit, and as it goes to infinity locking sets in, only to be broken once more without any sign of hysteresis when the coupling is reduced. When the beat period is long it exhibits clear signs of non-linearity, the envelope departing markedly from sinusoidal form, as the record in fig. 25 shows. This is not an isolated example – the other systems showing the Huygens phenomenon behave similarly. In this particular case the oscillators, when uncoupled, differed in frequency by 1 part in 41, but the beat period when coupled is not 41 but 67 cycles. This shows, as expected for resistive coupling, and in contrast to curve D of fig. 23, the attraction of the two frequencies towards the mean.

To conclude, let us return to Huygens himself and his observation of entrainment in pendulum clocks. It is quite easy to demonstrate the effect by extracting the mechanisms from two metronomes and mounting them on a light frame, which is suspended on threads so that its swinging can couple the two vibrations. With the adjustable weights removed the vibration frequency is about 100 cycles (200 clicks) a minute and strong

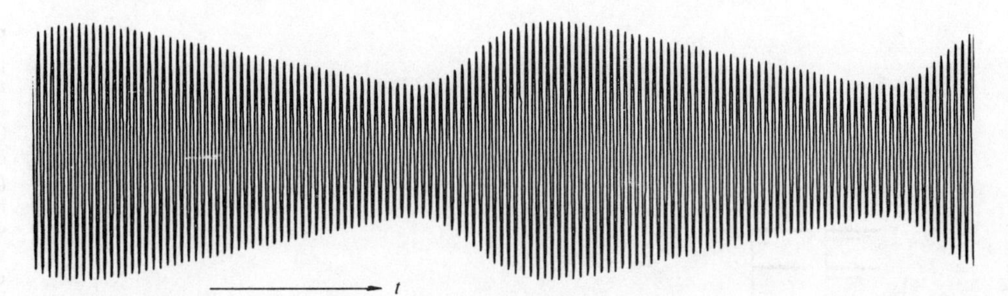

Fig. 25. Non-sinusoidal beat envelope for resistively-coupled oscillators not quite entraining.

coupling is observed. The records shown in fig. 26 are redrawn from a chart recording of the clicks picked up by a microphone. The records in (*a*) and (*b*) show how the metronomes vibrate independently, with a frequency difference of 2.5%, when the frame is clamped, but lock into a steady phase relationship when it is free to swing. The graph shows that the approach to the steady state is oscillatory, as expected for reactive (i.e. loss-free) coupling. An extra complication here, apart from the escapement mechanism, is the large amplitude of swing, up to $\pm 45°$ during the oscillatory approach to the steady state; changes in frequency with amplitude must be a significant factor in determining the details of the behaviour.

353 Frequency-locking of a multivibrator

The most striking difference between a class C oscillator and a relaxation oscillator, as regards their response to an injected sinusoidal disturbance, is that the former responds only to a nearly synchronous signal and is drawn into synchronism, while the latter can be controlled by a signal at a harmonic of its own oscillation. Perhaps the distinction is slightly exaggerated by this statement, in that a class C oscillator may indeed by locked by a powerful enough signal at the third or fifth harmonic, but this is a very mild affair compared with the way a multivibrator may be diverted, if only by a small amount, from its natural frequency into an exact submultiple of an injected frequency hundreds of times higher. It is this property that has given the multivibrator its place in the repertoire of electronic devices, for with two or three stages of subdivision one can convert the frequency of a standard source (e.g. a quartz crystal oscillator) from several megahertz down to a suitable frequency for running a clock.

The multivibrator is not only sensitive to harmonics; it can also be locked, though rather weakly, at a frequency which is a rational fraction of the injected frequency. An example is shown in fig. 27, where it can be seen that the frequency of the multivibrator is pulled in an extremely complicated way as the injected frequency is varied. The details depend critically on the amplitude of the perturbation, but the fundamental response is always the most marked. The odd harmonics pull the multivibrator more strongly than the even, and certain intermediate lockings are obvious. What is not obvious in the diagram is that every point marked on

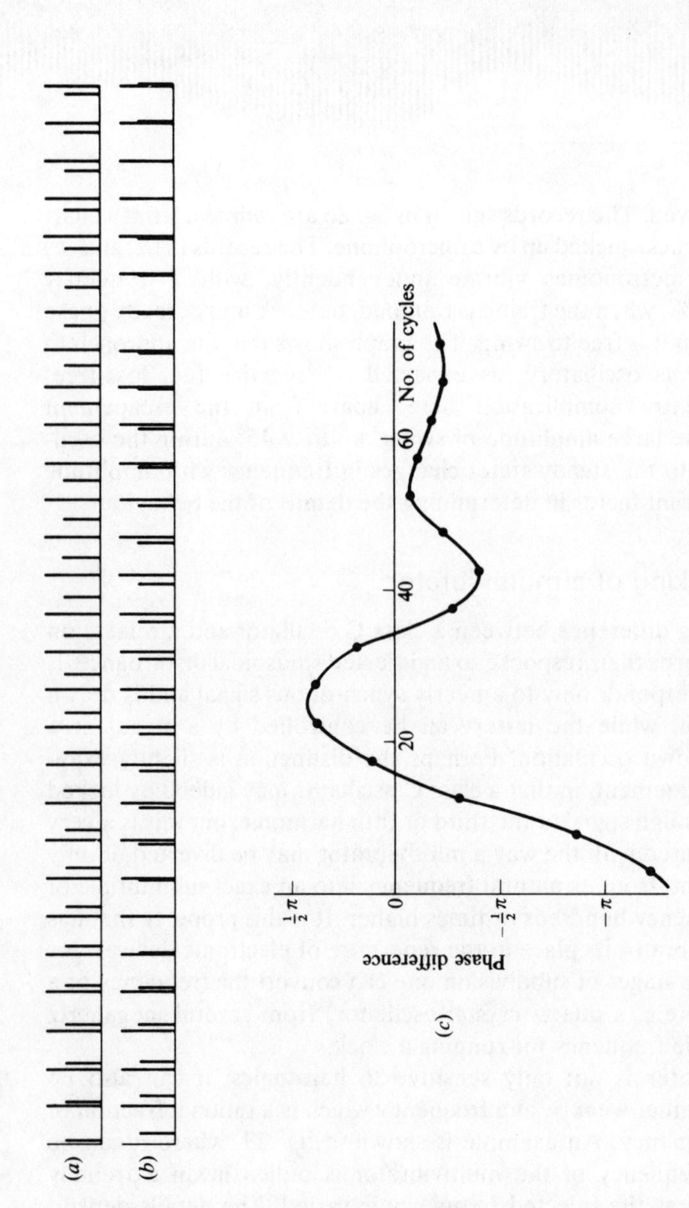

Fig. 26. The Huygens phenomenon with two metronome movements; (*a*) and (*b*) are transferred from a record of the clicks of the two metronomes (shown different in height here). In (*a*) the beam on which they were mounted was clamped, and they vibrated independently at different frequencies; in (*b*) the beam was swinging freely and entrainment occurred, as plotted in (*c*) which shows that the phase difference between the clicks settled down to a constant value. The oscillatory approach indicates a case of focal stability.

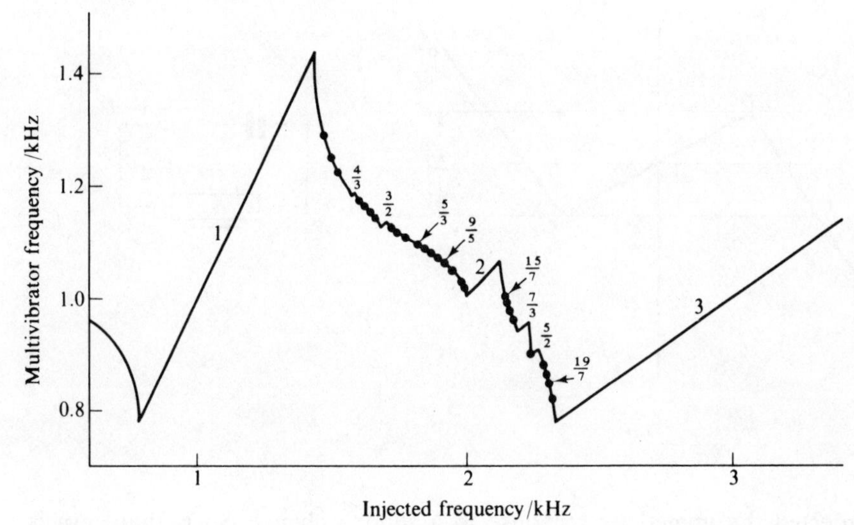

Fig. 27. Multivibrator as in fig. 11.38, entrained by a sinusoidal input of variable frequency. The labels show the ratio of the injected frequency to the entrained frequency of the multivibrator, i.e. n/m in the notation of p. 412.

the curve represents a locking, if only over a very short frequency interval (1 Hz or less), but nevertheless discernible; doubtless many more could be found by patient searching. Below the fundamental, however, it is hard to convince oneself that any locking occurs, though vestiges can be discovered, for example at the third subharmonic, by turning up the amplitude of the injected signal. We shall provide a rather superficial discussion of these observations, concentrating on generalities and not attempting a detailed theory of why some steps are long and others short; and one notable failure will be the absence of any explanation for the weakness of the sub-fundamental steps. This is not to imply that the model analysed is at fault, but only that elementary mathematics has failed to cope with a remarkably elusive phenomenon.

Let us begin with the contrast, already noted, between the class C oscillator and the multivibrator in their sensitivity to other than synchronous locking. Qualitatively, the insensitivity of the one and the sensitivity of the other to a harmonic disturbance is easily understood. The model we have adopted for a class C oscillator is essentially that of a free-running high-Q circuit kept going by a very weak impulse during each cycle. The injection of a third-harmonic signal can produce only the weakest perturbation compared with the effect of a signal near the resonance frequency. By contrast, the multivibrator is non-resonant, spending almost all the time relaxing, like an RC circuit, towards an unattainable point of rest ($\pm V_0$ in fig. 28(a)), only to be regularly tripped into a different state at B or D. The circuit can be idealized as in fig. 28(b), the device labelled N being the feedback network and a series resistor

354

(10 Ω in fig. 11.39), which together have the characteristic shown in (a); a locking e.m.f. v may be injected as shown. If $V_0 \gg \Delta V$, i and V move along AB or CD at a very nearly constant rate, $\dot{V} = \pm V_0/\tau$, τ being the time-constant R_sC. This is what happens when $v = 0$, but when a signal is

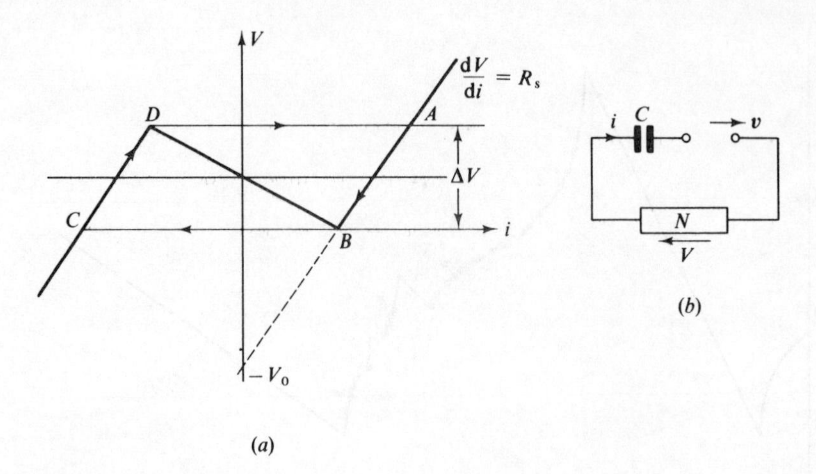

Fig. 28. (*a*) Assumed negative-resistance characteristic of the device N in (*b*) which represents the multivibrator circuit with sinusoidal input v.

injected the immediate response to a sudden change Δv is that $-\Delta v$ is developed across N, since the p.d. across C cannot change suddenly. Only over a time of the order of τ does C respond to the stimulus, and if v oscillates as $v_0 \cos \omega t$, with $\omega\tau \gg 1$, we may ignore everything but the initial effect; on the steady change of V is superposed an oscillation $-v_0 \cos \omega t$:

$$V = \text{const} \pm V_0 t/\tau - v_0 \cos \omega t. \tag{63}$$

Here, and in what follows, the upper sign refers to motion along CD, the lower to motion along AB; the constant changes its value at every switch from one side to the other, but it will soon become clear that it is of no importance to the argument.

It is convenient to introduce V' as a variable that increases steadily at the same rate as $|V|$, rather than oscillating to and fro; $\dot{V}' = |\dot{V}|$, and

$$V' = A + V_0 t/\tau \mp v_0 \cos \omega t. \tag{64}$$

Every time V' reaches a value that is a multiple of ΔV, as defined in the diagram, the sign changes, and so does the value of A to keep V' continuous. The switches occur at successive times, t_0, t_1 etc. such that

$$\Delta V = (V_0 t_1/\tau - v_0 \cos \omega t_1) - (V_0 t_0/\tau - v_0 \cos \omega t_0),$$
$$\Delta V = (V_0 t_2/\tau + v_0 \cos \omega t_2) - (V_0 t_1/\tau + v_0 \cos \omega t_1),$$

and so on with alternating signs. The process is illustrated in fig. 29(*a*), and the actual variation of V' in (*b*) where each element of vertical height ΔV has been shifted by appropriate choices of A so as to give a continuous curve. This is irrelevant, however, since all the interest lies in the placing of the successive segments on the sinusoids, to see whether it shows any sign of that regularity which must accompany entrainment.

In this example there is no obvious relationship between the mean switching frequency and ω the injected frequency though possibly, by continuing the diagram for long enough, a periodic pattern would be

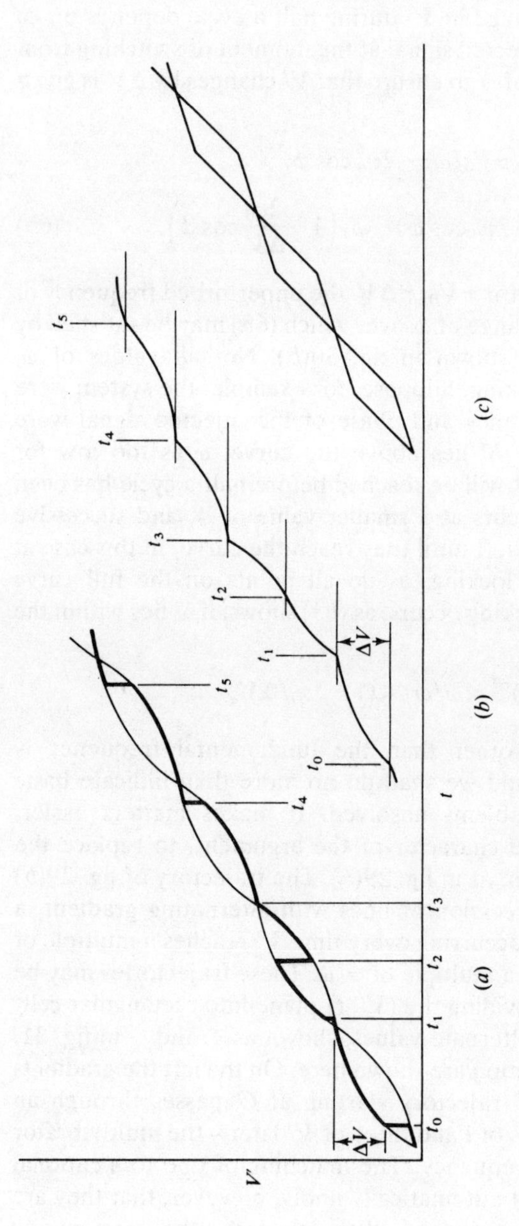

Fig. 29. (a) Each successive heavily-drawn segment has vertical height ΔV, and at the end of each the point is transferred to the other sinusoid to start the next segment; in (b) the successive segments are joined to show the variation of V' with time, as given by the solution of (64). (c) is a simplified form of the sinusoids in (a), used in constructing fig. 31.

revealed. It is much easier to see how locking arises in the specially simple case when switching is synchronized with the fundamental of the injected signal. If this has been achieved, the switch occurs at the same phase in each half-cycle, as fig. 30 shows for several different cases. For a given choice of ω, however, the change in V' during half a cycle depends on ϕ, defined as the phase of the injected signal at the moment of switching from D to A; from (64), the value of ω to ensure that V' changes by ΔV is given by the equation:

$$\Delta V = \pi V_0/\omega\tau - 2v_0 \cos\phi,$$

or

$$\omega = \pi V_0/\tau(\Delta V + 2v_0\cos\phi) \sim \omega_1\left(1 - \frac{2v_0}{\Delta V}\cos\phi\right), \tag{65}$$

if $v_0 \ll \Delta V$; here ω_1 is written for $\pi V_0/\tau\,\Delta V$, the unperturbed frequency of the multivibrator. There is a range of ω over which (65) may be satisfied by appropriate choice of ϕ, as shown in fig. 30(*b*). Not all values of ϕ, however, represent stable locking. Suppose, for example, the system were started at A when the frequency and phase of the injected signal were given by the point X. Since X lies above the curve, ω is too low for synchronous switching, and B will be reached before half a cycle has been executed; the next switch occurs at a smaller value of ϕ, and successive positions of X migrate to the left until they reach the curve, in this case at X', which represents stable locking, as do all points on the full curve $(0 < \phi < \pi)$. Consequently locking occurs, as (65) shows, if ω lies within the range

$$(1 + 2v_0/\Delta V)^{-1} < \omega/\omega_1 < (1 - 2v_0/\Delta V)^{-1}.$$

Discussion of locking at other than the fundamental frequency is considerably more difficult, and we shall do no more than indicate basic principles, leaving many problems unsolved. It makes matters easier, without seriously altering the character of the argument, to replace the sinusoidal input by a sawtooth, as in fig. 29(*c*). The trajectory of fig. 29(*b*) then takes the form of a succession of lines with alternating gradient, a switch from one to the other occurring every time V' reaches a multiple of ΔV, and every time t reaches a multiple of π/ω. These trajectories may be drawn without difficulty by dividing the (V', t)-plane into rectangular cells in which the gradients take alternate values, shown as + and − in fig. 31. Two examples of the construction are shown here. On the left the gradients have been chosen so that a trajectory starting at O passes through an equivalent point, P, two cycles of t and three of V' later – the multivibrator is oscillating at $\frac{3}{2}$ the input frequency. The matching of one to a rational multiple of the other does not automatically imply, however, that they are locked; in this particular case, indeed, they are not – they are merely adjusted to coincidence, as the following argument makes clear. Between the identical trajectories OP and O_1P_1, one cycle apart in time, there are five others with the same periodicity, obtained by starting the construction

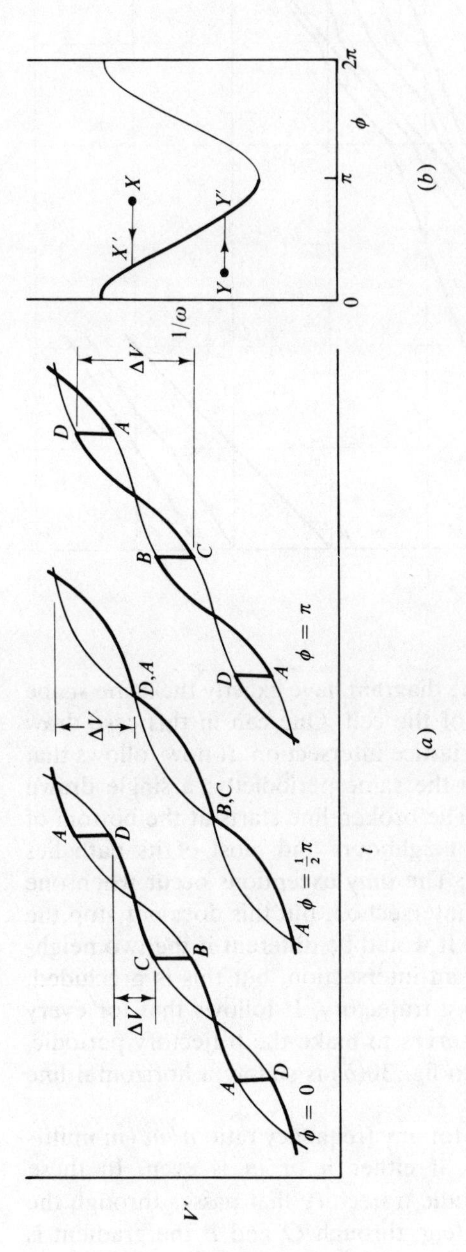

Fig. 30. Synchronization (entrainment) of a multivibrator at the fundamental, the curves in (*a*) showing how different values of ϕ demand a different value of ΔV for a given injected frequency, ω. The heavy line in (*b*) indicates the range of ϕ and ω over which synchronization occurs, and illustrates the process by the two phase migrations XX' and YY'.

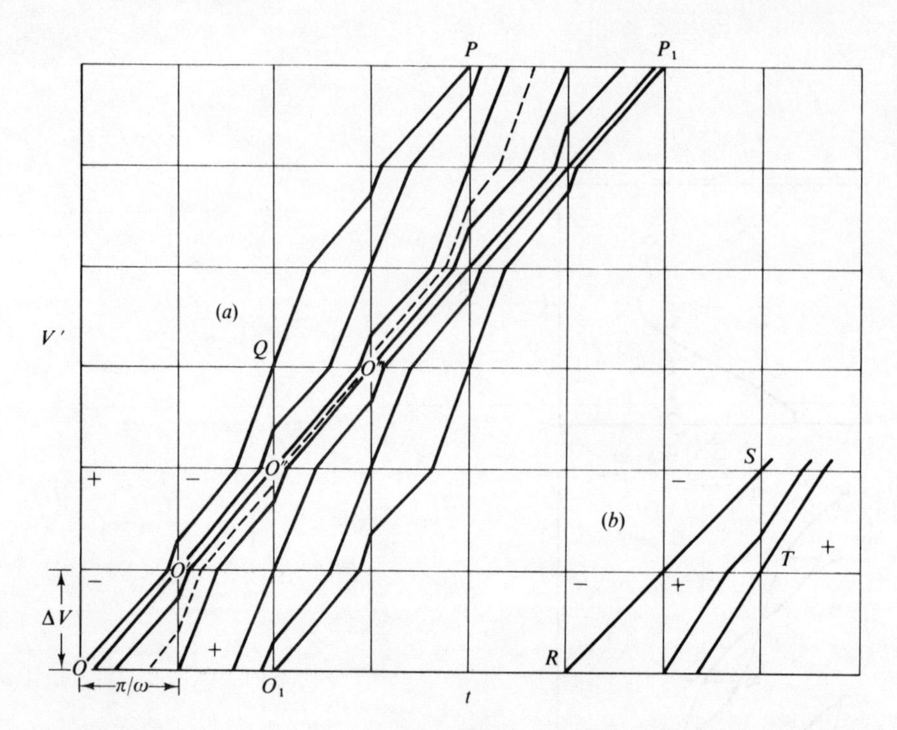

Fig. 31. Illustrating entrainment at other than the fundamental frequency.

at O', O'' etc.; all those shown in the diagram have exactly the same shape but are displaced by the diagonal of the cell. One can in this case draw identical trajectories through every lattice intersection. It now follows that intermediate trajectories also have the same periodicity; a single drawn example elucidates the reasoning. The broken line starts at the bottom of the diagram, halfway between its neighbours and most of its path lies between them as they run parallel. The only exceptions occur when one neighbour passes through a lattice intersection, but this does not stop the broken line lying halfway between. It would be different if the two neighbours could pass on either side of an intersection, but this is precluded, every intersection already lying on a trajectory. It follows that for every starting point the same frequency serves to make the trajectory periodic, and that the curve corresponding to fig. 30(*b*) is simply a horizontal line with no locking properties.

This argument holds equally well for any frequency ratio n/m (m multivibrator cycles to n input cycles), if either n or m is even. In these circumstances one can draw a periodic trajectory that passes through the two different types of intersection (e.g. through O and P the gradient is low, while through Q it is high), and this allows every intersection to lie on an identical trajectory. By contrast, if both n and m are odd there are two different trajectories, as illustrated in fig. 31(*b*); that through R and S is periodic while that through T is not, nor are intermediate trajectories.

Consequently, to force periodicity with a given choice of gradients, ω must depend on ϕ, providing, as in fig. 30(*b*), the right conditions for locking. It is apparent in fig. 27 that locking does in practice occur even if n or m is even, but the weakness of the effect when $n/m = 2$, compared with what happens at 1 and 3, suggests that small asymmetries in the multivibrator waveform may be invoked as an explanation.

The argument given above, although illustrated by means of a sawtooth waveform, does not depend on this assumption, but can be applied to any waveform that lacks even harmonics. In particular a sinusoidal input should behave very similarly, with no locking for frequency ratios n/m if either n or m is even, and locking in principle for every other rational fraction, whether greater or less than unity. It is worrying, therefore, that signs of subharmonic locking should be so elusive, while such a wealth of examples show up when the input frequency is greater than that of the multivibrator. One would expect in such a case to find a mathematical identity guaranteeing for subharmonic frequencies that property which we have noted for n or m even. But this has not been found, for the good reason that it is not true. Careful computation of the frequencies needed to give coincidence at the third subharmonic (i.e. $n/m = \frac{1}{3}$) starting from different values of ϕ, reveals very small variations. In principle subharmonic locking can occur, but in practice it seems to be too weak to be discerned.

Locking of multivibrator to a resonant circuit

The problem presented by a passive resonant circuit coupled to a multivibrator is similar to that of the coupled class C oscillator and resonant circuit, though in one respect it is simpler. The class C oscillator acts on the passive circuit and itself suffers a reaction that can change both the amplitude and frequency of its oscillation; the multivibrator is less sensitive – its frequency can be altered, but the waveform and amplitude are determined by the amplifier characteristic and do not change. If the coupling is reactive, by means of a mutual inductance, say, the square waveform of the multivibrator current induces impulsive e.m.f.s, of alternating sign, in the passive circuit, which will be strongly excited if its natural frequency is near an odd harmonic of the multivibrator. If it is not so well tuned it will respond less strongly, but always in the end will come into synchronism with the nearest odd harmonic (we shall suppose the Q of the resonant circuit to be high enough, and tuning close enough, for one harmonic to dominate). In this sense the multivibrator is always entrained by the resonant circuit, and the problem is to determine at what frequency the combination will oscillate. We shall consider only locking at the fundamental, the natural frequencies of the multivibrator, ω_1, and of the resonant circuit, ω_2, being close together.

According to (65) and the arguments leading to it, the multivibrator frequency is affected only by that component of injected e.m.f. which is in phase quadrature ($\phi = 0$ or π) with the fundamental of i_1; if $\phi = 0$ the frequency is reduced. In this particular phase relationship the e.m.f. leads the current by $\pi/2$, as if an extra inductance had been added. The response of the multivibrator is thus qualitatively similar to that of a class C oscillator, in that extra inductance lowers the frequency, while extra resistance can be compensated for and is of no consequence. There are, of course, differences in detail which are readily supplied. Thus in determining the possible frequencies for steady oscillation one must remember that if the multivibrator current oscillates, as we shall assume, between $\pm i_1$ with square waveform, its fundamental Fourier component has magnitude $4i_1/\pi$. Since the impedance at frequency ω presented by the passive circuit, through the mutual inductance, is $-\mathrm{j}\omega^2 M^2/2L(\omega - \omega_2)$, ω_2 being complex, the reactive component of the e.m.f., which must be substituted for $v_0 \cos\phi$ in (65), is $4i_1/\pi$ times the imaginary part. Hence, from the approximate form of (65), ω is a possible frequency for the combination if

$$\omega - \omega_1 = \kappa^2 \frac{\omega - \omega_2'}{(\omega - \omega_2')^2 + \omega_2''^2}, \tag{66}$$

in which $\kappa^2 = 4\omega^2\omega_1 M^2 i_1/\pi L \Delta V$. This is exactly the same as (33), only the meaning of κ^2 being changed. The general pattern of locking must clearly be similar to that shown in fig. 14, but it still remains to determine the stability range of the various branches. The treatment in terms of normal modes is inapplicable here, on account of the highly non-linear behaviour of the multivibrator, and we shall proceed in a manner more closely analogous to what was used for the Huygens phenomenon.

We shall suppose the oscillation in the passive circuit to be not perfectly in synchronism with the multivibrator and to have a magnitude slightly different from its equilibrium value; we shall then be able to determine whether it tends towards or away from equilibrium. At any instant let the (real) frequencies of multivibrator and passive circuit be Ω_1 and Ω_2, with the phase difference ϕ defined as before, so that $\dot\phi = \Omega_2 - \Omega_1$. When the amplitude of current oscillation in the passive circuit is i_2, the induced e.m.f. in the multivibrator circuit is $\Omega_2 M i_2$, and (65) takes the form:

$$\Omega_1 = \omega_1\left(1 - \frac{2\Omega_2 M i_2}{\Delta V}\cos\phi\right). \tag{67}$$

The passive circuit at the same time receives an induced e.m.f. of magnitude $4\Omega_1 M i_1/\pi$, leading the current i_2 by $\frac{1}{2}\pi - \phi$. It therefore responds as to an additional impedance δZ, where

$$\delta Z = 4\mathrm{j}\Omega_1 M i_1\, \mathrm{e}^{-\mathrm{j}\phi}/\pi i_2.$$

The imaginary part of δZ alters the natural frequency of the circuit, and the real part its decrement, so that

$$\Omega_2 = \omega_2\left(1 - \frac{2Mi_1}{\pi L i_2}\cos\phi\right),$$

and $$\mathrm{d}i_2/\mathrm{d}t = -i_2(R + 4\Omega_1 M i_1/\pi i_2 \sin\phi)/2L.$$

With the substitutions $\Delta = \omega_2 - \omega_1$, $\lambda = \frac{1}{2}R/L$, $x = (\pi\Omega_2 L)^{\frac{1}{2}} i_2/(\Delta V i_1)^{\frac{1}{2}}$ and κ as already defined, these equations, together with (67), take the form:

$$\dot{\phi} = \Delta + \kappa(x - 1/x)\cos\phi, \tag{68}$$

and $$\dot{x} = -\lambda x - \kappa\sin\phi. \tag{69}$$

Hence $$\frac{\mathrm{d}x}{\mathrm{d}\phi} = \frac{-\lambda x - \kappa\sin\phi}{\Delta + \kappa(x - 1/x)\cos\phi}, \tag{70}$$

from which trajectories of the same type as those in figs. 18 and 21 may be constructed. Examples are given in fig. 32. There is always one stable focus with x less than unity, shown as F_1 in the diagram. As Δ is raised from a negative value to a positive, F_1 moves up to cross the line $x = 1$ and eventually merges with the saddle point S and disappears, leaving F_2 as the only stable solution. This behaviour provides an immediate association with curve C of fig. 14 – the upper and lower heavily drawn curves represent F_1 and F_2, while the lightly drawn curve is S. If F_1 and F_2 are stable up to the point of merger with S, the full range of hysteresis is attained, but as with the class C oscillator it may prove that F_1 and F_2 become unstable before this point is reached.

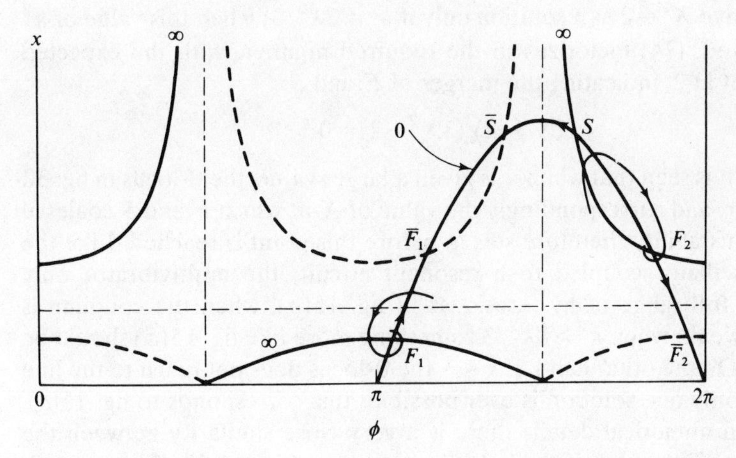

Fig. 32. 0-loci, ∞-loci and sketched trajectories for solutions of (70). As Δ goes from negative to positive, the ∞-loci change progressively from the form shown by full lines to that shown by broken lines; F_1 moves to $\bar{F}_1$, F_2 to $\bar{F}_2$, S to $\bar{S}$.

To investigate this we proceed exactly as before. An intersection of the loci occurs when $x = X$ and $\phi = \Phi$, such as to cause (68) and (69) to vanish together. Hence

$$\Delta = -\kappa(X - 1/X)\cos\Phi \quad \text{and} \quad \lambda X = -\kappa\sin\Phi. \tag{71}$$

Writing $\phi = \Phi + \xi$, $x = X + \eta$ in the neighbourhood of such a point, we have that

$$\frac{\mathrm{d}\eta}{\mathrm{d}\xi} = \frac{\alpha_3\xi + \alpha_4\eta}{\alpha_1\xi + \alpha_2\eta},$$

where

$$\alpha_1 = \lambda(X^2-1), \qquad \alpha_2 = -\frac{\Delta}{X}\frac{X^2+1}{X^2-1}, \qquad \alpha_3 = \frac{\Delta X}{X^2-1}, \qquad \alpha_4 = -\lambda. \tag{72}$$

The disappearance of the intersection above the line $x = 1$ is marked by the vanishing of the determinant D; what we have to discover is whether, as $|\Delta|$ is increased, this occurs before or after T passes from negative to positive and causes F to turn into an unstable focus. That is to say, we need to find Δ and κ such that D and T vanish together. The value of κ is then the critical coupling at which the full range of hysteresis ceases to be observable.

255

From (72), $T=0$ when $X^2=2$, and $D=0$ when $\lambda^2(X^2-1)^3 = \Delta^2(X^2+1)$. This is satisfied when $X^2=2$ if $\Delta^2=\frac{1}{3}\lambda^2$. All that remains is to find the value of κ such that when $\Delta^2=\frac{1}{3}\lambda^2$ an intersection occurs at $X^2=2$. By eliminating Φ from (71) an equation is obtained for the values of X^2 at the intersections:

$$\lambda^2X^6-(2\lambda^2+\kappa^2)X^4+(\lambda^2+\Delta^2+2\kappa^2)X^2-\kappa^2=0. \tag{73}$$

Substituting $\Delta^2=\frac{1}{3}\lambda^2$ we find:

$$\lambda^2X^6-(2\lambda^2+\kappa^2)X^4+(\tfrac{4}{3}\lambda^2+2\kappa^2)X^2-\kappa^2=0, \quad \text{if } \Delta^2=\tfrac{1}{3}\lambda^2. \tag{74}$$

This can have $X^2=2$ as a solution only if $\kappa^2=8\lambda^2/3$; when this value of κ^2 is substituted, (74) factorizes in the required manner, with the expected double root at 2, indicating the merger of F and S:

$$(X^2-2)^2(3X^2-2)=0.$$

From (69) it is seen that when κ is given a larger value the 0-locus in fig. 32 rises higher, and correspondingly the value of X at which F and S coalesce is higher; instability therefore sets in before this point is reached. Like the class C oscillator coupled to a resonant circuit, the multivibrator only shows the full range of hysteresis, as in fig. 15(b), when the coupling is relatively weak; when $\kappa^2>8\lambda^2/3$ something more like fig. 15(a) should be observed. On the other hand, if $\kappa<\lambda$ the 0-locus does not reach to the line $x=1$ and only one solution is ever possible; this corresponds to fig. 15(c). Apart from numerical details there is a very close similarity between the two systems. Examples of the behaviour are given in fig. 33, the approach to a vertical tangent at the point of instability being clearly marked in the case of weaker coupling, and entirely absent when it is stronger.

Superconducting weak links (Josephson junctions)[7]

It is a far cry from multivibrators to the essentially quantum-mechanical Josephson effect in superconductors, but their behaviour shows enough common features to justify a short exposition. Let us admit at the outset that any account of the basic theory is precluded; we shall do little more than state the fundamental equation and proceed to develop mechanical

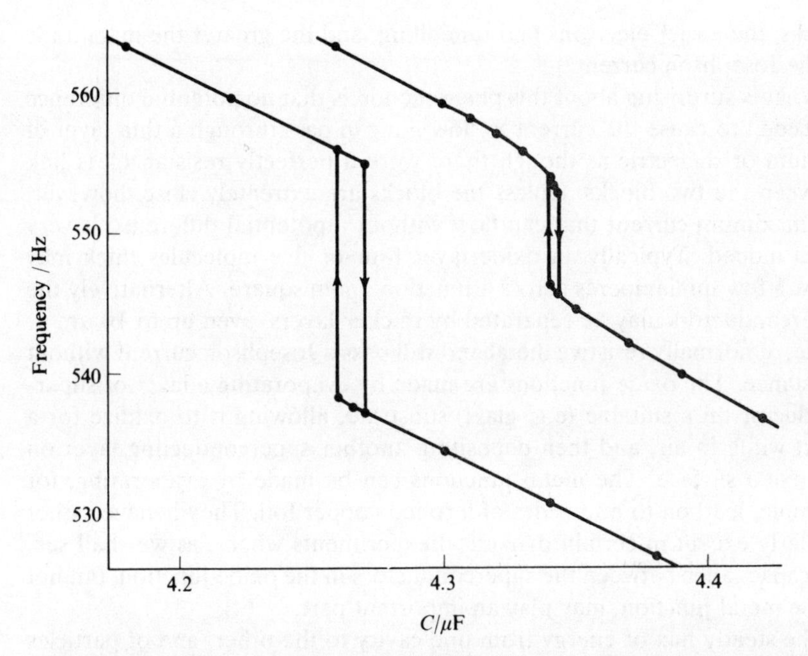

Fig. 33. Experimental observations of the locking of a multivibrator to a resonant circuit, the curve on the left being obtained with stronger coupling than for that on the right. The frequency of the multivibrator was altered by changing its capacitance, C.

models obeying the same rules. It is through these models, and mechanical models of the multivibrator, that the similarity will be revealed.

The electrons responsible for the characteristic properties of a block of metal which has been cooled below the temperature (a few K at most) at which it becomes superconducting are collectively in a single isolated quantum state having a well-defined energy. The wave-function describing this state is something like the resonant mode of a wave in a cavity, though far more complicated – the analogy is, however, good enough to be helpful. When another superconducting block is brought so near to it that it becomes possible for electrons to pass through the intervening space (by quantum tunnelling as in the tunnel diode), the two are coupled together in a similar manner to the coupling of two resonant electromagnetic cavities which can exchange energy. The coupled cavities, which we suppose to have been originally tuned to the same frequency, can only oscillate steadily if they are in phase or in antiphase, these two phase relationships being characteristic of the normal modes, and in either normal mode there is no energy exchange between the cavities. So too with the superconductors; there are two normal modes in which the wave-functions oscillate either in phase or in antiphase. These are stationary states; when the system is in either there is no exchange of electrons between the blocks. But if the cavities are set with a phase difference ϕ that is not 0 or π, energy flows from one to the other at a rate proportional to $\sin\phi$. And if the superconducting blocks have a phase difference that is not 0 or π a current of electrons flows, the magnitude being proportional to $\sin\phi$.[(8)] In both cases the maximum current that can flow, when $\phi = \pm\frac{1}{2}\pi$, is determined by the strength of the coupling – the closer the superconducting blocks, the easier electrons find tunnelling, and the greater the magnitude of the Josephson current.

What is surprising about this phenomenon is that no potential difference is needed to cause the current to flow – it can pass through a thin layer of vacuum or dielectric as though there were a perfectly resistanceless link between the two blocks. Unless the blocks are extremely close, however, the maximum current that can flow without a potential difference is very small indeed. Typically an oxide layer four or five molecules thick may allow a few milliamperes across a junction 1 mm square. Alternatively the superconductors may be separated by thicker layers, even up to 10 μm or more, of normally resistive metal and still pass a Josephson current without resistance. The oxide junctions are made by evaporating a layer of superconductor on a suitable (e.g. glass) substrate, allowing it to oxidize for a short while in air, and then depositing another superconducting layer on the oxide surface. The metal junctions can be made by evaporating, for example, lead on to both sides of a rolled copper foil. They behave rather similarly except in certain dynamical experiments where, as we shall see, the capacitance between the superconductors in the oxide junction, but not in the metal junction, may play an important part.

The steady flux of energy from one cavity to the other, and of particles from one superconductor to the other, continues only so long as the phase difference is constant, and this is only possible in Josephson junctions when there is no potential difference between the two superconductors. There is a basic association of energy with frequency in quantum mechanics, and two superconducting blocks, identical except for their electrostatic potential, will differ also in the frequency of their wave-functions; ϕ therefore changes with time, and the current oscillates to and fro at the difference frequency. Something very similar happens with resonant cavities; we may imagine the analogous behaviour to be what takes place when two cavities differing in resonant frequency by $\delta\omega$ but excited equally (and therefore in a mixture of normal modes) are coupled weakly; as the phase difference shifts steadily, so the energy flux oscillates at the difference frequency. One may describe this phenomenon, somewhat artificially perhaps, in quantum terms as follows: when one quantum of electromagnetic energy, $\hbar\omega$, is transferred the energy of the whole system must be changed by reason of the different quantum energy in the two cavities, $\delta E = \hbar\delta\omega$, and the frequency to be assigned to the process is $\delta E/\hbar$, which is just $\delta\omega$. The argument may be applied to the superconducting case by noting that the unit in which matter is transferred, the analogue of the quantum $\hbar\omega$, is an electron-pair, since in the superconducting ground state all electrons are associated in pairs, and one alone cannot be taken out or inserted. If a potential difference V is applied between the superconductors, an electron-pair changes its energy by $2eV$ in crossing the gap, and the frequency of oscillation of current in the gap is $2eV/\hbar$. This leads immediately to Josephson's expression for the dependence of phase on voltage:

$$\dot{\phi} = 2eV/\hbar, \tag{75}$$

to which must be added the relation between phase and Josephson current, i_s,

$$i_s = i_J \sin \phi, \tag{76}$$

where i_J is the maximum Josephson current that the particular junction can support, and is determined by the nature and dimensions of the junction.

Let us apply these equations to two special cases, the SIS junction in which superconducting layers are separated by a thin insulating layer, and the SNS junction in which they are separated by a metal foil having normal resistive properties. The former, SIS, in addition to its superconducting properties is also a capacitor. If its capacitance is C, when a current i enters one superconducting layer and leaves the other, charge builds up on the capacitor plates at a rate $i - i_s$, so that

$$C\dot{V} = i - i_s,$$

and, from (75) and (76),

$$\frac{\hbar C}{2e}\ddot{\phi} + i_J \sin \phi = i. \tag{77}$$

The SNS junction, on the other hand, has no capacitance between the superconducting layers, since no charges can remain within the metal at the interface between superconductor and normal metal, but if a potential difference is established a normal ohmic current will flow as well as the Josephson current. For a junction having ohmic resistance R,

$$i = V/R + i_s,$$

or

$$\frac{\hbar}{2eR}\dot{\phi} + i_J \sin \phi = i. \tag{78}$$

These equations, (77) and (78), describe how ϕ is affected by the injection of current i. The laboratory observer cannot observe ϕ directly, but (75) shows that $\dot{\phi}$ is uniquely associated with a potential difference, and therefore that observations of $V(t)$ in the presence of $i(t)$ provide information that can be compared directly with the theoretical predictions.

The appearance of $\sin \phi$ in both equations renders them non-linear and therefore difficult to solve in general. For this reason it is helpful to devise mechanical models that obey the same equations, so that physical intuition may replace mathematics up to the moment when exact solutions are required, and computing must take over. Fig. 34 shows some models. The pendulum (a), to which a torque Fa can be applied, obeys an equation of the same form as (77):

$$ml^2\ddot{\phi} + mgl \sin \phi = Fa,$$

while if it is immersed in so viscous a fluid as to be overdamped to the point where the initial term $\ddot{\phi}$ is negligible in comparison with the damping term $\lambda\dot{\phi}$, its equation of motion takes the form of (78):

$$\lambda\dot{\phi} + mgl \sin \phi = Fa.$$

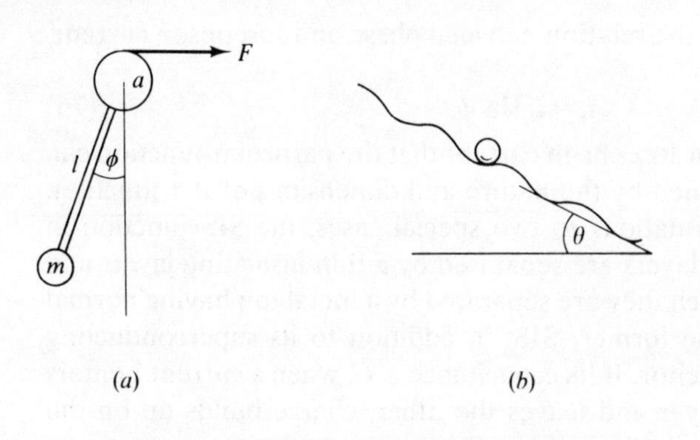

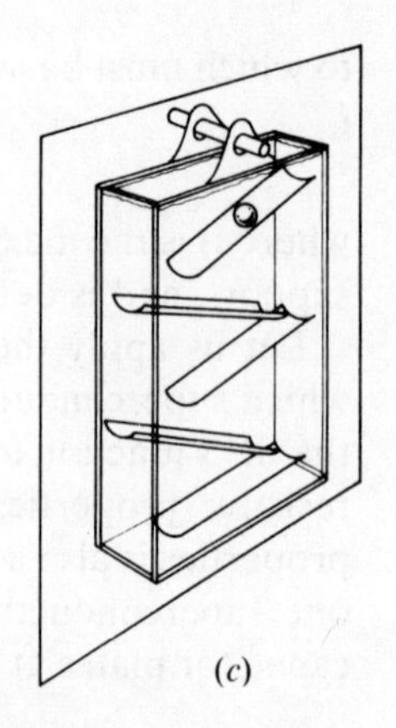

Fig. 34. Models of (a) and (b) Josephson junction, (c) multivibrator.

The same equations apply to the washboard model (b). If x represents distance measured down the slope, and the ripples have wavenumber k, the slope varies as $\theta - \alpha \sin kx$ if θ and α are both much less than unity. A ball rolling down in the absence of friction obeys the equation:

$$m\ddot{x} + g\alpha \sin kx = g\theta,$$

while immersing it in a viscous fluid, so that it always moves at its terminal velocity, causes it to obey the equation

$$\lambda\dot{x} + mg\alpha \sin kx = mg\theta.$$

In this model kx plays the part of ϕ, the maximum slope of the ripples ($k\alpha$) is related to i_J, and θ is related to i.

With the aid of these models the solutions of (77) and (78) under various conditions may be visualized. We shall devote more attention to (78), which is considerably easier to treat, though in the devices which employ Josephson effects it is the undamped SIS system, governed by (77), which is more important. If a steady current i is injected pure Josephson current, with no p.d., flows so long as $i < i_J$, for a solution $\phi = \text{constant}$ is possible.† In the models this corresponds to Fa being too weak to raise the pendulum beyond the horizontal position, or the slope θ being less than $\kappa\alpha$ so that the ball can sit at rest in any of the troughs. When $i > i_J$, however, the torque Fa pulls the pendulum over the top and it continues to rotate; with heavy damping the angular velocity is less on the way up than on the way down, so that the voltage, $\dot{\phi}$, has an oscillatory component superposed on a steady mean. In this case (78) is directly integrable:

$$t = \frac{\hbar}{2eR}\int \mathrm{d}\phi/(i - i_J \sin\phi).$$

† But not unique. By injecting a current greater than i_J the junction is set in a mode analogous to the pendulum going over the top in a continuous, but irregular, rotation. When damping is weak the motion can be maintained by a force F much weaker than that required to initiate it. In contrast to an SNS junction, the SIS junction is hysteretic.

In particular T, the period for one cycle, is obtained by setting limits 0 and 2π to the integral, and is equal to $\pi\hbar/eR(i^2-i_J^2)^{\frac{1}{2}}$. The mean value of $\dot{\phi}$ is $2\pi/T$, and the mean voltage follows from (75):

$$\bar{V}=R(i^2-i_J^2)^{\frac{1}{2}}.$$

This hyperbolic $\bar{V}$–i characteristic is shown in fig. 35(a).

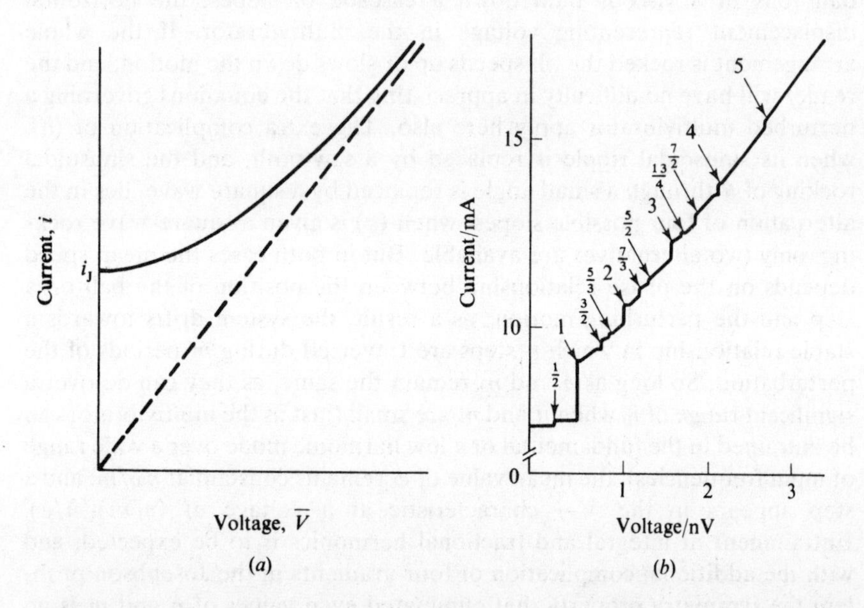

Fig. 35. V–i characteristics of SNS junction; (a) without a.c. injection, (b) with a.c. injection (unpublished measurements by Prof. J. Clarke). The frequency of the injected current was 260 kHz, and the numbers against each step are the ratio of the 'step frequency', $2e\bar{V}/\hbar$, to the injected frequency.

When an alternating current is added to the steady current something very interesting occurs; the $\bar{V}$–i characteristic develops a series of steps, such as in fig. 35(b). The precise size of each step depends in a complicated way on the strength of the alternating current. The equation governing ϕ follows from (78):

$$\frac{\hbar}{2eR}\dot{\phi}+i_J\sin\phi=i_0+i_1\sin\omega t$$

or $$\mathrm{d}\phi/\mathrm{d}x=I_0+I_1\sin x-I_J\sin\phi, \tag{79}$$

where $x=\omega t$ and $I=2eRi/\hbar\omega$. As soon as one thinks of sketching solutions of (79) on the (x,ϕ)-plane, the similarity to the entrainment problem of the multivibrator becomes obvious. For the slope of the trajectory of ϕ is periodic in x and ϕ, much like the periodicity shown in fig. 31. To emphasize this point, let us replace $\sin x$ and $\sin\phi$ by square waves of the same amplitude, as if the Josephson current were of a given magnitude, either positive or negative, and the alternating current had a square waveform. Then (79) may be written in shorthand notation

$$\mathrm{d}\phi/\mathrm{d}x=I_0\pm I_1\pm I_J. \tag{80}$$

and the square lattice into which the (x,ϕ)-plane is divided has four different types of cell, in each of which the gradient is constant; though a little more complicated than the simple alternating-gradient pattern in fig. 31, it is hardly different in principle.

The similarity of the two problems may be illustrated in a different way by comparing the washboard model of the SNS junction with a mechanical model simulating the multivibrator, fig. 34(b) and (c). In the latter model a ball rolls in a viscous fluid down a cascade of slopes, the horizontal displacement representing voltage in the multivibrator. If the whole arrangement is rocked the tilt speeds up or slows down the motion, and the reader will have no difficulty in appreciating that the equations governing a perturbed multivibrator apply here also. The extra complication of (b), when its sinusoidal ripple is replaced by a sawtooth, and the sinusoidal rocking of θ through a small angle is replaced by a square wave, lies in the alternation of four possible slopes; when (c) is given a square-wave rocking, only two alternatives are available. But in both cases the mean speed depends on the phase relationship between the position of the ball on a step and the perturbing motion; as a result, the system drifts towards a stable relationship in which n steps are traversed during m periods of the perturbation. So long as n and m remain the same, as they can do over a significant range of i_0 when n and m are small (just as the multivibrator can be entrained in the fundamental or a low harmonic mode over a wide range of input frequencies), the mean value of $\dot{\phi}$ remains constant at $n\omega/m$, and a step appears in the V–i characteristic at a voltage of $(n/m)(\frac{1}{2}\hbar/e)$. Entrainment at integral and fractional harmonics is to be expected, and with the additional complication of four gradients in the Josephson problem the symmetry property that eliminated even values of n and m is no longer present – all possible integral fractions should be candidates for entrainment. That they do not all appear in practice, but only a fairly rich selection, is probably a thermal effect, Johnson and other noise being adequate to disrupt the very weak phase linkage accompanying entrainment at a fractional harmonic ratio with large integers. As with the multivibrator, subharmonic entrainment is rarely observed.

See note on opposite page.

When a steady current is injected into the SNS junction, the mean voltage is accompanied by an oscillatory voltage at a frequency $2e\bar{V}/\hbar$. A resonant circuit connected across the junction should therefore be excited when $\bar{V}$ takes something near the correct value, and its response should react back on the junction, changing the behaviour of $\dot{\phi}$. The locking of junction and resonator might be expected to introduce steps, possibly with hysteresis, into the V–i characteristic, analogous to the frequency-locking of a multivibrator to a resonant circuit. Unfortunately this experiment seems not to have been tried but there are many published results to show the variety of ways in which the behaviour of an SIS junction may be modified by coupling a resonator to it. These are all, however, much more difficult to analyse than anything involving an SNS junction. The second derivative in (77) makes discussion by graphical methods virtually useless. We shall therefore leave the topic at this point, having made clear, it is to be hoped, that the study of classical systems can still cast light on essentially quantum-mechanical problems, and that some of these modern problems present the same sort of difficulties as their classical counterparts, demanding of the physicist an outlook that does not reject old attitudes but welcomes and improves on them.

Additional note (see opposite page)
A confusion of notation has made nonsense of the comparison between the locking of a multivibrator and of an SNS junction. In the former, locking is well marked (fig. 27) when the injected frequency is higher than the natural frequency of the multivibrator; in the latter (fig. 35) it is when the injected frequency is lower than the natural frequency, $2e\bar{V}/\hbar$. To reconcile this difference consider the mechanical model of the multivibrator depicted in fig. 34(c). A rapid sinusoidal rocking has small effect so long as the ball is not near the end of a channel; locking occurs because at the end the phase of the rocking motion can influence the precise moment at which the ball drops to the next channel. Very slow rocking is much less effective. But if we substitute a square wave for the sinusoidal rocking, the sudden changes of tilt, however infrequent, can hasten or retard the rolling towards the next drop; locking should therefore occur when n/m is both greater and less than unity. One may think of the process, loosely, as an interaction between any nearly coincident frequencies present in the two oscillatory processes. The multivibrator has a rich spectrum of harmonics to couple to a high injected sinusoid, but if the rocking follows a square wave, its own

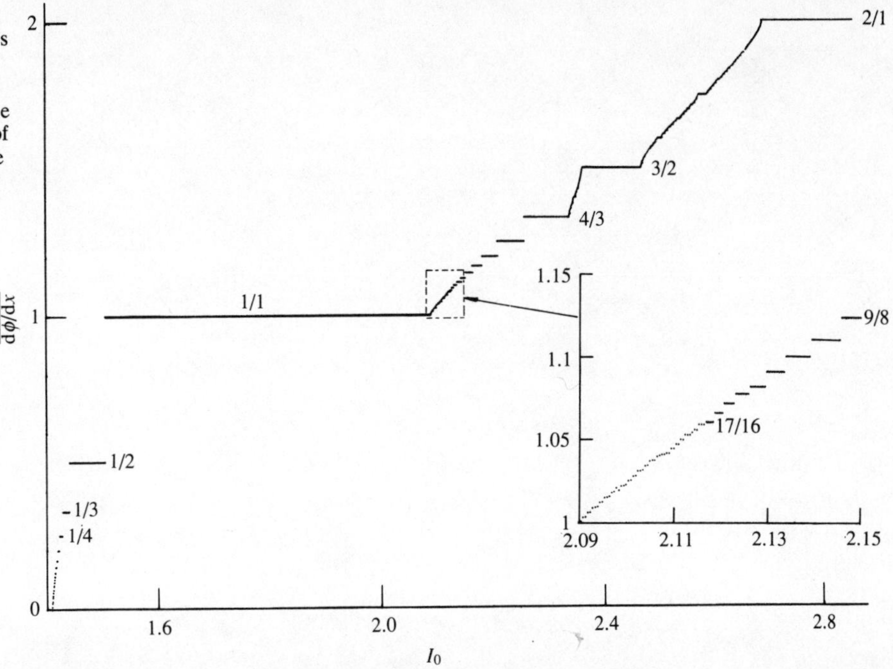

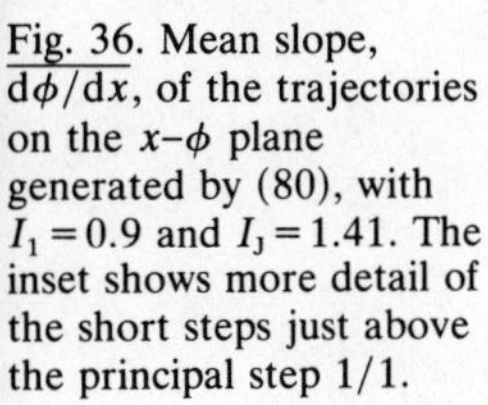
Fig. 36. Mean slope, $\overline{d\phi/dx}$, of the trajectories on the x–ϕ plane generated by (80), with $I_1 = 0.9$ and $I_J = 1.41$. The inset shows more detail of the short steps just above the principal step 1/1.

high-frequency components may couple to a lower multivibrator component to give subharmonic locking.

This intuitive picture is somewhat strained to account for fractional harmonic locking of high order in n and m, and for the locking of an SNS junction as modelled by fig. 34(b) where both natural and injected vibrations are nearly pure sinusoids. A more exact non-linear theory is then needed. But if we change the model to that represented by (80), a saw-toothed washboard rocked by a square wave, not only are the locking steps enhanced but the rich harmonic content of both vibrations allows locking at frequency ratios both larger and smaller than unity. The computed behaviour of this model is shown in fig. 36 and reveals all the expected complexity. Between the step labelled 2/1 at the top and the long step at 1/1 a sequence of the form $(n+1)/n$ is readily followed as far as $n=16$, and with patience could be taken still further. The same goes for the subharmonic sequence, of the form $1/n$, on the left.

Part 2

The simple vibrator in quantum mechanics

13 The quantized harmonic vibrator and its classical features

By the time a student of physics is ready to tackle quantum mechanics he has become familiar with the classical harmonic vibrator through many examples, and knows the crucial role it played in the development of Planck's ideas. It is natural then to concentrate on the mathematical aspects of the harmonic oscillator equation in quantum mechanics, the solution of Schrödinger's equation, normalization of the wave-function, calculation of mean values and of matrix elements, leaving the physics to look after itself. At a more advanced level the harmonic vibrator provides the entry into field theories, second quantization etc., and as a general rule it tends to be viewed more as a vehicle for instruction in more interesting matters than as a physical system having its own considerable interest and importance. Here we shall seek to redress the balance and study the vibrator as a thing in itself, without losing sight of the variety of physical problems to which the results can be applied. One especially important set of applications, however, will get little attention at this stage – the vibrations of compound bodies and of extended physical systems provide such wealth of interest as to justify a volume to themselves; and it is to such matters that the whole of the third part of this work will be devoted. As soon as one begins to contemplate the harmonic vibrator one becomes aware of the exceptional nature of its behaviour, in that it conforms more closely than any other system to the classical rules. It is surprising how many problems involving harmonic vibrators are found to have the same solution in classical and quantum mechanics, and we shall pay considerable attention to this question with a view to defining as clearly as possible the limits of validity of the classical approach.

xiii

There is always a danger in a venture like this of being misunderstood as attempting to obviate the need for quantum mechanics. Undoubtedly life would be simpler if the classical solution to every problem were correct; we need look no further than the next chapter for examples of non-linear vibrators which yield to classical methods without trouble but are distinctly less tractable in their quantized guise. But the classical solutions are not quite right, or sometimes are entirely wrong, so that however much we may value the insight that classical methods provide they can never be regarded as valid alternatives. Indeed, we have only to glance back once more to Planck to recognize that it was the non-classical features of that most nearly classical system, the harmonic vibrator, that started the whole quantum revolution on its course. Nevertheless there is another side to the coin; if it had been necessary at every stage to solve Schrödinger's equation before being able to understand in qualitative terms the transport of electricity in metals and semiconductors, solid state physics would not be the highly developed science it is nowadays. It was the insight provided by classical visualization of the processes which gave a firm structure to detailed quantum-mechanical analyses; and it is to make available these classical insights, and to delineate their range of validity, that we undertake this study of classical correspondences. All the same, to allay any residual fears of a reactionary attack on modern physics, we shall begin with an outline of the conventional approach to the quantized oscillator through the formal solution of Schrödinger's equation. No more than an outline is needed since the details are available in countless textbooks of quantum mechanics.[(1),(2)] We shall then have available the elementary results needed for the next stages of the analysis. It is worth remarking at this point that the treatment throughout will be elementary, based on the direct solution of Schrödinger's equation. This is not a textbook of quantum mechanics, but a study of real physical vibrators, and there is no call to employ sophisticated mathematical procedures that were developed either for the sheer love of elegance or for their direct descriptive power when applied to more deeply quantal problems having little or no point of contact with classical ideas. Even so, it is possible that the unconventionally thorough discussion of elementary matters will also enlighten the reader who is familiar with advanced methods.

One example of the general attitude to be adopted is provided by the theory of dissipative processes, such as a charged harmonic vibrator radiating electromagnetic energy. We shall show how the classical theory of coupled oscillators may be applied to this problem in such a way as to give confidence in the interpretation of processes which are commonly presented as rather mysterious, but essentially quantal: spontaneous and stimulated transitions.

In so far as the emphasis in this chapter leans towards classical processes, it is to indicate the value of classical reasoning where it yields the correct answers. The following three chapters also draw on classical methods, but with a rather different end in view. There are a few standard problems – the harmonic oscillator, the hydrogen atom, the square well – for which complete analytical solution of Schrödinger's equation is feasible and useful. There are many more whose analysis involves unfamiliar series expansions or other difficulties, such that the labour of solution seems disproportionate to the desired end. For these, and anharmonic vibrators come into this category, it is often possible to derive rather good approximations to the correct solutions by the semi-classical approach associated with the names of Bohr, Wilson and Sommerfeld; this is essentially a classical procedure supplemented by quantization rules which define the acceptable members of the continuous set of classical solutions. Between this approach and the solution of Schrödinger's equation lies the approximate method of Wentzel,

456

461

Kramers, Brillouin and Jeffreys. The next chapter illustrates the application of these techniques to a number of anharmonic vibrators which can also be treated exactly without too much difficulty. The power of the approximations, which is considerable, is thus revealed explicitly.

A third application of classical methods is developed in chapter 17; this is the replacement of a real physical system by a quite different classical system whose behaviour models certain features of the quantal behaviour of the real system. In particular the impulse response function of a quantized system, for instance an atom, can be modelled in the linear approximation by a set of classical harmonic oscillators. Consequently one may derive the response of the system to any weak time-dependent force from one's knowledge of the response of a harmonic oscillator. The possibility of designing such a model accounts for the success of classical interpretations of anomalous dispersion, and a similar modelling justifies the classical treatment of nuclear magnetic resonance.

510

The remainder of the volume builds on this identification of quantum transitions with classical harmonic oscillators, always with an eye to those features that have relevance to real vibrating physical systems. Thus the last two chapters deal with masers, lasers (but only very briefly) and bunching processes which form a conceptual link between masers and certain maintained oscillators, of which the klystron is selected for more detailed discussion. To reach this point it is necessary to examine the interaction of radiation with quantized systems in a number of different ways until, in chapter 20, there emerges a rather precise picture of a radiation-damped transition which, apart from a strictly quantal noise source, corresponds remarkably closely with the primitive classical model of a viscous-damped harmonic vibrator.

588

At various points in this development we have to face the question – if classical methods work so well for so many aspects of the behaviour of harmonic vibrators, wherein lies the essential quantal character? The answer we shall give is that the classical approach, however successful in describing a vibrator acted upon by a well-defined force, fails when the force is itself provided by a system that must be treated quantally. To understand this point is to possess something like an operational criterion for deciding whether classical methods are safe.

Solution of Schrödinger's equation

A particle of mass m, moving on a line under the influence of a linear restoring force, has potential energy V equal to $\frac{1}{2}m\omega_0^2x^2$ if its classical angular frequency of vibration is ω_0, and is governed by the time-independent Schrödinger equation:

$$(\hbar^2/2m)\,\mathrm{d}^2\psi/\mathrm{d}x^2+(E-\tfrac{1}{2}m\omega_0^2x^2)\psi=0. \tag{1}$$

In almost everything that follows we shall use reduced variables, setting

$\xi = (m\omega_0/\hbar)^{1/2}x$ and $\varepsilon = 2E/\hbar\omega_0$, so that (1) takes the form:

$$d^2\psi/d\xi^2 + (\varepsilon - \xi^2)\psi = 0. \qquad (2)$$

Note that the energy quantum, $\hbar\omega_0$, corresponds to $\varepsilon = 2$ and the unit of length, $(\hbar/m\omega_0)^{1/2}$, is the amplitude of the corresponding classical oscillator when its energy is unity, or $\frac{1}{2}\hbar\omega_0$ in laboratory units.

One may choose ε at will and, having also chosen values of ψ and ψ' at some ξ, proceed to integrate (2) step by step. The resulting solution will almost certainly diverge rapidly at both positive and negative ξ, and hence cannot be regarded as physically meaningful. Whether it diverges or not, however, this solution can be used to generate another equally good solution for the same value of ε, simply by reflecting it in the origin; for the occurrence of ξ in (2) only in quadratic form shows that if $\psi(\xi)$ is a solution, so also is $\psi(-\xi)$.

Since (2) is a linear equation, all solutions may be cast in symmetric or antisymmetric form by taking $\psi(\xi) \pm \psi(-\xi)$ as the standard pattern. Instead of starting the integration with arbitrary choices of ψ and ψ' we may ensure one or other of those forms by starting always at the origin, and either taking $\psi'(0)$ as zero and $\psi(0)$ as non-zero for a symmetric solution, or $\psi'(0)$ as non-zero and $\psi(0)$ as zero for an antisymmetric solution. For every choice of ε there is one symmetric and one antisymmetric solution, and in general both diverge at large $|\xi|$. For when $\xi^2 > \varepsilon$ the sign of ψ'' is the same as that of ψ, and from this point on divergence is almost inevitable. For if, as integration proceeds, we find ψ' taking the same sign as ψ, the curve can only move steadily away from the axis, since the sign of ψ' can never be reversed. If, on the other hand, the curve is pointing towards the axis, with ψ and ψ' opposite in sign, the slope may not decrease fast enough to prevent the curve crossing the axis; as soon as this happens divergence is inevitable, for ψ and ψ' have now the same sign. Convergence to zero at high values of ξ can only be achieved by the correct choice of ε so that the curve comes down to the axis at a grazing angle.

When ξ is very large the value of ε is almost irrelevant and by selecting convenient values the asymptotic behaviour can be readily determined. Thus when $\varepsilon = 1$ one of the two independent solutions of (2) is $e^{-\frac{1}{2}\xi^2}$, and when $\varepsilon = -1$ one of the solutions is $e^{+\frac{1}{2}\xi^2}$. Both are very nearly solutions for other values of ε, and as ξ increases the solution for any ε approaches a linear combination of the two:

$$\psi \sim A\,e^{\frac{1}{2}\xi^2} + B\,e^{-\frac{1}{2}\xi^2}.$$

Only by choosing ε correctly can the symmetric or antisymmetric solution for small ξ join smoothly to the asymptotic form containing B only, and therefore converging to zero as required. This is the background for proceeding to discover solutions of (2) in the form of $e^{-\frac{1}{2}\xi^2}$ multiplying a power series in ξ. The details are well presented in most standard textbooks, and we need only note here that the resulting power series in general diverges roughly as e^{ξ^2}, showing that in general $A \neq 0$ in the asymptotic form. By choosing ε correctly, however, the series may be terminated so that ψ is $e^{-\frac{1}{2}\xi^2}$ times a finite polynomial, $H_n(\xi)$, which can never increase so rapidly at large ξ as to overwhelm the exponential factor. The required values of ε are $2n+1$, corresponding to $E=(n+\frac{1}{2})\hbar\omega_0$ as conjectured by Planck except for the extra $\frac{1}{2}\hbar\omega_0$. We have then as acceptable solutions

$$\psi_n = \mathcal{N}_n H_n(\xi)\, e^{-\frac{1}{2}\xi^2}, \tag{3}$$

in which the $\mathcal{N}_n$ are normalizing coefficients to ensure that $\int_{-\infty}^{\infty} \psi_n^2\, d\xi = 1$:

$$\mathcal{N}_n = \pi^{-\frac{1}{4}}(2^n n!)^{-\frac{1}{2}}. \tag{4}$$

The $H_n(\xi)$ are Hermite polynomials, shown bracketed in (5), which is a sample of some of the lower-order wave-functions.[3]

$$\left.\begin{aligned}
n=0 \quad & \psi_0 = 0.75113\,(1)\, e^{-\frac{1}{2}\xi^2}\\
1 \quad & \psi_1 = 0.53113\,(2\xi)\, e^{-\frac{1}{2}\xi^2}\\
2 \quad & \psi_2 = 0.26556\,(4\xi^2-2)\, e^{-\frac{1}{2}\xi^2}\\
3 \quad & \psi_3 = 0.10842\,(8\xi^3-12\xi)\, e^{-\frac{1}{2}\xi^2}\\
\vdots \quad & \\
10 \quad & \psi_{10} = 1.2321\times 10^{-5}(1024\xi^{10} - 23\,040\xi^8\\
& \quad +161\,280\xi^6 - 403\,200\xi^4 + 302\,400\xi^2 - 302\,40)\, e^{-\frac{1}{2}\xi^2}
\end{aligned}\right\} \tag{5}$$

The Hermite polynomials are related by the equation

$$H_{n+1} = 2\xi H_m - 2nH_{n-1}, \tag{6}$$

so that it is easy to generate successive polynomials and hence as long a sequence of ψ_n as may be required. The examples given show the alternation of symmetric wave-functions (even powers of ξ) and antisymmetric (odd powers).

The form of a few wave-functions is shown in fig. 1, together with $\psi\psi^*$; since ψ is normalized, $\psi\psi^*\, d\xi$ is the probability of finding the oscillating particle in the range $d\xi$. For a classical oscillator this is the function $d\xi/[\pi(\xi_0^2-\xi^2)^{\frac{1}{2}}]$, strongly peaked at the extremities, $\pm\xi_0$, and the envelope of $\psi\psi^*$ approximates to this form when n is large. In the lowest states, however, there is no sign of this, and in particular the ground state probability distribution retains a certain width, comparable to that of a classical oscillator with energy $\frac{1}{2}\hbar\omega_0$, the zero-point energy of the quantized system.

When n is large the extreme peak becomes sharper, as can be seen by examining the form of (2) in the vicinity of the turning-point, $\xi^2=\varepsilon$. Substituting $\varepsilon^{\frac{1}{2}} + z/2^{\frac{1}{3}}\varepsilon^{\frac{1}{6}}$ for ξ, we find (2) transformed into Airy's equation:

$$d^2\psi/dz^2 - z\psi = 0, \tag{7}$$

if a second-order term in z is neglected. This is equivalent to treating the parabolic curve for V as a straight line in the immediate neighbourhood of the turning-point. The dimensionless form of (7) indicates that the

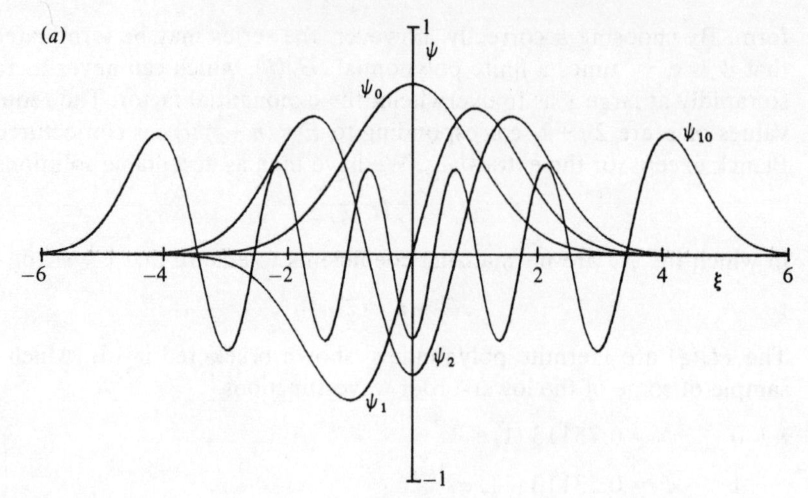

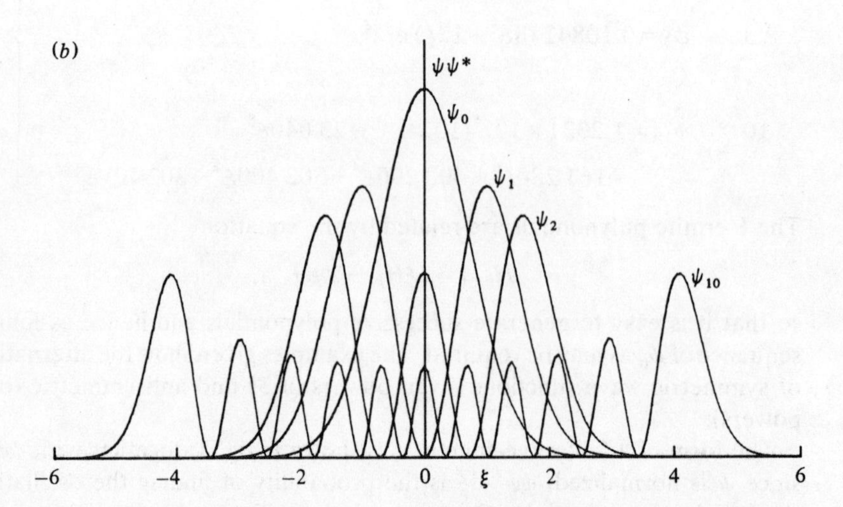

Fig. 1. (*a*) Wavefunctions of the harmonic vibrator stationary states ψ_0, ψ_1, ψ_2 and ψ_{10}, as given in (5). (*b*) $\psi\psi^*$ for the same wave-functions.

separation of the last two peaks is always the same when z is the co-ordinate, and that therefore in terms of ξ or the laboratory co-ordinate x the separation varies as $\varepsilon^{-\frac{1}{6}}$ – not a strong dependence but significant when the quantum number is 10^6 or more, such as we shall find when we come to discuss masers. By the same argument it follows that the tail of the wave-function, beyond the turning-point, narrows with increasing n. From a table of Airy functions[4] we find that the last peak of $\psi\psi^*$ occurs at $z=-1.019$ and that the value of $\psi\psi^*$ has dropped by a factor e just outside the turning-point at $z=0.116$. Taking $\Delta z \times 1.135$ as a measure of the width of the tail, we have that corresponding $\Delta\xi_n = 0.80/n^{\frac{1}{6}}$, or 0.08 when $n=10^6$. This may be compared with the width $\Delta\xi_0 = 1$ for the decay of $\psi_0\psi_0^*$ by a factor e from its peak

Momentum distribution

An alternative representation of the wave-function of a stationary state is as the superposition of travelling waves:

$$\psi_n(\xi) = \int_{-\infty}^{\infty} \varphi_n(\kappa)\, e^{i\kappa\xi}\, d\kappa. \tag{8}$$

Then $\varphi_n\varphi_n^* \, d\kappa$ represents the probability of finding the particle moving with wavenumber in the range $d\kappa$; and since $p \propto \kappa$, $\varphi_n\varphi_n^*$ describes the probability distribution for the momentum, p. By the Fourier transform theorem,

72

$$\varphi_n(\kappa) = \frac{1}{2\pi}\int_{-\infty}^{\infty} \psi_n(\xi)\, e^{-i\kappa\xi}\, d\xi. \tag{9}$$

We now demonstrate one of the peculiarities of the harmonic oscillator, that the differential equation obeyed by φ_n has the same form as that obeyed by ψ_n. We multiply each term in (2) by $e^{-i\kappa\xi}$ and integrate over ξ, making use of the following results obtained by differentiating within the integral signs of (8) and (9):

$$d^2\psi_n/d\xi^2 = -\int_{-\infty}^{\infty} \kappa^2\varphi_n\, e^{i\kappa\xi}\, d\kappa,$$

so that
$$\int_{-\infty}^{\infty} e^{-i\kappa\xi}(d^2\psi_n/d\xi^2)\, d\xi = -2\pi\kappa^2\varphi_n;$$

and
$$\int_{-\infty}^{\infty} \xi^2\psi_n\, e^{-i\kappa\xi}\, d\xi = -2\pi\, d^2\varphi_n/d\kappa^2.$$

Hence, from (2),

$$d^2\varphi_n/d\kappa^2 + (\varepsilon_n - \kappa^2)\varphi_n = 0. \tag{10}$$

The fact that this is the same as (2) shows that the diagrams in fig. 1 equally well represent the momentum distribution, just as with the classical vibrator, for which the peaked distribution is equally good for position and momentum, if the co-ordinates are suitably scaled. Our present choice of variables has this property. It may be noted that, since obviously $\langle\xi^2\rangle = \langle\kappa^2\rangle$, the mean values of kinetic and potential energy are the same. For just as $\xi = (m\omega_0/\hbar)^{\frac{1}{2}}x$, so $\kappa = (\hbar/m\omega_0)^{\frac{1}{2}}k$, k being the wavenumber measured in ordinary units, and

$$\left.\begin{aligned} \langle T\rangle &= (\hbar^2/2m)\langle k^2\rangle = \tfrac{1}{2}\hbar\omega_0\langle\kappa^2\rangle,\\ \text{while}\quad \langle V\rangle &= \tfrac{1}{2}m\omega_0^2\langle x^2\rangle = \tfrac{1}{2}\hbar\omega_0\langle\xi^2\rangle. \end{aligned}\right\} \tag{11}$$

This equality of mean kinetic and potential energy is also a classical property, and we shall show presently that it remains true even when the oscillator occupies a non-stationary mixture of states. For this purpose, however, we must consider the representation of non-stationary states by the superposition of stationary states.

Non-stationary states

The stationary state wave-functions, ψ_n, are the spatial parts of solutions of Schrödinger's time-dependent wave equation which, in one dimension, takes the form:

$$(\hbar^2/2m)\partial^2\Psi/\partial x^2 - V\Psi = -i\hbar\partial\Psi/\partial t. \tag{12}$$

When $V = \frac{1}{2}m\omega_0^2 x^2$ the appropriate form for the harmonic vibrator in reduced variables may be written

$$\partial^2\Psi/\partial\xi^2 - \xi^2\Psi = -2i\partial\Psi/\partial\tau, \tag{13}$$

in which $\xi = (m\omega_0/\hbar)^{\frac{1}{2}}x$ as before, and $\tau = \omega_0 t$. The full expression for the wave-function Ψ_n of the nth stationary state is $\psi_n\,\mathrm{e}^{-\mathrm{i}(n+\frac{1}{2})\tau}$ where ψ_n satisfies (2) if $\varepsilon = 2n+1$. Since (13) is linear in Ψ, arbitrary superpositions of various Ψ_n are also solutions:

$$\Psi = \sum_n a_n\Psi_n = \mathrm{e}^{-\frac{1}{2}i\tau}\sum_n a_n\psi_n\,\mathrm{e}^{-in\tau}. \tag{14}$$

When Ψ and all ψ_n are normalized to unity, as is customary,

$$1 = \int_{-\infty}^{\infty}\Psi\Psi^*\,\mathrm{d}\xi = \sum_n\sum_m a_n a_m^*\int_{-\infty}^{\infty}\psi_n\psi_m^*\,\mathrm{d}\xi = \sum_n a_n a_n^*, \tag{15}$$

since the ψ_n are orthogonal. One may interpret $a_n a_n^*$ as the probability of finding the system in the nth state, in an experiment designed to yield such information.† Or if one seeks to determine the energy of the system the mean value $\langle E\rangle$ for a large number of repetitions of the measurement is $\Sigma\, a_n a_n^* E_n$, the weighted average of the energy levels over the occupation probability.

So long as the system is left undisturbed, with V independent of time, Ψ keeps the form (14) with all a_n constant. Once the form of Ψ at any moment has been expressed in terms of a sum of ψ_n, the subsequent development of Ψ follows inexorably. Moreover the ψ_n obtained by finding all stationary solutions of Schrödinger's time-independent equation, with any chosen variation of V with position, constitute a complete set of orthogonal functions, and may be used to express as a Fourier sum any other function whose behaviour is not too pathological. If, therefore, Ψ is known at any instant, the recipe for determining its subsequent development is to express it as $\Sigma_n\, a_n\psi_n$ and then to supplement each ψ_n with its appropriate time-variation, as in (14).

† The Stern–Gerlach experiment is such an experiment, atoms in different states being spatially segregated by an inhomogeneous magnetic field; it is not easy to devise a corresponding experiment for a harmonic vibrator.

The coherent state

As an example consider a function which can easily be verified as a normalized solution of (13):

$$\Psi = \pi^{-\frac{1}{4}} \mathrm{e}^{-\frac{1}{2}(\xi - b\cos\tau)^2 - \mathrm{i}\theta(\xi,\tau)} \tag{16}$$

in which b is a real constant and

$$\theta = \tfrac{1}{2}\tau + b\xi\sin\tau - \tfrac{1}{4}b^2\sin 2\tau. \tag{17}$$

The phase θ disappears, of course, from the probability distribution:

$$\mathscr{P}(\xi) = \Psi\Psi^* = \pi^{-\frac{1}{2}} \mathrm{e}^{-(\xi - b\cos\tau)^2} \tag{18}$$

The wave-function (16) is a Gaussian wave-packet of unchanging shape, oscillating to and fro with amplitude b. It is known as a *coherent state* and is of special interest as it is the form of Ψ which results from applying a uniform time-varying force to a vibrator initially in its ground state. We shall show this in due course; meanwhile let us note some of its characteristics by resolving it into its stationary components at time $\tau = 0$, when $\theta = 0$ and the spatial part of Ψ is $\pi^{-\frac{1}{4}}\mathrm{e}^{-\frac{1}{2}(\xi-b)^2}$. Then, if

445

$$\pi^{-\frac{1}{4}}\mathrm{e}^{-\frac{1}{2}(\xi-b)^2} = \sum_n a_n\psi_n,$$

a_n is determined by multiplying by ψ_n and integrating:

$$a_n = \int_{-\infty}^{\infty} \pi^{-\frac{1}{4}}\mathrm{e}^{-\frac{1}{2}(\xi-b)^2}\psi_n\,\mathrm{d}\xi = (2^n\pi n!)^{-\frac{1}{2}}\mathrm{e}^{-\frac{1}{4}b^2}\int_{-\infty}^{\infty} \mathrm{e}^{-\xi^2}H_n(\xi)\,\mathrm{e}^{\xi^2-(\xi-\frac{1}{2}b)^2}\,\mathrm{d}\xi, \tag{19}$$

from (3) and (4). The rearrangement of the exponent shown in the last integral allows the use of a valuable identity concerning Hermite polynomials:

$$\mathrm{e}^{\xi^2-(\xi-\eta)^2} = \sum_l H_l(\xi)\eta^l/l! \tag{20}$$

The integral in (19) may therefore be written:

$$\begin{aligned}\int_{-\infty}^{\infty} \mathrm{e}^{-\xi^2}H_n\,\mathrm{e}^{\xi^2-(\xi-\frac{1}{2}b)^2}\,\mathrm{d}\xi &= \sum_l (\tfrac{1}{2}b)^l/l!\int_{-\infty}^{\infty}\mathrm{e}^{-\xi^2}H_nH_l\,\mathrm{d}\xi\\ &= \sum_l (\tfrac{1}{2}b)^l/l!\int_{-\infty}^{\infty}(\psi_n\psi_l/\mathcal{N}_n\mathcal{N}_l)\,\mathrm{d}\xi\\ &= (\tfrac{1}{2}b)^n/\mathcal{N}_n^2 n!,\end{aligned}$$

on account of the orthogonality of the ψ_n. Hence, by use of (4),

$$a_n = b^n\,\mathrm{e}^{-\frac{1}{4}b^2}/(2^n n!)^{\frac{1}{2}}, \tag{21}$$

and, from (14), the full form of $\Psi(\xi, \tau)$ follows:

$$\Psi(\xi, \tau) = e^{-\frac{1}{4}b^2 - \frac{1}{2}i\tau} \sum_n [b^n/(2^n n!)^{\frac{1}{2}}]\psi_n\, e^{-in\tau}. \tag{22}$$

It is readily verified that (15) holds, for

$$\sum a_n a_n^* = e^{-\frac{1}{2}b^2} \sum_n (\tfrac{1}{2}b^2)^n/n!,$$

and the summation is just the series expansion of $e^{\frac{1}{2}b^2}$; hence $\Sigma\, a_n a_n^* = 1$.

A classical oscillator whose ξ-amplitude is b has energy $\frac{1}{2}b^2(\hbar\omega_0)$, and if we write $\frac{1}{2}b^2$ as n_c, n_c is to be interpreted as the number of quanta needed to supply the energy of the corresponding classical oscillator. The probability of finding the quantized oscillator in its nth state, with n quanta in addition to the zero-point energy, is $a_n a_n^*$ or $n_c^n\, e^{-n_c}/n!$, and the mean number of quanta then follows:

$$\langle n \rangle = e^{-n_c} \sum_n n n_c^n/n! = n_c.$$

When n_c is large a good approximation to $a_n a_n^*$ is obtained by use of Stirling's formula, $\ln n! \sim n \ln n - n$. Then if x is written for $n - n_c$,

$$\ln (a_n a_n^*) \sim -(n_c + x) \ln (1 + x/n_c) + x,$$

and the desired approximation follows by expanding $\ln (1 + x/n_c)$ as $x/n_c - x^2/2n_c^2$. To this order

$$a_n a_n^* \propto e^{-(n-n_c)^2/2n_c}. \tag{23}$$

The energy levels that are significantly occupied lie in a range of about $n_c^{\frac{1}{2}}$, much smaller than n_c when the oscillator is highly excited. The energy of the oscillator in this non-stationary state is a constant of the motion and takes the same value as for a classical particle moving with the centroid of the wave-packet, apart from the half-quantum of zero-point energy which may be thought of as that required to confine the particle in the wave-packet; in fact the function $\pi^{-\frac{1}{4}}\, e^{-\frac{1}{2}\xi^2}$ is just the ground state wave-function as given in (5). Other close parallels with the classical solution may also be noted – the energy of excitation (i.e. $E - \frac{1}{2}\hbar\omega_0$) is periodically exchanged between potential and kinetic forms, and the mean displacement $\langle \xi \rangle$ oscillates with simple harmonic motion. These two results are not peculiar to the particular solution chosen here, but are quite general as we now demonstrate. They arise from the specially simple form of the energy level structure (evenly spaced) and of the wave-functions, and are matched in virtually no other physical system.

Potential and kinetic energy

To show the exchange of energy consider the wave-function at time τ_0 expressed as a Fourier sum:

$$\Psi(\xi, \tau_0) = \sum_n a_n \psi_n.$$

Then one-quarter of a cycle later, at $\tau_0+\frac{1}{2}\pi$, each component will be multiplied by a phase-factor $e^{-\frac{1}{2}in\pi}$, so that

$$\Psi(\xi, \tau_0+\tfrac{1}{2}\pi)=\sum_n (-i)^n a_n\psi_n.$$

The potential energy in reduced units is $\langle\xi^2\rangle$ and hence

$$V(\tau_0)=\sum_m\sum_n\int_{-\infty}^{\infty} a_m^*\psi_m\xi^2 a_n\psi_n\,d\xi.$$

Similarly $$V(\tau_0+\tfrac{1}{2}\pi)=\sum_m\sum_n\int_{-\infty}^{\infty}(-i)^{n-m}a_m^*\psi_m\xi^2 a_n\psi_n\,d\xi.$$

Now of all the terms in these double summations, the alternating even and odd symmetries of the ψ_n ensure that the only terms that matter are those for which $n-m$ is even, but in fact out of all the integrals only those for which $n-m=0$ or ± 2 are non-vanishing:

$$M_{n,n-2}^{(2)}\equiv\int_{-\infty}^{\infty}\psi_n\xi^2\psi_{n-2}\,d\xi=\tfrac{1}{2}[n(n-1)]^{1/2}$$

and $$M_{n,n}^{(2)}\equiv\int_{-\infty}^{\infty}\psi_n\xi^2\psi_n\,d\xi=n+\tfrac{1}{2}. \tag{24}$$

Hence $$V(\tau_0)=\sum_n\{a_na_n^*(n+\tfrac{1}{2})+2\,\mathrm{Re}\,[a_n^*a_{n-2}]M_{n,n-2}^{(2)}\},$$

while $$V(\tau_0+\tfrac{1}{2}\pi)=\sum_n\{a_na_n^*(n+\tfrac{1}{2})-2\,\mathrm{Re}\,[a_n^*a_{n-2}]M_{n,n-2}^{(2)}\}.$$

Now the total energy ε is $\Sigma_n\,a_na_n^*(2n+1)$, just twice the first term, from which it immediately follows that

$$V(\tau_0+\tfrac{1}{2}\pi)=\varepsilon-V(\tau_0)=T(\tau_0), \tag{25}$$

where $T(\tau_0)$ is the kinetic energy at τ_0. Every quarter-cycle the partitioning of the energy between potential and kinetic is interchanged. It may also be noted that the alternation of odd and even ψ_n ensures that $\Psi(\xi)\rightleftharpoons\Psi(-\xi)$ every half-cycle.

This is equally true for a classical harmonic vibrator, but there is one significant difference. It is possible to choose τ_0 for the classical vibrator so that the energy is entirely potential or entirely kinetic; there is complete interchange during each cycle. With the quantized vibrator, on the other hand, there are no states in which either potential or kinetic energy is entirely absent. Thus the function (16) at time $\tau=0$ takes the form $\pi^{-\frac{1}{4}}\,e^{-\frac{1}{2}(\xi-b)^2}$ in which the displacement is greatest; V has its maximal value of b^2 and T its minimal value of $\frac{1}{2}$, the zero-point kinetic energy associated with the constant-phase Gaussian wave-packet. One quarter-cycle later, Ψ is an undisplaced wave-packet and V is minimal, at $\frac{1}{2}$, while the phase-variation across the wave-packet contributed by θ now confers on T its maximal value of b^2. There is complete interchange between T and V except for the zero-point energy, $\frac{1}{2}$. By contrast, in a stationary state $T=V$ at all times, and no interchange takes place. The restricted interchange permitted by (25) may be modelled in a classical system by an ensemble of identical vibrators rather than a single one. If the vibrators have different phases each exhibits complete interchange, but there is never a moment when all are purely potential or purely kinetic; and if the phases are evenly distributed over all possible values the total potential and kinetic energies remain unchanged with time, just as in the stationary state of a quantized vibrator. As we shall see in a later section, this is by no means the only way an ensemble of classical vibrators can be made to model the quantized vibrator.

It is characteristic of the total potential and kinetic energies of the classical ensemble not simply that they obey (25) but that each oscillates sinusoidally at twice the natural frequency:

e.g. $$V = \bar{V} + \Delta V \cos 2\tau, \qquad T = T_0 - \Delta V \cos 2\tau.$$

The result quoted in (24) leads to the same result for the quantized vibrator. For

$$\Psi(\xi, \tau_0 + \tau) = \sum_n a_n \psi_n \, \mathrm{e}^{-\mathrm{i}(n+\frac{1}{2})\tau},$$

and therefore $$V(\tau_0 + \tau) = \sum_m \sum_n \int_{-\infty}^{\infty} a_m^* \psi_m \xi^2 a_n \psi_n \, \mathrm{e}^{-\mathrm{i}(n-m)\tau} \, \mathrm{d}\xi.$$

On multiplying out and performing the integrations the only surviving terms are constants, when $m = n$, or have time-dependence as $\mathrm{e}^{\pm 2\mathrm{i}\tau}$ when $n - m = \pm 2$. It is, of course, the regular spacing of the levels that ensures that all values of n generate the same fundamental frequency, 2 in this case; and it is the vanishing of $M_{n,m}^{(2)}$ for any $|n-m|$ greater than 2 that eliminates all harmonics.

Classical behaviour of $\langle \xi \rangle$

A similar result holds for $\langle \xi \rangle$ as for $\langle \xi^2 \rangle$:

$$\langle \xi \rangle = \sum_m \sum_n \int_{-\infty}^{\infty} a_m^* \psi_m \xi a_n \psi_n \, \mathrm{e}^{-\mathrm{i}(n-m)\tau} \, \mathrm{d}\xi.$$

Now symmetry permits only odd values of $|n-m|$ to contribute, and in fact only when $|n-m| = 1$ does the integral not vanish:

$$M_{n-1,n}^{(1)} = M_{n,n-1}^{(1)} \equiv \int_{-\infty}^{\infty} \psi_n \xi \psi_{n-1} \, \mathrm{d}\xi = (n/2)^{\frac{1}{2}}. \tag{26}$$

Consequently $\langle \xi \rangle$ contains terms only in $\mathrm{e}^{\pm \mathrm{i}\tau}$ and oscillates sinusoidally, without a constant term, at the natural frequency.

This last result may be proved, without Fourier analysis, directly from Schrödinger's equation by use of Ehrenfest's theorem[5], which is valid for any one-dimensional system:

$$m\frac{\mathrm{d}}{\mathrm{d}t}\langle x\rangle=\langle p\rangle \quad \text{and} \quad \frac{\mathrm{d}}{\mathrm{d}t}\langle p\rangle=-\langle \text{grad } V\rangle,$$

so that
$$m\,\mathrm{d}^2\langle x\rangle/\mathrm{d}t^2=-\langle \text{grad } V\rangle. \tag{27}$$

It is not necessary that the potential V should be constant – it may also include the potential of some time-varying force. The close similarity between (27) and the classical expression of Newton's laws of motion is rather deceptive, in that the mean position $\langle x\rangle$ responds not to $F(\langle x\rangle)$, the force at this one point, but to $\langle F(x)\rangle$, the weighted average value of the force over all possible positions of the particle. Thus (27) describes perfectly well the tunnelling of a particle through a barrier, an impossible process in classical mechanics. If, however, grad V is either constant or contains in its series expansion no term of higher than first order, $\langle x\rangle$ does indeed follow the classical law. For when $-\text{grad } V=f_0(t)+xf_1(t)$, (27) takes the form

$$m\,\mathrm{d}^2\langle x\rangle/\mathrm{d}t^2=f_0(t)+\langle x\rangle f_1(t)=F(\langle x\rangle), \tag{28}$$

so that $\langle x\rangle$ behaves exactly like the displacement of a classical particle. In particular, if $f_0=0$ and f_1 is a negative constant, $\langle x\rangle$ executes harmonic motion in the Hooke's law potential, as proved above. And if now a uniform force, represented by non-vanishing $f_0(t)$, is applied $\langle x\rangle$ develops in exactly the same way as a classical oscillating particle subjected to $f_0(t)$. It may also be remarked that if $f_0=0$ and f_1 includes a time-varying component, (28) describes an oscillator parametrically excited by variations of the force constant, and $\langle x\rangle$ in this case also develops along classical lines. This strict parallelism between quantal and classical does not extend to variables other than $\langle x\rangle$, but the idea can be carried somewhat further by introducing an ensemble of classical oscillators which collectively represent the probability distribution of x and its response to a uniform force or to parametric excitation. For this development we shall need to know how a quantized oscillator responds to a sharp impulsive force, and as this is a matter of wider significance than the present application we shall establish the required result in general terms, though leaving the wider discussion until a later chapter.

445
510

Impulse-response of a quantized system

A particle moving in a constant three-dimensional potential $V(\boldsymbol{r})$ and subjected to a varying force $F(t)$ in the x-direction, independent of $\boldsymbol{r}$, obeys the Schrödinger equation

$$(\hbar^2/2m)\nabla^2\Psi-(V-xF)\Psi=-\mathrm{i}\hbar\partial\Psi/\partial t. \tag{29}$$

Now let F be a sharp impulse $P\delta(t-t_0)$ applied at time t_0. During the infinitesimal interval of its application the term in F completely dominates the left-hand side, so that the immediate effect of the impulse is obtained by integrating the equation:

$$\partial\Psi/\partial t = \mathrm{i}xF\Psi/\hbar.$$

If Ψ_- and Ψ_+ are the wave-functions immediately before and after the impulse,

$$\Psi_+ = \Psi_- \,\mathrm{e}^{\mathrm{i}x\int F\,\mathrm{d}t/\hbar} = \Psi_- \,\mathrm{e}^{\mathrm{i}xP/\hbar}. \tag{30}$$

The magnitude of Ψ is unchanged, but a phase shift proportional to x is imposed. This result holds for any strength of impulse, but when the impulse is weak it may be convenient to use the approximate form:

$$\Delta\Psi = \Psi_+ - \Psi_- \sim \mathrm{i}xP\Psi/\hbar \tag{31}$$

It is worth noting that (31) follows by direct integration of (29) over a short interval Δt, in which F is taken as constant:

$$\Delta\Psi = (\mathrm{i}x/\hbar)F\Psi\Delta t + \frac{\mathrm{i}}{\hbar}\left(\frac{\hbar^2}{2m}\nabla^2\Psi - V\Psi\right)\Delta t.$$

The second term represents the development of Ψ in the absence of a disturbing force F, while the first term is the change due to the impulse $F\Delta t$. The linearity of the differential equation for Ψ ensures that the two processes take place independently.†

106

To determine the impulse response function let us start with the system in a non-stationary state represented by $\Sigma_l a_l\psi_l$ at time $t=0$. If a weak impulse P is applied at t_0, when $\Psi(t_0) = \Sigma_l a_l\psi_l\,\mathrm{e}^{-\mathrm{i}E_lt_0/\hbar}$, the resulting change in the wave-function follows from (31):

$$\Delta\Psi(t_0) = (\mathrm{i}xP/\hbar)\sum_l a_l\psi_l\,\mathrm{e}^{-\mathrm{i}E_lt_0/\hbar}. \tag{32}$$

This change in Ψ can be represented as a change Δa_l in each a_l, and if the system remains undisturbed thereafter the new set of a_l stays constant. We therefore evaluate the coefficients at t_0, writing

$$\Delta a_m = \int \psi_m^*\,\mathrm{e}^{\mathrm{i}E_mt_0/\hbar}\,\Delta\Psi(t_0)\,\mathrm{d}\boldsymbol{r} = (\mathrm{i}P/\hbar)\sum_l a_l M_{m,l}^{(1)}\,\mathrm{e}^{\mathrm{i}(E_m-E_l)t_0/\hbar}, \tag{33}$$

in which $M_{m,l}^{(1)}(=M_{l,m}^{(1)*})$ is written for the matrix element $\int \psi_m^* x\psi_l\,\mathrm{d}\boldsymbol{r}$.

† The linearity in Ψ of Schrödinger's equation, for any $V(\boldsymbol{r})$, implies that if any two functions, Ψ_1, and Ψ_2, are solutions then so is $a_1\Psi_1 + a_2\Psi_2$. This should not be confused with the linearity in x exhibited by the equation of motion of a classical harmonic vibrator, $m\ddot{x} + \mu x = F(t)$. The latter has the property that the response to two forces, $F_1(t)$ and $F_2(t)$, applied simultaneously, is the sum of the responses to each separately. It is not true that the development of Ψ can be similarly dissected into the responses to different components of the applied force, even for the harmonic vibrator (though in this case $\langle x\rangle$ has this property, as (28) shows).

Hence after a further interval t,

$$\Psi_+(t_0+t) = \sum_m (a_m + \Delta a_m)\psi_m \, e^{-iE_m(t_0+t)/\hbar}$$

$$= \Psi_-(t_0+t) + (iP/\hbar) \sum_m \sum_l a_l M^{(1)}_{m,l} \psi_m \, e^{-i(E_l t_0 + E_m t)/\hbar}, \quad (34)$$

where $\Psi_-(t_0+t) = \Sigma_l a_l \psi_l \, e^{-iE_l(t_0+t)/\hbar}$, representing the development of Ψ which would have taken place without the impulse.

We may now calculate the variation of any observable, and shall concentrate on $\langle x\rangle$, evaluated as $\int_{-\infty}^{\infty} \Psi_+^* x \Psi_+ \, dx$. From (34) it is seen that $\langle x\rangle$ contains a term in Ψ_- independent of P, which is the variation in the absence of an impulse, a term proportional to P which is the required impulse-response of $\langle x\rangle$, and a term in P^2 which we ignore since (31) is valid only to first order. Representing the middle term as $\langle \Delta x\rangle$, we have for the impulse-response function

$$h(t) \equiv \langle \Delta x\rangle / P = (i/\hbar) \sum_l \sum_m \sum_n a_n^* a_l M^{(1)}_{n,m} M^{(1)}_{m,l} \, e^{-i[(E_l - E_n)t_0 + (E_m - E_n)t]/\hbar} + \text{c.c.} \quad (35)$$

Clearly there may be many frequencies present in the oscillations of $h(t)$, a typical example being $(E_m - E_n)/\hbar$ arising from the beating of the wave-functions of the mth and nth states. Such a term may well have been present before the impulse, but of course $h(t)$ contains only the changes in its amplitude resulting from the impulse. If the lth state was originally present, its perturbation by P results in a change Δa_m in the amplitude of the mth state, to an extent determined by $a_l M^{(1)}_{m,l}$; the beating of the mth and the nth states, which yields an amplitude in $h(t)$ proportional to $a_m a_n M^{(1)}_{m,n}$, is to first order changed by $\Delta a_m \cdot a_n M^{(1)}_{m,n}$, so that the change in the amplitude at frequency $(E_m - E_n)/\hbar$ involves three levels, as indicated by the triple summation in (35).

Let us now take a special case, where the system is initially in the nth stationary state, so that $a_n = 1$ and all other a_l vanish. Then the triple summation collapses to a single summation, t_0 may be taken as zero (since for a stationary state all times are equivalent), and

$$h(t) = (i/\hbar) \sum_m |M^{(1)}_{m,n}|^2 \, e^{-i(E_m - E_n)t/\hbar} + \text{c.c.} \quad (36)$$

Writing ω_{mn} for $(E_m - E_n)\hbar$ and $f_{m,n}$ for $(2m\omega_{mn}/\hbar)|M^{(1)}_{m,n}|^2$, we have†

107

$$h(t) = \sum_m (f_{mn}/m\omega_{mn}) \sin \omega_{mn} t. \quad (37)$$

In this form the response of the system bears a strong resemblance to that of a set of classical oscillators of different frequencies, ω_{mn}, and masses

† Note that m is used both for the mass of the particle and as an index of the stationary state. In the latter sense it is not used except as a subscript.

m/f_{mn}; and this in spite of the real system having no resemblance to an oscillator – its character has indeed not been specified. This result will form the basis of a later general discussion of the transitions between states induced by a time-varying force. For the present we confine our attention to applying it to a harmonic vibrator. 510

It is well to note, however, one general point arising from (35) as a warning against using the impulse-response function too casually. The presence of t_0 in the expression for $h(t)$ shows that, in contrast to a classical vibrator, the quantized system will respond differently at different times if it is not in a stationary state. This is clearly because each Δa_m has phase as well as amplitude, so that when it beats with a_n the outcome will depend on the relative phase of a_m and the particular a_l whose perturbation generated Δa_m; hence the presence of $(E_l - E_n)t_0/\hbar$ in the phase. The only occasions when a time-independent response function may be used safely is when the system is never driven significantly far from a pure stationary state; or when it is in a mixture of stationary states but with unknown phase relationships, and all that is required is the average response taken over all possible phases. Phase-averaging of (35) eliminates all terms for which $m \neq l$, yielding a slightly generalized form of (37):

$$h(t) = \sum_m \sum_n a_n^* a_n (f_{mn}/m\omega_{mn}) \sin \omega_{mn} t. \tag{38}$$

Each occupied level in the initial state responds independently in proportion to its occupation probability $a_n^* a_n$, and the response to a time-varying force may be expressed as an integral over impulse-responses of this standard form provided the $a_n^* a_n$ are not greatly changed. This is the condition for treating the response as linear. It happens, as (28) shows, that even for strong perturbations the harmonic vibrator behaves linearly so far as $h(t)$, the response of $\langle x \rangle$, is concerned; but this is not obvious from the present argument.

The form of $h(t)$ for a harmonic vibrator initially in the nth state is especially simple, since according to (26) $M_{m,n}^{(1)}$ vanishes unless $m = n \pm 1$. In laboratory variables $|M_{n-1,n}^{(1)}|^2 = (\hbar/2m\omega_0)n$ and $|M_{n+1,n}^{(1)}|^2 = (\hbar/2m\omega_0)(n+1)$, so that $f_{n-1,n} = -n$ and $f_{n+1,n} = n+1$. Hence 504

$$h(t) = [(n+1)/m\omega_0] \sin \omega_0 t - [n/m\omega_0] \sin \omega_0 t = (\sin \omega_0 t)/m\omega_0, \tag{39}$$

which is, as expected, the same as the classical expression for the impulse-response function. It is the absence from (39) of the initial value of n that makes possible (though it does not prove) the application of this result to any strength of perturbation. The oscillatory response (39) is undamped, since we have introduced no damping mechanism. The problem of damping will be raised in chapter 16 and subsequently. 492

To proceed from the impulse-response to the response to a time-varying force, we refer to (33) to write down how the amplitudes a_n develop, treating the force $F(t)$ as a sequence of impulses:

$$\dot{a}_n = (\mathrm{i}F(t)/\hbar)\sum_l a_l M^{(1)}_{n,l}\,\mathrm{e}^{-\mathrm{i}\omega_{nl}t}. \qquad (40)$$

This is a general result, which we now apply to the harmonic vibrator, returning at this point to reduced variables, with Schrödinger's time-dependent equation in the form

$$\partial^2\Psi/\partial\xi^2 - (\xi^2 - 2\xi\phi)\Psi = -2\mathrm{i}\partial\Psi/\partial\tau,$$

in which $\phi(\tau) = F(t)/(m\hbar\omega_0^3)^{1/2}$. Then (40) takes the form

$$\mathrm{d}a_n/\mathrm{d}\tau = (\mathrm{i}\phi/\sqrt{2})[n^{\frac{1}{2}}a_{n-1}\,\mathrm{e}^{\mathrm{i}\tau} + (n+1)^{\frac{1}{2}}a_{n+1}\,\mathrm{e}^{-\mathrm{i}\tau}]. \qquad (41)$$

In particular, if ϕ is a sinusoidally varying force, $\mathrm{Re}\,[\phi\,\mathrm{e}^{-\mathrm{i}\omega_0(1+\delta)t}]$, we write it as $\frac{1}{2}(\phi^*\,\mathrm{e}^{\mathrm{i}(1+\delta)\tau} + \phi\,\mathrm{e}^{-\mathrm{i}(1+\delta)\tau})$ to give four terms in (41):

$$\mathrm{d}a_n/\mathrm{d}\tau = (i/2\sqrt{2})[n^{\frac{1}{2}}a_{n-1}(\phi^*\,\mathrm{e}^{\mathrm{i}(2+\delta)\tau} + \phi\,\mathrm{e}^{-\mathrm{i}\delta\tau})$$
$$+ (n+1)^{\frac{1}{2}}a_{n+1}(\phi^*\,\mathrm{e}^{\mathrm{i}\delta\tau} + \phi\,\mathrm{e}^{-\mathrm{i}(2+\delta)\tau})].$$

When $\delta \ll 1$, close to resonance, the rapidly alternating terms at frequency $2+\tau$ provide only a faint ripple on a_n and may be ignored, so that

$$\mathrm{d}a_n/\mathrm{d}\tau = (\mathrm{i}/2\sqrt{2})[\phi n^{\frac{1}{2}}a_{n-1}\,\mathrm{e}^{-\mathrm{i}\delta\tau} + \phi^*(n+1)^{\frac{1}{2}}a_{n+1}\,\mathrm{e}^{\mathrm{i}\delta\tau}]. \qquad (42)$$

As δ goes to zero the variations in a_n become steadily larger; the wave-packet, following the classical motion, oscillates with an amplitude b that beats at frequency δ, the peak amplitude varying inversely as δ. It does, in fact, take the form (16) with b now a function of time. Correspondingly, as (23) shows, the values of n which are most sharply excited oscillate at frequency δ; a plot of $a_n^* a_n$ against n at various times would show the system starting, say, in its lowest state, climbing up the ladder of levels until a rather wide range ($\sim n_c \pm n_c^{\frac{1}{2}}$) was sparsely occupied, and then collapsing back again to the ground state, to repeat the pattern unchanged and indefinitely so long as the same sinusoidal force was applied. Only at resonance, when $\mathrm{e}^{\pm\mathrm{i}\delta\tau} = 1$, does the climb proceed unabated.

145

Critical phenomena

It has been possible to describe the solution of the set of equations represented by (42) in these terms without a detailed calculation because the full solution (21) is already known. We therefore know, among other things, that so long as $\delta \neq 0$ the system does not diverge – there is always some value of n above which virtually no excitation is found at any time. In general, however, there are more than two terms in (42), although most will be far from resonance with any single excitation frequency. It is then hard to be sure that the response will be convergent, or that the pattern of development of the a_n will vary smoothly with the applied frequency. There are, indeed, simple examples exhibiting critical phenomena, which would be difficult to discern from the structure of the equations corresponding to (42). Such a case occurs with parametric excitation of a harmonic vibrator. **285**

As already noted, following (28), under the influence of parametric excitation by periodic variation of the force constant, $\langle x \rangle$ develops in the classical way. This means that for a given strength of excitation there is a band of frequencies around $2\omega_0$ in which an initial small vibration is amplified, growing exponentially without limit if the system is truly harmonic. Thus as the frequency is moved towards $2\omega_0$ the variations of a_n remain within bounds, but at a certain critical frequency they become divergent. We may ask whether the equations corresponding to (42) in this case exhibit any feature that allows the critical phenomenon to be discerned, and the actual critical frequency to be calculated. Unhappily the answer seems to be no, but let us derive the equations.

If the force constant is modulated at a frequency of $2r\omega_0$, r being close to unity, (13) takes the form:

$$\partial^2\Psi/\partial\xi^2 - \xi^2(1+\beta\cos 2r\tau)\Psi = -2\mathrm{i}\partial\Psi/\partial\tau, \tag{43}$$

β being a measure of the strength of the parametric excitation. We seek a solution in the form (14) (with l replacing n) where now the a_l may vary with time. Substitute (14) in (43), multiply by $\psi_n\,\mathrm{e}^{\mathrm{i}n\tau}$ and integrate over ξ; then

$$\mathrm{d}a_n/\mathrm{d}\tau = -\tfrac{1}{2}\mathrm{i}\beta\cos r\tau \sum_l a_l M_{l,n}^{(2)}\,\mathrm{e}^{\mathrm{i}(n-l)\tau},$$

the matrix elements $M_{l,n}^{(2)}$ being given in (24); only those for which $l-n=0$ or ± 2 are non-vanishing. When their values are substituted and the rapidly varying terms eliminated, as was done with (42), the remainder take the form:

$$\mathrm{d}a_n/\mathrm{d}\tau = -\tfrac{1}{8}\mathrm{i}\beta[(n+2)^{\frac{1}{2}}(n+1)^{\frac{1}{2}}a_{n+2}\,\mathrm{e}^{2\mathrm{i}\delta\tau} + n^{\frac{1}{2}}(n-1)^{\frac{1}{2}}a_{n-2}\,\mathrm{e}^{-2\mathrm{i}\delta\tau}], \tag{44}$$

in which $\delta = r-1$. This differs from (42) in that the coefficients in the square brackets vary as n, rather than $n^{\frac{1}{2}}$, at large values of n. Thus high on the ladder of levels there is greater mobility and one need not be surprised to learn that when δ is small enough (44) may represent an uncontrolled escalation, while (42) only shows this behaviour when $\delta = 0$.

It only requires a small modification in (42) to create a comparable situation. A slightly anharmonic vibrator, if nearly lossless, shows a discontinuous response as the driving force is tuned through resonance. In one direction the amplitude climbs to a high value and suddenly plunges, while in the reverse direction it remains low until with equal suddenness it soars. Presumably the form of (42) for this case is almost unchanged, except that the levels are not quite evenly spaced, and δ must be regarded as varying with n. The discontinuous switches are associated with δ going through zero at some non-zero value of n, and the system either climbing or descending catastrophically through this region of the ladder of energy **250** levels. However, as with parametric excitation, it is no easy matter to quantify the critical phenomenon when (41) or its equivalent is taken as the starting point. Examples of such critical behaviour may be only rarely encountered in quantum phenomena, but the very possibility of their undetected presence is worrying. This example perhaps demonstrates more clearly than any other the difference between the power of classical and quantum mechanics to cope with problems out of the ordinary. If for no other reason, classical treatments should be valued for the warning they give of potential dangers lurking in the mathematical jungle of quantum mechanics.

The equivalent classical ensemble (σ-representation)

We return to the undisturbed vibrator in a non-stationary state, with a wave-function that obeys (13). The particular solution (16) provides the simplest example of a general property of solutions of (13): if $\Psi_0(\xi, \tau)$ is any solution, then $\Psi(\xi, \tau)$ is also a solution, where

$$\Psi(\xi, \tau) = \Psi_0(\xi - b \cos \tau, \tau) \exp[-\mathrm{i}(b\xi \sin \tau_1 - \tfrac{1}{4}b^2 \sin 2\tau_1 + c)]. \quad (45)$$

Here τ_1 is written for $\tau + \tau_0$; τ_0, b and c are arbitrary real constants. The truth of this proposition may be tested by substitution in (13). The probability distribution, $\mathscr{P}(\xi, \tau)$, defined as $\Psi\Psi^*$, contains nothing of the rather complicated phase variation of (45):

$$\mathscr{P}(\xi, \tau) = \mathscr{P}_0(\xi - b \cos \tau, \tau), \quad (46)$$

where $\mathscr{P}_0 = \Psi_0\Psi_0^*$. The probability distribution corresponding to Ψ develops periodically in the same way as that corresponding to Ψ_0, but also vibrates bodily with amplitude b. We now show that the effect of a time-varying uniform force on a system initially described by Ψ_0 is to change Ψ_0 into Ψ, with the form (45), by progressive changes in the parameters b, c and τ_0. It is enough for this purpose to show that a system described by (45), when acted upon by an arbitrary impulse, merely suffers certain changes in these parameters. It will then follow that after the impulse, and therefore through the sequence of changes consequent upon applying a varying force, $\mathscr{P}(\xi, \tau)$ continues to go through exactly the same periodic variation while vibrating to and fro with a continuously changing amplitude, b.

To demonstrate this result, consider (45) as representing Ψ_- just before the impulse P; then immediately after, according to the exact result (30),

$$\Psi_+(\xi, \tau) = \Psi_0(\xi - b \cos \tau, \tau) \exp[-\mathrm{i}(b\xi \sin \tau_1 - \tfrac{1}{4}b^2 \sin 2\tau_1 + c - p\xi)],$$

in which p is the reduced form of $P (Px/\hbar = p\xi)$. This expression for Ψ_+ may now be rewritten in the same form as (45) by changing the constants to b', c' and τ_0', satisfying the relations:

$$b\cos\tau_1 = b'\cos\tau_1', \tag{47}$$

$$b\sin\tau_1 - p = b'\sin\tau_1', \tag{48}$$

and
$$c - \tfrac{1}{4}b^2\sin 2\tau_1 = c' - \tfrac{1}{4}b'^2\sin 2\tau_1'. \tag{49}$$

The change in c given by (49) does not concern us, though it may be remarked that it is represented in fig. 2(a) by the area of the shaded triangle. In the same diagram are shown the changes in b and τ_1, exactly the same as if they defined the clockwise-rotating vector describing a classical oscillator; the real part, the displacement, is unchanged by the impulse, but the imaginary part, the momentum, is increased by p.

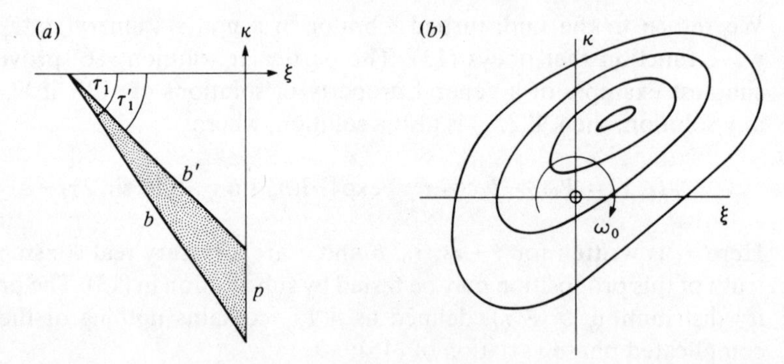

Fig. 2. (a) Illustrating the solution of (47)–(49). (b) Schematic diagram of the σ-representation, showing hypothetical contours of σ on the (ξ, κ)-plane.

We are now in a position to construct the classical model which will reproduce the behaviour of $\mathscr{P}(\xi, \tau)$ under the influence of a varying uniform force. Let us for the moment assume, as will be proved directly, that when no force is applied the periodic variations of $\mathscr{P}_0$ may always be represented by an ensemble of classical vibrators, defined by a two-dimensional density distribution $\sigma(\boldsymbol{R})$, where $\boldsymbol{R} = (\xi, \kappa)$, such as is shown schematically in fig. 2(b). The contours of σ are to be imagined drawn on a rigid lamina which spins at the oscillation frequency about the origin, so that at time τ the density is $\sigma(\xi\cos\tau - \kappa\sin\tau, \xi\sin\tau + \kappa\cos\tau)$. At any instant $\mathscr{P}_0$ is the projection of σ on to the ξ-axis:

$$\mathscr{P}_0(\xi, \tau) = \int_{-\infty}^{\infty} \sigma(\xi\cos\tau - \kappa\sin\tau, \xi\sin\tau + \kappa\cos\tau)\,d\kappa. \tag{50}$$

What the argument of the last paragraph showed was that the effect of an impulse may be represented by suddenly displacing the pattern of σ (i.e. the lamina) vertically through p, and then allowing the rotation to continue about the origin. This leaves the development of $\mathscr{P}$ unchanged, but superposes a bodily oscillation as described by a change of b and τ_1 in (46). We may think of σ as defining an ensemble of classical vibrators, each represented by a rotating vector, the heads of vectors having a density distribution σ. Since the ξ-component of each vector defines the instantaneous displacement of the corresponding oscillator, $\mathscr{P}_0(\xi, \tau)$ as defined by (50) is just the distribution of displacements at τ. And it may be remarked incidentally that the κ-components of the ensemble equally well describe the momentum-distribution in Ψ_0, and its variation with time; this explains the choice of κ to denote this coordinate. Finally, to complete the story, fig. 2(a) shows that if each member of the classical ensemble is subjected to the same impulse p, it responds in the required way to produce the modified form of σ; and what holds for an arbitrary impulse automatically holds for a time-varying force. Hence the σ-representation, which is the term we use from now on to refer to the classical ensemble set up to describe the initial probability distribution $\mathscr{P}_0$, and its cyclical evolution, will continue to give a correct description when the quantized system and its ensemble analogue are disturbed by an arbitrary uniform force.

Nothing has been said about the conditions that $\sigma(\boldsymbol{R})$ must satisfy in order to describe a possible distribution $\mathscr{P}(\xi, \tau)$. Obviously not every σ has this property (e.g. a strongly localized peak cannot describe the distribution arising from any real wave-function), but the restrictions on σ are hard to derive. For our purpose this is irrelevant, but it is not irrelevant that every possible $\mathscr{P}(\xi, \tau)$ can be represented by the appropriate σ. The proof is easy. Let σ at some instant be expressed as a two-dimensional Fourier integral,

$$\sigma = \iint A(k_\xi, k_\kappa)\, \mathrm{e}^{\mathrm{i}(k_\xi \xi + k_\kappa \kappa)}\, \mathrm{d}k_\xi\, \mathrm{d}k_\kappa.$$

Then $\mathscr{P}(\xi)$ at the same instant is $\int \sigma\, \mathrm{d}\kappa$ to which only Fourier components having $k_\kappa = 0$ contribute:

$$\mathscr{P}(\xi) \propto \int A(k_\xi, 0)\, \mathrm{e}^{\mathrm{i}k_\xi \xi}\, \mathrm{d}k_\xi.$$

Hence by the Fourier transform theory the instantaneous value of $\mathscr{P}(\xi)$ may be used to define A along the ξ-axis, and at any other time, when σ has turned to a new orientation, the variation of A along another radius is again uniquely determined by $\mathscr{P}$. It is therefore always possible to represent any $\mathscr{P}(\xi, \tau)$ by one, and one only, spinning σ. The fact that $\mathscr{P}$ reverses itself every half-cycle is an essential requirement.

For a stationary state σ has circular symmetry and its spinning produces no change. Thus for the unnormalized ground state in which $\Psi = \mathrm{e}^{-\frac{1}{2}(\xi^2 + \mathrm{i}\tau)}$, $\mathscr{P}(\xi)$ is $\mathrm{e}^{-\xi^2}$ and σ is e^{-R^2}, an isotropic two-dimensional Gaussian peak. If this narrow peak is displaced far from the origin and left to spin unchanged round a large circle it describes a coherent state of excitation. A stationary state of high energy, specified by a single n rather than the range expressed in (21), is by contrast represented by something like a narrow ring of constant amplitude all round. The mean level of $\mathscr{P}(\xi)$ follows closely the classical distribution, $1/\pi(\xi_0^2 - \xi^2)^{\frac{1}{2}}$, for an exact description of which σ must be wholly concentrated on a circle of radius ξ_0 and zero width. The lack of perfect sharpness at the edge, which we have already noted, means that the ring to which σ is confined must have a width of the order of $1/n^{\frac{1}{6}}$.

435
432

But this is not the whole story since $\mathscr{P}(\xi)$ oscillates, with n zeroes. To reproduce these oscillations the ring that generates the mean value must be supplemented by an extended pattern of relatively low-amplitude rings, on which σ alternates between positive and negative. We have no need to ask about the details of the pattern, but it is worth remarking that the presence of zeroes in $\mathscr{P}(\xi)$ makes negative values of σ inevitable. These have no classical interpretation; a negative density of vectors at ξ does not produce in $\mathscr{P}(\xi)$ the same contribution as a positive density at $-\xi$. The σ-representation is a convenient fiction, incapable of realization. We shall make considerable use of it later. 587

Energy imparted by an applied force

We shall now show that a uniform force with arbitrary time variation changes the energy of the harmonic vibrator in an essentially classical manner. It is sufficient to establish the result for an impulse acting on any state that can be represented on the $\sigma(\xi, \kappa)$ diagram. It follows from (11) that the potential energy $\langle V\rangle$ at any instant, measured in terms of $\frac{1}{2}\hbar\omega_0$, is $\langle\xi^2\rangle$, while $\langle\kappa^2\rangle$ is the value $\langle V\rangle$ will acquire one-quarter of a cycle later when the pattern has spun through $\pi/2$. Hence, from (25), $\langle\kappa^2\rangle$ is the kinetic energy $\langle T\rangle$, and the total energy E is $\langle\xi^2+\kappa^2\rangle$, in accordance with the earlier statement associating κ in this representation with momentum. 433
Writing $\xi^2+\kappa^2$ as R^2 we have that

$$\langle R^2\rangle = \bar{R}^2 + \langle(R-\bar{R})^2\rangle,$$

$\bar{R}$ being the co-ordinate of the centroid of the distribution. We have seen that an impulse shifts the distribution bodily in the κ-direction, so that the second term is unchanged. Thus we reach the required result, that the total energy changes by the same amount as if the impulse were applied to a classical oscillator at $\bar{R}$.

An alternative proof, of which the details need not be given, involves calculating $\langle T\rangle$ as $-(\hbar^2/2m)\int\psi^*(\mathrm{d}^2\psi/\mathrm{d}x^2)\,\mathrm{d}x$, and determining the change $\Delta\langle T\rangle$ when an impulse P replaces ψ by $\psi\,\mathrm{e}^{\mathrm{i}xP/\hbar}$, as in (30). Then $\Delta\langle T\rangle$ is found in general to be $P\langle\dot{x}\rangle+P^2/2m$, the same as the classical result in the special case of a harmonic vibrator for which we know $\langle x\rangle$ evolves classically under the influence of a uniform force.

At this point one may begin to worry that everything seems to conform too closely to the classical model, although we know that there are highly significant differences, which were responsible for the original discovery of the quantum. The mean energy of a vibrator in equilibrium with a gas at temperature T is not the classical equipartition energy $k_\mathrm{B}T$ but is given by Planck's expression $\hbar\omega_0/(\mathrm{e}^{\hbar\omega_0/k_\mathrm{B}T}-1)$, which is less than $k_\mathrm{B}T$ at all T.† If we suppose the gas atoms, colliding with the vibrator, to exert on it

† Planck's expression does not contain the zero-point energy $\frac{1}{2}\hbar\omega_0$, and is therefore the mean excitation energy, not the total energy. If the zero-point energy is included, the mean total energy is $\frac{1}{2}\hbar\omega_0\coth(\hbar\omega_0/2k_\mathrm{B}T)$, which is greater than $k_\mathrm{B}T$, its asymptotic value as $T\to\infty$.

a time-varying force we might take the argument just developed as good reason to expect the transfer of energy to be indifferent as to the classical or quantal character of the vibrator. The mistake lies in thinking that it reacts to the collision in the same way as to a specified impulse. This is not to imply that it is the non-uniformity of the force that is the sole cause of the difference; something more fundamental is involved. When two atomic systems interact, the consequences can only be determined by solving the Schrödinger equation for the two together; it is invalid to imagine that each acts on the other in some specified way, and that the two-body problem can be thus dissected into two one-body problems. An obvious illustration is provided by a vibrator in its ground state being struck by an atom moving so slowly that its energy is less than $\hbar\omega_0$. There is no objection to some of this energy being transferred in a classical collision, nor to $\langle E\rangle$ increasing if the dissection of the quantum system is permitted, and the impact treated as a weak time-dependent force; but we know that in the correct solution there must be no chance whatever of finding the vibrator in an excited state afterwards. This fundamental change in the dynamics is enough to dispel any expectation that equipartition of energy should apply.

The objection on the grounds of energy conservation is less serious when the colliding body is massive, for it may be moving very slowly and yet carry energy much in excess of $\hbar\omega_0$. Presumably, though we have not proved this, when the energy transfer calculated classically is much less than the energy of the colliding body it may be assumed that the motion of the body is virtually unperturbed and that the vibrator will react as to 491 a prescribed force. We shall meet a similar problem in chapter 16, where we shall find it impermissible to pretend that when two quantized vibrators are coupled together each responds independently to the time-varying force exerted by the other. Yet if one of the vibrators is very small and at a low level of excitation, while the other is large and in a high coherent state of oscillation, so that it approximates closely to a classical vibrator, it is surely a good approximation to suppose the former to be acted upon by a prescribed force.

Parametric excitation

We have already noted that Ehrenfest's theorem leads to classical behaviour 439 for $\langle x\rangle$ under the influence of parametric excitation, and the classical ensemble just developed may be used in this case also to illustrate how $\mathcal{P}(\xi, \tau)$ develops during parametric excitation. Rather than giving a complete demonstration we shall be content with a single example, worked through in detail to show the essential characteristics of the quantal behaviour. Let the vibrator be initially in its (unnormalized) ground state, $\Psi_0 = \mathrm{e}^{-\frac{1}{2}(\xi^2 + i\tau)}$, and let us seek a solution of (43) of the form

$$\Psi = \mathrm{e}^{-(y\xi^2 + k)}, \tag{51}$$

y and k both being functions of τ. For this solution to apply the following conditions must hold:

$$k = \mathrm{i}\int^{\tau} y\,\mathrm{d}\tau, \tag{52}$$

and

$$2\mathrm{i}\,\mathrm{d}y/\mathrm{d}\tau - 4y^2 + 1 + \beta\cos 2r\tau = 0. \tag{53}$$

The standard procedure[6] for an equation of Riccati type, such as (53), is to introduce $u(\tau)$, such that $2\mathrm{i}uy = \mathrm{d}u/\mathrm{d}\tau$. Then

$$\mathrm{d}^2u/\mathrm{d}\tau^2 + (1+\beta\cos 2r\tau)u = 0, \tag{54}$$

which is just the classical equation of motion for a parametrically excited vibrator. The analysis in chapter 10 shows that if the excitation is weak ($\beta \ll 1$) and perfectly in tune ($r = 1$), an adequate solution of (54) takes the form:

$$u = \mathrm{e}^{\Delta_c\tau}\cos(\tau + \tfrac{1}{4}\pi) + C\,\mathrm{e}^{-\Delta_c\tau}\cos(\tau - \tfrac{1}{4}\pi) \tag{55}$$

294

in which $\Delta_c = \frac{1}{4}\beta$. This case of perfect tuning is enough to illustrate the behaviour and we shall not go beyond it. The corresponding form of y follows immediately:

$$y(\tau) \doteqdot \tfrac{1}{2}\mathrm{i}[\mathrm{e}^{\Delta_c\tau}\sin(\tau+\tfrac{1}{4}\pi) + C\,\mathrm{e}^{-\Delta_c\tau}\sin(\tau-\tfrac{1}{4}\pi)]/[\mathrm{e}^{\Delta_c\tau}\cos(\tau+\tfrac{1}{4}\pi) + C\,\mathrm{e}^{-\Delta_c\tau}\cos(\tau-\tfrac{1}{4}\pi)].$$

In writing this we have assumed the excitation to be so weak ($\Delta_c \ll 1$) that the amplitude does not change appreciably in one cycle, and have accordingly ignored terms arising from derivatives of the exponential factors. If C is assigned the value i, $y(0)$ takes the ground state value $\frac{1}{2}$, the appropriate starting condition if the excitation is switched on when $\tau = 0$. Subsequently y oscillates at twice the natural frequency and the wave-function (51), while retaining its Gaussian profile, oscillates in width. The function $k(\tau)$ is a rounded staircase function, becoming squarer as time goes on, but fortunately there is no need to evaluate it. Its imaginary part contributes phase changes across the wave-function that disappear from $\mathscr{P}(\xi, \tau)$, while its real part serves to ensure conservation of $\int \mathscr{P}\,\mathrm{d}\xi$ throughout the oscillations of width. In this knowledge we write

$$\mathscr{P}(\xi,\tau) \propto (\overline{\xi^2})^{-\frac{1}{2}}\,\mathrm{e}^{-\xi^2/2\overline{\xi^2}}, \tag{56}$$

where

$$\overline{\xi^2} = 1/\mathrm{Re}\,[4y] = \tfrac{1}{2}\left(p^2\cos^2\theta + \frac{1}{p^2}\sin^2\theta\right), \tag{57}$$

$$p = \mathrm{e}^{\frac{1}{4}\beta\tau} \quad \text{and} \quad \theta = \tau - \tfrac{1}{4}\pi.$$

As p grows exponentially with time the pulsations of width become ever stronger, and the wave-packet oscillates vigorously twice a cycle between p and $1/p$ times its initial width.

This is what the classical ensemble predicts for the same excitation applied to all its members. If the vector representing any one classical vibrator is resolved at time $\tau = 0$, when the excitation is applied, into orthogonal components at $\pm\frac{1}{4}\pi$ to the real axis, each component develops independently, one growing exponentially as $e^{\frac{1}{4}\beta\tau}$ and the other decaying as $e^{-\frac{1}{4}\beta\tau}$. An observer rotating at the natural frequency would see the same scale transformation applied to each vector, its component at $-\frac{1}{4}\pi$ being increased and that at $\frac{1}{4}\pi$ decreased by a factor p. The initial isotropic σ, of the form $e^{-(\xi^2+\kappa^2)}$, becomes distorted into an elliptical Gaussian distribution, as in fig. 3, with axial ratio p^2, and in the laboratory framework this ellipse sweeps round at the natural frequency, its projection on the ξ-axis generating $\mathscr{P}(\xi, \tau)$ in accordance with (56), including the pre-exponential factor.

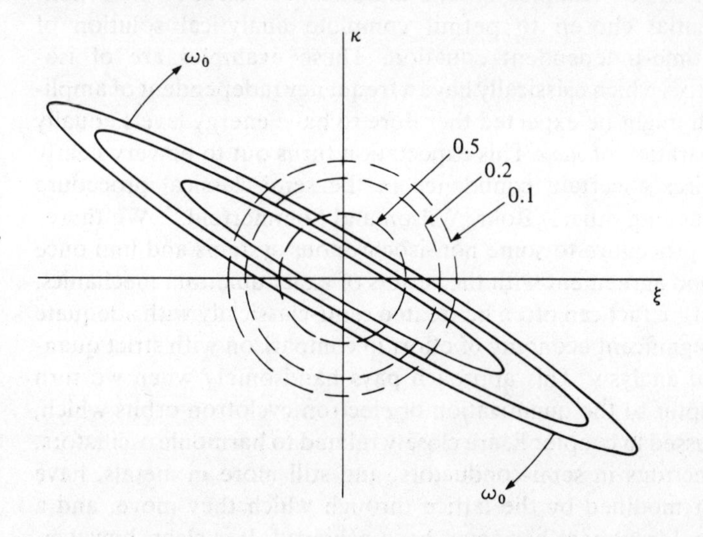

Fig. 3. The ground state, shown in the σ-representation as a circular Gaussian distribution (broken contours), is deformed by parametric excitation into an elliptical Gaussian distribution (full contours). In this case $p = 3$ and the axial ratio is 9.

It will be observed that $\mathscr{P}(\xi)$ remains even at all times in this case, and that $\langle x \rangle = 0$. This must be true, by symmetry, whenever the parametrically excited vibrator starts in a stationary state, and of course is also true for a classical vibrator initially at rest. In the classical case there may be a latent period before the advent of some disturbance to initiate the exponential growth, but with the quantized vibrator there is no latent period for the symmetrical evolution of Ψ; even in the ground state the zero-point distribution provides something for the parametric excitation to seize on, but the resulting symmetrical oscillation of $\mathscr{P}$ has no classical analogue except for the ensemble described here.

14 Anharmonic vibrators

We begin with some examples of one-dimensional vibration in a non-parabolic potential chosen to permit complete analytical solution of Schrödinger's time-independent equation. These examples are of isochronous vibrators which classically have a frequency independent of amplitude, and which might be expected therefore to have energy levels equally spaced at a separation of $\hbar\omega_0$. This expectation turns out to be very nearly right and inspires a certain confidence in the semi-classical procedure developed by (among others) Bohr, Wilson and Sommerfeld.[1] We therefore apply this procedure to some non-isochronous systems and find once more rather good agreement with the results of exact quantum mechanics. Periodic systems in fact can often be treated semi-classically with adequate accuracy, and significant economy of effort in comparison with strict quantum-mechanical analysis. This approach pays handsomely when we turn in the next chapter to the quantization of electron cyclotron orbits which, as already discussed in chapter 8, are closely related to harmonic oscillators. **237** Conduction electrons in semi-conductors, and still more in metals, have their behaviour modified by the lattice through which they move, and a complete quantal treatment has never been achieved. It is clear, however, from approximate calculations, often of great complexity, that the semi-classical method describes most of the interesting physical processes correctly and very simply. In chapter 15 we shall describe in outline some of the effects which can be treated quite well enough for most purposes without even writing down Schrödinger's equation.

Isochronous vibrators

It was shown in chapter 2 that the potential ax^2+b/x^2 is isochronous, with **16** ω_0 equal to $(8a/m)^{\frac{1}{2}}$, independent of b. It might appear that when $b=0$ the frequency is $(2a/m)^{\frac{1}{2}}$, but however small b may be the potential rises to infinity at $x=0$; when $b=0$ the parabola ax^2 is bisected by an impenetrable δ-function at the origin, so that the frequency is doubled. It is when b is larger that the well becomes more symmetrical and parabolic in form, $V\sim 2(ab)^{\frac{1}{2}}+4a[x-(b/a)^{\frac{1}{4}}]^2$. To avoid the base of the well changing height with b, let us redefine V as $(Ax-B/x)^2$, where $A^2=a$, $B^2=b$ and the minimum value of V is zero, at $x=(B/A)^{\frac{1}{2}}$. With the same reduced coordinates as for the harmonic oscillator, $\xi=(m\omega_0/\hbar)^{\frac{1}{2}}x$ and $\varepsilon=2E/\hbar\omega_0$, **430** and with $\beta = B(2m)^{\frac{1}{2}}/\hbar$, Schrödinger's equation takes the form:

$$\psi'' + [\varepsilon - (\tfrac{1}{2}\xi - \beta/\xi)^2]\psi = 0. \tag{1}$$

This can be solved in a power series once the singular behaviour at $\xi = 0$ and the asymptotic behaviour at infinity have been taken out. As for the latter, the term β/ξ becomes negligible and like the harmonic oscillator the solution that vanishes at infinity is of the form $e^{-\frac{1}{4}\xi^2}$. At the origin it is β/ξ that dominates the behaviour and, approximately,

$$\psi'' \approx \beta^2\psi/\xi^2.$$

If $\psi = \xi^s$, s being a positive number such that $s(s-1) = \beta^2$, the equation is satisfied and ψ vanishes at the origin. We therefore write

$$\psi = \xi^s\, e^{-\frac{1}{4}\xi^2} F(\xi), \tag{2}$$

where
$$s = \tfrac{1}{2}[1 + (1 + 4\beta^2)^{\frac{1}{2}}], \tag{3}$$

and seek for a solution $F(\xi) = \sum_0^\infty u_n \xi^n$, in which $u_0 \neq 0$. Substitution of this trial solution in (1) gives the recurrence relation:

$$\frac{u_{n+2}}{u_n} = \frac{n + s + \frac{1}{2} - \varepsilon - \beta}{(n + s + 2)(n + s + 1) - \beta^2} = \frac{n + s + \frac{1}{2} - \varepsilon - \beta}{(n + 2)(n + 2s + 1)}, \tag{4}$$

which tends to $1/n$ as $n \to \infty$. Now this asymptotic recurrence relation matches that of the series expansion of $e^{\frac{1}{2}\xi^2}$, so that if the series does not terminate, $F(\xi) \sim e^{\frac{1}{2}\xi^2}$ at large ξ; then ψ, varying as $e^{+\frac{1}{4}\xi^2}$, does not fall to zero at infinity. This is exactly the same behaviour as makes it necessary to terminate the infinite series in the solution of the harmonic oscillator problem. In this case ε must be chosen so that the numerator in (4) vanishes for some value of n; then $u_n\xi^n$ is the last term in the series, and $F(\xi)$ can grow no faster than ξ^n. The exponential $e^{-\frac{1}{4}\xi^2}$ now ensures the vanishing of ψ at infinity. Only even powers of ξ appear in $F(\xi)$, and if we write n as 2ν, the νth level has energy ε_ν given by the expression:

$$\varepsilon_\nu = 2\nu + s + \tfrac{1}{2} - \beta. \tag{5}$$

Hence
$$E_\nu = (\nu + \gamma)\hbar\omega_0, \tag{6}$$

where, from (3),
$$\gamma = \tfrac{1}{2}[1 - \beta + (\beta^2 + \tfrac{1}{4})^{\frac{1}{2}}]. \tag{7}$$

The energy levels are exactly spaced at $\hbar\omega_0$, as with the harmonic oscillator, but the fractional correction, γ, determining the zero-point energy $\gamma\hbar\omega_0$, is not $\frac{1}{2}$ except when β is very large and the well approximates to a parabolic potential. When $\beta = 0$, $\gamma = \frac{3}{4}$ as would be expected for a parabola bisected by a vertical barrier. For if the barrier were removed the natural frequency of the resulting harmonic oscillator would be $\frac{1}{2}\omega_0$ and the energy levels would be $(n + \frac{1}{2})\frac{1}{2}\hbar\omega_0$; however, only odd values of n generate wave-functions that vanish at the origin and hence can be fitted into the bisected potential. Writing $n = 2\nu + 1$ gives energy levels $(\nu + \frac{3}{4})\hbar\omega_0$, as required by (7). Other values of β span the range $\frac{1}{2}$ to $\frac{3}{4}$ for γ, but otherwise leave the level structure unaltered. Of course, the wave-function itself is different from that for a harmonic oscillator, being necessarily asymmetric to conform to the shape of the potential well, but perhaps less distorted than one might have guessed. Fig. 1 presents a comparison for the case $\nu = 3$, the horizontal scales having been adjusted so as to make the classical turning-points coincide.

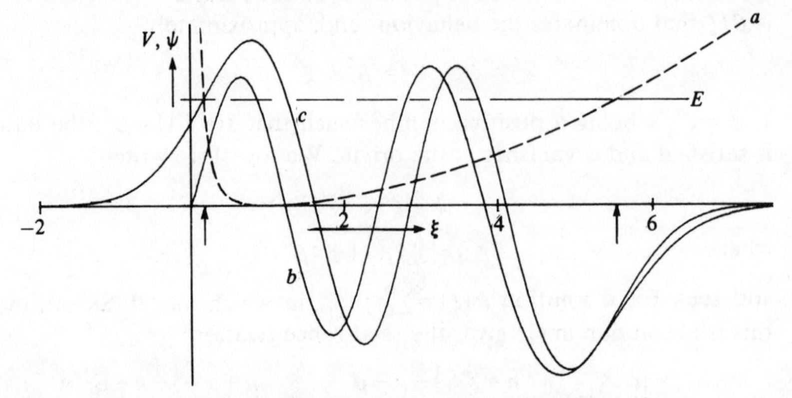

Fig. 1. The stationary state wave-function ψ_3 (curve b) for the isochronous potential $(\frac{1}{2}\xi - \beta/\xi)^2$, with $\beta = \frac{1}{2}$ (curve a). The arrows show the classical turning-points, $V = E$, and curve c is the corresponding solution for a harmonic vibrator having the same turning points.

To avoid any suspicion that all isochronous vibrators have strictly even level spacings, let us analyse a counter-example in which the potential is formed from two parabolic half-wells joined at their lowest points:

$$\left.\begin{aligned} V &= \tfrac{1}{2}m\omega_0^2x^2 \quad \text{for } x > 0 \\ &= \tfrac{1}{2}\beta^2 m\omega_0^2 x^2 \quad \text{for } x < 0, \end{aligned}\right\} \tag{8}$$

leading to a frequency $2\beta\omega_0/(1+\beta)$. On either side of the origin wave-functions that vanish at great distances can be found for all values of E, and quantization now arises from the necessity of making ψ and ψ' continuous at the origin. Let us first write down the general solution for $x > 0$, for which (13.2) applies with the same choice of reduced variables. Writing ψ as $e^{-\frac{1}{2}\xi^2}\sum_n u_n\xi^n$ we have the recursion formula

$$\frac{u_{n+2}}{u_n} = \frac{2n+1-\varepsilon}{(n+1)(n+2)}, \tag{9}$$

but we are not at liberty to choose ε so that the series terminates. Instead we recognize that (9) generates two independent series, one of even and the other of odd terms only, and that for each, since $u_{n+2}/u_n \to -2/n$ as $n \to \infty$, the behaviour at large ξ is as e^{ξ^2}. To make the solution acceptable we must arrange that the two series cancel each other at large ξ. Now the gamma function, $\Gamma(z)$, has the property[2] that for any z, $\Gamma(z+1) = z\Gamma(z)$, and this enables the coefficients u_n to be expressed compactly:

If n is even $\qquad u_n = 2^n \dfrac{\Gamma(\frac{1}{2}n - y)}{\Gamma(-y)n!}u_0;$

and
$$u_{n+1} = 2^n \frac{\Gamma[\frac{1}{2}(n+1)-y]}{\Gamma(-y+\frac{1}{2})(n+1)!} u_1,$$

where $y = \frac{1}{4}(\varepsilon - 1)$. The terms in each series take the same form if $u_1/u_0 = 2\Gamma(-y+\frac{1}{2})/\Gamma(-y)$, and they cancel each other asymptotically if the signs are opposite, i.e.

$$u_1/u_0 = -2\Gamma(-y+\tfrac{1}{2})/\Gamma(-y) = 2y \tan \pi y \cdot \Gamma(y)/\Gamma(y+\tfrac{1}{2}).$$

Now at the origin $(\mathrm{d}\psi/\mathrm{d}\xi)/\psi = u_1/u_0$, and therefore in laboratory coordinates

$$(\psi'/\psi)_0 = \tfrac{1}{2}(m\omega_0/\hbar)^{\frac{1}{2}}(\varepsilon-1)\tan[\tfrac{1}{4}\pi(\varepsilon-1)]\Gamma[\tfrac{1}{4}(\varepsilon-1)]/\Gamma[\tfrac{1}{4}(\varepsilon+1)]. \quad (10)$$

This is shown as curve 1 in fig. 2; the curve crosses the axis at those values of E which are even stationary states ($\psi' = 0$) of the harmonic oscillator, and rises to ∞ at the odd stationary states ($\psi = 0$).

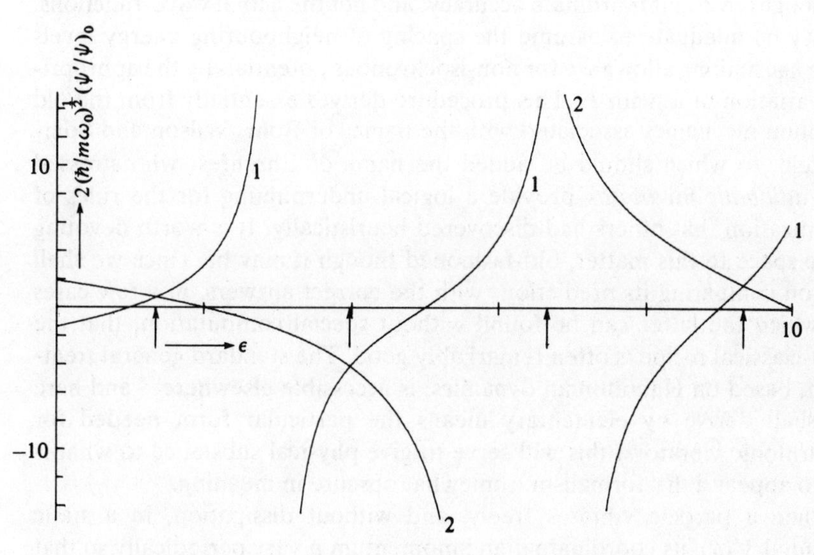

Fig. 2. Illustrating the energy levels for a particle in an asymmetric parabolic potential as the intersections (shown by arrows) of curves 1 and 2; see text for a full description.

On the negative side, $x < 0$, the same behaviour holds except that the sign of (ψ'/ψ) is reversed and ω_0 is replaced by $\beta\omega_0$. Thus to apply to the same values of E curve 2 must be derived from curve 1 by scaling horizontally by β, vertically by $\beta^{\frac{1}{2}}$, and reflecting in the horizontal axis. In this case β has been taken as 2. The intersections give those values of $E/\hbar\omega_0$ at which ψ and ψ' may be matched at the origin, and are therefore the stationary energy states.

Since the frequency is $\frac{4}{3}\omega_0$, one might expect eigenvalues of $(n+\frac{1}{2})\frac{4}{3}\hbar\omega_0$, and this is very nearly realized. When the energies are expressed as $(n+\gamma_n)\frac{4}{3}\hbar\omega_0$, computed values of the intersections give $\gamma_0 = 0.50895$, $\gamma_1 = 0.49497$, $\gamma_2 = 0.49937$, $\gamma_3 = 0.49890$, $\gamma_4 = 0.50012$ etc. Departures from strictly even spacing of the levels undoubtedly occur, but they are extremely small. The somewhat capricious deviations of γ from $\frac{1}{2}$ are probably attributable to the discontinuity in d^2V/dx^2 at $x=0$, whose effect will depend on whether ψ is large or small there. We shall see similar examples presently.

Arbitrary potential well; the semi-classical method

The potentials just analysed happen to submit to analytical treatment without difficulty, but an arbitrarily chosen form of $V(x)$ is unlikely to yield to anything but a computer. If $V(x)$ is symmetrical about the origin the fact that ψ is either odd or even simplifies the computation, for one can start a trial integration of Schrödinger's equation at the origin with either ψ or ψ' equal to zero, and seek a value of E that causes convergence to zero at a great distance. In the absence of symmetry the search is more tedious, for in addition to E the value of ψ'/ψ at some point must be correctly adjusted. This process may be obviated if only the energy levels are sought, without inordinate accuracy, and not the actual wave-functions. It may be adequate to assume the spacing of neighbouring energy levels to be $\hbar\omega$, making allowance for non-isochronous potentials by the appropriate variation of ω with E. This procedure derives essentially from the old quantum mechanics associated with the names of Bohr, Wilson and Sommerfeld, to which should be added the name of Ehrenfest who stressed that *adiabatic invariants* provide a logical underpinning for the rules of quantization that others had discovered heuristically. It is worth devoting some space to this matter, old-fashioned though it may be, since we shall find on comparing its predictions with the correct answers, in a few cases for which the latter can be found without special computation, that the semi-classical recipe is often remarkably good. The standard general treatment, based on Hamiltonian dynamics, is accessible elsewhere(3) and here we shall derive by elementary means the particular form needed for anharmonic vibrators; this will serve to give physical substance to what is apt to appear a dry formalism, somewhat obscure in meaning.

When a particle vibrates freely, and without dissipation, in a static potential $V(q)$, its coordinate q and momentum p vary periodically so that they may be represented by a closed curve on the q–p phase plane. A harmonic vibrator, with q and p varying sinusoidally in phase quadrature, executes elliptical trajectories, but in general all we can say is that there will be a family of closed curves, one for each energy of vibration, as in fig. 3, with mirror symmetry about the q-axis since reversing the sign of p leaves E unchanged. We now imagine the potential V to be very slowly deformed, allowing many cycles of oscillation for any appreciable change to occur. Any quantity which is unchanged by this process is an adiabatic invariant. The total energy is certainly not an adiabatic invariant, since at any instant the raising of V at the point where the particle happens to be increases its potential energy without immediately affecting its kinetic energy; i.e. $\dot{E}=\dot{V}$, in which $\dot{V}$ must be understood as the rate of change of V at the location of the particle. The change of E and the deformation of

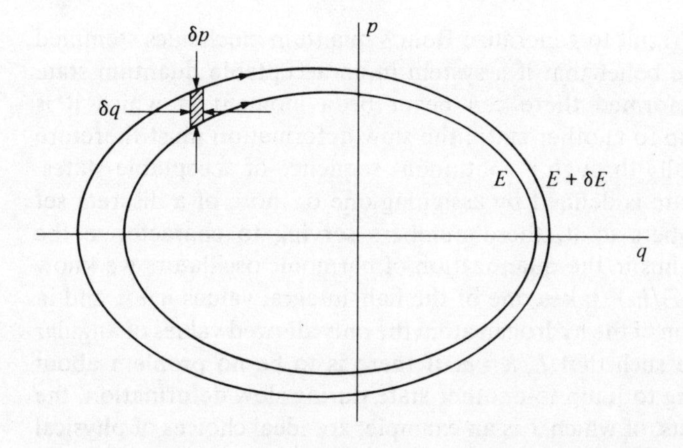

Fig. 3. Phase-plane trajectories for a classical anharmonic vibrator.

V both result in modifications of the shape of the phase-trajectory, but its area J, which may also be written $\oint p\,\mathrm{d}q$, remains invariant. To prove this result, let us consider the process in two stages – first we slowly make a small deformation, $\delta V(q)$, to V and calculate the resulting change in area, assuming E to remain constant; then we add the effect of the change in E that accompanies $\delta V(q)$. For the first stage, note that if (q, p) is a point on the original trajectory, where the velocity p/m is v, then $(q, p-\delta V(q)/v)$ is a point on the final trajectory of the same energy; the increase $\delta V(q)$ in potential energy must be compensated by a decrease in kinetic energy, resulting in a decrease of momentum δp, such that $v\delta p = \delta V$. In this process the area of the cycle is diminished by $\oint \delta p \cdot \mathrm{d}q$,

i.e. $$\delta J_1 = -\oint \delta V \cdot \mathrm{d}q/v.$$

For the second stage, note that if T is the periodic time of the vibration, the time $\mathrm{d}q/v$ spent by the particle in the element $\mathrm{d}q$ is a fraction $\mathrm{d}q/vT$ of the cycle; it therefore picks up this fraction of the change in V as it passes through $\mathrm{d}q$. Overall, $\delta E = \oint \delta V \cdot \mathrm{d}q/vT$ and we have for the corresponding change in area,

$$\delta J_2 = \left(\oint \delta V \cdot \mathrm{d}q/vT\right) \mathrm{d}J/\mathrm{d}E.$$

If we can show that $\mathrm{d}J/\mathrm{d}E = T$ it will follow that $\delta J_1 + \delta J_2 = 0$ and J is invariant under slow deformation of V. To achieve this final stage, consider two trajectories δE apart and imagine the representative point as it traverses the cycle sweeping out the area δJ of the enclosed annulus. Since a typical element of area is $\delta p\mathrm{d}q$, as shown in fig. 3, the rate of sweeping out of the annular area, $\mathrm{d}(\delta J)/\mathrm{d}t$, is $v\delta p$ which is constant and equal to δE; hence, integrating round the whole cycle, we arrive at the required result, $\delta J/T = \delta E$ or, in the limit,

$$\mathrm{d}J/\mathrm{d}E = T. \tag{11}$$

Consequently, J is an adiabatic invariant.

The use of this result to generalize Bohr's quantum mechanics stemmed from the plausible belief that if a system in an acceptable quantum state is very slowly deformed there can never be a moment at which it is stimulated to jump to another state; the slow deformation must therefore take it automatically through a continuous sequence of acceptable states. Each quantum state is defined by assigning one or more of a discrete set of quantum numbers to it, these numbers serving to characterize the dynamical state; thus in the quantization of harmonic oscillators we know that the quantity $E/\hbar\omega_0$ takes one of the half-integral values $n+\frac{1}{2}$, and in Bohr's quantization of the hydrogen atom the only allowed values of angular momentum L are such that $L/\hbar = n$. If there is to be no problem about the system needing to jump to another state during slow deformation, the adiabatic invariants, of which J is an example, are ideal choices of physical variables to which quantum numbers may be assigned. The two examples just mentioned illustrate this point. A harmonic oscillator has an elliptical phase trajectory with semi-axes q_0 (the amplitude) and $m\omega_0 q_0$; its area J is $\pi m\omega_0 q_0^2$, which is $2\pi E/\omega_0$. Quantization of J as $(n+\frac{1}{2})h$ quantizes E as $(n+\frac{1}{2})h\omega_0/2\pi$. As for an electron in a central orbit, the angular momentum is invariant under changes of the force law and is therefore a natural candidate for quantization. For the moment, however, we restrict the discussion to systems of one variable, for which the quantization rule of Sommerfeld and Wilson may be written:

$$J = \oint p\,\mathrm{d}q = (n+\gamma)2\pi\hbar, \tag{12}$$

the fractional correction γ being a later refinement.

For an isochronous vibrator $T = 2\pi/\omega_0$ and is constant, so that (11) gives J as $2\pi E/\omega_0$ and Planck's quantum law follows from (12) in the form $E = (n+\gamma)\hbar\omega_0$. Although the determination of γ is beyond the range of semi-classical quantum mechanics, it is a tribute to the power of the method that for the whole set of isochronous potentials in (1) γ was found to be constant for each choice of β, and for the whole range of β only varied between $\frac{1}{2}$ and $\frac{3}{4}$. Let us test it further by determining γ and its variation with energy for some non-isochronous potentials where the true answer can be found by reference to tabulated functions. First, the square well, with V equal to zero when $-\frac{1}{2}a < x < \frac{1}{2}a$ and infinite otherwise, so that the classical particle bounces back and forth at constant speed between hard walls. The phase trajectory is a rectangle of width $\Delta q = a$, and height $\Delta p = 2(2mE)^{\frac{1}{2}}$; hence $J = 2a(2mE)^{\frac{1}{2}}$. The energy of the nth level then follows from (12):

$$E_n = (n+\gamma)^2\pi^2\hbar^2/2ma^2, \tag{13}$$

which is to be compared with the correct result, $(n+1)^2\pi^2\hbar^2/2ma^2$, resulting from the requirement that the de Broglie wavelength shall be $2a/(n+1)$. In this case $\gamma = 1$, independent of energy. It may be noted that when $\beta = 0$ in (1), and the potential is parabolic on one side and hard on the other, the true value of $\gamma, \frac{3}{4}$, is the mean of $\frac{1}{2}$ for the former and 1 for the latter. It will become apparent in the following discussion of the WKB method that γ represents a correction arising from the boundary conditions satisfied by ψ at the ends of the trajectory, and that the value for the complete oscillation is always an arithmetic mean of contributions from the two ends. In saying this, we ignore the weak effect noted earlier, where slight fluctuations in γ were attributed to lack of smoothness of the potential.

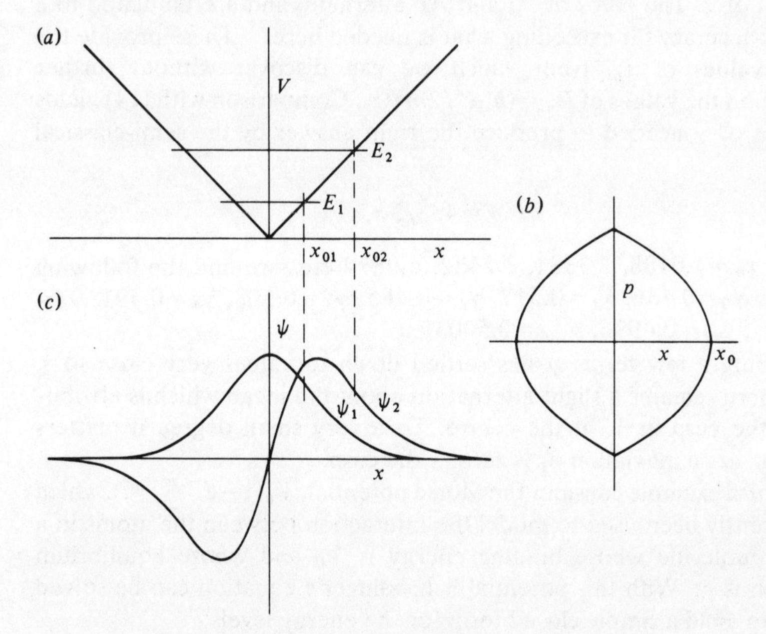

Fig. 4. A particle vibrating in the potential (a) has the classical phase-plane trajectory (b); the lowest two solutions of Schrödinger's equation are shown in (c), with the classical turning points indicated by vertical broken lines.

Next let us consider a symmetrical potential well with sloping sides, $V=\alpha|x|$, as in fig. 4(a). The classical particle experiences the same steady acceleration back towards the origin whichever way it is displaced. If the amplitude is x_0, $p^2=2m\alpha[x_0-|x|]$, giving phase trajectories in the form of two parabolic arcs, as in fig. 4(b). The area is $\frac{8}{3}x_0^{\frac{3}{2}}(2m\alpha)^{\frac{1}{2}}$ and therefore

$$J_n=(8/3\alpha)(2mE_n^3)^{\frac{1}{2}}=(n+\gamma)2\pi\hbar, \text{ from (12);}$$

consequently,
$$E_n=[\tfrac{3}{4}\pi\alpha\hbar(n+\gamma)]^{\frac{2}{3}}/(2m)^{\frac{1}{3}}. \tag{14}$$

To compare this with the true solution we note that V is symmetrical about the origin and that the solutions will have either $\psi(0)$ or $\psi'(0)$ equal to zero. It is necessary therefore to consider only solutions for positive x, with Schrödinger's equation in the form:

$$\hbar^2\psi''/2m+(E-\alpha x)\psi=0; \qquad x>0. \tag{15}$$

Now move the origin to the turning-point, $x_0=E/\alpha$, as shown in fig. 4(a), and write z for $(2m\alpha/\hbar^2)^{\frac{1}{3}}(x-x_0)$. Then

$$\mathrm{d}^2\psi/\mathrm{d}z^2-z\psi=0; \qquad z>-(2m/\hbar^2\alpha^2)^{\frac{1}{3}}E. \tag{16}$$

This is Airy's equation, whose solutions are extensively tabulated.[4] In the present case we need only note that the physically sensible solution is $\mathrm{Ai}(z)$, which decays smoothly to zero when $z>0$ and is oscillatory for $z<0$. The quantized energies are those that cause $\mathrm{Ai}(z)$ or its derivative to vanish when $-z=z_n=(2m/\hbar^2\alpha^2)^{\frac{1}{3}}E_n$; then the complete wave-function, of which fig. 4(*c*) shows examples, is constructed from this function and its mirror image, with reversed sign if necessary so that ψ and ψ' are continuous at the origin of x. The zeros of Ai and Ai′ alternate, and are tabulated to a degree of accuracy far exceeding what is needed here.[5] These provide the allowed values of z_n, from which we can discover without further computation the values of $E_n=(\hbar^2\alpha^2/2m)^{\frac{1}{3}}z_n$. Comparison with (14) yields the values of γ needed to produce the right answer by the semi-classical process:

$$n+\gamma=4z_n^{\frac{3}{2}}/3\pi,$$

and with $z_n=1.0188$, 2.3381, 3.2482, 4.0879 etc., we find the following sequence: $\gamma_0=0.436$, $\gamma_1=0.517$, $\gamma_2=0.485$, $\gamma_3=0.508$, $\gamma_4=0.491$, $\gamma_5=0.505,\ldots,\gamma_{99}=0.4996$, $\gamma_{100}=0.5003$ etc.

After only a few terms γ has settled down to values very close to $\frac{1}{2}$, though there remains a slight alternation about the mean which is attributable to the cusp in V at the centre. To a very small degree it matters whether ψ has a maximum or is zero at the cusp.

For a third example consider the Morse potential, $V_0(1-\mathrm{e}^{-a(r-r_0)^2})$, which has frequently been used to model the interaction between the atoms in a diatomic molecule whose binding energy is V_0 and whose equilibrium separation is r_0. With this potential Schrödinger's equation can be solved in series to yield a simple closed form for the energy levels[6]:

$$E_n=(n+\tfrac{1}{2})\hbar\omega_0-\tfrac{1}{4}(n+\tfrac{1}{2})^2\hbar^2\omega_0^2/V_0, \qquad (17)$$

in which ω_0 is $a(2V_0/m)^{\frac{1}{2}}$, the frequency for small-amplitude vibrations of the corresponding classical vibrator. The general expression for the variation of frequency with energy is readily derived by integrating (2.11): $\omega(\varepsilon)=\omega_0(1-\varepsilon)^{\frac{1}{2}}$, where $\varepsilon=E/V_0$, and the phase integral is equally readily evaluated:

14

$$\oint p\,\mathrm{d}q=(2\pi\omega_0 m/a^2)[1-(1-\varepsilon)^{\frac{1}{2}}].$$

Semi-classical quantization by use of (12) agrees perfectly with (17) if $\gamma=\frac{1}{2}$.

These examples suffice to illustrate the merit of the semi-classical approach, and if we now proceed to yet another it is more to draw attention to the perturbation technique which is of value when the form of V is close to one for which Schrödinger's equation is soluble in terms of well-understood functions. The mathematical procedure is described in standard texts,[7] and we need only quote a typical result, choosing a symmetrical, slightly anharmonic, potential in which the quadratic term is supplemented by a quartic:

$$V = \tfrac{1}{2}m\omega_0^2 x^2 + \beta x^4. \tag{18}$$

The perturbation approach, which expresses ψ as a Fourier series, using the harmonic oscillator wave-functions as basis, yields the energy levels as a power series in β. We shall assume β small enough to allow the series to be terminated after the first-order correction; then

$$E_n \approx \hbar\omega_0[(n+\tfrac{1}{2})+(n^2+n+\tfrac{1}{2})\Delta], \tag{19}$$

where $\Delta = 3\beta\hbar/2m^2\omega_0^3$. Let us see what value γ must take in the semi-classical approach to give the same answer. If $\beta = 0$, the trajectory for each E is an ellipse of area $2\pi E/\omega_0$, and the first-order correction due to β is easily calculated. At any value of p, (18) shows that x must be reduced by δx, equal to $\beta x^3/m\omega_0^2$, if V and hence E are to be unchanged. The resulting reduction in area is $\oint \delta x \cdot \mathrm{d}p$, and for the first-order correction p can be written in its unperturbed form, $[2m(E-\tfrac{1}{2}m\omega_0^2x^2)]^{\frac{1}{2}}$. Hence we find that

$$J = 2\pi E/\omega_0 - 3\pi\beta E^2/m^2\omega_0^5 = 2\pi E(1-E\Delta/\hbar\omega_0)/\omega_0.$$

If E_n is given by (19),

$$J_n \approx (n+\tfrac{1}{2}+\tfrac{1}{4}\Delta)2\pi\hbar, \text{ to first order in } \Delta,$$

which accords with (12) if $\gamma = \tfrac{1}{2}+\tfrac{1}{4}\Delta$, independent of n. The semi-classical solution is perfectly correct as far as it goes – the true energy levels (at least to first order in β) correspond to exactly equal increments of J; since there is no pretence of calculating γ, the solution cannot be faulted for not predicting the extent to which it differs from $\tfrac{1}{2}$. It will be observed that the correction to γ is positive, as might be expected from the increased steepness of the potential resulting from the quartic terms. To obtain some idea of the magnitude of the correction, let us suppose that the quartic term is large enough to raise the 5th level by $\hbar\omega_0$, i.e. in (19) let $(n^2+n+\tfrac{1}{2})\Delta$ be unity when $n=5$. This is a substantial anharmonicity, but Δ is still only 3.3×10^{-2}, and $\gamma = 0.508$, very little affected in fact.

WKB approximation

In almost all these tests the semi-classical approach, with γ put equal to $\tfrac{1}{2}$, has shown itself remarkably successful in predicting the energy levels, but of course it does not yield the wave-functions and is therefore powerless for any other quantum-mechanical calculation such as the expectation values of physical quantities or transition probabilities under the influence of time-dependent perturbations. Direct integration by computer of Schrödinger's equation in one dimension presents no problems and is always available if a certain numerical result is all that is wanted. If, however, an analytical form is desired it is worth bearing in mind that the WKB approach[8] offers an intermediate stage of approximation which may be good enough, especially for high quantum numbers. In its most elementary form, which is the form that is most generally useful, the solution of the equation

$$\psi'' + \psi f(x) = 0$$

is written as
$$\psi \approx C f^{-\frac{1}{4}} \exp\left[i \int_{x_0}^{x} f^{\frac{1}{2}}\,\mathrm{d}x\right] \tag{20}$$

This result is not peculiar to wave mechanics, but is a useful approximate form that applies to any problem of wave propagation along a line where the properties of the medium vary with x; $f^{\frac{1}{2}}$ is the local value of the wavenumber, and $\int_{x_0}^{x} f^{\frac{1}{2}}\,\mathrm{d}x$ is the phase difference between x_0 and x, on the assumption that the local wavenumber, in spite of varying with x, still describes the local rate of change of phase with x. Since the phase and group velocities of the wave change with x one must expect the amplitude also to change in order to conserve the flux of whatever property is being transported by the wave, and the coefficient $f^{-\frac{1}{4}}$ takes care of this.† The solution may conveniently be illustrated and checked by reference to Airy's equation (16), for which $f(x) = -x$ as in fig. 4(a), and for which (20) may be written in the form

$$\psi = C\bar{x}^{-\frac{1}{4}} \exp\left(\tfrac{2}{3}\mathrm{i}\bar{x}^{\frac{3}{2}}\right). \tag{21}$$

Here x_0 is incorporated in the complex constant C, and $\bar{x}$ is written for $-x$. This describes the oscillatory solution for negative x, the region in which $V < E$ and a classical particle can be found. As written, (21) describes a travelling wave accompanying the particle as it descends the potential slope, but for a particle bound in a V-shaped potential a standing wave is appropriate, as described by Re $[\psi]$. Whatever the phase of C, Re $[\psi]$ is a sinusoid of varying wavelength and with an amplitude enclosed within an envelope that varies as $\bar{x}^{-\frac{1}{4}}$; but the phase of C determines the placing of the zeros. This raises a question that cannot be answered within the framework of the elementary WKB theory. It is necessary to continue the solution through the point $x = 0$ into the classically inaccessible region where the wave-function must decay rapidly and vanish, and this decay occurs only if the zeros are correctly placed. The mathematical analysis of this problem by development of asymptotic expansions for the solution of (16) is subtle and historically important, but now that extensive tabulations of Airy functions are available we may appreciate the physical point merely by looking at graphs. Fig. 5 shows the physically valid function, Ai (x), which vanishes at $+\infty$, and for comparison the WKB solutions on either side of the origin, C being chosen for the latter so as to make the match

† $\psi\psi^* \propto f^{-\frac{1}{2}}$, and since $f = 2m(E-V)/\hbar^2$, $f^{\frac{1}{2}}$ is proportional to the velocity of a classical particle at x. The WKB solution thus reproduces in its amplitude variations the classical expectation of finding the particle at any point.

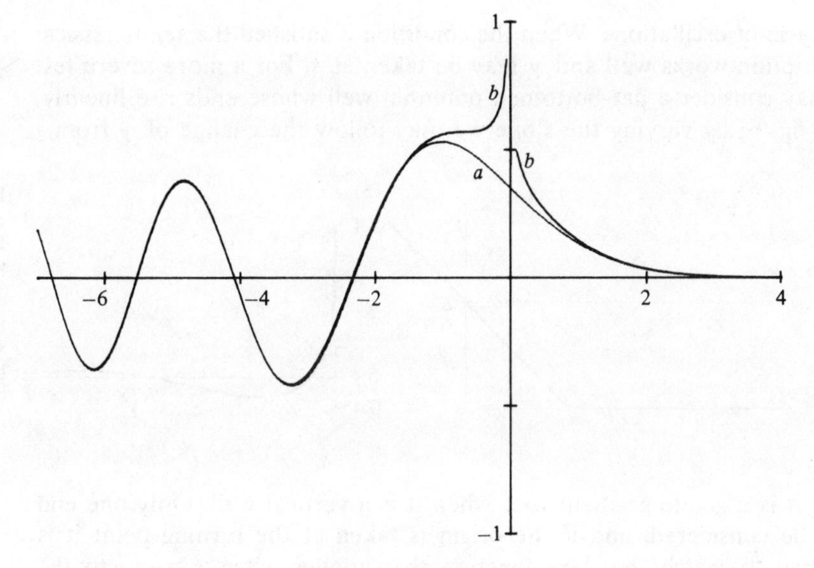

Fig. 5. Solution of Airy's equation: (*a*) exact, (*b*) WKB approximation fitted asymptotically on both sides.

as good as possible far from the origin:

$$\begin{aligned} \psi_{\mathrm{WKB}} &= \pi^{-\frac{1}{2}}\bar{x}^{-\frac{1}{4}} \sin\left(\tfrac{2}{3}\bar{x}^{\frac{3}{2}}+\tfrac{1}{4}\pi\right); \qquad x<0 \\ &= \tfrac{1}{2}\pi^{-\frac{1}{2}}x^{-\frac{1}{4}} \exp\left(-\tfrac{2}{3}x^{\frac{3}{2}}\right); \qquad x>0. \end{aligned} \tag{22}$$

These expressions are in fact just the first terms in the asymptotic expansions of Ai (x).

It is clear from the diagram that the WKB solution is extremely good except near the origin, where the local wavenumber (or extinction coefficient for positive x) suffers a relatively large change in the course of a wavelength. The analysis has shown that to a very good degree of approximation one may calculate the energy levels by treating the de Broglie wave as a wave on a resonant line, using the local wavenumber to determine the phase length and adding an eighth of a wavelength to each end (the $\frac{1}{4}\pi$ in the first equation of (22)) as end corrections. This prescription is indeed exactly the semi-classical quantization process, the value of $\frac{1}{2}$ assigned to γ being this end correction. We have already seen from the tabulated values of γ how good the prescription is, and now in fig. 5 we see how to proceed beyond this point and obtain by the WKB process a generally good picture of the wave-function itself.

All this, of course, has been developed by reference to one equation only, but there is no difficulty in believing that different smooth variations of V will be equally well served by the WKB approximation. Only at the classical turning point, where $E=V$, does it break down completely, and then it may well be possible to treat V as varying linearly with x through this critical range, so that the WKB solutions on either side are matched to the Airy function in the range itself. For this to work satisfactorily it is only necessary that V shall be nearly linear for a distance covering at least one cycle of oscillations. When the condition is satisfied the semi-classical prescription works well and γ may be taken as $\frac{1}{2}$. For a more severe test we may consider a flat-bottomed potential well whose ends rise linearly, as in fig. 6; by varying the slope we may follow the change of γ from $\frac{1}{2}$

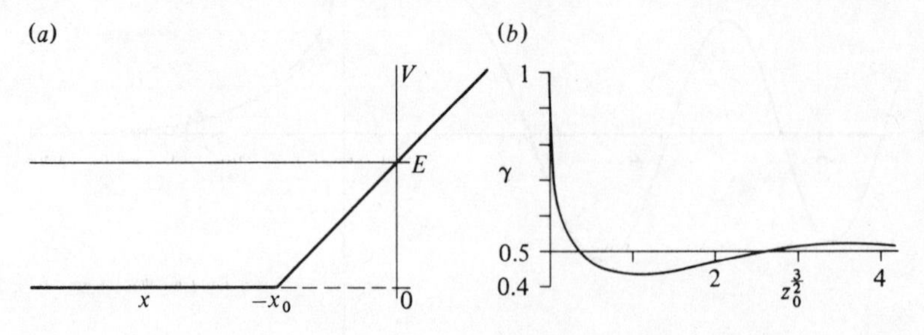

Fig. 6. Illustrating the end-correction γ for a flat-bottomed potential well with sloping ends (curve a); $z_0^{\frac{3}{2}}$ in curve b is the phase length of the sloping section of V between $-x_0$ and the classical turning point, 0.

when it is a gentle gradient to 1 when it is a vertical wall. Only one end need be considered, and if the origin is taken at the turning-point it is required to match the Airy function that applies when $x>-x_0$ to the sinusoidal standing wave that applies when $x<-x_0$. Writing α for the slope of the end, E/x_0, and z for $-(2m\alpha/\hbar^2)^{1/3}x$, we have

$$\left.\begin{aligned}\psi &= \mathrm{Ai}\,(z);\ z>-z_0\\ &C\cos(pz+\theta);\ z<-z_0\end{aligned}\right\} \tag{23}$$

where $z_0=(2m\alpha/\hbar^2)^{\frac{1}{3}}x_0\propto E^{\frac{1}{3}}x_0^{\frac{2}{3}}$, and $p=(\hbar^2/2m\alpha)^{\frac{1}{3}}(2mE)^{\frac{1}{2}}/\hbar=z_0^{\frac{1}{2}}$. In order that ψ and ψ' shall be continuous at $-z_0$,

$$z_0^{\frac{1}{2}}\tan(-z_0^{\frac{3}{2}}+\theta)=-\mathrm{Ai}'\,(-z_0)/\mathrm{Ai}\,(-z_0).$$

This equation determines θ and hence the placing of the standing wave in relation to the turning-pont, $z=0$:

$$\theta=z_0^{\frac{3}{2}}-\tan^{-1}[\mathrm{Ai}'\,(-z_0)/z_0^{\frac{1}{2}}\,\mathrm{Ai}\,(-z_0)]. \tag{24}$$

Now the asymptotic form (22) of Ai (z) at large negative values is $\cos[\frac{1}{4}\pi-\frac{2}{3}(-z)^{\frac{3}{2}}]$, from which it follows that when the slope of V is extremely gentle,

$$\theta\approx\tfrac{1}{4}\pi+\tfrac{1}{3}z_0^{\frac{3}{2}}.$$

The term $\frac{1}{3}z_0^{\frac{3}{2}}$ is the shift of the standing wave caused by the sloping potential, relative to the position it would adopt if V remained level right up to the end, $z=0$; that is to say, the difference between $z_0^{\frac{3}{2}}$ for constant V and $\frac{2}{3}z_0^{\frac{3}{2}}$ from (21). The first term, $\frac{1}{4}\pi$, is the phase correction we have come to expect for a linear potential variation. When z_0 is not so large the phase correction is $\theta-\frac{1}{3}z_0^{\frac{3}{2}}$. Since a phase correction of π brings the standing wave back to its original position, and since γ is defined to allow for both ends of the potential well, γ is $2/\pi$ times this phase correction:

$$\gamma=\frac{2}{\pi}\{\tfrac{2}{3}z_0^{\frac{3}{2}}-\tan^{-1}[\mathrm{Ai}'\,(-z_0)/z_0^{\frac{1}{2}}\,\mathrm{Ai}\,(-z_0)]\}. \tag{25}$$

Tables of the Airy function[4] allow γ to be calculated very readily, with the result shown in fig. 6(*b*); the abscissa, $z_0^{\frac{3}{2}}$, is the phase length of the sloping section for a wave whose wavenumber is $(2mE)^{\frac{1}{2}}/\hbar$, as on the level potential, and it is at first sight rather surprising how short a sloping section ($\frac{1}{20}$ of a wavelength) is needed to bring γ down from unity to something close to $\frac{1}{2}$. The explanation lies in the ease with which the wave-function spreads into the classically forbidden region. Since the extinction coefficient depends on the square root of the energy deficit, i.e. $\psi \sim e^{-\mu x}$ where $\mu = [2m(V-E)]^{\frac{1}{2}}/\hbar$, a very steep gradient of V is needed to force ψ down so sharply that a node of the standing wave lies close to the turning-point. It is easy to show that, for small z_0, $\gamma \approx 1 - 2z_0^{\frac{1}{2}}/\pi a$, so that for a given choice of E, $(1-\gamma)$ begins to rise as $x_0^{\frac{1}{3}}$. It is evident that the WKB method, with γ fixed at $\frac{1}{2}$, is considerably more successful than one would have dared hope.

Perhaps the most important physical problems involving anharmonic vibrators arise from the spectroscopy of diatomic molecules. For small amplitudes of vibration these behave reasonably well as harmonic oscillators, and the level spacing, revealed by their spectra, is nearly constant (if we leave out of account the centrifugal stretching effects due to rotation). At higher excitations, however, the potential well is wider than required for isochronous oscillation, as shown in fig. 7, and the frequency and level

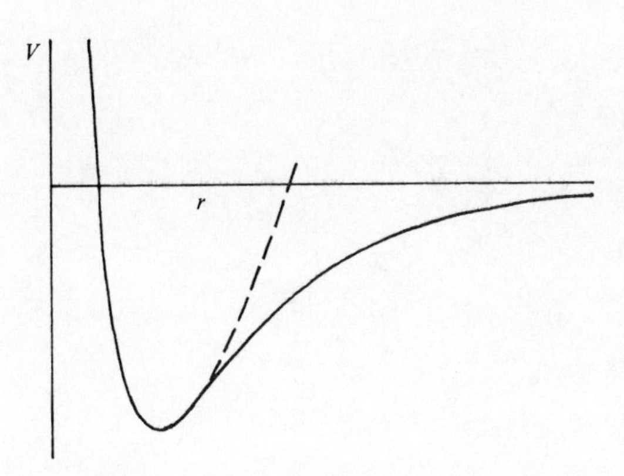

Fig. 7. Typical potential energy of interaction between two atoms (full curve) compared with an isochronous potential (broken curve).

spacing decrease. At a certain energy of excitation the molecule dissociates, and just below this the vibrational frequency is quite low. In principle it should be possible to derive a great deal of information on the shape of the potential well from the level spacing – not everything since, as with the classical oscillator, shearing of the potential curve has little or no effect on the levels. But the repulsive potential at close separations of the atoms is so steep that any plausible form for this leads to virtually the same shape at great distances, and the difficulty is not very serious. A good account of the processes available to determine $V(r)$ has been given by Le Roy,[9] and the reader will be struck by the central role played by extensions of the WKB method which, in contrast to a full wave-mechanical treatment, are manageable with fairly modest resources for computation. As a result of this work, $V(r)$ is known with considerable precision for a number of diatomic molecules; without it almost everything one could say about the details of $V(r)$ would be guesswork.

15 Vibrations and cyclotron orbits in two dimensions

For a particle moving in the quadratic potential, $V=\frac{1}{2}\mu_x x^2+\frac{1}{2}\mu_y y^2$, Schrödinger's equation takes the form, in Cartesian co-ordinates,

$$\hbar^2(\partial^2\psi/\partial x^2+\partial^2\psi/\partial y^2)/2m+(E-\tfrac{1}{2}\mu_x x^2-\tfrac{1}{2}\mu_y y^2)\psi=0, \tag{1}$$

which is separable by writing ψ as a product, $\psi(x, y)=X(x)Y(y)$; X and Y each obey the oscillator equation (13.1):

$$\hbar^2 X''/2m+(E_x-\tfrac{1}{2}m\omega_x^2 x^2)X=0; \qquad \hbar^2 Y''/2m+(E_y-\tfrac{1}{2}m\omega_y^2 y^2)Y=0,$$

in which $\omega_{x,y}^2=\mu_{x,y}/m$, and E_x+E_y must equal E. Thus each direction of motion is independently quantized, and the energy levels follow the rule:

$$E=(n_x+\tfrac{1}{2})\hbar\omega_x+(n_y+\tfrac{1}{2})\hbar\omega_y. \tag{2}$$

A typical wave-function for a stationary state, being the product of two oscillator functions, has its general form defined by a rectangular lattice of unequally spaced nodal lines, parallel to the axes, with the sign of ψ at the antinodes alternating from one cell to the next. Fig. 1(*a*) is a typical example in which $\omega_y=\frac{3}{4}\omega_x$, $n_x=5$ and $n_y=2$. The energy levels may be represented as in fig. 1(*b*), with each level n_x serving as the base for a ladder of n_y. Alternatively, as fig. 1(*c*) shows, they may be represented as a rectangular grid of points having co-ordinates $(n_x+\frac{1}{2})\hbar\omega_x$ and $(n_y+\frac{1}{2})\hbar\omega_y$; states of the same total energy lie on lines at 45°, and the linear increase in density of states with E is clearly seen in the increased length of these lines.

435, 445

The parallel between classical and quantum–mechanical behaviour discussed in chapter 13 carries through from one to two dimensions, and indeed to three or more in a potential which is a positive definite quadratic form. If the system is excited from its ground state by the application of a uniform, time-dependent force, the resulting wave-function (not a stationary state, of course) takes the form of a wave-packet of the same elliptical Gaussian shape as the ground state wave-function, moving in the classical trajectory; in general this will be an evolving Lissajous figure, if ω_x and ω_y are not simply related. In the special case of an isotropic oscillator, when $\omega_x=\omega_y$, the wave-packet, which now has circular symmetry, executes a steady elliptical trajectory if left alone after excitation. Unlike a wave-packet moving in a constant potential, which gradually spreads as it travels, this particular solution of Schrödinger's equation for a harmonic potential (not necessarily isotropic) keeps its compact form indefinitely.

28

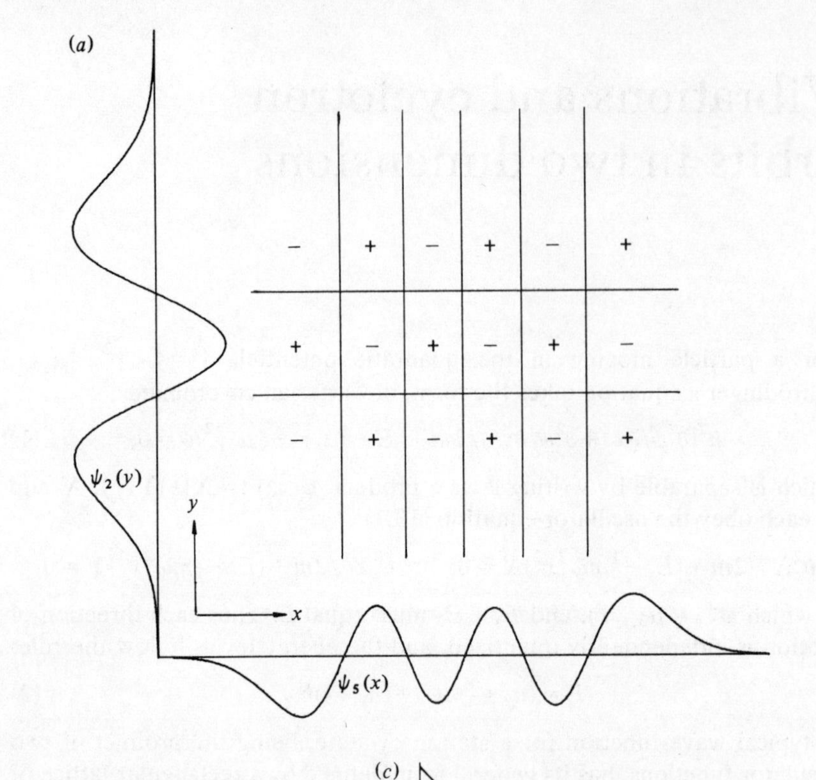

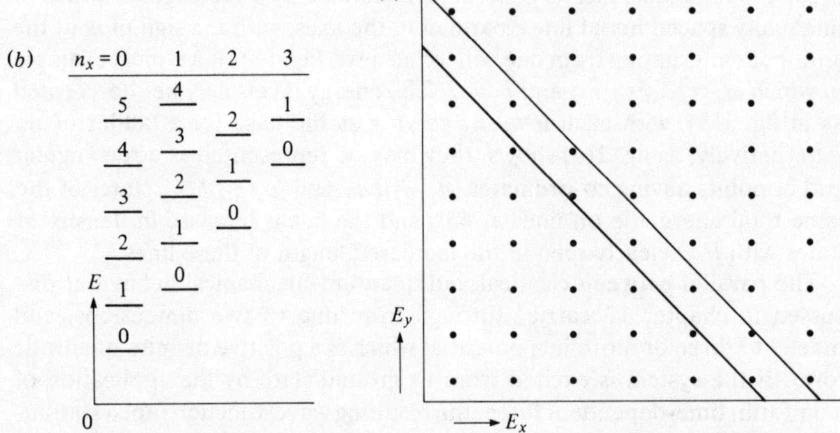

Fig. 1. (a) Wave-function for a two-dimensional harmonic vibrator. The stationary state is represented by a product wave-function $\psi_5(x)\psi_2(y)$, whose nodal lines form a rectangular grid; the antinodes take opposite signs in alternate cells. (b) Energy level diagram for the case $\omega_y = \frac{3}{4}\omega_x$; each level corresponds to a wave-function $\psi_{n_x}(x)\psi_{n_y}(y)$. Levels for each different value of n_x are shown as separate ladders, with the values of n_y attached to the levels themselves. (c) The same set of levels as points on a two-dimensional diagram. All points on a line at 45° have the same total energy, and the density of states is equal to the number of points between two lines corresponding to unit difference in total energy.

To look at this point from another aspect let us consider building up the wave-packet from stationary states. Since we seek illustration of the principle rather than useful results, it is sufficient to take the especially simple case of a circular trajectory, for which purpose the solutions of the equation in polar co-ordinates are appropriate. In the isotropic case $V = \frac{1}{2}\mu r^2$, and the equation to be solved has the form:

$$(\hbar^2/2m)\left[\frac{1}{r}\frac{\partial}{\partial r}\left(r\frac{\partial\psi}{\partial r}\right) + \frac{1}{r^2}\frac{\partial^2\psi}{\partial\phi^2}\right] + (E - \tfrac{1}{2}\mu r^2)\psi = 0. \qquad (3)$$

The co-ordinates are separable, and it is convenient to write the solutions in the form

$$\psi = \frac{1}{r^{\frac{1}{2}}} f(r)\, e^{il\phi},$$

in which l is a positive or negative integer,† and to introduce a new measure of radial distance by writing ξ^2 for $2(m\mu)^{\frac{1}{2}}r^2/\hbar$. Then (3) becomes an equation for $f(\xi)$ alone:

$$f'' + [\varepsilon - (\tfrac{1}{2}\xi - \beta/\xi)^2]f = 0,$$

in which $\varepsilon = E/\hbar\omega_0 - \beta$ and $\beta^2 = l^2 - 1/4$. This is the same as (14.1), which need cause no surprise since the potential function in (14.1) arises in the classical treatment of radial motion of a conical pendulum, being the sum of the restoring potential $\frac{1}{2}\mu r^2$ and the centrifugal potential proportional to r^{-2}. The energy levels follow immediately from (14.5):

$$E = (2\nu + 1 + |l|)\hbar\omega_0, \tag{4}$$

with $\nu = 0$ or any positive integer. As expected from the solution in Cartesian co-ordinates, the levels are equally spaced with total zero-point energy $\hbar\omega_0$, $\frac{1}{2}\hbar\omega_0$ for each degree of freedom. All but the lowest level exhibit degeneracy, the level for which $E = n\hbar\omega_0$ being n-fold degenerate, since there are n ways in which $n-1$ units of $\hbar\omega_0$ may be distributed between the two Cartesian co-ordinates. It is this that allows different representations of the stationary-state wave-functions to be constructed by linear combination of the members of a degenerate set.

For the present argument the polar representation is suitable, providing a basis for synthesizing a wave-packet moving on a circular classical orbit. An excited state with $\nu = 0$ has the highest value of $|l|$ compatible with the energy, and therefore the highest value of β, so that the wave-function is kept as far from the origin as possible. The solution then follows the form (14.2) with $F(\xi)$ a constant:

$$\psi = C\, e^{-\frac{1}{4}\xi^2}\xi^{|l|}\, e^{il\phi}. \tag{5}$$

When $|l|$ is large, and the energy is $(|l|+1)\hbar\omega_0$, ψ is large only in the vicinity of $\xi_c = (2|l|)^{\frac{1}{2}}$, which is the classical radius for circular motion with energy $|l|\hbar\omega_0$, the energy of excitation above the ground state at $\hbar\omega_0$. Approximately, $\psi \sim e^{-\frac{1}{4}(\xi-\xi_c)^2}\, e^{il\phi}$, the Gaussian form of the radial variation being the same as in the ground state but centred on ξ_c. Already, therefore, we have proceeded halfway towards constructing the required wave-packet, having achieved a narrow Gaussian radial spread but with no angular restrictions; at this stage the wave-function may be likened to a model railway track arranged in a circle, with narrow radial wavefronts like the

† The axially symmetric case, $l = 0$, needs special treatment, but as it presents no difficulties and is not required for the present argument we shall ignore it.

sleepers. Angular confinement is effected by combining similar wave-functions having $\nu = 0$ and a range of values for l, so that at any instant constructive interference occurs only near some special value of ϕ. We may imagine the superposition of waves sweeping round very nearly the same circular track, but with a range of wavenumber, k, and of frequency, Ω. Since $k \propto l$ while $\Omega = E/\hbar$ and is linear in l, the group velocity $\mathrm{d}\Omega/\mathrm{d}k$ is independent of k. This is just the required condition for a wave-packet to propagate unchanged in form, which is what we set out to demonstrate.

Extension of the theory of a harmonic vibrator from one dimension to two has added virtually nothing that is not intuitively obvious, once it is understood how closely the classical and quantal treatments agree. Where classical arguments work in one dimension they work in two; where they fail, the extra dimension does nothing to redeem the failure.

Fermi resonance[1]: non-linear coupling[2]

It was noted in chapter 12 that vibrational energy levels in some molecules are significantly perturbed by parametric coupling, for example when a stretching mode has something like twice the frequency of a bending mode, as in CO_2. The effect is simply modelled as the motion of a particle in the potential well of fig. 2(a). If the valley were a parabolic minimum (elliptical contours, as in the preceding section) there would be independent normal modes of vibration parallel to the x and y axes; and if the frequency along y were exactly twice that along x the Lissajous figures for arbitrary initial conditions would be closed curves like those in fig. 2.17(a)–(e). Curvature of the valley, however, couples the x and y motions, and there are no longer any normal modes; they cannot be expected except with quadratic potentials, and now the potential contains cubic and quartic terms. For small curvature we may choose as our model potential

378

28

$$V = \tfrac{1}{2}\mu x^2 + 2\mu(y - cx^2)^2. \tag{6}$$

The valley, defined as the line on which $\partial V/\partial y = 0$, is the parabola $y = cx^2$. The classical equations of motion can be expressed in the form

$$\left.\begin{aligned} &\ddot{\xi} + \xi - 8\xi(\eta - \xi^2) = 0 \\ \text{and} \qquad &\ddot{\eta} + 4(\eta - \xi^2) = 0, \end{aligned}\right\} \tag{7}$$

in which $(m/\mu)^{\frac{1}{2}}$ is taken as the unit of time, and $1/c$ as the unit of length.

The behaviour resulting from (7) is complicated, with exchange of energy between the ξ and η vibrations, and it is very helpful in appreciating the problem to make and study a real model. A simple pendulum in which the string is either replaced by, or incorporates, a spring capable of great extension is loaded until the extension of the spring is one-quarter the total length of the pendulum; then the vertical vibration has twice the frequency of the swing. No great care need be wasted on obtaining an exact ratio of

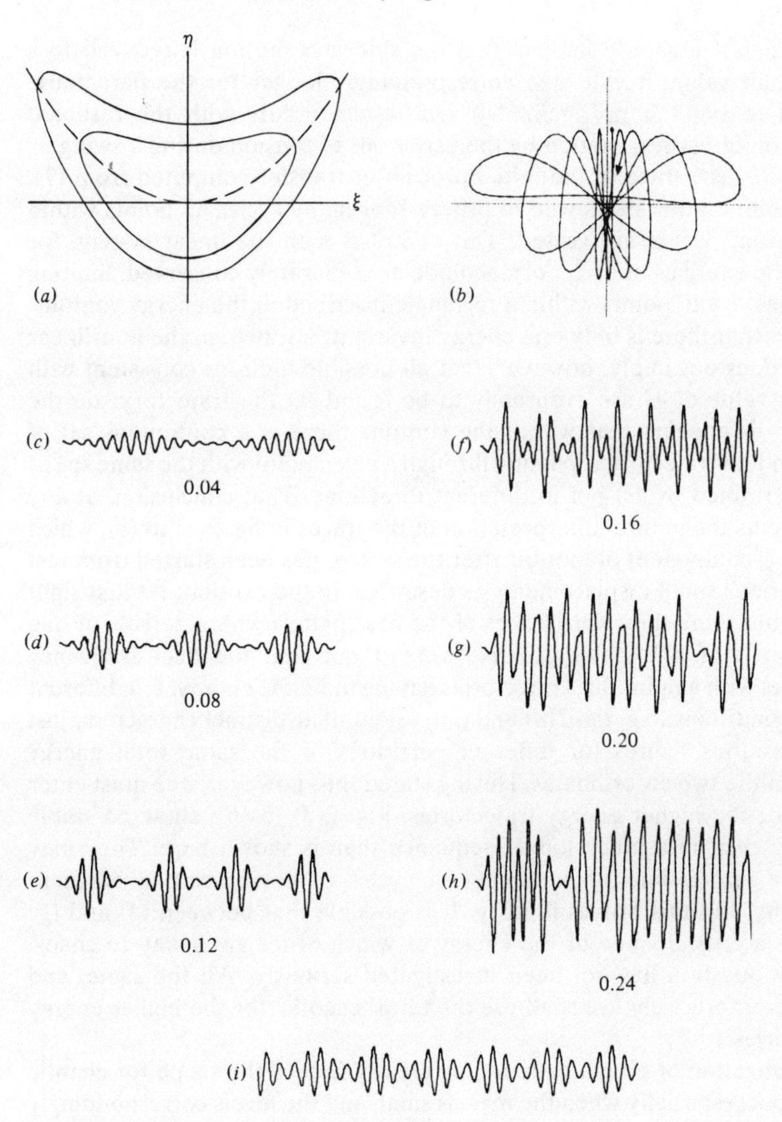

Fig. 2. (*a*) The anharmonic potential (6); the full lines are equipotentials and the broken line indicates the bottom of the valley. (*b*) Computed trajectory for the particle starting from rest at (0.01, 0.12). (*c*)–(*h*) $\xi(t)$ for the particle starting from rest at (0.01, η_0); values of η_0 are given below the curves. (*i*) $\xi(t)$ for the particle starting at the origin with initial velocity $(\dot{\xi}_0, \dot{\eta}_0) = (0.16, 0)$; the total energy is the same as in (*d*), but the beat period is noticeably shorter.

two between the frequencies – the behaviour is not affected by small departures from ideality. When the bob of the pendulum is pulled down and released, there is likely to be a component of the initial horizontal motion, albeit small, that is in the right phase to be amplified parametrically; amplification proceeds until the vertical motion has almost completely been converted into a horizontal swing. The variations of tension, at twice the frequency of swing, in their turn begin to re-establish the vertical vibration, and this resonant process proceeds until nearly all the sideways motion is lost again. The process then repeats itself almost in the same form; only the fact that each interchange of energy from one mode to the other is not quite complete gives rise to a slight variability from one 'beat' cycle to the next. Thus if it should happen that the sideways motion is reduced to a very small value, it will take correspondingly longer for the parametric process to build it up again. No such effect occurs with the resonant excitation of vertical motion by the variations of tension during a swing.

Fig. 2(*b*) is a trace of the initial process of transfer computed from (7). If the computation is allowed to proceed for many cycles all points within the contour $V = E$ are visited. This contrasts with the linear system, for which the energies of each normal mode are separately conserved, limiting the trajectory to points within a rectangle inscribed in the energy contour. The fact that there is only one energy invariant, not two, in the non-linear system does not imply, however, that all possible motions consistent with a given value of E are eventually to be found on the trajectory; on the contrary it appears that within the contour there is a continuous set of independent trajectories, passing through a given point with the same speed (as determined by E) but in different directions. That conclusion, at any rate, seems the natural interpretation of the traces in fig. 2(*c*) to (*e*), which are the ξ-component of motion after the system has been started from rest with various small displacements as described in the caption. At first sight one would think they were traces of the beat pattern characteristic of one component of a Lissajous figure, were it not that the beat frequency increases with amplitude. Trajectories having the same energy, but different initial conditions (e.g. fig. 2(*d*) and (*i*)), retain their distinct characters, just like Lissajous figures for different partitions of the same total energy between the two co-ordinates. Having stated this, however, one must enter a caveat; the higher energy trajectories, Fig. 2(*f*) to (*h*), show no stable pattern, even in a much longer sequence than is shown here. They may give the impression for a number of cycles of having settled down to regularity, but this proves illusory. It is possible that between (*f*) and (*g*) there is a critical value of the energy at which order gives way to chaos, but this question has not been investigated seriously. All the same, and however unprecisely, we shall use the term 'chaotic' for the higher energy trajectories.

270

Quantization of such a system may greatly reduce the scope for chaotic behaviour, especially when the mass is small and the levels correspondingly far apart. On the molecular scale, in fact, quite low excited levels already have energies that would lead to chaotic behaviour in a classical system. The quantal treatment is now markedly simpler than the classical and involves only elementary perturbation theory. It may be noted that the semi-classical approach offers no help here, since the motion is aperiodic and outside the scope of the Bohr–Wilson–Sommerfeld method, while the WKB solution runs into the same problems as the classical, that there is no explicit expression for the trajectories generated by (7).

456
461

The Hamiltonian derived from (6) has the form:

$$\mathscr{H} = (p_x^2 + p_y^2)/2m + \tfrac{1}{2}\mu x^2 + 2\mu(y - cx^2)^2,$$

and leads to Schrödinger's equation:

$$\partial^2\psi/\partial X^2+\partial^2\psi/\partial Y^2+[\varepsilon-X^2-4(Y-\gamma X^2)^2]\psi=0, \tag{8}$$

in which the reduced co-ordinates are not the same as in (7). Here

$$\left.\begin{aligned} X=\alpha x=\xi/\gamma,\quad Y=\alpha y=\eta/\gamma,\quad \varepsilon=2E/\hbar\omega_0=(2E/\hbar)(m/\mu)^{\frac{1}{2}},\\ \gamma=c/\alpha\quad\text{and}\quad \alpha=(\mu/\hbar\omega_0)^{\frac{1}{2}}.\end{aligned}\right\} \tag{9}$$

In this form γ is seen as the coupling constant between the otherwise independent harmonic vibrations of X and Y. To relate the magnitudes in the classical and quantal treatments, note the suggestion that chaotic behaviour sets in when η_0, the initial value of η, is about 0.16. The energy of vibration is then $2\mu\eta_0^2/c^2$, so that $\varepsilon=4\mu\eta_0^2/\hbar\omega_0c^2=4\eta_0^2/\gamma^2 \doteqdot 0.1/\gamma^2$. Thus if $\gamma=0.1$ the state $\varepsilon=10$, which is the second excited level of the Y-vibration at a frequency $2\omega_0$, has enough energy to cause trouble in the classical system. Let us bear this in mind when interpreting the outcome of the following first-order perturbation treatment of (8).

When $\gamma=0$, (8) is separable and ψ is the product of harmonic oscillator functions:

$$\psi(X,Y)=\psi_n(X)\psi_m(Y);\qquad \varepsilon=2n+4m+2. \tag{10}$$

The level $\varepsilon=2(N+1)$ has a degeneracy of $\frac{1}{2}N+1$ if N is even, and $\frac{1}{2}(N+1)$ if N is odd. The effect of a small non-vanishing γ is primarily to break this degeneracy, creating new states which are linear combinations of degenerate states. To first-order the admixture of other non-degenerate states may be neglected, and we may write as a trial solution

$$\psi_N(X,Y)=\lambda_0\psi_N(X)\psi_0(Y)+\lambda_1\psi_{N-2}(X)\psi_1(Y)+\cdots. \tag{11}$$

On substituting this in (8) we find

$$[\varepsilon-2(N+1)+8\gamma X^2Y](\lambda_0\psi_N\psi_0+\lambda_1\psi_{N-2}\psi_1+\cdots)=0, \tag{12}$$

if the term in γ^2 is dropped as irrelevant to the first-order calculation. Now multiply the left-hand side by $\psi_{N-2m}\psi_m$ and integrate over both variables:

$$\lambda_m\Delta+8\gamma\sum_l M^{(2)}_{N-2l,N-2m}M^{(1)}_{l,m}\lambda_l=0, \tag{13}$$

in which Δ is written for the displacement of the energy level, $\varepsilon-2(N+1)$, and $M^{(1)}$ and $M^{(2)}$ are the matrix elements as defined in (13.26) and (13.24). It should be noted, however, that $M^{(2)}$ relates here to an oscillator of frequency $2\omega_0$, and that correspondingly $M^{(2)}_{n,n-2}=\frac{1}{4}[n(n-1)]^{\frac{1}{2}}$. The only non-vanishing terms in (13) are those for which $l=m\pm1$, and hence

$$\lambda_{m-1}\gamma a_m+\lambda_m\Delta+\lambda_{m+1}\gamma a_{m+1}=0, \tag{14}$$

where $a_m=[2m(N-2m+1)(N-2m+2)]^{\frac{1}{2}}$. This set of equations for the unknowns λ_m, with m running from 0 to $\frac{1}{2}N$ or $\frac{1}{2}(N-1)$, has a solution only if the determinant of the coefficients vanishes:

$$\begin{vmatrix} \Delta & \gamma a_1 & 0 & 0 & \cdots \\ \gamma a_1 & \Delta & \gamma a_2 & 0 & \cdots \\ 0 & \gamma a_2 & \Delta & \gamma a_3 & \cdots \\ 0 & 0 & \gamma a_3 & \Delta & \cdots \\ \vdots & \vdots & \vdots & \vdots & \end{vmatrix} = 0. \tag{15}$$

Hence the energy levels are determined: $\varepsilon = 2(N+1)+\Delta$. The splitting of the lowest levels is easily calculated. The lowest two, $N=0$ and 1, are non-degenerate and unperturbed to first order in γ. The next two, $N=2$ and 3, involve a 2×2 determinant and hence a quadratic equation for Δ; the levels split into doublets, while $N=4$ and 5 involve a cubic equation and split into triplets, and so on. The splitting is symmetrical and proceeds as in the following table:

N	ε (unperturbed)	Δ
0	2	0
1	4	0
2	6	$\pm 2\gamma$
3	8	$\pm 2\sqrt{3}\gamma$
4	10	0 and $\pm 4\sqrt{2}\gamma$
5	12	0 and $\pm 8\gamma$
6	14	$\pm 2.516\gamma$ and $\pm 10.662\gamma$
7	16	$\pm 4.059\gamma$ and $\pm 13.547\gamma$

Thenceforth the solution of (15) becomes progressively more tedious. One may guess, however, at the behaviour when N is large by assuming that the levels continue, as in the table, to be more or less evenly spaced; then the sum of the Δ^2 gives the spacing and overall spread. If N is even, multiplying out (15) yields an algebraic equation of the Nth order containing only even powers, and the sum of the Δ^2 is minus the coefficient of Δ^{N-2}, which is $\gamma^2\sum a_m^2$, or $\gamma^2 N^2(N+2)(N+4)/24$. For convenience let us take N to be a multiple of 4, so that evenly spaced levels would lie at $\Delta = 0$, $\pm\delta$, $\pm 2\delta$ etc. Then

$$\delta^2[1+4+9+\cdots+(N/4)^2] = \gamma^2 N^2(N+2)(N+4)/24.$$

Since the left-hand side sums to $\delta^2 N(N+2)(N+4)/192$, it follows that

$$\delta = (8N)^{\frac{1}{2}}\gamma. \tag{16}$$

Correspondingly the whole band of levels lies in the range $2(N+1)\pm\gamma N^{\frac{3}{2}}/\sqrt{2}$.

The assumption of evenly spaced levels is consistent with the regular amplitude modulation exhibited by fig. 2(c)–(e). To make a quantitative comparison, we note that an initial displacement η_0 starts the system vibrating with ε equal to $4\eta_0/\gamma^2$, so that $N \sim 2\eta_0^2/\gamma^2$; hence from (16) the expected value of δ is $4\eta_0$ and the energy separation of the levels, i.e. $\frac{1}{2}\hbar\omega_0\delta$, is $2\hbar\omega_0\eta_0$. The resulting beat frequency should then be $2\omega_0\eta_0$, and the beat cycle should contain $1/2\eta_0$ cycles of the x-vibration. In the trace (f) in fig. 2 there are very nearly 3 cycles to the beat, entirely consistent with the initial value of 0.16 for η_0; and if the beats seem a little too long at the smaller values of η_0, the explanation probably lies in the very low amplitudes at the minima. Parametric amplification is a slow process in these circumstances but, as we have noted already in chapter 13, when the system is quantized the zero-point energy eliminates the early stages that are found in the classical system. The last trace, fig. 2(i), for which the energy is the same as in (d), shows that with a lesser degree of amplitude modulation the beat period is shorter; now, indeed, it is rather shorter than (16) predicts, $4\frac{1}{2}$ cycles rather than $6\frac{1}{4}$. The implication of this variation of beat period with depth of modulation is that the levels are not quite evenly spaced; and when a wave-packet is constructed to correspond to the classical solution different levels presumably are selected according to the initial conditions.

The conclusion of this comparison is that first-order perturbation theory works reasonably well at least until the beat frequency is about $\omega_0/3$, as in fig. 2(f), i.e. until the spacing of the levels is about one-third of the original separation of the degenerate levels. Clearly when N is large each band will have spread out to overlap many neighbouring bands, and it is perhaps surprising that they do not interact together so as to destroy all regularities. It may be noted, however, that even with the full perturbation expressed by $4(Y-\gamma X^2)^2$ in (8), the only components of the wave-function (11) that are coupled to components of $\psi_M(X, Y)$, belonging to another energy level, are those for which $|N-M|$ is 0, 2 or 4 and the m's differ by unity. The interaction between levels is consequently totally absent or not very strong, and the explanation for the modest success of the first-order calculation must surely lie here. The system deserves fuller investigation, both in its classical and its quantal aspects, than it has received so far.

See note on p. 488.

Cyclotron orbits

A classical non-relativistic charged particle, moving freely in a uniform magnetic field $\boldsymbol{B}$, executes helical orbits at a frequency eB/m, independent of its energy and thus of the orbit radius $eB/(2mE)^{\frac{1}{2}}$. Its motion in the plane normal to $\boldsymbol{B}$, which is all we shall concern ourselves with, closely resembles that of a conical pendulum in a circular orbit and it is not surprising to find the quantum mechanics also very similar. Schrödinger's equation now contains the vector potential $\boldsymbol{A}$, defined by the equation curl $\boldsymbol{A} = \boldsymbol{B}$:

237

$$[(\hbar/i)\nabla - e\boldsymbol{A}]^2\psi/2m - E\psi = 0, \tag{17}$$

V being taken as zero everywhere. We shall return to the form of this equation later, but for the present let us take it as given. If $\boldsymbol{A}$ is chosen to

477

be non-divergent,

$$\hbar^2\nabla^2\psi/2m - ie\hbar\boldsymbol{A}\cdot\nabla\psi/m + (E - e^2A^2/2m)\psi = 0. \qquad (18)$$

The sensible choice of $\boldsymbol{A}$ depends on the co-ordinate system to be employed, and we shall consider two examples, $\boldsymbol{A} = (-By, 0, 0)$ appropriate to Cartesian co-ordinates and $\boldsymbol{A} = \frac{1}{2}\boldsymbol{B}\wedge\boldsymbol{r}$ appropriate to polar co-ordinates. Then (18) takes the alternative forms:

$$\hbar^2(\partial^2\psi/\partial x^2 + \partial^2\psi/\partial y^2)/2m + (ie\hbar By/m)\partial\psi/\partial x + (E - e^2B^2y^2/2m)\psi = 0, \qquad (19)$$

or

$$\hbar^2\left[\frac{1}{r}\frac{\partial}{\partial r}\left(r\frac{\partial\psi}{\partial r}\right) + \frac{1}{r^2}\frac{\partial^2\psi}{\partial\phi^2}\right]\Big/2m + (ie\hbar B/m)\partial\psi/\partial\phi + (E - e^2B^2r^2/8m)\psi = 0 \qquad (20)$$

In both cases the variables are separable. In (19) we substitute $\psi = Y(y)\,e^{ikx}$ and find that Y must obey the equation:

$$\hbar^2Y''/2m + [E - e^2B^2(y + \hbar k/eB)^2/2m]Y = 0, \qquad (21)$$

which is the harmonic oscillator equation with the natural frequency $\omega_c = eB/m$ and with the attractive centre at the point $y = -\hbar k/eB$. For a large rectangular slab k can take many values to allow e^{ikx} to satisfy periodic boundary conditions in the x-direction, while still allowing the oscillator function to fit within the bounds of the sample in the y-direction. Each energy level $(n+\frac{1}{2})\hbar\omega_c$ is thus highly degenerate, so that infinitely many alternative representations of the wave-functions are possible. It is not to be expected that any one choice, determined by the mathematical technique of solving the equation, will bear a close resemblance to the classical orbit pattern. In this particular case the y-variation reproduces the characteristic pattern of harmonic motion, as if there had been superimposed a lot of wave-functions closely confined to the classical orbit, but centred at different points along the line $y = -\hbar k/eB$, with their phases adjusted to give a progressive phase variation e^{ikx}.

Wave-functions resembling a classical orbit are not merely hypothetical; one such arises from the solution of (20), from which the ϕ-variable may be separated by the substitution $\psi = (1/r^{\frac{1}{2}})f(r)\,e^{il\phi}$. Once more (14.1) emerges, with $\xi = |eB/\hbar|^{\frac{1}{2}}r$, $\varepsilon = 2E/\hbar\omega_c \pm l - \beta$ and $\beta^2 = l^2 - \frac{1}{4}$. Hence

469

$$E = (2\nu + 1 + |l| \mp l)\tfrac{1}{2}\hbar\omega_c, \qquad (22)$$

the sign being that of the charge on the particle. The same level structure emerges, of course, but the curious appearance of $|l| \mp l$ reflects the basic difference between cyclotron motion and a conical pendulum, in that only one sense of rotation is possible in the former. When $\nu = 0$, which gives a wave-function most nearly confined to the classical orbit, only positive values of l are permitted for positively charged particles, i.e. only anticlockwise circulation as viewed along the direction of $\boldsymbol{B}$, and only negative l, clockwise circulation, for negative particles. Otherwise the solution is like that for the conical pendulum, with the particle closely confined to the classical orbit when $|l|$ is large. Moreover, like the pendulum, stable wave-packets may be constructed to localize the particle on its orbit, and keep it so localized. This should give encouragement to those who are accustomed to think of the motion of particles in classical terms; when dealing with the dynamical behaviour of conduction electrons in metals and semiconductors, for example, the classical mode of thought is virtually a necessity on account of the complexities introduced by the crystal lattice. Quantum mechanics serves to show how the electron can travel through a perfect lattice without suffering collisions with individual atoms, and it can produce expressions for the modifications to the dynamical properties caused by the lattice. Once it has shown the character of the equivalent classical particle which would simulate these dynamical properties there is some merit in forgetting quantum mechanics and thinking of the classical trajectories of these particles as they move in electric and magnetic fields, and suffer scattering from defects of the lattice. The fact that one can construct stable wave-packets that follow the classical trajectory helps one to believe that the undoubted success of classical thinking applied to these problems may not be merely a lucky accident.

This is not to suggest of course, that it is always safe to neglect the discrete energy level structure when a magnetic field is present, but this is something that can often be supplied as an afterthought through the semi-classical process of phase-integral quantization, already shown to be a great value for anharmonic oscillators. If a free particle in its cyclotron orbit is allowed to present close analogies with a harmonic vibrator, the modified dynamics resulting from the crystal lattice may be seen as analogous to anharmonicity, often of extraordinary complexity; but this proves to present no obstacle to semi-classical quantization. A number of interesting and important points arising in the pursuit of this line of thought are all traceable to a rather strange effect that is already to be found in the behaviour of a free particle. The nearly-classical orbit, in which $\nu = 0$ and l is large has a perimeter of $2\pi v/\omega_c$, if v is the velocity of a classical particle of the same energy, $l\hbar\omega_c$. The wave that sweeps round it has l wavelengths fitting into the circuit. Hence the wavelength, measured round the orbit, is $2\pi v/\omega_c l$, i.e. $2h/mv$, since $l = \frac{1}{2}mv^2/\hbar\omega_c$. This wavelength is *twice* the value that one would naively expect for a de Broglie wave.

456

Quantization in a magnetic field

The explanation lies in the classical mechanics of a charged particle in a magnetic field. The non-dissipative Lorentz force can be incorporated in the Lagrangian formalism by redefining the momentum conjugate to the position variable; instead of being merely $m\boldsymbol{v}$, $\boldsymbol{p}$ must be taken as $m\boldsymbol{v} + e\boldsymbol{A}$, so that the Hamiltonian form of the energy is $(\boldsymbol{p} - e\boldsymbol{A})\cdot(\boldsymbol{p} - e\boldsymbol{A})/2m + V(\boldsymbol{r})$. The transition to Schrödinger's equation, by the substitution $\boldsymbol{p} = \hbar\nabla/\mathrm{i}$, leads to (17). In other words, the de Broglie wavelength is determined by $m\boldsymbol{v}+e\boldsymbol{A}$ rather than by $m\boldsymbol{v}$ alone, and in the case just discussed $e\boldsymbol{A}$ is antiparallel to $m\boldsymbol{v}$ and half its magnitude; for a positively charged particle moving anticlockwise in an orbit of radius r, $mv = eBr$, while the vector potential runs clockwise and has magnitude $\frac{1}{2}Br$.

Since the addition to $\boldsymbol{A}$ of $\nabla\beta$, β being any differentiable scalar function of $\boldsymbol{r}$, leaves curl $\boldsymbol{A}$ and therefore $\boldsymbol{B}$ unaltered, the local de Broglie wavelength can be modified at will, but this has no effect on the quantization of the orbits since the total phase change around the orbit is not changed. The two choices of gauge for $\boldsymbol{A}$ already discussed in connection with (19) and (20) illustrate this point. A detailed appreciation of the motion of electrons in a metal under the influence of $\boldsymbol{B}$ also depends on recognizing the contribution of $\boldsymbol{A}$ to the phase. If the periodic field of the lattice is very weak electrons moving in almost any direction hardly notice its existence, but it matters considerably if it can cause Bragg reflection; for this to occur there must be a Fourier component in the periodic lattice field having wavenumber $\boldsymbol{g}$ such that an electron with wavenumber $\boldsymbol{k}$ can be reflected into a state of the same energy and of wavenumber $\boldsymbol{k}+\boldsymbol{g}$. Very commonly, as in X-ray diffraction, the satisfying of the Bragg condition is achieved by making the magnitudes of $\boldsymbol{k}$ and $\boldsymbol{k}+\boldsymbol{g}$ the same, but this is not an essential requirement, and we have a particular case here where the more general statement of Bragg's law is required. An electron in its orbit may find itself reflected by a set of lattice planes into another that is differently centred (fig. 3), and under the right conditions one or more

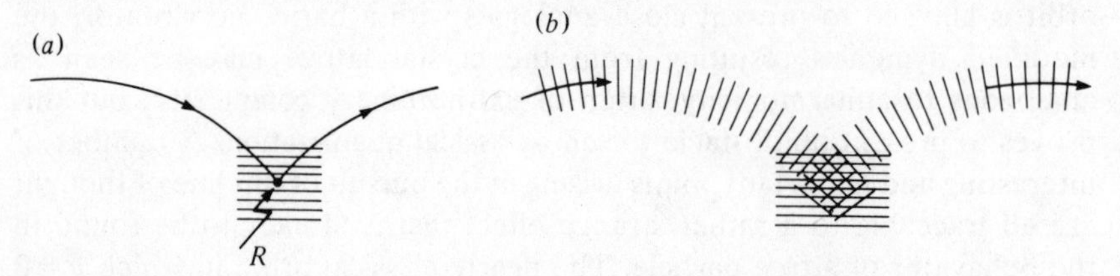

Fig. 3. (*a*) An electron moving in a uniform magnetic field suffers Bragg reflection at R from the lattice planes, shown as horizontal lines. (*b*) Schematic, but incorrect (see text), representation of Bragg reflection in a magnetic field.

further reflections may bring it back to the original orbit; an example will be found in the quasi-classical diagram, fig. 6. In line with our discussion of localized orbit wave-functions we might imagine the corresponding quantum-mechanical picture to take a form like that in fig. 3(*b*), and the process of Bragg reflection to act on a wavefront of limited extent, but still much wider than the lattice spacing. There seems little objection to this view except that it is incorrect to represent both sets of wavefronts as radial. For this is to imply that each orbit is gauged with its own centre as gauge centre,† while in practice we ought to choose the same gauge centre for both. This leads us to derive the rule for phase variation around an orbit which is gauged with respect to a point not its centre.

† If $\boldsymbol{A}$ is written as $\frac{1}{2}\boldsymbol{B}\wedge(\boldsymbol{r}-\boldsymbol{R})$, $\boldsymbol{R}$ is the position of the gauge centre round which $\boldsymbol{A}$ circulates.

If ψ_0 is a solution of (17) for a given value of E, the addition of $\nabla\beta$ to $\boldsymbol{A}$ can be compensated by replacing ψ_0 by $\psi_o\, e^{ie\beta/\hbar}$; it is readily verified that the latter is now a solution for the same E. If then we express ψ_0 as $f(\boldsymbol{r})\, e^{i\theta(\boldsymbol{r})}$, $f(\boldsymbol{r})$ being real, the change of gauge alters the phase factor without changing the amplitude, the new solution being $f(\boldsymbol{r})\, e^{i(\theta+e\beta/\hbar)}$. This result can be expressed otherwise in terms of the local wavenumber $\boldsymbol{k}$, defined as the gradient of phase. Originally $\boldsymbol{k}_0=\nabla\theta$, and the change of gauge converts $\boldsymbol{k}_0$ to $\boldsymbol{k}_0+e\nabla\beta/\hbar$, or $\boldsymbol{k}_0+e\Delta\boldsymbol{A}/\hbar$, where $\Delta\boldsymbol{A}$ is the change in $\boldsymbol{A}$ at the point considered. The significance of this rule is immediately apparent when applied to a wave-function which locally has the form of a plane wave, $e^{i\boldsymbol{k}_0\cdot\boldsymbol{r}}$, and which describes a particle of momentum $\hbar\boldsymbol{k}_0$. Now this momentum is not $m\boldsymbol{v}$ but the conjugate of the position vector, i.e. $\hbar\boldsymbol{k}_0=m\boldsymbol{v}+e\boldsymbol{A}$. The simultaneous addition of $\Delta\boldsymbol{A}$ to $\boldsymbol{A}$ and $e\Delta\boldsymbol{A}/\hbar$ to $\boldsymbol{k}_0$ leaves $m\boldsymbol{v}$ unchanged, so that the same dynamical behaviour is successfully described in the new representation.

For the particular gauge change in which the gauge centre is shifted from the origin to $\boldsymbol{R}$, $\Delta\boldsymbol{A}=\frac{1}{2}\boldsymbol{B}\wedge(\boldsymbol{r}-\boldsymbol{R})-\frac{1}{2}\boldsymbol{B}\wedge\boldsymbol{r}=-\frac{1}{2}\boldsymbol{B}\wedge\boldsymbol{R}$, which is constant. Thus every $\boldsymbol{k}_0$ is increased by the same vector increment, $-\frac{1}{2}e\boldsymbol{B}\wedge\boldsymbol{R}/\hbar$. For an orbit whose centre is the origin, $\boldsymbol{k}_0$ at a point $\boldsymbol{r}$ is originally $e\boldsymbol{r}\wedge\boldsymbol{B}/2\hbar$, and if the gauge centre is moved to $\boldsymbol{r}$ itself the increment to $\boldsymbol{k}_0$ is equal to $\boldsymbol{k}_0$ itself, so that now $\boldsymbol{k}_0=e\boldsymbol{r}\wedge\boldsymbol{B}/\hbar$. This is just $m\boldsymbol{v}/\hbar$, as if there were no magnetic field present, but it applies only to that point on the orbit which is the new gauge centre. At the opposite end of the diameter $\boldsymbol{k}$ is reduced to zero and, of course, $\oint\boldsymbol{k}\cdot d\boldsymbol{r}$ round the orbit is invariant with respect to gauge changes, since $\oint\Delta\boldsymbol{A}\cdot d\boldsymbol{r}=0$. It is now clear that the condition for Bragg reflection is essentially unchanged by the presence of $\boldsymbol{B}$. For if we choose the point of reflection, R in fig. 3(a), as the gauge centre, the local wavenumber on each orbit is determined by $m\boldsymbol{v}$ as if no field were present. Reflection occurs when the electron in its orbit happens to move in such a direction as would cause it to suffer reflection when $\boldsymbol{B}=0$. This result is naturally independent of the choice of gauge centre; since shifting the centre by $\boldsymbol{R}'$ adds the same vector increment to the wave-vectors on the two orbits, their difference remains equal to $\boldsymbol{g}$ whatever may happen to them individually.

Quantization of non-circular orbits in real metals

It is possible to proceed from this point to calculate the phase length round an orbit composed of a number of circular arcs joined by Bragg reflections, and hence to derive the quantization rule, which turns out to be more conveniently expressed in terms of the area enclosed by the orbit than in terms of its perimeter. The argument, which demands rather careful exposition, will not be given here.[3,4] It is sufficient to remark that it agrees with an earlier treatment due to Onsager,[5] of great elegance and economy, which by-passes the more awkward problems by going back to the semi-classical method of phase-integral quantization. We have seen how well this works with highly anharmonic oscillators, and equal success attends its application to the cyclotron orbits of electrons, even when they move in lattice fields strong enough for the assumption of nearly free motion between Bragg reflections to be quite inadequate. Onsager assumed that the electron could be treated as a classical particle executing a well-defined orbit, but with modified dynamical properties. This has been to a certain degree, though far from rigorously, justified by later analyses.(6) In the absence of a magnetic field the energy is taken to be a function, possibly very complicated, of momentum $\boldsymbol{p}$; $E = E(\boldsymbol{p})$ generalizes the Newtonian form $E = p^2/2m$. Surfaces of constant E plotted in a three-dimensional $\boldsymbol{p}$-space serve to represent the form of this function, and in real metals these energy surfaces are normally very different indeed from the concentric spheres that describe a Newtonian particle. A particular example is shown in fig. 4 to illustrate the need for an approach to these problems that is

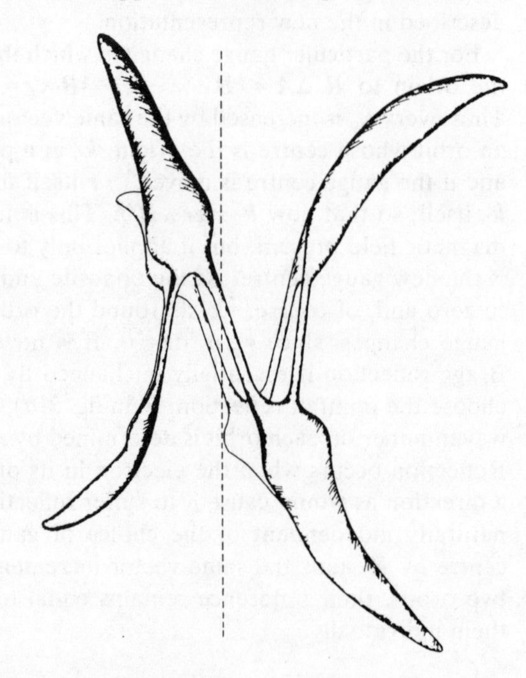

Fig. 4. Fermi surface of arsenic, as derived by Lin and Falicov(7).

not too strongly dependent on simplified analytical representations. The analogue to the de Broglie wave associated with a particle of momentum $\boldsymbol{p}$, and wavenumber $\boldsymbol{k} = \boldsymbol{p}/\hbar$, is a Bloch wave to which also wavenumber $\boldsymbol{k}$ may be assigned. Thus the surfaces representing $E(\boldsymbol{p})$ serve equally for $E(\boldsymbol{k})$ if the scale is altered. Since in fact the meaning of $\boldsymbol{k}$ for a Bloch wave is rather clearer than the meaning of $\boldsymbol{p}$, it is customary to proceed in terms of $\boldsymbol{k}$ rather than $\boldsymbol{p}$; but one must remember that although, when a magnetic field is applied, the same function $E(\boldsymbol{k})$ still defines the energy, $\boldsymbol{k}$ no longer defines the local wavenumber, which must now be taken as $\boldsymbol{k} + e\mathbf{A}/\hbar$. This does not, however, affect the group velocity which is still that of a wave of frequency $\omega = E(\boldsymbol{k})/\hbar$, and serves to determine the velocity $\boldsymbol{v}$ of a particle in the state $\boldsymbol{k}$. For one-dimensional motion the group velocity is $\mathrm{d}\omega/\mathrm{d}k$, i.e. $(\mathrm{d}E/\mathrm{d}k)/\hbar$, and the same applies to any component of the three-dimensional velocity; $\boldsymbol{v}$ therefore has the Cartesian form $\hbar^{-1}(\partial E/\partial k_x, \partial E/\partial k_y, \partial E/\partial k_z)$, i.e. $\boldsymbol{v} = \hbar^{-1}\nabla_k E$, where ∇_k is the vector derivative (gradient) with respect to position in $\boldsymbol{k}$-space. There is no component of $\nabla_k E$ lying in the constant energy surfaces, and $\boldsymbol{v}$ is therefore directed normal to these surfaces with a magnitude varying inversely as the separation of neighbouring surfaces, E and $E + \delta E$.

In a uniform magnetic field $\boldsymbol{B}$ the classical equation of motion for an electron under the influence of the Lorentz force takes the form $\dot{\boldsymbol{p}} = e\boldsymbol{v} \wedge \boldsymbol{B}$, and this result still holds, so that $\boldsymbol{k}$ changes according to the equation:

$$\dot{\boldsymbol{k}} = (e/\hbar)\boldsymbol{v} \wedge \boldsymbol{B}. \tag{23}$$

All changes in $\boldsymbol{k}$ are in the plane normal to $\boldsymbol{B}$ and, being also normal to $\boldsymbol{v}$, cause the point representing $\boldsymbol{k}$ for the electron to move round a constant energy surface. If we cut a plane section of the surface normal to $\boldsymbol{B}$ the resulting curve is the $\boldsymbol{k}$-trajectory of the electron. In general there will be a component of $\boldsymbol{v}$ parallel to $\boldsymbol{B}$, but we shall ignore this longitudinal motion and concentrate on the transverse component $\boldsymbol{v}_t$, normal to $\boldsymbol{B}$. Let $\boldsymbol{v}_t$ in (23) be written as $\dot{\boldsymbol{r}}_t$; then $\dot{\boldsymbol{r}}_t$ is related to $\dot{\boldsymbol{k}}$ by the process of multiplying by $\hbar/eB$ and turning through $\pi/2$. So too, on integrating, $\boldsymbol{r}_t$ is related to $\boldsymbol{k}$ except for an arbitrary integrating constant. As illustrated in fig. 5, when

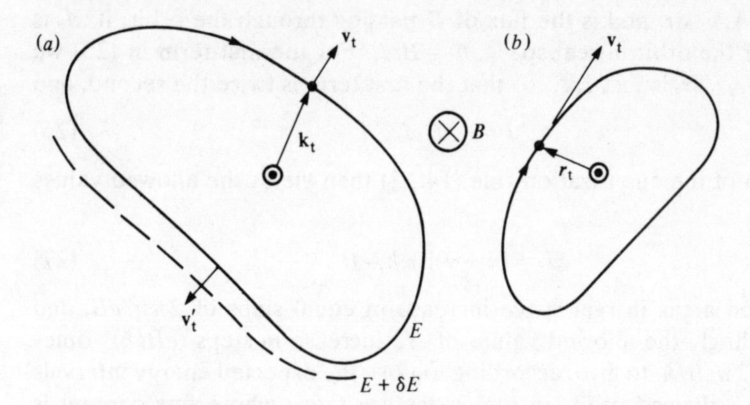

Fig. 5. Motion of an electron in a uniform magnetic field: (*a*) in $\boldsymbol{k}$-space (*b*) in real space. The curves are similar in shape, but oriented at 90° to one another.

the section of the constant energy surface is a closed curve, the electron whose $\boldsymbol{k}$ moves round this will describe in real space a similar closed orbit, which may be located anywhere and whose size is inversely proportional to the magnetic field strength. Strictly, when we take motion along $\boldsymbol{B}$ into account, this orbit is the projection on a plane normal to $\boldsymbol{B}$ of an irregular helical trajectory which repeats itself exactly every time round.

The cyclotron frequency ω_c of an orbit may be expressed in geometrical terms by calculating how long it takes to sweep out the annular area between the two neighbouring sections, E and $E + \delta E$, in fig. 5. The representative

237

point and its vector $\boldsymbol{v}'_t$ sweep around at a rate $\dot{k}=eBv'_t/\hbar$, from (23); the width of the annulus is $\delta E/\hbar v'_t$ and the annular area is therefore swept out at the constant rate $eB\delta E/\hbar^2$. If the area of the section of the constant energy surface is $\mathscr{A}_k(E)$, the annulus has area $\mathrm{d}\mathscr{A}_k/\mathrm{d}E\cdot\delta E$ and is swept out in a time $T=(\hbar^2/eB)\,\mathrm{d}\mathscr{A}_k/\mathrm{d}E$.

Hence $$\omega_c\equiv 2\pi/T=(2\pi eB/\hbar^2)/(\mathrm{d}\mathscr{A}_k/\mathrm{d}E). \qquad (24)$$

For the special case of free electrons moving in the plane normal to $\boldsymbol{B}$, $E=\hbar^2k^2/2m$, so that $\mathscr{A}_k=2\pi mE/\hbar^2$ and (24) gives eB/m for ω_c, as expected. If the quantized levels are separated in energy by $\hbar\omega_c$, they are represented on the $\boldsymbol{k}_t$-plane by a set of nesting curves, like that in fig. 5(a), each the section of a constant energy surface. The areas enclosed by the curves increase in equal steps of $2\pi eB/\hbar$, independent of any parameters relating to the particular metal. This is reminiscent of the phase-integral for vibrators, and is indeed confirmed by the following direct evaluation of the phase-integral.

458

Let us choose corresponding points in fig. 5 as origins for $\boldsymbol{k}_t$ and $\boldsymbol{r}_t$. Then when the electron is at a point $\boldsymbol{r}_t$, $\boldsymbol{k}_t$ has a value $e\boldsymbol{r}_t\wedge\boldsymbol{B}/\hbar$. This must be supplemented by $e\boldsymbol{A}/\hbar$ to give $\boldsymbol{k}'_t$ determining the phase variations. Consequently the phase length of the orbit is expressed in the form

$$J=\hbar\oint\boldsymbol{k}'_t\cdot\mathrm{d}\boldsymbol{r}_t=e\oint(\boldsymbol{r}_t\wedge\boldsymbol{B}+\boldsymbol{A})\cdot\mathrm{d}\boldsymbol{r}_t=e\left\{-\boldsymbol{B}\cdot\oint\boldsymbol{r}_t\wedge\mathrm{d}\boldsymbol{r}_t+\Phi\right\}, \qquad (25)$$

where $\Phi=\oint\boldsymbol{A}\cdot\mathrm{d}\boldsymbol{r}_t$ and is the flux of $\boldsymbol{B}$ passing through the orbit; if $\mathscr{A}_r$ is the area of the orbit in real space, $\Phi=B\mathscr{A}_r$. For the first term in (25) we note that $\oint\boldsymbol{r}_t\wedge\mathrm{d}\boldsymbol{r}_t$ is just $2\mathscr{A}_r$, so that the first term is twice the second, and

$$J=-eB\mathscr{A}_r, \qquad (26)$$

application of the quantization rule (14.12) then yields the allowed values of $\mathscr{A}_r$:

$$\mathscr{A}_r=(n+\gamma)2\pi\hbar/eB. \qquad (27)$$

The allowed areas in real space increase in equal steps of $2\pi\hbar/eB$, and correspondingly the allowed values of $\mathscr{A}_k$ increase in steps $(eB/\hbar)^2$ times larger, i.e. $2\pi eB/\hbar$, to give, according to (24), the expected energy intervals of $\hbar\omega_c$. The allowed orbits in real space are those whose flux content is $(n+\gamma)2\pi\hbar/e$. The unit of flux, $2\pi\hbar/e$, is sometimes referred to as the flux quantum, and in many respects it behaves as its name suggests, in that when it is possible to isolate a bundle of flux lines the quantum conditions applying to the system that effects this isolation are rather likely to impose a value on Φ that is an integral multiple of $2\pi\hbar/e$. The easiest way of achieving this isolation of Φ, however, is by enclosure within a superconducting ring, but in this case it is a multiple of the half-quantum $\pi\hbar/e$ that may be trapped. The concept of a flux quantum therefore needs to be treated with caution – it is not so absolute an entity as most quanta.

Onsager's demonstration of the quantization of cyclotron orbits in terms of their area alone, whatever the details of their shape, has had profound consequences for the study of metals. The de Haas–van Alphen effect,[8] whose essential feature is the oscillatory variation with magnetic field strength of the magnetization of a metal single crystal at low temperatures, is a direct consequence on the orbit quantization. By measuring the period of the oscillation the areas of certain dominant sections of the Fermi surface (the constant energy surface marking the division between filled and unfilled states) may be determined with great accuracy, and as the direction of $\boldsymbol{B}$ is changed other sections come under study. Enough information has been amassed for many metals[9] to allow the detailed shape of their Fermi surfaces to be constructed; fig. 4 is an example. This is the first step towards acquiring a full specification of the dynamical properties of the conduction electrons, and the systematic pursuit of this goal has been rewarded by a great advance in knowledge of metallic behaviour. Further, it has brought about greatly improved calculations from first principles of the shape to be expected for the Fermi surface in a given metal. Solution of Schrödinger's equation for electrons moving in the potential of the ionic lattice has proved too complicated to accomplish without making approximations, or introducing simplified models in place of the real ionic potentials. Knowledge of the required answer has provided an essential test of the theoretical simplifications, as a result of which these approximate methods may be applied confidently to more advanced problems where direct experimental confirmation is harder to acquire.

Magnetic breakdown[10]

The periodic field of the ionic lattice, through which the conduction electrons move, does not merely modify their orbits in matters of detail; it produces significant changes in the nature of the orbits, such as reversing their sense. The hexagonal two-dimensional pattern of fig. 6 illustrates this point. If the lattice potential is negligible, any of the circular orbits shown is one out of a myriad of possible free-electron orbits, with their centres arbitrarily located. As the potential is increased in strength, however, it is convenient to select for examination a subset, as in the diagram, chosen so that at all the intersections and only there, the Bragg condition is satisfied; the electrons cannot now complete their circular orbits but instead, depending on where they started, they may either traverse small triangular orbits ($\mathscr{T}$) in the same sense, or larger hexagonal orbits ($\mathscr{H}$) in a retrograde sense. The Onsager quantization rule (27) determines at which electron energies orbits of each kind are allowed. It may be noted that for a given magnetic field strength the free-electron orbits which are coupled by Bragg reflection are defined by the same lattice of orbit centres for all energies;† as E

† The lattice of orbit centres is the same as the reciprocal lattice, but scaled by $\hbar/eB$.

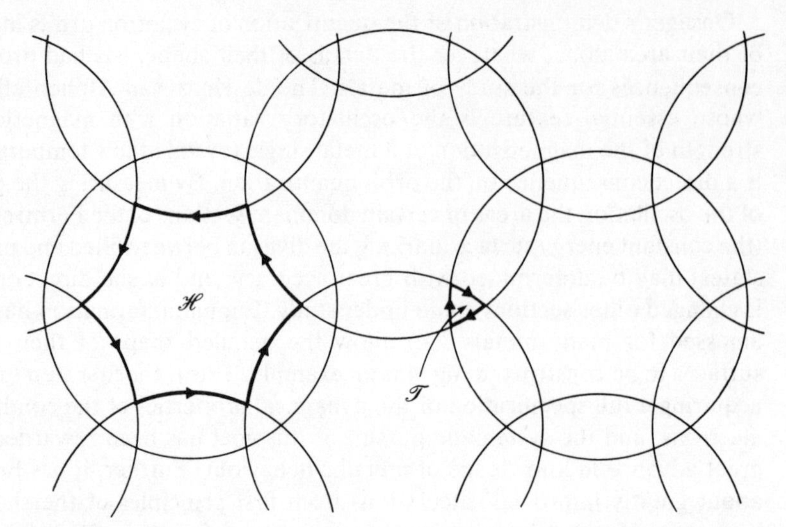

Fig. 6. Free electron orbits coupled by Bragg reflection from a hexagonal lattice, showing the network of paths made available by magnetic breakdown.

increases, therefore, and the circles get larger, the triangular orbits expand while the hexagonal orbits contract. The permitted energy levels then fall into two sets, not evenly spaced like free-electron levels (since the areas are not linear in energy) and as B is increased the levels for $\mathscr{T}$ move upwards in energy while those for $\mathscr{H}$ move downwards.

After this brief preliminary, let us ask how the transition from the free-electron orbits to $\mathscr{T}$ and $\mathscr{H}$ is effected as the lattice potential is gradually strengthened. The answer is simple in principle – with a weak lattice potential Bragg reflection is incomplete, and at every junction an electron has a certain probability of reflection, or alternatively may carry on round the free-electron orbit to the next junction, where the same choice arises. The process of Bragg reflection may be pictured as the production of a weak reflected wavelet by each lattice plane as the incident wave passes. At the correct angle of incidence all reflected wavelets add in phase to give a strong resultant – indeed if the incident wave is not deflected by a magnetic field reflection is total for angles of incidence within a certain narrow range round the Bragg angle. But the magnetic field, by bending the wave path, limits the length within which the Bragg condition is nearly enough satisfied, and if there is not enough reflected amplitude from the lattice planes within this length reflection will be only partial. The stronger the field the more likely the electron is to get through without reflection. This is the phenomenon of *magnetic breakdown.*

Under ideal conditions of purity and crystalline perfection the electrons may travel considerable distances without loss of phase information, and it is now essential to take account of interference between waves that have traversed different paths. The whole network in fig. 6 may be thought of as a system that can guide waves in the direction of the arrows and which will show a multitude of resonances, since each possible closed path has something of the character of a circulating resonator. Suppose, for example, that Bragg reflection is nearly complete; then a wave circulating anticlockwise on a hexagonal orbit will excite a small amplitude in each neighbouring triangular orbit, and the wave fed into the triangle will run round inside it, with only slight leakage into its contiguous hexagons. If the phase length of the triangle is an exact multiple of 2π it will resonate – the weak amplitude fed in from the hexagon will build up to such a resultant that ultimately there will be a significant extraction from the hexagon and transfer of wave amplitude into the other orbits in the vicinity. In this way the electron originally described as confined to a single hexagonal orbit can find its way to others and continue in a sort of diffusive motion to wander through the metal. As the effect depends on the electron's energy matching one of the permitted levels for the triangular orbits (this is the meaning of resonance in this context), the paths available to an electron may at some magnetic field strengths be closely confined to an orbit such as $\mathscr{H}$, and at others allow considerable excursions to distant regions. As a result, the conductivity of the metal, which is determined by how far an electron may travel, and in which directions, shows rapid oscillations with strength of magnetic field, as illustrated in fig. 7. The phenomenon of

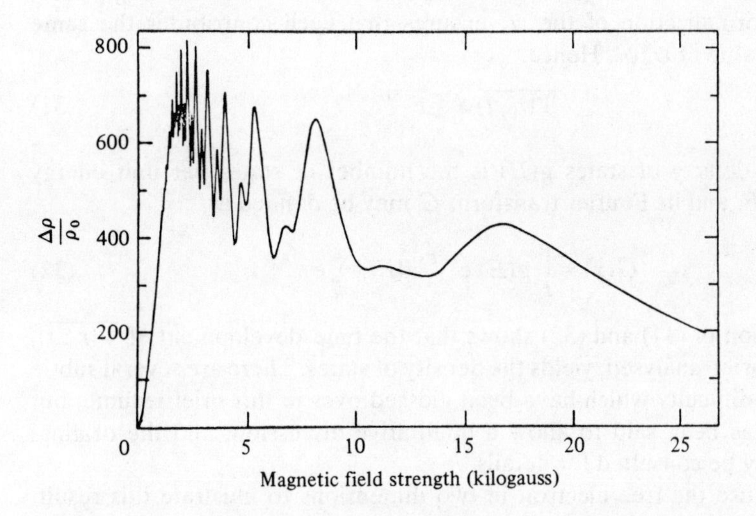

Fig. 7. Magnetoresistance in a pure zinc single crystal at 4 K (Stark[11]). Current flow is in the plane normal to the hexagonal axis, which is the direction of the applied magnetic field, $\boldsymbol{B}$; $\Delta\rho/\rho_0$ is the change in resistance of the sample expressed in terms of the resistance when $\boldsymbol{B} = 0$.

magnetic breakdown is not uncommon in metals; that is to say, there is a considerable number of examples where Bragg reflection is weak enough to allow breakdown in experimentally attainable magnetic fields. The details of what happens then depend critically on the arrangements of coupled orbits, and very few general rules have emerged from analyses of special cases.

There is one general result of some interest, however, and not solely in the context of magnetic breakdown, since it depends on a viewpoint that can be applied in other cases. We have seen that when the cyclotron orbits are well-defined the permitted energy levels form a discrete ladder. When, however, they are coupled by magnetic breakdown the levels spread into something more like a continuum, though it may still show gaps. A theorem due to Falicov and Stachowiak[12] provides a clue as to the character of the energy level diagram. Let us, at $t=0$, establish a wave-packet at the point $\boldsymbol{r}_0$ in the metal, and let this wave-packet be ideally small: $\Psi(\boldsymbol{r}_0, 0)=\delta(\boldsymbol{r}-\boldsymbol{r}_0)$. It will immediately begin to spread, and the manner of its development may be written down formally by expressing $\delta(\boldsymbol{r}-\boldsymbol{r}_0)$ as a Fourier sum over all stationary states of the electron in the magnetic field:

$$\delta(\boldsymbol{r}-\boldsymbol{r}_0)=\sum_n \psi_n^*(\boldsymbol{r}_0)\psi_n(\boldsymbol{r}), \tag{28}$$

the amplitude of the nth wave-function being simply $\psi_n^*(\boldsymbol{r}_0)$. At time t, therefore, the development of each wave-function gives rise to the resultant

$$\Psi(\boldsymbol{r}, t)=\sum_n \psi_n^*(\boldsymbol{r}_0)\psi_n(\boldsymbol{r})\,\mathrm{e}^{-\mathrm{i}E_n t/\hbar}. \tag{29}$$

In particular, at the point $\boldsymbol{r}_0$ where the wave-packet started,

$$\Psi(\boldsymbol{r}_0, t)=\sum_n \psi_n^*(\boldsymbol{r}_0)\psi_n(\boldsymbol{r}_0)\,\mathrm{e}^{-\mathrm{i}E_n t/\hbar}, \tag{30}$$

and if we take the average value of $\Psi(\boldsymbol{r}_0, t)$ over all initial starting points, $\boldsymbol{r}_0$, the normalization of the ψ_n ensures that each contributes the same average value of $\psi_n^*\psi_n$. Hence,

$$\overline{\Psi(\boldsymbol{r}_0, t)}\propto\sum_n \mathrm{e}^{-\mathrm{i}E_n t/\hbar}. \tag{31}$$

Now the density of states $g(E)$ is the number of states per unit energy range at E, and its Fourier transform G may be defined as

$$G(x)=\int g(E)\,\mathrm{e}^{-\mathrm{i}Ex}\,\mathrm{d}E=\sum_n \mathrm{e}^{-\mathrm{i}E_n x}. \tag{32}$$

Comparison of (31) and (32) shows that the time-development of $\overline{\Psi(\boldsymbol{r}_0, t)}$, when Fourier-analysed, yields the density of states. There are several subtle points of difficulty which have been glossed over in this brief résumé, but enough has been said to allow a qualitative discussion, and the original paper may be consulted for details.

Let us use the free electron in two dimensions to illustrate this result. The initial wave-packet spreads out in all directions as if a cluster of classical electrons had been released at $\boldsymbol{r}_0$. No matter what their original direction, they will all return to $\boldsymbol{r}_0$ after one cyclotron period, and we expect the wave-function to show a sharp peak as it reconstitutes itself, only to spread once more, but at subsequent equal intervals to return. Thus $\overline{\Psi(\boldsymbol{r}_0, t)}$ is expected to present an even sequence of pulses with a period of $2\pi/\omega_c$, and the Fourier transform of this is a sharp set of lines

$$g(E)\propto\int\overline{\Psi(\boldsymbol{r}_0, t)}\,\mathrm{e}^{\mathrm{i}Et/\hbar}\,\mathrm{d}t\propto\sum_\nu \mathrm{e}^{\mathrm{i}\nu E/\hbar\omega_c}\propto\sum_n \delta(E-n\hbar\omega_c).$$

Apart from the fractional correction γ, which our slapdash treatment has managed to lose, the result is the desired quantization of levels $\hbar\omega_c$ apart.

When the argument is applied to a breakdown network such as fig. 6, the form of $\overline{\Psi(\boldsymbol{r}_0, t)}$ becomes more complex, though it may still be expected to approximate to a sequence of pulses. But now the pulses do not arrive in a simple regular sequence, but at such times as are permitted by any of the closed paths through the network, and with such amplitudes as are determined by the probabilities of reflection and transmission at each junction. The density of states therefore contains periodicities characteristic of every possible closed orbit, and it is not surprising that it takes a complicated form. Indeed, the matter is a good deal more complicated than might be guessed, since the movement of levels both up and down, like those due to $\mathscr{T}$ and $\mathscr{H}$ in the hexagonal network, and at different speeds, means that although the Fourier amplitudes in $g(E)$ may be fairly

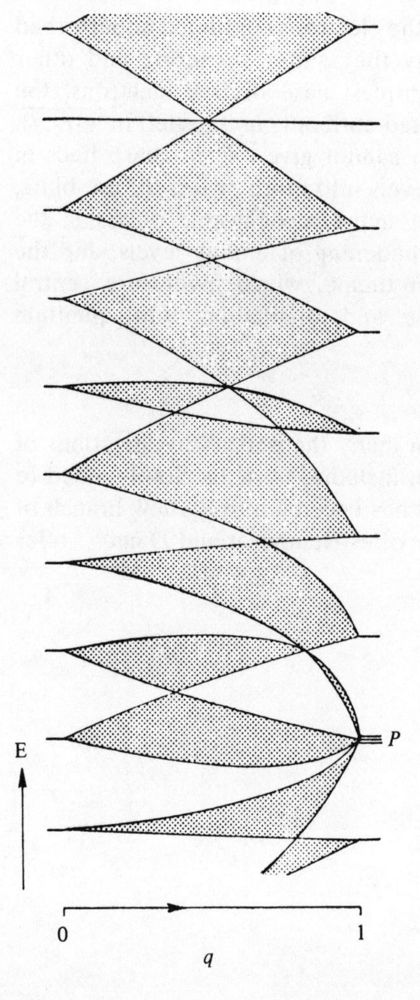

Fig. 8. Theoretical energy level structure for the network of fig. 6 (Pippard[3]). On the right, where Bragg reflection, as measured by q, is strong, the hexagonal orbit gives rise to evenly spaced sharp levels, with a second set of more widely spaced levels, having twice the degeneracy, due to the triangular orbits $\mathscr{T}$; at P levels due to $\mathscr{H}$ and $\mathscr{T}$ are shown in coincidence. On the left, when the lattice potential has been reduced to so small a value that magnetic breakdown is complete, the levels are also sharp and are characteristic of free electrons. In between there is partial reflection and partial transmission at each junction of the network, and the degenerate levels are spread into bands by the coupling of the orbits.

well defined, their relative phases develop in different ways with B, so that the resultant itself changes form with B. Such phenomena as the deHaas–van Alphen effect or the magneto-resistance oscillations are more strongly conditioned by the Fourier amplitudes than by the phases, and the extraordinary (indeed discontinuous) evolution of $g(E)$ as a function of B has few, if any, experimental consequences. Nevertheless it is worth noting, merely as a fact of nature, what variety is generated by the coupling of orbits into an extended network. As illustration of $g(E)$ in its simplest form, i.e. with B given a special value for the sake of the computation, fig. 8 shows how the discrete level structure due to $\mathscr{H}$ and $\mathscr{T}$, on the right, spreads into bands as the lattice potential is reduced and magnetic breakdown sets in; ultimately the bands collapse again to discrete free-electron levels on the left, when the lattice potential is completely removed. Falicov and Stachowiak have verified by direct computation, in a rather simpler case, that the Fourier transform of the energy level diagram conforms to the predictions of their theory.

One last point – it was assumed that the electrons continued unperturbed in their orbits for ever, but in reality they suffer scattering and other mishaps. As a result, even in the simplest case of free electrons, the succession of pulses is gradually disrupted until only noise is left in $\overline{\Psi(\boldsymbol{r}_0, t)}$. The Fourier transform of this pattern cannot give rise to sharp lines in $g(E)$ but must spread the individual levels into more or less diffuse blurs, according as there are only a few or many recognizable pulses in the sequence. Scattering results in the broadening of energy levels, and the next chapter takes up this important theme, which involves a central problem of introducing randomness and dissipation into quantum mechanics.

Additional note (see p. 475)
Since this was written there have been many theoretical investigations of chaotic systems and their quantization, including systems closely allied to the case of Fermi resonance. The field has become a lively new branch of mathematical physics, to whose complexities Reinhardt and Dana[13] offer a point of view.

16 Dissipation, level broadening and radiation

After this excursion into the field of non-linear vibrators, we now return to the harmonic vibrator and take up a point which had begun to reveal itself at the end of chapter 13, where we found that under the influence of a uniform, but arbitrarily time-varying, force the vibrator never forgets its initial state. If it started in the ground state, for ever afterwards its response can be described by the movement of a compact distribution in the σ-representation of equivalent classical vibrators; only the centroid of the distribution responds to the applied force. The harmonic vibrator is thus extraordinarily resistant to randomization. To be sure, if the force is not uniform, but depends on the displacement of the oscillating particle, the result just summarized is no longer true. Nevertheless, in the most important application, where an oscillator of atomic dimensions is influenced by electromagnetic vibrations, the force due to the electric field is as nearly uniform as makes no difference, since the wavelength of electromagnetic waves at a typical atomic resonant frequency is a thousand times the size of an atom. The disturbing aspect of the resistance of a vibrator to randomization is that in all theories of black-body radiation, before and after Planck, it is assumed that material oscillators and electromagnetic vibrations in a cavity will eventually share the chaotic state that allows statistical mechanics to be applied. We shall find, on looking into the matter from a consistently quantum-mechanical point of view, that the problem is not as serious as this too-classical discussion has suggested; at least it is no more serious than other fundamental questions in statistical mechanics that most physicists are content should be analysed, and if possible resolved, by mathematicians. We shall have to consider a material oscillator and a cavity vibration as a pair of coupled harmonic oscillators, whose behaviour is best described in terms of normal modes. It is the normal modes and not the individual oscillators that are quantized independently, and the recognition of this fact not only resolves the immediate difficulty but leads us on naturally to consider the quantum mechanics of energy exchange between a material oscillator and the cavity vibrations; in other words, the atomic processes involved in electromagnetic radiation and absorption. In the present chapter we shall only make a preliminary attack on the problem, mainly by the use of classical arguments where these are reasonably safe. In later chapters it will recur in various forms and the argument will be progressively refined.

449
594
366

Coupled harmonic vibrators

For the present purpose no deep understanding of the nature and patterns of electromagnetic vibrations in cavities is needed. It is enough to recognize that a large cavity whose walls are perfectly conducting can be stimulated, by an oscillating dipole within the cavity for instance, to resonate at a very large number of different frequencies. Each resonant mode exhibits its characteristic pattern of antinodes and nodal surfaces, and for each mode the strength of the electric field $\mathscr{E}$ at some specified point can serve to define the field strength everywhere; thus one parameter, analogous to the displacement x of a harmonic oscillator, is sufficient for the whole, while the magnetic field $\mathscr{B}$, similarly specified by a single parameter, plays a role analogous to $\dot{x}$. The energy in the cavity at any instant, being the space-integral of the energy density, $\frac{1}{2}\varepsilon_0\mathscr{E}^2+\frac{1}{2}\mathscr{B}^2/\mu_0$, can be written in a form that exactly parallels the energy, $\frac{1}{2}\mu x^2+\frac{1}{2}m\dot{x}^2$, of a material oscillator. The Schrödinger equation for the cavity mode then takes exactly the same form as (13.1), and can be cast in the reduced form (13.2) by assigning a suitable definition to ξ. Just as $\frac{1}{2}\xi^2$ is the potential energy and $\frac{1}{2}\dot{\xi}^2$, i.e., $\frac{1}{2}(\mathrm{d}\xi/\mathrm{d}\tau)^2$, the kinetic energy of the material oscillator, so $\frac{1}{2}\xi^2$ is the total cavity energy due to the electric field and $\frac{1}{2}\dot{\xi}^2$ that due to the magnetic field. In free oscillation the energy is interchanged between these two forms twice per cycle, in the same way as the energy of a harmonic vibrator is alternately potential and kinetic.

An oscillating dipole placed at some point inside the cavity, where its moment $\boldsymbol{p}$ is acted upon by the electric field $\mathscr{E}$ at that point, is coupled to the cavity mode by a coupling energy $\boldsymbol{p}\cdot\mathscr{E}$; this is proportional to $\xi_1\xi_2$ if ξ_2 is a measure of the electric field in the cavity and ξ_1 of the dipole moment. Dipole and cavity mode together behave like a pair of harmonic vibrators with potential coupling, and the classical and quantal analyses can be carried out for such a coupled pair without considering the actual nature of the vibrators.

376

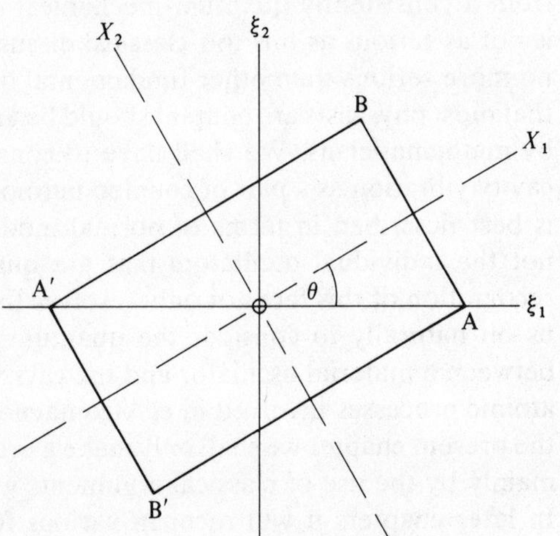

Fig. 1. Co-ordinates used to describe normal modes of two coupled harmonic vibrators (cf. fig. 12.3); ξ_1 and ξ_2 are co-ordinates of the individual vibrators, but when they are coupled the normal modes are represented by independent harmonic vibrations, at different frequencies, along X_1 and X_2.

There is no need to recapitulate the classical theory of coupled harmonic vibrators which was developed in chapter 12. The essential result is that if co-ordinates ξ_1 and ξ_2 are used, such that $\frac{1}{2}\dot{\xi}_{1,2}^2$ are the contributions to the kinetic energy of the uncoupled vibrators, then when they are coupled by a potential energy term proportional to $\xi_1\xi_2$ the normal modes are linear combinations obtained by rotating the axes in the (ξ_1, ξ_2)-plane, as in fig. 1 which differs from fig. 12.3 only in notation. Formally, the Hamiltonian for the coupled material vibrators takes the form

366

$$\mathcal{H}=p_1^2/2m_1+p_2^2/2m_2+\tfrac{1}{2}m_1\omega_1^2x_1^2+\tfrac{1}{2}m_2\omega_2^2x_2^2+Cx_1x_2. \tag{1}$$

If ω_1 and ω_2 are almost the same it is appropriate to define $\xi_{1,2}$ as $(m_{1,2}\bar{\omega}/\hbar)^{\frac{1}{2}}x_{1,2}$ and τ as $\bar{\omega}t$, where $\bar{\omega}=\frac{1}{2}(\omega_1+\omega_2)$. Then

$$\mathcal{H}=\tfrac{1}{2}\hbar\bar{\omega}(\dot{\xi}_1^2+\dot{\xi}_2^2+\omega_1^2\xi_1^2/\bar{\omega}^2+\omega_2^2\xi_2^2/\bar{\omega}^2+4\kappa\xi_1\xi_2/\bar{\omega}),$$

where $\kappa=C/[2\bar{\omega}(m_1m_2)^{\frac{1}{2}}]$. Writing $\xi_1=X_1\cos\theta+X_2\sin\theta$, $\xi_2=X_2\cos\theta-X_1\sin\theta$, such that

$$\tan 2\theta=2\kappa/(\omega_2-\omega_1), \tag{2}$$

we have that

$$\mathcal{H}=\tfrac{1}{2}\hbar\bar{\omega}(\dot{X}_1^2+\dot{X}_2^2+\Omega_1^2X_1^2/\bar{\omega}^2+\Omega_2^2X_2^2/\bar{\omega}^2), \tag{3}$$

the cross term having been eliminated. The normal mode frequencies, Ω_1 and Ω_2, are the solutions of the equation already derived as (12.11),

$$(\Omega_{1,2}-\bar{\omega})^2=\kappa^2+(\Delta\omega)^2, \tag{4}$$

in which $\Delta\omega=\frac{1}{2}(\omega_1-\omega_2)$. By working in normal mode co-ordinates the Schrödinger equation corresponding to (3) is made separable, and $\Psi(X_1,X_2)=\Psi_1(X_1)\Psi_2(X_2)$, Ψ_1 and Ψ_2 being solutions of harmonic oscillator equations for particles of the same mass but subject to different force constants. Alternatively (3) may be taken as the Hamiltonian of a particle moving on a plane in an anisotropic quadratic potential well. Its free motion, as a classical particle, would be a Lissajous figure traversed in the beat period, $2\pi/(\Omega_1-\Omega_2)$. As a quantized system, the component wave-functions would develop independently as already discussed, while the centroid of $\Psi\Psi^*$ would execute the classical Lissajous figure.

28

One can now see how randomization may enter. Let ξ_2 be the co-ordinate describing a cavity mode, and ξ_1 the dipole coupled to it. At the start of the process let us suppose both are in their ground state, so that a nearly circular Gaussian wave-packet sits at the origin in fig. 1. Excitation of the dipole by an impulsive electric field starts the wave-packet oscillating, unchanged in form, along the ξ_1-axis, say between A and A′; the subsequent Lissajous figure covers the rectangle ABA′B′, and the cavity oscillation is described by a Gaussian wave-packet whose amplitude of vibration slowly fluctuates at the beat period. So far no random element has appeared. Now let us imagine the dipole, uncoupled for the moment, to be perturbed by collisions with gas molecules; these interactions are strongly position-dependent and it is not unreasonable to suppose that the wave-function for the dipole is so disrupted that ultimately it is best described as the superposition of eigenstates whose amplitudes and phases are random. When the same process takes place while the dipole is coupled to a cavity vibration the randomization must include both. For any disturbance to the ξ_1 part of a normal-mode wave-function mixes in other normal-mode wave-functions, and therefore changes the ξ_2 part. The difference between this new treatment, which allows the transfer of randomness, and the old, which did not, lies in keeping both X_1 and X_2 as independent co-ordinates in (1) and thus in the Schrödinger equation; previously the equation for X_2 (or ξ_2) involved X_1 as a dependent variable, $X_1(t)$ defining the force that perturbed X_2. It is clear from this discussion that it is not enough to know how $\langle X_1 \rangle$ changes, or even the probability distribution for all possible changes of $\langle X_1 \rangle$; nothing of this sort can generate randomness in X_2. The mechanism of randomness is essentially non-classical, and this discussion serves to demarcate rather clearly the limit of classical reasoning about a harmonic oscillator – as soon as it interacts with other quantized systems, even another harmonic oscillator, great caution is needed if one hopes to avoid a fully quantal treatment of the coupled system. The argument applies, of course, equally well to the collision of a molecule with the oscillator; we implied above that the character of the interaction was in itself enough to produce a random outcome, but quantum mechanics makes this all the more inevitable.

Dissipation and level broadening

The foregoing analysis provides the basis for discussing the quantum mechanics of a vibrator subject to dissipation. The aim of this section is to clarify the meaning of the term *level broadening*. A classical lossy vibrator shows a Lorentzian resonance peak as its response to a force that may not be exactly tuned in frequency to the vibrator, and the quantal analogue is a slightly imprecise energy level which allows transitions to occur with emission or absorption of a quantum lying within a band of energy. In classical mechanics it is frequently permissible to introduce a loss term into the equation of motion without specifying its origin; thus a frictional force may be stated to be independent of velocity without a full treatment of the surface phenomena responsible for friction. The example we have just discussed, however, of a coupled dipole and cavity mode, illustrates how essential it is in quantum mechanics to be aware of the microscopic mechanisms at work, and to incorporate them in the analysis if necessary. There is no way of inserting rubbing friction into Schrödinger's equation, short of the prohibitively complicated expedient of writing down the many-body equation for every atom involved. We therefore base our discussion on realistic models of what may lead to dissipation in the microscopic world, and we must distinguish different types of dissipation, as well as recognize

136

that not all apparent level broadening is the result of dissipation. The following three examples will make the distinctions clear:

(1) *A particle in a potential well, through whose walls it may tunnel.* The solution of Schrödinger's equation in this case is virtually identical with the solution of a transmission line problem in which a resonator is defined by two reactances connected across the line, allowing weak transmission to the lines outside of the standing wave confined between them. It makes little difference, when the loss rate is slow, whether it arises from imperfect reflection of a wave incident on the barrier or by a process of absorption distributed throughout the well. The latter model crudely simulates the bound state of an unstable particle, or of one that by interaction with its surroundings may be scattered into a state irrelevant to the problem. The details of such processes should strictly be included in the treatment, but it is sometimes adequate to introduce particle non-conservation through the device of a complex potential. A particle constrained to move on a section of a line, of length a, on which the potential is $-\mathrm{i}V''$, is formally governed by Schrödinger's equation:

186

$$(\hbar^2/2m)\,\mathrm{d}^2\psi|\mathrm{d}x^2+(E+\mathrm{i}V'')\psi=0$$

and if $\psi=0$ at the two ends, $x=0$ and a, the nth eigenstate, $\psi_n=\sin(n\pi x/a)$, demands that $E_n=n^2\pi^2\hbar^2/(2ma^2)-\mathrm{i}V''$. The full wave-function, $\Psi_n=\psi_n\,\mathrm{e}^{-\mathrm{i}E_nt/\hbar}$, therefore has a decrement $\mathrm{e}^{-V''t/\hbar}$. The impulse-response function of this system, corresponding to the admixture after the impulse of other eigenstates, typically (from such products as $\Psi_m^*\Psi_n$) contains terms varying as $\mathrm{e}^{-\mathrm{i}(E_n-E_m^*)t/\hbar}$, an oscillation at the beat frequency but decaying as $\mathrm{e}^{-2V''t/\hbar}$. Correspondingly the frequency-response shows a Lorentzian resonance of half-width $2V''/\hbar$.

76

(2) *Doppler broadening of an otherwise sharp spectral line.* Here we have an ensemble of atoms moving at different speeds but all radiating identically from the point of view of an observer moving with the atom. The broadened line observed in a spectrometer is really composed of a multitude of sharp lines, unequally shifted by the relative motion, and does not constitute an example of line-broadening in the sense we are concerned with.

(3) *Radiative broadening.* This is the case to concentrate on, representing as it does a vibrating particle which, in contrast to the first example, is itself conserved but which is losing energy by radiation. We shall start with a strictly classical analysis, by a process which allows plausible extension to quantum mechanics.

The standard treatment of a Hertzian dipole(1) allows it to radiate into free space, but there are certain difficulties about incorporating unlimited free space into Schrödinger's equation, and we shall therefore return to the system already discussed, where a dipole is placed inside a large cavity, whose size may ultimately be increased without limit. There is now, to be sure, no mechanism of energy loss since the decay of the dipole oscillation accompanies the transfer of its energy to a huge number of loss-free normal modes of electromagnetic vibration in the cavity. It is possible that from time to time some of this energy will be returned to the dipole, but normally this will not be a perceptible effect, unless one makes the error of siting the dipole at the centre of a spherical cavity, when the radiated wave will be reflected and refocussed on the dipole. Even this is of small importance if the cavity is so large that the reflected wave only returns long after the dipole has radiated the greater part of its energy. And for this reason we need not worry about the inevitability of such a process taking place in the simplified model we shall treat first, in which the oscillating dipole is represented by a resonant circuit and the cavity is replaced by a one-dimensional analogue, a long transmission line, as in fig. 2.

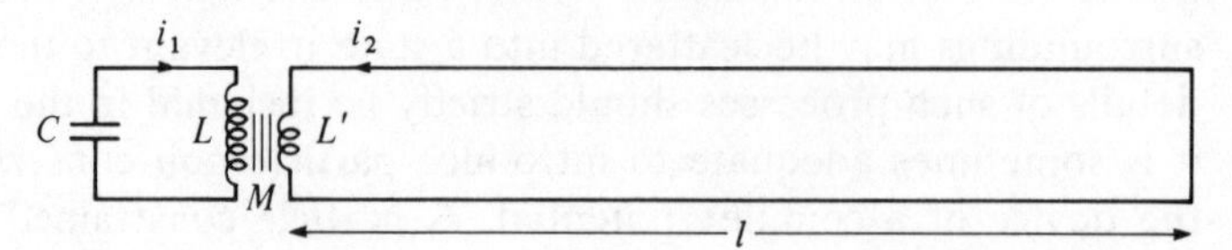

Fig. 2. Model of a radiating harmonic vibrator in the form of an LC circuit inductively coupled to a very long, lossless transmission line.

The line alone, uncoupled to the resonant circuit, behaves like one that is short-circuited at both ends, if L' is negligibly small. It has a closely spaced series of normal modes in which there is an integral number of half-wavelength standing wave loops; the nth mode has a frequency $n\pi c/l$, and the spacing is constant, $\pi c/l$. Hence the density of states $g(\omega)$, the number of modes in unit range of ω, is also constant with a value $l/\pi c$. When the circuit is coupled to the line the normal modes combine excitation of both line and circuit, and the coupling is especially evident for those modes whose frequency lies near the resonant frequency, ω_0, of the uncoupled circuit. As a result, the normal mode frequencies are displaced from their evenly spaced sequence, and the density of states is distorted. The first stage of the calculation is to show that a Lorentzian peak, centred on ω_0, is superimposed on the originally flat curve for $g(\omega)$.

Seen from the line, the resonant circuit presents an impedance Z_r, where

$$Z_r = \omega^2 M^2/(j\omega L - j/\omega C) \sim \omega_0^2 M^2/2j(\omega - \omega_0)L$$

when ω is very near ω_0.† The same Z_r would alternatively be produced by extending the length of the line by Δl, if $jZ_0 \tan(\omega \Delta l/c)$ is made equal to Z_r. The circuit therefore changes the frequency of any given mode by $\Delta\omega$, where $\Delta\omega/\omega = -\Delta l/l$; 165

i.e. $$\Delta\omega = (c/l)\tan^{-1}(\delta_0/\delta), \tag{5}$$

in which δ is written for $\omega - \omega_0$ and δ_0 for $\omega_0^2 M^2/2Z_0 L$. Modes well above and well below ω_0, for which $|\delta_0/\delta| \ll 1$, seem to be hardly affected, but this is a misleading interpretation. As δ goes from large positive to large negative, $\tan^{-1}(\delta_0/\delta)$ changes by π, so that on one side or other of the resonance the mode pattern slips by the mode spacing, to allow one extra

† For this circuit analysis, the electrical convention is adopted, as explained in chapter 3; j may be read as $-i$, if desired. 52

mode to appear, being that contributed by the resonant circuit. The rearrangement principally takes place in a frequency range of the order of δ_0, and if the line is so long that $c/l \ll \delta_0$ any shift is much smaller than the frequency range of interest. It is then permissible to treat the modes as forming virtually a continuum with a density $g(\omega)$ varying with δ; instead of the original value $l/\pi c$, the modified density of states has the form

$$g(\omega) = (l/\pi c)\left(1 - \frac{\mathrm{d}}{\mathrm{d}\delta}(\Delta\omega)\right) = l/\pi c + (\delta_0/\pi)|(\delta_0^2 + \delta^2), \tag{6}$$

from (5). The extra density of states appears as the second term at the right of (6), a Lorentzian peak of unit area, and width $2\delta_0$ at half-power, with a Q-value of $\omega_0/2\delta_0$.

The width of this peak is governed by the rate of radiation of energy from the resonant circuit into the transmission line. There is no doubt about what this rate is, since circuit theory allows the infinite line to be replaced by a resistance Z_0, appearing as ω^2M^2/Z_0 in the resonant circuit. Now the oscillation in an LCR circuit decays as $\mathrm{e}^{-\omega''t}$, with ω'' equal to $R/2L$; hence $\omega'' = \delta_0$ and Q for the circuit is $\omega_0/2\delta_0$, the same as characterizes the Lorentzian peak in the density of states. This argument, however, employs an infinite line and is not conveniently translated into quantal terms. Let us therefore derive the result by an alternative, if more roundabout, approach involving the transfer of energy to individual modes of the finite lossless line. The central idea to be brought out, and made quantitative, is that a state of affairs in which the circuit is excited into oscillation, while the transmission line is unexcited, is not a normal mode of the coupled system. Indeed its representation in terms of normal modes involves all of the latter, and it is the beating of these against each other that leads to the exponential decay of the circuit oscillation.

32

A synthetic approach to this question is the simplest; we write down a suitable combination of normal modes and then demonstrate that it shows the required properties. In this case it describes a sharp pulse incident on the unexcited resonant circuit and reflected as a pulse, but trailing behind it an exponentially damped wave that is immediately attributable to radiation from the circuit, following excitation by the pulse. A typical normal mode of amplitude A_n and phase ϕ_n has a standing wave pattern for $\mathscr{E}$, the field strength on the line:

$$\mathscr{E}(x, t) = A_n\,\mathrm{e}^{\mathrm{j}(\omega_n t + \phi_n)} \sin(\omega_n x/c - \theta_n), \tag{7}$$

in which ω_n must take the value $(n\pi + \theta_n)c/l$ in order that $\mathscr{E}$ shall vanish at the far end. At the near end $\mathscr{E}$ does not vanish at $x = 0$ but at $x = -\Delta l$ on account of the impedance presented by the circuit. Hence, from (5), $\tan\theta_n = \delta_0/\delta_n$.

The particular synthesis needed for the present purpose is that in which A_n is independent of n, and $\phi_n = \theta_n$, so that apart from a constant multiplier, which may be ignored,

$$\mathscr{E}(x, t) = \sum_n \{e^{j\omega_n(t+x/c)} - e^{j\omega_n(t-x/c)+2j\theta_n}\}.$$

Since the modes are almost equally spaced in frequency, summation may be replaced by unweighted integration; and the resultant variations of $\mathscr{E}(x, t)$ are expressed in the form

$$\mathscr{E}(x, t) = \int_0^\infty [e^{j\omega(t+x/c)} - e^{j\omega(t-x/c)+2j\theta(\omega)}]\, d\omega. \tag{8}$$

The first term describes a sharp pulse located at $x = -ct$ travelling towards the left-hand end and reaching it at time $t = 0$; thenceforth it lies to the left and has vanished from the picture. The presence of $\theta(\omega)$ in the second term, which describes the subsequent behaviour at positive x, shows that the reflected pulse is distorted. Let us evaluate this second term at the point $x = 0$, writing it in the form:

$$\mathscr{E}(0, t) = -e^{j\omega_0 t} \int_{-\infty}^{\infty} e^{j[\delta t + 2\tan^{-1}(\delta_0/\delta)]}\, d\delta, \tag{9}$$

in which $\delta = \omega - \omega_0$, as before. The lower bound has been extended from $-\omega_0$ to $-\infty$, a change which facilitates evaluation of (9) by contour integration without damaging the interesting part of the result which is the after-effect that follows the pulse. The integrand in (9) has one simple pole at $\delta = j\delta_0$. When $t < 0$ the term $e^{j\delta t}$ decreases exponentially as one proceeds in the negative imaginary direction of δ, and the negative semicircle at infinity makes no contribution to the contour integral which also includes the real axis. Since no poles are enclosed (9) vanishes at all negative t. When t is positive, however, the positive semicircle must be used to complete the contour, which now includes the pole. Thus if we disregard a constant factor in the residue, the integral is governed by the value of $e^{j\delta t}$ at the pole, i.e., $e^{-\delta_0 t}$, and $\mathscr{E}(0, t) \propto e^{(j\omega_0 - \delta_0)t}$ when $t > 0$. This decaying oscillation of $\mathscr{E}$ is of course induced by the circuit oscillation, and the analysis shows that it decays as $e^{-\delta_0 t}$, the result already known from circuit theory.

To conclude this section, it is worth stressing that the Lorentzian peak of width $2\delta_0$ is a peak in the density of very closely spaced sharp lines, each a loss-free normal mode of the complete system. If one likes to think of it as an extra, broadened, line characteristic of the radiation-damped oscillator alone, superimposed on the regular continuum of transmission line modes, no harm is done and it is probably a useful pictorial simplification. But fundamentally a broadened energy level has no meaning in quantum mechanics, and it is wise to bear in mind the interpretation in terms of a broad peak in the density of states.

See note on p. 509.

Electromagnetic radiation into free space

It is easy to extend the argument to the radiation of energy by an oscillating dipole into the modes of a large cavity.† Care is needed, however, in specifying the coupling of the dipole to the cavity modes. It must be remembered that of the many modes lying within any narrow frequency range some will have nodes and some antinodes of the standing waves of electric field $\mathscr{E}$ at the location of the dipole. The former will be uncoupled and the latter most strongly coupled, and all the intermediate cases will occur. Moreover all polarizations of $\mathscr{E}$ will be represented but only the component parallel to the dipole will be coupled to it. These effects are included in the argument by assigning the same, i.e. the average, coupling to each mode, simply by assuming the mean square value of the field responsible for coupling to be $\frac{1}{3}\overline{\mathscr{E}}^2$, where $\overline{\mathscr{E}}^2$ is the true mean square field in the cavity.

We now express δ_0 in the transmission line example in a form that allows its analogue in the cavity to be written down immediately:

$E_1 \equiv$ energy in oscillating circuit $= \frac{1}{2}Li_1^2$

$E_2 \equiv$ energy in a transmission line mode $= \frac{1}{4}Z_0 l i_2^2/c$

$E_{12} \equiv$ coupling energy $= Mi_1i_2$

$g(\omega) \equiv$ density of modes at circuit frequency $= l/\pi c$

The currents i_1 and i_2 are peak values in the circuit and on the line. In terms of these variables,

$$\delta_0 = \omega_0^2 M^2/2Z_0 L = \pi\omega_0^2 g(\omega_0)E_{12}^2/16E_1E_2. \quad (10)$$

Now for a dipole consisting of a mass m, carrying charge e and oscillating with amplitude a, the corresponding quantities for a given cavity mode are as follows:

$$E_1 = \tfrac{1}{2}m\omega_0^2 a^2, \qquad E_2 = \tfrac{1}{2}\varepsilon_0\overline{\mathscr{E}}^2 V = \tfrac{1}{2}\overline{\mathscr{E}}^2 V/\mu_0 c^2 \quad \text{and} \quad E_{12} = ea\mathscr{E}_\parallel,$$

where V is the volume of the cavity and $\mathscr{E}_\parallel$ is the peak electric field parallel to the dipole (we have not yet taken the average). Furthermore[(2)] $g(\omega_0) = \omega_0^2 V/\pi^2 c^3$. Substituting in (10), and writing $e^2a^2\overline{\mathscr{E}}_\parallel^2 = \frac{1}{3}e^2a^2\overline{\mathscr{E}}^2$ for the average value of E_{12}^2, we have

$$\delta_0 = \mu_0\omega_0^2 e^2/12\pi mc.$$

The time-constant for energy decay, τ_e, is $1/2\delta_0$;

i.e.
$$\tau_e = 6\pi mc/\mu_0\omega_0^2 e^2. \quad (11)$$

Alternatively expressed, the rate of energy loss by the dipole,

$$-\dot{W} = E_1/\tau_e = \mu_0\omega_0^4 e^2 a^2/12\pi c, \quad (12)$$

† Easy, but not obviously justifiable. A very serious problem of convergence arises here which is not present in the previous example. This will be discussed at the end of the chapter; it is a pervasive difficulty of electromagnetism but here, as in many other cases, it can apparently be ignored without invalidating the theory.

which is just Hertz's[1] expression for the rate of radiation by a dipole into free space. We have therefore satisfied ourselves that the dissipative radiative process can be represented by a multitude of non-dissipative energy exchanges with the normal modes of a cavity, provided the cavity is taken to be large enough for the spectrum of modes to be treated as a continuum. This is the same criterion as ensures that no reflected wave returns to the dipole until long after the decay is complete. Then at any instant the reaction of the cavity on the dipole is the same as if the cavity were completely unexcited until that very moment; since the system is linear and the rate of decay is always governed only by the instantaneous condition, it cannot be other than exponential. This argument carries over into the quantal treatment, and we shall take it for granted from now on. Thus, if we find the initial rate of energy loss immediately after coupling to the cavity to be αE_1, we shall infer a time-constant $1/\alpha$ for energy dissipation.

To illustrate this point, let us return to the transmission line model of fig. 2 and consider how the energy in the nth mode varies with time. At a frequency $\omega_n+\Delta$ the line presents at the input end an inductance $Z_0 \tan[(\omega_n+\Delta)l/c]$, i.e. $\Delta Z_0 l/c$ for small Δ. The same inductance would be presented by an LC circuit, resonant at ω_n, if $L=\frac{1}{2}Z_0 l/c$; it is convenient to replace the transmission line mode by this lumped circuit, so that the problem of energy transfer to the mode is reduced to the standard problem of energy transfer between coupled circuits. For the purpose of determining the initial decay of E_1 we assume the current i_1 to oscillate with constant amplitude and to excite the coupled LC circuit with an emf $\omega_0 M i_1 \sin \omega_0 t$. The response after switching on the excitation at $t=0$ takes the form (see 6.31):

146

$$i_2(t) = -i_2 \sin \bar{\omega} t, \qquad (13)$$

in which $$\bar{\omega} = \tfrac{1}{2}(\omega_0+\omega_n),$$

$$i_2 = (2\omega_0 M c i_1/Z_0 l)\sin(\tfrac{1}{2}\delta_n t)/\delta_n,$$

and $$\delta_n = \omega_n - \omega_0.$$

The energy E_2 in the mode is $\frac{1}{4}Z_0 l i_2^2/c$;

i.e. $$E_2 = (\omega_0^2 M^2 c i_1^2/Z_0 l)\sin^2(\tfrac{1}{2}\delta_n t)/\delta_n^2$$

$$= (4\delta_0 E_1/g\pi)\sin^2(\tfrac{1}{2}\delta_n t)/\delta_n^2. \qquad (14)$$

By writing the result in the form (14) it has been rendered independent of details of the particular model, and can be applied to other systems. So long as $\delta_n t \ll 1$, E_2 rises quadratically with time, but sooner or later the energy is restored to the first circuit, and from then on oscillates between the two with zero mean rate of increase. When all the other modes are considered to be similarly engaged, one sees that the number still involved in the quadratic increase are those in a frequency range $\pm\delta$ that decreases steadily as $1/t$, and the product of these two time variations causes the total energy in the modes to increase as t. From this the initial rate of decay of E_1 follows, and hence the time constant for the exponential radiation damping. To derive the result in detail, it is only necessary to integrate (14) over all modes:

$$E_1(0)-E_1(t)=\int_{-\infty}^{\infty}\dot{E}_2 g\,\mathrm{d}\delta_n=(2\delta_0 E_1 t/\pi)\int_{-\infty}^{\infty}\sin^2 x\,\mathrm{d}x/x^2=2\delta_0 E_1 t, \tag{15}$$

equivalent to a time constant $\tau_e = 1/2\delta_0$. This rate can be obtained by neglecting all the modes except those in a frequency range $2\pi/t$, for each of which perfect tuning is assumed, with $\delta_n = 0$. Thus each such mode makes a contribution to the initial decay of E_1 in the form (14) with $\delta_n = 0$:

$$E_1(0)-E_1(t)=\delta_0 E_1(0)t^2/g\pi = E_1(0)t^2/2\pi g\tau_e, \tag{16}$$

and the effective number of modes is $2\pi g/t$; the product yields (15).

Spontaneous radiation in quantum mechanics

We must now show how the essential features of these arguments can be carried over into quantum mechanics, and for this purpose we return to the basic model of an oscillating dipole interacting with a single cavity mode, the whole being represented by two coupled loss-free vibrators. As already noted, as Ψ evolves on the plane of fig. 1 the centroid $\langle \boldsymbol{r}\rangle$, where $\boldsymbol{r}$ represents the co-ordinates (ξ_1, ξ_2), executes a Lissajous figure of classical form. One cannot infer from this, however, that the energy exchange between the vibrators also behaves classically. The response of $\langle r^2\rangle$ to an applied force may follow classical rules, but we have already seen that adequate discussion of the energy transfer between quantized systems demands that the complete system be treated as a single quantized unit.

491

To appreciate what rules govern the exchange of energy let us first note some general points. If at any instant the wave-function for the two coupled vibrators, expressed in terms of normal mode co-ordinates, is $\Psi(X_1, X_2, t)$, this can be Fourier-analysed in terms of eigenfunctions which are themselves products of normal mode oscillator functions:

$$\Psi(X_1, X_2, t)=\sum_n\sum_m a_{nm}\psi_n(X_1)\psi_m(X_2)\,\mathrm{e}^{-\mathrm{i}[(n+\frac{1}{2})\Omega_1+(m+\frac{1}{2})\Omega_2]t}, \tag{17}$$

in which Ω_1 and Ω_2 are the normal mode frequencies, assumed to be fairly close together. If many terms are present in (17) the different frequencies are responsible for a complicated evolution of Ψ, but however complicated it may be the amplitude of Ψ must be strictly periodic at the beat frequency $\chi = |\Omega_1 - \Omega_2|$. Different physical quantities would show on evaluation different periodic patterns of behaviour, according to the harmonics of the beat frequency that appeared. The energy in either of the constituent vibrators, however, behaves in the simplest way possible, with only the fundamental playing a part. To evaluate E_1, for example, we must know

$\langle\xi_1^2\rangle$ and $\langle\dot{\xi}_1^2\rangle$; the point is sufficiently illustrated by considering $\langle\xi_1^2\rangle$, which in terms of normal modes is $\langle X_1^2\cos^2\theta-2X_1X_2\cos\theta\sin\theta+X_2^2\sin^2\theta\rangle$. Only the cross term in X_1X_2 generates the beat frequency χ or its harmonics when this is evaluated by use of (17):

$$\langle X_1X_2\rangle=\sum_n\sum_m\sum_{n'}\sum_{m'}\iint a_{nm}a^*_{n'm'}\psi_n(X_1)\psi_m(X_2)X_1X_2\psi^*_{n'}(X_1)\psi^*_{m'}(X_2)$$
$$\times\,\mathrm{e}^{\mathrm{i}[(n-n')\Omega_1+(m-m')\Omega_2]t}\,\mathrm{d}X_1\,\mathrm{d}X_2. \tag{18}$$

Now $\int\psi_nX_1\psi'_n\,\mathrm{d}X_1$ vanishes unless $n-n'=\pm1$, and similarly for X_2, so that the only frequencies to emerge from (18) are $\pm\Omega_1\pm\Omega_2$, and the only low frequency is χ. The energy exchange between the two vibrators is therefore purely sinusoidal.

Now in general if the two vibrators are set up in arbitrary non-stationary states and then coupled, this sinusoidal exchange of energy will not be caught at an extremal point, i.e. $\dot{E}_1=-\dot{E}_2\neq0$ initially. There are, however, certain situations in which initially $\dot{E}_1=0$, as for example when the vibrators are both in stationary states before coupling. More significant for the present discussion is for the cavity mode to be in its ground state while the dipole is arbitrarily excited. If, after coupling, E_2 is to oscillate at the beat frequency it must start at the minimum for the cycle, since there is no possibility that sampling at any later stage will reveal it to have an energy content less than its zero-point energy. To determine the range of oscillation of E_1 and E_2 we shall, for the sake of future reference, present the argument in a more general form than is needed for the immediate purpose.

At the moment of coupling, let the two vibrators have energies E_{10} and E_{20}:

$$E_{10}=\langle\xi_1^2+\dot{\xi}_1^2\rangle_0,\qquad E_{20}=\langle\xi_2^2+\dot{\xi}_2^2\rangle_0, \tag{19}$$

as follows from (13.11), if $\frac{1}{2}\hbar\omega_0$ is taken as the energy unit and all frequencies are close enough together for distinctions to be ignored, except when the beating effect is considered. Then the energies in the normal modes follow from rotating the co-ordinate axes, as in fig. 1:

$$\left.\begin{aligned}E_{n1}&=\langle X_1^2+\dot{X}_1^2\rangle=E_{10}\cos^2\theta+E_{20}\sin^2\theta+\langle\xi_{10}\xi_{20}+\dot{\xi}_{10}\dot{\xi}_{20}\rangle\sin2\theta\\E_{n2}&=\langle X_2^2+\dot{X}_2^2\rangle=E_{10}\sin^2\theta+E_{20}\cos^2\theta-\langle\xi_{10}\xi_{20}+\dot{\xi}_{10}\dot{\xi}_{20}\rangle\sin2\theta.\end{aligned}\right\} \tag{20}$$

Now let the cavity mode be initially in its ground state, so that E_{10} and E_{20} are extrema of E_1 and E_2. Then $\langle\xi_{20}\rangle$ and $\langle\dot{\xi}_{20}\rangle$ both vanish, and the third terms in (20) are absent. After coupling, E_{n1} and E_{n2} stay constant but the two normal modes have different frequencies, and consequently E_1 and E_2 oscillate about an easily determined mean. For on applying the reverse transformation we have that

$$\left.\begin{aligned}E_1&=E_{n1}\cos^2\theta+E_{n2}\sin^2\theta-\langle X_1X_2+\dot{X}_1\dot{X}_2\rangle\sin2\theta,\\E_2&=E_{n1}\sin^2\theta+E_{n2}\cos^2\theta+\langle X_1X_2+\dot{X}_1\dot{X}_2\rangle\sin2\theta,\end{aligned}\right\} \tag{21}$$

and the third terms oscillate with the beat frequency about a mean value of zero. Hence E_1 oscillates about a mean value:

$$(E_1)_{Av} = E_{n1}\cos^2\theta + E_{n2}\sin^2\theta = E_{10} + \tfrac{1}{2}(E_{20} - E_{10})\sin^2 2\theta \qquad (22)$$

from (20). Similarly

$$(E_2)_{Av} = E_{n1}\sin^2\theta + E_{n2}\cos^2\theta = E_{20} + \tfrac{1}{2}(E_{10} - E_{20})\sin^2 2\theta. \qquad (23)$$

Since E_{10} and E_{20} are extremal values for the oscillations of E_1 and E_2, it follows that E_1 oscillates between E_{10} and $E_{10}\cos^2 2\theta + E_{20}\sin^2 2\theta$, and E_2 between E_{20} and $E_{20}\cos^2 2\theta + E_{10}\sin^2 2\theta$.

This result may be compared with the beating behaviour of two classical vibrators. If at the moment of coupling, one is at rest with $E_{20} = 0$, the motion is initially represented on fig. 1 by a vibration between A and A′. After half a beat period the oscillation lies along BB′; E_2 has reached its maximum of $E_{10}\sin^2 2\theta$ and E_1 its minimum of $E_{10}\cos^2 2\theta$. The quantal behaviour is almost the same, except for the zero-point energy E_{20}; in fact the oscillations of energy are just as if the two initial values E_{10} and E_{20} indulged in independent beating in the classical manner, so that as a fraction $E_{10}\sin^2 2\theta$ was transferred from the first to the second vibrator, simultaneously $E_{20}\sin^2 2\theta$ was transferred in the reverse direction. Alternatively we may imagine only the excitation energy, $E_{10} - E_{20}$, as involved in an otherwise classical process. From this it is an obvious step to take over the whole classical description of the decay of a dipole oscillator in an empty cavity; it is not the total energy which is radiated to give exponential decay with time-constant $1/2\delta_0$, but only the excitation energy, $E - \frac{1}{2}\hbar\omega_0$, and the time-constant is the same as is given by the Hertzian theory of radiation.

Planck's radiation law; Einstein coefficients

The idea of independent energy exchange between two vibrators might be considered the most primitive form of Prévost's theory of exchanges, which is fundamental to the theory of heat transfer by radiation. The non-interaction of the traffic in both directions is readily justified in classical processes by the assumption of a random phase relationship between the bodies concerned. Thus when a vibrating dipole is placed in a cavity which is already excited, the initial power transfer results from the oscillating current of the dipole interacting with the oscillating electric field of the cavity modes; the mean value is zero if an average is taken over all possible phase relationships. It is only when each has had time to influence the vibration of the other, and introduce correlation into the phase relationship, that the mean power transfer is non-zero, and in fact is the same as if each were coupled to an unexcited vibrator. The proof of these statements is left as a simple exercise. The same is true when two quantized vibrators are coupled, and averages taken over every conceivable phase difference involved. Perhaps this is a fair representation of an assembly of dipoles placed in an excited cavity, but in fact it is not necessary to discard all phase relationships so wildly in order to achieve the effect of randomization. If the dipole and a cavity mode are established in some states of excitation, not necessarily stationary states, and are then coupled, in general there will be an immediate power transfer, with $\dot{E}_1 = -\dot{E}_2 \neq 0$ as already remarked. Let the experiment be repeated many times, however, with exactly the same initial conditions, but with varying intervals allowed to elapse before coupling, so that the relative phases of the two vibrators assume different initial values; it will be found that on the average the energy transfers in the two directions are independent.

The proof is straightforward. In each trial E_1 starts with the same value, E_{10}, but the oscillations vary in amplitude, though they are always sinusoidal with the same frequency, χ. Consequently when the average is found, it too will be sinusoidal with frequency χ. Moreover, although $\dot{E}_1$ does not initially vanish in most trials, the mean value $\dot{\bar{E}}_{10} = 0$ since positive and negative $\dot{E}_{10}$ are equally likely. Hence the oscillations of $\bar{E}_1$ start at an extremum, and are about an average value $(\bar{E}_1)_{\text{Av}}$ which is simply the mean of $(E_1)_{\text{Av}}$ taken over all the trials. Now when the vibrators are coupled after some arbitrary lapse of time there is no preferential phase correlation between the normal mode vibrations, so that the mean values of X_1X_2 and $\dot{X}_1\dot{X}_2$ are zero. Thus $(\bar{E}_1)_{\text{Av}}$ is obtained by dropping the last term in (21) and the argument proceeds exactly as if the cavity mode had been in a stationary state, with E_1 oscillating between E_{10} and $E_{10}\cos^2\theta + E_{20}\sin^2\theta$. And, as we have seen, this is the same as if the dipole was transferring its excitation energy to an unexcited cavity mode and conversely the cavity mode were transferring its excitation energy to an unexcited dipole. We may now be satisfied that when an excited dipole is placed in an excited cavity, there are enough modes of similar frequency and random phase for the mean power transfer to each to follow the same initial pattern, along the lines of (16), as if the cavity modes were all in their ground state. And similarly the average transfer from the cavity modes to the dipole starts in the same way as if the dipole were in its ground state.

It is clear from this that if the excitation energy of the dipole matches the average excitation energy for the cavity modes of the same frequency there will be no net transfer, so that thermal equilibrium between material oscillators and cavity modes is achieved when both exhibit the same dependence of mean energy on frequency. This, of course, was Planck's conclusion based on very careful, and obviously troubled, analysis of the process of interaction, and it led him to his black-body radiation law. The mean excitation energy of a quantized harmonic oscillator, and hence of a cavity mode, is $\hbar\omega/(e^{\hbar\omega/k_BT} - 1)$; and since the density of states per unit volume of cavity is ω^2/π^2c^3, the energy density (not counting zero-point energy) per unit range of ω takes the now familiar form:

$$u(\omega) = (\hbar\omega^3/\pi^2c^3)/(e^{\hbar\omega/k_BT} - 1). \tag{24}$$

Eighty years after the discovery one may have sufficient confidence in the result to point out that Planck's careful analysis was not strictly necessary.

We have seen how the choice of energy-based co-ordinates to describe coupled oscillators removes from the argument all parameters defining their magnitude or constitution. In these terms a material oscillator and a cavity mode at the same frequency become strictly equivalent, and the condition for statistical equilibrium could not be other than that their co-ordinates and hence their energy content should be the same. But to have expected Planck to see it in this light is to ignore the difficulties of a pioneer, and from the comfort of a limousine to belittle the hardships of those who drove the road through the jungle.

In 1916 Einstein published his justly famous analysis[(3)] of the black-body radiation problem in terms of spontaneous and stimulated processes. It might appear as if two rather different mechanisms were engaged, stimulated emission and absorption on the one hand, being the interaction of cavity modes with a material oscillator, and on the other hand spontaneous emission in which superficially it looks as if the cavity modes play no part. It is clear, however, from our discussion that it is exactly the same coupling between dipole and cavity modes that governs both processes, and that (at least when the dipole is a harmonic oscillator) everything may be described quite symmetrically in terms of spontaneous emission from either dipole or cavity mode into the other. Unfortunately many texts create an air of mystery about spontaneous emission by suggesting that it lies beyond the scope of Schrödinger's wave mechanics; it is even hinted sometimes that **593** the zero-point oscillations of the cavity are needed to initiate spontaneous emission, as it were by telling the otherwise isolated dipole that there is a suitable recipient for its energy near at hand. We have seen, however, that these difficulties evaporate once it is recognized that free vibrations of the dipole in an unexcited cavity do not constitute a normal mode.

As for the role of zero-point energy, what has emerged is that it is irrelevant to this discussion and is better ignored throughout – only excitations above the ground state enter the description of energy transfer.†

Nevertheless there is still a reconciliation to be effected between Einstein's vision and what has been developed here, and for this we apply his argument to an assembly of similar dipoles (harmonic oscillators) coupled to black-body radiation in a cavity. In the historical context of Bohr's quantum theory it is natural to suppose that each oscillator occupies at any one moment a well-defined energy level. Of the n_l oscillators occupying the lth level, some are making spontaneous transitions to the $(l-1)$th while some are being stimulated upwards and others downwards by the radiation. The rates, $-\dot{n}_l$, for the three processes are to be written, according to Einstein, as $A_l n_l$, $B_{l+1} n_l u(\omega_0)$ and $B_l n_l u(\omega_0)$. Only neighbouring levels are involved, and A_l and B_l are the coefficients for spontaneous and stimulated processes between l and $l-1$, while B_{l+1} is the coefficient for stimulated

† This should not be taken to imply, however, that zero-point effects are in principle unobservable; they have a deleterious influence on X-ray and neutron diffraction intensities, and have been invoked to explain the fluidity of helium down to 0 K.

processes between l and $l+1$. The mean rate of energy gain by each oscillator that happens to be excited in the lth level may therefore be written:

$$\dot{W}_l = (-A_l - B_l u + B_{l+1} u)\hbar\omega_0. \qquad (25)$$

Now we have already seen that the oscillator strength is proportional to the quantum number of the upper of the two levels involved, so that it is appropriate to write B_l as βl, β taking the same value for all levels. Hence

442

$$\dot{W}_l = (-A_l + \beta u)\hbar\omega_0, \qquad (26)$$

which is in exactly the form required to reconcile the different viewpoints. Each oscillator in the lth level is emitting energy at a rate $A_l\hbar\omega_0$, independent of u, while the cavity radiation is feeding it at a rate $\beta u\hbar\omega_0$, independent of l. It only remains to show that the calculated expressions for A_l and β make sense. For $A_l\hbar\omega_0$ we go back to (12), taking a to be the amplitude of a classical vibrator with energy $l\hbar\omega_0$, so that

$$A_l = \mu_0\omega_0^2 e^2 l/6\pi mc. \qquad (27)$$

For β we note that $\beta u\hbar\omega_0$ is the rate of absorption of energy from the radiation field by an oscillator in its ground state, $l=0$, for which $\beta_{l+1}=\beta$, and that this is the same rate as for the equivalent classical oscillator, starting from rest. Randomness of phase causes the increase in energy due to each mode to be additive, and the same argument as led to (16) applies here, that after time t the overall effect is as if only modes in the frequency range $2\pi/t$ around ω_0 had contributed. The energy density in these modes is $2\pi u/t$, so that the mean square of the component of electric field that may be considered to act on the dipole is $\frac{4}{3}\pi u/\varepsilon_0 t$. Since a field $\mathscr{E}$ oscillating in resonance imparts energy $\frac{1}{8}e^2\mathscr{E}^2t^2/m$ in time t, it follows that the dipole gains energy at a rate $\frac{1}{6}\pi e^2 u/\varepsilon_0 m$ or $\frac{1}{6}\pi\mu_0 e^2c^2u/m$.

Hence

$$\beta = \tfrac{1}{6}\pi\mu_0 e^2 c^2/\hbar m\omega_0, \qquad (28)$$

and

$$A_l/B_l = \hbar\omega_0^3/\pi^2c^3 = \hbar\omega_0 g/V. \qquad (29)$$

As is well known from standard treatments of this problem, (29) is the ratio of the coefficients for spontaneous to stimulated processes that is needed for Planck's radiation law to emerge from Einstein's theory.

The equation governing overall energy exchange between the cavity radiation and an assembly of oscillators follows from (26), which may be written by use of (29) in the form:

$$\dot{W}_l = \beta\hbar\omega_0(u - l\hbar\omega_0 g/V). \qquad (30)$$

Summing over all oscillators we find that their total excitation energy E obeys the equation:

$$\dot{E} = \sum_l n_l\dot{W}_l = \beta\hbar\omega_0(Nu - gE/V), \qquad (31)$$

since N, the total number of oscillators, is Σn_l and $E = \Sigma n_l l\hbar\omega_0$. Hence the mean energy of an oscillator, $\bar{E}(=E/N)$ relaxes towards its equilibrium value of uV/g according to the equation:

$$\dot{\bar{E}} = (\beta\hbar\omega_0 g/V)(uB/g - \bar{E}), \tag{32}$$

and the time-constant, τ_e, is $V/\beta\hbar\omega_0 g$, independent of u. This is exactly what we have been led to expect by the quasi-classical picture of cavity modes and oscillators exchanging energy as if each alone were excited. The value of τ_e is of course the same as the classical (Hertzian) value given by (11).

In placing so much emphasis on the strict reciprocity of the exchanges between a material oscillator and a cavity mode, there is some danger of hiding an important distinction between spontaneous and stimulated processes. Consider, for instance, radiation falling on a material surface and being partially reflected and partially absorbed. If the material is cold its response to the radiation is entirely to be ascribed to stimulated processes, and the reflection and absorption coefficients are controlled by the amplitude and phase of the dipole vibrations induced by the incident field. If the material is hot its own vibrators are excited quite independently of the incident field, and with no organized phase relationship to that of the field. In classical and quantum physics these processes, when they occur in harmonic vibrators, are uncoupled. One may therefore observe the spontaneous radiation of a red-hot sheet of metal without confusing it with the reflection of a bright light shone onto it; the extra dipole vibrations caused by the latter are phase-coherent with the source. It is not so easy to appreciate this point if one thinks of the material vibrators as existing only in pure quantum states, since these have no dipole moment. However, if we picture the incident radiation as possessing a well-defined phase, we know from chapter 13 that the additional motion excited in the vibrators will be essentially the same as if they were classical, and initially at rest. The energy transfer from field to vibrators is thus accompanied by coherent excitation, and conversely this reacts back on the radiation field also in a coherent fashion. These are the processes of stimulated absorption and emission. The spontaneous process is in no essential sense different from stimulated emission except that the emitting system is not phase-linked to the incident field.

445

We have made considerable progress without the need to set up quantum-mechanical equations of motion, but to go beyond this point demands more detailed analysis. Further discussion of radiative processes will therefore be postponed, to be resumed in chapters 18 and 20.

528, 588

Divergences in the theory of dipole radiation; mass enhancement

We discussed the level broadening resulting from spontaneous radiation in terms of the interaction of a material oscillator with the cavity modes having very similar frequencies. We did not enquire into the effect of high-frequency modes which unhappily lead to a divergent integral. It is convenient to reformulate the problem in terms of an ensemble of vibrators, each representing a cavity mode, coupled to the material oscillator but not among themselves. The normal modes of such an arrangement can be expressed as the solutions of a comparatively simple equation.

For simplicity let us ignore variations, from one mode to another, of polarization and location of the field maxima, treating all modes as coupled to the oscillator with the mean value of the coupling constant. When the energy of the ith mode is wholly electrical and of magnitude $\frac{1}{2}\varepsilon_0\overline{\mathscr{E}^2}V$, we equate this to the energy of the equivalent material vibrator, $\frac{1}{2}\mu_i x_i^2$, and take $(\frac{1}{3}\overline{\mathscr{E}^2})^{\frac{1}{2}}$ as the field acting on the oscillating charge. Thus the field has strength $(\mu_i/3\varepsilon_0 V)^{\frac{1}{2}}x_i$ and exerts a force $c_i x_i$ on the oscillating charge, where $c_i = e(\mu_i/3\varepsilon_0 V)^{\frac{1}{2}}$. When all modes are taken into account, the equation of motion of the material oscillator has the form:

$$m_0\ddot{x}_0 + \mu_0 x_0 - \sum_i c_i x_i = 0. \tag{33}$$

Correspondingly the dipole moment of the oscillator modifies the equation of motion of each mode in the expected reciprocal manner, as may easily be proved:

$$m_i\ddot{x}_i + \mu_i x_i - c_i x_0 = 0. \tag{34}$$

Let a normal mode solution of the set of equations consisting of (33) and the N equations analogous to (34) have frequency Ω. When this trial solution is substituted, the condition for the $N+1$ equations to be compatible is that the determinant of the coefficients shall vanish;

i.e.
$$\begin{vmatrix} \mu_0 - m_0\Omega^2 & -c_1 & -c_2\ldots \\ -c_1 & \mu_1 - m_1\Omega^2 & 0\ldots \\ -c_2 & 0 & \mu_2 - m_2^2\Omega^2\ldots \\ \vdots & \vdots & \vdots \end{vmatrix} = 0. \tag{35}$$

On multiplying out and dividing through by $(\mu_1 - m_1\Omega^2)$ $(\mu_2 - m_2\Omega^2)\ldots(\mu_N - m_N\Omega^2)$ the equation for Ω is obtained in the form:

$$\mu_0 - m_0\Omega^2 = \sum_i c_i^2/(\mu_i - m_i\Omega^2), \tag{36}$$

i.e.
$$\omega_0^2 - \Omega^2 = (e^2/3m_0\varepsilon_0 V)\sum_i \omega_i^2/(\omega_i^2 - \Omega^2). \tag{37}$$

The right-hand side of (37) oscillates with great rapidity, reaching $\pm\infty$ at every value of Ω that coincides with a cavity mode. The slope of the function is least at points lying roughly midway between successive modes, as shown in fig. 3(a), and it is clear that the solutions are bunched together most strongly where the intersections of the lines representing the left and right sides of (37) lie near the positions of minimal slope. We shall take

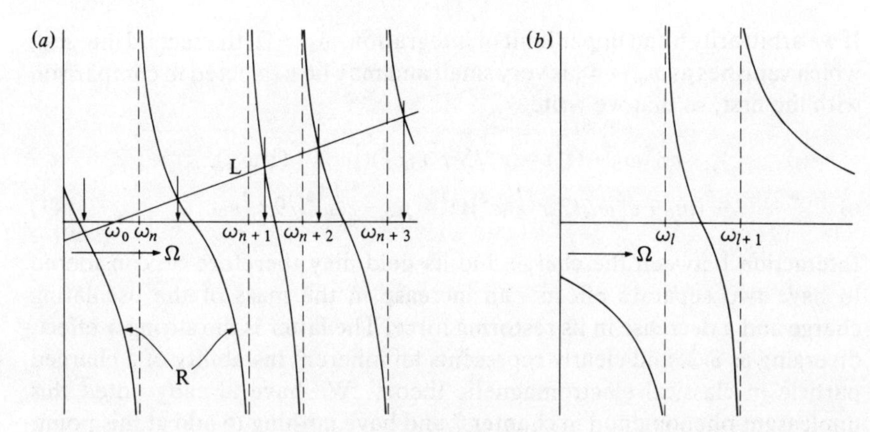

Fig. 3. (*a*) Schematic diagram of the left (L) and right (R) hand sides of (37) as functions of Ω. The normal modes are given by the intersections, as shown by the arrows, and it is the slope of L that causes the displacement of successive modes to vary progressively and leads to a slight bunching of the modes. In (*b*) is shown the contribution to R of two successive modes alone.

this as defining the centre of the broadened line. To determine its position, let us suppose it to lie between ω_l and ω_{l+1}, whose contribution to the right-hand side of (37) is shown in fig. 3(*b*). If that were all, the minimum slope would occur on the axis, and the line would have its centre at $\omega_0^2 - \Omega^2 = 0$. But all the other modes may be expected to provide a contribution to the right-hand side of relatively low slope, but not necessarily small magnitude; the centre of the line will occur when $\omega_0^2 - \Omega^2$ is equal to this contribution. We therefore evaluate (37), excluding the immediate vicinity of Ω from the sum; for large V and correspondingly dense distribution of modes, the sum may be replaced by the principal value of an integral:

$$m_0(\omega_0^2 - \Omega^2) = (e^2/3\varepsilon_0) ⨍ \frac{\omega^2 g(\omega)}{(\omega^2 - \Omega^2)} \, d\omega, \tag{38}$$

in which $g(\omega)$ is the density of states per unit volume of cavity, i.e. $\omega^2/\pi^2 c^3$. Hence

$$m_0(\omega_0^2 - \Omega^2) = (e^2/3\pi^2 \varepsilon_0 c^3) ⨍ \frac{\omega^4}{(\omega^2 - \Omega^2)} \, d\omega. \tag{39}$$

There is no reason to cut off the spectrum of cavity modes at any finite frequency, and the problem raised by the analysis is immediately obvious, since the integral in (39) diverges as ω^3 at high frequencies. The difficulty does not arise in the model represented in fig. 2. Analysis of inductively coupled circuits shows that the equation corresponding to (38) does not contain ω^2 in the numerator of the integral; and since for a transmission line $g(\omega) = \text{constant}$, the integral to be evaluated is $⨍_0^\infty d\omega/(\omega^2 - \Omega^2)$, which is zero. The earlier result is therefore confirmed, that the centre of the line is unshifted.

To return to the oscillating dipole, the integral in (39) may be conveniently rearranged

$$⨍ \frac{\omega^4}{(\omega^2 - \Omega^2)} \, d\omega = \int (\omega^2 + \Omega^2) \, d\omega + \Omega^4 ⨍ \frac{d\omega}{(\omega^2 - \Omega^2)}. \tag{40}$$

If we arbitrarily fix an upper limit of integration, $\omega_m \gg \Omega$, the second integral, which vanishes as $\omega_m \to \infty$, is very small and may be neglected in comparison with the first, so that we write

$$m_0(\omega_0^2 - \Omega^2) = (e^2/3\pi^2\varepsilon_0 c^3)(\tfrac{1}{3}\omega_m^3 + \Omega^3\omega_m),$$

or

$$(m_0 + e^2\omega_m/3\pi^2\varepsilon_0 c^3)\Omega^2 = \mu_0 - e^2\omega_m^3/9\pi^2\varepsilon_0 c^3. \tag{41}$$

Interaction between the charge and its field may therefore be considered to have two separate effects – an increase in the mass of the oscillating charge and a decrease in its restoring force. The latter is the stronger effect, diverging as ω_m^3, and clearly represents an inherent instability of a charged particle in classical electromagnetic theory. We have already noted this unpleasant phenomenon in chapter 7 and have nothing to add at this point, except to remark that it is not readily eliminated from either the classical or the quantal treatment. The achievements of quantum electrodynamics have taught us that it is possible to live in the shadow of infinities such as this, and to devise systematic procedures for getting the right answer to a variety of problems by pretending they are not there.[(4)]

200

The weaker divergence, on the left-hand side of (41), can be controlled, as the stronger cannot, by assigning non-vanishing radius to the oscillating charge. Suppose, for instance, that the charge is uniformly distributed on the surface of a sphere of radius a. Then it is easy to show that a field varying as $\mathscr{E}_0 \mathrm{e}^{ikx}$ acts on the charge with a force that is not $e\mathscr{E}_0$ but $e\mathscr{E}_0 \sin(ka)/ka$. In the light of this, we may replace the relevant part of (40), which is $\int \Omega^2\,\mathrm{d}\omega$, by $c\int_0^\infty \Omega^2 \sin ka\,\mathrm{d}k/a$, since $\omega = ck$; and in (41) the extra mass is altered by replacing ω_m by $\pi c/2a$. The effective mass is thus $m_0 + e^2/6\pi\varepsilon_0 c^2 a$. Now the electrostatic field energy outside a charged sphere is $e^2/8\pi\varepsilon_0 a$, and if we apply Einstein's mass–energy relation we expect this energy to have a mass $e^2/8\pi\varepsilon_0 c^2 a$, $\frac{3}{4}$ of the amount we have just derived. The history of electrodynamics from about 1880 till Einstein's work (1905) is full of attempts to resolve the precise relationship between electrostatic energy and mass, and is littered with numerical factors of the order of unity. In view of the fact that we have agreed to ignore the gross divergence on the right-hand side of (41), it would have been astonishing for the right answer to have emerged from any subsequent analysis of the problem. Nevertheless the form of the correction to the mass gives confidence in ascribing the mass correction to the field energy, even while we recognize that this is a palliative rather than a cure for the weakness. It is a basic conundrum in particle theory to reconcile the apparent point-character of the electron with its finite mass, and it is here that renormalization theory is particularly successful in sidestepping the issue.

The divergence problem does not arise when an oscillator embedded in a solid interacts with the lattice vibrations (*phonons*) rather than with electromagnetic cavity vibrations (*photons*), since the atomicity of the solid imposes a natural cut-off frequency at the upper end of the vibration spectrum. For a given energy of a vibrational mode, the electric field associated with a lattice vibration is normally very much less than that associated with the same mode in a cavity of the size and shape of the solid, and to this extent the coupling is weaker. The density of states, however, is now $\omega^2/\pi^2 v_s^3$ rather than $\omega^2/\pi^2 c^3$, where v_s is the velocity of sound, perhaps 10^4–10^5 times less than c, and this vastly greater density of states commonly makes phonons dominate photons as the mechanism of energy exchange. By the same token, the enhancement of mass by lattice polarization may be very significant.[5] In this case there is no need to invoke Einstein's mass–energy relation – the charged particle elastically deforms the lattice around itself, and enhances the local density; when the particle moves, the excess mass must also move with it. This at any rate is true for slow vibrations of the particle, or slow motion of a charged particle through the solid, to which an expression analogous to (41) applies. But if the particle is caused to vibrate at a frequency much greater than ω_m the neglected last term in (40) is important; indeed, when $\Omega \gg \omega_m$ it cancels the mass-enhancement. Obviously, if one forces the particle to vibrate at a frequency much higher than any possible lattice frequency, the lattice will not follow the motion and there will be no contribution from it to the mass of the particle. An alternative way of visualizing the process is in terms of the response of the particle to an impulsive force. At the instant of application of the impulse, P, the particle jumps forward with velocity P/m_0, responding as though it were a free particle, as already noted in chapter 13. It momentarily leaves behind its clothing of virtual phonons (i.e. the local lattice distortion), but after a time of the order of $1/\omega_m$ these begin to respond and as the lattice reacts in this way it drags the particle back so that eventually particle and lattice distortion proceed together at a lower speed, $P/(m_0 + m_{ph})$, where m_{ph} is the phonon mass-enhancement.

This discussion has perhaps deflected attention from the principal point, which is the description of radiative processes and the meaning of the resulting line broadening. The phenomenon of mass-enhancement can be treated as a quite separate issue, associated especially with the high-frequency end of the vibration spectrum, which is at least partially responsible for whatever mass the oscillating particle happens to exhibit. The line broadening, on the other hand, is something which concerns only those cavity modes whose frequencies lie very close to the oscillator frequency, and there is no reason to doubt the treatment of this process because of flaws in the treatment of other, essentially disconnected, processes.

Additional note (see p. 496)
The method of this section can readily be applied to calculate the contribution of a harmonic vibrator to the thermodynamic properties (entropy, free energy etc.) even when the vibrator is heavily damped or over-damped.[6]

17 The equivalent classical oscillator

This chapter may be regarded as a prologue to the next or as a brief gathering together of some threads from chapter 13, where it was shown that, under certain not too restrictive conditions, a quantum system responds to an applied force in a way that can be modelled by a set of classical harmonic oscillators, one for every possible transition between energy levels. The impulse-response function, $h(t)$, derived in (13.37), applies either to a system which, before the impulse, is known to be in a stationary state, or to an assembly of identical systems whose previous history has rendered them chaotic in phase. In either case the behaviour of $\langle x(t)\rangle$, considered either as the response of a simple system in the nth quantum state or that part of the average for the response of the assembly that is attributable to the occupation of the nth state, takes the form:

439

$$h_n(t) = \sum_l (f_{ln}/m\omega_{ln}) \sin \omega_{ln} t, \tag{1}$$

in which

$$f_{ln}/m\omega_{ln} = 2|\langle l|x|n\rangle|^2/\hbar = 2|M_{ln}^{(1)}|^2/\hbar, \tag{2}$$

and $\omega_{ln} = (E_l - E_n)/\hbar$. A set of classical oscillators, of which a typical member has mass m/f_{ln} and natural frequency ω_{ln}, would give in sum the same response.

The equivalence of the quantum system and the set of classical oscillators holds when a time-varying force is applied, provided the response is linear, to allow the force to be resolved into a sequence of impulses, each inducing a response of the form (1). One must not apply the argument if the perturbing force is so strong that the occupation of the nth quantum state is significantly altered or (what amounts to much the same) if the assumption of random phase is caused to fail, for $h_n(t)$ will thereby be changed. To estimate how serious this restriction is let us consider a single transition, between the nth and the lth levels, when an electron is perturbed by an electric field $\mathscr{E} \cos \omega t$ and ω is close to ω_{ln}. We put $a_n = 1$ and $a_l = 0$, and use (13.39) to find how a_l develops:

$$\dot{a}_l = (ie\mathscr{E}M_{ln}^{(1)}/\hbar)\, \mathrm{e}^{-\mathrm{i}\omega_{ln}t} \cos \omega t \sim (\tfrac{1}{2}i\mathscr{E}p/\hbar)\, \mathrm{e}^{i\delta t}, \tag{3}$$

in which $\delta = \omega - \omega_{ln}$ and $p = eM_{ln}^{(1)}$; a rapidly oscillatory term is neglected. The resulting amplitude of oscillation of a_l at the beat frequency δ is

$\frac{1}{2}\mathscr{E}p/\hbar\delta$, and the criterion for the continued application of (1), which is the criterion of linear response, is simply that a_l must never approach unity:

$$\mathscr{E}p \ll \hbar\delta. \tag{4}$$

Now for an atomic system p is typically the dipole moment resulting from shifting an electron through a distance comparable to the atomic radius, say 3.3×10^{-30} Cm (1 Debye unit). With gross mistuning, e.g. δ equal to the frequency of the line ($\sim4\times10^{15}\ \text{s}^{-1}$ for an optical transition) $\mathscr{E}$ must approach $10^{11}\ \text{V m}^{-1}$ to initiate non-linear processes. This is an enormous field strength, such as would be found in a beam of radiation whose power flux was about $10^{19}\ \text{W m}^{-2}$. Huge as this is, it is not beyond the

604

flux obtainable in short pulses from the largest lasers, and at this power level virtually every physical system will respond non-linearly, even if it is not immediately destroyed. With much less power and better tuning non-linear effects are relatively easy to obtain; thus with δ 10^5 times smaller than ω_{ln}, the required flux is only $10^8\ \text{W m}^{-2}$ – out of the question with monochromatic radiation before the invention of lasers, but readily available since. If there were no radiative loss, and the spectral line were consequently perfectly sharp, (4) would hold for all δ, however small; under exact tuning conditions any value of $\mathscr{E}$ would suffice to produce non-linear effects in the response. Such a state of affairs is not very different from the experimental conditions prevailing in nuclear magnetic resonance, where very low power levels can cause saturation of the line. In spectroscopy at optical frequencies, however, (4) begins to fail when δ is comparable to the line width and quite sizeable values of $\mathscr{E}$ are allowable. Loss of energy by radiative transitions then limits the build-up of a_l and encourages linear response.

The orders of magnitude just arrived at explain the interest, since the invention of lasers, in the theory of non-linear optical processes, which we shall not consider here;[(1)] but they also show that for many purposes the assumption of linearity is fully justified, and with it the use of the equivalent classical oscillator as an aid to description of processes and for problem-solving. Within the linear assumption the changes in the occupation probabilities $a_n a_n^*$ are so small that one need not worry about transitions involving three or more levels, such as the climbing of the ladder of oscillator levels discussed in chapter 13. Indeed the harmonic oscillator is an especially sensitive system from this point of view on account of the even spacing of the levels. More usually the significant excitation of a system from the nth level by radiation closely tuned to ω_{nl} will find it in the lth level with nowhere else to go except back again, all other potentially accessible levels being at the wrong spacing for near-resonance. Systems with equally spaced levels, like the infinite set for a harmonic oscillator, or the $2s+1$ levels for a particle of spin s, respond in a way which has marked classical analogies, even when strongly excited, and thereby are not inaccessible to exact theory. With other systems non-linearity only begins to involve more than two levels with a much stronger perturbing force, such that (4) fails for a whole chain of connected levels. The theoretical problem then begins to be formidable except to the sledgehammer approach of a computer. The intermediate case, however, can be approximated by considering only the two levels which are near resonance with the perturbing force, and this provides the incentive for detailed analysis of the two-level system in the next chapter. To take things in proper order, however, let us first develop some consequences of the equivalent classical oscillator theory.

The f-sum rule

It was noted in chapter 13 that when an impulse P is applied to a quantum system the immediate response may be derived by neglecting V in the time-dependent Schrödinger equation and consequently treating the particle as free. Moreover it is easily shown that the result (13.30) implies that the expectation value of the momentum is increased by P, while the position is unchanged, just as for a classical particle. Now, according to (1), when unit impulse is applied to the system in its nth state, the instantaneous response of the velocity may be written

439

$$\dot{h}_n(0) = \sum_l f_{ln}/m,$$

and it is this that must match the response of a classical particle, which is $1/m$. It follows that when a real system, in which only one particle is affected by the applied force, is replaced by a set of equivalent oscillators, each acted upon by the same force, the oscillator strength, f_{ln}, involved in transitions from a given level, n, to any other level, l, must obey the f-sum rule:

$$\sum_l f_{ln} = 1. \tag{5}$$

We have already had an example in the harmonic oscillator, where a uniform force can induce transitions from n only to $n \pm 1$, and where $f_{n+1,n} = n+1$ and $f_{n-1,n} = -n$, so that (5) is obviously true. Let us explicitly verify the rule in another case which is easy to treat, a particle confined to a limited stretch of a line, $0 < x < a$, in which range $V = 0$. Then the nth wave-function (normalized) is $(2/a)^{\frac{1}{2}} \sin(n\pi x/a)$ and its energy is $n^2(\pi^2\hbar^2/2ma^2)$. Consequently $\omega_{ln} = \pi^2\hbar(l^2 - n^2)/2ma^2$ and

442

$$\langle l|x|n\rangle = (2/a)\int_0^a x \sin(l\pi x/a)\sin(n\pi x/a)\,\mathrm{d}x$$

$$= -8nla/[\pi^2(l^2-n^2)^2] \text{ if } l-n \text{ is odd,} \tag{6}$$

$$= 0 \text{ if } l-n \text{ is even.}$$

Hence $f_{ln} = 64n^2l^2/[\pi^2(l^2-n^2)^3]$ if $l-n$ is odd. Let us evaluate f_{ln}, for transitions from the state $n = 4$:

$l=$	1	3	5	7	9	11	13	15
$f_{l4}=$	−0.0307	−2.7234	3.5581	0.1415	0.0306	0.0108	0.0049	0.0026

The sum of the f_{l4} up to this point is 0.9954 and the higher terms account for the missing fraction. If we had considered transitions from any other state, the f-values would of course have all been different, but the sum is always unity.

It will be noted that, as with the harmonic oscillator, some of the terms are greater than unity, indicating a stronger response than a classical oscillator could provide, and that downward transitions have negative strength. The contribution of these to the impulse response function is opposite in sign to that of a classical bound particle having positive mass. There are consequences which will play an important role in what follows. For example, upward and downward transitions contribute oppositely to the polarization of a molecular system by a steady field, upward transitions giving rise to a positive electrical susceptibility, downward to a negative. And when the response function exhibits decay as a result of radiative transitions or collisions, the upward transitions reveal this as dissipation, but the downward as negative dissipation, or gain. The action of masers and lasers depends on this sign reversal.

347
574

The f-sum rule expressed in (5) applies to a single particle, but a generalization is straightforward. We consider the commonest case, that of electric dipole transitions, where the force is applied by a uniform electric field, $\mathscr{E}$; different particles in a complex system experience different forces, $q\mathscr{E}$, according to their charge q. A free particle of charge q_i and mass m_i, acted on by an impulsive field, $P_e = \int \mathscr{E}\, \mathrm{d}t$, responds with an instantaneous velocity change $P_e q_i/m_i$, and an instantaneous change in electric current $P_e q_i^2/m_i$. If the response of the nth state of the system is to be simulated by a set of classical oscillators, all carrying charge e but with masses m/f_{ln}, then

$$\sum_l f_{ln} e^2/m = \sum_i q_i^2/m_i,$$

$$\text{i.e.} \qquad \sum_l f_{ln} = \sum_i (q_i/e)^2/(m_i/m). \tag{7}$$

For a system composed of N identical particles, electrons for instance, it is natural to choose e and m to be the electronic charge and mass, whereupon (7) takes the form $\sum_l f_{ln} = N$.

It often happens that in the system to which the f-sum rule is applied some of the particles are more or less free, such as conduction electrons in a metal, some are loosely bound (electrons near the top of the valence band in a semiconductor), some tightly bound (inner shell electrons) and some very tightly bound indeed by these standards (individual nucleons in the nuclei). The corresponding oscillator frequencies cover a huge frequency range, but if they fall into well-separated bands it may be possible to neglect the high-frequency oscillators when considering low-frequency phenomena; nuclear structure is of small consequence to the conduction properties of solids, and indeed the nucleus and the inner electron shells are often treated as a single structureless charged particle, the ion core. There is, however, no blanket rule to cover neglect of tightly bound structures; the polarization of an atom or an insulating solid may be dominated by the equivalent oscillators of low frequency, but inner shells often make a far from negligible contribution. Care is always needed when the oscillator spectrum is truncated.

Static polarizability

A typical equivalent oscillator, of strength f_{ln}, and characteristic frequency ω_{ln}, must be assigned a restoring force constant μ_{ln} equal to $m\omega_{ln}^2/f_{ln}$. Under the influence of a steady field $\mathscr{E}$ the displacement is $e\mathscr{E}/\mu_{ln}$ and the resulting dipole moment $e^2\mathscr{E}/\mu_{ln}$. Hence the total dipole moment acquired by the system in its nth state may be written

$$p = (e^2\mathscr{E}/m)\sum_l f_{ln}/\omega_{ln}^2. \tag{8}$$

If the system considered is a single atom or molecule the molecular polarizability α is defined by the equation $p = \alpha\varepsilon_0\mathscr{E}$. With N molecules per unit volume the polarization per unit volume $P = N\alpha\varepsilon_0\mathscr{E} = \kappa\varepsilon_0\mathscr{E}$, where κ is the volume susceptibility, $N\alpha$†. From (8),

$$\alpha_n = (e^2/m)\sum_l f_{ln}/\omega_{ln}^2. \tag{9}$$

The example already studied, the particle moving on a limited stretch of line, serves to illustrate this result, and we shall compute the polarizability in the ground state, $n = 1$. From (6), $f_{l1} = 64l^2/[\pi^2(l^2-1)^3]$ for even l, and $\omega_{l1} = (l^2-1)(\pi^2\hbar^2/2ma^2)$; hence

$$\begin{aligned}\alpha_0 &= \frac{256me^2a^4}{\pi^6\hbar^2}\sum_{l\,\mathrm{even}} l^2/(l^2-1)^5\\ &= \frac{me^2a^4}{\pi^4\hbar^2}(42\,696\,663 + 54\,652 + 1778 + 167 + 27 + 6 + 2 + 1)\\ &\quad\times 10^{-8} = 0.42753296\,me^2a^4/\pi^4\hbar^2.\end{aligned} \tag{10}$$

To calculate α_0 directly Schrödinger's equation must be solved for the potential $V = -\mathscr{E}x$, but the wave-function need be correct only to first order in $\mathscr{E}$. It may be checked by substitution that the following solution is adequate:

$$\psi = (1-\beta x)\sin kx + \beta kx^2\cos kx, \tag{11}$$

where $\hbar^2k^2/2m = E$ and $\beta = e\mathscr{E}ma^2/2\pi^2\hbar^2$. The zeros of ψ are not exactly

† Strictly $\boldsymbol{\mathscr{E}}$ and $\boldsymbol{P}$ are vectors and κ is a second-rank tensor. In the three-dimensional case one must recognize that the matrix element $\langle l|x|n\rangle$ must be treated as a vector $\langle l|x_i|n\rangle$ and that f_{ln} is a second-rank tensor, $(2m\omega_{ln}/\hbar)\langle l|x_i|n\rangle\langle l|x_j|n\rangle^*$, conferring tensorial anisotropy on the reciprocal mass $1/m_{ln} = f_{ln}/m$. This is a complication which must be taken account of in many real computations, but does not affect the physical principles discussed here.

at 0 and a, but at 0 and $X(=a(1-\beta a))$; this does not matter, however, provided the same stretch of line is used to calculate $\langle x\rangle$ before and after $\mathscr{E}$ is applied, the difference yielding α correct to first order in $\mathscr{E}$. Now, from (11), $\int_0^X \psi^2\,dx = \frac{1}{2}a(1-2\beta a)$ and $\int_0^X x\psi^2\,dx = \frac{1}{4}a^2[1-(10/3-5/\pi^2)\beta a]$. It follows that after $\mathscr{E}$ is applied

$$\langle x\rangle = \int_0^X x\psi^2\,dx \Big/ \int_0^X \psi^2\,dx = \tfrac{1}{2}a[1-(4/3-5/\pi^2)\beta a],$$

while when $\mathscr{E}=0$, $\langle x\rangle = \frac{1}{2}X$. The shift is $\frac{1}{2}\beta a^2(5/\pi^2-1/3)$, and hence

$$\alpha = (me^2a^4/\pi^4\hbar^2)(5/4-\pi^2/12) = (me^2a^4/\pi^4\hbar^2)\times 0.42753297,$$

in agreement with (10).

It is tempting to conclude from the appearance of (10) that the transition to the next level so dominates α for all others to be negligible unless a very precise value is needed. And the temptation is compounded when we apply this principle to helium, for which the resonance line at a wavelength of 59 nm corresponds to the excitation of one electron only from $1s$ to $2p$ (by symmetry the excitation $1s-2s$ has an f-value of zero). If we assume that this provides the only significant equivalent oscillator, with f near unity, the resulting estimate of the volume susceptibility κ at NTP is 8.4×10^{-5}, comparing very favourably with the measured value of 7.0×10^{-5}. The same argument applied to xenon, however, whose resonance line is at 124 nm, would suggest that κ should be about 4 times as great as in helium, whereas in fact it is 20 times. Obviously the large number of electrons in the atom leads to a much higher density of excited states which, even if contributing little individually, add up to an overwhelming total.

Another warning is provided by calculating the f-values for the transitions from the ground state of hydrogen, $1s$, to the higher p-states, $2p$, $3p$ etc., which are the only transitions for which f does not vanish. In this case the wave-functions are well known and the integrations straightforward, if tedious.[2] It is found that $f_{21}=0.4162$, $f_{31}=0.0791$, $f_{41}=0.0290\ldots$, with a sum $\Sigma_{n=2}^{\infty} f_{n1}=0.5641$. The f-sum is not unity because we have not taken into account the continuum of unbound states with positive energy, and it is clear that they contribute nearly half the oscillator strength. These examples serve to emphasize once more how cautious one must be in truncating the oscillator set. The lowest lying states may allow the order of magnitude of α to be estimated, but for anything approaching exactitude there are no short cuts.

It will be appreciated that in the calculation of the static polarizability there is no preferential weighting of certain equivalent oscillators by exciting them near resonance, except in so far as the presence of ω_{ln}^2 in the denominator of (9) makes low-energy transitions particularly important. As soon as we concern ourselves with resonant responses, however, we may narrow our vision, as already discussed, to exclude all but those levels to which there is an enhanced transition probability.

18 The two-level system

We now specialize the results of the last chapter, to consider in some detail the behaviour of a system in which only two levels are significantly occupied. In proton spin resonance, for example, all the translational states of the proton are irrelevant under ideal conditions since they are not affected by the oscillating magnetic field that induces transitions between the spin states; and in the case of a particle tunnelling between two wells the lowest two states are much closer in energy than they are to other excited states, so that at the low frequencies required to change the occupations of these lowest states the others remain uninvolved. In neglecting all others we do not imply that they do not exist – we do not, for example, assign an oscillator strength of unity to the transition between the states of interest – we simply ignore them and define the state of the system by the two complex amplitudes, a_1 and a_2, of the eigenfunctions, $\chi_1(\boldsymbol{r})$ and $\chi_2(\boldsymbol{r})$, taken as real. At any instant the wave-function takes the form

$$\Psi(\boldsymbol{r}, t) = a_1\chi_1\,\mathrm{e}^{-\mathrm{i}E_1t/\hbar} + a_2\chi_2\,\mathrm{e}^{-\mathrm{i}E_2t/\hbar}, \tag{1}$$

in which $E_2 > E_1$. If $E_2 - E_1 = 2\Delta_0$ the resonance frequency is $2\Delta_0/\hbar$. Under the influence of disturbing forces a_1 and a_2 change with time, but if the wave-function is initially normalized $a_1a_1^* + a_2a_2^* = 1$ always.

A convenient model for developing the argument is a one-dimensional two-well system for which χ_1 and χ_2 are functions of y only. We shall find that most of the results are expressible in terms of three characteristic dipole moments:

520

$$\begin{aligned} p_1 &= \langle 1|ey|1\rangle \\ p_2 &= \langle 2|ey|2\rangle \\ p_{12} &= \langle 2|ey|1\rangle \end{aligned} \tag{2}$$

By using these, the results will appear to depend very little on the details of the model, as indeed is the case; we use a particular two-well system to make the derivations explicit and elementary, but shall not hesitate to apply the results in a wider context, even to spin resonance. It should be noted, however, that for all its practical importance we shall have little to say on that topic, other than to give some justification for the classical treatment in chapter 8.

218

General theory

When a time-varying electric field, $\mathscr{E}(t)$, is applied parallel to y, a_1 and a_2 change in accordance with (13.40):

$$\left.\begin{aligned}\dot{a}_1 &= (\mathrm{i}\mathscr{E}/\hbar)(p_1 a_1 + p_{12} a_2\,\mathrm{e}^{-2\mathrm{i}\Delta_0 t/\hbar})\\ \text{and}\qquad \dot{a}_2 &= (\mathrm{i}\mathscr{E}/\hbar)(p_2 a_2 + p_{12} a_1\,\mathrm{e}^{2\mathrm{i}\Delta_0 t/\hbar})\end{aligned}\right\} \tag{3}$$

A field varying as $\mathscr{E}\cos\omega t$ can be written as $\frac{1}{2}\mathscr{E}(\mathrm{e}^{\mathrm{i}\omega t}+\mathrm{e}^{-\mathrm{i}\omega t})$, and each equation then contains terms involving three frequencies, $|\omega|$ and $|\omega \pm 2\Delta_0/\hbar|$. Close to resonance the slowest vibration, $|\omega - 2\Delta_0/\hbar|$, dominates the evolution of a_1 and a_2, and the others only superimpose rapid oscillations of small amplitude. If these are ignored the resulting simplified versions of (3) are readily solved. Writing δ for $\hbar\omega/2\Delta_0 - 1$, δ_0 for $p_{12}\mathscr{E}/2\Delta_0$ and τ for $2\Delta_0 t/\hbar$, we have in this approximation:

$$\mathrm{d}a_1/\mathrm{d}\tau = \tfrac{1}{2}\mathrm{i}\delta_0 a_2\,\mathrm{e}^{\mathrm{i}\delta t} \quad\text{and}\quad \mathrm{d}a_2/\mathrm{d}\tau = \tfrac{1}{2}\mathrm{i}\delta_0 a_1\,\mathrm{e}^{-\mathrm{i}\delta\tau}. \tag{4}$$

On substituting the trial solution

$$a_1 = c_1\,\mathrm{e}^{\mathrm{i}\gamma\tau}, \qquad a_2 = c_2\,\mathrm{e}^{\mathrm{i}(\gamma-\delta)\tau}, \tag{5}$$

it is found that

$$\left.\begin{aligned}\gamma(\gamma-\delta) &= \tfrac{1}{4}\delta_0^2, \quad \text{or } \gamma = \tfrac{1}{2}[\delta \pm (\delta^2+\delta_0^2)^{\frac{1}{2}}]\\ \text{and}\qquad c_2/c_1 &= 2\gamma/\delta_0.\end{aligned}\right. \tag{6}$$

The two solutions (5) are conveniently written in the form:

$$\left.\begin{aligned} a_1 &= \mathrm{e}^{\mathrm{i}\gamma_1\tau}, \qquad a_2 = (2\gamma_1/\delta_0)\,\mathrm{e}^{\mathrm{i}\gamma_2\tau}\\ \text{and}\qquad a_1 &= \mathrm{e}^{-\mathrm{i}\gamma_2\tau}, \qquad a_2 = -(2\gamma_2/\delta_0)\,\mathrm{e}^{-\mathrm{i}\gamma_1\tau},\end{aligned}\right\} \tag{7}$$

in which $\gamma_1 = \frac{1}{2}[(\delta^2+\delta_0^2)^{\frac{1}{2}}+\delta]$ and $\gamma_2 = \frac{1}{2}[(\delta^2+\delta_0^2)^{\frac{1}{2}}-\delta]$. The general solution is a linear combination:

$$\left.\begin{aligned} a_1 &= C_1\,\mathrm{e}^{\mathrm{i}\gamma_1\tau} + C_2\,\mathrm{e}^{-\mathrm{i}(\gamma_2\tau+\phi)}\\ \text{and}\qquad a_2 &= (2/\delta_0)[C_1\gamma_1\,\mathrm{e}^{\mathrm{i}\gamma_2\tau} - C_2\gamma_2\,\mathrm{e}^{-\mathrm{i}(\gamma_1\tau+\phi)}].\end{aligned}\right\} \tag{8}$$

Since the absolute phases of a_1 and a_2 are irrelevant C_1 and C_2 may be taken as real, and ϕ as an arbitrary phase. In the normalized solution $a_1 a_1^* + a_2 a_2^* = 1$, and hence

$$\left.\begin{aligned}\gamma_1 C_1^2 + \gamma_2 C_2^2 &= \delta_0^2/4\delta',\\ \text{where}\qquad \delta' = \gamma_1+\gamma_2 &= (\delta^2+\delta_0^2)^{\frac{1}{2}}.\end{aligned}\right. \tag{9}$$

The dipole moment associated with the general solution is readily evaluated:

$$\langle p\rangle = e\langle y\rangle = a_1 a_1^* p_1 + a_2 a_2^* p_2 + (a_1 a_2^*\,\mathrm{e}^{\mathrm{i}\tau} + a_2 a_1^*\,\mathrm{e}^{-\mathrm{i}\tau})p_{12}. \tag{10}$$

The first two terms are non-oscillatory except as a result of changes in a_1 and a_2 caused by external influence, and they are present only if the stationary states themselves have permanent dipole moments. They are absent in proton resonance but present when a particle tunnels between two unequal potential wells. The latter case will arise later in the chapter, but for the moment we shall concentrate on the oscillatory dipole resulting from superposition of the states. By using (9), the last term of (10) may be evaluated for the general solution (8):

$$\langle p\rangle/p_{12}=\mathrm{Re}\,[\mathrm{e}^{\mathrm{i}(1+\delta)\tau}\{[(\delta_0/\delta')^2-(4C_1C_2)^2]^{\frac{1}{2}} + (4C_1C_2/\delta_0)(\gamma_1\,\mathrm{e}^{-\mathrm{i}(\delta'\tau+\phi)}-\gamma_2\,\mathrm{e}^{\mathrm{i}(\delta'\tau+\phi)})\}]. \qquad (11)$$

This result has the simple geometrical interpretation shown in fig. 1. The quantity in braces, { }, represents on the complex plane an elliptical trajectory which is the projection of uniform motion round a tilted circle; the

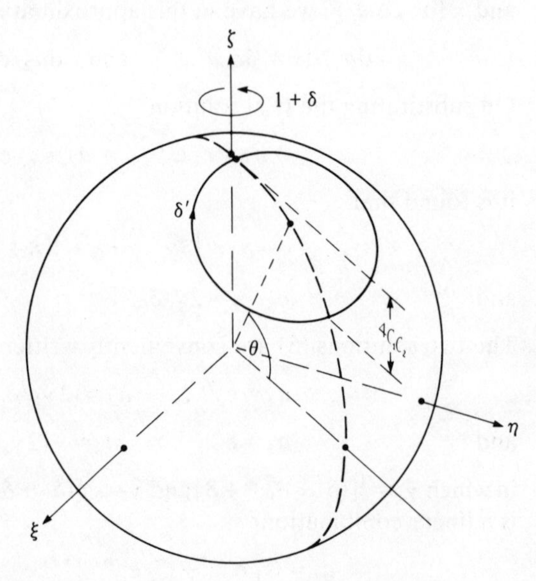

Fig. 1. Geometrical representation of (11) as the projection, onto the horizontal (ξ, η) plane, of motion round a circular orbit on the surface of a sphere.

radius and tilt of the circle can be inferred from the major and minor semi-axes $(4C_1C_2/\delta_0)(\gamma_1\pm\gamma_2)$, while the centre of the ellipse is at $[(\delta_0/\delta')^2-(4C_1C_2)^2]^{\frac{1}{2}}$. It is easily shown that the circle is a plane section of a sphere of unit radius, at an angle $\theta=\tan^{-1}(\delta/\delta_0)$, and that the vertical excursion is $4C_1C_2$. The representative point moves relatively slowly round this circle, with angular velocity δ', while the whole diagram spins at the driving frequency, $1+\delta$. The ξ-co-ordinate of the point is $\langle p\rangle/p_{12}$, and it executes a harmonic oscillation at something like the beat frequency $(E_2-E_1)/\hbar$ with an amplitude which is modulated at something like the detuning frequency, δ, or more precisely δ'. This behaviour is exactly the same as that described in chapter 8 and illustrated in fig. 8.7 – the motion of a gyromagnetic top in a steady vertical magnetic field, when perturbed by a weak field oscillating or rotating in the horizontal plane. It is the exactness

220

of the analogy which allows us, just as with the harmonic oscillator, to discuss in classical pictorial terms what would seem to be the strictly quantal phenomenon of spin resonance.

The diagrammatic representation of fig. 1 acquires new significance when described in terms of Cartesian co-ordinates. At any instant let Ψ be $a_1\chi_1 + a_2\chi_2$, the phase factors $e^{-iE_{1,2}t/\hbar}$ being incorporated in a_1 and a_2, and let this wave-function be represented on the unit sphere by a point defined as follows:

$$\zeta = a_2 a_2^* - a_1 a_1^*, \qquad \rho = \xi + i\eta = 2a_1 a_2^*. \tag{12}$$

Thus the upper pole of the sphere ($\zeta = 1$) represents the pure excited state, the lower ($\zeta = -1$) the pure ground state, while the azimuthal angle of any other point defines the instantaneous phase difference between the two components; the wave-function is automatically normalized, $a_1a_1^* + a_2a_2^* = 1$. Now from the last term in (10) it is clear, since $e^{i\tau}$ is incorporated in $a_1a_2^*$, that ξ so defined represents $\langle p\rangle/p_{12}$, and that the evolution of a_1 and a_2 causes the point to move in the manner required by (11). What is rather special about spin resonance, as compared with the double-well system that is also described by a simple two-component wave-function, is that all the Cartesian components are meaningful, and not just the ξ-component. In fact we have done no more than take over from the quantum theory of spin a pictorial representation that serves a much wider range of interest. The Pauli matrices:[1]

$$\sigma_\xi = \frac{\hbar}{2}\begin{pmatrix} 0 & 1 \\ 1 & 0 \end{pmatrix}, \qquad \sigma_\eta = \frac{\hbar}{2}\begin{pmatrix} 0 & -i \\ i & 0 \end{pmatrix}, \qquad \sigma_\zeta = \frac{\hbar}{2}\begin{pmatrix} 1 & 0 \\ 0 & -1 \end{pmatrix},$$

generate expectation values for the three components of angular momentum of the same form as (12):

$$\langle L_\xi\rangle = \frac{\hbar}{2}(a_1^* \quad a_2^*)\begin{pmatrix} 0 & 1 \\ 1 & 0 \end{pmatrix}\begin{pmatrix} a_1 \\ a_2 \end{pmatrix} = \frac{\hbar}{2}(a_1^*a_2 + a_1a_2^*)$$

$$\langle L_\eta\rangle = \frac{\hbar}{2}(a_1^* \quad a_2^*)\begin{pmatrix} 0 & -i \\ i & 0 \end{pmatrix}\begin{pmatrix} a_1 \\ a_2 \end{pmatrix} = -\frac{i\hbar}{2}(a_1^*a_2 - a_1a_2^*)$$

$$\langle L_\zeta\rangle = \frac{\hbar}{2}(a_1^* \quad a_2^*)\begin{pmatrix} 1 & 0 \\ 0 & -1 \end{pmatrix}\begin{pmatrix} a_1 \\ a_2 \end{pmatrix} = \frac{\hbar}{2}(a_1a_1^* - a_2a_2^*)$$

In the light of this there is little need to justify in greater detail the use of classical arguments in chapter 8 to describe proton resonance, and we shall continue to rely on their validity on those occasions when the discussion of that chapter needs extension. We shall not consider particles of spin other than $\frac{1}{2}$, which have more than two equally spaced levels and which bridge the gap between spin-$\frac{1}{2}$ particles (two levels) and harmonic oscillators (unlimited number of levels).

The double-well model

It is now convenient to keep in mind a model system, having two closely spaced levels, which is closer in form than is the spin-$\frac{1}{2}$ particle to the systems with which most of this and the next chapter will be concerned. We consider first the system shown in fig. 2(a) in which the one-dimensional

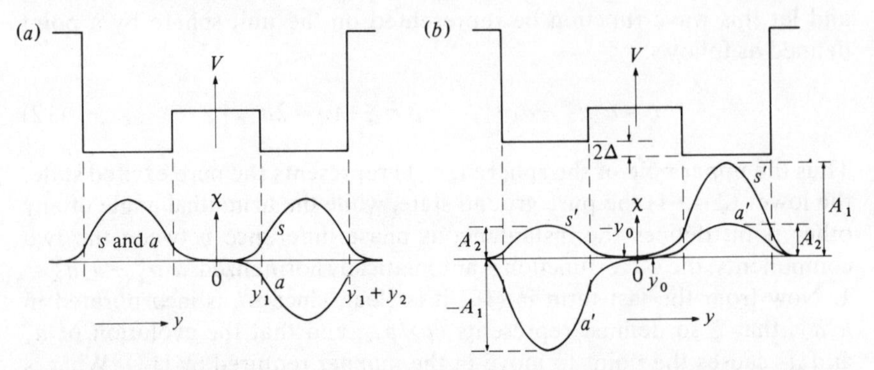

Fig. 2. The double-well model: (a) equal wells (above) giving rise to symmetrical and antisymmetrical wavefunctions (below); (b) unequal wells for which the lower state (s′) has larger amplitude in the deeper well, and the upper state (a′) larger amplitude in the shallower well.

potential wells are equal in depth and width. If the barrier is high and wide enough to preclude tunnelling, the wells are uncoupled and the ground state wave-function may have arbitrary amplitudes in each of them, but once they are coupled by the possibility of tunnelling there are two low-lying stationary states, one symmetrical (even) and the other antisymmetrical (odd), the former having lower energy than the latter. Inside the wells the wave-function is a sinusoid, at the outside edges it is exponential, and in the barrier it takes the form cosh μy in the even case, sinh μy in the odd, μ being determined by the height of the barrier. Because of the spread of χ beyond the sides of the well, the half-wavelength of the sinusoid in the well is slightly more than the width y_2. It is a very good approximation to take this small excess to be the sum of the values of χ/χ' at the edges, i.e. to extrapolate the sinusoid linearly to zero and take the effective edge to lie where χ hits the axis. At the outside edges the excess is constant and can be neglected, but the variations of the excess on the inside edges, depending on the tunnelling probability, are responsible for the shift of energy levels, which is the essential feature of the model. Thus if $\chi = \cosh \mu y$ in the barrier, χ/χ' at the barrier edge ($y = y_1$) takes the value $(\coth \mu y_1)/\mu$, rather than the value $1/\mu$ which it would take if y_1 were so large as to make the barrier impenetrable. Tunnelling therefore increases the effective width by $(\coth \mu y_1 - 1)/\mu$ and correspondingly lowers the kinetic energy of the even state, which is $\hbar^2 k^2/2m$ for a sinusoidal wave-function of wavenumber k. Hence

$$\Delta E/E = 2\Delta k/k = -2(\coth \mu y_1 - 1)/\mu y_2,$$

y_2 being the width of the well. Conversely, in the odd state the effective width is narrowed, sinh μy being steeper than $e^{\mu y}$, and the energy is raised by nearly the same amount. The levels are thus at $\pm\Delta_0$ relative to the uncoupled wells, where

$$\left.\begin{array}{ll} & \Delta_0 = C(\coth \mu y_1 - 1) \\ \text{or} & C(1 - \tanh \mu y_1) \end{array}\right\} \sim 2C\,\mathrm{e}^{-2\mu y_1} \tag{13}$$

$$\text{and} \qquad C = \pi^2\hbar^2/\mu m y_2^3.$$

When the system is set up in a mixture of the two states, the beating of the wave-functions results in a sinusoidal exchange of the particle between the wells at a frequency of $2\Delta_0/\hbar$, i.e. $(\pi^2\hbar/\mu m y_2^3)\,\mathrm{e}^{-2\mu y_1}$. If one finds a classical picture helpful, the pre-exponential factor is, apart from a constant of order unity, the frequency at which a particle of the same energy strikes the inner edge of its cell, and the exponential represents its chance of penetrating the barrier at each attempt (strictly, the amplitude reduction across the barrier). But this is little more than a mnemonic.

When the potential wells are unequal, as in fig. 2(*b*), the centre of the wave-function in the barrier is shifted from the origin to some new position y_0. It is now convenient to synthesize a solution and then to determine what problem has been solved. For example, let the even solution be modified so that the wave-function in the barrier is $\cosh \mu(y + y_0)$, the minimum being moved from the centre to $-y_0$. The same approximation as leads to (13) gives a different kinetic energy shift in the two wells, $-2C\,\mathrm{e}^{-2\mu(y_1 \pm y_0)}$ or $-\Delta_0\,\mathrm{e}^{\pm 2\mu y_0}$, the upper sign referring to the left-hand well in which the kinetic energy is lower than in the right-hand well. The difference in kinetic energy must be compensated by the difference in potential energy, 2Δ, from which it follows that this solution obtains when $2\Delta = \Delta_0(\mathrm{e}^{2\mu y_0} - \mathrm{e}^{-2\mu y_0})$. Hence

$$\Delta = \Delta_0 \sinh(2\mu y_0). \tag{14}$$

The overall lowering of the energy is obtained by summing kinetic and potential contributions in either well; thus from the right-hand well the energy is found to be $-E$, where

$$E = \Delta + \Delta_0\,\mathrm{e}^{-2\mu y_0} = \Delta_0 \cosh(2\mu y_0) = (\Delta_0^2 + \Delta^2)^{\frac{1}{2}}. \tag{15}$$

In the upper state, for which the wave-function in the barrier is $\sinh \mu(y + y_0)$, the energy is $+E$.

The wave-function in the wells is so nearly sinusoidal that a single number serves to define the amplitude in each well. Since the gradient of the sinusoid as it approaches zero is determined by its amplitude, it follows that the values of χ' at the inner edges of the wells fix the relative amplitudes. If these are A_1 and A_2 in the right and left wells when the system is in its lower state, they are A_2 and $-A_1$ in the upper state, and

$$\left.\begin{array}{ll} & A_1/A_2 = \sinh \mu(y_1 + y_0)/\sinh \mu(y_1 - y_0) \\ \text{or} & \cosh \mu(y_1 + y_0)/\cosh \mu(y_1 - y_0) \end{array}\right\} \sim \mathrm{e}^{2\mu y_0} = \frac{E + \Delta}{\Delta_0} = \frac{\Delta_0}{E - \Delta}. \tag{16}$$

Hence $$A_1^2/A_2^2=(E+\Delta)/(E-\Delta), \tag{17}$$

or $$A_1=(1+\Delta/E)^{\frac{1}{2}}/\surd 2 \quad \text{and} \quad A_2=(1-\Delta/E)^{\frac{1}{2}}/\surd 2, \tag{18}$$

if the amplitudes are normalized so that $A_1^2+A_2^2=1$. We may now relate the dipole moments defined in (2) to a standard moment p_0 obtained by placing the charge centrally in one of the wells. Then

$$\left.\begin{aligned} p_1=-p_2=p_0(A_1^2-A_2^2)&=p_0\Delta/E \\ \text{and} \qquad p_{12}=2p_0A_1A_2&=p_0\Delta_0/E. \end{aligned}\right\} \tag{19}$$

For equal wells p_1 and p_2 vanish and $p_{12}=p_0$. When the wells are unequal

$$p_1^2+p_{12}^2=p_0^2. \tag{20}$$

The results expressed in (15), (16) and (17) are of the same form as those describing coupled resonant circuits. Thus (15) is to be compared with (12.11) and fig. 12.2, frequencies in the circuits being replaced by energies in the double-well system, while the tunnelling probability as measured by Δ_0 is the analogue of the coupling parameter κ for the circuits. Similarly the ratio of the energy content of the two circuits when a pure normal mode is excited, as given by (12.10), finds an exact parallel in the ratio of probabilities of finding the particle in the two wells, as given by (17). Neither for the circuits nor for the double-well are the results sensitive to the details of the system. The shape of the potential wells does not matter – it is enough that they should take the same form apart from a vertical displacement, and that the coupling should be weak.[(2)] Even the shapes of the wells need not be identical provided care is taken in defining the amplitudes $A_{1,2}$ so that their squares represent probabilities of finding the particle in each well.

366

Response functions

Application of a uniform electric field $\mathscr{E}$ superposes a linear variation of potential on the double well. The tilt of the bottom of each well is of secondary importance compared to the shift of the mean levels, and the effect of the field may be adequately described by a change of Δ by $p_0\mathscr{E}$. If the system was initially in its lower state and is able after the application of $\mathscr{E}$ to revert to this state its dipole moment is changed by δp_1 i.e., from (19) and (15), $p_0^2\Delta_0^2\mathscr{E}/E^3$. In the upper state the result is the same except for a change of sign. The pure states therefore have polarizabilities

$$\alpha=\mp p_0^2\Delta_0^2/E^3, \tag{21}$$

the upper sign referring to the upper state. When the wells are equal (21) becomes $\mp p_0^2/\Delta_0$, and this is the condition of maximum polarizability.

The arguments of the last chapter find a very simple application here. There is only one f-value of significance and the force constant for the equivalent oscillator is $E/\langle 2|y|1\rangle^2$, i.e. e^2E/p_{12}^2, giving a polarizability of

p_{12}^2/E; (19) shows that this is the same as (21). As the coupling between equal wells is reduced E falls while p_{12} is constant, and α increases without limit. The value of f_{12}, according to (17.2), is $4m\Delta_0 p_{12}^2/\hbar^2 e^2$, roughly equal to the tunnelling factor $\mathrm{e}^{-2\mu y_1}$, and therefore very small. Here is an example of how the overwhelmingly most significant equivalent oscillator may account for only a tiny fraction of the oscillator strength; its low frequency and hence its weak force constant are responsible for its importance.

The result expressed by (21) holds only for a pure state. An assembly of identical systems in thermal equilibrium has its mean polarizability $\bar{\alpha}$ determined by the excess of systems in the lower state relative to those in the upper state, for the two states have equal and opposite polarizabilities. When Δ_0 is small the effect of the high individual values of α is largely compensated by the small excess in the lower state, and in fact $\bar{\alpha}$ tends to become independent of Δ_0. Let us suppose that before $\mathscr{E}$ is applied half the systems have an inequality of their wells of one sign, Δ, and half are reversed in sign, $-\Delta$, so that the resultant total dipole moment is zero. Consider the first half only, of which fractions $\mathrm{e}^{\mp E/k_BT}/(\mathrm{e}^{E/k_BT}+\mathrm{e}^{-E/k_BT})$ will be in the two states, giving a mean moment

$$\bar{p}/p_0=(\Delta/E)\tanh(E/k_\mathrm{B}T). \tag{22}$$

The mean polarizability follows by differentiating and setting $\mathrm{d}\Delta/\mathrm{d}\mathscr{E}$ equal to p_0:

$$\bar{\alpha}=\mathrm{d}\bar{p}/\mathrm{d}\mathscr{E}=p_0^2\,\mathrm{d}(\bar{p}/p_0)/\mathrm{d}\Delta=p_0^2[(\Delta_0^2/E^3)\tanh(E/k_\mathrm{B}T)$$
$$+(\Delta^2/E^2k_\mathrm{B}T)\operatorname{sech}^2(E/k_\mathrm{B}T)]. \tag{23}$$

The other half of the assembly gives an identical result, so that (23) is correct for the whole assembly. The two terms in the square brackets reveal two distinct sources of polarization by the field. The first reflects the polarizability of each state (21), reduced by the factor $\tanh(E/k_\mathrm{B}T)$ which is the excess fraction of systems in the lower state. The second results from the change in E, and therefore of the Boltzmann factor, when $\mathscr{E}$ is applied. Whereas the dipole moment due to the first term appears virtually instantaneously, since the energy levels adjust themselves immediately, the second depends on systems making transitions between states and thereby approaching a new condition of statistical equilibrium; if the systems are isolated from extraneous disturbances and have to rely on radiative processes, a long time may elapse before the second term makes itself felt.

We have, however, oversimplified in suggesting that the first process is instantaneous, for it is clear that the sudden application of a weak field will not immediately change the occupation of the wells, and the total dipole moment must therefore suffer no sudden change. When $\mathscr{E}$ is applied those systems that were in the lower state, with certain values of A_1 and A_2 as given by (18), now find themselves with the same $A_{1,2}$ which are no longer appropriate to a stationary state with altered Δ. In their new mixed state their dipole moments have a small oscillatory component at frequency

$2E/\hbar$; and the systems initially in the upper state also acquire an oscillatory moment, at the same frequency but vibrating in antiphase. The difference in numbers of systems in the two states leads to an excess contribution from the lower state. The dipole moment after the application of $\mathscr{E}$ therefore begins to oscillate as shown in fig. 3(*a*), starting at the same value as before

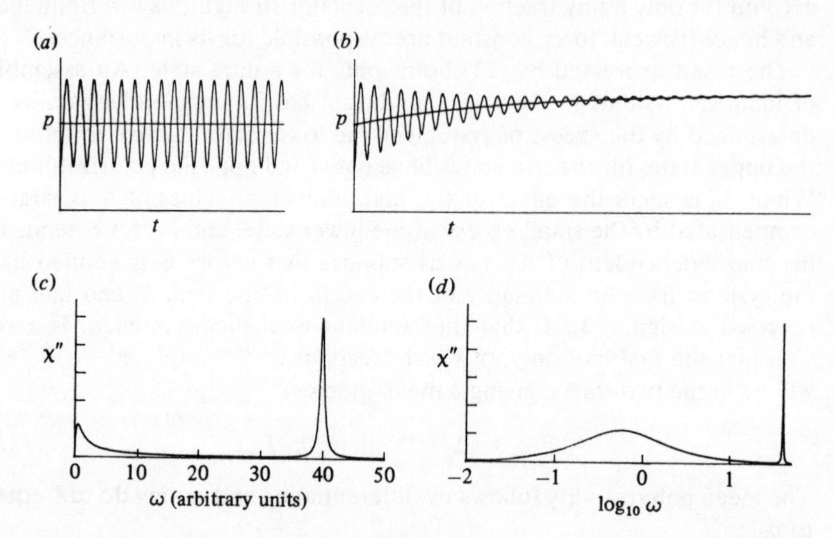

Fig. 3. (*a*) Step-response function of dipole moment of double-well system in the absence of dissipation. (*b*) Step-response function for unequal double-well system with dissipation. (*c*) Lossy part of $\chi(\omega)$ derived from (*b*) according to (64); the amplitudes of the two peaks are not in the correct proportion. (*d*) The same as (*c*) but plotted against $\log \omega$ to show symmetrical form (but markedly different widths) for the relaxation peak on the left and the resonance peak on the right.

$\mathscr{E}$, but now oscillating about a new mean value, which is given by the first term of (23). As time proceeds the oscillation is damped out by radiative processes or randomized by other disturbances, and at the same time the transitions occur to re-equilibrate the assembly and bring in the second term. The latter process, as we shall discuss later, does not normally take place at the same rate as the damping of the oscillations, and the behaviour shown in fig. 3(*b*) may typify the dipole moment response to a step in $\mathscr{E}$ – an exponentially decaying oscillation about a baseline that itself tends exponentially to a constant level.

528

The physical picture of some systems entirely in one stationary state and the rest in the other is a valuable simplification that was perforce taken for granted in the early days of quantum theory. It gives the correct result when there is no correlation in phase between the occupation amplitudes, a_1 and a_2, since the mean value of cross-terms, e.g. $a_1a_2^*$, is zero and all observables are defined by $a_1a_1^*$ and $a_2a_2^*$. It can lead to error, however, if successive perturbations are close together and insufficient time has elapsed between them for randomization. We shall therefore not use this approximation more than necessary and shall proceed to make the argument of the last paragraph quantitative without invoking its aid until the very end. Let us derive the response of a system, in an arbitrary state, to a step-function $\mathscr{E}$ and to an impulse, the latter being the derivative of the former if the response is linear. For this purpose, and for other problems that follow, the geometrical representation of the state of the system, as in fig. 1, is convenient but needs generalization to include such cases as the unequal double well. For if the poles of the sphere are to continue representing the pure stationary states, account must be taken of their permanent dipole moments.

The interpretation of ζ and ρ in (12) still stands, but the motion of the representative point must be recalculated from (3), with p_2 put equal to $-p_1$, according to (19). Then

$$\dot{a}_1 = (\mathrm{i}\mathscr{E}/\hbar)(p_1 a_1 + p_{12} a_2\, \mathrm{e}^{-\mathrm{i}\omega_0 t})$$

and
$$\dot{a}_2 = (\mathrm{i}\mathscr{E}/\hbar)(-p_1 a_2 + p_{12} a_1\, \mathrm{e}^{\mathrm{i}\omega_0 t}),$$

in which $\omega_0 = 2E/\hbar$. It follows that

$$\dot{\zeta} = \dot{a}_2 a_2^* - \dot{a}_1 a_1^* + \text{c.c.} = -2p_{12}\mathscr{E}\eta/\hbar. \tag{24}$$

In evaluating $\dot{\rho}$ it must be remembered that in (12) the phase factors $\mathrm{e}^{\pm Et/\hbar}$ are included in $a_{1,2}$, so that even in the absence of $\mathscr{E}$ the representative point spins at frequency ω_0 round the ζ-axis, i.e. $\dot{\rho} = \mathrm{i}\omega_0\rho$ if $\mathscr{E} = 0$. In the presence of $\mathscr{E}$, (12) shows that

$$\dot{\rho} = \mathrm{i}\omega_0\rho + 2(\dot{a}_1 a_2^* + a_1\dot{a}_2^*)\,\mathrm{e}^{\mathrm{i}\omega_0 t} = \mathrm{i}(\omega_0 + 2p_1\mathscr{E}/\hbar)\rho + 2\mathrm{i}p_{12}\mathscr{E}\xi/\hbar. \tag{25}$$

The first term reflects the change in rotational speed as E is changed by the interaction of $\mathscr{E}$ with $\pm p_1$. If this rotation is now taken for granted, the remaining changes in the position of the representative point may be collected together from (24) and (25) to give the new equations of motion:

$$\dot{\zeta} = -2p_{12}\mathscr{E}\eta/\hbar, \qquad \dot{\eta} = 2p_{12}\mathscr{E}\xi/\hbar, \qquad \dot{\xi} = 0. \tag{26}$$

These simply describe rotation about the ξ-axis at an angular velocity $2p_{12}\mathscr{E}/\hbar$. The permanent moments $\pm p_1$ in the stationary states play no part in this motion except for the replacement of ω_0 by $\omega_0 + 2p_1\mathscr{E}/\hbar$ in the overall spinning of the pattern about the ρ-axis, and the reduction of p_{12} to a value less than p_0.

Where the interpretation is significantly altered is in the value of $\langle p\rangle$ to be derived from the diagram. According to (10), with the phase factors incorporated in $a_{1,2}$,

$$\langle p\rangle = p_1(a_1 a_1^* - a_2 a_2^*) + p_{12}(a_1 a_2^* + a_1^* a_2) = p_{12}\xi - p_1\zeta. \tag{27}$$

Now (20) allows p_{12} to be written as $p_0 \cos\varepsilon$ and p_1 as $p_0 \sin\varepsilon$, and (19) shows that $\tan\varepsilon = \Delta/\Delta_0$. Hence

$$\langle p\rangle = p_0(\xi\cos\varepsilon - \zeta\sin\varepsilon),$$

and is the ξ-component of the projection of the representative point onto a plane tilted about the y-axis at an angle ε to the horizontal, as in fig. 4(a). The point moves at angular velocity $2E/\hbar$ in a horizontal circular orbit $\mathscr{O}$, when the system is undisturbed; its projection on the tilted plane describes, through its ξ'-component which is the only significant component, an oscillatory dipole moment superposed on a steady component. The

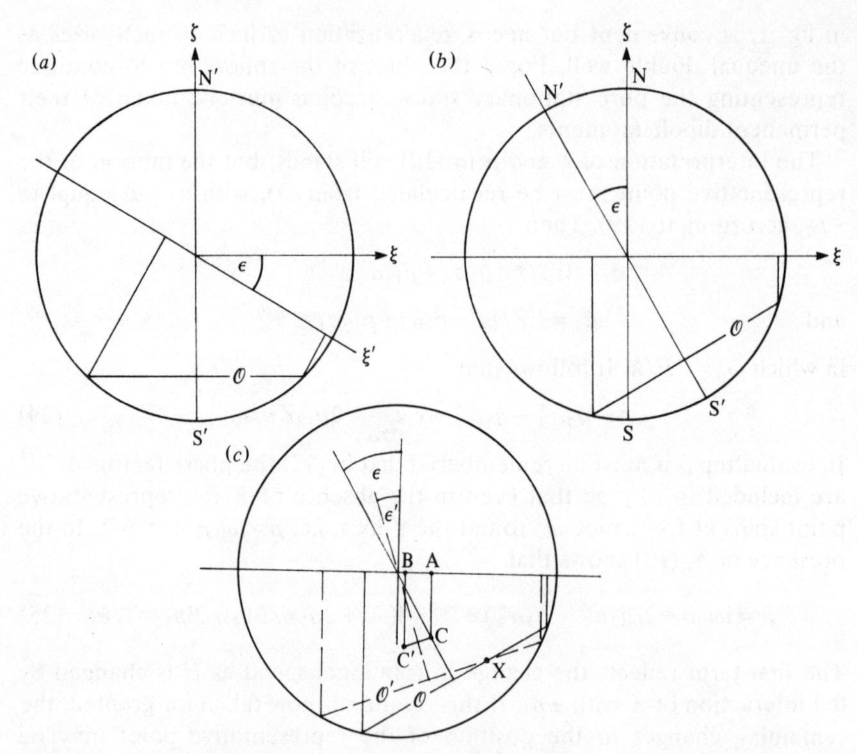

Fig. 4. (a) Modification of fig. 1 appropriate for unequal wells; only the $\xi-\zeta$ section is shown, with $\tan\varepsilon = \Delta/\Delta_0$, and the observed dipole moment is represented by the ξ'-component of the representative point. (b) An alternative form of (a), tilted through ε so that the ξ-component represents the dipole moment, but the orbit of the representative point for the undisturbed system is now tilted. (c) Illustrating the calculation of the step-response function.

latter arises from the permanent moments in the stationary states, still represented by the poles, N' and S'; and is positive if the amplitude of the lower state exceeds that of the upper.

An alternative representation, which has certain advantages, is shown in fig. 4(b), which is just 4(a) tilted through ε so that $\langle p\rangle$ is once more determined by projection onto the horizontal plane. The plane of the orbit $\mathcal{O}$ is now tilted at ε, and N' and S' are the stationary states. This representation amounts to describing each state of the system in terms of the stationary states $\chi_1^{(0)}$ and $\chi_2^{(0)}$ of the equal-well system, in which $\Delta = 0$ and $A_1^2 = A_2^2 = 1/2$. Thus the lower stationary state S', in which A_1 and A_2 are given by (18), is a superposition $a_1\chi_1^{(0)} + a_2\chi_2^{(0)}$ with

$$a_1 = \tfrac{1}{2}[(1+\Delta/E)^{\frac{1}{2}} + (1-\Delta/E)^{\frac{1}{2}}]$$

$$a_2 = \tfrac{1}{2}[(1+\Delta/E)^{\frac{1}{2}} - (1-\Delta/E)^{\frac{1}{2}}],$$

so that, according to (12), $\zeta = -\Delta_0/E = -\cos\varepsilon$ and $\xi = \Delta/E = \sin\varepsilon$; these, of course, are the co-ordinates of S'. It may be noted in passing that in a non-stationary state the amplitudes A_l and A_r in the left-hand and right-hand wells are simply determined by the position of the representative point, since

$$\xi = A_\mathrm{r}A_\mathrm{r}^* - A_\mathrm{l}A_\mathrm{l}^* \quad \text{and} \quad -\zeta + \mathrm{i}\eta = 2A_\mathrm{r}^*A_\mathrm{l}. \tag{28}$$

The ξ-coordinate is the difference in occupation of the two wells, which confirms that it represents the dipole moment, while the azimuth angle around the ξ-axis, measured from a point directly below the axis, gives the phase difference between A_r and A_l.

In the representation of fig. 4(*b*) the response to a step-function $\mathscr{E}$ is immediately obvious. If Δ is the inequality of the wells before $\mathscr{E}$ is applied the point is orbiting at an angle $\tan^{-1}(\Delta/\Delta_0)$; A_l and A_r are not immediately changed by $\mathscr{E}$, but the tilt of the orbit is now altered to $\tan^{-1}(\Delta'/\Delta_0)$, where $\Delta' = \Delta + p_0\mathscr{E}$. In fig. 4(*c*) the original orbit $\mathscr{O}$ is shown switching to $\mathscr{O}'$ at X, the position of the point at the instant $\mathscr{E}$ is applied. The change in the mean value and in the oscillatory amplitude of $\langle p \rangle$ can be seen in the projections of the extremes of the orbits onto the horizontal plane. The difference is the step response function, and clearly depends on the position of X. The impulse response is similarly calculated; a strong field $\mathscr{E}$ applied for a short time δt momentarily raises Δ' to a very high value $p_0\mathscr{E}$, and ε rises to $\frac{1}{2}\pi$. The point then rotates about the ξ-axis through $2p_0\mathscr{E}\delta t/\hbar$ before the impulse ceases and the original motion is resumed, but on a slightly displaced orbit. An impulse P_e, defined as $\int \mathscr{E}\, \mathrm{d}t$, causes a rotation $2p_0P_e/\hbar$, and the resulting difference in the dipole oscillation, which is P_e times the impulse response function, again clearly depends on the moment of application of the impulse.

In an assembly of randomly phased systems, therefore, different systems will respond differently. But the response of the mean is easily discovered. Randomness of phasing implies that before a step is applied the centroid of all points representing the different members of the assembly lies on the ζ'-axis, N′S′ in fig. 4(*b*), and that the mean of all the oscillatory dipole moments vanishes, though there will in general be a steady moment since except at a very high temperature the centroid lies below the centre of the sphere. When application of $\mathscr{E}$ causes each point to orbit about a new axis, the centroid does the same. Thus if C in fig. 4(*c*) is the original centroid, the mean response will be the orbit CC′, whose projection gives the response $\bar{p}(t)$; it is clear that, as in fig. 3, $\bar{p}$ starts at one extreme of its cycle of oscillation. If $\mathscr{E}$ is small, the angle $\varepsilon' - \varepsilon$ is $(\mathrm{d}/\mathrm{d}\Delta)\tan^{-1}(\Delta/\Delta_0)\cdot p_0\mathscr{E}$, i.e. $\Delta_0 p_0\mathscr{E}/E^2$; a short calculation shows that if f is the length OC, the fractional excess in the lower state, the oscillatory amplitude, $\frac{1}{2}\mathrm{AB}\times p_0$, is $(\Delta_0 p_0^2\mathscr{E}f/E^2)\cos\varepsilon$, or $\Delta_0^2 p_0^2\mathscr{E}f/E^3$, in agreement with the first term of (23) when f takes the thermal equilibrium value $\tanh(E/k_\mathrm{B}T)$. The step-response function, when $\mathscr{E} = 1$, is written down by inspection:

$$S(t) = (\Delta_0^2 p_0^2 f/E^3)[1 - \cos(2Et/\hbar)], \tag{29}$$

and its derivative $\mathrm{d}S/\mathrm{d}t$ is the impulse-response function:

$$h(t) = (2\Delta_0^2 p_0^2 f/\hbar E^2)\sin(2Et/\hbar). \tag{30}$$

This is just what results from turning the point C through $2p_0/\hbar$ about the ξ-axis, as prescribed above. So long as the displacements of C are small, linearity holds and the effects of successive impulses are additive. These expressions take no account of dissipative processes and therefore offer no explanation of the second term in (23). This will arise out of the next development of the theory.

Radiative decay of the two-level system

We shall begin by solving a model from which the required results can be derived without much additional calculation, by building on the ideas developed in chapter 16. The model consists of a two-level system for a particle of mass m_1 tunnelling between equal wells, and a harmonic oscillator of mass m_0 and frequency ω_0, close to $2\Delta_0/\hbar$. The two are weakly coupled as if an electric field proportional to x, the displacement of the oscillator, acted to polarize the double well; conversely the double well acts on the oscillator with a force of constant magnitude but whose sign depends on which well the particle occupies. The coupling term is incorporated in the Hamiltonian for the combined system:

490

$$\mathscr{H}(x, y) = T_0(\dot{x}) + V_0(x) + T_1(\dot{y}) + V_1(y) + Cx \operatorname{sgn}(y), \qquad (31)$$

in which T_1 and V_1 refer to the double well, T_0 and V_0 to the oscillator, and $\operatorname{sgn}(y) = \pm 1$ according as y is positive or negative. Correspondingly Schrödinger's time-independent equation takes the form

$$(\hbar^2/2m_1)\partial^2\Phi/\partial y^2 + (\hbar^2/2m_0)\partial^2\Phi/\partial x^2 + [W - V_1 - V_0 - Cx \operatorname{sgn}(y)]\Phi = 0. \qquad (32)$$

If $C = 0$ the solutions, of energy W_0, are product wave-functions, $\chi(y)\psi(x)$, of which the components are of either even parity ($\chi_1(y)$, $\psi_{2l}(x)$) or odd ($\chi_2(y)$, $\psi_{2l+1}(x)$). The coupling term is odd in both x and y, and the matrix element connecting two product wave-functions vanishes unless one contains χ_1 and the other χ_2; moreover, since the $\psi(x)$ are oscillator functions, the two appearing in any non-vanishing matrix element must be neighbours, ψ_l and ψ_{l+1}. With weak coupling, then, every $\chi_1\psi_n$ suffers an admixture of $\chi_2\psi_{n\pm1}$, and every $\chi_2\psi_n$ an admixture of $\chi_1\psi_{n\pm1}$. The mixing is strongest when the energy levels involved are close together; thus when $\hbar\omega_0 \approx 2\Delta_0$, $\chi_1\psi_{n+1}$ is most strongly modified by the admixture of $\chi_2\psi_n$, and $\chi_2\psi_n$ has the strongest admixture of $\chi_1\psi_{n+1}$. We therefore seek real solutions of (32) with the form:

442

$$\Phi_n = \alpha_n\chi_1\psi_{n+1} + \beta_n\chi_2\psi_n; \qquad \alpha_n^2 + \beta_n^2 = 1. \qquad (33)$$

Substituting in (32) we have that

$$\alpha_n[W + \Delta_0 - E_{n+1} - Cx \operatorname{sgn}(y)]\chi_1\psi_{n+1}$$
$$+ \beta_n[W - \Delta_0 - E_n - Cx \operatorname{sgn}(y)]\chi_2\psi_n = 0, \qquad (34)$$

in which $E_n = (n + \frac{1}{2})\hbar\omega_0$. Small terms omitted through the approximations in (33) give a negligible contribution when (34) is multiplied by either $\chi_1\psi_{n+1}$ or $\chi_2\psi_n$ and integrated over both variables to yield the first-order approximations:

$$\alpha_n(\Omega_n+\omega_e)=\beta_n CM^{(1)}_{n,n+1}/\hbar \quad\text{and}\quad \beta_n(\Omega_n-\omega_e)=\alpha_n CM^{(1)}_{n,n+1}/\hbar, \tag{35}$$

in which $\Omega_n = W/\hbar-(n+1)\omega_0$, $\omega_e=\Delta_0/\hbar-\frac{1}{2}\omega_0$, and $M^{(1)}_{n,n+1}$ is the same matrix element as in (13.26) except that in laboratory units its value is $[\hbar(n+1)/2m_0\omega_e]^{\frac{1}{2}}$.

366

Once again we have a close analogy to the coupled circuits of chapter 12, the energy level structure obtained by eliminating α_n and β_n from (35) having the same hyperbolic form as the normal mode frequency structure of (12.11):

$$\Omega_n^2=\omega_e^2+\gamma^2(n+1), \quad \text{where } \gamma=C/(2\hbar\omega_0 m_0)^{\frac{1}{2}}. \tag{36}$$

When the systems are uncoupled the mismatch between ω_0 and $2\Delta_0/\hbar$ causes the levels to appear in pairs, as shown on the left of fig. 5. Coupling

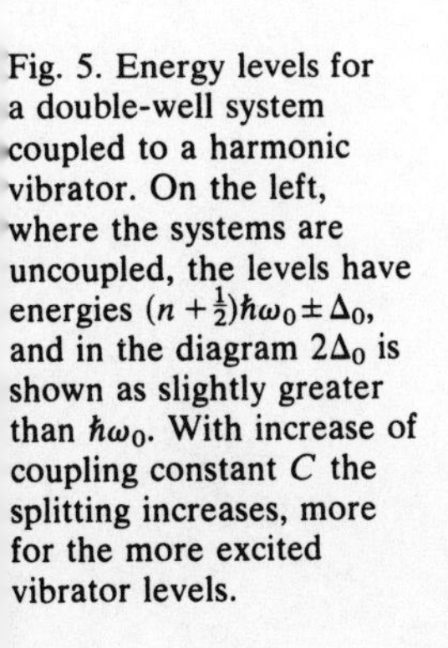

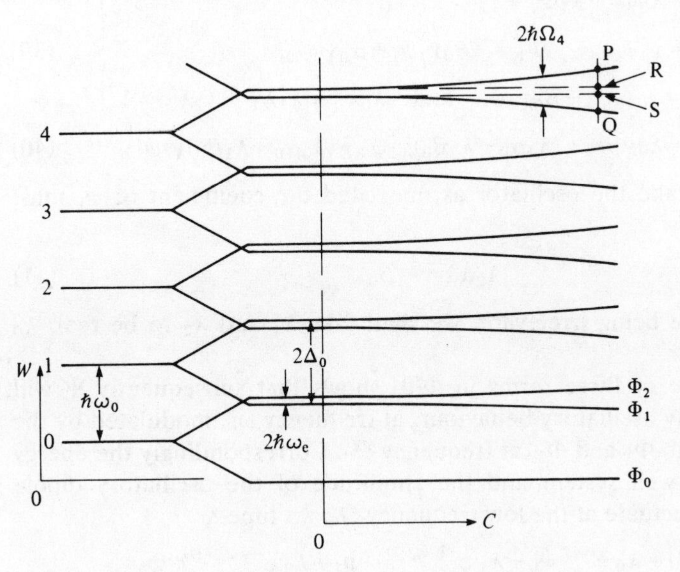

Fig. 5. Energy levels for a double-well system coupled to a harmonic vibrator. On the left, where the systems are uncoupled, the levels have energies $(n+\frac{1}{2})\hbar\omega_0\pm\Delta_0$, and in the diagram $2\Delta_0$ is shown as slightly greater than $\hbar\omega_0$. With increase of coupling constant C the splitting increases, more for the more excited vibrator levels.

enlarges the separation, and the presence of $(n+1)$ in (36) enhances the effect at the higher levels; the solitary lowest level is unperturbed. It should be remembered that this analysis goes only as far as first-order effects of C on the wave-function (second-order in energy), so that by the time the splitting is as large as on the right of the diagram higher-order effects will certainly begin to play a part and change the details. We shall treat only first-order effects. From (35),

$$\alpha_n^2/\beta_n^2=(\Omega_n-\omega_e)/(\Omega_n+\omega_e). \tag{37}$$

In the diagram ω_e is taken to be positive, and the state P has $(\Omega_n-\omega_e)$ represented by PR, $(\Omega+\omega_e)$ by PS. Hence $\alpha_n/\beta_n=(\text{PR}/\text{PS})^{\frac{1}{2}}$ and (35) shows that α_n and β_n have the same sign. The state Q has the ratio $|\alpha_n/\beta_n|$ inverted, and opposite signs for α_n and β_n. This behaviour exactly parallels what has already been noted in analogous cases. Since $\alpha_n^2+\beta_n^2=1$, it follows from (37) that in the state P

$$\alpha_n^2=\tfrac{1}{2}(1-\omega_e/\Omega_n), \qquad \beta_n^2=\tfrac{1}{2}(1+\omega_e/\Omega_n). \tag{38}$$

Let us use this result to follow the behaviour of the two-well system very weakly coupled to the unexcited oscillator. This will serve as a model for the radiation of energy into one mode of a cavity. As in chapter 16 we shall suppose the cavity to be so large that at any instant after coupling the chance of finding any one mode excited is extremely small. This places a helpful restriction on the form of the wave-function which will contain only the lowest three states of the combined system, Φ_0, Φ_1, and Φ_2 as shown in fig. 5. At the instant of coupling, taken as $t=0$ for convenience, the spatial part Ψ may be written

497

$$\left.\begin{aligned} &\Psi(0)=\lambda_0\Phi_0+\lambda_1\Phi_1+\lambda_2\Phi_2,\\ \text{where}\quad &\Phi_0=\chi_1\psi_0, \qquad \Phi_1=-\beta_0\chi_1\psi_1+\alpha_0\chi_2\psi_0,\\ &\Phi_2=\alpha_0\chi_1\psi_1+\beta_0\chi_2\psi_0 \quad\text{and}\quad \lambda_0\lambda_0^*+\lambda_1\lambda_1^*+\lambda_2\lambda_2^*=1. \end{aligned}\right\} \tag{39}$$

Hence $\Psi(0)=\lambda_0\chi_1\psi_0+(\lambda_1\alpha_0+\lambda_2\beta_0)\chi_2\psi_0+(\lambda_2\alpha_0-\lambda_1\beta_0)\chi_1\psi_1, \qquad (40)$

and since we take the oscillator as unexcited the coefficient of ψ_1 must vanish:

i.e.
$$\lambda_2\alpha_0=\lambda_1\beta_0. \tag{41}$$

Absolute phase being irrelevant, we shall take λ_1 and λ_2 to be real; λ_0 may be complex.

The presence of three terms in $\Phi(0)$ shows that subsequently Ψ will exhibit a rapidly oscillatory behaviour, at frequency ω_0, modulated by the slow beat due to Φ_1 and Φ_2, at frequency Ω_0. Correspondingly the energy of the double-well system and the amplitude of the oscillatory dipole moment will fluctuate at the low frequency Ω_0. At time t,

$$\Psi(t)=\lambda_0\,e^{i\omega_e t}\,\Phi_0+\lambda_1\,e^{-i(\omega_0-\Omega_0)t}\,\Phi_1+\lambda_2\,e^{-i(\omega_0+\Omega_0)t}\,\Phi_2$$

which may be written, like (40):

$$\begin{aligned}\Psi(t)=\lambda_0\,e^{i\omega_e t}\chi_1\psi_0&+(\lambda_1\alpha_0\,e^{i\Omega_0 t}+\lambda_2\beta_0\,e^{-i\Omega_0 t})\,e^{-i\omega_0 t}\chi_2\psi_0\\ &+(\lambda_2\alpha_0\,e^{-i\Omega_0 t}-\lambda_1\beta_0\,e^{i\Omega_0 t})\,e^{-i\omega_0 t}\chi_1\psi_1.\end{aligned} \tag{42}$$

The probability $\mathscr{P}_2$, of finding the double-well system in its excited state is determined by the second term, the only one to contain χ_2:

$$\begin{aligned}\mathscr{P}_2(t)=|\lambda_1\alpha_0\,e^{i\Omega_0 t}+\lambda_2\beta_0\,e^{-i\Omega_0 t}|^2&=(\lambda_1\alpha_0+\lambda_2\beta_0)^2-4\lambda_1\lambda_2\alpha_0\beta_0\sin^2\Omega_0 t\\ &=\mathscr{P}_2(0)[1-(\gamma^2/\Omega_0^2)\sin^2\Omega_0 t],\end{aligned} \tag{43}$$

by use of (36), (38) and (41).

The exchange of energy implied by (43) has exactly the same form as the radiative exchange between an oscillator and a cavity mode, as described by (16.14), if the coupling is so weak that Ω_0 may be put equal to ω_e. As in the earlier example, we have made here no assumptions about the initial state, and its parameters have vanished from (43); it follows that coupling to all modes in an empty cavity causes $\mathcal{P}_2$ and the excitation energy of the system to decay exponentially. Since (43) shows that the initial energy transfer to a single mode follows the same behaviour as (16.16), i.e. $E(0)-E(t)=E(0)\gamma^2t^2$, we may proceed immediately to determine τ_e, the time constant for energy loss into a cavity. Equating γ^2 to $1/2\pi g\tau_e$, we have that

$$\tau_e = 1/2\pi g\gamma^2, \tag{44}$$

497 where $g=4\Delta_0^2V/\pi^2\hbar^2c^3$, the density of states for a cavity of volume V at a frequency of $2\Delta_0/\hbar$.

To find the value of γ we go back to the Hamiltonian (31). Consider a typical mode at frequency $2\Delta_0/\hbar$, and at some instant when the field $\mathscr{E}$ is at its peak write $\overline{\mathscr{E}^2}$ for the mean square field, averaged over the cavity. Then the mode energy, $\frac{1}{2}\varepsilon_0V\overline{\mathscr{E}^2}$, must be equated to $\frac{1}{2}m_0(2\Delta_0/\hbar)^2x^2$, the potential energy of the equivalent oscillator; and the square of the coupling energy, C^2x^2 in (31), must be equated to $\frac{1}{3}p_0^2\overline{\mathscr{E}^2}$, the factor 1/3 arising as in chapter 16 from the different polarizations of different modes. Hence

$$C^2 = 4m_0\Delta_0^2p_0^2/3\hbar^2\varepsilon_0V, \tag{45}$$

and

$$\gamma^2 = \Delta_0p_0^2/3\hbar^2\varepsilon_0V, \tag{46}$$

from (36), with $\hbar\omega_0$ put equal to $2\Delta_0$.

From (44), then,

$$\tau_e = 3\pi\hbar^4c/8\mu_0\Delta_0^3p_0^2. \tag{47}$$

Now a Hertzian dipole of mass m_{12}, oscillating at $2\Delta_0/\hbar$, has a time-constant 497 for energy loss of $3\pi m_{12}\hbar^2c/2\mu_0\Delta_0^2e^2$. These time-constants coincide if f_{12}, 523 i.e. m/m_{12}, is $4m\Delta_0p_0^2/\hbar^2e^2$, which is the value already calculated directly. The double-well system not only responds to weak stimulation as if it were a harmonic oscillator, but it also radiates spontaneously like the same equivalent oscillator. Of course, if we had not reached this conclusion it would have been very disturbing, for the Einstein argument involving detailed balance between the radiation field and a material oscillator (to 536 which we revert shortly) demands that a unique relation, as expressed by (16.29), between stimulated and spontaneous processes must hold if the Boltzmann distribution is to apply universally.

When the motion of a classical oscillator decays exponentially the time-constant for energy decay is half that for the decay of amplitude, i.e. dipole moment, and the same holds for the double-well system. To show this, we return to (42) and evaluate $\langle y(t)\rangle$, i.e. $\iint \Psi^*(t)y\Psi(t)\,dy\,dx$. Of the nine terms in the product $\Psi^*\Psi$, by symmetry only two survive, those which have the form $\chi_1\chi_2\psi_0^*\psi_0$; with the help of (41) we find:

$$\langle p\rangle/p_0 = 2\,\mathrm{Re}\,[(\lambda_0^*\lambda_1/\alpha_0)\,e^{-i(\omega_0+\omega_e)t}(\alpha_0^2\,e^{i\Omega_0t}+\beta_0^2\,e^{-i\Omega_0t})]. \tag{48}$$

The first part, including $e^{-i(\omega_0+\omega_e)t}$, is a fast oscillation whose amplitude is slowly modulated by the following term in brackets, and it is the absolute magnitude of this latter term, and its variation with time, that carries the information needed to determine the decay of the dipole oscillation. If we write Amp (p) for $|\alpha_0^2 e^{i\Omega_0 t}+\beta_0^2 e^{-i\Omega_0 t}|$, we have that

$$[\text{Amp}(p)]^2 = 1-4\alpha_0^2\beta_0^2 \sin^2 \Omega_0 t = 1-(\gamma^2/\Omega_0^2)\sin^2 \Omega_0 t,$$

which has the same form as (43). We conclude that the square of the oscillatory dipole moment decays exponentially with time-constant τ_e, and consequently the moment itself decays with time-constant τ_p equal to $2\tau_e$, like a classical dipole.

Decay by spontaneous radiation is readily incorporated in the geometrical representation of the double-well system. In the absence of external perturbations the representative point spins at angular frequency $2E/\hbar$ about the tilted axis of fig. 4(b)†, but now its height above the lower state S′, measured along S′N′, decreases exponentially with time-constant τ_e, since this is a measure of its excitation energy, while the radius of its circular orbit, which determines the dipole oscillations, decreases at half the rate. The necessity of allowing the point to leave the surface of the sphere illustrates a basic feature of quantum mechanics, that while the wave-function of two or more uncoupled systems may be written as a simple product, once they are coupled the function evolves so that in general it can never again be written thus, even after the coupling is reduced to zero; only by making observations on the state of each component separately can the product wave-function be restored. In this case the representative point, once located on the sphere, must remain there so long as the system is truly isolated from other quantal systems, since the energy and dipole amplitude are then uniquely related. Coupling to the cavity destroys the uniqueness of the relation.

Equations of motion of a damped double-well

When the double-well is subject to weak external forces as well as to radiative damping the two processes may be treated as independent in any short interval. The proviso that the force shall be weak is necessary because τ_e and τ_p vary with Δ. To find the variation, note that according to (16.11) a Hertzian dipole of mass m_{12} has a decay time proportional to m_{12}/ω_{12}^2. Now (17.2) and (2) show that f_{12} here is $2m\omega_{12}p_{12}^2/\hbar e^2$, varying as $1/E$ according to (19), since $\omega_{12} \propto E$. Hence τ_e and τ_p, being proportional to $1/\omega_{12}^2 f_{12}$, vary as $1/E$. Relative to the equal well system, the decay times must be reduced by the factor Δ_0/E, i.e. $\Delta_0/(\Delta_0^2+p_0^2\mathscr{E}^2)^{\frac{1}{2}}$. Moreover when $p_0\mathscr{E}$ is comparable to Δ_0 the representation of fig. 4(b) and (c) shows that decay takes place along a tilted axis. If the force is imagined to be applied

† Here we generalize to an unequal double-well the results derived only for the equal wells.

as a sequence of sharp impulses, decay is fast during the application of the impulses. It is readily shown, however, that provided the mean value of $p_0\mathscr{E}$ is much less than Δ_0 these effects may be neglected, and the decay assumed to take place only between the impulses, and at the normal speed. We shall now make this assumption in deriving the impulse-response function of a radiatively damped double-well.

At any instant the wave-function may be expressed as a sum over stationary state solutions for the uncoupled systems, which are product wave-functions of the type $\chi_1\psi_n(x_i)$ or $\chi_2\psi_n(x_i)$, $\psi_n(x_i)$ being a shorthand notation for product wave-functions comprising one oscillator function for each cavity mode. Immediately before the impulse

$$\Psi = \sum_n [\lambda_n\chi_1\psi_n(x_i) + \mu_n\chi_2\psi_n(x_i)]. \tag{49}$$

A weak impulse, $\int \mathscr{E}\,\mathrm{d}t = P$, acting on the double-well only, adds on to Ψ an increment $ieyP\Psi/\hbar$, according to (13.31). Now χ_1 has amplitudes $+1/\sqrt{2}$ in each well when they are equal (the only case we shall consider) and χ_2 has amplitudes $\pm 1/\sqrt{2}$. The impulse adds $\mathrm{i}p_0P/\sqrt{2}\hbar$ to the amplitude in the right-hand well of χ_1, and subtracts the same from the left-hand well – that is to say, χ_1 is converted into $\chi_1 + (\mathrm{i}p_0P/\hbar)\chi_2$; similarly χ_2 is converted into $\chi_2 + (ip_0P/\hbar)\chi_1$. Every λ_n in (49) is thus converted into $\lambda_n + \mathrm{i}\kappa\mu_n$ and every μ_n into $\mu_n + \mathrm{i}\kappa\lambda_n$, κ being written for $p_0P/\hbar$.

The initial probability $\mathscr{P}_1$ of finding the double-well system in its ground state, whatever the oscillators may be doing, is $\Sigma_n \lambda_n\lambda_n^*$, and the probability $\mathscr{P}_2$ that it will be found excited is $\Sigma_n \mu_n\mu_n^*$. The change in each λ_n and μ_n due to the impulse results in a first-order change to $\mathscr{P}_1$ and $\mathscr{P}_2$:

$$\delta\mathscr{P}_1 = -\delta\mathscr{P}_2 = \mathrm{i}\kappa \sum_n (\mu_n\lambda_n^* + \mu_n^*\lambda_n). \tag{50}$$

By forming $\langle\Psi^*|y|\Psi\rangle$ the dipole moment is calculated:

$$\langle p\rangle/p_0 = \sum_n (\mu_n\lambda_n^* + \mu_n^*\lambda_n). \tag{51}$$

These results immediately suggest the appropriate modification to the geometrical representation expressed by (12):

$$\zeta = \sum_n (\mu_n\mu_n^* - \lambda_n\lambda_n^*) = \sum_n (1 - 2\lambda_n\lambda_n^*), \qquad \rho = \xi + \mathrm{i}\eta = 2\sum_n \lambda_n\mu_n^*. \tag{52}$$

It is not difficult to verify that no point specified in this manner can lie outside the unit sphere, and that to lie on the sphere $\mu_n = c\lambda_n$, c being a complex constant so that (49) takes the form appropriate to uncoupled system and cavity, $\Psi = (\chi_1 + c\chi_2)\Sigma_n\lambda_n\Psi_n$. Otherwise the point lies within the sphere, as we have come to expect from the analysis of coupled systems. The prescription (52) matches the requirement that $\frac{1}{2}(\zeta+1)$ shall describe the probability of finding the double-well system excited, and that ξ shall describe the dipole moment. Further, when unperturbed the representative point executes an orbit round the ζ-axis with angular velocity $2\Delta_0/\hbar$, since each term in ρ has its phase angle increasing at this rate. We may be satisfied, therefore, that the dissipative process will take place as already determined, with time-constant τ_e for $1+\zeta$, $2\tau_e$ for ρ, and we can also write down the response to an impulse:

$$\delta\rho = 2\sum_n (\lambda_n \delta\mu_n^* + \mu_n^* \delta\lambda_n) = 2i\kappa \sum_n (\mu_n\mu_n^* - \lambda_n\lambda_n^*) = 2i\kappa\zeta.$$

$$\therefore \quad \delta\xi = 0, \qquad \delta\eta = 2\kappa\zeta, \quad \text{and similarly } \delta\zeta = -2\kappa\eta.$$

This describes rotation through 2κ, i.e. $2p_0P/\hbar$, about the ξ-axis, exactly the same as found in (30) for the isolated double-well.

We may now write the equations of motion, under the influence of a varying field $\mathscr{E}(t)$, in Cartesian co-ordinates:

$$\left.\begin{aligned} \dot{\xi} &= 2\Delta_0\eta/\hbar - \xi/\tau_p \\ \dot{\eta} &= -2\Delta_0\xi/\hbar - \eta/\tau_p + 2p_0\mathscr{E}\zeta/\hbar \\ \dot{\zeta} &= (\zeta_0 - \zeta)/\tau_e - 2p_0\mathscr{E}\eta/\hbar. \end{aligned}\right\} \qquad (53)$$

The three contributions, rotation about ζ at frequency $2\Delta_0/\hbar$, rotation about ξ due to $\mathscr{E}$, and dissipation with two separate time-constants, are immediately recognizable in the structure of these equations. The opportunity has been taken to replace $-(1+\zeta)/\tau_e$ in the third equation by $(\zeta_0 - \zeta)/\tau_e$, so that the ζ-coordinate is permitted to relax exponentially towards a point which is not the pure lower state. This, as will be justified shortly, allows for stimulated as well as spontaneous processes at a non-zero temperature. **536**

These equations of motion were developed for the equal-well case and are virtually the same as the Bloch equations for nuclear spin resonance, (8.18) and (8.19). The two time-constants in (53) are counterparts of T_1 (longitudinal relaxation time $\equiv \tau_e$) and T_2 (transverse relaxation time $\equiv \tau_p$). In general there are other dissipative mechanisms besides radiative transitions, and τ_p need not be $2\tau_e$ – in fact, as we discuss later, energy-conserving dephasing processes can lower τ_p without changing τ_e; it is not possible for τ_p to exceed $2\tau_e$. When the wells are unequal or $\mathscr{E}$ is strong, or both together, modifications must be made to the axes and speeds of rotation, and to the decay times; in the representation of fig. 4(*b*) the tilt, ε, is determined by the instantaneous value of E, to which both Δ and $p_0\mathscr{E}$ contribute, and the tilted axis is also the direction of energy relaxation with τ_e replaced by $\Delta_0\tau_e/E$. Clearly if $\mathscr{E}$ is both strong and variable the equations become unpleasantly non-linear, and we shall not even write them down. **227** **552**

When τ_e and τ_p are long and $\mathscr{E}$ oscillates at a frequency close to $2\Delta_0/\hbar$, even a weak field can cause considerable perturbation. The behaviour illustrated in fig. 1 is characteristic of a non-dissipative system, and is essentially transient when dissipation is present. Let us look for the steady-state solution of (53), once this transient has died away and the variations of ξ, η and ζ are synchronized to the driving frequency, ω. If $\mathscr{E} = \mathscr{E}_0 \cos\omega t = \frac{1}{2}\mathscr{E}_0(e^{i\omega t} + e^{-i\omega t})$, the first two equations show that ξ and η are proportional to $\mathscr{E}$ in the lowest order of approximation, so that the third equation contains $\mathscr{E}$ only as a quadratic term. Let us assume, then, that ζ settles down to a constant value, $\bar{\zeta}$, which is not ζ_0 since there is a steady term contained in $\mathscr{E}\eta$. With ζ constant the first two equations are conveniently written in terms of $\rho(=\xi+\mathrm{i}\eta)$:

$$\tau_p\dot{\rho}+(1+2\mathrm{i}\Delta_0\tau_p/\hbar)\rho=(\mathrm{i}p_0\tau_p\bar{\zeta}\mathscr{E}/\hbar)(\mathrm{e}^{\mathrm{i}\omega t}+\mathrm{e}^{-\mathrm{i}\omega t}). \qquad (54)$$

Near resonance the second exponential dominates the solution if $2\Delta_0\tau_p\gg 1$, and we may neglect the positive exponential. Then $\rho=\rho_0\,\mathrm{e}^{-\mathrm{i}\omega t}$, where

$$\rho_0=(\mathrm{i}p_0\mathscr{E}_0\tau_p\bar{\zeta}/\hbar)/[1+\mathrm{i}(\omega_{12}-\omega)\tau_p]; \qquad \omega_{12}=2\Delta_0/\hbar. \qquad (55)$$

The imaginary part of ρ_0 is the amplitude of η, from which it follows that $\eta=(p_0\mathscr{E}_0\tau_p\bar{\zeta}/\hbar)[\cos\omega t-(\omega_{12}-\omega)\tau_p\sin\omega t]/[1+(\omega_{12}-\omega)^2\tau_p^2]$, and the mean value of $\mathscr{E}\eta$ is $\frac{1}{2}p_0\mathscr{E}_0^2\tau_p\bar{\zeta}/\hbar[1+(\omega_{12}-\omega)^2\tau_p^2]$. When this is substituted in the third equation of (53), and $\dot{\zeta}$ put equal to zero, the mean value of ζ emerges as the solution:

$$\bar{\zeta}=\zeta_0\bigg/\left[1+\frac{p_0^2\mathscr{E}_0^2\tau_e\tau_p/\hbar^2}{1+(\omega_{12}-\omega)^2\tau_p^2}\right]. \qquad (56)$$

If ζ_0 is the equilibrium position of the unperturbed system maintained 523 in contact with black-body radiation at temperature T, $\zeta_0=-\tanh(\Delta_0/k_BT)$ and the reduction in $|\zeta_0|$ expressed by (56) can be interpreted as a raising of the effective temperature, T_{eff}. The energy input by virtue of $\mathscr{E}$ acting on the oscillating dipoles can only be dissipated at a rate $(\bar{\zeta}-\zeta_0)/\tau_e$ times the excess. The reason why τ_p appears in (56) as well as τ_e is that it helps determine the magnitude of the oscillating dipole and through this the energy input. If τ_e and τ_p are long enough, and $\omega_{12}-\omega$ small enough, $\bar{\zeta}$ may be reduced practically to zero ($T_{\text{eff}}\to\infty$) while $\mathscr{E}$ is still small enough for the linear approximation to apply.

The strength of the dipole response follows from (55):

$$|\rho_0|=(p_0\mathscr{E}_0\tau_p\zeta_0/\hbar)[1+(\omega_{12}-\omega)^2\tau_p^2]^{\frac{1}{2}}/[1+(\omega_{12}-\omega)^2\tau_p^2+p_0^2\mathscr{E}_0^2\tau_p\tau_e/\hbar^2]. \qquad (57)$$

So long as $\mathscr{E}_0$ is weak enough to cause insignificant 'heating' $|\rho_0|\propto\mathscr{E}_0$, but as $\mathscr{E}_0$ is increased $|\rho_0|$ ultimately falls to zero as $1/\mathscr{E}_0$. Naturally the saturation of $|\rho_0|$ is most readily achieved at resonance, when $\omega=\omega_{12}$.

At resonance $\bar{\zeta}$, according to (56), is reduced by a factor $1+2p_0^2\mathscr{E}_0^2\tau_e^2/\hbar^2$, 531 if $\tau_p=2\tau_e$. Writing $\hbar e^2 f/2\omega_{12}m$ for p_0^2, $6\pi mc/f\mu_0\omega_{12}^2e^2$ for τ_e and $2\Phi/\varepsilon_0 c$ for $\mathscr{E}_0^2$, Φ being the power flux in the irradiation, we have that for a transition at a wavelength λ,

$$2p_0^2\mathscr{E}_0^2\tau_e^2/\hbar^2=9m\lambda^5\Phi/4\pi^3\mu_0c^2e^2\hbar\doteqdot 2\times 10^{29}\lambda^5\Phi/f.$$

A typical optical transition, with $f\sim 1$ and $\lambda\sim 5\times 10^{-7}$ m, requires a flux of about 160 $W\,\mathrm{m}^{-2}$ to reduce ζ_0 by a factor 2 and thus go some way towards saturating the transition. This is a flux density that is readily available in a monochromatic beam by use of a continuous gas laser. At lower frequencies, as for example in the microwave range of the spectrum, the dependence of Φ on λ^{-5}, to give a certain degree of saturation, would suggest that even the weakest sources would saturate the transitions. It should be remembered, however, that this calculation takes note only of radiative processes in dissipation, and it is the inefficiency of these at long wavelengths that produces this result. In practice collisions between the molecules dominate the loss mechanisms, but even so it is not hard to saturate the transitions with very modest irradiating power. The same applies to nuclear magnetic resonance where very low power levels are normally needed.

Stimulated and spontaneous transitions

The introduction of ζ_0 in (53), instead of -1 as required by the theory of spontaneous transitions, is a fairly obvious generalization, but one which deserves discussion. Let us imagine the double-well system immersed in a bath of cavity radiation in equilibrium at temperature T. Following our usual practice we consider the interaction with one cavity oscillator, but now it is not initially in its ground state, but in some mixture of excited states, so that at the instant of coupling the product wave-function takes the form:

$$\Phi(0) = (a_1\chi_1 + a_2\chi_2) \sum_n b_n\psi_n. \tag{58}$$

If we are concerned with the average behaviour of the energy contained in an assembly of similar systems we may assume not only that the cavity oscillators have random phases for their ψ_n, but that the only results of interest are those obtained by averaging over all phases of a_1 and a_2. This allows us to follow the development of each term in (58) separately (which the linearity of Schrödinger's equation permits) and to ignore phase correlations when combining the results. Thus, we start with a typical term, $a_1\chi_1 b_n\psi_n$, and express it as a sum of stationary states (33):

$$\chi_1\psi_n = \alpha_{n-1}\Phi' + \beta_{n-1}\Phi'', \tag{59}$$

in which $$\Phi' = \alpha_{n-1}\chi_1\psi_n + \beta_{n-1}\chi_2\psi_{n-1}$$

and $$\Phi'' = \beta_{n-1}\chi_1\psi_n - \alpha_{n-1}\chi_2\psi_{n-1}.$$

The subsequent development of the wave-function results from the beating of the two terms in (59) with frequency $2\Omega_{n-1}$. After half a cycle the relative phases are reversed, and $\chi_1\psi_n$ is turned into $(\alpha_{n-1}^2 - \beta_{n-1}^2)\chi_1\psi_n + 2\alpha_{n-1}\beta_{n-1}\chi_2\psi_{n-1}$. The probability $\mathscr{P}_2$ of finding the system excited is now $4\alpha_{n-1}^2\beta_{n-1}^2$, i.e., $\gamma^2 n/\Omega_{n-1}^2$ from (38) and (36). At any other time

$$\mathscr{P}_2(t) = (\gamma^2 n/\Omega_{n-1}^2)\sin^2(\Omega_{n-1}t) \propto nt^2 \text{ for small } t.$$

Proceeding along the same lines as for the unexcited cavity we conclude that if the system is initially unexcited $\mathscr{P}_2$ will start to rise linearly with time, the rate being determined by the mean value of n for those oscillators which are in resonance with the system.

If, however, we proceed likewise with an initial state $a_2\chi_2 b_n\psi_n$ the argument is similar except that the analogue of (59) contains ψ_n and ψ_{n+1} rather than ψ_{n-1} and ψ_n. The probability $\mathscr{P}_1$, of finding the system unexcited again rises linearly, but the rate is controlled now by the mean value of $n+1$, rather than n.

The development of $\mathscr{P}_1$ and $\mathscr{P}_2$ for the assembly is now clear, for randomness of phase-relationships eliminates any interference effects, so that the two processes make independent contribution to the evolution of $\mathscr{P}_1$ and $\mathscr{P}_2$;

$$-\dot{\mathscr{P}}_2 = \dot{\mathscr{P}}_1 = -\bar{n}\mathscr{P}_1/\tau_e + \overline{(n+1)}\mathscr{P}_2/\tau_e = (\bar{\mathscr{P}}_1 - \mathscr{P}_1)/\tau_e', \tag{60}$$

in which $\tau_e' = \tau_e/(2n+1)$ and τ_e is the decay time for spontaneous radiation into the empty cavity. The equilibrium value of $\mathscr{P}_1$ is $\bar{\mathscr{P}}_1$, where

$$\bar{\mathscr{P}}_1 = \frac{\bar{n}+1}{2\bar{n}+1} \quad \text{and} \quad \bar{\mathscr{P}}_2 = 1 - \bar{\mathscr{P}}_1 = \frac{\bar{n}}{2\bar{n}+1}. \tag{61}$$

Since according to Planck's law, $\bar{n} = (e^{2\Delta_0/k_B T} - 1)^{-1}$,

$$\bar{\mathscr{P}}_1/\bar{\mathscr{P}}_2 = 1 + 1/\bar{n} = e^{2\Delta_0/k_B T},$$

as required by the Boltzmann distribution. This is the justification for placing ζ_0 instead of -1 in (53), but it should also be noted that the stimulated processes resulting from the excited cavity modes speed up the process of equilibration, and that τ_e in (53) should be multiplied by $1/(2\bar{n}+1)$, i.e. $\tanh(\Delta_0/k_B T)$. The same holds for τ_p which remains equal to $2\tau_e$, although to prove this it would be necessary to take account of phase correlation between a_1 and a_2, since otherwise there would be no dipole oscillation to decay.

The dependence of τ_e on $\bar{n}$, and thus on the ambient temperature, contrasts with the constancy of τ_e for a harmonic vibrator. In fact the transition rate between any two levels is affected by temperature in exactly the same way in both systems, but as T is raised the number of occupied vibrator levels increases proportionately, and the consequent increased number of transitions needed to reach equilibrium compensates for the greater speed. Nothing of this sort can of course occur with the two-level system.

503 The outcome of the argument in this section, so far as the energy is concerned, confirms Einstein's analysis in which the terms proportional to $\bar{n}$ in (60) represent stimulated processes while the additional $\mathscr{P}_2/\tau_e$ describes the spontaneous deexcitation. At the time of Einstein's work, which took place in the context of Bohr's quantum theory, virtually nothing could be said concerning such linear response functions as the dipole oscillations. It would be tempting to suppose that the impulse response would continue

oscillating undamped until a sudden transition, as envisaged by Bohr, stopped it dead. The random occurrence of such transitions would lead in the average, for an assembly of systems, to exponential decay, not with time-constant τ_p but with the same time-constant, τ_e or τ'_e, as for the energy. One must therefore be careful, when using Einstein's model for visualization of the process, not to apply it blindly. Given care, however, it is a very real aid to thought and, of course, was historically of the greatest importance.

The frequency-dependent susceptibility

The Bloch equations (53), with their solution (55) for $\rho_0 \mathrm{e}^{-\mathrm{i}\omega t}$ when the applied field is $\mathscr{E}_0 \mathrm{e}^{-\mathrm{i}\omega t}$, yield a complex susceptibility $\chi(\omega) = p_0\rho_0/\mathscr{E}_0$ of characteristically resonant form for the case of equal wells. When the wells are unequal various modifications must be made to the Bloch equations, the details of which are left to the reader. One must bear a number of points in mind. It follows from (26) that in the representation of fig. 4(*a*) the equations of motion of the loss-free system are unchanged in form, though ω_0 is now $2E/\hbar$ and p_0 must be replaced by p_{12}, i.e. $p_0\Delta_0/E$, from (19). Proceeding to (53), we note that τ_p and τ_e (better, τ'_p and τ'_e) must be multiplied by Δ_0/E, and that $\zeta_0 = -\tanh(E/k_\mathrm{B}T)$. This last point is significant, since ζ now suffers a first-order oscillation as a result of the dependence of E on $\mathscr{E}$; and since $\langle p \rangle$ is obtained by projecting onto a tilted plane, this oscillation of ζ makes a contribution to the susceptibility. This contribution is indeed the same as that attributable to the varying base-line in the step-response function, fig. 3(*b*). The reason for not displaying this approach to the problem in detail is that it is easier to incorporate the various modifications into the response function, and derive $\chi(\omega)$ therefrom. Thus (29), with f put equal to $\tanh(E/k_\mathrm{B}T)$ and the oscillatory term appropriately damped, supplies the first term of (62). The second term rises smoothly from 0 to its final value $(\Delta^2 p_0^2/E^2 k_\mathrm{B}T)\,\mathrm{sech}^2(E/k_\mathrm{B}T)$ which is $-(\Delta p_0^2/E)\,\mathrm{d}\zeta_0/\mathrm{d}\Delta$; unit field changes the equilibrium value of ζ_0 by $p_0\,\mathrm{d}\zeta_0/\mathrm{d}\Delta$, whose projection onto the tilted plane gives a change in $\langle p \rangle$ of $-p_0\Delta/E$ times this. Hence

$$S(t) = (\Delta_0^2 p_0^2/E^3)\tanh(E/k_\mathrm{B}T)[1 - \mathrm{e}^{-Et/\Delta_0\tau'_p}\cos(2Et/\hbar)] + (\Delta^2 p_0^2/E^2 k_\mathrm{B}T)\,\mathrm{sech}^2(E/k_\mathrm{B}T)(1 - \mathrm{e}^{-Et/\Delta_0\tau'_e}). \qquad (62)$$

The impulse-response is $\mathrm{d}S/\mathrm{d}t$:

$$h(t) \doteqdot (2\Delta_0^2 p_0^2/\hbar E^2)\tanh(E/k_\mathrm{B}T)\,\mathrm{e}^{-Et/\Delta_0\tau_p}\sin(2Et/\hbar) + (\Delta^2 p_0^2/E\Delta_0 k_\mathrm{B}T\tau_e)\,\mathrm{sech}^2(E/k_\mathrm{B}T)\,\mathrm{e}^{-Et/\Delta_0\tau'_e}, \qquad (63)$$

if $E\tau_p/\hbar \gg 1$ and the oscillations are weakly damped. Finally, according to

(5.4) the susceptibility $\chi(\omega)$ is the Fourier transform of $h(t)$:

$$\chi(\omega)=\int_0^{\infty} h(t)\,\mathrm{e}^{\mathrm{i}\omega t}\,\mathrm{d}t$$
$$\doteqdot (4\Delta_0^2 p_0^2/\hbar^2 E)\tanh(E/k_{\mathrm{B}}T)/(\omega^2-4E^2/\hbar^2+2\mathrm{i}\omega E/\Delta_0\tau_p)$$
$$+(\Delta^2 p_0^2/E^2 k_{\mathrm{B}}T)\,\mathrm{sech}^2(E/k_{\mathrm{B}}T)/(1-\mathrm{i}\omega\Delta_0\tau_e/E). \qquad (64)$$

524

The imaginary part of χ is shown in fig. 3(c), with two peaks corresponding to the two terms of (64). The first gives the peak on the right of the diagram, a typical Lorentzian resonance, centred on $2E/\hbar$, whose width at half power is $2E/\Delta_0\tau_p$; to the left of the diagram is the Debye-type relaxation peak[3] contributed by the second term. There is some risk of regarding the peaks as having similar shapes, especially in the conventional plot such as fig. 3(d), where χ'' is displayed against log ω rather than against ω itself. Since $\chi''=\chi'(0)\omega\tau/(1+\omega^2\tau^2)$, we derive the form of the logarithmic plot by writing λ for ln ω and λ_0 for ln $(1/\tau)$:

$$\chi''(\lambda)=\chi'(0)\,\mathrm{e}^{\lambda-\lambda_0}/(1+\mathrm{e}^{2(\lambda-\lambda_0)})=\chi'(0)\,\mathrm{sech}\,(\lambda-\lambda_0), \qquad (65)$$

a symmetrical bell-shaped curve superficially resembling a Lorentzian resonance if we overlook its width – between the half-peak points there is a factor of 13.9 in frequency, the two half-points being defined by $\omega\tau=2\pm\sqrt{3}$. On the logarithmic plot of fig. 3(d) the relaxation peak has a width of 1.14, whatever the value of τ which only determines the position of the peak ($\omega\tau=1$), while the resonant peak has a width of about $1/Q$, much less than the relaxation peak even for a low-Q resonance.

The area under each curve on the logarithmic plot, i.e. $\int_0^{\infty}\chi''\,\mathrm{d}\lambda$, is a convenient measure of the strength of the process. According to the

110

Kramers–Kronig relations (5.9) the area is $\frac{1}{2}\pi$ times $\chi'(0)$, the contribution of the process to the static polarizability. Hence, from (64), we have the two strengths which should be compared with the two terms in (23):

$$\text{Resonance strength}=(\pi p_0^2\Delta_0^2/2E^3)\tanh(E/k_{\mathrm{B}}T), \qquad (66)$$
$$\text{Relaxation strength}=(\pi p_0^2\Delta^2/2E^2 k_{\mathrm{B}}T)\,\mathrm{sech}^2(E/k_{\mathrm{B}}T). \qquad (67)$$

These strengths are independent of the relaxation times and are determined by the tunnelling probability, as measured by Δ_0, and the asymmetry Δ of the potential wells in relation to $k_{\mathrm{B}}T$. The resonance has its greatest strength of $\pi p_0^2/2\Delta_0$ at low temperatures and when $\Delta=0$, but the relaxation strength is zero when Δ and $T=0$, having a maximum value of $0.27p_0^2/\Delta_0$ when $\Delta=\sqrt{2}\Delta_0$ and $k_{\mathrm{B}}T=2.24\Delta_0$. Under the same conditions the resonance strength is reduced to $0.20p_0^2/\Delta_0$, principally because of E^3 replacing Δ_0^3 in the denominator of (66). When the wells are markedly uneven relaxation can still proceed by the particle tunnelling from the higher well to the lower, but there is very little of an oscillatory dipole associated with this process. When the temperature is high enough for tanh $(E/k_{\mathrm{B}}T)$ to be replaced by $E/k_B T$ and $\text{sech}^2 (E/k_B T)$ by unity, the two expressions, (66) and (67), sum to $\pi p_0^2/2k_B T$, independent of Δ_0 or Δ. A static susceptibility varying inversely with temperature is just what is predicted by the simplest theories of orientable permanent dipoles, as in Curie's law for paramagnetic materials and Debye's analysis of dipolar dielectrics. We shall comment further on this point in due course, with specific reference to the ammonia molecule.

547

The relaxation peak that arises by quantum-mechanical tunnelling between potential wells is very much rarer than the loss peaks of the same form which are found in dielectrics,[3] magnetic materials,[4] elastic materials[5] and many others, where the exponential approach to equilibrium after application of a step-function perturbation is by thermal activation over a potential barrier. These processes are unaccompanied by any resonant behaviour and will be considered no further here, except to remark that, through the operation of the Boltzmann factor, the time-constant for crossing the barrier increases rapidly at low temperatures, usually as $\exp (T_0/T)$, where T_0 is a characteristic temperature commonly in the range around 10^4 K. With quantum-mechanical tunnelling, however, we may expect a much slower temperature variation. According to (60), τ_e' varies inversely as the mean energy (including zero-point energy) of the modes at frequency $2E/\hbar$ which are the agents of the relaxation process; because of spontaneous emission, τ_e' tends to a constant as T falls to zero, rather than rising indefinitely as it does when thermal excitation is responsible.

Dielectric loss in polyethylene – an example of tunnelling relaxation

Although pure polyethylene, consisting of very long chains of $—CH_2—$ units, is remarkably free from dielectric loss at low temperatures, when lightly oxidized by heating in air it develops a pattern of loss that conforms very well to the Debye model and can be explained in detail as arising from a sparse distribution of double-well potentials between which a proton can tunnel. The chemical configuration of these wells is a matter of uncertainty which is, however, irrelevant to our purpose. What emerges from the study of this material is that the two wells are reasonably similar in depth – though not exactly equal, for then they would show no relaxation peak. The experimental curves, fig. 6(a), of the frequency-variation of χ''/χ' exhibit peaks superimposed on a low background loss whose exact value is somewhat conjectural but certainly consistent with the observed loss in unoxidized material. After allowance for the background, the peaks conform to the Debye pattern (65) very well indeed; the highest peak, for example, has a width at half-peak amounting to a factor of 14.5 in frequency, only slightly larger than the theoretical 13.9. This could not happen if there were a substantial spread in relaxation times among the different double-well systems, and it may be inferred that they are all closely similar in

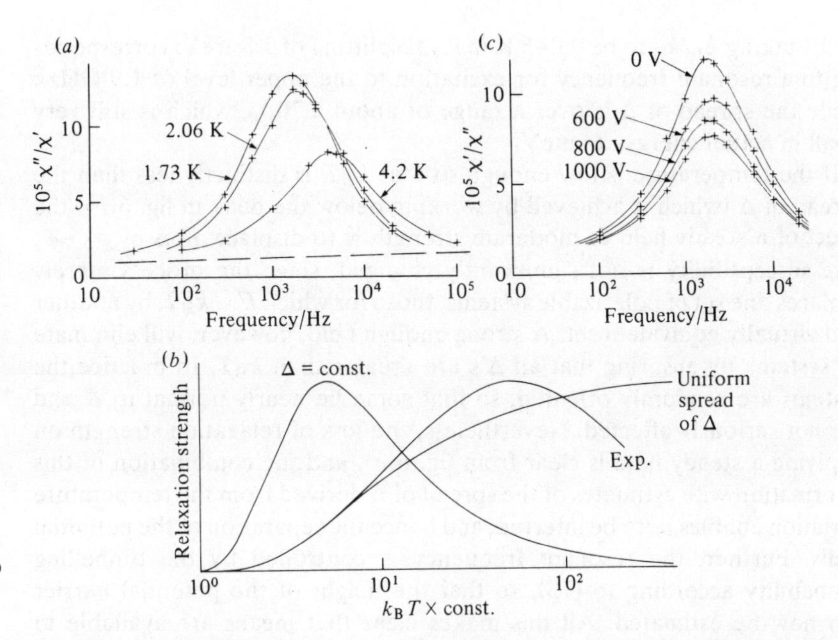

Fig. 6. (a) Dielectric relaxation loss in polyethylene (Phillips[2]); the temperatures at which the measurements were made are shown beside the curves. (b) Theoretical curves for the temperature-variation of relaxation strength for two models – an assembly of identical unequal double-wells (Δ = const) and an assembly having a uniform wide spread of Δ – and an experimental curve (Frassati and Gilchrist[6]). (c) Application of a steady strong electric field polarizes the double-wells and lowers the relaxation strength without affecting the relaxation frequency.[2]

structure and environment. Too much, however, must not be read into this; the very width of the Debye peak has the consequence that a moderate spread of relaxation times, and hence of peak positions, produces little extra breadth. A spread over a factor of 1.5 is not inconsistent with the observations.

The curves in fig. 6(a) show how the relaxation strength increases as the temperature is lowered, but this cannot continue down to zero temperature and the form of the variation is revealing. Let us suppose, as is fully justified by detailed experimental study, that the double-well systems differ from one another principally in their asymmetry. Ideally, Δ would be zero and the wells identical, but local strains and other environmental factors disturb the potential so that Δ may be taken to vary randomly over what turns out to be a relatively narrow range. It is easy to compute from (67) how the area under the relaxation loss curve will vary with temperature for a given choice of the spectrum of Δ, and two extreme examples are given in fig. 6(b), one for constant Δ and one for a uniform spread of Δ between $\pm\infty$; the curves are scaled to the same peak height, as are the experimental observations. The fall-off of the strength at higher temperatures is a sure indication that the spread of Δ is finite, though the behaviour at the lowest temperatures, which is dominated by the systems for which Δ is near zero, shows that a limited spread around zero is the model to be preferred over that of a single non-zero Δ. The temperature range in which the relaxation strength disappears as T is lowered is the most obvious indicator of Δ_0, while the fall-off at higher temperatures gives a good measure of the range of Δ. For this sample the disappearance at low temperatures is well under way at 0.1 K, and the entire variation with temperature can be accounted for by taking Δ_0/k_B to be 0.045 K (a level splitting of 7.8 μeV, corresponding to a resonant frequency for excitation to the upper level of 1.9 GHz), while the spread of Δ is over a range of about $\pm 70\Delta_0$, which is still very small in actual energy, $\pm\frac{1}{2}$ meV.

If the temperature is low enough so that $k_B T$ is distinctly less than the spread of Δ (which is achieved by working below the peak in fig. 6(*b*)) the effect of a steady field of moderate strength is to displace all Δ by $\boldsymbol{p}_0 \cdot \boldsymbol{\mathscr{E}}$. The susceptibility is not significantly changed, since the process merely replaces one set of polarizable systems, those for which $E \lessdot k_B T$, by another and virtually equivalent set. A strong enough field, however, will eliminate all systems by ensuring that all Δ's are greater than $k_B T$. In practice the systems are randomly oriented, so that some lie nearly normal to $\boldsymbol{\mathscr{E}}$ and are not seriously affected. Nevertheless, the loss of relaxation strength on applying a steady field is clear from fig. 6(*c*), and the combination of this information with estimates of the spread of Δ derived from the temperature variation enables p_0 to be inferred, and hence the separation of the potential wells. Further, the resonant frequency is controlled by the tunnelling probability according to (13), so that the height of the potential barrier can now be estimated. All this makes clear that means are available to determine the parameters of the system experimentally; the outcome is a model potential which not only has the required properties to account for all the data quantitatively, but is also a very reasonable double-well of such dimensions and with such a barrier as might be expected for a proton (i.e. hydrogen ion) in a hydroperoxide group, —OOH, which is one conjectured source of double wells.

The frequency at which the relaxation peak occurs is determined by the strength of the coupling of the double-well to the thermally maintained vibrations of its environment, for it is these that supply or remove the energy involved in the transitions between levels that restore and maintain the state of equilibrium. With the low frequency (1.9 GHz) and modest dipole moment (6×10^{-30} Cm) the time constant for equilibration by spontaneous electromagnetic radiation is far too long (~40 years) to be significant. We know, however, from the spread of Δ among otherwise identical systems that the potential is influenced by strain, as if there were a polarizing electric field associated with the deformed lattice. The double-well is thus coupled to the lattice vibrations (phonons) of the host polymer, and the tunnelling process is in consequence referred to as phonon-assisted tunnelling. For non-dispersive waves such as lattice waves or light the density of states g at a given frequency is inversely proportional to the cube of the wave velocity v_s, which in polyethylene is about 2×10^5 times less than c. If a single elastic mode were to generate the same Δ as an electromagnetic mode with the same energy density, the two would interact identically with the double-well system, and the change in g would reduce the time constant by $(v_s/c)^3$, from 40 years to 2×10^{-7} s. In fact the coupling to the lattice strain must be markedly less, not only because the time constant at the lowest temperatures (where spontaneous processes dominate) is 10^3 times larger than this, but because it would be hard to understand otherwise the limited spread of Δ. As a result of anisotropic contraction on cooling, strains of the order of 1% must exist in the polymer, giving rise to strain energy densities of about 10^5 J m^{-3}. An electric field having this energy density, $\mathscr{E} = 1.5 \times 10^8$ V m^{-1}, would generate a value of Δ equal to 6 meV, more than 10 times the observed spread. Since the time-constant varies inversely as the square of the coupling constant, there is at least a factor of 100 here towards the 1000 needed. In view of the rough arguments involved, this is near enough to suggest that a consistent picture is emerging.

One last point is worthy of notice. In view of the spread of E, and hence the resonance frequency, from Δ_0 to $70\Delta_0$, the very narrow range of relaxation times needs explanation. This in fact comes about automatically for the suggested coupling process. We have seen that τ_e for the equal-well system is reduced to $\tau_e\Delta_0/E$ by the introduction of inequality, and that the relaxation time in the presence of thermal radiation (phonons in this case) is reduced by a further factor $1/(2\bar{n}+1)$. Hence we expect τ to have the form $\tau_e\Delta_0/E(2\bar{n}+1)$. Since $2\bar{n}E$ is just k_BT for oscillators of level spacing $2E$, $\tau \sim \tau_e\Delta_0/k_BT$; Δ has vanished, and τ takes the same value for all.

Other examples of double-well systems; ammonia

The hydrogen molecule-ion, H_2^+, with one electron moving in the field of two protons, is as simple an example as one could hope to find, and it was the subject of many detailed calculations in the early days of quantum mechanics. The energy can be lowered by sharing the electron between the protons rather than having it bound to one only and, as expected, the electron wave-function in the ground state is symmetric and describes the bound state of the ion. In this case the barrier is very transparent and in the equilibrium configuration the minimum of $\psi\psi^*$ between the protons is shallow.

A considerable number of molecules[7] show a variety of stable configurations and can transform from one to another by tunnelling or by thermal excitation over the barrier. In ethane $H_3C{\cdot}CH_3$ the two configurations fig. 7(*a*) and (*b*) differ in energy by about 1/8 eV, corresponding to a temperature of 1500 K, and the CH_3 groups can rotate through 120° from one stable position of the form (*b*) to another by passing over or tunnelling through this potential hump. The tunnelling frequency (~10 GHz) is about 100 times lower than the frequency of torsional oscillations around a stable position, so that the ground state is slightly split, but not by enough to make a great difference to the bonding energy. In dimethyl-acetylene $H_3C{\cdot}C{:}C{\cdot}CH_3$ the CH_3 groups are further apart and the potential hump much smaller; in this case a good approximation to the energy results from assuming that the groups rotate freely, with a hardly significant correction for the correlation of their angular positions.

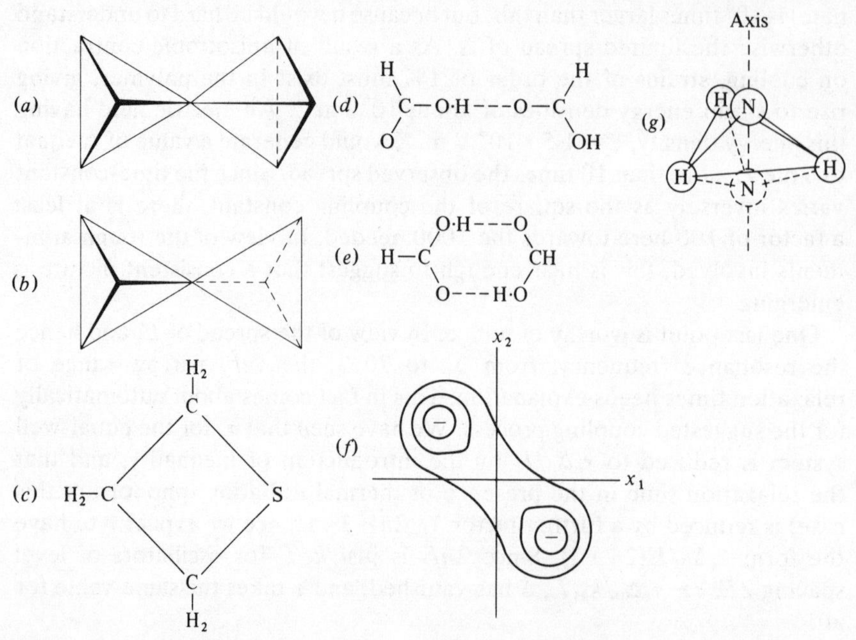

Fig. 7. (*a*) and (*b*) two configurations of ethane $(CH_3)_2$, shown as tetrahedra with carbon atoms at their centres, single-bonded at points in contact, and having H-atoms at the free points. In (*a*), the cis-configuration, the H_3 triangles at either end are similarly oriented; in (*b*), the trans-configuration of lower energy, they are 60° apart. (*c*) Trimethylene sulphide; the potential minima for the S-atom lie just above and just below the plane of the paper. (*d*) Two formic acid molecules bonded by a single hydrogen bond. (*e*) Two formic acid molecules bonded by two hydrogen bonds. (*f*) Schematic representation of the potential energy surface for (*e*). (*g*) Tetrahedral structure of ammonia, showing the two equivalent energy minima for the N-atom relative to the H_3-triangle.

The hindered rotation of CH_3 groups such as those, or of a polymer chain such as $(CH_2)_n$ about any one of the C—C bonds, makes possible a large variety of stable molecular configurations differing rather little in energy and separated by potential barriers. Important though these are to the polymer chemist, they are too complicated for our present purpose, as well as showing very little in the way of tunnelling; to switch from one configuration to another involves moving too many massive atoms too far, and is normally accomplished only by thermal excitation. On the other hand ammonia and trimethylene sulphide exhibit readily discernible tunnelling behaviour in the splitting of their ground states by the coupling of two alternative equivalent configurations. In view of its importance in the development of the maser, to which we shall turn in chapter 20, we shall pay special attention to ammonia, and in a few words dispose of trimethylene sulphide. The ring of 3 carbon atoms and 1 sulphur (fig. 7*c*) has its lowest energy when bent (*puckered*), and we may imagine the carbon atoms held fixed while the sulphur moves normal to the diagram in a symmetrical double well with the potential maximum lying in the plane of the carbons. This exaggerates the difficulty experienced by the sulphur in tunnelling from one side to the other; when the molecule is free the carbons also move and the sulphur behaves as if its mass were considerably less than its real mass. We shall not derive the required correction, which is exactly analogous to the reduced mass correction required for nuclear motion in the theory of the energy levels of hydrogen-like atoms, and is straightforward to calculate. Since the molecule has a component of dipole moment normal to the carbon plane, which reverses when the sulphur changes sides, the tunnelling process carries an oscillatory dipole moment and transitions between the symmetrical ground state and the antisymmetrical state, 34 μeV above, can be detected as an absorption line in the microwave spectrum at 8.2 GHz. By contrast, when the sulphur atom is replaced by oxygen the barrier is so low that the ground state lies above it; its presence is, however, readily apparent in the microwave spectrum around 24 GHz since the vibrations of the oxygen atom are highly anharmonic.

One more example is worth mention, though the experimental evidence is not so clear. Carboxylic acids tend to form dimers, of which the dimer of formic acid HCOOH is the simplest example. If the molecules were brought together as in fig. 7*d* the hydrogen bond would have the proton much closer to its original carbon atom than to the other, and it would be sitting in a simple potential well. Electrostatic repulsion prevents it transferring to the other molecule, but this difficulty vanishes when two hydrogen bonds are present, as in fig. 7*e*. There are two equivalent configurations, mirror images, and it is probable that they, rather than a configuration with both protons placed midway, are the states of lowest potential energy. This provides a model of a tunnelling process in which two atoms tunnel simultaneously, and serves to illustrate what may be even more complex in other cases, with more than two atoms tunnelling and more than one route connecting the stable states. A simplified representation is probably adequate to illustrate the principal points. Since the timescale for proton tunnelling is slow in comparison with other vibrations and with electron rearrangements, the Born–Oppenheimer (adiabatic) approximation[8] allows us to imagine that the rest of the molecule has plenty of time to adjust itself to changes of the probability distribution of the protons; they are thus to be thought of as moving in a potential which is a well-defined function of their position co-ordinates. And we may further simplify the problem by making the proton motion one-dimensional, along the line joining its own oxygen atoms. In this way the problem is reduced to motion of a particle of protonic mass on a plane in which the potential is $V(x_1, x_2)$, x_1 and x_2 being the co-ordinates of the two protons. The general form of $V(x_1, x_2)$ is easily guessed – there must be minima at points corresponding to the stable configurations, e.g. fig. 7*e*, and nowhere else; any departure from the line $x_1+x_2=0$ means a different net charge on the two molecules, with extra electrostatic energy, so that the central position, $x_1=x_2=0$, is probably a saddlepoint of V, as shown in fig. 7*f*. The simultaneous tunnelling of the two protons is now described by a simple two-well model, the wells being further apart by a factor $\surd 2$ than the stable positions of the individual protons; this may well reduce the tunnelling probability by a large factor, the same as would arise if we kept the original well separation and doubled the mass of the tunnelling particles. The ground state energy is lowered a little by the tunnelling, and corresponds to a symmetrical wave-function, with the antisymmetrical state slightly above. In both states $\psi\psi^*$ has equal maxima in the two wells, so that one may say that *if* one proton is found to be displaced to the right, there is a high probability that the other will be displaced to the left; and *vice versa*. Both configurations are equally likely. In a mixture of the two states the configurations interchange periodically, but there is no oscillatory dipole associated with the process, only a quadrupole which will be very weakly coupled to an electromagnetic field and correspondingly difficult to detect.

Let us now turn to the especially important case of ammonia, to which is traditionally ascribed the tetrahedral structure of fig. 7*g*. The centroid of negative charge is not compelled by symmetry to coincide with the centroid of positive charge, and one must expect to find a dipole moment oriented along the axis of the molecule. In fact the dipole moment is measured to be 1.48 Debye units, or 4.94×10^{-30} Cm, equivalent to one electron displaced through 0.03 nm (about half the Bohr radius). This model is, however, too classical since an equivalent structure is obtained by taking the nitrogen atom through the hydrogen plane to a mirror position opposite, and there is ample evidence that this inversion proceeds quite readily by tunnelling. In the spirit of the last paragraph we may suppose the only co-ordinate of significance to be the position of the nitrogen atom on a line normal to the hydrogen triangle, passing through its centroid. If a nitrogen atom has mass m_N and a hydrogen atom mass m_H, the effective mass of the tunnelling particle is m^*, where $1/m^* = 1/m_N + 1/3m_H$, i.e. $m^* = 2.5$ hydrogen masses. The chemical evidence which makes the tetrahedral structure so natural must now be interpreted as evidence for two equivalent potential wells separated by a barrier. The ground state is described by a symmetrical wave-function, with the antisymmetric state lying a little above it, and although neither state in itself has a dipole moment the superposition of the two produces an oscillatory moment exactly along the lines already discussed. The inversion phenomenon had been inferred from the line-splitting in the infra-red spectrum of ammonia, and extensively studied theoretically, before it was observed in the plainest manner as an absorption line in the microwave region, about 24 GHz, resulting from the direct transition between the split levels. This was in 1934, with the most primitive of equipment, and it gave microwave spectroscopy its first success. After 1945, with better equipment, the absorption at lower pressures could be studied, and it then became clear that there were many lines in the spectrum; more than 60 have been detected, of which about half are to be seen in fig. 8. They arise because the many rotational states of the molecule that are excited at room temperature suffer different centrifugal distortions of the bond lengths. In consequence the potential barrier and the level splitting are different in each rotational state. In addition, the nitrogen nucleus has a quadrupole moment that interacts with the field of the hydrogen atoms, and there are other internal effects as well. These complications fortunately need not be allowed to interfere with the essential simplicity of the following discussion, but we shall return to them briefly in the next chapter.

517

550

There would seem at first sight to be a contradiction between the idea of a permanent dipole, as measured, and the symmetry of the charge

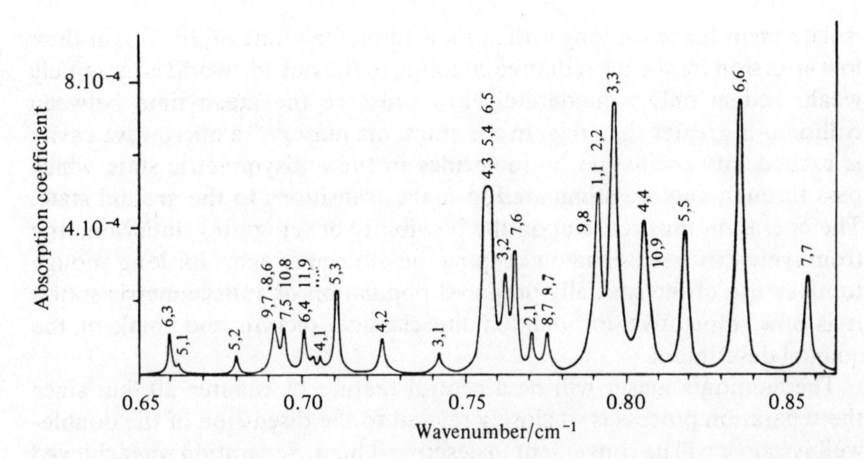

Fig. 8. Microwave absorption in ammonia gas at room temperature and a pressure of 1.2 Torr (Bleaney and Penrose[9]).

distribution in the stationary states, but this is resolved by a calculation which shows that the polarization induced by an electric field in the initially unpolarized stationary states simulates the polarization by orientation of permanent dipoles, provided $k_B T$ is greater than the level splitting. The necessary calculation has already been carried out for the double-well system; according to (23), if the wells are equal $\bar{\alpha} = (p_0^2/\Delta_0)\tanh(\Delta_0/k_B T) \approx p_0^2/k_B T$ if $k_B T \gg \Delta_0$. And this is just what is found for a permanent dipole which is allowed two orientations, either along or against the field. The two states will be occupied in the proportion $e^{\pm p_0 \mathscr{E}/k_B T}$, or $1 \pm p_0\mathscr{E}/k_B T$ when $\mathscr{E}$ is small; the excess fraction, $p_0\mathscr{E}/k_B T$, along the field gives a mean polarizability of $p_0^2/k_B T$. In the case of ammonia, with $\Delta_0/k_B \sim 0.6$ K, both models should yield virtually the same value of $\bar{\alpha}$ at all temperatures above 2 K.

This example illustrates the irrelevance, in many practical applications, of an apparently basic distinction between the classical and quantal views of symmetry. The wave-function in the ground state of ammonia is typical of many ground states which reflect the symmetry of the potential. If the potential is a symmetrical double well, $\psi\psi^*$ is symmetrical in all stationary states, but classically the particle must rest in either well, not simultaneously in both. With a potential barrier so high that it would take a year for tunnelling to occur it would be pedantic to insist on interpreting the observation of the particle in one well in terms of the superposition of two states differing in energy by 10^{-41} J, unless one had means of separating the two states, and sufficient freedom from thermal disturbances for a time exceeding the tunnelling time. In this case the energy splitting corresponds to a temperature of 10^{-18} K – only in fantasy could one pretend that the 'correct' symmetry-maintaining quantal picture has any advantage over the symmetry-breaking classical picture. By the same argument, so long as we are concerned with the behaviour of ammonia molecules in reasonably close contact with an environment well above 2 K, the classical picture is adequate and easier to use than the quantal. But now it is not difficult to isolate them for much longer than their tunnelling time of 10^{-11} s; at their low inversion frequency radiative coupling to the outside world is extremely weak, and at only a moderately low pressure the mean time between collisions is greater than this. In the ammonia maser[10] a microwave cavity is excited into oscillation by molecules in the antisymmetric state which pass through and are stimulated to make transitions to the ground state. The operation thus relies upon the possibility of separating antisymmetric from symmetric states and on keeping the molecules apart for long enough to make use of the specially prepared population of antisymmetric states. It is now imperative to abandon the classical picture and think of the quantal description.

The ammonia maser will be a central feature of chapter 20, but since the separation process is so closely related to the discussion of the double-well system it will be convenient to describe it here. Separation was achieved in the original maser by an electrostatic lens, making use of the positive polarizability of the symmetric state and the equal, but negative, polarizability of the antisymmetric, $\alpha = \pm p_0^2/\Delta_0$ from (21). In an inhomogeneous electric field the symmetric state can lower its energy still further by moving towards a region of stronger field, and is attracted in this direction, while the antisymmetric state has its energy raised and is therefore repelled. A field of magnitude $\mathscr{E}$ induces a dipole moment $\alpha\mathscr{E}$ and the energy is thereby reduced by $\frac{1}{2}\alpha\mathscr{E}^2$; consequently the force is grad $(\frac{1}{2}\alpha\mathscr{E}^2)$. By means of an array of four cylindrical electrodes with the cross-sectional form shown in fig. 9, a transverse electric field of quadrupole symmetry is established.

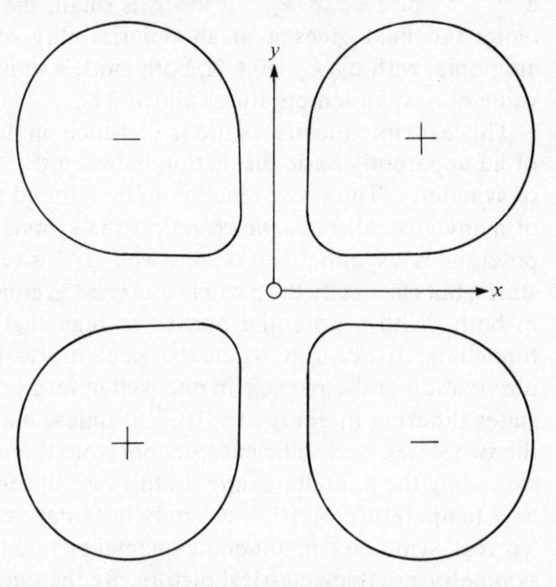

Fig. 9. Cross-section of quadrupole electrostatic lens used to focus ammonia molecules in the antisymmetric state (Gordon, Zeiger & Townes[10]).

The inner cheeks of the electrodes have the form of rectangular hyperbolas, xy = constant, compatible with the desired potential near the axis,

$$V = axy,$$

a being a constant. The field $\mathscr{E}$ is $-\text{grad}\, V$, i.e. $-a(y, x)$ and $\mathscr{E}^2$ is just $a^2(x^2+y^2)$, i.e. a^2r^2. The force on a molecule is therefore radial, drawing the antisymmetric molecules towards the axis and repelling the symmetric. If an axial but slightly divergent stream of atoms emerges from a source into the evacuated space of the lens, only atoms in the upper state are focussed back towards the axis and, passing through a hole in a diaphragm, are rather thoroughly separated from the atoms in the ground state. To appreciate the technical problem let us calculate the distance between the source and the point of focus. The equation of radial motion of a molecule is the equation of simple harmonic motion, $m\ddot{r} = -\text{grad}\,(\frac{1}{2}p_0^2a^2r^2/\Delta_0) = -p_0^2a^2r/\Delta_0$, and the period of half a cycle is $(\pi/p_0a)(m\Delta_0)^{\frac{1}{2}}$. A molecule moving along the axis with velocity v_z travels $(\pi/p_0a)(2E\Delta_0)^{\frac{1}{2}}$ in this time, where E is $\frac{1}{2}mv_z^2$, the molecule's kinetic energy, about $\frac{1}{2}k_BT$. In the original experiment a took the value $6\times10^8\ \text{V m}^{-2}$ and the ammonia molecules issued from a reservoir at room temperature. If we use for p_0 the measured dipole moment and for $2\Delta_0/\hbar$ the inversion frequency, 24 GHz, the focussing distance is calculated to be about 20 cm, much the same as was found in practice. The value of a deserves comment – 1 cm from the axis the field strength would be $6\times10^6\ \text{V m}^{-1}$, so that rather high potentials are needed on the electrodes, but not so high as to ruin the experiment by arcing.

It should be noted that the Maxwellian distribution of velocities gives different molecules different focussing distances, so that there is a measure of velocity selection. Nevertheless the velocities of those that actually enter the cavity, and hence the time they take to pass through, are liable to spread over a rather wide range; moreover, without detailed knowledge of the geometry of the arrangement, including the sizes of orifices, no estimate can be made of this distribution. This places certain restrictions on comparisons between theory and experimental performance of a maser, on the basis of published information, but it will become clear in chapter 20 that more interest attaches to the general physical principles than to details of behaviour.

19 Line broadening

The natural line broadening resulting from electromagnetic or acoustic radiative processes, the only causes of broadening discussed so far, by no means exhausts the mechanisms available and indeed is usually so minor an effect as to be of small practical importance. We may distinguish broadening due to different behaviour on the part of different members of an ensemble from broadening exhibited by each member on its own. In the first class are Doppler broadening and broadening due to variations in environment, to which may be added the effects of slight differences between superficially similar systems (e.g. different isotopes). In the second class, in addition to radiative processes, must be counted anything, especially collision with other atoms, which interrupts or distorts the wave-train emitted by a single system so as to widen its spread of Fourier components. There is a very large literature[1] on these effects, whose detailed analysis is both taxing and controversial. No attempt will be made here to go beyond an elementary discussion and illustration of some of the leading ideas, with examples of how the line-width may be reduced or its effects mitigated for the purpose of high-precision measurements of the central frequency. We start with the second class of processes, and for our purpose the two-level system provides an adequate model, with ammonia as a practical realization.

493

The absorption spectrum of ammonia at a rather low pressure, 1.2 torr, in the wavelength range from 1.1 to 1.5 cm is shown in fig. 18.8, each line resulting from transitions between pairs of levels in different rotational states of the molecule, as defined by the pairs of quantum numbers above each line. The first, J, may be visualized as defining the total angular momentum, while K defines that component of J along the axis, normal to the plane of hydrogen atoms, as indicated in the vector diagram of fig. 1. When $K = J$ the molecule is essentially spinning about this axis, and the hydrogens are thrown our centrifugally, to lower the potential barrier and increase the splitting of the pair of levels. The magnitude of the effect is determined roughly by J^2, and this explains the systematic progression from 1,1 to 7,7 at the right of fig. 18.8. Where K is less than J there is also a component of angular momentum about an axis orthogonal to the former axis, and the centrifugal effect of this elongates the barrier, so that lines with large J and small K appear at the left, low-frequency, side. The widths of the lines are real, not instrumental, and orders of magnitude greater than can be explained by radiative broadening or the Doppler effect. They are proportional to gas pressure; at 100 torr, 80 times higher than was used for fig. 18.8, the lines are so broadened as to overlap, and the resulting absorption spectrum shows no fine detail.

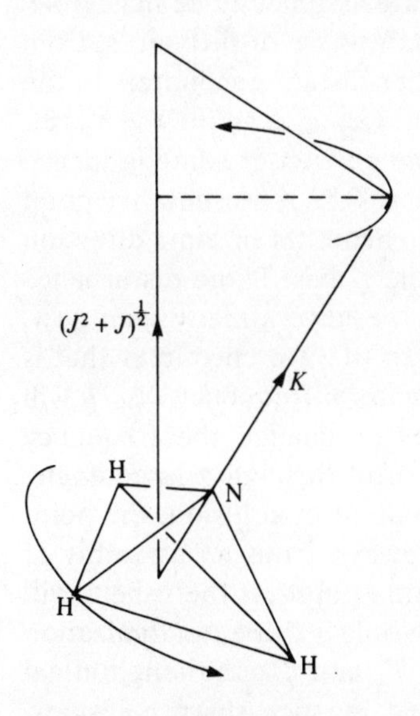

Fig. 1. Vector diagram of angular momentum of the ammonia molecule. The total angular momentum, $(J^2+J)^{\frac{1}{2}}$ units of $\hbar$, is drawn vertical, and $\boldsymbol{K}$ represents the component along the axis of the molecule, whose instantaneous position is drawn at the foot of the diagram; $\boldsymbol{K}$ precesses about the axis of total angular momentum.

The line-broadening is not usually so marked when the experimental chamber contains only a small amount of ammonia and the pressure is made up with a non-polar gas. Thus to obtain the same broadening in a mixture which is predominantly helium, with a trace of ammonia, 16 times more pressure is needed than in pure ammonia. The extent of the broadening observed with helium is close to what would be expected if every impact of a helium atom on an ammonia molecule disturbed the latter to such effect that its subsequent vibration was uncorrelated in phase to what had been taking place before. We may suppose the step-function response of a single molecule to proceed as in fig. 18.3(a) with negligible radiative decay, until a collision introduces an arbitrary phase shift; averaging over an ensemble of such molecules, all suffering their first collision at the same instant, we shall find that the wave-train stops abruptly at this point. But, of course, different molecules are hit at different times and the impacts are random, so that the probability of the wave-train lasting beyond a given time t decays exponentially, as e^{-t/τ_c}, where τ_c is the mean time between impacts of helium atoms on a given ammonia molecule. The theory is exactly analogous to the theory of the free path distribution in a gas. With an exponentially decaying response function the line shape, being the Fourier transform, is Lorentzian about the mean frequency, ω_1: $I(\omega) \propto [1+(\omega-\omega_1)^2\tau_c^2]^{-1}$. The line-width at half the peak intensity is $2/\tau_c$, and is proportional to the collision rate, i.e. to the pressure, as observed.

524

76

Phase diffusion

The description just given of line-broadening by collisions is perhaps a little cavalier in the sharp distinction made between catastrophic processes, which randomize the phase, and others which are assumed to be negligible. In reality one must expect a continuous range of phase disturbances, from a multitude of small changes arising from rather distant encounters to the truly catastrophic effects of a head-on collision. Let us examine the matter in a little more detail, taking as a model for an encounter some influence (e.g. the field of the intrusive molecule) that causes the representative point in fig. 18.4 to move during the duration of the encounter in some direction other than in its undisturbed motion round the ζ-axis. If the disturbance acts in any way like an applied electric field, we have already seen how, as in 18.4(b) this causes the axis of the orbit to tilt. An encounter that is completed during a fraction of a cycle at the natural frequency $2\Delta_0/\hbar$ will be effective only if the field is so strong as to change the frequency substantially, and this automatically tilts the orbit through a large angle. According to the phase in the cycle at the moment of collision, the point will be displaced upwards, downwards or sideways. Thus an assembly of molecules all originally represented by the same point on the sphere will soon appear as a spreading cloud of points; not only is there randomization of phase but also of ζ-co-ordinate, and both T_1 and T_2, the longitudinal and transverse relaxation times, are shortened by such sharp collisions. This is what must be expected in ammonia if the colliding molecule approaches as near as 1 nm. With a typical molecular velocity at room temperature of 700 m s^{-1}, the most effective period of the encounter lasts for something like 2 ps, during which the oscillation achieves only $\frac{1}{20}$ cycle; closer collisions have their main effect even more quickly. On the other hand, in optical transitions at 10^4 times the frequency, there is time for 500 cycles during the collision; a very small change of frequency, with the orbit correspondingly inclined at a very slight angle, can then seriously alter the phase, with hardly any perceptible shift in the ζ-direction. In such collisions T_2 is shortened but T_1 is hardly affected.

526

534

As the cloud of representative points is thus caused to diffuse over the sphere, its centroid moves towards the centre of the sphere and correspondingly the mean oscillatory dipole strength per molecule decreases. We shall now show that the decrease is exponential, not simply for catastrophic collisions but for any mixture of weak and strong collisions. For simplicity the argument will be confined to the situation more appropriate to optical lines, when the spread of points is around a line of constant ζ, but it can readily be generalized to include diffusion over the whole surface of the sphere. The only assumption needed is that the position of a point on this line is immaterial to the scattering process: if $P(\phi, \phi')\,\mathrm{d}\phi'$ is the chance that in a certain time interval a point at ϕ will be scattered into the range $\mathrm{d}\phi'$ at ϕ', we assume P has the form $P(\phi-\phi')$, and no explicit dependence on ϕ alone. This is a natural assumption when the collision lasts for so many cycles that the precise moment at which it begins is irrelevant. We do not assume that $P(\phi-\phi')=P(\phi'-\phi)$; indeed, since any electric field increases, never decreases, the orbiting frequency it would be mistaken to suppose forward and backward phase shifts to be equally likely. We now express any distribution of points, having density $A(\phi)$ on the line of constant ζ as a Fourier sum:

$$A(\phi)=\sum_{-\infty}^{\infty} a_n\,\mathrm{e}^{\mathrm{i}n\phi},\ \text{with}\ a_{-n}=a_n^*\ \text{since}\ A\ \text{is real}. \tag{1}$$

Consider how any one term in this sum is affected by scattering; a density that began as $a_n\,\mathrm{e}^{\mathrm{i}n\phi}$ has become $a'_n\,\mathrm{e}^{\mathrm{i}n\phi}$, where

$$a'_n\,\mathrm{e}^{\mathrm{i}n\phi'}=\int_0^{2\pi} a_n\,\mathrm{e}^{\mathrm{i}n\phi}P(\phi-\phi')\,\mathrm{d}\phi, \tag{2}$$

and if $P(\phi-\phi')$ is also expressed as a Fourier sum $\Sigma_{-\infty}^{\infty}\,p_m\,\mathrm{e}^{\mathrm{i}m(\phi-\phi')}$, only one term survives integration:

$$a'_n=2\pi p_{-n}a_n. \tag{3}$$

The form of the nth component is unchanged, but its amplitude is multiplied by $2\pi p_{-n}$, and since subsequent intervals produce the same relative effect, a_n simply varies exponentially: $a_n\propto \mathrm{e}^{-\alpha_n t}$, in which α_n may be complex and different for every Fourier component of $A(\phi)$. This does not, however, prevent the centroid from relaxing exponentially, since its position is determined by one component only. Thus

$$\bar{\rho}=\bar{\xi}+\mathrm{i}\bar{\eta}=\int_0^{2x} A(\phi)\,\mathrm{e}^{\mathrm{i}\phi}\,\mathrm{d}\phi\Big/\int_0^{2\pi}A(\phi)\,\mathrm{d}\phi=(a_{-1}/\bar{A})\,\mathrm{e}^{-\alpha_{-1}t},$$

the appropriate exponential decrement having been supplied in the last expression. The imaginary part of α_{-1} modifies the frequency of the centroid as it spins round the ζ-axis (that is to say, the centre of the resonance line is shifted by α''_{-1}) while the Lorentzian broadening is determined by the real part of α_{-1}.

It must be emphasized that the model employed here, of a two-level system perturbed by a transient electric field, is altogether inadequate to describe the complexities of collisions between real molecules. The purpose of this section has been to indicate by a specific example how the exponential character of the response functions, and in consequence the Lorentzian form of the broadened line, are not accidents of a special collision mechanism, but rather to be expected as the norm. It is therefore permissible to replace the continuous range of interactions by an equivalent model in which all close collisions are treated as catastrophic and the rest ignored. The precise definition of a close collision must depend on the spectrum of phase changes, especially if diffusion on the sphere is mainly accomplished in small steps, but for rough estimating one will not usually go far wrong in assuming that scattering of the representative point through more than 1 radian is catastrophic. When the pressure is high, however, so that the spectral lines are considerably broadened and shifted, they begin to suffer shape changes as well, for which this crude modelling is inadequate and which can only be understood by a very thorough analysis of the collision mechanism.

The rather general result that diffusive spread of the representative points over the sphere leads to exponential decay of the oscillatory dipole prompts the question whether the exponential radiative decay discussed in the last chapter may not be described analogously as a progressive loss of phase information when an excited system is coupled into a cavity. Can one, in other words, retain the representation appropriate to an isolated system, or one subject to a determinate imposed force, when the perturbation results from interaction with other quantized systems? The answer has already been given – no; once two systems have been coupled their wave-functions can only be separated again by making observations, and for the purpose of predicting the probability of a given observation the combined wave-function must be retained.† The last paragraphs may legitimately be criticized on this score, but probably no great error has been incurred in what was only intended as a very rough analysis. We shall be obliged in the next section to take a little more care, but not to be so distrustful of elementary procedures as to throw away the classical models which have so far proved very helpful. As we have seen with the Bloch equations, the original representation involving motion on the surface of a sphere needed extension, when cavity modes were coupled to the double well, so that the representative point could leave the surface. In a somewhat analogous fashion we shall find in the next chapter that when a harmonic oscillator (a resonant cavity) is coupled to a host of other quantized systems (ammonia molecules) the equivalent classical ensemble of chapter 13 does not lose its usefulness, but simply needs to be extended to take account of new circumstances.

528

532

587

445

This does not mean that we may not use the idea of diffusion on a sphere as a convenience to describe the evolution of the expectation value of the dipole moment or the energy, and in this way put the dissipative mechanisms of radiation and collisions on an equivalent footing. One must always remember, however, that the ζ-co-ordinate of the representative point defines the *mean* value of p as a result of many identical trials; in any one trial the particle will be found in one well or the other, never anywhere else, and $\pm p_0$ are the only possible outcomes, with probabilities $\frac{1}{2}(1\pm\xi)$. In the same way a statistical distribution of points representing an assembly of identical systems is only significant as an aid to calculating the probability of finding a given number of systems with moment $+p_0$ and the rest with $-p_0$. As the assembly increases in size the probability distribution becomes sharply peaked; if the centroid of N points is at $\bar{\xi}$, $\frac{1}{2}N(1+\bar{\xi})$ systems will

† It is the instinctive reluctance to accept this proposition, especially as applied to systems which were once coupled but have since been mutually isolated, that has generated most of the paradoxes, such as that of Einstein, Podolsky and Rosen,[(2)] by which the foundations of quantum mechanics have been (unsuccessfully) assailed.

be found in the right-hand well, with an RMS deviation from this value of $\frac{1}{2}[N(1-\bar{\xi}^2)]^{\frac{1}{2}}$. For most purposes, when macroscopic assemblies are being considered, deviations from the mean are negligible, and only the position of the centroid is relevant to the observations.

Stark broadening and resonance broadening

When a spectral line is observed in the presence of a gas of polar molecules, the broadening tends to be rather larger than with non-polar molecules on account of the electric field of the dipoles, which can cause significant phase shifts even when the approach is not close enough to be regarded as a direct collision. The Stark effect,[3] the shift of a spectral line by a uniform strong electric field, was first detected by the use of fields of about 10^7 V m^{-1}, and a dipole of 1 Debye unit produces this order of field strength at a distance of 1 nm, several times a typical atomic radius. A free electron or proton produces an equal field from ten times as far away, and line-broadening effects in plasmas present a real problem in the use of spectral analysis for diagnosing the molecular processes at work. We shall not attempt to discuss this very difficult field and shall even be content to pass lightly over the slightly less difficult Stark broadening by polar molecules. To obtain a rather more quantitative measure of its magnitude, consider a dipole p passing at velocity v, with its closest distance of approach equal to r, so that its field, say $p/4\pi\varepsilon_0 r^3$, may be supposed to act for a time of order r/v. Let us see how close r must be for the resulting phase shift of the resonance line ($1s \rightarrow 2p$) of a hydrogen atom to be 1 radian. The change of frequency, $\Delta\omega$, of this line in a field $\mathscr{E}$ is $3\hbar\varepsilon_0\mathscr{E}/2\pi me$, so that the critical distance is $(3\hbar p/8\pi^2 mev)^{\frac{1}{2}}$, i.e. about $\frac{1}{2}$ nm if $p = 1.5$ Debye units and $v = 700$ m s^{-1}. This is a rather stronger effect than we should have calculated by use of the double-well model, for which at this sort of field strength the Stark effect is quadratic in $\mathscr{E}$ and rather weak. The Stark effect is, to tell the truth, a good deal too complicated for simple models to be satisfactory, and we shall leave it at that in order to return to a line broadening effect which may be usefully discussed in terms of the double-well model. This is *resonance broadening*, resulting from the interaction between two identical, or nearly identical, molecules, and it can be very strong.

511

Let us first look at a completely classical model, with the molecules replaced by identical harmonic oscillators, and the pressure assumed low enough for only binary collisions to matter. Suppose that a step function electric field has set them all vibrating parallel to $\boldsymbol{\mathscr{E}}$ with the same amplitude and phase; the measured response function $S(t)$ is the sum of all individual responses. When any two approach closely enough to be appreciably coupled by their dipole fields, the normal mode co-ordinates are the sum and difference of the individual displacements, and these continue to vibrate with unchanged amplitude. In particular the sum, which is the measured quantity, is unchanged in amplitude. The phase of the vibration is, however, changed since the normal mode frequency is altered by the coupling, and

368

if this frequency change is $\Delta\omega$ (i.e. if $2\Delta\omega$ is the frequency difference between the two normal modes) the phase shift for an encounter of duration t_0 is $t_0\Delta\omega$. We take $1/\Delta\omega$ as a measure of the length of encounter needed for scattering through 1 radian, i.e. for the contributions of the molecules concerned to be effectively removed from the response; an effective collision cross-section may be defined in terms of the distance of approach required to produce this phase change. A molecule with an optical transition of f-value unity may be modelled by an electron undergoing harmonic oscillations. When the displacements of the electrons in two such molecules are x_1 and x_2, the potential energy of coupling is of the order of $e^2x_1x_2/4\pi\varepsilon_0r^3$ and this leads to normal modes displaced by $\pm e^2/8\pi\varepsilon_0mr^3\omega_0$ from the mean, ω_0. An assumed encounter time of r/v leads to a phase shift of 1 radian if r is $e/(8\pi\varepsilon_0m\omega_0v)^{\frac{1}{2}}$, about 6 nm for sodium atoms at a temperature of 1000 K. If each atom presents a target area πr^2 for collisions and there are n atoms per unit volume, the mean time, τ, between collisions is $1/(\pi r^2n\sigma)$ and the line width is $2/\tau$:

441

$$\text{Line width} = nfe^2/4\varepsilon_0m\omega_0, \tag{4}$$

the oscillator strength f having been added in this expression. Various attempts to calculate the line width in this particular case have led to similar expressions, but with multiplying factors ranging between 1 and 8/3. Experimentally a factor of about 2 is indicated. The target radius must therefore be about 9 nm, much greater than the atomic radius. It is interesting to note that when sodium at low pressure is diluted with helium it requires 100 times the pressure of helium to reproduce the self-broadening resonance effect – the target radius is more like 0.9 nm. This is still rather larger than the distance of approach, say 0.3 nm, at which a direct collision could be said to occur, and one must infer that the interaction responsible for the van der Waals forces[9] between non-polar molecules is responsible; the random fluctuating dipole moment of the helium atom, though changing on a time scale ten times as fast as the radiation frequency, can at these close distances perturb the sodium atom enough to cause a significant phase shift.

The model of identical oscillators just analysed is not really appropriate to the resonance line-broadening in ammonia at such low pressures that the lines in fig. 18.8 are well resolved. Most collisions will be between molecules in different rotational states, and the very fact that the lines are resolved implies that enough time elapses before a given molecule is dephased for the others to have gone through several cycles at least. Thus it may be assumed that the colliding molecules are different and have random relative phase, and that it is the change of phase suffered by each separately that matters, rather than by the sum as in the previous calculation. On the other hand the duration of the encounter is much less than a cycle, and rather strong mutual perturbations are needed to generate effective dephasing. The small differences in frequency between the colliding molecules are therefore unimportant when it comes to calculating the phase

547

shift, and we may safely assume them identical during the encounter. Consider then the two harmonic oscillators whose states of vibration are initially represented by the complex amplitudes A and B. The normal modes have amplitudes $(A+B)/\sqrt{2}$ and $(A-B)/\sqrt{2}$, and during the encounter they change phase at different rates, so that the upper mode finishes up θ ahead of, and the lower mode θ behind, the phase which the unperturbed oscillators would have reached. With normal mode amplitudes $(A+B)\,\mathrm{e}^{\mathrm{i}\theta}/\sqrt{2}$ and $(A-B)\,\mathrm{e}^{-\mathrm{i}\theta}/\sqrt{2}$ at the end of the encounter the two oscillators are left with amplitudes $\frac{1}{2}[(A+B)\,\mathrm{e}^{\mathrm{i}\theta} \pm (A-B)\,\mathrm{e}^{-\mathrm{i}\theta}]$, i.e. $A\cos\theta + \mathrm{i}B\sin\theta$ and $B\cos\theta + \mathrm{i}A\sin\theta$. So far as the assembly of A–molecules is concerned, the random phase of B means that the mean amplitude after the encounter is $A\cos\theta$, the term $\mathrm{i}B\sin\theta$ vanishing in the average. Similarly B is reduced to $B\cos\theta$ by the collision. It appears that $\theta = 1$ is a fair criterion for an effective collision. Since $\theta = t\Delta\omega$, the encounter must last for a time $1/\Delta\omega$, which is the same result as was found appropriate for the dephasing of the total moment of identical molecules. The details of the model used seem unimportant.

We may now look at the same problem from the point of view of two double-well systems, treated quantally, rather than classical oscillators. The encounter is modelled by laying the two systems side by side and allowing their interactions to be switched on for a certain time. If the co-ordinates of the two systems are y_1 and y_2 the arrangement is now represented, as in fig. 2(a), by a single particle in a potential consisting of four square wells

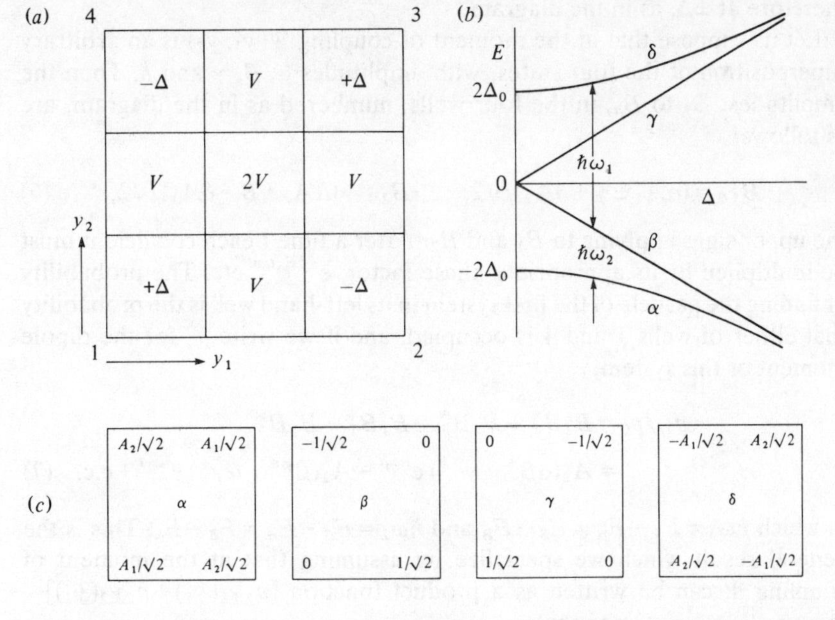

Fig. 2. (a) Potential energy for two double-wells coupled by dipole interaction, as measured by Δ; the wave-function is characterized by the amplitude of ψ in the wells at the four corners of the diagram. (b) Perturbation of energy levels by Δ. (c) Amplitudes of ψ in the four wells of (a) for the four stationary states of (b).

separated round the sides of the square by slightly penetrable barriers, the same as for the isolated double well, and with a central impenetrable region where the barrier is twice as high. The wells are taken as equal when the systems are isolated, but when they interact it is more favourable for opposite, rather than adjacent, wells to be occupied; each system may be supposed when polarized to produce a field disturbing the equality of the wells in the other. This is expressed by the potential $+\Delta$ in wells 1 and 3, and $-\Delta$ in 2 and 4. The energy levels are shown in fig. 2(*b*) and the corresponding amplitudes in the four wells in fig. 2(*c*). When $\Delta=0$ the levels β and γ represent one system in its symmetrical ground state and the other in the antisymmetrical excited state; the degeneracy of those two states of the combined system is removed by the coupling, while at the same time the outer levels, which in the uncoupled condition represent both in the lower or both in the upper state, are further separated by the interaction. In the latter states the cosh and sinh functions in the barriers are displaced by the same y_0, and the ratio A_1/A_2 depends on y_0 exactly as in the isolated systems; the presence of a tunnelling barrier on two sides of each well, however, doubles the kinetic energy shift. Thus instead of (18.15), we have that

$$E_{\alpha,\delta} = \pm(4\Delta_0^2+\Delta^2)^{\frac{1}{2}} \tag{5}$$

but A_1^2/A_2^2 is still given by (18.17). In the states β and γ the wave-function can be assumed to decay exponentially in going away from an occupied well; what minute amplitude reaches the next well is nullified by what arrives from the opposite corner. Since the wave-function is so nearly exponential there is no correction to the kinetic energy, and the levels are therefore at $\pm\Delta$, as in the diagram.

Let us suppose that at the moment of coupling $\Psi(y_1, y_2)$ is an arbitrary superposition of the four states, with amplitudes α, β, γ and δ. Then the amplitudes, B_1 to B_4, in the four wells, numbered as in the diagram, are as follows:

$$B_{1,3} = (\alpha A_1 \pm \gamma + \delta A_2)/\surd 2, \qquad B_{2,4} = (\alpha A_2 \pm \beta - \delta A_1)/\surd 2, \tag{6}$$

the upper signs applying to B_1 and B_2. After a time t each coefficient must be multiplied by its appropriate phase factor, $e^{-iE_\alpha t/\hbar}$ etc. The probability of finding the particle of the first system in its left-hand well is the probability that either of wells 1 and 4 is occupied, and if we write p_1 for the dipole moment of this system,

$$\begin{aligned}\langle p_1\rangle/p_0 &= B_2B_2^* + B_3B_3^* - B_1B_1^* - B_4B_4^* \\ &= A_2(\alpha\beta^* - \gamma\delta^*)\,e^{i\omega_2 t} - A_1(\beta\delta^* + \alpha\gamma^*)\,e^{i\omega_1 t} + \text{c.c.}\end{aligned} \tag{7}$$

in which $\hbar\omega_1 = E_\gamma - E_\alpha = E_\delta - E_\beta$ and $\hbar\omega_2 = E_\beta - E_\alpha = E_\delta - E_\gamma$. This is the general result which we specialize by assuming that at the moment of coupling Ψ can be written as a product function $[a_1\chi_1(y_1) + a_2\chi_2(y_1)]$ $[b_1\chi_1(y_2) + b_2\chi_2(y_2)]$, so that

$$B_{1,3} = \tfrac{1}{2}(a_1 \mp a_2)(b_1 \mp b_2), \qquad B_{2,4} = \tfrac{1}{2}(a_1 \pm a_2)(b_1 \mp b_2). \tag{8}$$

By comparing (6) and (8) the amplitudes α etc. are determined in terms of a_1, b_1 etc. After a certain amount of algebra and use of (18.12), (7) emerges in the form:

$$\langle p_1\rangle/p_0 = \text{Re}\,[\{A_2^2(\xi_1+\text{i}\eta_2\zeta_1)+A_1A_2(\xi_2\zeta_1+\text{i}\eta_1)\}\,\text{e}^{\text{i}\omega_2 t} + \{A_1^2(\xi_1-\text{i}\eta_2\zeta_1)-A_1A_2(\xi_2\zeta_1-\text{i}\eta_1)\}\,\text{e}^{\text{i}\omega_1 t}], \quad (9)$$

in which (ξ_1, η_1, ζ_1) and (ξ_2, η_2, ζ_2) are the positions of the representative points in fig. 18.1 for the two systems at the start of the interaction. In the case where the second system arrives with its dipole oscillation randomly phased, the average effect on $\langle p_1\rangle$ is obtained by putting $\bar{\xi}_2=\bar{\eta}_2=0$, and

$$\langle \bar{p}_1\rangle/p_0 = \text{Re}\,[(A_2^2\xi_1+\text{i}A_1A_2\eta_1)\,\text{e}^{\text{i}\omega_2 t}+(A_1^2\xi_1+\text{i}A_1A_2\eta_1)\,\text{e}^{\text{i}\omega_1 t}]. \quad (10)$$

When the interaction is weak, A_1 and A_2 differ only slightly from $1/\sqrt{2}$ and the principal effect arises from the beating of the two terms in (10), whose frequencies are $(2\Delta_0\pm\Delta)/\hbar$ to first order in Δ. Then

$$\langle \bar{p}_1\rangle/p_0 \approx \text{Re}\,[\rho_1\,\text{e}^{2it\Delta_0/\hbar}]\cos(t\Delta/\hbar). \quad (11)$$

This shows the same form of behaviour as do two classical oscillators; the amplitude of the oscillatory dipole is altered by a factor $\cos(\Delta\omega t)$, where in this case $\Delta\omega=\Delta/\hbar$. The results are in fact identical, since for two oscillators lying side by side $\Delta\omega = fe^2/8\pi\varepsilon_0 mr^3\omega_0$, and if f is given the value $4m\Delta_0 p_0^2/\hbar^2e^2$ appropriate to the double-well system, $\Delta\omega = p_0^2/4\pi\varepsilon_0\hbar r^3$ which is $\Delta/\hbar$ for two double wells in the same configuration. The classical equivalent oscillator is therefore, as we have now come to expect, a satisfactory model for this case of weak interaction.

This result does not hold when the interaction is strong, which is hardly surprising since we have already seen that the equivalent oscillator does not describe strong perturbations of the two-level system. The extreme case of strong interaction occurs when the mutual polarization causes Δ to become much greater than Δ_0, so that $A_1\sim 1$, $A_2\sim 0$; then $\omega_1\sim 2\Delta/\hbar$ and (10) reduces to the expression

$$\langle \bar{p}_1\rangle/p_0 = \xi_1\cos(2t\Delta/\hbar). \quad (12)$$

Since $\Delta \gg \Delta_0$ the perturbation needs to act for only a fraction of a cycle to be effective, and $\hbar/2\Delta$ may be taken as a measure of the interaction time required. This differs by only a factor of 2 from the criterion that applies when the interaction is weak.

It is the strong interaction result (12) that is most appropriate to the broadening of the microwave lines in ammonia. Let us ignore for the moment the rotation of the molecules and use the same approximations as before to estimate the critical distance of approach. Since $\Delta = p_0^2/4\pi\varepsilon_0 r^3$ and the encounter time is r/v, a value for r of $(p_0^2/2\pi\varepsilon_0\hbar v)^{\frac{1}{2}}$ is required to make $2t\Delta/\hbar$ equal to unity. Hence the target area is $p_0^2/2\varepsilon_0\hbar v$ and the collision rate when there are n molecules per m^3 is $p_0^2 n/2\varepsilon_0\hbar$, i.e. $1.3\times 10^{-14}\,n$. This may be compared with the experimental observations on the lines in fig. 18.8(c) for which $K = J$ and which approximate most closely to the model of non-rotating molecules. The line width is proportional to pressure and at a pressure of 0.5 torr is about 1.5×10^7 Hz, equivalent to a collision rate of $4.7 \times 10^7\ \mathrm{s}^{-1}$. At this pressure n is about $1.7 \times 10^{24}\ \mathrm{m}^{-3}$, so that the experimentally determined relationship is that the collision rate is $2.8 \times 10^{-14}\ n$. As with the estimate for the optical line of sodium the theoretical estimate is too low, but only by a factor of about 2, which is certainly within the limits of error of the estimate.

The discrepancy is really rather greater than has been suggested, perhaps by another factor of 2, because the rotation of the molecules has been ignored. When the quantum numbers are J and K, the total angular momentum is $\hbar[J(J+1)]^{\frac{1}{2}}$, of which $\hbar K$ is the component along the axis, as shown in fig. 1. The rotational speed is much higher than the inversion frequency, so that only a fraction $K/(J^2+J)^{\frac{1}{2}}$ of the dipole moment along the axis plays a significant role in the coupling between two molecules. The mean value of the steady dipole strength of the molecules colliding with a given one is therefore less by perhaps a factor of 2, and the line broadening to be expected is similarly reduced. The same argument implies that molecules having K equal, or nearly equal, to J should be especially sensitive to collisions, and those with small K less so; in fact the simple arguments given here suggest that the line width should be proportional to $K/(J^2+J)^{\frac{1}{2}}$. The measurements of Bleaney and Penrose are plotted in fig. 3 to show the strong correlation of line-width with $K/(J^2+J)^{\frac{1}{2}}$, but it

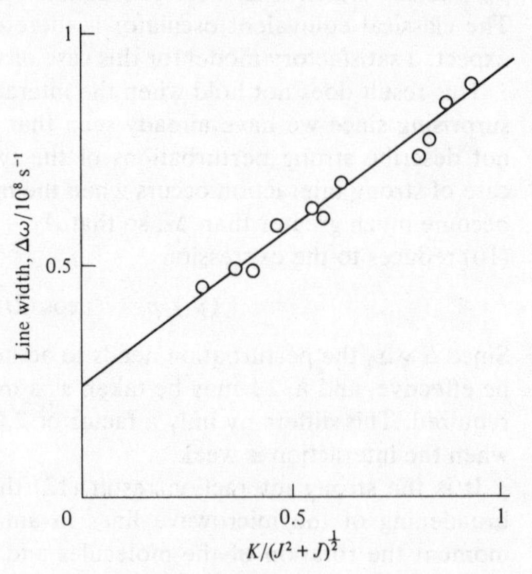

Fig. 3. Linear variation of line-width in the microwave spectrum of ammonia (fig. 18.8) when plotted against $K/(J^2+J)^{\frac{1}{2}}$.

is only fair to remark that they expected, on the basis of an equally rough-and-ready theory,[4] that the width should vary as $[K/(J^2+J)^{\frac{1}{2}}]^{\frac{2}{3}}$, and found in their results quite as satisfying a confirmation. It is not necessarily a refutation of the interpretation given here that the straight line does not pass through the origin. This may only mean that there are residual interactions between two ammonia molecules even when one of them has no steady dipole component; and the fact that some non-polar gas molecules can produce considerable broadening of the ammonia lines testifies to this. Let us therefore recognize that the line broadening processes can be fairly well understood in terms of simple models, even simple classical models, but that to take the argument further is no trivial matter. We therefore leave the question of what causes line broadening and proceed to the practical problem of how to get round it.

Doppler broadening; saturation spectroscopy

493

There is no need to say more about the origin of Doppler broadening. What is of interest here is the ingenious technique of saturation spectroscopy[5] by which its effects may be largely eliminated and the line-width reduced to something like the natural width, with consequent improved accuracy in wavelength measurement. It may be taken for granted that in such a measurement, aiming at the highest precision, the gas pressure is kept low enough for collision broadening to be unimportant, and that stray electric and magnetic fields are eliminated as far as possible, so that the natural width and the Doppler effect are all that are left to be contended with. There is not much to be done about the natural width, but the Doppler broadening is well worth removing. For example, the wavelengths of atomic hydrogen lines, such as $2P_{\frac{1}{2}} - 3D_{\frac{3}{2}}$ at a wavelength of about 656 nm, are needed to very high precision for the determination of fundamental constants. This particular line has a theoretical natural life-time of 16 ns, corresponding to a Q-value of 4.5×10^7, so that a line-of-sight velocity for the atom of c/Q, or 7 ms^{-1}, would shift the line by as much as its width. With typical atomic velocities of 2 kms^{-1} the advantage to be gained is obvious.

535

The technique depends on the availability of tunable lasers giving enough power to cause appreciable saturation of the line by heating. As (18.56) shows, the value of $|\zeta_0|$, which determines the excess of unexcited over excited atoms, is reduced by irradiation, especially if the radiation is tuned to the resonance frequency of the transition. When the laser frequency is close to the transition frequency those atoms having the correct line-of-sight velocity will resonate exactly to the Doppler-shifted light and will be strongly excited; with strong irradiation, indeed, ζ_0 may be reduced nearly to zero, and the excited and unexcited populations will be nearly equal. Atoms moving with other line-of-sight velocities will be less affected. Another, and weaker, light beam at the same frequency, injected in the reverse direction, will for its part be absorbed most strongly by those atoms which are moving with the opposite velocity to those most affected by the first beam. If, then, the laser is detuned from the resonance frequency the atoms which are strongly affected by the first beam are not those that will absorb the second, and the absorption coefficient for the second will be hardly affected. On the other hand, when the laser is perfectly tuned it is the atoms at rest, or moving in the plane normal to the light, that are saturated by the first beam and are exactly those atoms that would absorb the second beam if they were not saturated. It is the disappearance of absorption that signals the condition of exact tuning.

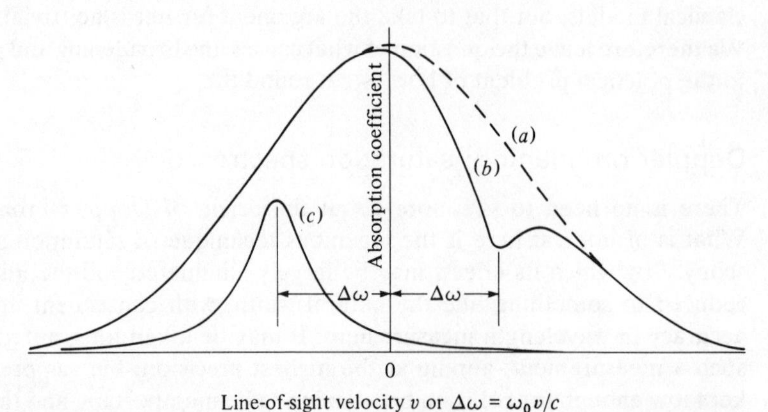

Fig. 4. Illustrating the principle of saturation spectroscopy; for description see text.

In fig. 4, curve (a) shows the line-of-sight velocity distribution, $f(v) \propto e^{-mv^2/2k_BT}$. When irradiated by light at a frequency $\Delta\omega$ above the resonance frequency, only those atoms moving away from the source with velocity close to $c\,\Delta\omega/\omega_0$ will absorb. If the laser power is weak, curve (a) represents the absorption coefficient of the gas as a function of frequency when the abscissae are scaled by ω_0/c to convert v into $\Delta\omega$. With more power, atoms moving in a band of velocities equivalent to something like the natural line-width suffer saturation and their absorption is reduced, as shown by curve (b) for $\Delta\omega > 0$. Now a weak beam from the same laser, but travelling in the opposite direction, is absorbed by atoms whose velocity is $c\,\Delta\omega/\omega_0$, also in the opposite direction. Since atoms which are not exactly in resonance, but lie within the natural line width, can absorb according to a Lorentzian formula, the total absorption for the weak beam is obtained by multiplying curve (b) by the Lorentzian (c), centred on $-\Delta\omega$, and determining the area under the product curve. Its variation with $\Delta\omega$ follows the general shape of the Doppler-broadened curve except when $\Delta\omega$ is so small that the dip in (b) and the hump of (c) overlap. It is this narrow dip that now constitutes the measured line-shape. The Doppler width, being so much greater, is virtually irrelevant except as a small correction factor in the analysis of the results; to develop the theory quantitatively we shall assume curve (a) to be flat except for the dip due to irradiation.

For an optical experiment conducted at room temperature ζ_0 may be taken to be -1, i.e. only spontaneous emission matters, and the representative point for a typical irradiated atom has co-ordinates ρ and ζ as given by (18.55) and (18.56); it describes a circular orbit of radius ρ_0 at a level

$\bar{\zeta}$, where

$$\bar{\zeta} = -1/[1 + I/(1 + \delta_+^2 \tau_p^2)]. \tag{13}$$

Here I is written for $p_0^2 \mathscr{E}_0^2 \tau_e \tau_p / \hbar^2$ and δ_+ for $\Delta\omega + \omega_0 v/c$, the mistuning of the Doppler-shifted radiation as seen by the moving atom. The spinning co-ordinate ρ is phase-linked to the radiation but this is irrelevant to the second beam which, travelling in the reverse direction, catches atoms in all phases with equal probability. Any individual atom experiences the second beam at a frequency mistuned by δ_-, i.e. $\Delta\omega - \omega_0 \sigma/c$, and the two beams combine to give an oscillatory field slightly modulated at the beat frequency $2\omega_0\sigma/c$. If the beams are both strong the modulation will cause periodic variations of $\bar{\rho}$ and lead to non-linear mixing, but with a weak second beam it is correct, to first order, to treat $\bar{\zeta}$ as constant at the value (13) determined by the first beam, and to use (18.55) to describe the response to the second beam:

$$\rho_0 = -\mathrm{i}C\mathscr{E}_2/[1 + I/(1 + \delta_+^2 \tau_p^2)][1 + \mathrm{i}\delta_- \tau_p], \tag{14}$$

where $C = p_0 \tau_0 / \hbar$ and $\mathscr{E}_2$ is the electric field strength in the second beam. The loss per atom is determined by $\mathrm{Im}\,[\rho_0/\mathscr{E}_2]$ and we must find the total contribution by atoms moving at different speeds. The maximum value of $\mathrm{Im}\,[\rho_0/\mathscr{E}_2]$ is $-C$, and occurs when the laser is perfectly tuned, the intensity weak, and $v = 0$; then $\delta_+ = \delta_-$, and $I \ll 1$. If we form W, defined by the expression,

$$W(\Delta\omega) = -\frac{1}{C}\int_{-\infty}^{\infty} \mathrm{Im}\,[\rho_0/\mathscr{E}_2]\,\mathrm{d}v,$$

W is a measure of the attenuation of the second beam at a mistuning of $\Delta\omega$. It is expressed as an equivalent velocity band-width – the attenuation is the same as if all atoms in the range $\pm W$ absorbed at the maximum value and the rest not at all. From (14),

$$W(\Delta\omega) = \int_{-\infty}^{\infty} \mathrm{d}v (1 + \delta_+^2 \tau_p^2)/(1 + \delta_+^2 \tau_p^2/\alpha^2)(1 + \delta_-^2 \tau_p^2), \text{ where } \alpha^2 = 1 + I,$$

$$= \frac{\pi c}{\omega_0 \tau_p}\left\{1 - \frac{\alpha^2 - 1}{\pi \alpha^2}\int_{-\infty}^{\infty} \mathrm{d}x/[1 + (x_0 + x)^2/\alpha^2][1 + (x_0 - x)^2]\right\},$$

in which $x = \omega_0 \tau_p v/c$ and $x_0 = \tau_p \Delta\omega$. The integral is easily evaluated by taking a contour along the real axis and back round the semicircle at infinity. Then

$$W(\Delta\omega) = \frac{\pi c}{\omega_0 \tau_p}\{1 - (1 - 1/\alpha)/(1 + \Delta\omega^2 \tau_1^2)\}, \tag{15}$$

in which $\tau_1 = 2\tau_p/(1 + \alpha)$. At low intensities of the first beam, $(1 - 1/\alpha) \ll 1$, the dip in attenuation is small and the width is the natural line-width determined by τ_p. As I is increased the dip becomes stronger but at some cost in terms of line-width. The physical reason is obvious – the stronger beam affects $\bar{\zeta}$ over a wider range of frequency or velocity, and a really strong beam can saturate the transition over as wide a range as one may choose.

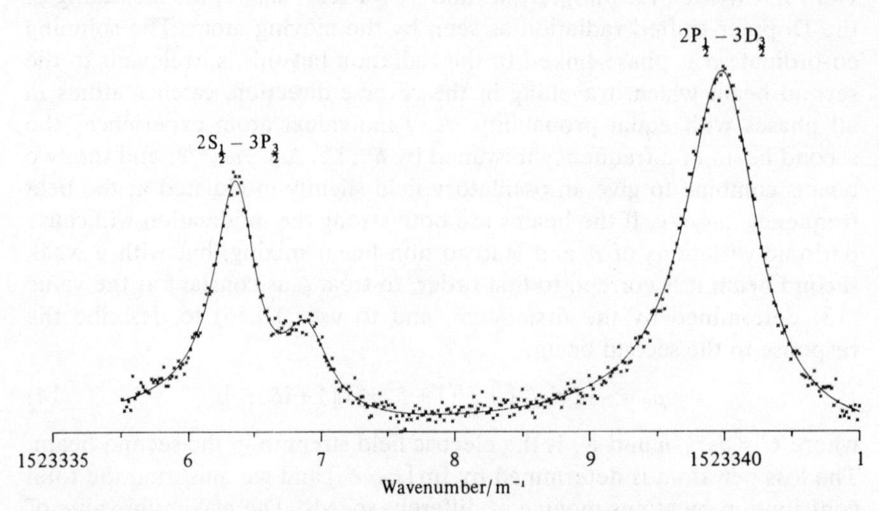

Fig. 5. Fine structure in the hydrogen spectrum resolved by saturation spectroscopy (Petley and Morris[6]). The full curve is theoretical, with the adjustable parameters chosen to give the best fit; from these parameters the wavenumber of each component can be determined with a precision of 0.01 ppm. The separation of the principal peaks (*Lamb shift*) is about 2.4 ppm, less than one-tenth of the Doppler width of each line.

The power of the technique is well illustrated in fig. 5, showing resolution of three fine-structure components of the hydrogen spectrum that would be lost in the Doppler profile.[6] The two principal components, separated by the Lamb shift, differ in frequency by 2 parts in 10^6; the Doppler profile for atomic hydrogen at room temperature is something like ten times wider than this.

Spin echoes and related effects

Our final example concerns the elimination of line broadening due to inhomogeneities, and we shall discuss two very different cases: nuclear spin resonance where the natural width of the lines is extremely small so that very small variations of magnetic field cause appreciable broadening; and double-well systems in glasses where the inhomogeneity is so great that no lines are normally detectable. The origin and use of echoes in such circumstances is conveniently discussed in terms of proton spin resonance, making use of the classical representation of the spin orientation as a point on a sphere. Since this carries over directly to any two-level system, the double well in particular, the second type of problem can be understood in principle as a straightforward extension of the first.

We shall be concerned with thermally equilibrated assemblies, every member of which is represented by a point on the sphere, with the centroid at some point ζ_0 on the vertical axis. In the absence of dissipation, and so long as the systems are identical, they respond to external forces in such a way that their centroid remains on a sphere of radius $|\zeta_0|$. It is convenient to replace the unit sphere of figs 18.1 and 18.4 by this smaller sphere, so that the lower pole represents the point to which the centroid tends as a result of dissipative effects; as before, the time-constant T_1 for relaxation in the ζ-direction is not normally the same as T_2 for relaxation in the ξ-η plane. The centroid provides all the information needed to interpret observations of the magnetic moment of a strictly homogeneous assembly of spins. If, however, the magnetic field varies with position in the sample, the spins precess at different rates and a cloud of points is now needed, each point representing the centroid of the spins in a region small enough to be treated as homogeneous.

224 Let us consider the effect of inhomogeneity on a typical nuclear induction experiment.[7] At the start all the points lie at the lower pole, but irradiation at the resonance frequency causes them to rotate about some horizontal axis, which we take as the ξ-axis, and in the ideal experiment irradiation is stopped when they have moved through 90° and reached the equator. In the frame of reference spinning at the resonant frequency they would all remain at the same point if the field were uniform; in the laboratory frame they would spin about a vertical axis to give a nuclear induction signal, constant in amplitude until dissipative effects supervened. Inhomgeneity of the steady field, however, causes different points to spin at different speeds; in a reference frame spinning at the mean speed the points fan out along the equator, forwards and backwards, causing their centroid to move in towards the axis. Without any dissipative process, therefore, the signal diminishes until, when the points are spread evenly around the equator, it disappears entirely. It can, however, be recovered by irradiation for a second period, twice as long as the first, so as to cause each point to turn through 180° about the ξ-axis. After an interval equal to that between the two irradiations the fan has closed up again and the induction signal is restored, only to fade away again as the fanning process continues. The reconstructed signal is the spin echo, and the diagram of fig. 6 illustrates

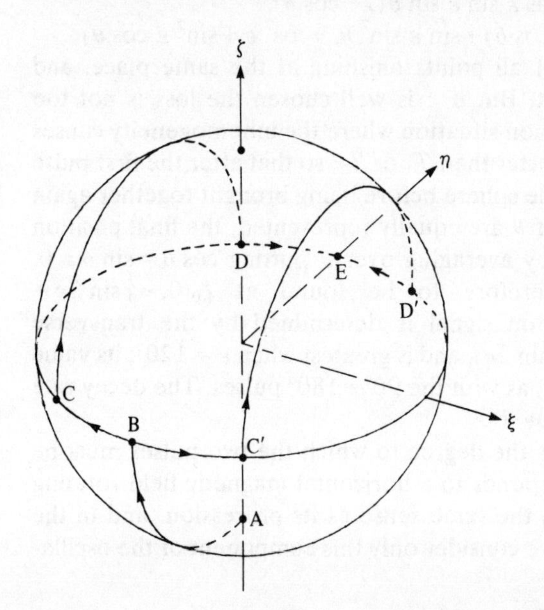

Fig. 6. Illustrating spin echo. A 90°-pulse takes the centroid of the spin assembly from A to B, where the representative points fan out round the equatorial plane; C and C′ show where two such points are at the instant a 180°-pulse is applied to take them to D and D′. After a time all points momentarily coincide at E. Note that the axes in this diagram are turned through 90° relative to fig. 7 and to fig. 18.1

the successive stages; this may also be expressed in Cartesian co-ordinates by carrying through the various rotations for a typical point which drifts through an angle θ during the fanning process:

A Initial state: $\zeta_0(0, 0, -1)$
B After 90° rotation about ξ-axis: $\zeta_0(0, -1, 0)$
C After fanning through θ about ζ-axis: $\zeta_0(-\sin\theta, -\cos\theta, 0)$
D After 180° rotation about ξ-axis: $\zeta_0(-\sin\theta, \cos\theta, 0)$
E After fanning through θ about ζ-axis: $\zeta_0(-\sin\theta\cos\theta + \cos\theta\sin\theta, \cos^2\theta + \sin^2\theta, 0)$, i.e. $\zeta_0(0, 1, 0)$.

So long as there is no random spreading of the points by dissipative effects, every point finishes up at E, as illustrated by the two trajectories in the diagram. At its momentary peak the echo is as strong as if there were no field inhomogeneity. Decay of the echo strength when longer intervals are allowed to elapse between the first 90° pulse and the second 180° pulse may be attributed to real dissipation, and thus the true time-constant may be separated from the artefacts of inhomogeneity. In an induction experiment it is the transverse component that is measured, and therefore T_2 controls the rate of echo decay.

Two matters are worth comment to supplement this brief account. First, it is technically easier to use identical pulses rather than have the second twice as long as the first. Let us follows the fortunes of a typical point when each pulse turns it through ε about the ξ-axis:

A Initial state: $\zeta_0(0, 0, -1)$
B After the first pulse: $\zeta_0(0, -\sin\varepsilon, -\cos\varepsilon)$
C After fanning through θ: $\zeta_0(-\sin\varepsilon\sin\theta, -\sin\varepsilon\cos\theta, -\cos\varepsilon)$
D After the second pulse:
$\zeta_0(-\sin\varepsilon\sin\theta, -\sin\varepsilon\cos\theta\cos\varepsilon - \cos\varepsilon\sin\varepsilon, -\cos^2\varepsilon + \sin^2\varepsilon\cos\theta)$
E After fanning through θ:
$\zeta_0(-\sin\varepsilon\sin\theta\cos\theta - \cos\varepsilon\sin\varepsilon\sin\theta(1+\cos\theta),$
$-\cos\varepsilon\sin\varepsilon(\cos^2\theta + \cos\theta) + \sin\varepsilon\sin^2\theta, -\cos^2\varepsilon + \sin^2\varepsilon\cos\theta)$.

There is no question now of all points finishing at the same place, and inevitably some signal is lost. But if ε is well chosen the loss is not too great. Let us consider a common situation where the inhomogeneity causes signal loss in a time much shorter than T_1 or T_2, so that after the first pulse the points spread all round the sphere before being brought together again by the second. If all values of θ are equally represented, the final position of the centroid is obtained by averaging over θ putting $\cos\theta = \sin\theta = 0$, $\cos^2\theta = \sin^2\theta = \frac{1}{2}$, it is therefore to be found at $\zeta_0(0, -\frac{1}{4}\sin 2\varepsilon + \frac{1}{2}\sin\varepsilon, -\cos^2\varepsilon)$. The induction signal is determined by the transverse component, i.e. $\zeta_0(\frac{1}{2}\sin\varepsilon - \frac{1}{4}\sin 2\varepsilon)$, and is greatest when $\varepsilon = 120°$; its value is then $0.65\,\zeta_0$, rather than ζ_0 as with the $90° + 180°$ pulses. The decay rate of the echo is still governed by T_2.

The second point concerns the degree to which the two pulses must be phase-coherent. The spin responds to a horizontal magnetic field rotating at the resonant frequency in the same sense as its precession, and in the rotating frame of reference we consider only this component of the oscillatory applied field, representing it as pointing in a direction in the horizontal plane that is determined by the phase of the oscillation. It is about this axis, which was taken as the ξ-axis in the above account, that the point rotates as a result of the irradiation. The assumption that the same axis applies to the effect of both pulses is tantamount to supposing that the oscillator remains switched on, at exactly the right frequency, between the two pulses, which are applied by gating the oscillator output at the appropriate moments. In fact, this rather delicate technique is not normally required, and it is enough to switch the oscillator on for each pulse separately, not worrying about the phase relationship. The two rotations through ε are now about different horizontal axes, and the effect of this is most readily seen graphically. In fig. 7 the original process is shown in perspective and

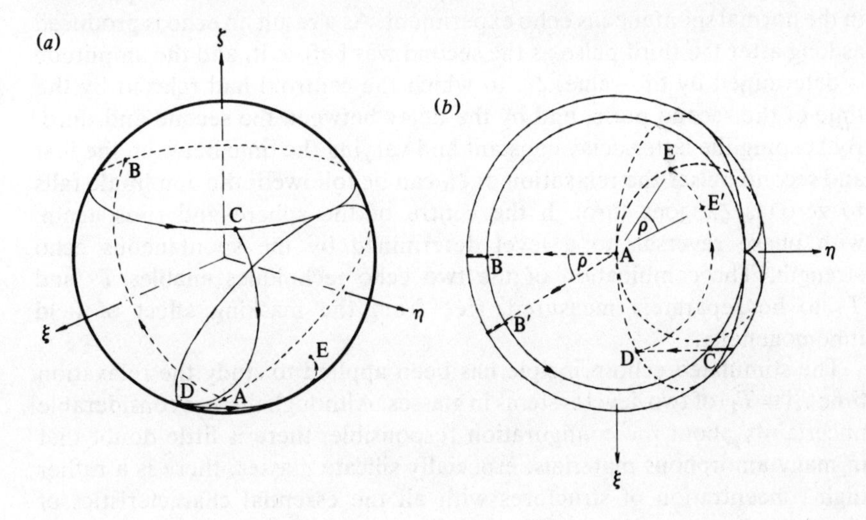

Fig. 7. Trajectory of a representative point in a spin echo experiment with two 120° pulses. (*a*) perspective view, (*b*) plan view. The trajectory ABCDE is typical for coherent pulses, while AB′CDE′ is typical for pulses with a relative phase shift of ϕ.

in plan, with $\varepsilon = 120°$, and as in fig. 6 the path followed by one system (with $\theta = 135°$) is shown in detail. When enough time is allowed for the points to make many revolutions and to spread evenly round the horizontal circle through B the second pulse, which tips the circle through 120°, gives rise to an induction signal since there is now a non-vanishing mean transverse moment, but this disappears as all the points fan out at different speeds, only to come together at the echo which arrives as the distribution collapses momentarily onto the curve which in plan resembles a cardioid. It is this that takes the place of the single point E in fig. 6.

Now let us suppose that the first rotation was about a different axis, so that points originally at A were shifted to B′. Then the system whose fortunes we followed from B to C, to D and ultimately to E is now most nearly matched in behaviour by another system whose fanning angle is $\theta - \phi$. This goes from A to B′ and then to C, whence it follows the same pattern as the original in going to D. However, in the last stage it does not fan round to E, but only to E′, since its fanning angle is smaller by ϕ. And the same may be said of every point, so that at the moment of echo all points lie on an identical cardioid, but rotated through ϕ. So long as there was time for the points to cover the circle completely, the cardioid will be covered in exactly the same way for all ϕ, so that only the phase, not the amplitude, of the echo is affected by lack of coherence between the pulses.

The echo produced as just described, by the use of two pulses, is known as the *spontaneous echo* to distinguish it from the *stimulated echo* for which three unequal pulses are required. The first is of such strength and duration as to turn the representation through 180°, so that the centroid which started at the south pole, $-\zeta_0$, is now at the north pole, $+\zeta_0$. There is no induction signal since no mean rotating dipole is created. The centroid relaxes back towards $-\zeta_0$ with the longitudinal relaxation time T_1, and at some stage before this process is complete two more pulses are applied as in the normal spontaneous echo experiment. As a result an echo is produced as long after the third pulse as the second was before it, and the amplitude is determined by the value, ζ_0', to which the centroid had relaxed by the time of the second pulse, and by the delay between the second and third. By keeping the latter delay constant and varying the time between the first and second pulses the relaxation of ζ_0' can be followed; the amplitude falls to zero as ζ_0' goes through the centre of the sphere and rises again, with phase reversal, to a level determined by the spontaneous echo strength. The combination of the two echo techniques enables T_1 and T_2 to be separately measured, free from the masking effect of field inhomogeneity.

The stimulated echo principle has been applied to study the relaxation time $\tau_e (\equiv T_1)$ of two-level systems in glasses. Although there is considerable uncertainty about the configuration responsible, there is little doubt that in many amorphous materials, especially silicate glasses, there is a rather high concentration of structures with all the essential characteristics of double-well systems. Most likely there is some commonly occurring atomic arrangement which provides single protons with the choice of two similar potential wells, close enough together to permit tunnelling. But these are not sparsely distributed and highly symmetrical like the proton systems in polyethylene; all inequalities of wells are to be found so that, even if they had the same Δ_0, the spread of Δ and hence of E would smear out any resonance line that might otherwise be detected. The ensemble of systems thus bears a certain resemblance to the spin system just discussed, and can be represented in the same way, but in place of a small degree of field inhomogeneity and consequent small spread of E there is here a very wide spread. The echo technique, however, allows systems whose values of E lie in a narrow band to be selected and their relaxation time measured. Since this is related to the tunnelling probability and the coupling to lattice vibrations the information that is yielded in principle is the same as can be obtained from dielectric loss studies in polyethylene and similar materials where the confusing effects of inhomogeneity are almost absent. In glasses the hypothetical Debye relaxation peaks are smoothed out into an almost

540

frequency-independent dielectric loss, from which little of interest can be gleaned.

A *photon-echo* experiment, analogous to the stimulated spin echo, has been performed, at very low temperatures, on samples of silica glass in a microwave cavity.[8] The electric field pulses used to shift the representative points round the sphere were applied as bursts at the cavity frequency 720 MHz, at which some at least of the double-well systems resonate. Let us suppose that all the systems have the same Δ_0, with a wide spread of Δ, and (for convenience) that the oscillator has been tuned to $2\Delta_0/\hbar$. Then the systems with $\Delta=0$ will ideally be inverted by the first pulse, their representative points being turned through 180° about a horizontal axis. Those, however, with non-zero Δ will be mistuned by an amount $\Delta_0/E-1$, which we denote by δ as in (18.4), and their points will turn at a rate $\delta'=(\delta_0^2+\delta^2)^{\frac{1}{2}}$ about an axis tilted $\tan^{-1}(\delta/\delta_0)(=\theta)$ from the horizontal,

517

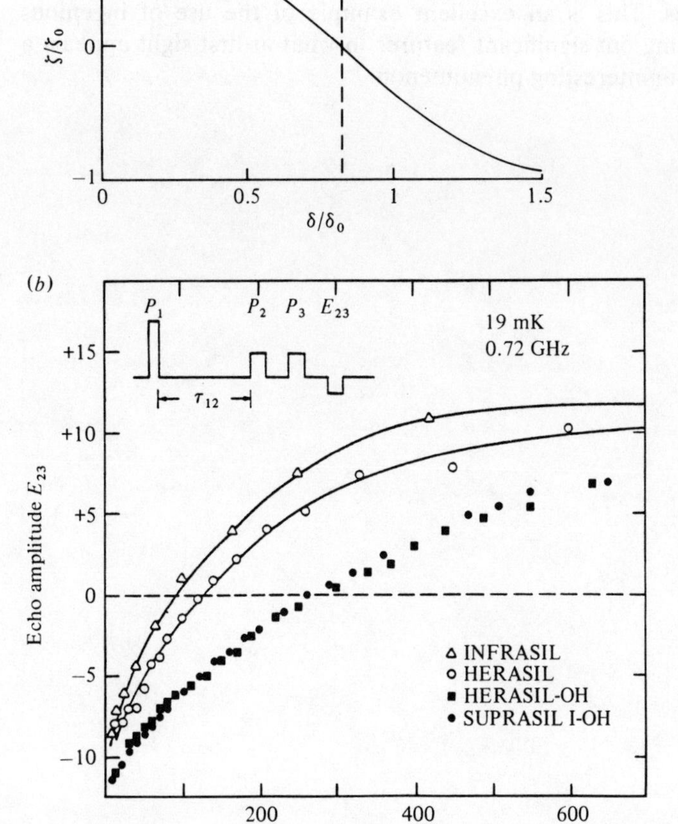

Fig. 8. (*a*) The centroid of representative points, initially at $-\zeta_0$, is raised to ζ by a 180° pulse; the curve shows how ζ is affected by mistuning between the frequency of the double-well and the pulse. The area under the curve is indicated by the broken-line rectangle of width 0.8. (*b*) Showing how T_1 may be determined in a photon echo experiment by the variation of echo amplitude with the time delay, τ_{12}, between the 180° pulse and the first 90° pulse[8]. The different points refer to different proprietary brands of silica glass.

as in fig. 18.1; here δ_0, is a measure of the strength of the oscillatory field. If the length of pulse is π/δ_0, a typical off-tune point is turned through $\pi(\delta_0^2+\delta^2)^{\frac{1}{2}}/\delta_0(=\beta)$, giving the ideal of π when $\delta=0$. Now a point initially at $-\zeta_0$, after turning through β about an axis tilted θ upwards from the ξ-axis, finishes at $\zeta_0(-\frac{1}{2}\sin 2\theta(1-\cos\beta), -\cos\theta\sin\beta, \cos^2\theta(1-\cos\beta)-1)$. The spread of E ensures that in practically no time the points have fanned out round the sphere, and we shall concern ouselves only with the ζ-component, which depends on δ/δ_0 as shown in fig. 8(a). The centroid of the assembly has been raised as much as if all those systems with $|\delta|<0.8\delta_0$ had been perfectly inverted and the rest left untouched. If the pulse is weak and appropriately long, δ_0 is small and there is sharp selectivity of systems for promotion. It is on these that the subsequent echo experiment is performed, the second and third pulses being close together since it is T_1 and not T_2 that is of interest. The experimental curves in fig. 8(b), taken at a temperature of 19 mK, show clearly the decay of ζ_0' from positive to negative, with the echo amplitude passing through zero. It is interesting that these experiments revealed clearly two different double-well systems, with different relaxation times (140 μs and 410 μs in the examples shown here) and requiring different pulse strengths to bring them to their maximum echo amplitudes. This is an excellent example of the use of ingenious technique to bring out significant features in what at first sight appears a featureless and uninteresting phenomenon.

20 The ammonia maser

The original maser of Gordon, Zeiger and Townes,[1] driven by a focussed stream of ammonia molecules in their antisymmetrical state, provides a conveniently explicit example on which to base a discussion of the principles underlying coherent excitation of a vibrator by stimulated emission. It was shown in chapter 18 how a quadrupole electrostatic lens served to separate symmetric from antisymmetric states, and we shall assume that separation is perfect; it is easy to extend the argument to include a proportion of molecules in the symmetric state. In addition we shall ignore any complications arising from the multitude of rotational states leading to the fine structure shown in fig. 18.8, and shall assume that only one line contributes, for example the strong 3,3 line at 23.9 GHz. Since the microwave cavity resonator, if it is to be excited by the molecules, must normally be very closely tuned to their natural frequency this assumption is realistic.

548

The simplest intuitive approach to the maser is by way of Einstein's treatment of radiation in terms of stimulated and spontaneous processes.† Excited molecules passing through the resonator, when it is already in an oscillatory condition, are stimulated by the field; if the resonator frequency is well matched to the molecular levels they may make a transition down to the ground state and on leaving the resonator have $2\Delta_0$ less energy than when they entered. The radiated energy is phase-coherent with the cavity vibration, whose amplitude is thereby increased; every molecule making a transition increases the quantum number of the vibration by one. The lifetime for spontaneous decay of the excited state of the molecule is so long in relation to the residence time in the resonator that spontaneous processes do not play a large role and may be ignored for the moment. The resonator walls are normally dissipative, and if the maser is used as a microwave source the extraction of power augments the natural dissipation. Since the rate of power transfer from a molecule to the resonator is proportional to the energy density in the resonator, and since the dissipation rate is also proportional to energy density, there is a critical flux of excited molecules below which the excitation is insufficient to overcome damping; for higher than critical flux, however, the excess power provided by the

503, 537

† If we refer to this approach as the Einstein argument, it is not meant to imply that he had any part in it except for the introduction of the radiation coefficients into the vocabulary of physics. Certainly he is not responsible for the fundamental flaws in the argument as applied to the maser.

molecules causes the level of oscillation to rise until the assumptions of the argument are invalidated, and the resonator settles down to a steady level. These processes have the appearance of being readily understood and formulated mathematically, as we shall show, but considerable refinement is called for if correct quantitative results are sought. A more classical approach is much less open to criticism, and we shall develop the response-function argument of chapter 18 for this purpose. The Einstein argument enables the threshold flux to be calculated and gives a fairly good picture of the steady state. It also, by an elementary application of the uncertainty principle, indicates how well tuned the cavity must be – if the residence time of a molecule is T, an error in frequency amounting to less than $1/T$ will hardly be noticed before the molecule has left. The response-function approach supplies everything in the Einstein argument, while showing much more explicitly by how much the critical flux increases with detuning, and hence it provides a figure for the frequency range within which maser action will occur for a given flux. Furthermore it shows that when the system is mistuned the frequency of steady oscillation is a compromise between that of the resonator and that of the molecule, with the latter strongly favoured. It does, however, fail to reveal the existence of a noise source which the fully quantal treatment shows to be present, not just as a concomitant of the dissipative process (Johnson noise) but equally as inseparable from the excitation. The effect of noise is principally to cause the phase of oscillation to drift randomly and thus to limit the purity of the output. In practice quantum noise is considerably less important than thermal noise for a maser operating in the microwave frequency range, but the way it emerges from the quantum mechanics is so intellectually satisfying that we shall develop the argument in detail.

Stimulated emission

From now on we shall adopt a simplified version of the resonator, assuming that along the line travelled by a molecule as it passes through, not only the phase but also the amplitude of electric field oscillation is constant rather than varying sinusoidally with position as befits a half-wavelength loop of a standing wave. If the energy in the resonator is E when this field reaches its maximum value $\mathscr{E}_0$, we write $\frac{1}{2}\varepsilon_0\mathscr{E}_0^2 V_{\text{eff}}$ for E and thus define the effective resonator volume, V_{eff}. The molecule moves in an electromagnetic field of energy density $u=\frac{1}{2}\varepsilon_0\mathscr{E}_0^2$, and if we take the Einstein-inspired arguments of chapters 16 and 18 at their face value we shall assume that an excited molecule will be stimulated to emit a quantum at a rate proportional to u, while an unexcited molecule will be stimulated to absorb at the same rate. In fact, when the resonator supports only one mode of interest this is not a valid point of view, for after an excited molecule has lost a quantum its wave-function is still coherent with the regular oscillation of the electromagnetic field, so that its subsequent behaviour is not independent of its past; and this is not the only failure of physical reasoning, as

503
532

will become apparent shortly. Nevertheless, we ignore such niceties and proceed in naive hope. If at a time t after entering the resonator a fraction f of the molecules are still excited, we write the de-excitation rate $-\dot{f}$ as $B_{12}uf$; while as a result of re-excitation there is a contribution $B_{12}u(1-f)$ to $\dot{f}$. Hence

$$\dot{f} = B_{12}u(1-2f),$$

so that f relaxes exponentially towards the steady-state value $\frac{1}{2}$. In particular, if $f = 1$ when the molecules enter, f varies during passage as $\frac{1}{2}(1+\mathrm{e}^{-2B_{12}ut})$, and the mean value of this expression at the moment of exit gives the fraction that have not communicated their energy to the resonator. Because of the inevitable spread of molecular velocities, the residence times vary widely, and it is not a very satisfactory approximation to insert the mean residence time T in the exponential, as we now do. However, the general form of the behaviour will not be seriously falsified.

Let us write Φ for the molecular flux, the number of excited molecules entering in unit time; then $\frac{1}{2}\Phi(1+\mathrm{e}^{-2B_{12}uT})$ emerge excited, the rest contributing $\hbar\omega_0$ to the resonator energy, or $\hbar\omega_0/V_{\mathrm{eff}}$ to u. Hence

$$\dot{u} = \tfrac{1}{2}\Phi(1-\mathrm{e}^{-2B_{12}uT})\hbar\omega_0/V_{\mathrm{eff}} - u/\tau_{\mathrm{e}}, \tag{1}$$

where τ_{e} is the decay time for energy in the freely oscillating resonator. By writing x for $2B_{12}uT$ and $F\Phi_{\mathrm{c}}$ for Φ, where $\Phi_{\mathrm{c}} = V_{\mathrm{eff}}/B_{12}T\hbar\omega_0\tau_{\mathrm{e}}$, (1) is cast in the form:

$$\tau_{\mathrm{e}}\dot{x} = F(1-\mathrm{e}^{-x}) - x, \tag{2}$$

from which the development, if any, of the energy of oscillation can be determined. When x is small, $\tau_{\mathrm{e}}\dot{x} \sim (F-1)x$, and it is clear that no growth can occur unless $F > 1$; Φ_{c} as defined above is therefore the critical flux. In terms of Φ_{c}, $x = 2E/\hbar Q_{\mathrm{r}}\Phi_{\mathrm{c}}$, where $Q_{\mathrm{r}} = \omega_0\tau_{\mathrm{e}}$, the natural quality factor of the resonator.

In the steady state $\dot{x} = 0$ in (2). When $F - 1 \ll 1$, x settles down to the value $x_{\mathrm{s}} = 2(F-1)$ and as F is increased x_{s} tends asymptotically to F. One obtains a clear picture of the scale of operation of a maser by calculating the mean quantum number $\bar{n}_{\mathrm{s}}$ of the resonator oscillation corresponding to a given value of x_{s}:

$$\bar{n}_{\mathrm{s}} = V_{\mathrm{eff}}u_{\mathrm{s}}/\hbar\omega_0 = V_{\mathrm{eff}}x_{\mathrm{s}}/2B_{12}T\hbar\omega_0 = \tfrac{1}{2}x_{\mathrm{s}}\tau_{\mathrm{e}}\Phi_{\mathrm{c}}. \tag{3}$$

In the original maser Φ_{c} was about 10^{13} molecules per second and τ_{e} about 10^{-7} s ($Q_{\mathrm{r}} \sim 10^4$), so that with $F = 2$ and $x_{\mathrm{s}} = 1.6$, $\bar{n}_{\mathrm{s}} \sim 10^6$. In everything that follows it will be assumed that $\bar{n}_{\mathrm{s}}$ is very high, and this will permit occasional mathematical simplifications.

It is easy enough to integrate (2) numerically to show how x builds up from a small value when F is brought above unity – exponentially at first with a time-constant of the order of τ_{e}, then slower, and finally an exponential approach to x_{s} with a different time-constant, also of the order of τ_{e}. Unfortunately this argument overlooks the fact that the residence time T

is about 10^{-4} s, considerably greater than τ_e. It is therefore quite wrong to assume that u is sensibly constant during the passage of a molecule, and the rate is in fact determined more by T than by τ_e; this does not invalidate the expressions for Φ_c and x_s, which depend on the nature of the steady state. There is so much wrong with the basic arguments leading to (2) that it is not worth attempting to derive a better treatment of the non-stationary state. The next approach we shall develop is much better in principle and will be taken as far as studying fluctuations from the steady state where the conflict of two very different characteristic times has important consequences.

Dielectric response of a molecular beam

The second approach to the ammonia maser has already been adumbrated in chapter 11, but we are now in a position to make the argument quantitative. We have seen that the response functions of the unexcited and excited states are opposite in sign, so that if the former gives rise to dielectric loss the latter conversely gives rise to dielectric gain. The critical flux is such that the beam of molecules, considered as a dielectric rod in the resonator, is just able by its gain to neutralize the resonator losses. Linear response theory will take one thus far, but to reach a steady state of oscillation non-linearity must play a part. We are concerned to find the mean dipole moment of a typical molecule as it traverses the resonator, whose vibration frequency is not necessarily the same as the natural resonator frequency. Since a number of different frequencies and related quantities enter the theory, we shall list the most important before proceeding further:

347
513

- ω_0 natural frequency (real) of resonator, whose natural decrement is described by τ_e or Q_r $(=\omega_0\tau_e=\frac{1}{2}\omega_0\tau_a)$.
- ω_m natural frequency of the molecular transition $=2\Delta_0/\hbar$; the Q associated with spontaneous decay is enormous and will not concern us, but we shall introduce an effective Q related to the residence time T rather than to the lifetime: $Q_m=\frac{1}{6}\omega_0 T$.
- ω the actual frequency of the maser in its steady state.
- ω_e the mistuning of empty resonator and molecule, defined as $\frac{1}{2}(\omega_m-\omega_0)$.
- ω_r the resultant mistuning of maser and molecule, defined as $\frac{1}{2}(\omega_m-\omega)$.
- Ω_n a measure of the normal mode frequency shift when a molecule is coupled to the resonator in its nth excited state; according to (18.36), $\Omega_n^2=\omega_e^2+\gamma^2(n+1)$ where γ is the coupling constant.

529

To relate γ to the resonator characteristics, note that C in (18.31) gives Cx as the interaction energy between an oscillator with displacement x and the double-well system with the particle in one well. Hence Cx is equivalent to $\mathscr{E}_0 p_0$, while by considering energy we see that $m_0\omega_0^2x^2$ is equivalent to $\varepsilon_0 V_{\text{eff}}\mathscr{E}_0^2$. Hence C^2 is to be interpreted as $m_0\omega_0^2p_0^2/\varepsilon_0 V_{\text{eff}}$, and γ^2, which is $C^2/2\hbar\omega_0 m_0$, is $\omega_0 p_0^2/2\hbar\varepsilon_0 V_{\text{eff}}$.

We now introduce three dimensionless quantities:

$\Gamma = 2\omega_r T$,

$\Gamma_0 = p_0\mathscr{E}_0 T/\hbar = 2\gamma n^{\frac{1}{2}} T$,

and $\Gamma' = (\Gamma^2 + \Gamma_0^2)^{\frac{1}{2}}$. For future reference it may be noted that $\Gamma' = 2\Omega_n T$ if ω_e is replaced by ω_r and the difference between n and $n+1$ is ignored.

The analysis accompanying fig. 18.1 contains what is needed to find the mean dipole moment of a molecule that enters in the excited state and experiences a very large number of cycles of resonator oscillation at frequency ω during residence. The point representing the momentary amplitude, and phase relative to the resonator oscillation, starts at the upper pole and moves for time T in a circular orbit on the sphere at angular velocity $(4\omega_r^2 + p_0^2\mathscr{E}_0^2/\hbar^2)^{\frac{1}{2}}$, i.e. Γ'/T, so that it traces out an arc of length Γ'. The inclination of the orbit, θ, is $\tan^{-1}(\Gamma/\Gamma_0)$ and the orbit radius is $\cos\theta$, i.e. Γ_0/Γ'. The centroid of the arc is easily found, and its projection on to the horizontal plane gives both the real $(\bar{\xi})$ and imaginary $(\bar{\eta})$ parts of the mean dipole moment in terms of p_0:

$$\bar{\xi} = -\Gamma\Gamma_0(\Gamma' - \sin\Gamma')/\Gamma'^3, \qquad \bar{\eta} = -\Gamma_0(1 - \cos\Gamma')/\Gamma'^2. \tag{4}$$

There are ΦT molecules present at any instant, so that the total dipole moment P_{tot} is $\Phi T p_0(\bar{\xi} + i\bar{\eta})$. This is not spread throughout the resonator in proportion to the local electric field, as the polarization of a uniform linear dielectric would be, but an energetic argument is readily devised to show that the effective volume susceptibility $\kappa_e\,(=\kappa_e' + i\kappa_e'')$ is $P_{\text{tot}}/V_{\text{eff}}\varepsilon_0\mathscr{E}_0$. Hence, from (4),

$$\kappa_e = -(\Phi T^2 p_0^2/\varepsilon_0\hbar V_{\text{eff}}\Gamma'^2)[\Gamma(\Gamma' - \sin\Gamma')/\Gamma' + i(1 - \cos\Gamma')]. \tag{5}$$

The sign of κ_e' is opposite to that of Γ, negative when $\omega < \omega_m$. The lower state of the molecule has positive polarization at low frequencies but the upper state has negative. Similarly the sign of κ_e'' describes dielectric gain rather than the loss shown by the lower state.

Since $\mathscr{E}_0$ is involved in Γ', κ_e is obviously not independent of $\mathscr{E}_0$, except in weak fields; then Γ' can be replaced by Γ without significant error or, when the system is perfectly tuned, $1 - \cos\Gamma'$ can be replaced by $\frac{1}{2}\Gamma'^2$. In the latter case $\kappa_e' = 0$ and $\kappa_e'' = -\Phi T^2 p_0^2/2\varepsilon_0\hbar V_{\text{eff}}$. The resonator behaves as if filled with a dielectric of relative permittivity $\varepsilon = 1 + \kappa_e$, and its resonant frequency is shifted from ω_0 to $\omega_0/\varepsilon^{\frac{1}{2}}$, i.e. $\omega \sim \omega_0(1 - \frac{1}{2}\kappa_e)$ when $\kappa_e \ll 1$. The imaginary part of ω is $-\frac{1}{2}\kappa_e''\omega_0$ and consequently the decay time for energy is $1/\kappa_e''\omega_0$. It follows that if the natural decrement of the resonator is described by a certain τ_e, the critical flux needed to overcome this is such that $\kappa_e'' = -1/\omega_0\tau_e = -1/Q_r$. Hence

$$\Phi_c = 2\varepsilon_0\hbar V_{\text{eff}}/Q_r p_0^2 T^2 = 1/\tau_e\gamma^2 T^2, \tag{6}$$

and (5) may be written:

$$\kappa_e = -(2F/Q_r\Gamma'^2)[\Gamma(\Gamma' - \sin\Gamma')/\Gamma' + i(1 - \cos\Gamma')], \tag{7}$$

with $F = \Phi/\Phi_c$, as in (2).

Before proceeding to discuss the consequences of this result, let us note that comparison of the two expressions for Φ_c, (6) and that following (1), shows what meaning should be ascribed to the stimulated emission coefficient B_{12}:

$$B_{12}=p_0^2T/2\varepsilon_0\hbar^2.$$

The appearance of the residence time T in this expression casts grave doubt on the argument, since B_{12} was introduced to define the de-excitation rate of a molecule in a field of given energy density, with no thought of how long it was to remain in that field. What we have unjustly called the Einstein argument involved the assumption that the probability of de-excitation is initially proportional to the elapsed time after exposure to resonator field. This is only tenable when a large number of incoherent modes are simultaneously acting on the molecule. If there is only one mode the probability is initially proportional to t^2, and the intrusion of another time-like quantity into the expression can be seen to be required by dimensional considerations. Unfortunately, however appealing the Einstein argument may be to the physical intuition, it is deeply flawed as a quantitative procedure.

498

Returning to the dielectric approach, when the amplitude of oscillation is low enough for Γ' to be replaced by Γ, κ_e has the form shown in fig. 1.

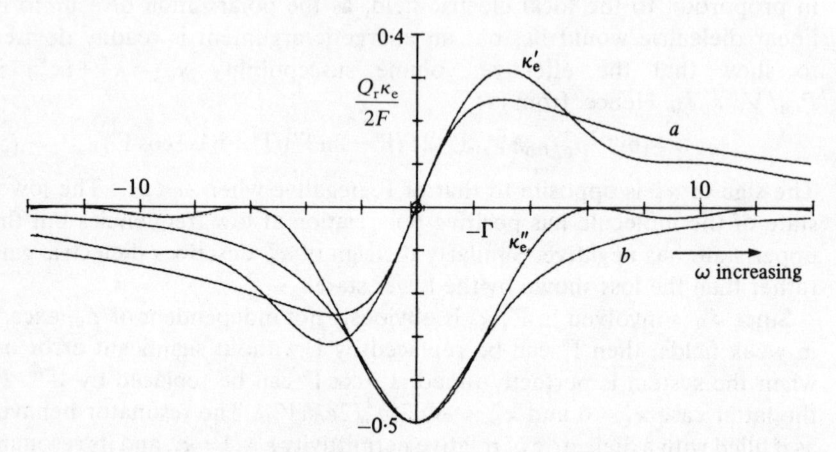

Fig. 1. Real (κ_e') and imaginary (κ_e'') parts of the susceptibility of a beam of excited ammonia molecules, compared with the corresponding curves (a and b) for a Lorentzian resonance adjusted to fit at $\Gamma=0$.

The ripples on the wings, which result from the assumption that T is the same for all molecules, would be smoothed out in any real device. Even without smoothing, the general form of the curve is fairly close to the Lorentzian response of a simple resonant system, and agreement would probably be still better after smoothing. The Lorentzian curves shown for comparison are the real and imaginary parts of $-(\mathrm{i}F/Q_r)/(1+\frac{1}{3}\mathrm{i}\Gamma)$, which is chosen to match the slope of κ_e' and the magnitude of κ_e'' at $\Gamma=0$. The width of this curve, $\Delta\Gamma$, is 6, so that the frequency width is $6/T$, corresponding to a Q-value of $\frac{1}{6}\omega_0 T$, which is what we have chosen to define Q_m, the quality factor of the molecular transition; Q_m is typically of the order of

10^6. It is the residence time, not spontaneous radiation, that determines the effective width of the quasi-resonant response. It will be observed that the unsmoothed form of κ_e'' goes to zero when Γ (or in general Γ') $= 2\pi$; this carries the implication that the oscillation cannot build up to the point where $\Gamma' = 2\pi$, and all points outside $\pm 2\pi$ are irrelevant. This would not be true for the smoothed curve.

As the amplitude builds up, non-linearities bring the system to a steady state in which κ_e'' is just sufficient to overcome dissipation in the resonator. The argument developed for linear response is still valid since κ_e'' has been defined, even in the non-linear situation, in such a way that $-\frac{1}{2}\kappa_e''\varepsilon_0\mathscr{E}_0^2 V_{\text{eff}}$ is $-\frac{1}{2}\varepsilon_0\mathscr{E}_0 \operatorname{Im}[P_{\text{tot}}]$, the power supplied by the molecular beam. Hence in the steady state $\kappa_e'' = -1/Q_r$ and, from (7),

$$2F(1 - \cos\Gamma') = \Gamma'^2, \text{ or } F = (\tfrac{1}{2}\Gamma'/\sin\tfrac{1}{2}\Gamma')^2. \tag{8}$$

The value of Γ' is thus uniquely determined by F. It follows then from (7) that κ_e' in the steady state is uniquely determined by F and the mistuning; thus by carrying this argument through we are able to discover ω, the frequency of oscillation in the steady state. For the effect of κ_e', the real part of the susceptibility, is to change ω_0 to $\omega_0(1 - \frac{1}{2}\kappa_e')$, and since $\omega - \omega_0 = 2(\omega_e - \omega_r)$ we have that

$$\omega_e - \omega_r = \tfrac{1}{4}\omega_0\kappa_e' = (F\omega_0\omega_r T/Q_r\Gamma'^3)/(\Gamma' - \sin\Gamma'), \tag{9}$$

from (7), with Γ replaced by $\omega_r T$. Hence

$$\omega_r/\omega_e = [1 + F\omega_0 T(\Gamma' - \sin\Gamma')/Q_r\Gamma'^3]^{-1}, \tag{10}$$

or, since the second term dominates,

$$\omega_r/\omega_e \sim (Q_r/Q_m) \times (1 - \cos\Gamma')\Gamma'/3(\Gamma' - \sin\Gamma') = (Q_r/Q_m)R(F), \tag{11}$$

in which F is related to Γ' by (8). With a typical residence time of 10^{-4} s, $Q_m \sim 2.5 \times 10^6$; it is the large value of Q_m/Q_r that justifies the approximation (11) under normal circumstances. The function $R(F)$ defined by (11) is equal to unity when $F - 1$ is small, and falls as F increases, but rather slowly, as shown in fig. 2. With the typical figures quoted, ω_r is of the order of 250 times smaller than ω_e – the molecule, having a much sharper resonance than the resonator, determines the resultant frequency almost entirely, though not so effectively when F is large. If, however, there is plenty of molecular flux in hand Q_r may be reduced, for instance by extracting more of the power from the resonator, so that the maser is only just maintained; as Q_r is thus reduced, the control of the frequency by the molecules is enhanced. It is this very high measure of control that is one factor making the maser a reliable frequency standard. Though good, the ammonia maser is not the best for this purpose, and at the end of this chapter we shall describe an even better system, the hydrogen maser devised by Ramsey. 601

This analysis also provides a measure of the detuning that is allowable with a given excess flux. Since (8) must have a solution for steady oscillation to occur, the maximum detuning is that which ascribes the whole of Γ' to

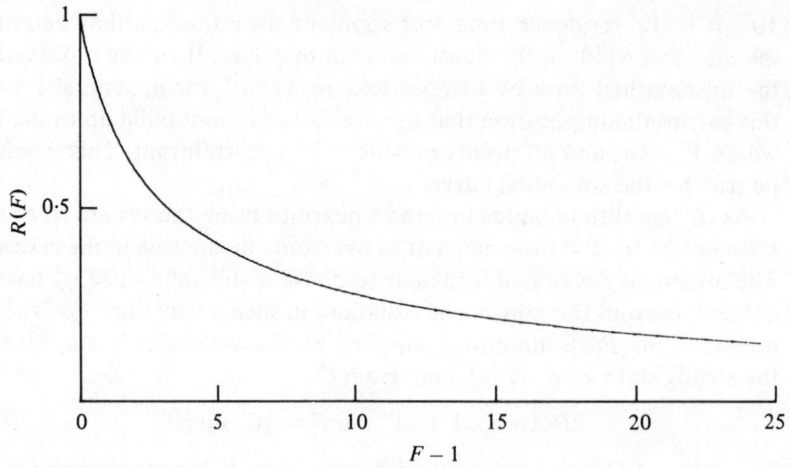

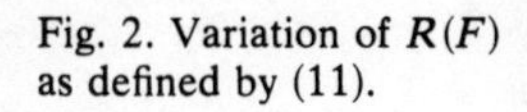
Fig. 2. Variation of $R(F)$ as defined by (11).

Γ, and none to Γ_0;

i.e $$F = (\tfrac{1}{2}\Gamma_{max}/\sin\tfrac{1}{2}\Gamma_{max})^2. \quad (12)$$

With this unsmoothed model, Γ_{max} cannot exceed 2π, so that $|\omega_r| < \pi/T \sim \omega_0/Q_m$, roughly what we have already inferred by an uncertainty-principle argument. 572

Fluctuations in amplitude and phase

In chapter 11 the effect of white noise on a maintained oscillator was shown to be controlled by Q_{sat}, defined in terms of the rate of return to the steady state amplitude after a disturbance; in the case discussed Q_{sat} was the same as the natural Q_r of the unexcited resonant circuit, but this is not always so. In what follows reasonable familiarity with the discussion in chapter 11 will be assumed. The essence of the result derived there is that if the state of the oscillator is represented by a vector defining the amplitude and phase, the end of the vector performs a random walk; but while the excursions of the amplitude are restrained by a radial restoring force, there is no such restraint on tangential motion, and the phase of the oscillation drifts unchecked. Something similar is found in the maser subjected to noise, but the details are different in several important respects. At the root of the difference lies the fact that maser oscillations are maintained by a source that is much more sharply resonant than the system it is exciting, in contrast to the oscillatory circuit in which the amplifier has a very flat frequency response. This makes the analysis rather lengthier – indeed unless the model is carefully devised the mathematical complications are fairly troublesome – and what emerges is that the meaning to be ascribed to Q_{sat} is much closer to Q_m, which is of the order of $\omega_0 T$, than to Q_r: and that white noise does not result in a purely diffusive process for the amplitude and phase. 361

To reduce the labour of analysis we shall assume that not every molecule enjoys the same residence time T, but that there is an exponential distribution so that after time t only a fraction $e^{-t/T}$ are still reacting coherently to the field; the rest have either left the cavity or have suffered catastrophic collision such as to make them on the average ineffective thereafter. This is in fact a rather good approximation to what happens in the hydrogen maser, and it is in connection with this maser, considered as a frequency standard, that the theory becomes most relevant. We shall concern ourselves only with the case of perfect tuning, when in the representation of fig. 18.1 the electric field $\mathscr{E}_0$ continues to point along η until disturbed by noise. On entering the resonator an excited molecule is represented by a point at the north pole, N, having co-ordinates (0, 0, 1); so long as $\mathscr{E}_0$ is constant and the molecule remains effective the point moves clockwise about the η-axis with angular velocity $p_0\mathscr{E}_0/\hbar$. To find the steady-state value of P_{tot} we need to know $\boldsymbol{R}$, the vector sum of the displacements from the origin of all those molecules that are still effective. Then P_{tot} is p_0 times the projection of $\boldsymbol{R}$ (OP in fig. (3(*a*)) onto the (ξ, η)-plane. There are three

601

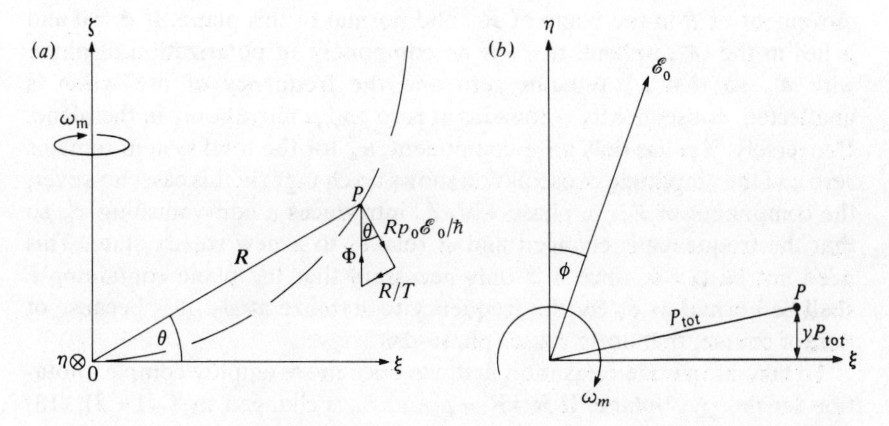

Fig. 3. (*a*) $\xi-\zeta$ section of fig. 18.1, showing the competing processes which are balanced in the steady state of the double-well. (*b*) Illustrating the calculation of phase drift due to noise.

processes moving P, which in the steady state must balance – the field causes tangential movement at a speed $Rp_0\mathscr{E}_0/\hbar$: attrition with time-constant T causes movement towards the origin at a speed R/T: and the arrival of new excited molecules causes vertical movement at a speed Φ. These processes are conveniently expressed by treating $\boldsymbol{R}$ as a complex number, writing $R = \xi + i\zeta$. Then

$$\dot{R} = -i(p_0\mathscr{E}_0/\hbar)R - R/T + i\Phi, \tag{13}$$

and in the steady state, when $\dot{R} = 0$,

$$R = R_0 = i\Phi T/(1 + i\Gamma_0), \tag{14}$$

in which Γ_0 has virtually the same significance as before, being defined as $p_0\mathscr{E}_0T/\hbar$. As $\mathscr{E}_0$ is increased, R_0 travels round the circular arc in the diagram, and $\cot\theta = \Gamma_0$.

The total polarization P_{tot} is $p_0 \operatorname{Re}[R_0]$, i.e. $\Phi p_0 T \Gamma_0/(\Gamma_0^2+1)$, and is in phase quadrature with $\mathscr{E}_0$, so that

$$\left.\begin{aligned}\kappa_e'' &= -P_{tot}/\varepsilon_0 \mathscr{E}_0 V_{eff} = -\Phi p_0^2 T^2/\hbar\varepsilon_0 V_{eff}(\Gamma_0^2+1) = -F/Q_r(\Gamma_0^2+1),\\ \text{where}\quad F &= \Phi/\Phi_c \text{ as before,}\quad \text{and}\quad \Phi_c = \hbar\varepsilon_0 V_{eff}/Q_r p_0^2 T^2.\end{aligned}\right\} \quad (15)$$

The expression for Φ_c arises from the requirement that at the threshold, when $F=1$ and Γ_0 is small, κ_e'' shall be equal and opposite to the dissipative susceptibility, $1/Q_r$, due to resonator losses. When $F>1$ the steady state has $\Gamma_0^2+1=F$, i.e. $\sin\theta = 1/F^{\frac{1}{2}}$.

Let us return for a moment to vector notation in order to consider the effect of a small arbitrary disturbance in which P is moved from $\boldsymbol{R}_0$ to a point $\boldsymbol{R}_0+\boldsymbol{\rho}$, not necessarily in the same plane. At the same time let $\mathscr{E}_0$ be changed to $\mathscr{E}_0(1+\delta)$ and the phase of oscillation changed by ϕ. As a result $\mathscr{E}_0$ does not now point along the η-axis but has been rotated through ϕ about the ζ-axis. All these perturbations are assumed small enough for linear approximations to hold. The subsequent relaxation of the system back to the steady state can be analysed as two independent processes, movement of P in the plane of $\boldsymbol{R}_0$, and normal to this plane. If $\phi=0$ and $\boldsymbol{\rho}$ lies in the (ξ,ζ)-plane, there is no component of polarization in phase with $\mathscr{E}_0$, so that κ_e' remains zero and the frequency of oscillation is unaffected; consequently ϕ remains at zero and $\boldsymbol{\rho}$ moves only in the plane. Conversely, if $\boldsymbol{\rho}$ has only an η-component, κ_e'' for the total system remains zero and the amplitude of oscillation shows no change; in this case, however, the component of P_{tot} in phase with $\mathscr{E}_0$ introduces a non-vanishing κ_e', so that the frequency is changed and ϕ relaxes to a new steady state. This need not be $\phi=0$, since it is only necessary that the plane containing P shall be normal to $\mathscr{E}_0$ for the frequency to stabilize at ω_0. It is because of this, of course, that noise causes phase-drift.

To take amplitude relaxation first, we once more employ complex notation for the (ξ,ζ)-plane. If $R=R_0+\rho$ and $\mathscr{E}_0$ is changed to $\mathscr{E}_0(1+\delta)$, (13) takes the form, to first order in ρ and δ,

$$\dot{\rho} = -[\rho(1+\mathrm{i}\Gamma_0)+\mathrm{i}\Gamma_0 R_0\delta]/T. \quad (16)$$

This shows how ρ responds to a given time-variation of $\mathscr{E}_0$, as expressed by $\delta(t)$, and it must be supplemented by an equation showing how $\mathscr{E}_0$ responds to a given time-variation of $p_0 \operatorname{Re}[\rho]$, the change in P_{tot}. Since the steady-state value of P_{tot} suffers a fractional change $\operatorname{Re}[\rho]/\operatorname{Re}[R_0]$, while the field suffers a fractional change δ, the consequent change of the molecular contribution to κ_e'' may be written

$$\Delta\kappa_e''/\kappa_e'' = p_0 \operatorname{Re}[\rho]/P_{tot} - \delta = x - \delta, \quad (17)$$

in which x is the real part of $\rho p_0/P_{tot}$, i.e. of $(\Gamma_0^2+1)\rho/\Phi\Gamma_0 T$. The resonator losses alone would cause the field to decay with time-constant τ_a, a process that would be neutralized by κ_e'' at its steady-state value. According to

(17), however, the time-constant for gain is $\tau_a/(1+\Delta\kappa_e''/\kappa_e'')$, or $\tau_a/(1+x-\delta)$. Consequently

$$\frac{\mathrm{d}}{\mathrm{d}t}[\mathscr{E}_0(1+\delta)] = \mathscr{E}_0(1+\delta)[-1/\tau_a+(1+x-\delta)/\tau_a],$$

so that $$\dot{\delta} = (x-\delta)/\tau_a, \text{ to first order in } \delta \text{ and } x. \tag{18}$$

If z is now written for the imaginary part of $\rho p_0/P_{\text{tot}}$, the real and imaginary parts of (16) provide two further equations to supplement (18):

$$\dot{x} = -(x-\Gamma_0 z-\delta)/T \tag{19}$$

and $$\dot{z} = -(z+\Gamma_0 x+\Gamma_0\delta)/T. \tag{20}$$

The coupling of x and z in these equations arises from the presence of $1+\mathrm{i}\Gamma_0$ in (16). In consequence, P does not relax back to equilibrium along a straight line but along a spiral path, which may converge quite slowly if F and hence Γ_0 are large; correspondingly the amplitude of $\mathscr{E}_0$ may also show oscillatory relaxation, though the effect is of small importance, as we shall now show.

Equations (18)–(20) yield a third-order equation for δ:

$$\alpha T^3\dddot{\delta}+(1+2\alpha)T^2\ddot{\delta}+[1+\alpha(\Gamma_0^2+1)]T\dot{\delta}+2\Gamma_0^2\delta=0,$$

in which $\alpha=\tau_a/T$ and is very small. The three solutions have the form $\mathrm{e}^{-\lambda t}$, with

$$\lambda \doteqdot [1\pm\mathrm{i}(8\Gamma_0^2-1)^{\frac{1}{2}}]/2T \text{ or } 1/\tau_a. \tag{21}$$

It is the first two solutions that give oscillatory relaxation if $F>9/8$. These describe a much slower process than the third; (18) shows that δ approaches its momentary steady-state value x with the time-constant τ_a characteristic of the resonator, while (19) and (20) show that x and z themselves are limited by the effective residence time of the molecules to much slower variations. This enables an adequate solution to be constructed without the tedious process of solving the cubic equation exactly. Suppose an impulsive noise-source, acting on the resonator, causes $\mathscr{E}_0$ to be changed slightly, so that at $t=0$, $\delta=\delta_0$ while $x=z=0$. Then in a short interval of a few times τ_a, before x and z have had time to change significantly, δ relaxes to x, i.e. something very close to zero:

$$\delta \doteqdot \delta_0\,\mathrm{e}^{-t/\tau_a}. \tag{22}$$

It is now reasonable to solve (19) and (20) on the assumption that x and z are zero during this interval – i.e. δ provides effectively an impulsive source in these equations, immediately after which $x(+0)=\alpha\delta_0$ and $z(+0)=-\Gamma_0\alpha\delta_0$. Subsequently only the slow solutions in (21) matter, which we match to those initial conditions by writing

$$\left.\begin{aligned} x=\delta &= (\alpha\delta_0 \operatorname{cosec}\varepsilon)\,\mathrm{e}^{-t/2T}\sin(\chi t+\varepsilon) \\ \text{and}\quad \Gamma_0 z &= (\alpha\delta_0\operatorname{cosec}\varepsilon)\,\mathrm{e}^{-t/2T}[\chi T\cos(\chi T+\varepsilon)-\tfrac{1}{2}\sin(\chi t+\varepsilon)], \end{aligned}\right\} \tag{23}$$

in which χ is the frequency $(8\Gamma_0^2-1)^{\frac{1}{2}}/2T$ given by (21) and $\tan\varepsilon = -2\chi T/(2\Gamma_0^2-1)$.

The variation of δ following an initial impulse is thus made up of a rapidly decaying primary response (22) and a long slow secondary oscillation according to the first equation of (23). The Fourier transform of this impulse response determines the spectrum of the fluctuations of δ when the resonator is subjected to a white noise input. In particular, the low-frequency ($\omega T \ll 1$) components of the fluctuations are simply proportional to the time-integral of the impulse response. On performing the integration, we find that the secondary oscillation changes the integral of (22) alone, which is $\delta_0\tau_a$, to $\delta_0\tau_a(\Gamma_0^2+1)/2\Gamma_0^2$, or $\frac{1}{2}\delta_0\tau_a/(1-1/F)$. When $F>2$ the low-frequency fluctuations are reduced, but never to less than half the amplitude for the passive resonator. So far as amplitude fluctuations are concerned the maser differs little from a maintained oscillator circuit $Q_{\text{sat}} \sim Q_{\text{r}}$.

The same is not true of the phase-drift, which is greatly reduced in the maser. We now assume, as shown in fig. 3(b), that P is displaced from the (ξ, ζ)-plane by an amount represented by y so as to give an extra total moment of yP_{tot}. If there were no phase shift of $\mathscr{E}_0$ this would create a real susceptibility κ_{e}' equal to $yP_{\text{tot}}/\varepsilon_0\mathscr{E}_0V_{\text{eff}}$, i.e. y/Q_{r} as follows from (15). If, however, $\mathscr{E}_0$ is advanced in phase by ϕ the component of the moment parallel to $\mathscr{E}_0$ is increased, and

$$\kappa_{\text{e}}' = (y+\phi)/Q_{\text{r}} = 2(y+\phi)/\omega_0\tau_a. \tag{24}$$

There is no first-order change in the component of $\mathscr{E}_0$ normal to the plane of P, and if $\mathscr{E}_0$ were to remain pointing in the direction ϕ, P would relax back to the plane normal to $\mathscr{E}_0$ with time-constant T, as the molecules grew ineffective and were replaced. Hence

$$\dot{y} = -(y+\phi)/T. \tag{25}$$

On account of (24), however, the resonator frequency is reduced fractionally by $\frac{1}{2}\kappa_{\text{e}}'$, so that

$$\dot{\phi} = -\tfrac{1}{2}\kappa_{\text{e}}'\omega_0 = -(y+\phi)/\tau_a. \tag{26}$$

From (25) and (26),

$$\ddot{\phi} = -\mu\dot{\phi}, \quad \text{where} \quad \mu = 1/\tau_a + 1/T,$$

$$\left.\begin{aligned} &\text{so that} \quad \phi = \phi_1 + \phi_2\,\mathrm{e}^{-\mu t} \\ &\text{and, from (25),} \quad y = -\phi_1 + \alpha\phi_2\,\mathrm{e}^{-\mu t}, \quad \text{where} \quad \alpha = \tau_a/T \text{ as before.} \end{aligned}\right\} \tag{27}$$

If the effect of a noise impulse is to produce an initial phase shift ϕ_0 without immediately changing the polarization, $\phi(0)=\phi_0$ and $y(0)=0$. Hence $\phi_1 + \phi_2 = \phi_0$ and $\phi_1 = \alpha\phi_2$;

$$\text{i.e.} \qquad \phi = \phi_0(\mathrm{e}^{-\mu t}+\alpha)/(1+\alpha).$$

The immediate phase shift is rapidly undone by the molecules; after a few times τ_a it is reduced to $\alpha\phi_0$, if $\alpha \ll 1$. Now according to (11.44) the Gaussian spread of phase in time t, for a given voltage noise, is inversely proportional to the circuit inductance, and hence to Q_r (since the given noise source implies R is constant). Thus the result we have derived implies that the phase drift is controlled by the molecular resonance, having $Q_m = \omega_0 T$, not the resonator with $Q_r = \omega_0 \tau_e$. It is as if, roughly speaking, the resonator and the excited molecules were to be treated as a single system, with the latter contributing so much energy that the total energy of oscillation (as may easily be shown) is something like Q_m/Q_r times as great as the field energy alone; the losses and the noise input, however, are due to the resonator only and the combined system therefore behaves like a simple oscillatory circuit with decay time T, and correspondingly enhanced phase stability.

Random noise input is not the only source of fluctuations; statistical variations in the flux of molecules give rise to amplitude variations but not to phase drift in a well-tuned system. This process can be shown to simulate thermal noise corresponding to a temperature of about $\hbar\omega_0/k_B$, which is some hundreds of times less than room temperature for the ammonia maser. The effect is important in optical masers, where $\hbar\omega_0 \gg k_B T$, but we shall not analyse it here. In addition there is, as already mentioned, a purely quantal source of noise which the analysis in terms of dielectric response misses altogether; it operates on the phase as well as the amplitude to cause fluctuations in both. We shall find it emerging automatically, if only as an incidental feature, from a fairly rigorous quantal treatment which goes a long way, apart from this point, towards justifying the dielectric approach. It is reasonable to expect that each process will be buffered by the molecules in the same way as thermal noise. The molecules do not interact directly with one another, but only indirectly through their effect on the resonator field. Any perturbation of phase by one molecule, therefore, whether of quantum or statistical origin, first changes $\mathscr{E}$ and the change is then diminished by a factor τ_a/T as the other molecules react.

Quantum mechanics of the resonator–molecule interaction

Once a molecule has come into interaction with the resonator the wave-functions of the two remain inseparable until an observation of the state of the resonator enables a new start to be made. We are interested in the probable state of the resonator after the maser has been running freely for a time, and must therefore keep in the composite wave-function all the molecules that have passed through during that time. This proves a far less formidable proceeding than might be feared since the molecules, after leaving the resonator, are effectively uncoupled from it and from each other, and the stationary states of an assembly of uncoupled systems are represented simply as products of eigenfunctions of individual members of the assembly. We shall permit ourselves one simplification which certainly does not accord with the facts, by assuming only one molecule to be in interaction at any instant, the total effect on the resonator being the sum of all such binary interactions. In the light of the foregoing discussion, and somewhat in the spirit of the Hartree approximation, we may suppose that the cavity and all molecules except the one under consideration are so closely coupled as to behave as a single resonator, with decrement governed by T rather than τ_e. Thus we shall work out the simplified model on the assumption that only one molecule is actually in the resonator cavity at any time, but shall then replace Q_r by Q_m when it seems reasonable. Similarly, although the empty resonator may be mistuned by ω_e with respect to the molecules, ultimately this mistuning is reduced to ω_r by the combined action of all molecules present, and we shall assume that ω_r is the mistuning experienced by any one molecule during its passage. It is perhaps distasteful to have recourse to such adjustments during what pretends to be a rather rigorous treatment, but the alternative involves so much more analysis that the basic physics may well be lost. By the procedure adopted here the quantum mechanics is at all events reduced to a fairly simple operation – we follow one molecule through the resonator to see how the many-body wave-function is thereby changed. In particular we determine the evolution of $\mathcal{P}(x, t)$, the probability that at time t the harmonic oscillator representing the resonator has displacement x. We have no interest in the molecules as such, but only in the state of the resonator mode as a result of their interactions with it. From now on it will be assumed that each molecule spends the same time T in the resonator rather than treating the exponential distribution of residence times as in the last section.

The analysis differs in detail according to the state of the molecule at the moment of entry; we shall work through the analysis for an excited molecule and at the end indicate what changes are needed for one that is unexcited. Just before the new molecule enters the resonator the wave-function of the entire system can be expressed as a Fourier sum over many-body eigenfunctions, each of which, since all parts are uncoupled, is a product of individual eigenfunctions, e.g. an oscillator function $\psi_n(x)$, a multitude of two-level functions $\chi_{1,2}(q_i)$ for previous molecules, and the initial wave-function of the new molecule $\chi_2(y)$. The $\chi(q_i)$ remain uncoupled, and their product may be gathered under a single umbrella function $U_m(q_i)$. Then since the same $\chi_2(y)$ appears in all eigenfunctions making up the initial state we write for the wave-function of the whole at the outset $(t=0)$:

$$\Psi(x, y, q_i; 0) = \chi_2 \sum_n \sum_m A_{nm}\psi_n U_m, \tag{28}$$

all possibilities being encompassed in the choice of A_{nm}. Once the new molecule is coupled to the resonator, $\chi_2\psi_n$ ceases to be an eigenfunction for the two and must be rewritten as a sum of the true eigenfunctions, as given by (18.33): **528**

$$\Psi(x, y, q_i; 0) = \sum_n \sum_m A_{nm} U_m[\beta_n(\alpha_n\chi_1\psi_{n+1} + \beta_n\chi_2\psi_n) - \alpha_n(\beta_n\chi_1\psi_{n+1} - \alpha_n\chi_2\psi_n)], \tag{29}$$

in which α_n and β_n are given by (18.38) and both are positive. The two terms in (29) now oscillate at different frequencies. No time-variation that is common to all the eigenfunctions in (29) plays any role; we may therefore ignore the oscillations associated with U_m and take the two terms in square brackets to have frequencies $(n+1)\omega \pm \Omega_n$, where Ω_n is given by (18.36) with ω_e replaced by ω_r. Also it should be noted that the resultant frequency, ω, is used instead of ω_0.

Then
$$\Omega_n = [\omega_r^2 + \gamma^2(n+1)]^{\frac{1}{2}}, \tag{30}$$

and Ω_n has the same sign as ω_r. To find how Ψ develops, we must multiply the first term in (29) by $\exp \mathrm{i}[-(n+1)\omega - \Omega_n]t$ and the second term by $\exp \mathrm{i}[-(n+1)\omega + \Omega_n]t$. Then after some rearrangement

$$\Psi(x, y, q_i; t) = \mathrm{e}^{-\mathrm{i}\omega t} \sum_n \sum_m U_m \psi_n [A_{nm}\chi_2(\beta_n^2\, \mathrm{e}^{-\mathrm{i}\Omega_n t} + \alpha_n^2\, \mathrm{e}^{\mathrm{i}\Omega_n t})$$
$$-2\mathrm{i}A_{n-1,m}\alpha_{n-1}\beta_{n-1}\chi_1 \sin \Omega_{n-1}t]\,\mathrm{e}^{-\mathrm{i}n\omega t}. \tag{31}$$

Now
$$\mathscr{P}(x, t) = \int \cdots \int \Psi^*\Psi\, \mathrm{d}q_i\, \mathrm{d}y, \tag{32}$$

and if Ψ^* is written in the same form as (31), but with subscripts k, l instead of n, m, the integrals over q_i are separable in the form $\int \cdots \int U_k^* U_m\, \mathrm{d}q_i = \delta_{km}$. Hence the U's disappear from (32), which now becomes:

$$\mathscr{P}(x, t) = \sum_l \sum_m \sum_n \mathrm{e}^{\mathrm{i}(l-n)\omega t} \int \psi_l^* \psi_n\, \mathrm{d}y [A_{lm}^* \chi_2^*\, \mathrm{e}^{\mathrm{i}\omega t}(\beta_l^2\, \mathrm{e}^{\mathrm{i}\Omega_l t} + \alpha_l^2\, \mathrm{e}^{-\mathrm{i}\Omega_l t})$$
$$+2\mathrm{i}A_{l-1,m}^* \alpha_{l-1}\beta_{l-1}\chi_1^* \sin \Omega_{l-1}t][A_{nm}\chi_2\, \mathrm{e}^{-\mathrm{i}\omega t}(\beta_n^2\, \mathrm{e}^{-\mathrm{i}\Omega_n t} + \alpha_n^2\, \mathrm{e}^{\mathrm{i}\Omega_n t})$$
$$-2\mathrm{i}A_{n-1,m}\alpha_{n-1}\beta_{n-1}\chi_1 \sin \Omega_{n-1}t]. \tag{33}$$

Since χ_1 and χ_2 are orthogonal and normalized, integration over y is trivial, and

$$\mathscr{P}(x, t) = \sum_l \sum_n \mathrm{e}^{\mathrm{i}(l-n)\omega t} \psi_l^* \psi_n [\rho_{ln}(\beta_l^2\, \mathrm{e}^{\mathrm{i}\Omega_l t} + \alpha_l^2\, \mathrm{e}^{-\mathrm{i}\Omega_l t})(\beta_n^2\, \mathrm{e}^{-\mathrm{i}\Omega_n t} + \alpha_n^2\, \mathrm{e}^{\mathrm{i}\Omega_n t})$$
$$+4\rho_{l-1,n-1}\alpha_{l-1}\beta_{l-1}\alpha_{n-1}\beta_{n-1} \sin \Omega_{l-1}t \sin \Omega_{n-1}t], \tag{34}$$

in which ρ_{ln} is written for the density matrix:

$$\rho_{ln} = \sum_m A_{lm}^* A_{nm} = \rho_{nl}^*. \tag{35}$$

If the original wave-functions were normalized, $\Sigma_n \rho_{nn} = 1$ automatically. This concludes the quantum-mechanical analysis, which we now proceed to interpret.

At the moment the molecule enters $(t=0)$ let us write ρ_{ln} as $\rho_{\bar{l}\bar{n}}$; then

$$\mathscr{P}(x, 0) = \sum_l \sum_n \rho_{\bar{l}\bar{n}} \psi_l^* \psi_n, \tag{36}$$

and at the moment it leaves ($t = T$) (34) shows that

$$\mathscr{P}(x, t) = \sum_l \sum_n \mathrm{e}^{\mathrm{i}(l-n)\omega T} \overset{+}{\rho}_{ln} \psi_l^* \psi_n, \tag{37}$$

in which

$$\left.\begin{aligned} \overset{+}{\rho}_{ln} &= (1+\lambda_{ln})\overset{-}{\rho}_{ln} + \mu_{l-1,n-1}\overset{-}{\rho}_{l-1,n-1}, \\ \lambda_{ln} &= (\beta_l^2\,\mathrm{e}^{\mathrm{i}\Omega_l T} + \alpha_l^2\,\mathrm{e}^{-\mathrm{i}\Omega_l T})(\beta_n^2\,\mathrm{e}^{-\mathrm{i}\Omega_n T} + \alpha_n^2\,\mathrm{e}^{\mathrm{i}\Omega_n T}) - 1, \\ \text{and} \qquad \mu_{ln} &= 4\alpha_l\beta_l\alpha_n\beta_n \sin\Omega_l T \sin\Omega_n T. \end{aligned}\right\} \tag{38}$$

During the passage of one molecule the phase of each term in (37) advances at its characteristic rate, $(l-n)\omega$, while $\overset{-}{\rho}_{ln}$ is changed to $\overset{+}{\rho}_{ln}$. If no molecule is passing through, ρ_{ln} remains unchanged but the phase advance continues unabated. We may therefore proceed to the effect of a steady stream of molecules by writing

$$\mathscr{P}(x, t) = \sum_l \sum_n \mathrm{e}^{\mathrm{i}(l-n)\omega t} \rho_{ln}(t) \psi_l^* \psi_n, \tag{39}$$

and using (38) to define the change suffered by ρ_{ln} through interaction with a single excited molecule:

$$\delta\rho_{ln} = \lambda_{ln}\rho_{ln} + \mu_{l-1,n-1}\rho_{l-1,n-1}. \tag{40}$$

For a flux of Φ molecules in unit time,

$$\dot{\rho}_{ln} = \Phi(\lambda_{ln}\rho_{ln} + \mu_{l-1,n-1}\rho_{l-1,n-1}). \tag{41}$$

If the elements of ρ_{ln} are set out in a square array, each line of elements parallel to the diagonal develops independently, according to (41). On the diagonal itself, where $l = n$, (38) shows that

$$-\lambda_{nn} = \mu_{nn} = 4\alpha_n^2\beta_n^2 \sin^2\Omega_n T. \tag{42}$$

Hence $\Sigma_n \dot{\rho}_{nn} = 0$, and the trace of the matrix, $\Sigma_n \rho_{nn}$, is independent of time, as already noted; the elements form a conserved distribution along the diagonal which may evolve in form but certainly climbs steadily to higher n as each ρ_{nn}, according to (41), acquires some of the distribution lying next below it and passes on its own distribution to the level above. Since each molecule in its passage moves ρ_{nn} upwards by a fraction μ_{nn} of an integer, μ_{nn} may be interpreted as the probability that a molecule emerges unexcited.

The off-diagonal lines do not conserve their distribution but suffer gradual dissipation. To show this it is enough to expand $\lambda_{ln} + \mu_{ln}$ as a power series in $n - l (\equiv \nu)$, stopping at the leading real and imaginary terms. After some tedious evaluations involving the derivatives of (38) and (30) we find

$$\sum_n \dot{\rho}_{n-\nu,n} = \Phi \sum (\lambda_{n-\nu,n} + \mu_{n-\nu,n})\rho_{n-\nu,n},$$

and

$$\begin{aligned} \lambda_{n-\nu,n} + \mu_{n-\nu,n} \approx &-(\gamma^4 T^2\nu^2/8\Omega_n^2)[1 + (\omega_r^2/\gamma^2 n\Omega_n^2 T^2)\sin^2\Omega_n T] \\ &+ \mathrm{i}(\gamma^2\omega_r\nu/4\Omega_n^3)[2\Omega_n T - \sin 2\Omega_n T]. \end{aligned} \tag{43}$$

Only on the diagonal, $\nu = 0$, does $\Sigma_n \dot{\rho}_{n-\nu,n}$ vanish; elsewhere it is negative, and the sum decays at a rate that depends on n and is also proportional to ν^2. We shall defer discussion of this and of the frequency shift that results from the imaginary part of (43).

598

The equation (41) describing the development of the density matrix was worked out for the case of excited molecules entering the resonator. If instead we had assumed unexcited molecules the step from (28) to (29) would have been different, with χ_1 instead of χ_2 in (28) and correspondingly $[\alpha_{n-1}(\alpha_{n-1}\chi_1\psi_n + \beta_{n-1}\chi_2\psi_{n-1}) + \beta_{n-1}(\beta_{n-1}\chi_1\psi_n - \alpha_{n-1}\chi_2\psi_{n-1})]$ in (29); the two eigenfunctions in the square brackets have frequencies $n\omega \pm \Omega_{n-1}$. As a result (38) must be rewritten in the form

$$\left.\begin{aligned} \overset{+}{\rho}_{ln} &= (1+\lambda'_{ln})\overset{-}{\rho}_{ln} + \mu'_{l+1,m+1}\overset{-}{\rho}_{l+1,n+1}, \\ \text{where} \quad \lambda'_{ln} &= \lambda^*_{l-1,n-1} \quad \text{and} \quad \mu'_{l+1,n+1} = \mu_{ln}. \end{aligned}\right\} \tag{44}$$

Hence (41) now has the form

$$\dot{\rho}_{ln} = \Phi(\lambda'_{ln}\rho_{ln} + \mu'_{l+1,n+1}\rho_{l+1,n+1}), \tag{45}$$

which describes a steady progression of the distribution downwards towards lower l, n.

Graphical representation of density matrix

445

The σ-representation, the ensemble of classical oscillators that was introduced in chapter 13, is readily extended to describe the more versatile function $\mathscr{P}(x, t)$ in (39). If the cavity is left undisturbed, so that each ρ_{ln} is constant, the probability $\mathscr{P}(x, t)$ of finding it displaced by x at time t is made up of the superposition of sinusoidally oscillating real functions, of which $\rho_{n-\nu,n}\,\mathrm{e}^{i\nu\omega t}\psi^*_{n-\nu}\psi_n + \rho_{n,n-\nu}\,\mathrm{e}^{-i\nu\omega t}\psi^*_n\psi_{n-\nu}$ is a typical example, the two terms being complex conjugates by virtue of (35). This represents a certain spatial variation of $\mathscr{P}$ with x which oscillates harmonically at a frequency of $\nu\omega$. It is always possible to find a real radial density $\sigma(r)$ such that the two-dimensional distribution $\sigma(r)\cos(\nu\theta + \varepsilon)$ when spun at angular velocity ω projects, as explained in chapter 13, onto the x-axis the desired $\mathscr{P}(x)$ oscillating at a frequency $\nu\omega$.† Therefore any $\mathscr{P}(x, t)$ described by ρ_{ln} has its counterpart in the form of an ensemble of classical oscillators of which the number displaced by an amount x at time t is proportional to $\mathscr{P}(x, t)$. The distribution $\sigma(r, \theta)$ defines the number of oscillators at any instant with amplitude r and phase θ, and in an undisturbed system $\sigma(r, \theta)$ spins unchanged. The diagonal terms of the density matrix are represented by axially symmetric forms of $\sigma(r, \theta) = \sigma_0(r)$, while all terms on the off-diagonal lines defined by $l - n = \pm\nu$ are represented collectively by $\sigma_\nu(r)\cos(\nu\theta + \varepsilon_\nu)$.

† If $\sigma(r)$ is the Bessel (Hankel)$^{(2)}$ function $H_\nu(kr)$, the projection of $\sigma(r)$ has the form e^{ikx}. Any periodic $\mathscr{P}(x, t)$ may therefore be Fourier-analysed in x and t, and the amplitude at frequency $\nu\omega$ and wavenumber k used as the amplitude of the corresponding $H_\nu(kr)$ to synthesize the required $\sigma(r)$.

The coherent state (13.16) and (13.18), represented by a narrowly confined hump of σ running around a ring, requires a certain spread in the values of n to provide tangential confinement. We know, in fact, from (13.23), that the wave-function of the isolated oscillator in the coherent state centred on n_c contains each ψ_n with amplitude a_n roughly proportional to $e^{-(n-n_c)^2/4n_c}$, and from the definition of A_{nm} in (28) it is clear that m is now irrelevant and that A_{nm} is simply the amplitude of ψ_n. The density matrix is then a product of the separate amplitudes,

436

$$\rho_{ln} = \alpha_l^* a_n \propto e^{-\Delta^2/2n_c} \tag{46}$$

where $\Delta^2 = (l - n_c)^2 + (n - n_c)^2$. The coherent state appears as an axially symmetrical Gaussian hump in both the equivalent classical ensemble and the density matrix.

The restriction of ρ_{ln} to a product, as in (46), is of course not peculiar to the coherent state, but is characteristic of any pure state, that is to say, one in which the behaviour of the oscillator is expressible in terms of oscillator eigenfunctions alone. When the oscillator is coupled to other systems it is in a mixed state, necessitating eigenfunctions of all the systems involved to describe it, and it is then that the more general form (35) for ρ_{ln} makes its appearance. It is clear from the dissipative property of (43) that as time goes on the coupling inevitably results in ρ_{ln} becoming more strongly confined to the diagonal (in which form it cannot be expressed as a product). Correspondingly in the σ-representation much of the tangential detail fades away until in the end, when ρ_{ln} is purely diagonal, σ is axially symmetric, though it may still be strongly peaked around a certain radius r. Such a distribution, in which the oscillator may have a well-defined amplitude (or energy) but no preference as to phase, cannot be expressed by any combination of oscillator wave-functions, but is a characteristic example of the greater generality allowed by the density matrix. It is interesting to observe how the coupling of quantum systems leads to new effects such as this, without immediately ruling out the use of a classical model to represent the behaviour. We shall even find that the rules governing the development of the model are visualizable in classical terms. Let us derive these rules.

The dissipative harmonic oscillator

The essential results can be demonstrated without stimulating the resonator into self-sustained maser action, and at the same time the earlier, rather sketchy, discussions of dissipative processes can be rounded out. It was pointed out by Scully and Lamb[3] that a stream of unexcited molecules passing through a resonator served to damp its oscillation, and that this mechanism was considerably easier to work out completely than damping by radiation, since each element of the dissipative process is a two-level system rather than a cavity mode with its ladder of equally spaced levels. There is no need for the molecules to be well tuned to the resonator, since

499, 528

even if poorly tuned they still have a chance, albeit small, of emerging excited.† It is positively advantageous, indeed, to model the dissipation with a very high flux Φ_d of very weakly coupled molecules, since the fluctuations in the flux can thereby be made negligibly small. All random effects can then be attributed to the interaction of the quantized systems and, as we shall see, it is easy to include thermal noise as well. Moreover, if we assume the residence time T to be much shorter than τ_e, all the problems discussed earlier about the appropriate value of Q disappear. We suppose that there are equal fluxes of molecules with positive and negative mistuning, ω_r, so that the imaginary part of (43) disappears. In the real part the second term in the square brackets can be made large by reducing the coupling constant γ, so that only this term need be retained.

† As described here, the process appears not to conserve energy – an unexcited molecule, whose frequency ω_m is not the same as the cavity frequency ω_0, may gain $\hbar\omega_m$ while the cavity loses $\hbar\omega_0$. It is not enough to take refuge in the Uncertainty Principle with the observation that a transition is only likely if $|\omega_m - \omega_0|T < 1$; this may be true, but the fact remains that the initial and final states of both molecule and cavity can be ascertained at leisure, with arbitrarily small uncertainty. To resolve the problem it is necessary to bring in the translational kinetic energy of the molecule, which does not remain constant during the interaction. The physical mechanism that causes it to change may be traced to the fringing field at the entrance and exit of the cavity; as the molecule crosses the threshold it begins to be polarized, and the gradient of field acts on the dipole moment to exert a translational force. In the light of this let us see how a stationary-state wave-function might be constructed to describe the complete process, involving the cavity and the whole path of the molecule from before its entry until after its exit. Initially we might have the molecule and cavity in well-defined states, $\chi_1\,e^{ik_0x}$ for the molecule and ψ_m for the cavity; the term e^{ik_0x} describes the initial translational motion. If the total energy, including kinetic energy, is E, the complete wave-function is $(\psi_n\chi_1\,e^{ik_0x})\,e^{-iEt/\hbar}$. This must be matched, at the entrance to the cavity, to the wave-function inside, where because of the interaction the relevant eigenstates will have the form $\alpha\chi_1\psi_n + \beta\chi_2\psi_{n-1}$ as in (18.33), with two choices of α and β. The complete wave-function will be a superposition of the two eigenstates, each multiplied by its appropriate e^{ik_1x} or e^{ik_2x} so as to give the same total energy E; only then can the matching persist at all times. The 'internal' energy difference $2\hbar\Omega_n$ is compensated by the difference in kinetic energy $\hbar^2(k_1^2 - k_2^2)/2m$, or $\hbar^2 k_0\delta k/m$ if the coupling is weak. Because of the wavenumber difference, at the exit the two waves will have suffered a relative phase shift of $L\delta k$, and will not recombine outside to form the original wave-function once more. Instead there will be a mixture of the original $(\psi_n\chi_1\,e^{ik_0x})\,e^{-iEt/\hbar}$ and $(\chi_{n-1}\chi_2\,e^{ik_0'x})\,e^{-iEt/\hbar}$, with k_0' different from k_0 so as to maintain the same total energy E. Clearly, if an electrostatic lens is used to separate these two states it will be found that whichever path any given molecule takes it will emerge with exactly the right energy. In summary, when in the text it is argued that the two eigenfunctions in the cavity suffer a phase shift of $2\Omega_n T$ because their energies differ, we now attribute the resulting transition probability to the difference in wavenumber, and write the phase shift as $L\delta k$, i.e. $2\Omega_n mL/\hbar k_0$. Since the molecular speed is $\hbar k_0/m$, these two expressions are equivalent. We may therefore continue to use the argument in the text, secure in the knowledge that energy is conserved in spite of appearances.

Then, since $\sin(\Omega_n T) \approx \Omega_n T$ when T is small, and $\Omega_n \approx \omega_r$ for large mis-tuning,

$$\lambda'_{n-\nu,n} + \mu'_{n-\nu,n} \approx -\gamma_d^2 T_d^2 \nu^2/8n, \tag{47}$$

in which γ_d and T_d are used to distinguish the coupling constant and residence time of the dissipative molecules from those of the molecules responsible for maser amplification. When $\nu = 0$, on the diagonal,

$$\begin{aligned} \mu'_{nn} = -\lambda'_{nn} &= 4\alpha_{n-1}^2 \beta_{n-1}^2 \sin^2(\omega_r T_d), \text{ from (38) and (44)}, \\ &\approx \gamma_d^2 T_d^2 n, \text{ from (18.38) and (18.36)}. \end{aligned} \tag{48}$$

If we write B_d for $\Phi_d \gamma_d^2 T_d^2$, the development of ρ_{ln} may be summarized thus. On the diagonal, from (45) and (48),

$$\dot{\rho}_{nn} = B_d[-n\rho_{nn} + (n+1)\rho_{n+1,n+1}]. \tag{49}$$

Off the diagonal, $-\lambda_{n-\nu,n}$ and $\mu_{n-\nu,n}$ are almost the same as when $\nu = 0$, so that (49) very nearly describes the development of ρ_{ln}. However, the difference between μ' and $-\lambda'$, which is smaller than either by something like ν^2/n^2, leads to non-conservation since, from (45),

$$\sum_n \dot{\rho}_{n-\nu,n} = \Phi_d \sum_n (\lambda'_{n-\nu,n} + \mu'_{n-\nu,n})\rho_{n-\nu,\nu} = -\tfrac{1}{8} B_d \nu^2 \sum_n \rho_{n-\nu,n}/n. \tag{50}$$

A reasonably narrow distribution, of width $n^{\frac{1}{2}}$ say, allows n to be taken outside the summation in (50);

$$\sum_n \dot{\rho}_{n-\nu,n} \approx -(B_d \nu^2/8\bar{n}) \sum_n \rho_{n-\nu,n}, \tag{51}$$

indicating that while this line of the matrix develops on the whole in the same way as the diagonal, the distribution decays steadily, but not exponentially since the instantaneous time-constant varies as $\bar{n}$, which is steadily falling.

We discuss the two aspects of the development separately; first the diagonal terms and then the decay of the off-diagonal terms. Since (49) is linear in ρ_{nn}, any solution may be synthesized once we know how a single term develops. The appropriate Green's function has the form

$$\rho_{rr}(t') = {}^N C_r (e^{t'} - 1)^{N-r} e^{-Nt'}, \tag{52}$$

which will be found by substitution to satisfy (49) if $t' = B_d t$. To find the initial state, let $t' \to 0$; then $\rho_{rr} \to 0$ for all r except $r = N$, for which $\rho_{NN} = 1$. The subsequent conservation of $\Sigma_r \rho_{rr}$ follows by considering the generating function $(1 - e^{-t'})^N [1 + 1/(e^{t'} - 1)]^N$; on expanding the square brackets as a binomial series, the rth term is seen to be ρ_{rr}, and since the function itself is obviously equal to unity it follows that $\Sigma_r \rho_{rr} = 1$. A convenient approximation to (52) results from applying Stirling's approximation to (52) followed by Taylor expansion:

$$\rho_{rr} \sim \mathcal{N}_r \, e^{-N(r-r_m)^2/2r_m(N-r_m)}, \tag{53}$$

in which r_m, which defines the position of the peak, is $N e^{-t'}$, and $\mathcal{N}_r$ is a time-dependent normalizing coefficient to conserve ρ_{rr}. The decay of r_m, and hence of energy, as $e^{-t'}$ shows that $t = t/\tau_e$, i.e. $B_d = 1/\tau_e$. The mean square width of the Gaussian peak follows from (53):

$$\overline{(r-r_m)^2} = r_m(1-r_m/N) = N\,e^{-t'}(1-e^{-t'}). \tag{54}$$

As the centre of the peak falls exponentially with time from N to zero, the R.M.S. width expands from zero to a maximum of $\frac{1}{2}N^{\frac{1}{2}}$, when $r_m = \frac{1}{2}N$, and then decreases, ultimately as $r_m^{\frac{1}{2}}$. Stirling's approximation renders (54) unreliable when r_m is small, but (52) shows that, as expected, the system ends up in the ground state.

We have just determined the development of the density matrix starting from a pure eigenstate, ψ_N. If we had started with a diagonal ρ_{nn} in the form of a narrow Gaussian peak centred on N and of mean square width w_0, $\rho_{nn} \propto e^{-(n-N)^2/2w_0}$, each element of ρ_{nn} would have developed according to (53), so that after time t the distribution of centres would have collapsed to a peak of mean square width $w_0\,e^{-2t'}$ centred on $N\,e^{-t'}$. On the other hand, each element would now have spread into a peak of width given by (54), so that the resultant would be the convolution of two Gaussian curves; such a convolution is itself Gaussian, with a mean square width w that is the sum of the separate mean square widths:

$$w = w_0\,e^{-2t'} + N\,e^{-t'}(1-e^{-t'}). \tag{55}$$

In particular, if we start with $w_0 = N$, subsequently $w = N\,e^{-t'} = r_m$. This means that the dissipative decay of a distribution that has been set up to match the coherent state of the oscillator, with a mean square width of the n-distribution equal to the mean value of n itself, maintains the coherent state throughout.

This result was derived for the diagonal terms of ρ_{ln}, but it holds for the off-diagonal terms also. To verify this point, let us start with the coherent pure state in which ρ_{ln} is an isotropic Gaussian peak of mean square width N centred on the point (N, N),

$$\rho_{ln} \propto e^{-[(l-N)^2+(n-N)^2]/4N}. \tag{56}$$

Along any off-diagonal line $l = n - \nu$, the distribution is Gaussian, with the same width N, and height reduced by $e^{-\nu^2/8N}$. Now let this whole peak move down towards the origin, remaining isotropic while shrinking in width so that the mean square width is always equal to the value of n at the centre. When the centre has reached $\bar{n}$ the off-diagonal section has a peak height reduced from that at the centre by a factor $e^{-\nu^2/8\bar{n}}$. In order to maintain the coherent state the off-diagonal elements must develop in the same way as the diagonal elements, except that they must decay at the same time. Thus the sum of ρ_{ln} along a line of constant ν, which is constant if $\nu = 0$, must in general have the property that

$$\sum_n \rho_{n-\nu,n} \propto e^{-\nu^2/8\bar{n}}.$$

Then, by taking the logarithmic derivative, we have that

$$\sum_n \dot{\rho}_{n-\nu,n} = (\nu^2/8\bar{n}^2)\dot{\bar{n}} \sum_n \rho_{n-\nu,n} = (-\nu^2/8\bar{n}\tau_e) \sum_n \rho_{n-\nu,n}, \tag{57}$$

since $\bar{n}$ decays exponentially with time-constant τ_e. Now we have already found that $B_d = 1/\tau_e$, so that (51) and (57) are identical.

We have demonstrated that when an oscillator in a coherent state suffers dissipation, it remains in a coherent state. This result is immediately translatable into the σ-representation, since we already know what distribution $\sigma(\boldsymbol{R})$ describes the coherent state; it has the form of the ground state distribution e^{-R^2}, where $\boldsymbol{R}$ is the vector (ξ, κ), but displaced from the origin to a new centre, $\boldsymbol{R}_0$ say; $\sigma \propto \exp\{-|\boldsymbol{R}-\boldsymbol{R}_0|^2\}$. The result just derived shows that if we allow this coherent state to decay the distribution remains unchanged in form and size while its centre $\boldsymbol{R}_0$ relaxes exponentially back to the origin. The whole pattern is of course spinning at angular frequency ω_0 round the origin, but we take this for granted. A relaxation of this character is characteristic of a distribution of points which are subject to an organized drift towards the origin, at a speed proportional to displacement, superimposed on diffusive motion obeying Fick's law. If the concentration of points is $\sigma(\boldsymbol{R})$, the local flux $\boldsymbol{V}$, as determined by these processes, is $-\alpha\sigma\boldsymbol{R} - (D \operatorname{grad} \sigma)$. The first term is due to organized drift and the second to diffusion. In the steady state $\sigma \propto e^{-\alpha R^2/2D}$, for which distribution $\boldsymbol{V} = 0$ everywhere. Let us now shift this pattern to a new centre, $\boldsymbol{R}_0$. At any other point $\boldsymbol{R}$ the organized drift velocity is $-\alpha\boldsymbol{R}$ while the diffusive velocity is away from $\boldsymbol{R}_0$ and of magnitude $\alpha(\boldsymbol{R}-\boldsymbol{R}_0)$. The resultant $\boldsymbol{V} = -\alpha\sigma\boldsymbol{R}_0$, so that every part of the σ-distribution moves at the same speed, $\alpha\boldsymbol{R}_0$. If there was no random element associated with the dissipation the distribution would diminish in radius as every part relaxed towards the origin; but the random effect of the dissipative mechanism, represented by the diffusive term, counteracts this diminution. The angular spread of the distribution, i.e. the uncertainty in phase of the coherent oscillation, may be very small, say $\bar{n}^{-\frac{1}{2}}$, when $\bar{n}$ is large. Clearly, however, it increases as dissipation drives the distribution towards the origin; eventually the distribution is isotropic round the origin, and no phase information remains. This is a simple pictorial representation of the decay of the off-diagonal elements of ρ_{ln}.

445

We have exhibited a special case of the diffusive effect, but it can be proved quite generally by returning to ρ_{ln} and remembering that every line $l = n - \nu$ develops independently of the others. What follows is an outline of the argument, without details of the manipulations. Considering one such line whose contribution to $\mathscr{P}(\xi, t)$ may be written, following (39),

$$\mathscr{P}_\nu(\xi, t) = \sum_n \rho_{n-\nu,n} \Psi^*_{n-\nu} \Psi_n, \tag{58}$$

we take the ψ_n as real and ignore the oscillatory term, which is the same as ignoring the spinning of the σ-distribution. It is convenient to write ψ_n

431 in terms of Hermite polynomials, as in (13.3):

$$\psi_n = \pi^{-\frac{1}{4}}(2^n n!)^{-\frac{1}{2}} H_n(\xi)\,\mathrm{e}^{-\frac{1}{2}\xi^2},$$

and to make use of the following properties of the polynomials:[4]

$$\left.\begin{aligned} \mathrm{d}H_n/\mathrm{d}\xi &= 2nH_{n-1} \\ \text{and}\qquad \mathrm{d}^2H_n/\mathrm{d}\xi^2 &= 4n\xi H_{n-1} - 2nH_n. \end{aligned}\right\} \tag{59}$$

From (58) and (45) we have that

$$\partial\mathcal{P}_\nu/\partial t = \Phi_\mathrm{d}\sum_n \rho_{n-\nu,n}(\lambda'_{n-\nu,n}\psi_{n-\nu}\psi_n + \mu'_{n-\nu,n}\psi_{n-\nu-1}\psi_{n-1}).$$

In addition, the argument leading to (48) may be extended to show that $\Phi_\mathrm{d}\lambda'_{n-\nu,\nu} = -(n-\frac{1}{2}\nu)B_\mathrm{d} = -(n-\frac{1}{2}\nu)/\tau_\mathrm{e}$ and $\Phi_\mathrm{d}\mu'_{n-\nu,\nu} = [n(n-\nu)]^{\frac{1}{2}}/\tau_\mathrm{e}$.

Given this information, the reader may verify that

$$\partial^2\mathcal{P}_\nu/\partial\xi^2 + 2\partial(\mathcal{P}_\nu\xi)/\partial\xi = 4\tau_\mathrm{e}\partial\mathcal{P}_\nu/\partial t, \tag{60}$$

independent of ν and therefore true for any $\mathcal{P}(\xi, t)$.

Since this equation continues to hold as the σ-distribution turns round, and since $\mathcal{P}$ is the projection of σ onto the ξ-axis, (60) must represent one component of the development of σ itself, so that we may write

$$D\nabla^2\sigma + 2D\,\mathrm{div}\,(\sigma\boldsymbol{R}) = \partial\sigma/\mathrm{d}t, \tag{61}$$

where $D = 1/4\tau_\mathrm{e} = \omega_0/4Q_\mathrm{r}$. The left-hand side is the divergence of a flux vector $-D\,\mathrm{grad}\,\sigma + \sigma\boldsymbol{R}/2\tau_\mathrm{e}$, in which the second term describes exponential decay of the individual vectors with a time-constant $2\tau_\mathrm{e}$, i.e. τ_a.

This result looks remarkably classical. If $\sigma(\boldsymbol{R})$ is considered as representing an ensemble of real classical vibrators, subject to linear dissipative forces and to random noise, the latter causes each point in the distribution to execute a random walk which is described statistically by a diffusion coefficient. Such an equation of motion for σ is identical with (61), and the magnitude of the noise determines D. Now if the noise is thermal, we know that the ensemble will ultimately settle down to a Boltzmann distribution, $\sigma \propto \mathrm{e}^{-R^2/2w}$, with w chosen so that the mean potential energy is $\frac{1}{2}k_\mathrm{B}T$. Since according to (13.11) the mean potential energy is $\frac{1}{2}\hbar\omega_0\overline{\xi^2}$, i.e. $\frac{1}{2}\hbar\omega_0 w$, it follows that

$$\sigma \propto \mathrm{e}^{-R^2\hbar\omega_0/2k_\mathrm{B}T}.$$

For the quantized vibrator, isolated from thermal noise, the final distribution is e^{-R^2}, so that the intrinsic quantum noise has the same effect as if the vibrator were classical and subjected to noise from a source at temperatures $\hbar\omega_0/2k_\mathrm{B}$. If the losses were due to radiation we might be tempted

503 to ascribe this noise to zero-point motion of the cavity modes, each of which has mean energy $\hbar\omega_0/2$. There is, however, nothing obviously analogous to zero-point motion in the model lossy mechanism that has led

to this result, and one would be wise to resist the temptation to interpret zero-point energy in classical terms.

The existence of quantum noise reveals the limitations of the dielectric model which in almost all other respects works extremely well. The defect is much the same as we have noted before in attempts to extend classical arguments too far – the dielectric model treats two interacting systems (resonator and molecule) as if they were separate entities, each acted upon by a determinate force due to the other. Thus the molecules are imagined as responding to a well-defined resonator field, and likewise the resonator to well-defined polarization oscillations of the molecules. We may expect, by analogy with the earlier examples, that on treating the coupled systems in a consistently quantal way new statistical features will enter. This is indeed the case, with fluctuations of amplitude and phase appearing quite independently of any external cause.

574

449, 532

It should be remembered that the noise temperature derived here is valid only for a perfectly cold system, since the molecules responsible for dissipation are all unexcited. A lossy resonator with walls at temperature T_0 may be simulated by a stream of untuned molecules for which the flux of unexcited molecules is $\Phi_d\, e^{\hbar\omega_0/k_B T_0}/(e^{\hbar\omega_0/k_B T_0}-1)$ and of excited molecules $\Phi_d/(e^{\hbar\omega_0/k_B T_0}-1)$; there is a net excess Φ_d of unexcited molecules to give the required dissipation, and the ratio of the two fluxes is the Boltzmann factor for temperature T_0. All molecules contribute equally to the fluctuations, which are therefore increased in proportion to the total flux, $\Phi_d \coth(\hbar\omega_0/2k_B T_0)$, so that the equivalent noise temperature T_N takes the form:

$$k_B T_N = \tfrac{1}{2}\hbar\omega_0 \coth(\hbar\omega_0/2k_B T_0). \tag{62}$$

At low temperatures the residual noise is purely quantal, with $T_N = \hbar\omega_0/2k_B$, and as the temperature is raised it increases until $T_N = T_0$ and the noise can be considered to be entirely thermal and to have the magnitude expected from classical arguments (cf. Johnson noise). The form of (62) matches Planck's expression for the mean energy of a harmonic vibrator, with the zero-point energy added, and the transition from quantum noise at low temperatures to classical noise at high is precisely parallel to the transition from zero-point energy of the vibrator to the classical equipartition result. At any temperature the equilibrium form of the σ-distribution is Gaussian, with a mean energy of $k_B T_N$ which is just the result of Planck. It is satisfactory to find, albeit in a special case, confirmation of the view expressed in chapter 16, that randomness is communicated, and statistical equilibrium established, in the interaction of quantized systems.

158

489

Finally we note that when the vibrator is acted upon by a determinate force the response of the equivalent classical ensemble is exactly the same as described in chapter 13, every member responding independently. We therefore can give a complete prescription for finding how $\mathscr{P}(\xi, t)$ develops for a quantized vibrator acted upon by a determinate force and subject to dissipation from perfectly cold absorbers: $\mathscr{P}(\xi, t)$ is the projection on the

ξ-axis of a two-dimensional distribution $\sigma(\boldsymbol{R}, t)$ which develops according to (61), spins about the origin at angular velocity ω_0, and in this spinning state is displaced by a force $F(t)$ which gives rise to a bodily shift, always in the κ-direction, at a rate $\dot{\kappa} = F/(m\hbar\omega_0)^{\frac{1}{2}}$. In laboratory co-ordinates, x and $\dot{x}/\omega_0$ instead of ξ and κ, (61) takes the form:

$$\dot{\sigma} = (\hbar/4mQ_r)\nabla^2\sigma + (\omega_0/2Q_r)\,\text{div}\,(\sigma\boldsymbol{r}). \qquad (63)$$

Any applied force $F(t)$ moves the spinning σ-distribution bodily in the y-direction at a speed $\dot{y} = F/m\omega_0$. The projection of the distribution on to the x-axis gives $\mathscr{P}(x, t)$.

It is worth noting that when Q_r is large and F is a force applied at the resonant frequency, the force may be resolved into two rotating components, each of magnitude $F/2$, one of which is synchronized with the rotating σ-distribution, while the counter-rotating component produces only a tiny perturbation and may be disregarded. In the spinning frame, therefore, the distribution suffers a steady drift in a constant direction at a velocity $\boldsymbol{F}/2m\omega_0$. This adds a term $(-\boldsymbol{F}/2m\omega_0)\cdot\text{grad}\,\sigma$ to the right-hand side of (63), which is equivalent to replacing the second term in this equation by $(\omega_0/2Q_r)\,\text{div}\,[\sigma(\boldsymbol{r}-\boldsymbol{r}_0)]$, where $\boldsymbol{r}_0 = Q_r\boldsymbol{F}/m\omega_0^2$, the classical value for the amplitude at the resonance peak. The steady state is then a Gaussian peak, of width appropriate to the temperature, displaced bodily from the origin; and if $k_BT_0 \ll \hbar\omega_0$ this is exactly the coherent state which is thus seen to be less artificial in conception than might have appeared from the way it was introduced.

See note on p. 602.

Dissipation and fluctuations in the maser

The general ideas developed in the last section are immediately extendable to the spontaneously oscillating maser, in which two processes are at work independently, the regenerative action of excited, well-tuned molecules and dissipation, modelled by untuned, unexcited molecules. The excited molecules cause ρ_{ln} to climb upwards, according to (41), while the unexcited oppose the climb, according to (49); the two processes acting together result in an equation for the development of the diagonal terms:

$$\text{d}\rho_{nn}/\text{d}t' = -C_n\rho_{nn} + C_{n-1}\rho_{n-1,n-1} - n\rho_{nn} + (n+1)\rho_{n+1,n+1}, \qquad (64)$$

in which $t' = t/\tau_e$ as before; $C_n = \tau_e\Phi\mu_{nn}$ and represents the number of molecules that are de-excited in time τ_e. From (38),

$$C_n = \tau_e\Phi\gamma^2(n+1)\sin^2\Omega_nT/\Omega_n^2. \qquad (65)$$

575

The rate of climb is determined by C_n, which in essence is the same as the negative loss $-\kappa_e''$ of the dielectric treatment. At low values of n, $C_n \,\widetilde{\propto}\, n$ and the excited molecules respond linearly to the resonator field. The critical flux, at which maser action begins, is such that $C_n = n$ when n is small, so that the first and second pairs of terms in (64) have equal and opposite effects; hence $\Phi_c = 1/\tau_e\gamma^2T^2$ in agreement with (7). We shall not worry about the precise variation of C_n with n since this is influenced by the unknown spread of residence times. It is enough to recognize that the initial proportionality to n breaks down and eventually C_n reaches, or even goes through, a maximum. The steady state of oscillation lies around the value of n at which $C_n = n$.

Before discussing the oscillatory steady state, however, let us examine the behaviour before maser action begins, and immediately after F exceeds unity.† So long as linearity prevails, with $C_n = Fn$ and F constant, the procedure that led from (45) to (60) and (61) applies equally well to (64), and allows the development of $\sigma(\boldsymbol{R})$ to be expressed as a differential equation:

$$\frac{1+F}{4\tau_e}\nabla^2\sigma + \frac{1-F}{2\tau_e}\operatorname{div}(\sigma\boldsymbol{R}) = \partial\sigma/\partial t. \qquad (66)$$

The drift back to the origin is now governed by a time-constant $2\tau_e/(1-F)$ which exhibits clearly the opposition of the dissipative and the regenerative processes, balanced when $F=1$. On the other hand the diffusive process involves the cooperation of both, as shown by the extra factor $(1+F)$. If $F<1$ the steady state has the Gaussian form of the ground state of the oscillator, centred on the origin, but spread out by a factor $[(1+F)/(1-F)]^{\frac{1}{2}}$:

$$\sigma \propto e^{-R^2/2w}; \qquad w = \tfrac{1}{2}(1+F)/(1-F). \qquad (67)$$

When F is raised above unity, the distribution explodes:

$$\left.\begin{aligned} &\sigma \propto w^{-1}\,e^{-R^2/2w}, \\ \text{and} \qquad &w(t) = w_0\,e^{(F-1)t/\tau_e} - \tfrac{1}{2}(F+1)/(F-1). \end{aligned}\right\} \qquad (68)$$

So long as $C_n = Fn$ the exponential spread of the distribution continues, with the maximum of σ remaining at the origin. As n increases, however, and C_n rises less rapidly than n, the outer regions of σ begin to slow down, while the inner regions continue to spread. Ultimately, when $C_n \sim n$ at the outside, the leading edge stops and the inner regions pile up against it, until the steady state has σ in the form of a ring.

At this point the σ-representation shows a weakness. The derivation of (66) from (64) depends on the validity of the linear approximation; as soon as C_n ceases to be proportional to n it becomes impossible to eliminate n in deriving a differential equation for σ. This does not mean that (66) must be utterly discarded, but only that it must be treated with caution as at best approximate, resting on the hope that to allow the coefficients to vary with R will take care of most of the problems. We shall find that the

† It should be remembered that the model is unrealistic at this point, since in the real maser the rate of growth is limited by T rather than τ_e. But the picture of how the development proceeds is probably fairly reliable apart from the time scale and is, in any case, introduced primarily to exhibit certain limitations of the σ-representation.

steady-state solution is indeed consistent with slowly varying drift and diffusion coefficients. We shall, however, also find that different diffusion coefficients are needed for radial and tangential variations of the distribution.

To discuss the radial variation first, it is enough to note that in the steady state ρ_{ln} is diagonal and governed by (64). Let us define $\bar{n}$ as $\Sigma_n n\rho_{nn}$; then

$$\mathrm{d}\bar{n}/\mathrm{d}t' = \sum_n n\,\mathrm{d}\rho_{nn}/\mathrm{d}t' = \sum_n (C_n - n)\rho_{nn} = \bar{C}_n - \bar{n}. \tag{69}$$

On the assumption of a narrow distribution, $\bar{n}$ in the steady state $= n_0$, where $C_n(n_0) = n_0$. In the vicinity of n_0 let $C_n = n_0 - C'_n(n - n_0)$, C'_n being the (positive) value of $-\mathrm{d}C_n/\mathrm{d}n$ at n_0. Then (69) may be rewritten

$$\mathrm{d}\bar{n}/\mathrm{d}t' = (n_0 - \bar{n})(1 + C'_n), \tag{70}$$

showing that $\bar{n}$ approaches n_0 exponentially with time constant $\tau_e/(1 + C'_n)$.

578

In terms of Q_{sat},

$$Q_{\mathrm{sat}}/Q_{\mathrm{r}} = 1/(1 + C'_n). \tag{71}$$

The same Q_{sat} determines the spread of n in the steady state. If $\mathrm{d}\rho_{nn}/\mathrm{d}t' = 0$ and ρ_{nn} is expanded as a Taylor series about n_0, (64) takes the approximate form

$$(1 + C'_n)(\rho_{nn} + y\rho'_{nn}) + n_0\rho''_{nn} \doteqdot 0, \text{ where } y = n - n_0.$$

As usual, the solution is Gaussian, $\rho \propto \mathrm{e}^{-y^2/2w}$, and $w = n_0/(1 + C'_n)$; the mean square width $= n_0 Q_{\mathrm{sat}}/Q_{\mathrm{r}}$.

Let us translate this into a distribution for σ, which will be sharply peaked round a ring of radius $(2n_0)^{\frac{1}{2}}$. Each term in ρ_{nn} is represented by an axially symmetrical distribution which would be a perfectly sharp ridge at $(2n)^{\frac{1}{2}}$ if $\psi_n^2(\xi)$ followed the classical probability pattern exactly. As we have seen, however, the ridge is not perfectly sharp, and inside it there is a series of subsidiary rings, alternatively positive and negative. Now when we add all the patterns due to the ψ_n^2 in a range of about $n_0^{\frac{1}{2}}$, the radii of the ridges extend over a range of about $n^{\frac{1}{6}}$ times the widest separation of the subsidiary rings in any one pattern, say 10 times if $n = 10^6$. This is enough to ensure, first, that the subsidiary oscillations cancel each other and, secondly, that the resulting ridge is wide enough for the width of an individual constituent to be unimportant. That is to say, the ring in σ reflects very closely the distribution ρ_{nn}. Since each n appears as a ridge of radius $R = (2n)^{\frac{1}{2}}$, a range Δn is translated into a range $\Delta R = \Delta n/(2n)^{\frac{1}{2}}$; and if $\overline{\Delta n^2} = n_0/(1 - C'_n)$, $\overline{\Delta R^2} = 1/2(1 + C'_n)$, which is the mean square width of the coherent state, $\frac{1}{2}$, multiplied by the factor $Q_{\mathrm{sat}}/Q_{\mathrm{r}}$.

432

This is a good approximation to the solution of (64) in the steady state. Let us see what drift and diffusion coefficients are needed to reproduce it, for which purpose we return to (64). If only the last two terms were present the vibration would decay exponentially, and the σ-distribution would relax towards the origin with local velocity $-\boldsymbol{R}/\tau_a$. With the amplifying terms added the local velocity is reduced by a factor $(n - C_n)/n$, which of course vanishes at $n = n_0$. In the vicinity of n_0, where $C_n = n_0 - C'_n(n - n_0)$, the drift velocity is $-R_0(1 + C'_n)(n - n_0)/n_0\tau_a$, or $-2(1 + C'_n)(R - R_0)/\tau_a$ since $n \propto R^2$. The flux is σ times this, and represents a drift of the distribution towards R_0 from above and below. It is counteracted by the diffusion due to random noise. If $\sigma \propto \exp[-(R - R_0)^2/2w]$, the diffusive flux $-D\,\mathrm{d}\sigma/\mathrm{d}R$ is $(R - R_0)D\sigma/w$, i.e. $2(1 + C'_n)(R - R_0)D\sigma$ for w equal to $1/2(1 + C'_n)$ as determined above. In the steady state the fluxes must balance and therefore D must take the value $1/\tau_a$, or $1/2\tau_e$, which is twice as great as in the free vibrator. If we care to attribute the noise to sudden impulses occurring every time a quantum is transferred, it is clear that in the steady state the number of quanta communicated to the resonator by the molecules must equal the number removed, i.e. so far as noise is concerned the effective value of F is always unity in the neighbourhood of the steady state. It would be unwise to view this interpretation as anything but a convenience; there is nothing in the quantum mechanics to justify the semi-classical concept of sudden switches of state giving rise to sharp impulses. Nevertheless it works, as does the classical modelling on which we have laid so much emphasis, and there is no objection to using these ideas in solving problems provided one does not fall into the habit of believing that they describe what could really be observed, if only our senses were delicate enough to follow the processes – this is the slippery slope leading to hidden-variable interpretations of quantum mechanics, which as yet have made no useful contribution to advancing the progress of physics.

To evaluate the phase-diffusion caused by quantum noise, it is necessary to translate the non-conserved off-diagonal elements of ρ_{ln}, as expressed for example by (43), into a diffusion coefficient. Consider a narrow ring of radius R in the σ-representation, round which σ varies in an arbitrary manner. In a Fourier analysis of this variation, the term in $\mathrm{e}^{i\nu\theta}$ represents a linear phase variation of the form $\mathrm{e}^{i\nu s/R}$, if s is measured along the perimeter of the ring. Now such a variation decays exponentially as a result of diffusion, the diffusion equation,

$$D\partial^2\sigma/\partial s^2 = \partial\sigma/\partial t,$$

being satisfied by $\sigma \propto \exp[i\nu s/R - D\nu^2 t/R^2]$, so that the time-constant is $R^2/D\nu^2$. A time-constant varying as $1/\nu^2$ is exactly what the real part of (43) predicts, and comparison of the two expressions shows that when the system is perfectly tuned, so that $\Omega_n^2 = \gamma^2(n+1)$,

$$D = \Phi\gamma^2 T^2 R^2/8(n+1) = \tfrac{1}{4}\Phi\gamma^2 T^2 = F/4\tau_e, \text{ from (7).} \qquad (72)$$

This is the tangential diffusion coefficient due to the excited molecules alone; the unexcited molecules have already been treated by a slightly different method and have yielded in (57) a time-constant of $8\bar{n}\tau_e/\nu^2$, i.e. $4R^2\tau_e/\nu^2$, equivalent to $D = 1/4\tau_e$. The diffusive processes are independent, both making contributions of the same sign to the real part of (43), so that the resultant tangential diffusion coefficient is $(1+F)/4\tau_e$, the same as for both radial and tangential diffusion in the initial growth as governed by (66). Nothing has been said at this point, however, about n being small and, in contrast to radial diffusion, the tangential diffusion coefficient does not fall off as the steady state is approached. Here is additional reason for treating the σ-representation with caution; it gives an excellent general picture of the true quantal behaviour but needs tinkering here and there if it is to be used quantitatively – and the tinkering, however minor, is of such a nature as to make one sceptical about deriving any deep meaning from the model.

Measurement of maser frequency

When a maser is connected to a counter to measure its frequency, the counter will respond every time the signal entering it passes through zero. The act of responding tells the observer that, for example, $\mathscr{E}=0$ at some instant which is rather well defined in relation to the period of oscillation. Possessing such information, the observer must set up a wave-packet $\sigma(\xi,\kappa)$, with a well-marked peak, to describe his knowledge of the maser. So long as the observer is content to let the maser and counter run unregarded, his ability to predict the phase at a later moment is limited by the spread of the σ-peak round its ring. Suppose that the R.M.S. angular width of the peak has increased to $2\Delta\theta$; he must be uncertain to this extent of the phase of the oscillation when at last he notes the response of the counter. The spread $\pm\Delta\theta$ implies that the counter may respond any time within the interval $\pm\Delta\theta/\omega_0$; hence the observer must be uncertain of the frequency to the extent of $\Delta\theta/\omega_0 t$. If $\Delta\omega_0$ is the R.M.S. uncertainty in the frequency, as measured by counting cycles for time t,

$$\Delta\omega_0/\omega_0=\Delta\theta/\omega_0 t. \tag{73}$$

Now tangential diffusion leads to a Gaussian distribution $e^{-s^2/4Dt}$, with $\Delta s=(2Dt)^{\frac{1}{2}}$; therefore $\Delta\theta=(2Dt)^{\frac{1}{2}}/R=(Dt/\bar{n})^{\frac{1}{2}}$, and

$$\Delta\omega_0/\omega_0=(D/\bar{n}\omega_0^2 t)^{\frac{1}{2}}. \tag{74}$$

We shall discuss this result in the next section, but must first draw attention to a point which has been so far ignored, the imaginary term in (43) which vanishes in a perfectly tuned maser. If $\omega_r\neq 0$ every excited molecule induces a phase shift to the off-diagonal terms in ρ_{ln}, proportional to ν, and the stream of molecules gives rise to a rate of phase shift Φ times as great. This is equivalent to a change in the angular velocity of the σ-representation, which will produce ν times the change in the oscillation frequency of the νth Fourier component projected onto the ξ-axis. Self-consistency demands that ω, the resultant frequency in the steady state, is such that the shift from ω_0, which is $2(\omega_e-\omega_r)$, is correctly given by (43):

$$2(\omega_e-\omega_r)=(\Phi\gamma^2\omega_r/4\Omega_n^3)[2\Omega_n T-\sin(2\Omega_n T)],$$

so that

$$\omega_r/\omega_e=\{1+(\Phi\gamma^2/8\Omega_n^3)[2\Omega_n T-\sin(2\Omega_n T)]\}^{-1}.$$

Apart from a change of notation, this is the same as (11). The dielectric model of the molecular beam gives as satisfactory an account of the frequency control as it does of the amplifying process. Only in the question of fluctuations does it fail.

The maser as frequency standard

A frequency standard is in essence an oscillator whose frequency is as little subject to perturbation as possible, whether from internal or external causes. A pendulum clock is limited in constancy of time-keeping by variations of pendulum length (temperature, wear of support), of g if it is moved to a different location, and of amplitude. Electrical oscillators such as maintained quartz crystal vibrators similarly suffer from temperature changes and secular drifts of the components in the maintaining circuit. On top of these effects, which are minimized by good design and maintenance, there is the intrinsic limitation of phase drift resulting from internal noise. It was noted in chapter 11 that a feedback oscillator suffered random walk in its phase, θ; after time t, when (11.44) is rewritten terms of $\Delta\theta$, the R.M.S. deviation, **363**

$$\Delta\theta = (t/2\pi I_\omega)^{\frac{1}{2}}/\omega_0 L q_0,$$

where I_ω is the spectral intensity of voltage fluctuations in the circuit. If these fluctuations are due to Johnson noise in the resistor, at temperature T_0, $I_\omega = 2Rk_BT_0/\pi$ from (6.49), and **92**

$$\Delta\theta = (k_BT_0t/E\tau_e)^{\frac{1}{2}} \quad \text{or} \quad \Delta\omega_0/\omega_0 = (k_BT_0/Pt)^{\frac{1}{2}}/Q_r, \tag{75}$$

in which E is the energy of oscillation, $q_0^2/2C$, P is the power dissipation, E/τ_e, and $\tau_e = L/R$.

To compare this with the maser we must first examine the relevance of (72) as an expression for the diffusion coefficient. This only enhances the dissipative contribution, as in (61), by a factor $F+1$, so that the equivalent noise temperature is $\frac{1}{2}(F+1)\hbar\omega_0/k_B$, still much less than room temperature unless F is very large indeed. In saying this it is taken for granted, in the light of the earlier discussion, that the buffering action of the molecules operates with quantum as with thermal noise. The conclusion is that thermal noise is the only source worth considering and that D, instead of being $1/4\tau_e$, would be $(2k_BT_0/\hbar\omega_0)/4\tau_e$, i.e. $k_BT_0/2\hbar Q_r$, in the absence of buffering. But since every step in the random walk due to noise is in practice reduced by a factor τ_e/T, and since D is proportional to the square of the step length, we have that

$$D = k_BT_0Q_r/2\hbar Q_m^2, \tag{76}$$

and from (74),

$$\Delta\omega_0/\omega_0 \sim (k_BT_0Q_r/2Q_m^2E\omega_0t)^{\frac{1}{2}} \sim (k_BT_0/2Pt)^{\frac{1}{2}}/Q_m. \tag{77}$$

The approximate signs are intended to show that no great care has been taken in relating T to Q_m, so that a numerical factor of the order of unity may well appear in a more meticulous treatment.

In the ammonia maser the residence time is about 10^{-4} s and Q_m about 10^7. A flux of 10^{13} molecules per second carries about 10^{-10} W, which is an upper limit for P; then if $T_0 = 300$ K,

$$\Delta\omega_0/\omega_0 \sim 5\times10^{-13}/t^{\frac{1}{2}} \ (t \text{ in seconds}).$$

In ten seconds the maser executes 2.4×10^{11} cycles and the uncertainty in frequency is about 2 parts in 10^{13}, equivalent to a phase uncertainty of 1/20 cycle, say 20°. This is very good indeed, but the hydrogen maser is still better.

The hydrogen maser[5] makes use of the transition in atomic hydrogen in which the electron and proton spins change their coupling. Both particles have spin $\frac{1}{2}$ and couple to give either a triplet state with parallel spins or a singlet with antiparallel, the former being higher in energy. The energy difference corresponds to a frequency of 1.42 GHz – the celebrated 21 cm line which has played a key role in galactic studies by radio astronomy. In a magnetic field, as fig. 4(a) shows, the triplet is split into Zeeman components; but it does not need a very strong field to break the coupling – the

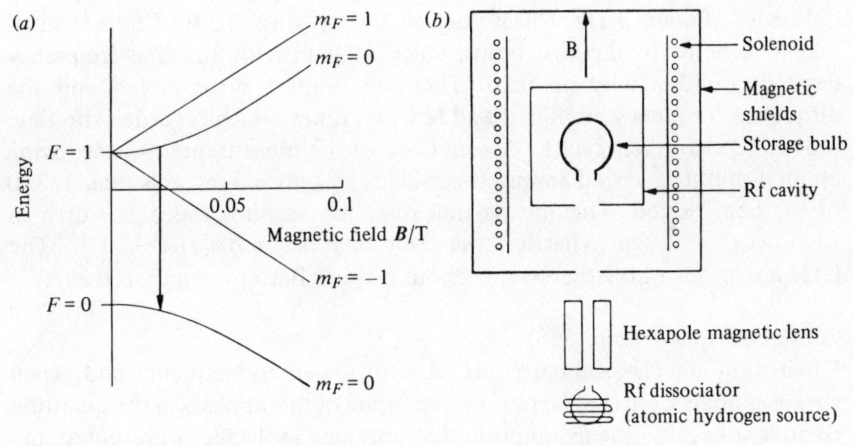

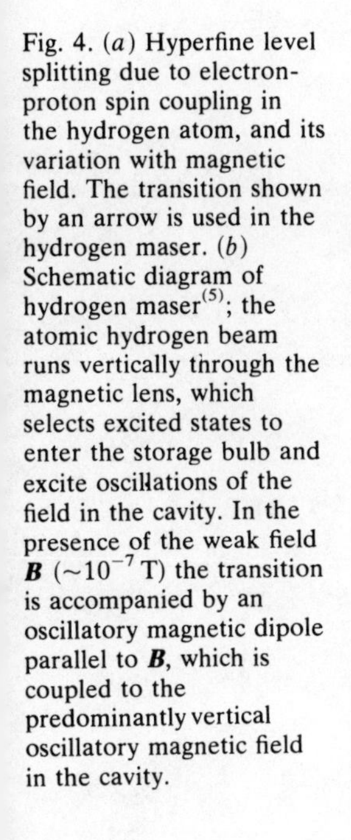
Fig. 4. (a) Hyperfine level splitting due to electron-proton spin coupling in the hydrogen atom, and its variation with magnetic field. The transition shown by an arrow is used in the hydrogen maser. (b) Schematic diagram of hydrogen maser[5]; the atomic hydrogen beam runs vertically through the magnetic lens, which selects excited states to enter the storage bulb and excite oscillations of the field in the cavity. In the presence of the weak field $\boldsymbol{B}$ ($\sim10^{-7}$ T) the transition is accompanied by an oscillatory magnetic dipole parallel to $\boldsymbol{B}$, which is coupled to the predominantly vertical oscillatory magnetic field in the cavity.

process is nearly complete at the right-hand side of the diagram. The details are unimportant for our discussion; what matters is that the upper two states ($F = 1$, $m_F = 0$ and 1), whose energy increases with the field strength, can be focussed by a magnetic lens (see fig. 4(b)) analogous to the electrostatic lens in the ammonia maser, while the lower two states are defocussed. In this way the transition shown can be used to stimulate oscillations of a resonant cavity by injecting a population that is largely in the excited state. What is special about the device is that the atomic beam does not pass straight through the cavity but into a silica bulb that is lined with teflon (polytetrafluoroethylene), in which the atoms bounce around until they manage to escape and are picked up by the vacuum pumps. Ramsey's intuition that such an arrangement could be made to work was brilliantly verified beyond all reasonable expectation – the coupling of the spins is so little perturbed by collision with the walls or with each other that the interaction between atoms and resonator field remains phase-coherent over very nearly one second, during which time an atom hits the walls several thousand times. Consequently Q_m may be as high as 5×10^9, perhaps 10^5 times the cavity Q. The obvious consequence is that temperature changes and other direct perturbations of the cavity affect the resonant frequency only 10^{-5} times as much as if the freely running cavity were used as a primary frequency standard. In this respect the maser has a great advantage over the quartz clock, but of course the atom or molecule which activates the maser must still be protected with the utmost care from disturbances to its environment. The very high Q_m also leads to high stability against noise; for a recent portable model[6] the theoretical estimate of $\Delta\omega_0/\omega_0$ is $4\times10^{-14}/t^{\frac{1}{2}}$ when t is measured in seconds, and this stability is realized over times ranging from 10 s to an hour. Extreme attention to detail in the design is needed to achieve this fine performance.

To appreciate the meaning of such a result it is helpful to consider a specific example. In an earlier test of the hydrogen maser principle, Vessot and Peters[7] constructed two masers which were tuned to a frequency difference of about 1 Hz. This is possible by applying slightly different small magnetic fields to the two bulbs, since at low fields the line frequency depends quadratically on field. The two outputs were mixed and the difference frequency amplified and fed to a timer which recorded the time (about 10 s) for ten cycles. A sequence of 19 measurements, occupying about 4 minutes, varied among themselves by only ±3 ms, less than 1/300 of the beat period. This means that over ten seconds the phase drift of each maser was somewhat less than one degree, or $\Delta\omega_0/\omega_0 \sim 10^{-13}$. The later model, mentioned above, is about 6 times better even than this.

Additional note (see p. 595)
Dissipation in a classical harmonic vibrator lowers its frequency and, when strong enough, destroys the periodicity. None of this appears in the quantum treatment because the assumption that only one molecule is present at any instant necessarily implies very weak dissipation. To achieve critical damping under these conditions γ would have to exceed ω_0.

21 The family of masers; from laser to travelling-wave oscillator

The ammonia and hydrogen masers are especially simple examples of maintained oscillators depending on the establishment of an *inverted population*. The general idea behind this expression can be understood by reference to an assembly of quantum systems in thermal equilibrium at some temperature T_0. The chance of finding any one system in a state of energy ε is given by the Boltzmann factor, $e^{-\varepsilon/k_B T_0}$, which is a monotonically decreasing function of ε. The radiation field due to the walls of the container, at T_0, passes through and interacts with the systems, but does not on the average disturb their equilibrium distribution between the energy levels. The upward transitions, stimulated by the radiation, from the more populous systems in lower-lying states are matched by the downward transitions from the less populous systems in higher states, which are enhanced in their activity since they emit spontaneously as well as by stimulation. If the distribution is modified by increasing the population of the higher levels at the expense of the lower, the coherent stimulated emission amplifies the radiation field more than stimulated absorption attenuates it; and the dominance of amplification is of course complete in the ammonia maser if only excited molecules are allowed to enter. This is an extreme example of an inverted population behaving as an active medium, that is, one that can amplify under appropriate conditions – in this case when the medium occupies a cavity resonator that is very closely tuned to the transition frequency of the molecules.

We have seen that the cavity plays only a minor role in controlling the frequency, the sharp molecular levels being far more effective for the purpose. If the active medium occupies a large volume, many wavelengths in extent, the problem of other dissipative mechanisms may vanish and any wave in the frequency bandwidth of the transition will be amplified as it passes through, by stimulated emission from the excited molecules. This is the most likely explanation of certain strong microwave emissions observed by radio astronomers, with brightness temperatures of the sources far exceeding that of the emitting gas cloud.(1) Such an excited gas cloud is intrinsically unstable, any spontaneously generated wave of the right frequency being amplified as it spreads out. Once the process has started, the amplitude soon rises above the ambient noise level, and the wave then provides the dominant mechanism for de-excitation, since the rate of de-excitation is proportional to the power level of the wave. By the same token any relatively strong signal entering the cloud from outside will assume the dominant role as it passes through. It is possible that the excited clouds are the product of a stellar explosion and that the expanding gas shell is not only maintained in an excited state by optical and ultra-violet radiation from the stellar remnant, but microwave radiation from the remnant may also be the primary source of the strongly amplified signal that ultimately emerges. The excitation will be effected by the higher energy photons, but only narrow bands of low energy radiation will be sufficiently well in tune with the relevant energy levels in the gas to be amplified.

There is no long-term coherence of the radiation produced in this way, since any primary source whose frequency lies in the bandwidth of the transition may be amplified; one expects to see an incoherent spectral line, possibly shifted and broadened by the Doppler effect in the expanding and turbulent gas cloud. If, however, the amplified wave after one passage through the cloud is fed back into it, it becomes the dominant source and will grow at the expense of other accidental sources, so long as the medium is replenished by whatever process is responsible for its excited state. Continuous cycling of the wave leads to coherence, though not necessarily to spectral purity. For a very strong wave may exhaust the medium so rapidly that only a short stretch behind the leading edge is amplified; in successive passages, then, the disturbance evolves into a short but intense pulse. This is a commonplace in high-power lasers[(2)] operating in the visible or near-visible spectral region, when Fabry–Perot interferometers or other optical resonators are used to cycle the radiation again and again through the active medium. For laser fusion machines fluxes of extraordinary intensity have been generated – 500 J in a pulse 500 ps long (15 cm in space), giving a flux of 10^{19} Wm^{-2} when focussed on to a target. This implies field strengths of 6×10^{10} Vm^{-1} and 200 T, and a radiation pressure not far short of 10^6 Atmospheres.†

In an optical laser the Fabry-Perot interferometer, or some other optical arrangement, replaces the resonant cavity of the maser. New problems arise from the very large number of resonant modes now available, some of them very closely spaced in frequency, so that the active medium may be able to support many. Even in low-power continuous lasers, where there is no sharpening of pulses by exhaustion of the medium, spectral purity is not automatic. This is but one aspect of laser design that will not be touched on here, and indeed the reader whose primary interest is in lasers should not seek enlightenment from the following pages, but from the multitude of specialist texts now available.[(3)] In the last chapter we examined the ammonia maser in some detail for its intrinsic interest as a soluble model of an important physical principle. At the end of the present chapter we shall look rather closely at a number of microwave devices and related

† It should be pointed out that in attempts at laser fusion by compression of pellets it is not the radiation pressure that does the work, but the still higher recoil pressure of the surface layers evaporated by the pulse.

processes whose analysis gives further substance to the argument, concerning the equivalence of classical and quantal treatments of the maser principle, which was emphasized in chapter 20. The following few pages are prescribed as an antidote to the sense of oppression which these slightly pedantic concerns may induce.

Mechanisms of population inversion

The direct separation by lens of excited from unexcited molecules is of very limited applicability. In this section we shall note some of the procedures for establishing an inverted population, with special reference to those that can be used for optical lasers. A large number of methods have been studied, with many specific examples of each put into practice, and an exhaustive catalogue will not be attempted. Instead we shall exhibit a few of the most important methods.

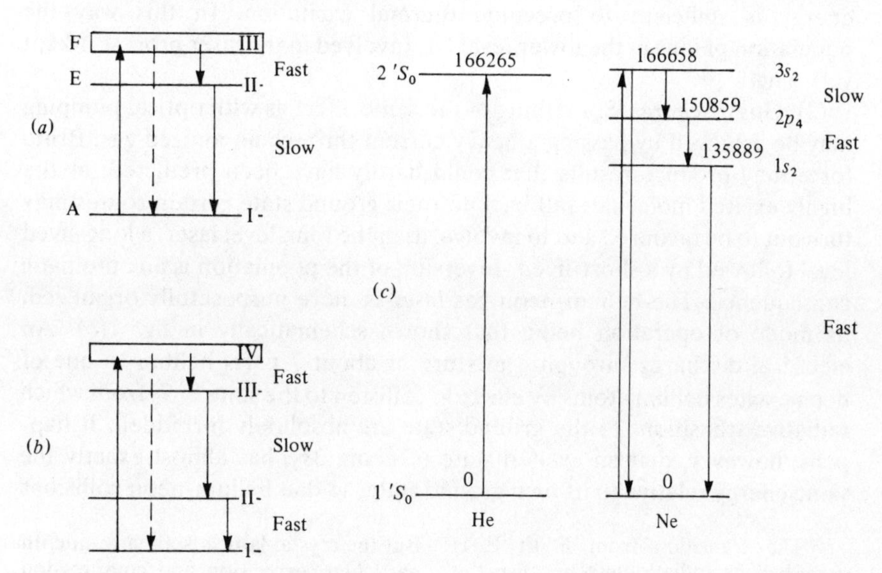

Fig. 1. (a) Schematic diagram of energy levels in a 3-level laser (ruby). (b) Energy levels for a 4-level laser (Nd in YAG). (c) Energy levels for the continuous He-Ne gas laser.

(1) *Optical pumping*. The active species is raised from its ground state, (I) in fig. 1(a), to an excited state (III) by irradiation at a higher quantum energy than will result from the laser action. The excited state must have a preference for decaying into a less highly excited state (II) from which spontaneous downward transitions are rare. The state (II) will then build up a substantial population; with strong enough pumping the ground state may be so depleted that its population falls below that of II and laser action becomes a possibility. Clearly a number of conditions must be satisfied, which may be summed up in the requirement that the active medium shall have a high fluorescence yield – the preferred route for de-excitation must involve an intermediate energy level. In ruby, the Cr^{3+} ions present as impurities have this property. The crystal absorbs green light to excite the

Cr^{3+} into the rather broad level labelled F in fig. 1(a); this is, however, not a very strong absorption and return to the ground state by re-radiation is much less likely than the emission of a number of phonons (quanta of lattice vibration) to attain the rather long-lived state labelled E. Intense pumping by a flash bulb can largely deplete the ground state and transfer a sizeable fraction of the ions to E in a time short compared with its lifetime of 3 ms. The mechanism for the fast radiationless† transition from F to E is somewhat obscure, but measurements of the number of red quanta (transitions from E to A) emitted for a given number of green quanta (A to F) absorbed leave no doubt about this route being highly preferred.

In contrast to the 3-level laser, severe depopulation of the ground state is not required for the 4-level laser (fig. 1(b)), of which Nd^{3+} in yttrium aluminium garnet or (more cheaply) a suitable silicate glass provide examples. The fluorescent level (III) here decays not to the ground state but to an excited state (II) which itself decays very rapidly, and whose energy is sufficient to preclude thermal excitation. In this way the population of ions in the lower level (II) involved in the laser process is kept very small.

(2) *Gas discharge.* Something of the same effect as with optical pumping may be achieved by passing a heavy current through an ionized gas. Brute force may produce results that could hardly have been predicted; as the highly excited molecules fall back to their ground state certain routes may turn out to be favoured and to involve, as in the four-level laser, a long-lived level followed by a short-lived. Inversion of the population is an automatic consequence. The helium-neon gas laser is more purposefully organized, its mode of operation being that shown schematically in fig. 1(c). An electrical discharge through a mixture of about 7 parts helium to one of neon excites helium atoms by electron collision to the state 2^1S_0 from which radiative transitions to the ground state are absolutely forbidden. It happens, however, that an excited state of neon, $3s_2$, has almost exactly the same energy relative to its own ground state, so that helium-neon collisions

† The transition from F to E is described as radiationless because the energy is emitted as lattice vibrations (phonons) rather than as a photon. Since the separation of the levels is many times the energy of any phonon that the lattice can support, a number of phonons must be emitted simultaneously. The probable mechanism may be visualized in quasi-classical terms by imagining it to have started, so that a Cr^{3+} ion is in a mixture of F and E states, with its charge distribution and the local surroundings vibrating rapidly. The frequency of vibration is, we know, much too high to generate an acoustic wave and so radiate a single phonon. But the crystal lattice is anharmonic. In particular, expansion and compression are not equivalent, and a process analogous to rectification takes place, the same process as is responsible for thermal expansion. In this case the vibrational motion around the ion has the effect of increasing the local pressure and initiating a shock wave. It is the shock wave that very quickly carries the energy away, and as there is no necessary limit to the intensity of shock, so there is no limit to the number of low frequency phonons that it can ultimately decompose into; the time taken for this last process is of no importance to the ion.

have a high cross-section for yielding this state of neon by exchange of excitation energy. From then on the neon behaves like a 4-level laser, making a transition fairly readily (10^{-7} s) to a lower level that decays considerably faster (2×10^{-8} s). It is the $3s_2$–$2p_4$ population that is inverted by this process. It should be noted that the $3s_2$ state is perfectly able to fall back to the ground state as well as to $2p_4$, but that trapping of the resonance radiation prevents depopulation of the excited state by this means. In this process the radiation emitted is reabsorbed by an unexcited neon atom which is thus raised to $3s_2$ to replace the originally excited atom. If there were no Doppler shift, the cross-section for resonant reabsorption would be $\frac{1}{2}\pi/k^2$, k being the wavenumber of the radiation; and with n neon atoms per unit volume the mean free path of the photon would be $2k^2/\pi n$. Typically, at a pressure of 0.1 torr the mean free path would be 1 μm, but the Doppler shift increases this in the ratio of Doppler to natural line width, since only those atoms whose Doppler shift does not take them out of resonance will be effective absorbers. Even so, a photon is likely to suffer many processes of absorption and re-emission before escaping from the discharge, and during this time the atom that is temporarily excited may promote laser action.

203

(3) *Chemical excitation.* When hydrogen and fluorine combine to form hydrogen fluoride the first state of the molecule is one of high excitation. Left to itself the molecule loses energy by collision with others and the resulting rise of temperature is interpreted as the heat of reaction. But if the de-excited molecules are swept away the steady production of excited molecules may generate and maintain an inverted population. A laser can be operated in a steady state condition with continuous gas flow, H_2 and F_2 being admitted to the mixing chamber and HF extracted. The chemical process proceeds by the collision of fluorine atoms with hydrogen molecules, or hydrogen atoms with fluorine molecules, the two reactions forming a self-perpetuating sequence:

$$F+H_2\rightarrow HF^*+H,$$

followed by

$$H+F_2\rightarrow HF^*+F.$$

The stars indicate excited states. A little nitric oxide added to the hydrogen ensures the production of fluorine atoms when the streams mix: $NO+F_2\rightarrow NOF+F$. The interaction between, for example, F and H_2 does not need anything like a head-on collision, and begins when the two are some distance apart, one of the hydrogens being attracted to the fluorine atom; thus HF and H may separate with an interatomic spacing in the HF considerably larger than the equilibrium separation. The molecule is left in a high vibrational state, and decays with a strong oscillatory dipole moment at the vibrational frequency which lies in the infra-red. It is this frequency which is coherently stimulated in the laser. For all its simplicity, however, the extreme chemical reactivity of the components makes this system unattractive.

No problem of removal of the de-excited molecule arises when excimers are used. There is no stable compound of krypton and fluorine, but KrF forms a bound molecule (excimer) in certain excited states. When an electrical discharge is passed through a mixture of krypton and fluorine, $Kr^* + F_2 \rightarrow KrF^*$, and the excimer decays very rapidly with the emission of a rather broad line in the ultra-violet, around 250 nm. The excimer can therefore be used to operate an ultra-violet laser which is tunable over the line-width. The excimer Xe_2 produces an even higher energy photon, at 170–176 nm; at this quantum energy the spontaneous de-excitation is so fast that the gas must be compressed to about 10 Atmospheres and very vigorously excited to create a sufficient population of excimers. The power dissipated by spontaneous emission is correspondingly high and demands excitation by a very intense (about 1 MW) electron beam. At such a level only pulsed operation is feasible, since most of the injected energy goes into heating the gas rather than into the coherent ultra-violet output. Nevertheless, very strong pulses of ultra-violet can be achieved.

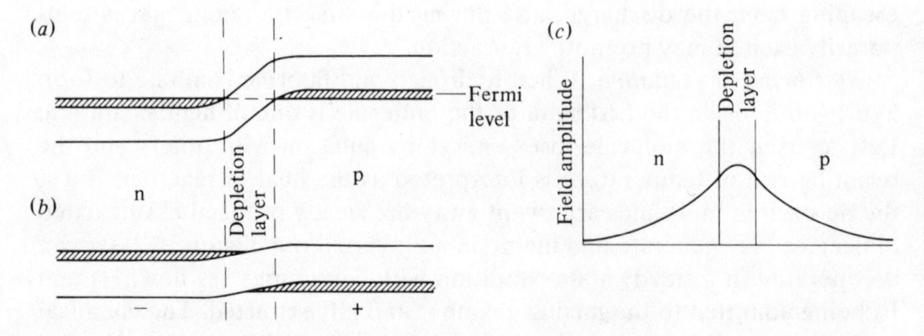

Fig. 2. (*a*) Impurity bands (shown shaded) of electrons in the conduction band of the n-type, and holes in the valence band of the p-type GaAs, separated by a depletion layer. In equilibrium the Fermi level is the same throughout the material. (*b*) A potential difference, as indicated by the + and − signs, allows electrons in the conduction band to fall into the valence band in the depletion layer. (*c*) Electrons and holes lower the dielectric constant except in the depletion layer. The amplitude profile shown speeds up the wave in the depletion layer and slows it down elsewhere, thus allowing a guided wave to travel with constant profile along the interface between n and p material.

(4) *The semiconductor laser.* Fig. 2(*a*) shows schematically the energy level structure of a p–n junction in a semiconductor such as GaAs. On the left there is a high enough doping level of tellurium donors for them to form a band of filled states just below the top edge of the energy gap, while on the right zinc atoms (acceptors) have been diffused in to overwhelm the donors so that the resulting p-type material has a band of vacant states just below the bottom edge of the gap. Without any external bias there is an electric field in the depletion layer between n and p, the Fermi level being the same on both sides. Application of a forward biasing voltage to reduce this field leads to the situation in fig. 2(*b*), and there is now a narrow region in which electrons in the conduction band coexist with vacancies in the valence band. This provides the population inversion, and de-excitation can occur by electrons falling into the vacancies and emitting a photon of wavelength 840 nm, in the near infra-red. The transition is rapid ($\sim 10^{-9}$ s) because in GaAs, unlike many other semiconductors, there is no need for a phonon to be emitted along with the photon to conserve crystal momentum. The material is transparent to light of the wavelength emitted, with a refractive index that is slightly lower in the p and n regions than in the

narrow active lamina between them. This allows confinement of the radiation in the vicinity of the lamina, as shown in the diagram; a wave travelling unchanged in form at a velocity intermediate between that of a plane wave in the lamina and that in the exterior regions must have a convex amplitude profile in the lamina to make it go faster, and a concave profile outside to make it go slower. This is the mode that can be caused to resonate by reflection to and fro between polished ends of the material, and to be amplified by stimulated transitions of electrons downwards into vacant states.

(5) *Surf-riding and bunching.* By these terms we describe a class of inversion procedures which operate through the interaction of a travelling wave with a stream of particles moving slightly faster than the wave, and being caused to slow down and thereby to transmit some of their energy to the wave. The amplification of acoustic waves, especially in conducting piezo-electric crystals, by passing a current exemplifies this process, as does the travelling-wave tube that is used to generate microwaves of millimetre wavelength. Such devices are normally analysed in a strictly classical manner, but their operation may equally well be considered as resulting from population inversion and stimulated emission. They therefore provide examples to link classical and quantal treatments, and for this reason deserve the fuller treatment to which we now turn.

Classical perspective on maser processes

572

The discussion of stimulated emission in chapter 20 showed that it can be considered as giving rise to an active dielectric medium such as, in a capacitor, would reveal itself as a negative resistance. To this extent the maser has an obvious resemblance to a negative-resistance oscillator and the parallelism was developed further in the analysis of fluctuations of amplitude and phase. One might attempt to reverse the process of thought and apply quantum mechanics to the oscillatory circuit, seeking a mechanism that will coherently refresh the circuit by providing photons to replace those dissipated by lossy processes; but a difficulty arises. In a typical maser the inverted population is relatively long-lived in the absence of stimulated transitions, spontaneous transitions being rather rare; on the other hand, in a negative-resistance circuit (or tunnel diode) there is steady dissipation at zero frequency and the stimulated transitions are commonly a rather minor perturbation of this continuing attempt to achieve equilibrium. Consequently, although it would probably not be out of the question to cast the theory of the oscillator into a form resembling that of a maser, it would amount to little more than an intellectual curiosity.

600

Electron–cyclotron oscillator

The same criticism does not apply to a class of oscillators exemplified by the klystron, the magnetron and the travelling-wave tube,[4] which can be understood in general terms from classical and quantal points of view with equal ease. For the purpose of illustration we shall examine the excitation of a microwave cavity by electrons executing cyclotron motion in a powerful magnetic field. Oscillators employing this mechanism have been realized,[5] as laboratory models rather than marketable devices, but we shall be little concerned with technical details, only with physical principles. The essential features are modelled by an electron cloud in two dimensions; each particle moves independently in a circular orbit in a plane normal to the magnetic field, $\boldsymbol{B}$, and is acted upon by a uniform electric field $\mathscr{E}$ rotating in the same sense as the cyclotron motion, with frequency ω. In practice the electrons, continuously injected with only a slight spread of energy around a mean of something like 20 kV, drift along $\boldsymbol{B}$ in helical paths, spending long enough in the resonant cavity, under the influence of $\mathscr{E}$, to execute many orbits. For mathematical convenience, on a point of no great physical significance, we shall assume an exponential distribution of residence times, with a probability $e^{-t/T}$ that an electron experiences $\mathscr{E}$ for a longer time than T.

237

Let us now calculate the impulse response function for the electrons distributed uniformly on the circular orbit of fig. 3 and consequently

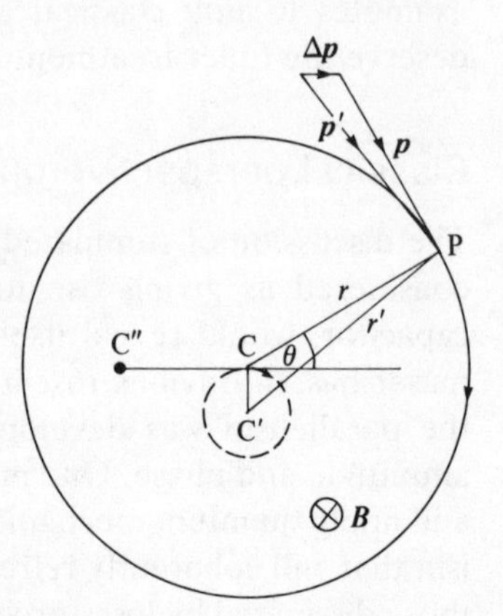

Fig. 3. Illustrating calculation of impulse response function for an assembly of electrons undergoing cyclotron motion.

exhibiting no resultant oscillatory dipole moment. The radius of the orbit, r, is p/eB for an electron having momentum $\boldsymbol{p}$, and this holds for relativistic as well as slow electrons. A typical electron, lying at P at the instant of the impulse (applied horizontally in the diagram), acquires momentum $\Delta\boldsymbol{p}$ from the impulse, so that $\boldsymbol{p}$ is changed to $\boldsymbol{p}'$ while the electron remains instantaneously at P. Its new orbit must be tangential to $\boldsymbol{p}'$ and have radius r' such that $r'/r = p'/p$; it follows by simple geometry that C′ lies vertically at a distance $\Delta p/eB$ below C, irrespective of the position of P. Hence the ring of electrons now begins to spin bodily about C, and the subsequent oscillations of dipole moment are as if the whole charge were initially concentrated at C and then set spinning at the cyclotron frequency in the orbit shown as a broken circle. In the absence of dissipation this impulse response leads to a loss-free contribution to the susceptibility, but decay

237 as $e^{-t/T}$ introduces a lossy component, exactly as in chapter 8. In reaching this conclusion, however, we have assumed that all electrons have the same cyclotron frequency, $\omega_c = eB/m$, and have ignored the relativistic change of frequency with momentum. The change, small though it may be for these electrons of rather low energy, is nevertheless an essential feature of the device since it is only through its presence that the electrons may impart energy rather than absorb it.

When the impulse is applied, electrons near the top of the orbit in the diagram have their momentum increased, and their cyclotron frequency decreased, while those at the bottom have their momentum decreased and their frequency increased. If one were to observe the motion from a frame rotating at the mean frequency one would see electrons at the top and the bottom slowly drifting towards the left-hand side of the orbit, bunching together at the left and leaving a deficiency on the right. The centroid of the electrons would thus be seen to move to the left from C at an easily calculated rate. An electron at P has its momentum increased in magnitude by $\Delta p \sin\theta$, and its frequency by $\omega_c' \Delta p \sin\theta$, where $\omega_c' = d\omega_c/dp$, and is negative. This is its angular velocity as seen in the rotating frame, and the horizontal component of its velocity is $r\omega_c' \Delta p \sin^2\theta$, with a mean value for the whole ring of $\frac{1}{2} r\omega_c' \Delta p$. After time t, therefore, the impulse response is represented not by the vector C′C, but by C′C″, where $CC'' = \frac{1}{2} r\omega_c' t \Delta p$; at the same time, in the laboratory frame the vector C′C″ has rotated clockwise through $\omega_c t$. It should be noted that for a small impulse CC′ is correspondingly small, and that in time CC″ may well grow to a considerably greater magnitude without getting anywhere near the orbit radius, so that the linear approximation is not invalidated.

The position of the centroid after time t, expressed as a complex number, determines the impulse response function. As a result of an impulsive unit electric field, $\mathscr{E} = \delta(t)$, the momentum change $\Delta p = e$, so that $CC' = 1/B$ and $CC'' = \frac{1}{2} r\omega_c' e t = \omega_c' p t / 2B$; if there are N electrons in the ring the response of the dipole moment takes the form:

$$h(t) = (Ne/B)(i + \tfrac{1}{2}\omega_c' p t)\, e^{-i\omega_c t - t/T}; \qquad t > 0. \tag{1}$$

By (5.4), the susceptibility follows:

$$\chi(\omega) = \int_0^\infty h(t)\, e^{i\omega t}\, dt = \frac{NeT}{B(1 - i\delta T)}[i + \tfrac{1}{2}\omega_c' p T/(1 - i\delta T)], \tag{2}$$

in which $\delta = \omega - \omega_c$. Gain or loss is controlled by the imaginary part of χ,

$$\chi''(\omega) = \frac{NeT}{B(1 + \delta^2 T^2)}[1 + \omega_c' p \delta T^2/(1 + \delta^2 T^2)]. \tag{3}$$

The first, non-relativistic, term is always lossy, but the loss may be converted to gain by the relativistic bunching provided the second term is large enough and negative; δ must be positive, since ω_c' is negative, and the optimum tuning is such as to make $\delta T = \frac{1}{2}$. The residence time T must be long enough for $|\frac{1}{2}p\omega_c' T| > 1$. Now $p = mv/(1-v^2/c^2)^{\frac{1}{2}}$ and $\omega_c = \omega_{c0}(1-v^2/c^2)^{\frac{1}{2}}$, where m is the rest mass of the electron and $\omega_{c0} = eB/m$, from which it follows that $\omega_c' = -\omega_c v/E$, in which E is the total energy of the electron, including its rest energy mc^2. The criterion for amplification is that $\omega_c T > 2E/pv$. For electrons of only modest energy, $2E/pv$ is roughly the ratio of rest energy to kinetic energy, i.e. about 25 for electrons of 20 keV energy. Amplification should become possible, then, when $\omega_c T > 25$ and the residence time is long enough for more than four orbits to be accomplished; this presents no technical problem. When $\omega_c T$ is much greater than this critical value the tuning range in which amplification occurs runs from just above ω_c to $\omega_c + |\omega_c' p|$, though the factor outside the brackets in (3) limits the usefulness of the process when δ is large. It is clear, however, that elaborate precautions are not necessary to ensure uniformity of energy for the electrons, provided that they are used to excite a cavity tuned to a frequency slightly above the highest cyclotron frequency present, that of the least energetic electrons.

The quantal analysis of the same system starts conveniently with a single electron in its nth energy state, n being very large. We know that the behaviour closely corresponds to that of a harmonic oscillator, with the oscillator strength f equal to $n+1$ for upward and $-n$ for downward transitions to neighbouring levels, the only allowed processes. Because of the finite residence time the levels are broadened, with Lorentzian profiles if the distribution of lifetimes is exponential. When the electron is irradiated at such a frequency as will stimulate downward transitions, upward transitions will also take place unless the level spacings are different. In the non-relativistic case of equal spacings the larger f-number for upward transitions will in fact ensure that stimulated absorption outweighs stimulated emission, and the system will be lossy. It is only when relativistic effects cause the level spacing to decrease at higher energies that an irradiating frequency can be found, higher than ω_c, which will sufficiently favour the downward transitions that the net result is gain rather than loss. If the electrons had a Boltzmann energy distribution it would be necessary for the frequency to be higher than ω_c for the slowest, and the majority of processes would then involve just those electrons for which the relativistic advantage is least. By using a fairly homogeneous beam of slightly relativistic electrons all are enabled to contribute more or less equally to the gain.

512

To make the argument quantitative we note, from (13.39), that the impulse response function is expressed as the difference between two terms, but it must now be remembered that the frequency ω_0 is not quite the same in each term, being $\omega_c(n)$ for transitions between the levels $n+1$ and n. Thus we should write for a linear oscillator in a high quantum state, unlimited by finite resident time:

442

$$h(t) = \frac{\mathrm{d}}{\mathrm{d}n}[(ne^2/m\omega_c)\sin\omega_c t] = \frac{\mathrm{d}}{\mathrm{d}n}[(ne/B)\sin\omega_c t],$$

a factor e^2 being supplied so that $h(t)$ is the dipole response to unit impulsive electric field. The orbiting electron is represented by a two-dimensional oscillator, and the appropriate modification allows $h(t)$ to rotate and be expressed as a complex quantity:

$$h(t) = \frac{\mathrm{d}}{\mathrm{d}n}[(\mathrm{i}ne/B)\,\mathrm{e}^{-\mathrm{i}\omega_c t}] = (e/B)[\mathrm{i} + nt\,\mathrm{d}\omega_c/\mathrm{d}n]\,\mathrm{e}^{-\mathrm{i}\omega_c t}. \quad (4)$$

We now multiply by $N\,\mathrm{e}^{-t/T}$ to allow for the number of electrons and their residence times; further we note that, since $n \propto p^2$, $n\,\mathrm{d}\omega_c/\mathrm{d}n = \frac{1}{2}p\,\mathrm{d}\omega_c/\mathrm{d}p = \frac{1}{2}p\omega_c' = \frac{1}{2}reB\omega_c'$. With these changes, (4) becomes identical to (1).

After all that has gone before it will not come as a surprise that the classical and quantal treatments agree so perfectly, since we are dealing in effect with a harmonic oscillator where this is to be expected. What is instructive is the demonstration that the process which preferentially encourages stimulated emission (in this case the unequal level spacing) is the same as induces bunching and permits the systematic development of a polarized state of the assembly, capable of energy transfer to the applied field. It is, of course, equally true that with a lower irradiation frequency, when δ in (3) is negative, the bunching process leads in both treatments to an enhanced absorption. Bunching is the classical expression of a mechanism of energy transfer that can be arranged to favour either absorption or emission. The conditions under which emission is favoured turn out to be such as to maximize the transfer rate at a point in a distribution where the population is inverted.

From the point of view of quantum mechanics there is little difference in principle between this system and a hydrogen atom which, under the stimulus of an oscillatory electric field, develops an oscillatory charge concentration by which energy transfer is effected. Unlike the hydrogen atom, however, the extended systems discussed here are equally amenable to classical analysis.

Bunching and maser action

A very similar bunching action can be seen as the classical mechanism by which acoustic waves in certain solids are absorbed, or in the right circumstances amplified, by interaction with conduction electrons;[6] and here too the process when viewed quantally is closely analogous to maser action. It is most marked with longitudinal compressional waves in piezo-electric semiconductors, such as cadmium sulphide, CdS, where the rarefactions and compressions give rise to an electric field pattern, parallel to the propagation direction and with sinusoidally varying amplitude. The wave

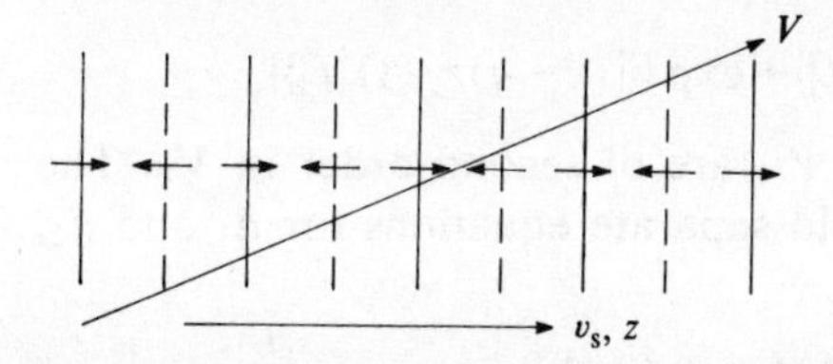

Fig. 4. Crests (full lines) and troughs (broken lines) of a compressional wave in a piezo-electric semiconductor and the resulting electric field pattern (short arrows); the path of an electron through this field pattern is also shown.

in the material lattice is thus accompanied by a wave of electrostatic potential, $V_0 \cos(qz - \omega t)$ if propagation is along the z-direction, as shown in fig. 4. Any electrons, excited thermally or optically, may interact with this potential and extract energy from, or transfer it to, the acoustic wave; but if the free path of the electrons between collisions is much longer than the acoustic wavelength, strong interaction is confined to those electrons whose velocity component v_z matches the sound velocity v_s, so that they move in a nearly constant field. This is obvious from a classical point of view, since any electron that travels through alternate crests and troughs is alternately accelerated and decelerated, with only small excursions of its energy from the mean. By contrast, for those electrons having v_z equal to v_s some are continuously accelerated and others continuously decelerated, and there is a real possibility of substantial energy transfer. The same condition arises naturally if one considers the wave as a quantized entity able to transfer energy in quanta of $\hbar\omega$ and momentum in quanta of $\hbar q$, provided these transfers are compatible with the conservation laws. When an electron receives a small increment, $\Delta \boldsymbol{p}$, of momentum its kinetic energy changes by $\boldsymbol{p}\cdot\Delta\boldsymbol{p}/m$, i.e. $\hbar q p_z/m$ if $\Delta\boldsymbol{p} = (0, 0, \hbar q)$. Equating the energy change to $\hbar\omega$ yields the required result that p_z/m, which is v_z, must equal ω/q, which is v_s.

To illustrate the analogy between the classical bunching process and the maser principle it is enough to consider a simple one-dimensional model in which an electron in the state $\Psi_0 = e^{i(kz-\Omega_k t)}$, where $\Omega_k = \hbar k^2/2m$, is perturbed by a potential, $V_0 \cos(qz - \omega t)$, switched on at $t = 0$. It is convenient to work in the frame in which the wave is at rest and the electron is moving at velocity $v = v_z - v_s$, having $k = mv/\hbar$. If $v > 0$, as assumed in the calculation, the electron runs ahead of the wave in the laboratory frame, this being the condition for wave amplification. We concern ourselves with interactions so weak that the solution need contain only terms linear in V_0. The time-dependent Schrödinger equation:

$$(\hbar^2/2m)\partial^2\Psi/\partial z^2 - eV_0\Psi\cos qz = -i\hbar\partial\Psi/\partial t, \tag{5}$$

has in this approximation the solution

$$\Psi_1 = \exp\{i(kz - \Omega_k t)\} + a_1(t)\exp\{i[(k+q)z - \Omega_{k+q}t]\} + a_2(t)\exp\{i[(k-q)z - \Omega_{k-q}t]\}. \tag{6}$$

Only at higher orders of V_0 are terms with wavenumber other than $k \pm q$ excited. When (6) is substituted in (5), and $\cos qz$ expressed as $\frac{1}{2}(e^{iqz} + e^{-iqz})$,

$$\begin{aligned}\dot{a}_1 \exp\{i[(k+q)z - \Omega_{k+q}t]\} + \dot{a}_2\exp\{i[(k-q)z - \Omega_{k-q}t]\} \\ = (eV_0/2i\hbar)(e^{iqz} + e^{-iqz})\Psi_1 \\ \sim (eV_0/2i\hbar)\{\exp\{i[(k+q)z - \Omega_k t]\} + \exp\{i[(k-q)z - \Omega_k t]\}\},\end{aligned}$$

since the terms arising from $a_{1,2}$ in Ψ_1, are of second order in V_0. The terms in $(k+q)z$ and in $(k-q)z$ yield separate equations for $\dot{a}_1$ and $\dot{a}_2$, from which, since $a_{1,2} = 0$ when $t = 0$,

$$a_1 = \alpha_1\{1 - \exp[i\hbar(2kq + q^2)t/2m]\}$$

and

$$a_2 = \alpha_2\{1 - \exp[-i\hbar(2kq - q^2)t/2m]\}, \tag{7}$$

where $\alpha_1 = meV_0/[\hbar^2(2kq+q^2)]$ and $\alpha_2 = -meV_0/[\hbar^2(2kq-q^2)]$. Hence from (6),

$$\Psi_1 = \Psi_0\{1+\alpha_1 e^{iqz}[e^{-i\hbar(2kq+q^2)t/2m}-1]+\alpha_2\, e^{-iqz}[e^{i\hbar(2kq-q^2)t/2m}-1]\}. \tag{8}$$

If we take $\Psi_0^*\Psi_0$ as defining the initially uniform electron density, $\bar{\rho}$, its development follows from (8):

$$\rho(t)/\bar{\rho} = \Psi_1\Psi_1^* \sim 1+2(\alpha_1+\alpha_2)[\cos q(z-vt)\cos(\hbar q^2 t/2m)-\cos qz] + 2(\alpha_1-\alpha_2)\sin q(z-vt)\sin(\hbar q^2 t/2m). \tag{9}$$

This expression shows how the electron stream develops a periodic bunching which increases with time and drifts along, relative to the potential, at the mean electron velocity v. The classical limit, which is reached by letting either $\hbar$ or q/k tend to zero, follows from (9) on substituting for α_1 and α_2:

$$2(\alpha_1+\alpha_2) \sim -eV_0/mv^2$$

and

$$2(\alpha_1-\alpha_2) \sim 2meV_0/\hbar^2 kq.$$

Hence

$$[\rho(t)/\bar{\rho}]_{\mathrm{Cl}} \sim 1-\frac{eV_0}{mv^2}[\cos q(z-vt)-\cos qz - qvt\sin q(z-vt)]. \tag{10}$$

This result is the same as is obtained by a classical calculation of the effect of the potential on a uniform stream of particles initially all moving with the same speed. The last term, with its coefficient proportional to t, grows without limit; it has its origin in the fact that when the potential is switched on, electrons at different points have the same kinetic energy but different potential energies, and therefore different total energies. As they migrate over the hills and valleys of potential their mean velocities are correspondingly different, so that in addition to the periodic ripple in ρ caused by electrons spending different times near crests and troughs of potential, there is a systematic growth in the bunching process as the faster electrons catch up the slower. This process, which will reappear in our discussion of the klystron, does not go on indefinitely, for as the faster electrons actually overtake the slower the bunches diffuse out again; but this is only apparent when higher orders of V_0 are included in the calculation. The smaller V_0 is, the longer the time before non-linear effects intervene.

Rearrangement of (10) shows more clearly how the peaks of ρ grow and shift relative to the potential:

$$[\rho(t)/\bar{\rho}]_{\mathrm{Cl}} \sim 1-(eV_0/mv^2)(4\sin^2\tfrac{1}{2}X+X^2-2X\sin X)^{\frac{1}{2}}\sin(qz-X-\theta), \tag{11}$$

where $X = qvt$ and $\tan\theta = 2\sin^2\frac{1}{2}X/(X-\sin X)$. The amplitude and phase are shown in figs 5(a) and (b). Initially the bunching of ρ rises as t^2 and is strongest where $qz = \pi$, 3π etc., i.e. at the potential minima, as would

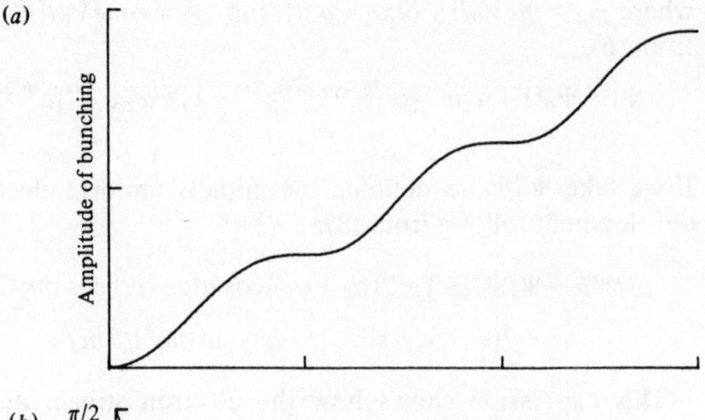

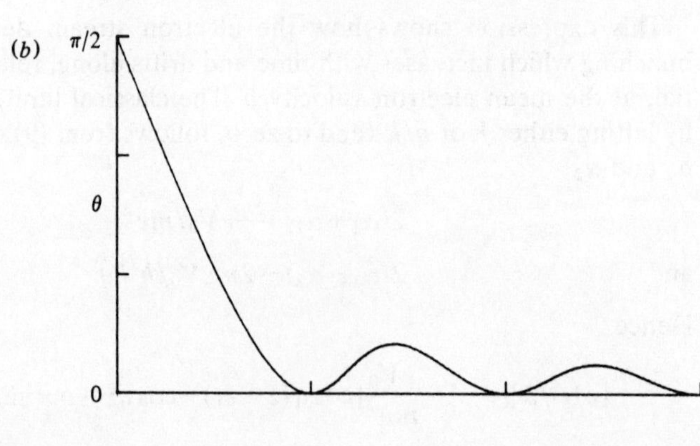

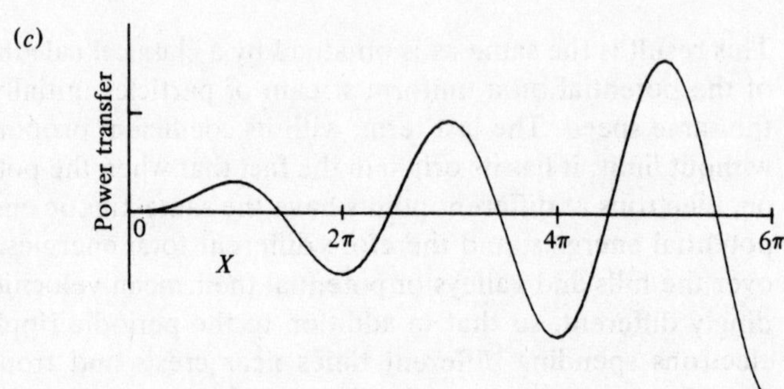

Fig. 5. Bunching of an initially uniform electron beam travelling slightly faster than a longitudinal wave field: (a) amplitude and (b) phase of density fluctuations at a distance X from the point of entry of the uniform beam; (c) rate of transfer of energy from electrons to wave.

be expected from elementary dynamical arguments. As time proceeds, the linear growth dominates and, as θ approximates to zero, the centres of the bunches move with those electrons that have the mean velocity v, having been at the outset at points of zero potential. Half these points act as foci of electron bunches, where those behind overtake and those ahead fall back; the other half conversely act as foci of rarefactions.

The rate of energy transfer, P, from the electrons to the wave is measured by the mean retarding force exerted by the fluctuating potential:

$$P \propto \langle -e\rho(\mathrm{d}V/\mathrm{d}z)\rangle \propto qe^2 V_0 \langle \rho \sin qz \rangle,$$

and is therefore determined by the magnitude and sign of the $\sin qz$

component of ρ. This follows immediately from (10):

$$P \propto \sin X - X \cos X, \tag{12}$$

and is shown in fig. 5(c). Obviously the oscillations of the curve reflect the migration of the bunches through alternate regions of retarding and accelerating force. In the initial stages the bunches, starting at potential minima, move forward to experience a retarding force, but the amplifying power of the electron beam is exhausted by the time $X = \pi$; if the electron velocity exceeds that of the wave by v, the useful time of interaction is not more than π/qv. This accords with the Uncertainty Principle, as the following argument shows, which holds in any frame of reference. For a real transition to occur there must be conservation of wavenumber and energy which, as we have seen, requires when $q \ll mv_z/\hbar$ that the electronic and acoustic velocities match. Now the wavenumber of the potential is well defined, but for an interaction time t there is an uncertainty in frequency, $\Delta\omega$, of roughly $1/t$; the perturbation experienced by the electron is therefore resolvable into Fourier components with velocities, ω/q, that spread over a range of $\Delta\omega/q$, roughly $1/qt$. Hence for a component to be present that satisfies the conservation rules, $1/qt$ must be at least comparable with v, the discrepancy in velocity, and the expected result follows: t must not be longer than something like $1/qv$. In oscillators that depend on an electron beam interacting with a wave the physical size of the device limits the interaction time, while with acoustic waves in solids it is the scattering of the electrons by defects that produces an equivalent effect, in this case more by rendering the electronic wavenumber slightly uncertain.

The argument has been presented for an electron travelling faster than the wave, so that the bunching runs ahead into a region of retarding force. Conversely, an electron moving more slowly than the wave lags behind and finds itself in a region of accelerating force; such an electron attenuates, rather than amplifies, the wave. Looked at from a quantum point of view the relative probabilities of exciting electronic states $k \pm q$ are governed by α_1^2 and α_2^2, i.e. $(2kq \pm q^2)^{-2}$ if $k = mv/\hbar$. When the relative velocity v is positive, k is positive and $\alpha_1^2 < \alpha_2^2$, the electron being more likely to lose than to gain energy; but the opposite is the case when $v < 0$. The difference between α_1 and α_2 becomes very marked when q, instead of being much less than k, is comparable. In particular, when $k = \frac{1}{2}q$, α_2 becomes infinite. An observer moving with the wave sees the electron suffer Bragg reflection. In the laboratory frame also the electron is reflected off the moving lattice formed by the wave, and it suffers an energy loss as a result of the Doppler shift of its frequency; correspondingly the wave benefits from the recoil. As we approach the classical limit the relative velocity v required for Bragg reflection goes to zero, and in any practical case $|k|$ may be taken as much greater than q. Then the probability of a downward transition exceeds that of an upward transition by $\alpha_2^2 - \alpha_1^2$, which now is proportional to $1/k^3$, i.e. $1/v^3$, becoming very strong as v approaches zero and changing rapidly from an excess of downward to an excess of upward excitations. If the wave is interacting with a stream of electrons of different velocities, it will be amplified only if the distribution of velocities provides more moving slightly faster than are moving slightly slower – in that range of distribution where $v_z \sim v_s$, the population must vary in the opposite way to a thermally equilibrated population, with more at higher energy than at lower. From the quantum viewpoint this is to ensure more downward than upward transitions; from the classical to ensure that the bunching is preferentially disposed towards retarding regions of the wave.

Having established once more the correspondence between bunching and maser action, let us note a few examples of waves interacting with electron streams, starting with the example that introduced the discussion.

Acoustic attenuation and amplification in solids[6]

In cadmium sulphide the electrons that are to interact with acoustic waves can be generated by the internal photoelectric effect, bound electrons in the valence band being excited optically into the conduction band where they soon become thermalized, with a Maxwell–Boltzmann distribution of velocities. The R.M.S. velocity at room temperature is something like $10^5\ \mathrm{ms^{-1}}$, perhaps 50 times the acoustic velocity. Let us make the (unfortunately unrealistic) assumption that the mean free path of the electrons is very long; then the only electrons capable of exchanging energy with a given acoustic wave are those moving so as to keep pace with the wave. A wave moving from left to right in fig. 6 can exchange energy with electrons whose $\boldsymbol{k}$-vector lies near the interaction surface, a plane normal to the wave-velocity and displaced $mv_s/\hbar$ from the origin. Since the normal distribution of $\boldsymbol{k}$-values, as in fig. 6(*a*), has its greatest density at the origin, falling off steadily as k increases, there is a greater density to the left of the surface than to the right, and hence more processes involve absorption

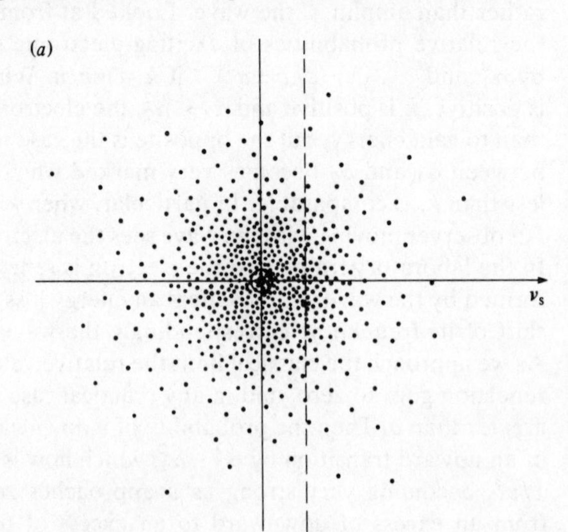

Fig. 6. (*a*) The dots represent a Maxwell–Boltzmann distribution of velocities (or $\boldsymbol{k}$) among the electrons; the broken line is the interaction surface on which the electrons stay on a wave-front of the acoustic wave. (*b*) An electric field has shifted the $\boldsymbol{k}$-distribution so that its centroid lies to the right of the interaction surface. (*c*) Attenuation in CdS as a function of longitudinal electric field strength[7]. At the left, where $\mathscr{E} > 700\ \mathrm{Vcm^{-1}}$, the wave is amplified.

Fig. 6 (*cont.*) (*b*)

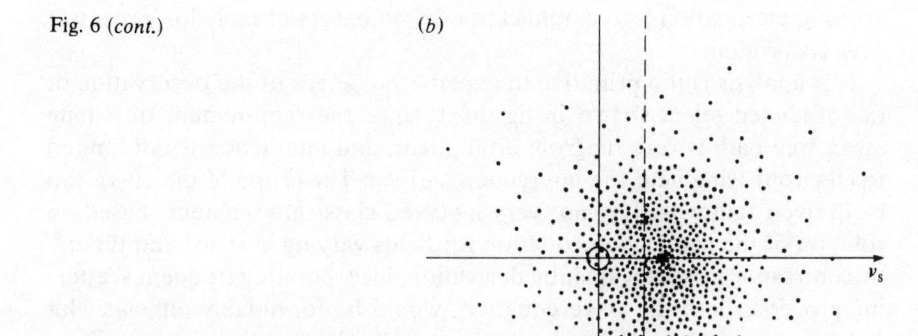

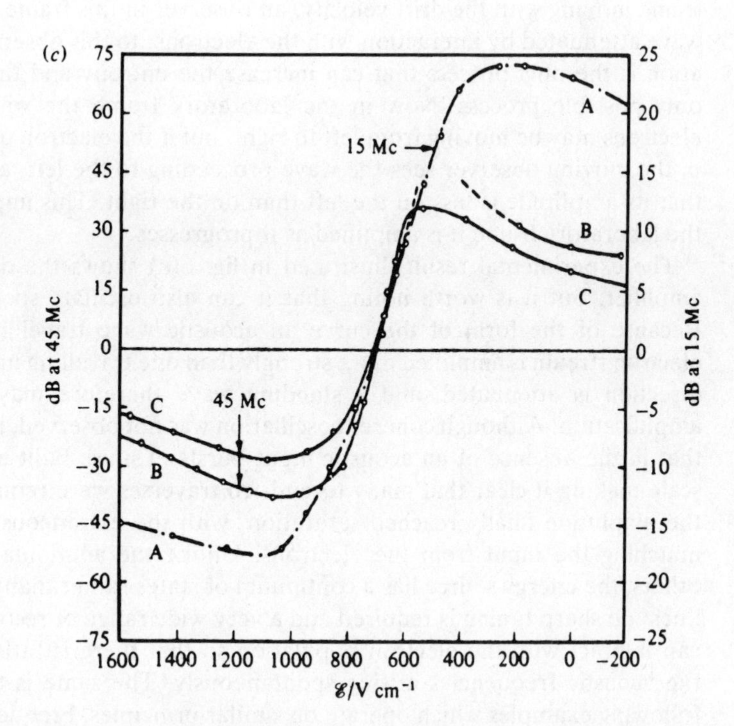

of energy from the wave than emission to it; this accounts for the acoustic attenuation. By applying a strong electric field, however, it is possible to shift the distribution very considerably, as indicated in fig. 6(*b*). This has been drawn to show a bodily shift, but it is likely that the distribution will be broadened asymmetrically at the same time. Leaving such a comparatively minor detail aside, it is clearly possible in principle to arrange that the density is greater on the right than on the left, and thus to convert acoustic attenuation into amplification by an excess of emission processes over absorption.

This analysis is too primitive to explain the details of the observation of the predicted effect shown in fig. 6(*c*), since the requirement of a long mean free path is very far from attainment, and interaction is not limited to electrons lying near the interaction surface. The shape of the curve can be derived rather well by not very involved classical arguments based on solution of the conduction equation for fields varying in space and time;[(8)] by contrast, a genuinely quantal derivation, incorporating frequent scattering processes into the wave equation, would be formidably difficult. Not only that, it might well hide the fundamental point that the transition from attenuation to amplification when the drift velocity exceeds the velocity of sound is an almost inevitable consequence of any theory in which the wave and the electron assembly are the only important constituents. So long as the steady electric field serves mainly to impart a drift velocity to the electrons, without grossly modifying their distribution as observed in a frame moving with the drift velocity, an observer in this frame will see the wave attenuated by interaction with the electrons; to this observer, attenuation is the only process that can increase the entropy and therefore the only possible process. Now in the laboratory frame the wave and the electrons may be moving from left to right, but if the electron drift exceeds v_s the moving observer sees the wave proceeding to the left, and will find that its amplitude is less on the left than on the right. This implies that in the laboratory frame it is amplified as it progresses.

The experimental result illustrated in fig. 6(*c*) shows the device as an amplifier, but it is worth noting that it can also oscillate spontaneously. Because of the form of the curve an acoustic wave travelling with the electron stream is amplified more strongly than one travelling in the reverse direction is attenuated, and a standing wave therefore may suffer net amplification. Although coherent oscillation was not observed, it was found that in the absence of an acoustic input bursts of noise built up, the time scale making it clear that many to-and-fro traverses were required before the amplitude finally reached saturation, with the extraneous dissipation matching the input from the electrons. Unlike the ammonia maser and others, the energy source has a continuum of states rather than a few sharp lines; no sharp tuning is required and a very wide range of resonant modes can interact with the electron population so that there is little control of the acoustic frequencies arising spontaneously. The same is true for the following examples which operate on similar principles. Frequency control is supplied by a resonant cavity and the devices may be tuned over a considerable range without losing the amplification from the non-resonant active medium.

Travelling-wave oscillator

A helix of wire, surrounded by a cylindrical metal sheath, as in fig. 7, acts as a coaxial transmission line on which electromagnetic waves travel rather

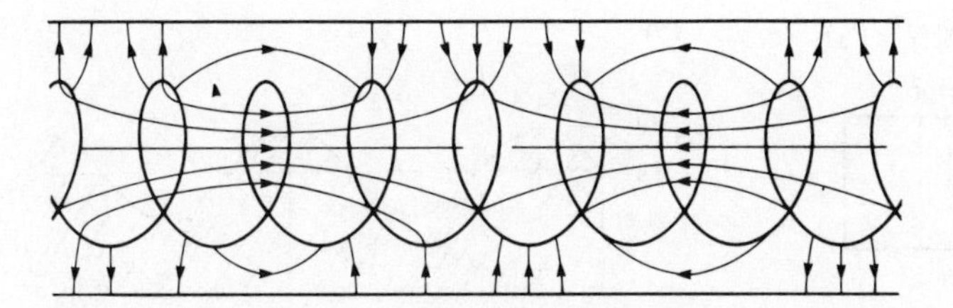

Fig. 7. Electric field distribution in a coaxial transmission line with helical inner conductor.

slowly on account of the inductance of the inner conductor. Moreover the electrical field-lines outside the helix terminate on surface charges which also generate a field component along the axis. This is one example of a wave which can interact with an electron stream moving faster than itself so as to be amplified. Unless we are to undertake a full analysis, such as is certainly desirable when the detailed design of an oscillator is the aim, what has already been said is enough to reveal the principle as virtually the same as for wave-amplification in CdS. Here, however, there is no problem arising from collisions (though space-charge forces may be significant) and the quantum theory of the interaction may be appreciated as having as good a status as the classical theory of bunching. Whereas in CdS the thermalized distribution of electron velocities spans a range much greater than the wave velocity, the reverse is true here; the electron gun can produce a rather well-defined stream of nearly mono-energetic electrons, so that by adjustment of the accelerating voltage the majority can be caused to move only slightly faster than the wave, and the regenerative interaction is correspondingly strong. In an actual device the amplified wave that emerges is fed back into the input so that the whole forms a ring resonator which may be imagined as spontaneously excited by the maser action of the inverted population of energy states in the electron beam.

The klystron

We shall devote more space to the klystron than to other oscillators, not because it is technically more important, but because in an idealized form its behaviour exemplifies both the bunching process and maser action with such simplicity that detailed analysis presents no difficulties. There are many variants of the design, according to the intended application, and we shall consider only one, the *reflex klystron* which for many years, until its supremacy was challenged by solid state devices, was the most convenient and reliable low-power continuous oscillator in the frequency range 10^9–10^{11} Hz. The resonator, shown in section in fig. 8(*a*), has axial symmetry and supports a fundamental mode which produces an oscillatory electric field in the gap at the centre. This region is in the vacuum tube (not shown) and an electron beam is accelerated from a gun at negative potential so that it passes through the gap and then is reflected back by the retarding field of the still more negative reflector electrode. It returns through the gap and subsequently is lost from the beam. We shall assume that the gap

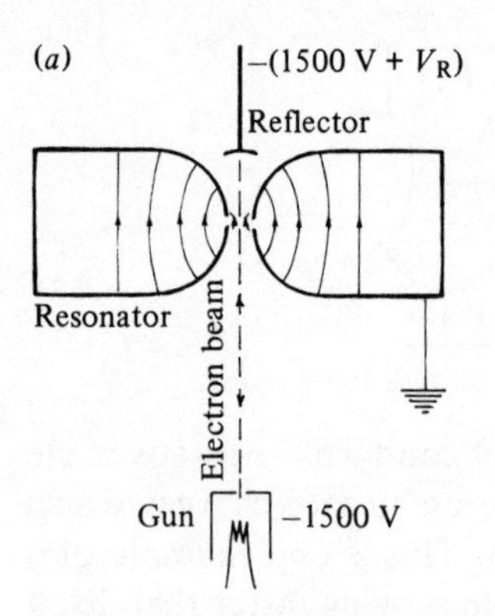

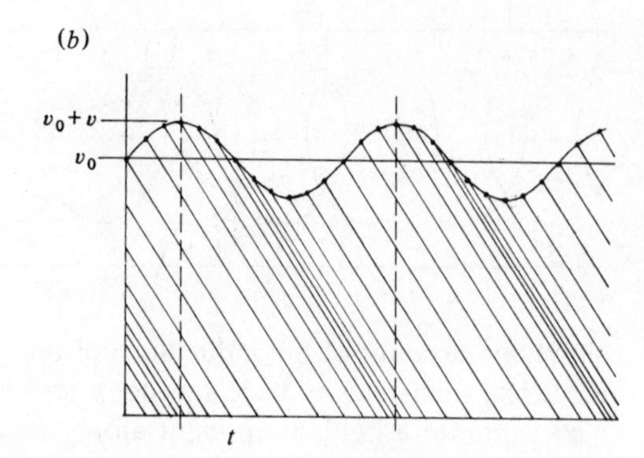

Fig. 8. (*a*) Schematic diagram of reflection klystron; the resonator, shown in section, is a torus with a gap round its inner surface through which the electric field spreads to interact with the electron beam. (*b*) Illustrating the bunching mechanism.

is very narrow and that the retarding field is constant between the resonator and the reflector. When the cavity is excited at frequency ω the electrons arriving at the gap with velocity v_0 leave with velocity $v_0 + v \cos \omega t$, being accelerated or retarded according to the phase of oscillation at the instant of passage. The same deceleration is then experienced by all, so that the fastest take longest to return, just like a ball thrown in the air. For values of ωt around $\pi/2$ successive electrons gain progressively less velocity from the gap and consequently return sooner so that the beam is locally more concentrated; half a cycle later the beam is correspondingly emaciated. A schematic representation is shown in fig. 8(*b*); an electron arriving at t returns to the gap after an interval $C(v_0 + v \cos \omega t)$, where $2/C$ is the deceleration. The lines, whose gradient is $-1/C$, show by their intersection with the horizontal axis the times of return for an originally equally spaced set of electrons, starting at values of ωt differing successively by $\pi/6$. It is clear that if the cavity excitation (i.e. v) is high enough, or the retarding field weak enough, the sinusoid may be steeper than the lines over part of its cycle, so that the electrons return with their order changed; where the straight lines are tangential to the sinusoid the local density on return is infinite (or would be if space charge did not prevent it). In principle, therefore, the bunching may lead to a highly non-sinusoidal distribution of charge in the beam, and indeed the klystron has proved a good device for frequency multiplication by virtue of its strong harmonic content.

Our concern, however, is with the self-excited oscillator, for which purpose we need only know the fundamental component of the bunching. This is determined by $\int_0^{2\pi/\omega} j(t')e^{i\omega t'}\,dt'$, where $j(t')$ is the current in the gap due to the reflected electrons. Now each electron, irrespective of its velocity, is responsible for a sharp impulse $j(t')$ such that $\int j(t')\,dt' = e$, its charge. To compute the Fourier integral, therefore, it is necessary only to sum $e^{i\omega t'}$ for each electron, with t' representing the instant of the return passage, and the difficulties that might have been anticipated from the peaked form of $j(t)$ are eliminated. Thus the electrons that made their first passage between t and $t + dt$ carry a total charge proportional to dt (we shall leave out all constants and look solely at the form of the behaviour), and arrive back when t' is $t+C(v_0+v\cos\omega t)$; it follows immediately that j_1, the fundamental component of the current, takes the form

$$j_1 \propto \int_0^{2\pi/\omega} e^{i\omega[t+C(v_0+v\cos\omega t)]}\,dt$$
$$\propto i\,e^{i\omega C v_0} J_1(\omega C v), \tag{13}$$

where $J_1(x)$ is the Bessel function of the first kind and first order:

$$J_1(x) = \tfrac{1}{2}x - \tfrac{1}{16}x^3 + \tfrac{1}{384}x^5 - \cdots.$$

The meaning of this result is clear from fig. 8(*b*); $J_1(\omega C v)$ is a measure of the degree of bunching, while $\omega C v_0$ represents the phase of the bunches relative to the oscillating field. In the diagram the bunches have been drawn with their maximum density coinciding with the peak of the cavity field – the current maxima in the reflected beam meet a strong decelerating field and thereby transfer energy to the resonator. The ratio of j_1 to the field in the gap (to which v is proportional) may be regarded as the effective impedance presented by the electron beam when the behaviour of the resonant cavity is modelled by a series LCR circuit. In addition to the fixed circuit components there is an impedance of the form

$$Z_{\text{beam}} = -iA\,e^{i\omega C v_0} J_1(\omega C v)/v, \tag{14}$$

where A is a constant determined by the beam density, gap configuration etc. In the figure $\omega C v_0 = 3\pi/2$ and Z_{beam} is real and negative, providing the required regeneration. More generally R_{beam}, the resistive part of Z_{beam}; is equal to $A\sin(\omega C v_0)J_1(\omega C v)/v$ and the reactive part $X_{\text{beam}} = -iA\cos(\omega C v_0)J_1(\omega C v)/v$. If oscillation can be maintained, it settles down at an amplitude of v such that $R_{\text{beam}} + R = 0$. Fig. 9 shows how the solution might appear under nearly optimal tuning conditions, $\omega C v_0 = 3\pi/2$, and the beam is four times stronger than is needed to initiate oscillation. The

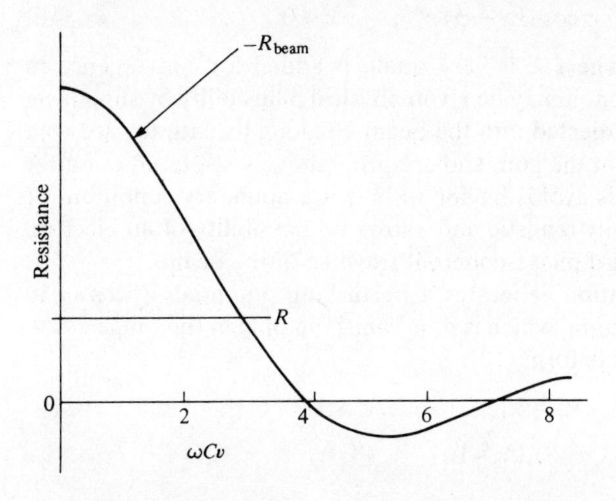

Fig. 9. Plot of the real part of (14).

amplitude builds up until v approaches the first zero of R_{beam}, which occurs when $\omega Cv = 3.83$, the first zero of $J_1(x)$. At this point $v \sim 0.8v_0$ and the beam is very heavily modulated. When there is, as in this example, plenty of amplification in hand, the reflector field may be varied over a considerable range, to reduce the magnitude of $\cos(\omega Cv_0)$, without stopping oscillation, and the consequent introduction of X_{beam} causes the frequency to shift. It is easily seen that the klystron is tunable by this means over a range several times the frequency band of the freely oscillating cavity.

Let us turn now to the quantum mechanics of the system, regarding the cavity field as a time-dependent perturbation of the electrons in the beam. This is a simple one-dimensional problem of a particle in a potential which, when the cavity is unexcited, is zero for $x < 0$ (between electron gun and gap) and equal to $2mx/C$ for $x > 0$ (between gap and reflector). Since the deBroglie wavelength is millions of times less than the dimensions, the WKB solution can be expected to provide a good approximation to the Airy function which is the solution for a linear potential variation, as in the region of positive x. For negative x a sinusoidal standing-wave solution applies, whose nodes are fixed by the need to join up with the Airy function. Now an electron of energy E has kinetic energy $E - 2mx/C$ at x, and wavenumber $(2mE - 4m^2x/C)^{\frac{1}{2}}/\hbar$; its phase change 2ϕ in going from gap to reflector and back is $2\int k\,\mathrm{d}x$, and may be written down immediately:

461

$$2\phi = [2(2mE)^{\frac{1}{2}}/\hbar] \int_0^{x_0} (1 - x/x_0)^{\frac{1}{2}}\,\mathrm{d}x,$$

where $x_0 = CE/2m$ and marks the classical reflection point. Hence

$$\phi = \sqrt{2}CE^{\frac{3}{2}}/3\hbar m^{\frac{1}{2}}. \tag{15}$$

Every increment of E that increases ϕ by π shifts the standing wave at $x = 0$ from one node to the next. The unnormalized wave-function in the region $x < 0$ may therefore be written

$$\psi = \cos(kx - \phi)\,\mathrm{e}^{\lambda x}; \qquad x < 0. \tag{16}$$

The exponential $e^{\lambda x}$, where λ is very small, is added for convergence in the following calculation; it may be given physical plausibility by supposing that the electrons are injected into the beam all along its path towards the resonator, rather than at the gun, and are progressively scattered out after the return passage. This avoids having to invent a boundary condition for the gun that is physically realistic and allows no possibility of an electron making a second or third phase-coherent traverse of the beam.

The cavity in oscillation generates a perturbing potential $V' \cos \omega t$ in the form of a step-function, which is non-vanishing only in the range $x < 0$, where ψ takes the simple form (16):

$$V' = V_0; \qquad x > 0$$

$$= V_0 + V_1; \qquad x < 0,$$

where $V_1 = \frac{1}{2}m[(v_0+v)^2 - v_0^2]$. When V_1 is weak, the only case we consider, the strict periodicity of $V' \cos \omega t$ ensures that a given state E, having wavenumber k, is significantly perturbed only by the introduction of the states $E \pm \hbar\omega$. The probability of a transition occurring to one of these states, k' say, is proportional to the square of the matrix element:

$$M_{kk'} = \int_{-\infty}^{\infty} \psi_{k'}^* V' \psi_k \,\mathrm{d}x = V_1 \int_{-\infty}^{0} \psi_{k'}^* \psi_k \,\mathrm{d}x,$$

since orthogonality ensures that $V_0 \int_{-\infty}^{\infty} \psi_{k'}^* \psi_k \,\mathrm{d}x = 0$. Hence, from (16), when $\lambda \to 0$,

$$M_{kk'} \approx \tfrac{1}{2} V_1 \sin(\phi' - \phi)/(k' - k), \tag{17}$$

the much smaller terms having $k + k'$ in their denominator being neglected. It is the difference in $|M_{kk'}|^2$ for upward and downward transitions that concern us. Binomial expansion of (15), with E replaced by $E \pm \hbar\omega$, shows that

$$\phi' = \phi(1 \pm \tfrac{3}{2}\varepsilon + \tfrac{3}{8}\varepsilon^2 \cdots), \qquad \text{where } \varepsilon = \hbar\omega/E$$

and the upper sign refers to the transition to $E + \hbar\omega$. Hence

$$\sin(\phi' - \phi) = \pm\sin(\tfrac{3}{2}\varepsilon\phi) + \tfrac{3}{8}\varepsilon^2\phi \cos(\tfrac{3}{2}\varepsilon\phi) + \cdots. \tag{18}$$

A similar expansion for k, which is proportional to $E^{\frac{1}{2}}$, shows that

$$1/(k' - k) = \pm 2/\varepsilon k + 1/2k + \cdots. \tag{19}$$

Hence

$$|M_{kk'}|^2 = (EV/m\omega v_0)^2[\sin^2\xi \pm \tfrac{1}{2}(\hbar\omega/E)\sin\xi(\sin\xi + \xi\cos\xi) + \cdots], \tag{20}$$

where $\xi = \frac{3}{2}\varepsilon\phi = \frac{1}{2}\omega C v_0$. Maser action is strongest when the negative sign gives the larger matrix element, i.e. when $\sin\xi(\sin\xi + \xi\cos\xi)$ takes its maximum negative value. The second term, proportional to $\xi \sin 2\xi$, gives the same condition as the classical calculation; if C is varied amplification is greatest when $C \sin(\omega C v_0)$ has its maximum negative value. The first term, which is relatively rather unimportant, is probably the result of inadequate approximations to the wave-functions. When the matrix elements are to be expanded to second order in ε the WKB approximation may require refinement; but we shall not attempt to achieve perfect agreement. At this degree of approximation the main point is clear enough, that bunching and maser action, as in the other examples, are alternative descriptions of a basic and essentially simple physical process of energy transfer.

Epilogue

As is appropriate in an exposition of the quantum mechanics of vibrators, we started with a quantal calculation and have finished with one; but we have never strayed far from the classical models. We began, indeed, in chapter 13 the task of establishing limits to the validity of classical reasoning, and the very last example of this chapter has been used to demonstrate, somewhat perversely it might be thought, that the method of quantum mechanics can occasionally be applied to systems which physicists and engineers would instinctively regard as classical. So long as the discussions give insight into physical processes and reveal the strengths and weaknesses of the analytical tools available, no apology is needed and no defence is offered except the evidence of the book itself. Even the rather simple systems which provide material for the whole of the volume demonstrate clearly the increased power available to those who can handle both classical and quantal reasoning. The reader who wishes to apply his skill to complex vibratory processes will find his tasks eased if he can use whichever seems advantageous at each stage, and be confident that his understanding of simple processes will enable him to recognize the dangers inherent in both approaches – the danger of carrying classical reasoning too far into the quantum domain, and the danger of forcing over-simplifications on physical systems to make them amenable to the unforgiving methodology of quantum mechanics.

References

Chapter 2: The free vibrator

1 C. Huygens, *Horologium Oscillatorium* (1673).
2 F. Hope-Jones, *Electrical Timekeeping*, 2nd edition (London: N.A.G. Press, 1940).
3 H. L. F. Helmholtz (trans. A. J. Ellis), *The Sensations of Tone*, p. 68 (London: Longmans, Green, 1875).
4 P. C. Wraight, 1971, *Phil. Mag.* **23**, 1261.
5 D. G. Fincham and P. C. Wraight, 1972, *J. Phys.* A **5**, 248.
6 F. Homann, 1936, *Forsch. Geb. Ing.* **7**, 1.
7 J. S. Lew, 1956, *Amer. J. Phys.* **24**, 46.
8 Ref. 7.3, vol. 1, chapter 4.

Chapter 3: Applications of complex variables to linear systems

1 E. C. Titchmarsh, *Introduction to the Theory of Fourier Integrals*, p. 305 (Oxford: Clarendon Press, 1948).
2 P. M. Morse, *Vibration and Sound*, p. 114 (New York: McGraw-Hill, 1936).

Chapter 4: Fourier series and integral

1 K. F. Riley, *Mathematical Methods for the Physical Sciences*, chapter 8 (Cambridge University Press, 1974).
2 S. G. Lipson and H. Lipson, *Optical Physics*, p. 152 (Cambridge University Press, 1969).
3 K. G. Budden, *Radio Waves in the Ionosphere*, p. 256 (Cambridge University Press, 1961).
4 L. M. Milne-Thomson, *Theoretical Hydrodynamics*, 4th edition, chapter 14 (London: Macmillan, 1960).
5 J. L. Lawson and G. E. Uhlenbeck (editors), *Threshold Signals*, chapter 3 (New York: McGraw-Hill, 1950).
6 J. C. Dainty, The Statistics of Speckle Patterns, in *Progress in Optics* vol. 14, ed. E. Wolf (Amsterdam: North-Holland, 1977).
7 C. Kittel, *Thermal Physics* (New York: Wiley, 1969).
8 H. Feder, 1929, *Ann. Phys.* (Germany) **1**, 497.

Chapter 5: Spectrum analysis

1 G. A. Vanasse and H. Sakai, Fourier Spectroscopy, in *Progress in Optics* vol. 6, ed. E. Wolf (Amsterdam: North-Holland, 1967).
2 Ref. 4.2, pp. 210 and 225.
3 Ref. 4.1, chap. 16. *See also* E. G. Phillips, *Functions of a Complex Variable*, p. 96 (Edinburgh: Oliver and Boyd, 1940).
4 R. Kronig, 1926, *J. Opt. Soc. Amer.* **12**, 527.
H. A. Kramers, 1927, *Atti congr. fis.* (Como) 545.

5 F. Abelès (editor) *Optical Properties and Electronic Structure of Metals and Alloys* (Amsterdam: North-Holland, 1966).
6 R. T. Schumacher and W. E. Vehse, 1963, *J. Phys. Chem. Solids* **24**, 297.
7 R. K. Potter, G. A. Kopp and H. C. Green, *Visible Speech* (New York: Van Nostrand, 1947).
8 A. H. Benade, 1973, *Scientific American* **229**, 24.
9 A. B. Pippard, 1947, *Proc. Roy. Soc.* A **191**, 370.
10 H. Bremmer, *Encylopaedia of Physics* **16**, 573 (Berlin: Springer, 1958).

Chapter 6: The driven harmonic vibrator

1 E. Kappler, 1931, *Ann. Phys.* (Germany) **11**, 233. *See also* R. B. Barnes and S. Silverman, 1934, *Rev. Mod. Phys.* **6**, 162.
2 R. V. Jones and C. W. McCombie, 1952, *Phil. Trans. Roy. Soc.* A **244**, 205.
3 J. B. Johnson, 1928, *Phys. Rev.* **32**, 97. *See also* ref. 4.5, chapter 4.
4 A. W. Hull and N. H. Williams, 1925, *Phys. Rev.* **25**, 148.

Chapter 7: Waves and resonators

1 B. I. Bleaney and B. Bleaney, *Electricity and Magnetism*, 3rd edition, chapter 9 (Oxford University Press, 1976).
2 Ref. 1, p. 72.
3 J. W. Strutt, Baron Rayleigh, *The Theory of Sound*, vol. 2, p. 196 (New York: Dover, 1945).
4 H. Lamb, *Hydrodynamics*, 6th edition, p. 74 (Cambridge University Press, 1924).
5 A. B. Pippard, Metallic Conduction at High Frequencies and Low Temperatures, in *Advances in Electronics and Electronic Physics*, vol. 6, ed. L. Marton (New York: Academic Press, 1954).
6 P. H. Ceperley, I. Ben-Zvi, H. F. Glavish and S. S. Hanna, 1975, *IEEE Trans. Nuc. Sci* **22**, 1153.
7 A. H. Cook, *Interference of Electromagnetic Waves*, chapter 6 (Oxford: Clarendon Press, 1971).
8 S. Tolansky, *Introduction to Interferometry* (London: Longmans, Green, 1955).
9 C. G. Montgomery (editor), *Technique of Microwave Measurements*, p. 319 (New York: McGraw-Hill, 1947).
10 L. I. Schiff, *Quantum Mechanics*, 3rd edition, chapter 5 (New York: McGraw-Hill, 1968).
11 J. D. Jackson, *Classical Electrodynamics*, 2nd edition, chapter 17 (New York: Wiley, 1975).
12 A. E. S. Green, *Nuclear Physics*, p. 454 (New York: McGraw-Hill, 1955).
13 D. J. Bennet, *The Elements of Nuclear Power*, p. 175 (London: Longmans, 1972).
14 A. B. Pippard, *Elements of Classical Thermodynamics*, p. 78 (Cambridge University Press, 1966).
15. J. W. Strutt, 1871, *Phil. Mag.* **41**, 107.
16. S. A. Korff and G. Breit, 1932, *Rev. Mod. Phys.* **4**, 471.

Chapter 8: Velocity-dependent forces

1 A. B. Pippard, *Forces and Particles*, pp. 33 and 223 (London: Macmillan, 1972).
2 Ref. 7.1, p. 750.
3 S. Timoshenko, *Vibration Problems in Engineering*, 2nd edition, p. 222 and chapter 5 (London: Constable, 1937).

4 Ref. 7.1, chapter 24.
5 E. Segré, *Nuclei and Particles*, p. 130 (Reading, Mass.: Benjamin, 1965).
6 Ref. 7.1, p. 729.
7 G. Dresselhaus, A. Kip and C. Kittel, 1955, *Phys. Rev.* **98**, 368.
8 N. W. Ashcroft and N. D. Mermin, *Solid State Physics*, p. 11 (New York: Holt, Rinehart and Winston, 1976).
9 Ref. 7.1, p. 236.
10 R. Bowers, C. Legendy and F. Rose, 1961, *Phys. Rev. Lett.* **7**, 339. *See also* ref. 4.3, chapters 5 and 6.
11 J. A. Delaney and A. B. Pippard, 1972, *Rep. Prog. Phys.* **35**, 677.

Chapter 9: The driven anharmonic vibrator; subharmonics; stability

1 N. Minorsky, *Non-linear oscillations* (Huntington, NY: Krieger, 1974).
J. J. Stoker, *Non-linear vibrations* (New York: Interscience, 1950).
2 R. Thom, *Structural Stability and Morphogenesis* (Reading, Mass.:Benjamin. 1975).
3 A. E. H. Love, *Mathematical Theory of Elasticity*, 4th edition, p. 401 (Cambridge University Press, 1927).
4 R. A. Cowley, 1964, *Phys. Rev.* **134**, A981.
5 P. Weiss and R. Forrer, 1926, *Ann. Phys.* (France) **5**, 153.
6 K. A. Müller and W. Berlinger, 1971, *Phys. Rev. Lett.* **26**, 13.
7 Ref. 8.1, p. 119.
8 P. H. Hammond, *Feedback Theory and its Applications*, p. 67 (London: English University Press, 1958).
9 Minorsky, Ref. 1, chapter 16.
10 R. M. May, 1976, *Nature* **261**, 459.
11 M. Ya. Azbel' and E. A. Kaner, 1956, *J. Exp. Theor. Phys.* **30**, 811.
W. M. Walsh, Resonances both Temporal and Spatial, in *Solid State Physics* (vol. 1 Electrons in Metals) ed. J. F. Cochran and R. R. Haering (New York: Gordon and Breach, 1968).
12 W. M. Walsh, ref. 11, p. 158.
13 A. B. Pippard, 1947, *Proc. Roy. Soc.* A **191**, 385.
14 A. B. Pippard, *Response and Stability* (Cambridge University Press, 1985).

Chapter 10: Parametric excitation

1 J. W. Strutt, Baron Rayleigh, 1887, *Phil. Mag.* **24**, 145.
2 E. T. Whittaker and G. N. Watson, *Modern Analysis*, 4th edition, p. 412 (Cambridge University Press, 1927).
3 Ref. 2, chapter 19.
4 B. J. Robinson, 1961, *Institute of Electrical Engineers Monograph* 480 E.
5 T. H. Beeforth and H. J. Goldsmid, *Physics of Solid State Devices*, p. 48 (London: Pion, 1970).

Chapter 11: Maintained oscillators

1 D. C. Gall, 1942, *J. Inst. Elec. Eng.* **89**, 434.
2 H. Nyquist, 1932, *Bell Syst. Tech. J.* **11**, 126. *See also* ref. 9.8, p. 71.
3 A. Hund, *Phenomena in High-frequency Systems*, chapter 2 (New York: McGraw-Hill, 1936).
4 G. B. Clayton, *Operational Amplifiers* (London: Butterworths, 1971).
5 Ref. 2.2.
6 W. J. Gazeley, *Clock and Watch Escapements*, p. 23 (London: Heywood, 1956).

7 J. H. Simpson and R. S. Richards, *Physical Principles and Applications of Junction Transistors*, p. 176 (Oxford: Clarendon, 1962).
Ref. 10.5, p. 45.
8 A. Liénard, 1928, *Rev. Gén. de l'Electricité*, **23**, 901.
9 J. R. Bristow, 1946, *Proc. Roy. Soc.* A **189**, 88.
10 Ref. 7.3, p. 232.
11 M. Sargent, M. O. Scully and W. E. Lamb, *Laser Physics*, chapter 5 (Reading, Mass.: Addison-Wesley, 1974).
12 J. P. Gordon, H. Z. Zeiger and C. H. Townes, 1954, *Phys. Rev.* **95**, 282.
13 H. Abraham and E. Block, 1919, *Ann. Phys.* (Germany) **12**, 237.
14 *Encyclopaedia Britannica*, 15th edition – Macropaedia **8**, 132; Geysers and Fumaroles (Chicago, 1974).
15 H. Kalmus, 1941, *Nature* **148**, 626.

Chapter 12: Coupled vibrators

1 Unpublished observation by Dr N. Elson.
2 P. J. Wheatley, *Molecular Structure*, 2nd edition, p. 66 (Oxford: Clarendon Press, 1968).
3 G. Herzberg, *Molecular Spectra and Molecular Structure*, vol. 2, p. 215 (New York: Van Nostrand, 1945).
4 Ref. 8.1, p. 37.
5 Minorsky, ref. 9.1, chapter 18.
6 Ref. 11.11, chapter 11.
7 J. R. Waldram, 1976, *Rep. Prog. Phys.* **39**, 751.
8 B. D. Josephson, 1962, *Phys. Lett.* **1**, 251.

Chapter 13: The quantized harmonic vibrator and its classical features

1 J. L. Powell and B. Crasemann, *Quantum Mechanics*, p. 127 (Reading, Mass: Addison-Wesley, 1961).
2 L. Pauling and E. B. Wilson, *Introduction to Quantum Mechanics*, p. 67 (New York: McGraw-Hill, 1935).
3 Ref. 2, p. 81.
4 M. Abramowitz and I. A. Stegun, *Handbook of Mathematical Functions*, p. 475 (New York: Dover, 1965).
5 Ref. 1, p. 98.
6 G. M. Murphy, *Ordinary Differential Equations and Their Solutions*, p. 15 (Princeton, NJ: Van Nostrand, 1960).

Chapter 14: Anharmonic vibrators

1 A. Sommerfeld (tr. H. L. Brose), *Atomic Structure and Spectral Lines*, p. 232 (London: Methuen, 1923).
2 Ref. 13.4, p. 256.
3 Ref. 1, p. 304.
D. ter Haar, *Elements of Hamiltonian Mechanics*, 2nd Ed., p. 129 (Amsterdam: North-Holland, 1964).
4 Ref. 13.4.
5 J. C. P. Miller, *The Airy Integral*, (B.A. Mathematical Tables, Cambridge University Press, 1946).
6 Ref. 13.2, p. 271.
7 Ref. 13.1, p. 387.

8 Ref. 13.1, p. 140.
9 R. J. Le Roy, *Molecular Spectroscopy, Vol. 1*, Ch. 3 (London: The Chemical Society, 1973).

Chapter 15: Vibrations and cyclotron orbits in two dimensions

1 G. Herzberg, *Molecular Spectra and Molecular Structure, Vol. 2*, p. 215 (New York: Van Nostrand, 1945).
2 N. Minorsky, *Non-linear Oscillations*, p. 506 (Huntington, NY: Krieger 1974). E. Breitenberger and R. D. Mueller, 1981, *J. Math. Phys.* **22**, 1196.
3 A. B. Pippard, 1964, *Phil. Trans. Roy. Soc.* A**256**, 317.
4 J. M. Ziman (ed.), *The Physics of Metals, 1. Electrons*, p. 118 (Cambridge University Press, 1969).
5 L. Onsager, 1952, *Phil. Mag.* **43**, 1006.
6 E. I. Blount, *Solid State Physics* (ed. Seitz and Turnbull), *Vol. 13*, p. 305 (New York: Academic Press, 1962).
7 P. J. Lin and L. M. Falicov, 1966, *Phys. Rev.*, **142**, 441.
8 Ref. 4, p. 145.
9 Ref. 4, p. 62.
10 Ref. 4, p. 129.
11 R. W. Stark, 1964, *Phys. Rev.*, **135**, A1698.
12 L. M. Falicov and H. Stachowiak, 1966, *Phys. Rev.*, **147**, 505.
13 W. P. Reinhardt and I. Dana, 1987, *Proc. Roy. Soc.* A**413**, 157.

Chapter 16: Dissipation, level broadening and radiation

1 B. I. Bleaney and B. Bleaney, *Electricity and Magnetism*, 3rd edition, p. 248 (Oxford University Press, 1976).
2 R. L. Sproull and W. A. Phillips, *Modern Physics*, 3rd edition, p. 653 (New York: Wiley, 1980).
3 Ref. 2, p. 258.
4 J. M. Jauch and F. Rohrlich, *The Theory of Photons and Electrons*, 2nd edition, p. 171 (New York: Springer-Verlag, 1976).
5 N. W. Ashcroft and N. D. Mermin, *Solid State Physics*, p. 519 (New York: Holt, Rinehart and Winston, 1976).
6 A. B. Pippard, 1987, *Eur. J. Phys.* **8**, 55.

Chapter 17: The equivalent classical oscillator

1 D. H. Auston, Topics in Applied Physics (No. 18, Ultrashort Light Pulses, ed. S. L. Shapiro) p. 123 (Berlin: Springer-Verlag, 1977).
2 E. U. Cordon and G. H. Shortley, *The Theory of Atomic Spectra*, p. 133 (Cambridge University Press, 1935).

Chapter 18: The two-level system

1 Ref. 13.1, p. 358.
2 W. A. Phillips, 1970, *Proc. Roy. Soc.* A**319**, 565.
3 H. Fröhlich, *Theory of Dielectrics*, p. 70 (Oxford: Clarendon Press, 1949).
4 C. J. Gorter, *Paramagnetic Relaxation* (New York: Elsevier, 1947).
5 C. Zener, *Elasticity and Anelasticity of Metals* (Chicago University Press, 1948).
6 G. Frassati and J. le G. Gilchrist, 1977, *J. Phys. C.* **10**, L509.
7 H. W. Kroto, *Molecular Vibration Spectra*, Ch. 9 (London: Wiley, 1975).
8 Ref. 16.5, p. 425.

9 B. Bleaney and R. P. Penrose, 1947, *Proc. Roy. Soc.* A**189**, 358.
10 J. P. Gordon, H. Z. Zeiger and C. H. Townes, 1954, *Phys. Rev.*, **95**, 282.

Chapter 19: Line broadening

1 S. Ch'en and M. Takao, 1957, *Rev. Mod. Phys.*, **29**, 20.
H. Margenau and M. Lewis, 1959, *Rev. Mod. Phys.*, **31**, 569.
G. Peach, 1975, *Contemp. Phys.* **16**, 17.
1981, *Ad. in Phys.* **30**, 367.
2 D. Bohm, *Quantum Theory*, p. 611 (London: Constable, 1951).
3 J. Stark, 1914, *Ann. Physik*, **43**, 965.
4 B. Bleaney and R. P. Penrose, 1948, *Proc. Phys. Soc.*, **60**, 540.
5 T. W. Hänsch, M. D. Levenson and A. L. Schawlow, 1971, *Phys. Rev. Lett.*, **26**, 946.
G. W. Series, 1974, *Contemp. Phys.*, **15**, 49.
6 B. W. Petley and K. Morris, 1979, *Nature*, **279**, 141.
7 A. Abragam, *The Principles of Nuclear Magnetism*, p. 58 (Oxford: Clarendon Press, 1961).
8 B. Golding, M. v. Schickfus, S. Hunklinger and K. Dransfeld, 1979, *Phys. Rev. Lett.*, **43**, 1817.
9 Ref. 13.2, p. 383.

Chapter 20: The ammonia maser

1 Ref. 18.9.
2 Ref. 13.4, p. 358.
3 M. O. Scully and W. E. Lamb, 1967, *Phys. Rev.*, **159**, 208.
4 Ref. 13.2, p. 79.
5 D. Kleppner, H. M. Goldenberg and N. F. Ramsey, 1962, *Phys. Rev.*, **126**, 603.
6 R. F. C. Vessot in *Radio Interferometry Techniques for Geodesy*, p. 203 (NASA conference publication 2115, 1980).
7 R. F. C. Vessot and H. E. Peters, 1962, *I.R.E. Trans.* (Instrumentation), 183.

Chapter 21: The family of masers; from laser to travelling-wave oscillator

1 A. H. Cook, *Celestial Masers*, Ch. 1 (Cambridge University Press, 1977).
2 D. J. Bradley, in *Ultrashort Light Pulses* (ed. Shapiro) p. 17 (Berlin: Springer-Verlag, 1977).
3 e.g. B. A. Lengyel, *Introduction to Laser Physics* (New York: Wiley, 1966), O. Svelto (tr. D. C. Hanna), *Principles of Lasers* (New York: Plenum, 1976).
4 A. H. W. Beck, *Thermionic Valves*, Chs 12, 14, 15 (Cambridge University Press, 1953).
5 I. B. Bott, 1965, *Phys. Lett.*, **14**, 293.
6 J. H. McFee in *Physical Acoustics* (ed. Mason), *Vol. IV A*, p. 1 (New York: Academic Press, 1966).
7 A. R. Hutson, J. H. McFee and D. L. White; 1961, *Phys. Rev. Lett.*, **7**, 236.
8 G. Weinreich, 1956, *Phys. Rev.*, **104**, 321.

Index